AF752458

Theoretical Models of Chemical Bonding

The Concept of the Chemical Bond

Theoretical Models of Chemical Bonding

Part 1: Atomic Hypothesis and the Concept of Molecular Structure

Part 2: The Concept of the Chemical Bond

Part 3: Molecular Spectroscopy, Electronic Structure and Intramolecular Interactions

Part 4: Theoretical Treatment of Large Molecules and their Interactions

Theoretical Models of Chemical Bonding

Part 2:

The Concept of the Chemical Bond

Editor:
Prof. Dr. Zvonimir B. Maksić
Zagreb

AKADEMIE-VERLAG BERLIN

Die Originalausgabe „Theoretical Models of Chemical Bonding“ (Part 1—4)
erscheint im Springer-Verlag
Berlin · Heidelberg · New York · London · Paris · Tokyo · Hongkong

Vertriebsrechte für die sozialistischen Länder:
Akademie-Verlag Berlin

Vertriebsrechte für alle Staaten mit Ausnahme der sozialistischen Länder:
Springer-Verlag Berlin · Heidelberg · New York · London · Paris · Tokyo · Hongkong

Alle Rechte vorbehalten
© Springer-Verlag Berlin · Heidelberg 1990

ISBN 3-540-51553-4 Springer-Verlag Berlin · Heidelberg · New York

Maksić: Theoretical 1 Models of Chemical Bonding (Part 1—4)
ISBN 3-05-500807-3 Akademie-Verlag Berlin
Part 2 ISBN 3-05-500810-3

Erschienen im Akademie-Verlag Berlin, Leipziger Straße 3—4, Berlin, DDR-1086
Printed in the German Democratic Republic
Gesamtherstellung: Druckhaus „Thomas Müntzer“ GmbH, DDR-5820 Bad Langensalza
LSV 1215
Bestellnummer: 764 116 3 (9296/2)

To the memory of my parents
Olivera and Branko Maksić

Preface

There are three fundamental concepts which provide the axiomatic framework of classical chemistry. These are: the notion of atoms as molecular building blocks, the idea that atoms are orderly arranged in molecules giving rise to some structural patterns and finally the concept of the chemical bond realized by two nearest neighbour atoms. It is a common knowledge that a microscopic description of chemical phenomena requires application of wave mechanics because atoms and molecules are quantum objects. It appears, however, that it is very difficult to derive the most basic principles of empirical chemistry from quantum laws. The state of the art concerning the concepts of modified atoms in chemical environments and of the molecular structure in its generic and nongeneric sense were discussed *in extenso* in the first book of this Series. In the present volume, manifestations of covalent chemical bonding are considered in detail. An emphasis on the conceptual and interpretive aspects of the quantum theory of molecules is made in keeping up with a general philosophy inaugurated in the Foreword of the Series "Theoretical Models of Chemical Bonding". Concomitantly, a great deal of attention is focused on the rationalization of experimental facts and observed regularities. It will turn out, as expounded in the first chapter, that a detailed physical description of chemical bonding involves quite sophisticated subtleties. However, trends of changes of many properties within families of related molecules and/or (bond) additivity rules can be reduced to and explained by somewhat crude but nevertheless practical, simple and pictorial concepts and models.

It is hoped that the presented text will help the practicing chemist in comprehending modern theory thus allowing a better amalgamation of theoretical and experimental knowledge. This is of paramount importance because overlapping of abstract theory, pictorial models and empirical findings is a *conditio sine qua non* in revealing and understanding the secrets of Nature.

Z. B. Maksić

Table of Contents

The Physical Origin of the Chemical Bond

W. Kutzelnigg
Lehrstuhl für Theoretische Chemie, Ruhr-Universität Bochum, 4630 Bochum, FRG

1 Introduction

In the world by which we are surrounded, the only free atoms are those of the rare gases. All other atoms exist chemically bound as constituents of molecules or of higher aggregates such as solids. Without doubt the chemical bond plays a very important role in nature, and to understand its mechanism is certainly of central scientific interest.

Bonding of physical entities to more complex composite entities is a widespread phenomenon in physics that arises on a whole hierarchy of levels. In fact, according to our present knowledge

(1) quarks bind to form nucleons
(2) nucleons bind to form nuclei
(3) nuclei and electrons bind to form atoms
(4) atoms bind to form molecules, clusters, metals or ionic crystals
(5) molecules bind via intermolecular forces to form higher aggregates such as molecular crystals

It is not certain whether (1) is really the first step and whether quarks are not made up from 'more elementary' particles, it is also not excluded that beyond (5) the concept of another type of bonding may turn out to be relevant for some further organization, e.g. in biology. If one likes, one may even regard the formation of planetary systems as a kind of bonding and proceed to still higher levels of bonding in astrophysics.

Usually, bonding implies that the composite system has lower energy than the sum of the energies of the constituents, though this is not absolutely necessary — it may be suffient that the desintegration into the constituents is inhibited somehow (in this case, one refers to metastable bonding).

The forces that are basically responsible for bonding in steps (1) and (2) are those of the gluon field, in steps (3) to (5) those of the electromagnetic (photon) field. In these last 3 steps the residual gluon forces are negligible due to their extremely short range, in steps (1) and (2) the photon field is also present, but relatively weak and of minor importance.

We are here only concerned with step (4): "The chemical bond". However, we cannot ignore step (5) completely, since there is not always a clear distinction between the chemical bond and intermolecular forces. We must also say a few words about step (3), since atoms are the building blocks of molecules and since bonding in atoms and in molecules are related to each other.

The bonding in atoms is relatively easy to understand, at least qualitatively. Nuclei and electrons have opposite electric charges and attract each other. That a nucleus with charge number Z can at least bind Z electrons is quite plausible, though a rather simple rigorous proof of this has been given only quite recently [1]. The Schrödinger equation of the simplest atom, hydrogen, can be solved in closed form. A proton and an electron even bind in classical mechanics (the Kepler problem). Stable orbits are characterized by a balance of the attractive Coulomb potential and the repulsive centrifugal energy. Quantum mechanics essentially adds that there is a ground state with lowest possible energy and that bound states are only possible for certain discrete energy levels. The simplest way to correct classical mechanics for quantum effects is to apply the Bohr-Sommerfeld quantization rules. On this level (the Bohr model) many features of the H atom are understood somehow, but not the fact that an H atom is able to bind another atom to form an H_2 molecule. So, in a way, the chemical bond is more fundamentally quantum mechanical than the existence of atoms. Of course, an important ingredient of quantum mechanics for many-electron systems is the Pauli principle. This plays a decisive role both for the structure of atoms and of molecules.

That neutral atoms attract other atoms is harder to understand on a quasiclassical level than the attraction between nuclei and electrons, because in an atom the Coulomb forces are, so to say, saturated and there is no net charge and hence no significant residual electrostatic force left. While at large distances between two neutral atoms the electrostatic interaction vanishes, there is some electrostatic interaction when the electron clouds of the atoms start to overlap. Then the repulsive forces between the nuclei and between the electrons no longer cancel the attractive forces between a nucleus and the electrons. This quasiclassical interaction (that is obtained in ignoring all quantum mechanical aspects of the interaction of the two atoms) is usually attractive at distances where chemical bonds take place. For H_2 the quasiclassical attraction energy is roughly 10% of the actual binding energy, in H_2^+ which is chemically bound, the quasiclassical energy is even repulsive everywhere. For most molecules larger than H_2 the quasiclassical energy is of the same order of magnitude as the binding energy [69]. We will come back to this oberservation in Sect. 9. Even if one could understand the chemical bond one purely classical terms, it would be illegal to ignore quantum effects, mainly the Pauli principle. On the quasiclassical level one also ignores that in different molecular states with a common dissociation limit the same atoms may either repel or attract each other.

One can circumvent the question of how, say, two H atoms bind to a H_2 molecule, if one does not build up H_2 from two atoms, but from two nuclei and two electrons. The numerical solution of the Schrödinger equation of the H_2 molecule has been possible to any desired level of accuracy [2]. If, however, one compares the wave function of H_2 with the wave functions of two H atoms one finds little similarity. In the H_2 molecule the two constituent H-atoms no longer exist. They have lost their individuality.

This loss of the individuality of the constituent atoms makes a theory of the chemical bond difficult. In order to describe the H_2 molecule as a system of two H atoms bound to each other, at least two formal steps (which are not physically observable) are necessary, one that involves the interaction of the atoms, the other their deformation under the interaction.

We have already mentioned that the formation of atoms from electrons and a nucleus involves a potential energy part which is attractive, and a kinetic energy part which is repulsive. One must consider both contributions, since without the kinetic energy the electrons would collapse into the nucleus. The kinetic energy — together with the quantum mechanical uncertainty principle — prevents the electrons from coming too close to the nucleus. In discussing the chemical bond one usually separates the electronic and nuclear motions in the spirit of the Born-Oppenheimer separation, which implies that the electronic potential curve plays the role of the potential energy for the motion of the nuclei. One then identifies a minimum of the potential curve (or potential hypersurface) with an equilibrium structure of the molecule, without worrying about the kinetic energy of the relative motion of the nuclei.

This different attitude towards the kinetic energy in the two cases is justified. The Hamiltonian of an atom is bounded from below only if the kinetic energy of the electrons is included, the potential alone is not bounded from below. The Hamiltonian for the motion of the nuclei on a potential hypersurface is bounded from below even if one ignores the kinetic energy. The kinetic energy of the relative motion of the nuclei raises the lower bound only slightly (from the minimum of the potential hypersurface to the level of the zero-point motion). Therefore a (global) potential minimum usually implies that the vibronic ground state is bound as well.

There are exceptions to this. The kinetic energy of the nuclear motion can either prevent bonding — if the potential well is too shallow as compared to the kinetic energy of the zero point motion (a possible candidate for this is He_2, although it now appears that at least $^4He_2$ is very slightly bound, witP a binding energy of less than 10^{-3} K to be compared with the well depth of 10.8 K [3], while $^3He_2$ with essentially the same well depth is very likely not bound) — or it can give rise to bonding in spite of the absence of a potential well, as in the so-called vibrational bond in systems like IHI [4]. We shall here not be concerned with these exceptions.

It must be pointed out, however, that although the kinetic energy of the relative motion of the nuclei is rather irrelevant for an understanding of the chemical bond, the kinetic energy of the electrons plays a decisive role. This is so because the electronic Hamiltonian of a molecule is, like that of an atom, only bounded from below if the kinetic energy of the electrons is taken care of.

It is an interesting question (that we shall not discuss here) whether a theory of the chemical bond is meaningful only in the context of the Born-Oppenheimer separation. Primas and Müller-Herold [5] have stressed the relation between the concept of a molecular structure and the Born-Oppenheimer separation.

The theory of the long-range interaction (of van-der-Waals type) between arbitrary atoms or molecules is, at least as far as its asymptotic expansion in powers of $1/R$ is concerned, not plagued by the problem that the interacting atoms or molecules loose their individuality, and is hence formally somewhat easier. These forces are due to the electrostatic interaction of instantaneous multipoles, i.e. to interatomic correlation effects [6]. The interatomic forces decay as R^{-k}, while forces related to chemical bonding (but also to interatomic short range repulsion) go as $\exp(-\alpha R)$. Van-der-Waals type forces are also present if a chemical bond is formed. In most cases they only give a minor contribution to the binding energy. Exceptional cases are the alkali dimers and their ions, due to the extremely high polarizability of the valence electrons of alkali atoms. So, although van-der-Waals forces are not the topic of this chapter, we cannot fully ignore them.

It is often said that the chemical bond is related to spin-pairing. However, spin only plays a role in relation to the Pauli principle. The direct interaction of the spins (an interaction of two magnetic dipoles) is entirely negligible. Other magnetic effects like spin-orbit interaction do not affect the chemical bond appreciably for light atoms. If, in heavy atoms, the spin-orbit interaction gets the same magnitude as binding energies, it reduces the latter significantly (as in Tl_2 or Pb_2 [7]).

Let us summarize some of the items which make a theory of the chemical bond so difficult, that it has not yet been formulated in a general and fully satisfactory way.

1) In forming a molecule, atoms loose their individuality to a large extent. Only the atomic cores remain relatively unaffected by the bond formation.
2) The mechanism of bond formation can, in spite of its electrostatic origin, not be understood in purely electrostatic terms. The kinetic energy (of the electrons) plays an important role, which is not easily visualized.
3) The role of the kinetic energy in bond formation is paradoxical insofar as the driving force is a lowering of the kinetic energy, while at the end the kinetic energy is raised.
4) The best-studied molecule, as far the mechanism of its chemical bond is concerned, is H_2^+ [8]. Unfortunately it is not in all respects representative for covalent bonds in general.
5) If one tries to analyse other bonds than that in H_2^+, this not only turns out to be more difficult, but one easily gets the impression that any bond differs from any other one, such that it is hard to extract what is common to all (covalent) bonds [9, 10].
6) The Hellmann-Feynman theorem [11] tells us that all forces in a molecule can be understood on purely classical grounds, provided that the exact electron density in the considered molecule is known (or at least a density derived from a wave function that satisfies the Hellmann-Feynman theorem). This led some authors to think that an analysis of the electron density would be sufficient to understand the chemical bond. The real problem is, however, not to understand what happens once one knows the electron density, but rather how to get the electron density in a molecule and to reationalize its origin. (See also the end of this section).

The key to an understanding of the chemical bond is found in the work of Ruedenberg [8, 12], which extends older studies by Mulliken [13], Hellmann [14] and others [15]. Ruedenberg's original method for the analysis of the chemical bond [12] is very complicated, since it involves a large number of contributions, that are often hard to visualize. In Ruedenberg's analysis of H_2^+ [8] only three contributions turned out to be important, namely a) the quasiclassical interaction, b) interference, and c) AO-deformation (promotion). Based on this study Driessler and Kutzelnigg [16] as well as Bitter and Schwarz [10] have tried a general analysis of the chemical bond in which the same three terms play the central role. The two approaches differ in the way in which the correlation energy is treated. Bitter and Schwarz start from SCF and regard the correlation energy as quasiclassical, while Driessler and Kutzelnigg start from an energy expression that approximately takes care of the left-right correlation and guarantees the correct dissociation of the bond. A further difference is that in ref. [16] all electrons are considered, while in ref. [10] only the valence electrons in a pseudopotential are treated explicitly. The pseudopotential approach makes the analysis somewhat more transparent. While in ref. [10] the roles of kinetic and potential energy are analyzed in detail, the interest of Ref. [16] is more related to the question to which extent the chemical bond is a one-electron effect, that is generally describable by Mulliken's reduced resonance integral β.

In fact the question as to the nature of the chemical bond is closely related to the problem of the justification of simple models of chemical bonding (e.g. of Hückel or extended Hückel type) and to the physical meaning of the parameters that appear in such models. It has turned out [16] that one-electron models are more easily justified than methods that take care of the electron interaction, but in which the overlap (or differential overlap) is formally neglected. The only model with neglect of overlap, but with inclusion of the electron interaction, that can be justified theoretically, is that of Pariser-Parr-Pople for π-electron systems [54, 55, 66].

Further important work towards an understanding of the chemical bond, mainly concerning the role of the kinetic energy, has been published by Wilson and Goddard [17] and Hurley [18]. The role of electron correlation for the chemical bond has been discussed by Lievin et al. [19] and in particular by Malrieu et al. [20].

We shall, in this chapter, mainly consider the bond between two partners, i.e. in diatomic molecules. It appears that all fundamental problems arise for and can be discussed with diatomic molecules. The new aspects in polyatomic molecules are rather well understood and of less fundamental nature [21, 22].

There are two main ways towards an analysis of the chemical bond, the 'energetic approach' and the 'force approach'. In the former, which we shall follow here, one searches for a mechanism which lowers the energy in a molecule as compared to the separated atoms. In the latter one studies the forces that act on the atoms in a molecule [23–27]. In principle, since a force is the gradient of the energy, the two approaches should be closely related. However, the energetic view is the more direct and less problematic one, since even with approximate wave functions the energy expectation value $E_\Phi = \langle\Phi| H |\Phi\rangle$, with H the electronic Hamiltonian is a meaningful quantity. In the force approach one considers $f_x = \partial E_\Phi/\partial x$, where x is some internal coordinate. The correct expression

$$f_x = \left\langle\frac{\partial\Phi}{\partial x} \,|\, H \,|\, \Phi\right\rangle + \left\langle\Phi \,|\, H \,|\, \frac{\partial\Phi}{\partial x}\right\rangle + \langle\Phi|\,\frac{\partial H}{\partial x}\,|\Phi\rangle \tag{1}$$

reduces to the Hellmann-Feynman force

$$f_x^{HF} = \langle\Phi|\,\frac{\partial H}{\partial x}\,|\Phi\rangle \tag{2}$$

only if Φ is exact, i.e. solution of the Schrödinger equation — or is at least optimized in a very special sense. All discussions become somewhat problematic if one uses the Hellman-Feynman force although Φ does not satisfy the Hellman-Feynman theorem.

Another drawback of the force approach is that it does (at variance with the energetic approach) not furnish the electron density which must be got from other sources. However, the construction of the density is more crucial than its subsequent analysis.

2 The Bond in H_2^+

The chemical bond in H_2^+ has been studied in detail by Ruedenberg et al. [8] and it can be regarded as entirely understood. We follow here, with some modifications, the present author's presentation [21, 22] of the Ruedenberg analysis.

As announced in the introduction we divide the bond formation somewhat artificially into various steps that lead us from the non-interacting atoms to the molecules.

We start by considering the 'quasiclassical' interaction of the undeformed partners, which in the case of H_2^+ are an H-atom and a proton at a distance R. The wave function of the H-atom is, in Hartree units

$$a(\vec{r}) = N \cdot e^{-r_a} \tag{3}$$

with N a normalization factor and r_a the distance of the electron from nucleus A (see Fig. 1). The electron density corresponding to (3) is

$$\varrho_a(\vec{r}) = \{a(\vec{r})\}^2 = N^2 e^{-2r_a} \tag{4}$$

The energy of the H-atom (in its ground state) is —1/2 hartree, which is eigenvalue and expectation value of the Hamiltonian

$$\hat{h}_a = -\frac{1}{2}\Delta - \frac{1}{r_a} \tag{5}$$

$$\langle a|\, \hat{h}_a \,|a\rangle = -\frac{1}{2}\langle a|\, \Delta \,|a\rangle - \langle a|\, \frac{1}{r_a} \,|a\rangle = \frac{1}{2} - 1 = -\frac{1}{2} \tag{6}$$

The quasiclassical interaction energy between the H-atom and the proton is the sum of the interactions between the two protons ($1/R$) and the attraction of the electron by the 'other' proton (from which is has the distance r_b, see Fig. 1)

$$\Delta E_{QK} = \frac{1}{R} - \langle a|\, \frac{1}{r_b} \,|a\rangle = e^{-2R}\left(1 + \frac{1}{R}\right) \tag{7}$$

This interaction (plotted in Fig. 2) is positive (repulsive), because the attraction between the electron and the other nucleus is smaller than the repulsion between the two nuclei. The reason is that the 'other' nucleus B only interacts with that part of the 'electron cloud' which is inside a sphere of radius R around nucleus A. As soon as the nucleus B 'penetrates' significantly into the H-atom, it feels a repulsion. In view of the factor e^{-2R} this repulsion is of very short range.

In the next step we take care of the symmetry of the problem. In a way, the electron does not know to which of the two nuclei it 'belongs'. We must hence allow that with the same probability it is in the atomic orbital a or b (b is like (3) but with

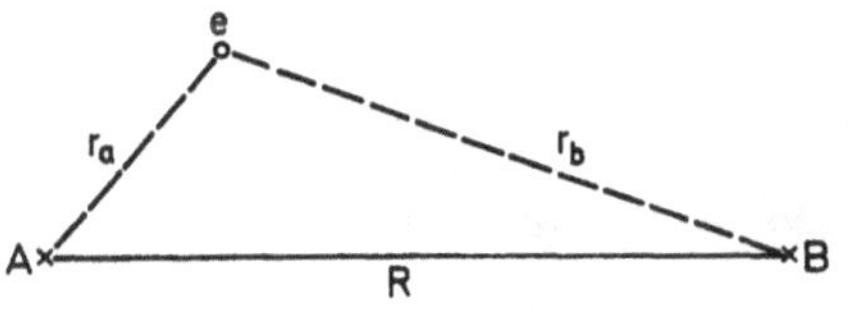

Fig. 1. Coordinates in H_2^+

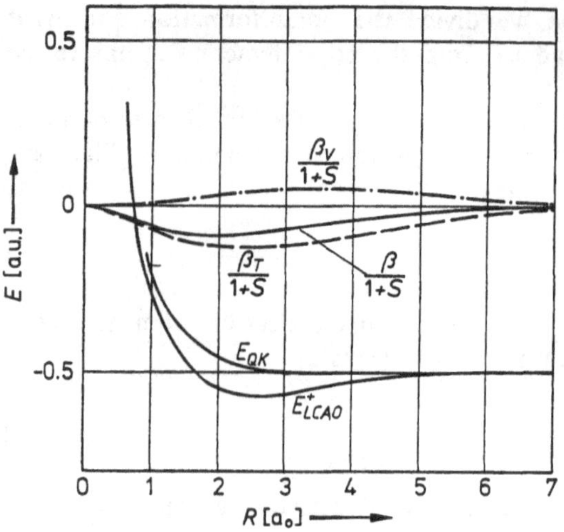

Fig. 2. Contributions to the binding energy in H_2^+: The quasiclassical interaction energy ΔE_{QK} and the interference energy $\beta/(1 + S)$ with the decomposition of the latter into kinetic (T) and potential (V) contributions

r_a replaced by r_b). The quasiclassical interaction energy is not changed by this symmetrization, since (it is part of the definition of 'quasiclassical' that) the two situations don't interfer. What changes is the electron density, which now becomes

$$\varrho_{QK} = \frac{1}{2}\{[a(\vec{r})]^2 + [b(\vec{r})]^2\} \tag{8}$$

In analogy we can construct the one-particle density matrix for this symmetrized quasiclassical structure

$$\gamma_{QK}(\vec{r}, \vec{r}') = \frac{1}{2}\{a(\vec{r})\, a(\vec{r}') + b(\vec{r})\, b(\vec{r}')\} \tag{9}$$

From the molecular Hamiltonian (see Fig. 1 for the coordinates used)

$$\hat{h} = -\frac{1}{2}\Delta - \frac{1}{r_a} - \frac{1}{r_b} + \frac{1}{R} \tag{10}$$

and the quasiclassical density matrix γ_{QK} we get the energy expectation value

$$\begin{aligned} E_{QK} = Tr\hat{h}\gamma_{QK} &= -\frac{1}{4}\{\langle a|\,\Delta\,|a\rangle + \langle b|\,\Delta\,|b\rangle\} \\ &\quad -\frac{1}{2}\left\{\langle a|\,\frac{1}{r_a}\,|a\rangle + \langle a|\,\frac{1}{r_b}\,|a\rangle + \langle b|\,\frac{1}{r_a}\,|b\rangle + \langle b|\,\frac{1}{r_b}\,|b\rangle\right\} + \frac{1}{R} \\ &= \langle a|\,\hat{h}_a\,|a\rangle + \Delta E_{QK} \end{aligned} \tag{11}$$

with ΔE_{QK} defined by (7) (where we have used that certain matrix elements are equal by symmetry).

So far we have argued in quasiclassical terms, i.e. we have ignored phase relations between the two situations (electron at nucleus A or nucleus B) and simply added densities. Now we take care of quantum mechanical interference, which implies that wave functions rather than densities have to be combined linearly. There are now two possibilities (phase relations)

$$\varphi_+ = \frac{1}{\sqrt{2(1+S)}}\{a+b\} \tag{12}$$

$$\varphi_- = \frac{1}{\sqrt{2(1-S)}}\{a-b\} \tag{13}$$

where

$$S = \langle a | b \rangle \tag{14}$$

and where the factors in front of {...} in (12, 13) care for normalization. The wave functions (12, 13) were first proposed by Pauling [28]. To each of the two choices (12, 13) corresponds a density (and a density matrix)

$$\varrho_\pm = \frac{1}{2(1 \pm S)}\{a^2 \pm 2ab + b^2\} = \varrho_{QK} \pm \frac{1}{1 \pm S}\{ab - S\varrho_{QK}\} \tag{15}$$

$$\gamma_\pm(\vec{r}, \vec{r}\,') = \gamma_{QK}(\vec{r}, \vec{r}\,') \pm \frac{1}{2(1 \pm S)}\{a(\vec{r})\, b(\vec{r}\,') + b(\vec{r})\, a(\vec{r}\,') - 2S\gamma_{QK}(\vec{r}, \vec{r}\,')\} \tag{16}$$

If we now evaluate the expectation value of the Hamiltonian (10) with the density matrix (16) we get

$$E^\pm_{LCAO} = E_{QK} \pm \Delta E^\pm_{int} \tag{17}$$

$$\Delta E^\pm_{int} = \frac{\beta}{1 \pm S}; \quad \beta = \langle a | \hat{h} | b \rangle - S\langle a | \hat{h} | a \rangle \tag{18}$$

where the subscript 'LCAO' stands for 'linear combination of atomic orbitals' and the subscript 'int' means interference, β is referred to as 'reduced resonance integral' [13].

The energy is now split into two values E^+_{LCAO} and E^-_{LCAO}. Both have the quasiclassical part E_{QK} in common, but they differ in the interference part $\Delta E^\pm_{int}$. Since β is negative (see Fig. 2), E^+_{LCAO} has the lower energy. If one plots E^+_{LCAO} as function of the distance one gets in fact a potential curve with a minimum near $2a_0$ that accounts hence for the chemical bond, while E^-_{LCAO} is repulsive everywhere. The two states φ_+ and φ_- have also different densities. In ϱ_+ there is an accumulation of electronic charge in the bond region (between the nuclei), while in ϱ_- the electronic charge in the bond region is diminished (see Fig. 3).

The interpretation of these results, found in many textbooks, that probably goes back to Slater [29] that the charge accumulated between the nuclei is under the

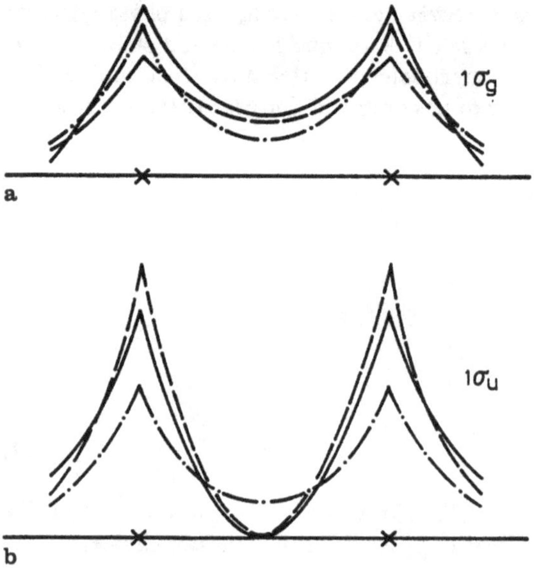

Fig. 3. The electron density along the molecular axis in the $1\sigma_g$ and $1\sigma_u$ state of H_2^+. —·—: quasiclassical, – – –: with interference, —: with interference and promotion. From Ref. 21 with kind permission of Verlag Chemie

attractive influence of the two nuclei, which lowers the potential energy of the electrons and causes bonding is, however, premature. To see what really happens, let us decompose β as defined by (18) into the contributions of kinetic and potential energy

$$\beta = \beta_T + \beta_V \tag{19}$$

$$\beta_T = -\frac{1}{2}\langle a|\Delta|b\rangle + \frac{1}{2}S\langle a|\Delta|a\rangle = -\frac{1}{3}R^2e^{-R} \tag{20}$$

$$\beta_V = -\langle a|\frac{1}{r_a} + \frac{1}{r_b}|b\rangle + S\langle a|\frac{1}{r_a} + \frac{1}{r_b}|a\rangle$$

$$= e^{-R}\left\{\frac{1}{R} - \frac{2}{3}R + \frac{1}{3}R^2\right\} - e^{-3R}\left\{\frac{1}{R} + 2 + \frac{4}{3}R + \frac{1}{3}R^2\right\} \tag{21}$$

A separate plot of $\beta_T/(1 + S)$ and $\beta_V/(1 + S)$ (Fig. 2) shows that β_T is everywhere negative with the minimum at $R = 2a_0$ which is very close to the minimum of the potential curve for the + state, while β_V is positive everywhere. This indicates that contrary to the current picture just mentioned, the potential energy is not lowered, but rather raised (though not very much), while the lowering of the kinetic energy is responsible for the chemical bond.

It is not hard to understand why the potential energy is raised. The electronic charge accumulated in the middle of the bond has been taken away from regions close to either nucleus which are more attractive than the region between the nuclei. Slater [29] failed to realize this. The lowering of the kinetic energy is less easily rationalized. The best explanation is probably that the wave function becomes flatter in the bond region in the + state, hence its gradient becomes smaller. In view of the alternative expression for the kinetic energy

$$T = \frac{1}{2}\langle \nabla\varphi \mid \nabla\varphi \rangle \tag{22}$$

a flatter wave function implies a smaller kinetic energy. Another rationalization that goes back to Hellmann [14] is to say that the uncertainty of the position is larger in H_2^+ than in an H atom, and that consequently the momentum uncertainty and hence the kinetic energy is smaller. One gets the same rationalization by assuming an analogy to the particle in a box [22, 30]. An interesting explanation of this lowering of the kinetic energy has been given by Wilson and Goddard [17] in terms of the 'contragradience' of the overlapping AOs.

It is not really important that β_V is positive, the essential message is that $|\beta_V|$ is much smaller than $|\beta_T|$, i.e. that the interference has a large effect on the kinetic energy (lowering), but a rather small one on the potential energy.

We must now perform the fourth step on the way from an H atom and a proton to the H_2^+ ion. We must deform the AOs a and b, from which we have constructed our LCAO wave function to some $\tilde{a}$ and $\tilde{b}$, in such a way that with a and b replaced by $\tilde{a}$ and $\tilde{b}$ the wave functions (12, 13) become exact H_2^+ wave functions (which requires of course, that the sets $\tilde{a}$ and $\tilde{b}$ are different for the two states). It is not possible to define $\tilde{a}$ and $\tilde{b}$ uniquely [31], but this is not important for the present discussion.

This fourth step is conveniently divided into two substeps. In the first substep (4a) we scale the atomic orbitals, i.e. we replace (3) by

$$\tilde{a}(\vec{r}) = \tilde{N}e^{-\eta r_a} \tag{23}$$

with η a variable parameter depending on R, that is determined such that it minimizes the molecular energy for fixed R.

The optimum value of η at the equilibrium distance of $2a_0$ is 1.25, while it approaches 1 for $R \to \infty$ and 2 for $R \to 0$ (see Fig. 4).

Replacing a by $\tilde{a}$ in the LCAO wave function [32] lowers the potential curve and leads to rather good agreement with the exact curve (the well depth is now 0.090 hartree below the energy of an H atom, the exact value being 0.103 hartree). Although this 'scaling' of the wave functions does not change the total energy by very much, namely by 0.04 hartree, it affects the ratio of the kinetic and the potential energy substantially. In order to understand better what happens, we change the order of the steps. Assuming that we know that $\eta_{opt} = 1.25$, for $R = 2a_0$, we take a change (promotion) of the AOs a and b to $\tilde{a}$ and $\tilde{b}$, as the new first step in bond formation and consider next the quasiclassical and interference energies.

The promotion changes the energy of an H atom from (6) to

$$\langle \tilde{a}|\, \hat{h}_a \,|\tilde{a}\rangle = \frac{1}{2}\eta^2 - \eta \tag{24}$$

While formerly the virial theorem held for the isolated atom, (which states that the expectation value of the kinetic energy is equal to minus the total energy)

$$\langle a|\, T \,|a\rangle = -\frac{1}{2}\langle a|\, \Delta \,|a\rangle = -\langle a|\, \hat{h}_a \,|a\rangle = \frac{1}{2}\,\text{hartree} \tag{25}$$

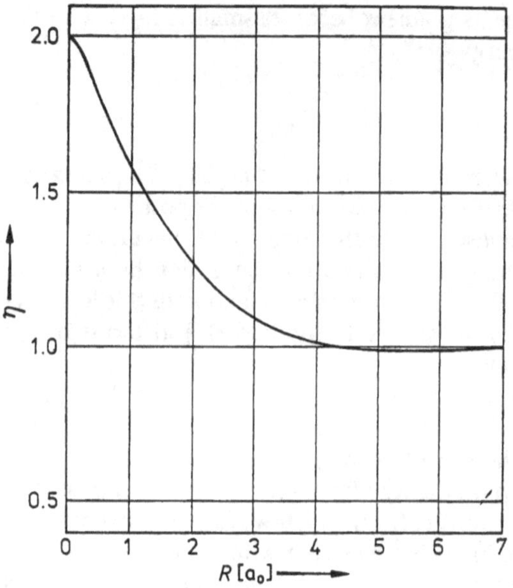

Fig. 4. Dependence of the variational parameter η in H_2^+ on the internuclear distance R. From Ref. 21 with kind permission of Verlag Chemie

we now have ($\eta = 1.25$)

$$-\frac{1}{2}\langle\tilde{a}|\,\Delta\,|\tilde{a}\rangle = \frac{1}{2}\eta^2 = 0.78125 \text{ hartree} \tag{26}$$

$$\langle\tilde{a}|\,\hat{h}_a\,|\tilde{a}\rangle = -0.46875 \text{ hartree} \tag{27}$$

The total energy (of the promoted isolated H-atom) is somewhat raised (by 0.03125 hartree) as compared to a real isolated H atom, but the kinetic energy is raised by much more (0.28125 hartree) and the potential energy is correspondingly lowered (by 0.25 hartree). There is also a change in the quasiclassical interaction energy (7). It now becomes

$$\Delta\tilde{E}_{QK} = \eta e^{-2\rho}\left(1 + \frac{1}{\rho}\right); \qquad \rho = \eta R \tag{28}$$

(One should not confuse ρ for the scaled internuclear distance with the electron density ϱ)

The change from (7) to (28) is not very large, since $|\Delta E_{QK}|$ is small anyway. Of course, the reduced resonance integral β and its components β_T and β_V are also changed. Instead of (20, 21) we now have

$$\tilde{\beta}_T = -\frac{1}{3}\eta^2\rho^2\, e^{-\rho} \tag{29}$$

$$\tilde{\beta}_V = \eta e^{-\rho}\left\{\frac{1}{3}\rho^2 - \frac{2}{3}\rho + \frac{1}{\rho}\right\} - \eta e^{-3\rho}\left\{\frac{1}{3}\rho^2 + \frac{4}{3}\rho + 2 + \frac{1}{\rho}\right\} \tag{30}$$

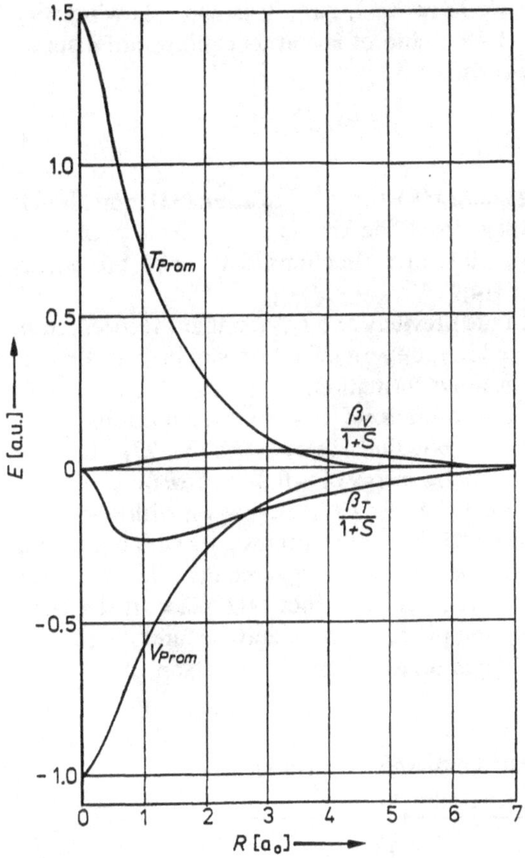

Fig. 5. The kinetic and potential energy parts of promotion and interference energy in the $1\sigma_g$ state of H_2^+. From Ref. 21 with kind permission of Verlag Chemie

The change with respect to (20,21) can be seen by comparison of Fig. 2 with Fig. 5. At distances close to the minimum of the potential curve $|\tilde{\beta}|$ is significantly larger then $|\beta|$, i.e. $\tilde{\beta} < \beta$, however, a large part of the difference between β and $\tilde{\beta}$ is compensated by the atomic promotion energy, such that the potential curve is only slightly lowered.

What has changed spectacularly, is the balance between kinetic and potential energy. While for $\eta = 1$ we had at $R = 2a_o$

$$\langle T \rangle = 0.39\,, \quad \langle V \rangle = -0.94\,, \quad E = \langle T + V \rangle = -0.55 \tag{31}$$

we now have

$$\langle T \rangle = 0.59 \quad \langle V \rangle = -1.18 \quad E = \langle T + V \rangle = -0.59 \tag{32}$$

One sees that for the scaled molecular wave function with $\eta = 1.25$ the virial theorem is satisfied, which was not the case for the simple LCAO function.

One could, of course, have argued simply in terms of the virial theorem without worrying about the various steps that we have used, namely in the following way. The exact wave functions of both one H åtom and of H_2^+ at its equilibrium distance satisfy the virial theorem in the simple form

$$E = -\langle T \rangle \tag{33}$$

Since bonding means lowering of the energy (E), the virial relation (33) immediately implies that bonding is accompanied by a rise in the kinetic energy. This holds both for the exact wave functions and any trial wave functions that make the energy stationary with respect to a scale factor [14].

This result seems to be in conflict with the previous observation that the mechanism which causes binding is a lowering of the kinetic energy. Ruedenberg [8] calls this the paradoxical role of the kinetic energy in bond formation.

On closer inspection one sees that this discrepancy is somewhat spurious. If we argue that the atoms are first prepared for bonding, we see (Eqs. 26, 27), that this step is accompanied by an increase of the kinetic energy of ~0.28 hartree (as compared to an 'unprepared' H atom), while the final rise of $\langle T \rangle$ in H_2^+ (again with respect to H) is ~0.09, such that the binding between the 'prepared' atoms goes with a lowering of $\langle T \rangle$ by ~0.19 hartree. The rise of the kinetic energy required by the virial theorem has essentially an intraatomic origin and does not take place in the bond region. In particular it does not support the picture that we have refuted, according to which the attraction of the charge accumulated in the bond region, by the two nuclei is responsible for bonding.

Table 1. The kinetic energy $\langle T \rangle$ of H_2^+ and its components $\langle T_\parallel \rangle$ parallel and $\langle T_\perp \rangle$ perpendicular to the molecular axis

	$\langle T_\parallel \rangle$	$\langle T_\perp \rangle$	$\langle T \rangle$
separated atoms $\eta = 1$	0.17	0.17	0.50
separated atoms $\eta = 1.25$	0.26	0.26	0.78
molecule $\eta = 1$	0.09	0.15	0.39
molecule $\eta = 1.25$	0.14	0.23	0.60

One can decompose the kinetic energy into components parallel and perpendicular to the bond (see Table 1) and one sees then that the rise of the kinetic energy takes only place perpendicular to the bond, while in bond direction there is a lowering of $\langle T \rangle$ even with $\eta = 1.25$.

We have, so far, only taken care of a radial deformation of the AOs (describable by a scale factor η in (23)). An angular deformation (describable by admixture of p-AOs to the s-AO) also takes place, but it is less important. It lowers the total energy by only 0.01 hartree. Still less important are radial deformations that go beyond the simple scaling.

The formation of the chemical bond in H_2^+ is obviously accompanied by orbital contraction. There has been some controversy about this point, it has especially been argued that this orbital contraction is an artifact of variational calculations [33, 34]. However, Kutzelnigg and Schwarz [35] have shown that one can define

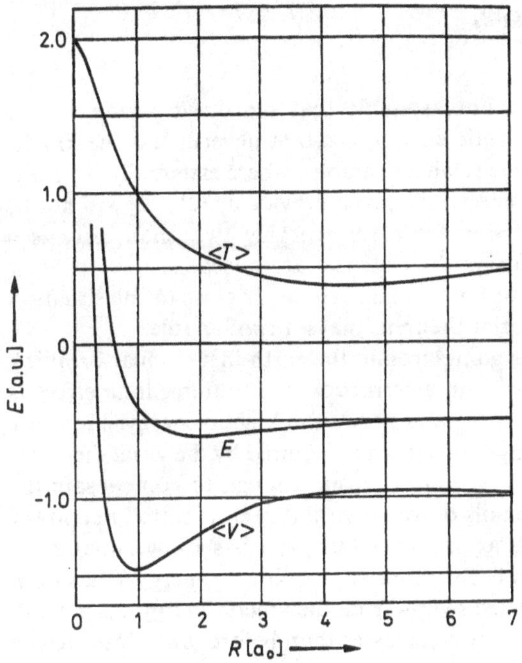

Fig. 6. The total energy (E) of H_2^+ and its decomposition into kinetic ($\langle T\rangle$) and potential energy ($\langle V\rangle$) contributions as functions of the internuclear distance R. From Ref. 21 with kind permission of Verlag Chemie

various criteria for orbital contraction and that according to all of these the exact wave function of H_2^+ displays orbital contraction.

In order to understand the mechanism of bond formation in H_2^+ one must appreciate that the kinetic energy is raised, although the driving mechanism is a lowering of the kinetic energy. In view of the importance of this we give in the next section a few examples of analogous mechanisms, to show that it is less uncommon than one might think.

The respective roles of kinetic and potential energy are even better understood if one displays them not only for the equilibrium distance, but for the whole potential curve. One sees from Fig. 6 that even for the exact H_2^+ wave functions the kinetic energy $\langle T\rangle$ is lower than for an isolated H atom for $R \gtrsim 2.5a_o$. The region of distances in which the virial theorem enforces that $\langle T\rangle$ is raised with respect to the isolated atom is rather limited. Hurley [18] has pointed out that the lowering of $\langle T\rangle$ for $R \geqq R_c (R_e < R_c < \infty)$ is a necessary and sufficient condition for the occurence of a covalent chemical bond. This result is based on the fact that for arbitrary R the virial theorem is [14]

$$2\langle T\rangle + \langle V\rangle + R\left(\frac{\partial E}{\partial R}\right) = 0 \tag{34}$$

which reduces to (33) if $(\partial E/\partial R) = 0$.

3 Other Paradoxical Mechanisms

To convince the reader that it is not unreasonable that the driving force of the bond formation is a lowering of the kinetic energy, but that nevertheless the kinetic energy is finally raised, it is useful to give related examples where essentially the same things happen. A paradigm from commercial life given previously [21, 22] will not be repeated here in detail. There the driving force for merging two enterprises is a lowering of expenses, although after fusion the expenses are raised.

Instead we will give three physical examples which are rather close to the situation in the chemical bond and where the virial theorem plays a similar role.

Let us first consider a satellite that is going around the earth in a circular orbit [8]. We want to slow down its motion by firing a retrorocket. The immediate effect is certainly a lowering of the angular velocity and hence the kinetic energy. However potential and kinetic energy have no longer the balance required by the virial theorem. In other words, the centrifugal force is no longer strong enough to compensate the attraction by the earth and the satellite falls down — until it gets stabilized in a lower orbit, in which the virial theorem holds again. Since this orbit has a lower (negative) total energy than originally, i.e. $|E|$ has been increased, the kinetic energy T has been increased as well, although we have started by lowering the kinetic energy. As a result of being slowed down the satellite finally rotates faster than before (and closer to the earth).

The next example is even closer to that of the chemical bond, namely an illustration of Hund's rule. We want to explain e.g. the energy difference between the two lowest ${}^1P$ and ${}^3P$ states of the configuration $1s2p$ of the He atom. We can define a quasiclassical energy of these two states (identical with the 'average-of-configuration' energy) as consisting of the energy of either electron in the field of the nucleus and the electrostatic repulsion of the electron clouds corresponding to the two electrons. (as to the meaning of the electron interaction integral $(1s1s \mid 2p2p)$ see Eq. (57)).

$$E_{QK} = \langle 1s| -\frac{1}{2}\Delta - \frac{2}{r}|1s\rangle + \langle 2p| -\frac{1}{2}\Delta - \frac{2}{r}|2p\rangle + (1s\,1s|2p\,2p) \quad (35)$$

we can e.g. optimize the AOs $1s$ and $2p$ such that they minimize the energy (35) (average-of-configuration Hartree-Fock). To simplify things we take simple Slater approximations to these AOs

$$1s = Ne^{-\alpha r} \quad (36)$$

$$2p = N're^{-\beta r}Y_1^m(\vartheta, \varphi) \quad (37)$$

where N and N' are normalization factors and where m can be 1,0 or -1, with E independent of the choosen m. With the choice (36, 37) we get for (35) (see e.g. ref. 52)

$$E_{QK} = \frac{1}{2}\alpha^2 - 2\alpha + \frac{1}{2}\beta^2 - \beta + \frac{\xi}{32}(1-\tau^2)(14 - 7\tau - \tau^2 + 3\tau^3 - \tau^4) \quad (38)$$

with

$$\tau = \frac{\alpha - \beta}{\alpha + \beta}; \qquad \xi = \frac{\alpha + \beta}{1} \tag{39}$$

Slater's rules [36] predict $\alpha = 2$, $\beta = (2 - 0.85)/2 = 0.575$ and hence $\xi = 1.2875$, $\tau = 0.5534$. Rather than to determine both ξ and r such that they minimize (38) we choose τ according to Slater's rules and minimize E with respect to ξ. The result is $\xi_{opt} = 1.2752$. With these values of ξ_{opt} and r we get

$$E_{QK} = -2.1242 \text{ hartree} \tag{40}$$

Taking care of interference (phase relations) we get a splitting between the singlet and the triplet states

$$E(^1P) = E_{QK} + (1s2p \mid 2p1s); \qquad E(^3P) = E_{QK} - (1s2p \mid 2p1s) \tag{41}$$

$$(1s2p \mid 2p1s) = \frac{7}{96}\xi(1 + \tau)^3 (1 - \tau)^5 = 0.00625 \text{ hartree} \tag{42}$$

The singlet/triplet splitting is hence 0.0125 hartree and the triplet has the lower energy since the exchange integral is positive. We get

$$E(^3P) = -2.13046 \text{ hartree}; \qquad E(^1P) = -2.11796 \text{ hartree} \tag{43}$$

The difference in the energy of the two states is here entirely due to a difference in the potential energy (more specifically the electron interaction energy). Both states have, on this level of approximation, the same kinetic energy. Of course, neither state satisfies the virial theorem. This only holds for E_{QK}. We can achieve that the virial theorem holds, by choosing a different ξ for two states. The result is

$$\xi_{opt}(^3P) = 1.277; \qquad \xi_{opt}(^1P) = 1.233 \tag{44}$$

The orbitals in the triplet state are somewhat more compact, hence closer to the nucleus. The corresponding improved energies agree with (43) to all significant figures. However, if we now want to analyse the difference of the energies of the two states in terms of the corresponding changes of the expectation values of the kinetic energy T, the electron nuclear potential V_{ne} and the electron-electron potential V_{ee}, we get (in hartrees)

$$\Delta\langle T\rangle = +0.01249, \quad \Delta\langle V_{ne}\rangle = -0.01333; \quad \Delta\langle V_{ee}\rangle = -0.01167 \tag{45}$$

According to this breakdown (45), it appears that all three terms contribute by about the same extent to the singlet-triplet splitting.

In the next step we allow for radial deformations of the AOs (independently for the two states), that go beyond simple scaling (this had little effect for bonding in H_2^+). Now the energies are further lowered to −2.13143 and −2.12246 hartree respectively and (45) is replaced by [37, 38]

$$\Delta\langle T\rangle = +0.0093\ ;\quad \Delta\langle V_{ne}\rangle = -0.0416\ ;$$
$$\langle V_{ee}\rangle = +0.0225 \tag{46}$$

i.e. the paradoxical result is obtained that $\langle V_{ee}\rangle$ is larger in the ^3P state than in the ^1P state although the basic mechanism at the origin of Hund's rule has been that $\langle V_{ee}\rangle$ is smaller in the triplet state. What happens is that in the ^3P state, because the Pauli principle prevents an electron in a $2p$ AO from coming too close to an electron in the $1s$-AO, the AOs can adjust in such a way that on the average the two electrons can come closer to each other and to the nucleus, while in the ^1P state the electrons must be more distant on the average from each other and from the nucleus, since the Pauli principle does not prevent them from coming close to each other at the same time.

On the last step — inclusion of electron correlation — qualitatively not much is changed. We now get [37, 38]

$$\Delta\langle T\rangle = +0.00932\ ;\quad \Delta\langle V_{ne}\rangle = -0.04024\ ;$$
$$\Delta\langle V_{ee}\rangle = +0.02162 \tag{47}$$

Some authors [39, 40] have concluded from these results that the traditional interpretation of Hund's rule is incorrect and that the energetic order of the states is determined by the contribution of $\langle V_{ne}\rangle$ i.e. the nuclear-electronic interaction term. The observation that $\Delta\langle V_{ne}\rangle$ is negative is, however, by no means helpful for an understanding of Hund's rule.

One has a chance to understand Hund's rule only if one accepts that its explanation is necessarily paradoxical. The Pauli principle prevents the electrons in the state with highest spin-multiplicity to come close to each other instantaneously. This however, allows them to get closer to each other on the average. So although the driving mechanism for the lower energy of this state is a reduction of the electron repulsion, the repulsion energy between the electrons is finally increased.

It is obvious that, if one wants to understand why of two states with the same configuration one of them has the lower energy, one must take the meaning of the 'same configuration' literally, i.e. describe both states by the same orbitals. On this level the traditional explanation of Hund's rule — in terms of the sign of the exchange integral only — is valid. As a second step one must realize that the two states have (necessarily) the same configuration only approximatively (and that even the concept of configuration is only an approximation for either state).

In the appendix we give an alternative rationalization of Hund's rule in terms of the concept of effective Hamiltonians. In this context an exchange-type interaction turns out to be decisive, as in the naive interpretation of Hund's rule.

The third example is the so-called relativistic contraction of the cores [10]. One observes in fact that the mean radius of 1s-AOs in heavy atoms is reduced with a relative shrinking $\sim Z^2/c^2$ where in Hartree units c is equal to the inverse of the fine structure constant ($c \approx 137$).

If one wants to find a simple explanation of this contraction, one must compare the relativistic and nonrelativistic Hamiltonians. To explain the relativistic contraction of the K shell the spindependent terms in the relativistic Hamiltonian are rather irrelevant and one can account for this contraction by considering simply the relativistic kinematics. This means the operator of the kinetic energy

$$T = \frac{p^2}{2m} \tag{48}$$

has to be replaced by

$$T_{\text{rel}} = \sqrt{m^2c^4 + p^2c^2} - mc^2 = mc^2\left\{\sqrt{1 + \frac{2T}{mc^2}} - 1\right\} = T - \frac{T^2}{2mc^2} + \ldots \tag{49}$$

It is easy to see that the operator of the relativistic kinetic energy is smaller than the nonrelativistic kinetic energy — this means that for any wave function belonging to the domain of both T and T_{rel}

$$\langle\varphi|\, T_{rel}\, |\varphi\rangle < \langle\varphi|\, T\, |\varphi\rangle \tag{50}$$

while the operator of the potential energy remains unchanged. This lowering of the expectation value of the kinetic energy allows the electron to come closer to the nucleus, since there is less centrifugal repulsion, and it causes also the electron to be more strongly bound. Although the operator of the kinetic energy has become smaller, the electron adjusts to this situation like in the previous examples in such a way that the kinetic energy is finally larger in the relativistic than in the non-relativistic case.

Bitter et al. [10] give a few more physically very interesting examples of the same type. They stress that one should clearly distinguish between the actual value of an observable and the same observable as function of a model parameter.

4 The Bond in H_2

We take the same steps as in H_2^+. The first of them involves the quasiclassical interaction between two H atoms. For this we need the quasiclassical density and density matrix

$$\varrho_{QK}(\vec{r}) = \{a(\vec{r})\}^2 + \{b(\vec{r})\}^2 \tag{51}$$

$$\gamma_{QK}(\vec{r}, \vec{r}') = a(\vec{r})\, a(\vec{r}') + b(\vec{r})\, b(\vec{r}') \tag{52}$$

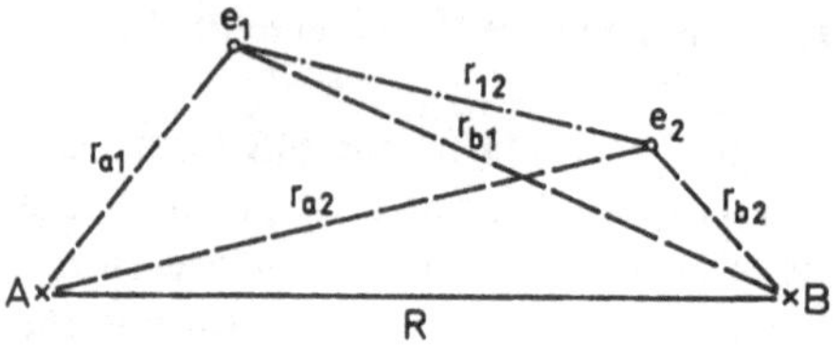

Fig. 7. Coordinates in H_2

and the pair density

$$\pi_{QK}(\vec{r}_1, \vec{r}_2) = \frac{1}{2}\{a(\vec{r}_1)\, b(\vec{r}_2)\}^2 + \frac{1}{2}\{b(\vec{r}_1)\, a(\vec{r}_2)\}^2 \tag{53}$$

The latter expresses the fact that either the first electron 'belongs' to atom A and the second to atom B or vice versa. The construction of (52) implies a symmetrization in analogy to (8).

With the Hamiltonian (see Fig. 7 for the coordinates)

$$H(1, 2) = -\frac{1}{2}\Delta_1 - \frac{1}{2}\Delta_2 - \frac{1}{r_{a1}} - \frac{1}{r_{b1}} - \frac{1}{r_{a2}} - \frac{1}{r_{b2}} + \frac{1}{r_{12}} + \frac{1}{R} \tag{54}$$

we get the following quasiclassical energy [21]

$$E_{QK} = 2\langle a|\, h_a\, |a\rangle + \Delta E_{QK} \tag{55}$$

$$\Delta E_{QK} = -2\,\langle a|\, \frac{1}{r_b}\, |a\rangle + \frac{1}{R} + (aa\,|\,bb) = \left(\frac{1}{R} + \frac{5}{8} - \frac{4}{3}R - \frac{1}{6}R^2\right)e^{-2R} \tag{56}$$

where $\langle a|\, h_a\, |a\rangle = -1/2$ is the energy of an H atom and where we have used the Mulliken-notation [13] for the electron interaction integral

$$(pq\,|\,rs) = \langle p(1)\, r(2)|\, \frac{1}{r_{12}}\, |q(1)\, s(2)\rangle \tag{57}$$

The quasiclassical interaction is, as expected, the sum of the attraction of one electron at one atom by the other nucleus, the repulsion of the nuclei and the repulsion of the two electrons. The total quasiclassical energy E_{QK} is the sum of ΔE_{QK} and the energy (−1 hartree) of two isolated H atoms. This is plotted in Fig. 8.

ΔE_{QK} is strongly repulsive for small R, but it becomes attractive for $R \gtrsim 1.4a_o$ with a minimum of −0.0196 hartree at $R \approx 1.9a_o$. Since the exact potential curve of H_2 has a well depth of −0.174 hartree (at $R = 1.4a_o$), the quasiclassical interaction accounts only for a minor part of the binding energy in H_2. The difference of ΔE_{QK} in H_2^+ (for which ΔE_{QK} is repulsive everywhere with $\Delta E_{QK} = +0.104$ hartree at $R = 1.4a_o$ and $\Delta E_{QK} = 0.034$ hartree at $R = 1.9a_o$) is, however, responsible in part for the much shorter equilibrium distance in H_2 ($1.4a_o$) as compared to H_2^+ ($2.0a_o$).

The next step is to take care of a phase relation and hence of interference. A straight-

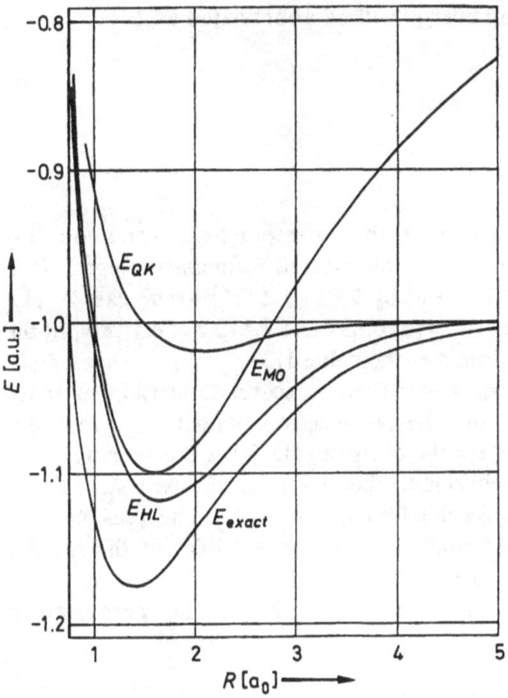

Fig. 8. The quasiclassical energy and the potential curve of H_2 in various approximations

forward way to do this (though not the only one possible) is to choose the Heitler-London (HL) wave function [41], which is the prototype case of the valence-bond (VB) approximation.

$$\psi^{\pm}_{HL}(1,2) = \frac{1}{\sqrt{2(1 \pm S^2)}} \{a(1)\, b(2) \pm b(1)\, a(2)\} \tag{58}$$

which leads to the density and density matrix

$$\varrho^{\pm}_{HL} = \varrho_{QK} \pm \Delta\varrho^{\pm} \tag{59}$$

$$\Delta\varrho^{\pm} = \frac{S}{1 \pm S^2} \left\{a(\vec{r})\, b(\vec{r}) - \frac{S}{2}\, [a(\vec{r})]^2\, \frac{S}{2}\, [b(\vec{r})]^2\right\} \tag{60}$$

$$\gamma^{\pm}_{HL}(\vec{r}, \vec{r}\,') = \gamma_{QK}(\vec{r}, \vec{r}\,') \pm \Delta\gamma^{\pm}(\vec{r}, \vec{r}\,') \tag{61}$$

$$\Delta\gamma^{\pm} = \frac{S}{2(1 \pm S^2)} \{a(\vec{r})\, b(\vec{r}\,') + b(\vec{r})\, a(\vec{r}\,') - S[a(\vec{r})\, a(\vec{r}\,') + b(\vec{r})\, b(\vec{r}\,')]\} \tag{62}$$

further the pair density

$$\pi^{\pm}_{HL} = \pi_{QK} \pm \Delta\pi^{\pm} \tag{63}$$

$$\Delta\pi^{\pm} = \frac{1}{1 \pm S^2} \{a(1)\, b(1)\, a(2)\, b(2) - S^2 \pi_{QK}\} \tag{64}$$

The contribution of $\Delta\pi^{\pm}$ to the binding energy will be abbreviated as $E^{(2)}$.

$$E_{HL}^{\pm} = E_{QK} + E_{int}^{\pm} + E^{(2)} \tag{65}$$

$$E_{int}^{\pm} = \pm \frac{2S\beta}{1 \pm S^2} \tag{66}$$

We first neglect $E^{(2)}$ and compare (66) with the corresponding expression (18) for H_2^+. Both expressions contain the factor β, the reduced resonance integral. It is multiplied by $(1 \pm S)^{-1}$ in the case of H_2^+ and by $2S(1 \pm S^2)^{-1}$ in the case of H_2. At $r = 2a_o$ these two factors are (for the + sign) 0.631 and 0.872 respectively, in the limit $R \to 0$ the ratio of the factors approaches the value 1/2.

The comparison of (18) and (66) reveals that there is no fundamental difference between the one-electron bond in H_2^+ and the two-electron bond in H_2. In either case the reduced resonance integral β plays the decisive role. That β is negative, and that hence binding takes place, is entirely due to the kinetic energy part β_T. Like in H_2^+ lowering of the kinetic energy is responsible for the bond in H_2. The quasiclassical energy stabilizes H_2 with respect to H_2^+ and is in part responsible for the smaller equilibrium distance in H_2 ($1.4a_o$) than in H_2^+ ($2.0a_o$).

Let us now consider the contribution of $\Delta\pi^{\pm}$ (63,64) to the binding energy, which we have so far neglected. It is

$$E^{(2)} = \pm \frac{1}{1 \pm S^2} \{(ab \mid ba) - S^2(aa \mid bb)\} \tag{67}$$

The exchange integral $(ab \mid ba)$ leads to a very complicated expression, that is hard to analyze. It has turned out, however, that the Mulliken approximation [13] is rather accurate for integrals between H-atom wave functions. This approximation consists of replacing a general two electron integral $(pq \mid rs)$ by

$$(pq \mid rs) \approx \frac{1}{4} S_{pq} S_{rs} \{(pp \mid rr) + (qq \mid rr) + (pp \mid ss) + (qq \mid ss)\} \tag{68}$$

If we apply the Mulliken approximation (68) to the expression (67) we get

$$E^{(2)} \approx \pm \frac{S^2}{2(1 \pm S^2)} \{(aa \mid aa) - (aa \mid bb)\} \tag{69}$$

For the + state (+ sign in (58)) this expression is positive (repulsive). In the interesting region it varies between 0.01 and 0.02 hartree.

Neglect of $E^{(2)}$ lowers the potential curve and makes it closer to the exact one, an example how by a fortunate cancellation of errors one can improve quantum chemical calculations.

As for H_2^+ we can now perform the *4th* step, namely preparation of the AOs for bonding by scaling and further deformation. Scaling has essentially the same (but quantitatively a smaller) effect as in H_2^+ and need not further be discussed (the optimum is now $\eta = 1.19$). While in the case of H_2^+ the further (angular) deformation (that was small anyway) leads to the exact wave function, this is not so in H_2. Here one can only arrive at the best possible wave function of the form (58) with a and b replaced by $\tilde{a}$ and $\tilde{b}$, so to say the optimized Heitler-London function. Such a function can be rewritten in an alternative so-called two configuration form [42]

$$\psi = c_1\chi_1(1)\,\chi_1(2) - c_2\chi_2(1)\,\chi_2(2) \tag{70}$$

The two orbitals χ_1 and χ_2 are orthonormal (while $\tilde{a}$ and $\tilde{b}$ are non-orthogonal, with the overlap integral S) and are related to $\tilde{a}$ and $\tilde{b}$ as

$$\chi_1 = \frac{1}{\sqrt{2(1+S)}}(\tilde{a}+\tilde{b}), \qquad \chi_2 = \frac{1}{\sqrt{2(1-S)}}(\tilde{a}-\tilde{b}) \tag{71}$$

$$c_1 = \frac{1+S}{\sqrt{2(1+S^2)}}; \qquad c_2 = \frac{1-S}{\sqrt{2(1+S^2)}} \tag{72}$$

Obviously χ_1 is an MO of σ_g symmetry and χ_2 of σ_u symmetry and (70) is the simplest case of a CI (configuration interaction) wave function. One sees from (72) that for $R \to \infty$, where $S \to 0$ the two coefficients c_1 and c_2 approach the same value $1/\sqrt{2}$, while for $R \to 0$ and $S \to 1$, c_1 approaches 1 and c_2 tends to zero.

At the equilibrium distance c_2 is sufficiently small as compared to c_1, such that a discussion of the chemical bond can also be based on a single MO configuration $\chi_1(1)\,\chi_1(2)$.

With the best single configuration (MO theory) one takes care of 77% of the binding energy, with the best two configuration wave function (70) of 87%. To account for the remaining 13% a configuration interaction ansatz (natural expansion) [42].

$$\psi = \sum_k c_k\chi_k(1)\,\chi_k(2) \tag{73}$$

is necessary. This means that an important part of the chemical bond in H_2 comes from electron correlation, i.e. a correlation of the motion of the two electrons, that is not describable in terms of an MO or a VB wave function. The electrons try to avoid each other and thus their repulsion is reduced as compared to that of two electron clouds.

We have here based the analysis of the bond in H_2 on the VB (Heitler-London) function, according to Driessler and Kutzelnigg [16]. This has, as compared to the alternative based on the single configuration MO function, favoured e.g. by Schwarz and Bitter [10], the advantage that the two-electron contribution to the binding energy is very small, so that its neglect can be justified. Further the part of the correlation energy that is not explicitly considered, amounts to only 13% of the binding energy (vs 23% in the MO approach). Moreover, neglect of $E^{(2)}$ almost compensates for the neglect of this part of correlation energy. A drawback of this analysis is that a generalization to many-electron molecules is difficult, though not impossible [16]. An appealing feature of an MO-based analysis is, at first glance that the interference contribution in H_2 is exactly twice that in H_2^+. Now the problem is to explain why the binding energy in H_2 is actually less than twice that in H_2^+. It turns out that this is due to a two-electron effect, called 'sharing penetration' [12] by Ruedenberg. The MO wave function allows both electrons to be in the same AO (admixture of ionic structures) and this raises the electronic repulsion. The change of the electronic repulsion with respect to the quasiclassical one is no longer negligible and this complicates the analysis. The main disadvantage of the MO approach is that the MO wave function does not describe the dissociation of H_2 to two H-atoms for $R \to \infty$ correctly. In the analysis of Driessler and Kutzelnigg the MO formalism is corrected so that the bonds dissociate correctly [16].

5 The Repulsion of Closed Shells

While two H atoms bind to an H_2 molecule, an He_2 molecule does not exist. There is a weak van-der-Waals type long-range attraction $\sim R^{-6}$ between two He atoms, that we will not study here, and a strong short-range repulsion. The van-der-Waals minimum arises where the attractive and repulsive forces cancel each other (at $5.6a_o$) with a well-depth of 10 meV.

To analyze the closed-shell repulsion between two He atoms we proceed as in the case of the chemical bond in H_2^+ or H_2. We start with the quasiclassical energy. To formulate this we need the quasiclassical density, density matrix and pair density. We cannot formulate this exactly as we did for H_2^+ or H_2, since the ground state wave function of an He atom is not known in closed form. We therefore use an independent particle wave function for the He atom, which is sufficient for a qualitative discussion

$$\varphi_A(\vec{r}_1, \vec{r}_2) = a(\vec{r}_1)\, a(\vec{r}_2) \tag{74}$$

where a can be the Hartree-Fock AO of He or a simple approximation to it, e.g. a wave function of the form (23) with $\eta = 1.6875$. The quasiclassical density etc. corresponding to this type of wave function for atom A and the analogous one for the atom B are

$$\varrho(\vec{r}) = 2[a(\vec{r})]^2 + 2[b(\vec{r})]^2 \tag{75}$$

$$\gamma(\vec{r}, \vec{r}') = 2a(\vec{r})\, a(\vec{r}') + 2b(\vec{r})\, b(\vec{r}') \tag{76}$$

$$\begin{aligned}\pi(\vec{r}_1, \vec{r}_2) = {} & [a(\vec{r}_1)\, a(\vec{r}_2)]^2 + [b(\vec{r}_1)\, b(\vec{r}_2)]^2 + [a(\vec{r}_1)\, b(\vec{r}_2)]^2 \\ & + [b(\vec{r}_1)\, a(\vec{r}_2)]^2\end{aligned} \tag{77}$$

The Hamiltonian is

$$H(1,2,3,4) = -\frac{1}{2}\sum_{k=1}^{4} \Delta_k - \sum_{k=1}^{4}\sum_{\mu=1}^{2} \frac{2}{r_{\mu k}} + \sum_{k<l=1}^{4} \frac{1}{r_{kl}} + \frac{4}{R} \tag{78}$$

where the labels k, l count electrons and μ nuclei.

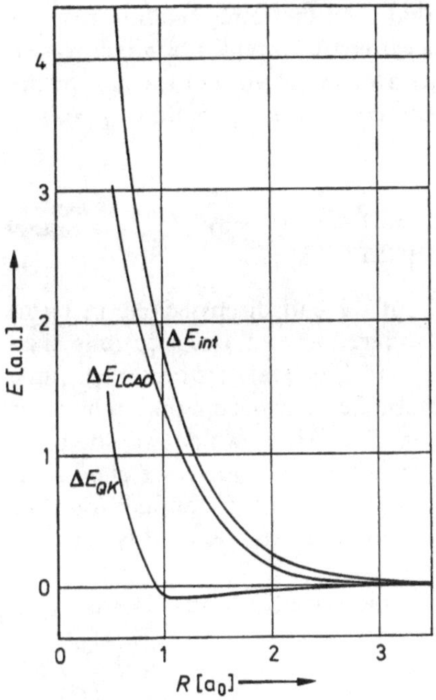

Fig. 9. The interference ΔE_{int} and quasiclassical (ΔE_{QK}) contributions to the MO-LCAO interaction energy (ΔE_{LCAO}) in He_2

The quasiclassical energy in terms of (75) to (77) is

$$E_{QK} = 2E(\mathrm{He}) + \Delta E_{QK} \tag{79}$$

$$\Delta E_{QK} = -\,8\langle a|\frac{1}{r_b}|a\rangle + 4(aa\,|\,bb) + \frac{4}{R} \tag{80}$$

The quasiclassical interaction energy ΔE_{QK} is 4 times that between two hypothetical H atoms — with the orbital exponent η valid for He. Like in the case of the interaction of two H atoms, ΔE_{QK} is slightly attractive, here for $R \gtrsim 0.9a_o$ and repulsive for shorter distances. ΔE_{QK} is plotted in Fig. 9.

The next step in the cases of H_2^+ and H_2 has been to add wave functions (rather than densities) with a definite phase relation. This has been possible, since one can formulate 1 and 2-electron problems spinfree and without worrying about the Pauli principle, i.e. the antisymmetry of the wave function. For a one-electron problem this is trivially so, while for a two-electron problem a function symmetric with respect to exchange of the coordinates of the electrons can be multiplied with an antisymmetric spin function (singlet), and a function antisymmetric in space with a symmetric spin function (triplet), to yield an overall antisymmetric function. In the case of two He atoms, described by independent particle wave functions, the total wave function can be formulated approximatively as a Slater determinant.

$$\Phi(1,2,3,4) = \frac{1}{\sqrt{4!}} \begin{vmatrix} \varphi_1\alpha(1) & \varphi_1\beta(1) & \varphi_2\alpha(1) & \varphi_2\beta(1) \\ \varphi_1\alpha(2) & \varphi_1\beta(2) & \varphi_2\alpha(2) & \varphi_2\beta(2) \\ \varphi_1\alpha(3) & \varphi_1\beta(3) & \varphi_2\alpha(3) & \varphi_2\beta(3) \\ \varphi_1\alpha(4) & \varphi_1\beta(4) & \varphi_2\alpha(4) & \varphi_2\beta(4) \end{vmatrix} \tag{81}$$

where φ_1 and φ_2 are two orthogonal functions that span the same function space as a and b. To use a and b directly in (81) — with a different normalization factor — is possible (and mathematically equivalent), but inconvenient in view of the non-orthogonality of a and b. The simplest choice of φ_1 and φ_2 are the normalized + and — linear combinations, i.e. the bonding and antibonding MOs

$$\varphi_1 = \frac{1}{\sqrt{2(1+S)}}(a+b); \qquad \varphi_2 = \frac{1}{\sqrt{2(1-S)}}(a-b) \tag{82}$$

The difference to the cases of H_2^+ and H_2 is (if we had discussed H_2 in terms of the MO rather than the VB function) that there we had less electrons than available spin-orbitals, such that we could choose which orbitals are occupied, while we now have as many electrons as spin orbitals, i.e. the occupation scheme is predetermined. There is only one possible state. (For He_2^+, which we shall not discuss, there are only three electrons and four spin orbitals, we can hence choose the occupation scheme as either $1\sigma_g^2 1\sigma_u$ or $1\sigma_g 1\sigma_u^2$. For the former configuration we get a rather strong bond, like in H_2^+, since the bonding MO is doubly and the antibonding MO singly occupied).

The density and density matrix corresponding to the wave function (81) are

$$\varrho(\vec{r}) = 2[\varphi_1(\vec{r})]^2 + 2[\varphi_2(\vec{r})]^2 = \varrho_{QK} + \Delta\varrho \tag{83}$$

$$\Delta\varrho = \left\{ab - \frac{S}{2}(aa+bb)\right\}\left\{\frac{1}{1+S} - \frac{1}{1-S}\right\} = -\frac{2Sab - S^2(aa+bb)}{1-S^2} \tag{84}$$

$$\gamma(\vec{r}, \vec{r}') = 2\varphi_1(\vec{r})\,\varphi_1(\vec{r}_1') + 2\varphi_2(\vec{r})\,\varphi_2(\vec{r}') = \gamma_{QK} + \Delta\gamma \tag{85}$$

$$\Delta\gamma = -\frac{S\{a(\vec{r})\,b(\vec{r}') + b(\vec{r})\,a(\vec{r}') - Sa(\vec{r})\,a(\vec{r}') - Sb(\vec{r})\,b(\vec{r}')\}}{1-S^2} \tag{86}$$

The contribution of the change of the pair density is small and we ignore it.

Although we have not explicitly imposed a phase relation between the wave function of the two He atoms, the Pauli principle leads to interference terms nevertheless. We get again that the reduced resonance integral β determines the interference part ΔE_{int} of the interaction energy.

$$E_\Phi = E_{QK} + \Delta E_{int} \tag{87}$$

$$\Delta E_{int} = -\frac{4S\beta}{1-S^2} \tag{88}$$

This term is positive (since β is negative) and it explains in fact the bulk of the repulsion between the two He atoms. The expression (88) is, by the way, formally the same as twice the interaction in the lowest triplet state ψ_{HL}^- of the H_2 molecule. In that state, in a MO description both φ_1 and φ_2 are singly occupied, while in He_2 they are doubly occupied.

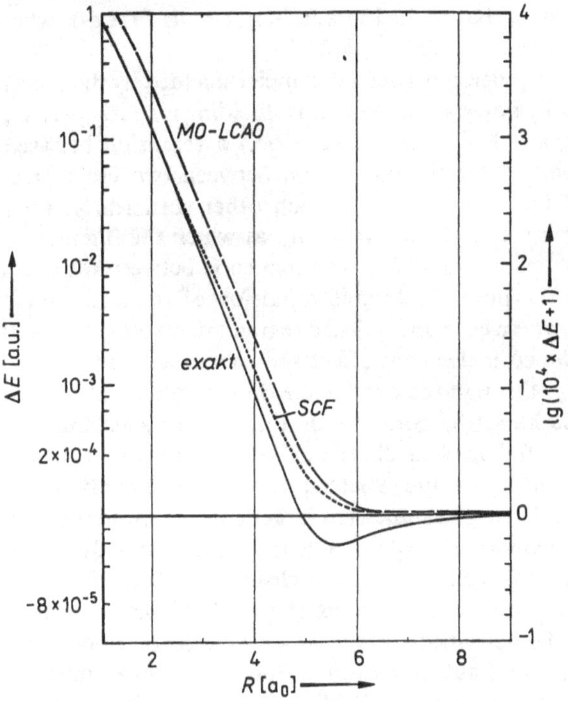

Fig. 10. The interaction energy between two He atoms in MO-LCAO, optimized SCF (Hartree-Fock) approximations and exact. From Ref. 21 with kind permission of Verlag Chemie

One can say that the repulsion between two He atoms results from the fact that the antibonding MO is — in view of the factor $(1 - S)^{-1}$ — more strongly antibonding than the bonding MO, with the factor $(1 + S)^{-1}$, is bonding.

Since the reduced resonance integral β is responsible for this repulsion, the underlying mechanism is formally the same as in H_2^+ or H_2. A rise of the kinetic energy (as in the antibonding state of H_2^+) is the driving force for the repulsion, but after a radjustment of the AOs of He (or promotion) the total kinetic energy is finally lowered.

We see in Fig. 9 that the dominant contribution to the MO-LCAO-interaction energy is ΔE_{int}, we see further on Fig. 10 that the simple MO-LCAO energy of He_2 represents qualitatively the repulsion between two He atoms. For small R an almost quantitatively correct description is obtained in the SCF approximation, i.e. by choosing the best possible wave functions of the form (81). To go beyond SCF, i.e. to account for electron correlation effects is mainly necessary in order to describe the van-der-Waals minimum. Note that the repulsion energy is very close to $\sim \exp(-\alpha R)$ over a wide range of distances.

6 The Bond in Li_2^+ and Li_2

In this section we follow in part unpublished work of Bitter, Schwarz et al. [10]. Li has a closed shell core ($1s^2$) and a valence electron ($2s$). It differs from H mainly in the presence of the core. A further difference is that the $2p$-AO is energetically quite close to $2s$, which makes the $2s$-AO highly polarizable. The dipole polarizability of the H atom ground state is 4.5 atomic units, that of Li 165 a.u. [43]. Let us first ask what consequences the presence of the core has.

There is, of course, the Coulomb repulsion between the nuclei shielded by the cores. Since the core electrons are practically non-overlapping, this shielding is nearly perfect, i.e. the cores have effective charges $+1$. One may also expect a repulsion between the cores like that discussed in Sect. 5 for the interaction between two He atoms. However this takes only place if the cores approach each other sufficiently, such that they start to overlap. This arises at distances $R \lesssim 1a_o$, at which the interaction between the valence-AOs is already highly repulsive. The repulsion between the cores is hence pratically identical with the pure Coulombic repulsion of point-like ionic cores. (Only for very heavy atoms where core and valence region are no longer so well separated, can the repulsion of the core electrons affect the chemical bond).

The cores only matter indirectly, through their influence on the valence electrons. This influence is twofold. On one hand the core shields the nucleus such that the valence electron does not 'see' the full nuclear charge ($Z = 3$). Very far from the nucleus it 'sees' the charge 1, but since the electron has some probability to penetrate into the core, it sees on the average a somewhat larger charge, approximately that given by Slater's rules [36], namely $Z_{\text{eff}} = 1.3$. On the other hand the Pauli principle prevents the valence electrons from coming too close to the core electrons. This is usually accounted for by orthogonalizing the valence-AOs to the core-AOs. Although this exclusion effect can be described strictly only by means of a non-local operator, a local approximation to the Pauli repulsion is possible, as has first been suggested by Hellmann [44]. One can combine the Pauli repulsion together with the incomplete shielding to a so-called 'pseudopotential', which is 'felt' by the valence electron.

Keeping this in mind we can proceed as in the case of H_2^+ or H_2, just replacing the nuclear potential $1/r_a$ by an effective core potential $V(r_a)$ that includes the pseudopotential. The quasiclassical step and the interference step are then formally very similar to those in H_2^+ or H_2. Again the reduced resonance integral β is decisive. What is changed, however, is the decomposition of β into the potential and kinetic contributions β_V and β_T. As for H_2^+ or H_2, β_T is negative, i.e. interference lowers the kinetic energy. However, while β_V was slightly positive for H_2^+ and H_2, for Li_2^+ and Li_2 even β_V is negative. The argument, refuted for H_2^+ and H_2, that accumulation of charge in the binding region is energetically favorable, since it is attracted by both nuclei, appears now valid. The difference is that in the H-atom as well as in H_2^+ or H_2 the 'best place' for a valence electron from the point of view of the potential energy is the region close to the nucleus, whereas this is no longer so for the valence electron in Li. Because of the Pauli repulsion it cannot approach the nucleus very closely, while the region where the pseudopotential has its minimum, is rather flat and not very pronounced. In Li_2^+ and Li_2 the 'best place' for the electrons is in fact the region between the two nuclei, both from the point of view of the potential and the kinetic energy.

The next step has been adjustment of the MO by scaling (or preparation for bonding, if one reverts the order of the steps). In the cases of H_2^+ and H_2 optimum scaling led to a wave function that satisfies the virial theorem, which implied that finally the kinetic energy was higher in the molecule than in the isolated atoms. This again is different now.

For an electron in a pseudopotential the virial theorem is no longer valid. More precisely, a different virial theorem holds [43] (at least in some one-electron cases) and not that which is valid for the Coulomb potential. Rather than to argue in terms of the virial theorem appropriate for the chosen pseudopotential, we go back to the all-electron wave function and note that it does satisfy the virial theorem. So binding must lead to an overall rise in the kinetic energy. However it is not necessary that this is achieved by a scaling (steepening) of the valence-AOs, it can as well happen (and this is usually more effective) in the core. So the virial theorem is no longer a severe restriction on the valence part provided that a core is present.

The genuine 'chemical' contribution to the bond in Li_2^+ and Li_2 is rather small compared to that in H_2^+ or H_2, since the 'diffuse' 2*s*-AOs overlap much less effectively than the compact 1*s*-AOs. Consequently at the respective equilibirium distances the overlap integral S as well as the absolute value $|\beta|$ of the reduced resonance integral β are much smaller in Li_2^+ and Li_2 than in H_2^+ or H_2. As Bitter et al. [10] have pointed out the small binding energy in Li_2^+ is not really surprising if one considers that in all X_2^+-ions with X = H, Li, Na etc. (even X = Cu, Au and in XY^+ ions like CH^+, BH^+) the binding energy is 20 to 25% of the ionisation potential of X. The 2*s*-AO in Li is diffuse, which also means that it is loosely bound and that it has a small kinetic energy (in the pseudopotential formulation; in an all-electron theory the orthogonality of 2*s* to 1*s* introduces a spurious kinetic energy that cancels the deep potential energy near the nucleus). It is not unreasonable to expect that the lowering of the kinetic energy, which leads to bonding, is somehow proportional to the kinetic energy of the electron that is involved.

Another peculiarity of Li_2^+ and Li_2 is due to the high polarizability of the 2*s*-electron. (It is near-degenerate with 2*p*). This implies that a type of interaction which is usually referred to as intermolecular (or interatomic) and not regarded in connection with the chemical bond, becomes now very important.

In the case of H_2^+ and H_2 we have not even mentioned these terms. They were implicitly contained in the last steps 'deformation of the AOs beyond scaling' and 'electron correlation' (the latter only in H_2). For H_2^+ and H_2 these effects are small at distances close to the equilibrium value, but they dominate at very large distances. All chemical forces (due to overlap or interference) decay exponentially with R, while the induction (interaction of a permanent charge with an induced dipole) present in H_2^+ (and Li_2^+) goes as $\sim R^{-4}$ and the dispersion (interaction of instantaneous dipoles) dominant in H_2 and Li_2 for large R goes as $\sim R^{-6}$.

The weak chemical bond in Li_2^+ and the large polarizability of Li imply that in Li_2^+ at its equilibrium distance almost 60% of the energy lowering is due to induction (only less than 10% in H_2^+).

Moreover the deformation (promotion) of the AOs is very different in H_2^+ and Li_2^+. While it is mainly radial in H_2^+ (describable by a scale factor η), it is mainly angular (describable by admixing 2*p* to 2*s*) in Li_2^+, which means that the contribution of this promotion to the binding energy is essentially of the 'induction' type.

This different polarization has the further consequence that in Li_2^+ (unlike in H_2^+) the final kinetic energy in bond direction is larger than that perpendicular to it [45].

While the binding energy of the two-electron bond in H_2 (4.75 eV) is a little less than twice as large as that of the one-electron bond in H_2^+ (2.78 eV), the comparison between Li_2^+ and Li_2 leads to an astonishing result. The binding energy in Li_2 (1.05 eV) is even smaller than that in Li_2^+ (1.29 eV). There are two reasons for this unexpected behaviour. If we take the expressions (18) and (66) for the interference contribution to the binding energy in a one-electron and a two electron bond, we realize that the two-electron bond is twice as strong as the one-electron bond only in the limit of strong overlap ($S \approx 1$), the two bonds become equally strong for $S = \sqrt{2} - 1 \approx 0.414$; while for smaller S the two electron bond is weaker than the one-electron bond. This discussion is not entirely correct, since the two-electron bond is further stabilized by electron correlation and since the equilibrium distance and hence both S and β differ somewhat for the neutral molecule and the ion. It explains nevertheless, why the 'chemical' contribution to the bond energy is smaller in Li_2 than in Li_2^+.

Of course, this is a VB type argument [16]. In an analysis based on MO wave functions one must argue that the sharing penetration (inclusion of ionic terms) is relatively larger in Li_2 than in H_2 and raises the energy [10].

A more important effect is that the strong polarization of the valence AO into the middle of the bond in Li_2^+ is less favourable in Li_2, where *two* electrons ought to be polarized into the bond region, where they repel each other. Another way of phrasing the same thing is to say that the positive core of one Li in Li_2^+ induces a dipole moment at the valence AO attached to the other core, while in Li_2 there is no positive core left and instead of induction only dispersion is possible.

The role of electron correlation is quite different in Li_2 and in H_2. While in H_2 the most important correlation effects is left-right correlation, describable by a two-configuration wave function ($1\sigma_g^2/1\sigma_u^2$) as (70), in Li_2 the angular correlation, describable by admixture of $1\pi_u^2$, is more important. This means that in H_2 the HL wave function is a rather good approximation, with little contribution of ionic terms. In Li_2 the weight of ionic terms is relatively larger. The increased probability to find two valence electrons near the same atom requires that one should take a wave function that describes these ionic contributions (including the intraatomic correlation) well.

One may wonder whether the bonds in Li_2^+ or Li_2 are more representative for 'ordinary' covalent chemical bonds than the bonds in H_2^+ or H_2. This is, in spite of the presence of a core in Li_2^+ and Li_2, certainly not the case. The bonds in the alkali dimers are very atypical both in view of their weakness and of the pronounced near-degeneracy effects. The fact that Li atoms prefer the metallic (many center) bond to a two-center covalent bond is another hint that the covalent bond in Li_2^+ or Li_2 is not a very 'good' chemical bond.

One must be aware that the best-understood bonds are those in H_2^+, H_2, Li_2^+ and Li_2 and that the comparison of H_2^+ with Li_2^+ and H_2 with Li_2 is supposed to give insight into what may happen in the more general case.

7 An Exceptional Bond: Be_2

That there is a stable Be_2 molecule was proved experimentally only a few years ago [46]. However, quantum chemical calculations had not only predicted [47] that Be_2 should exist, but had also furnished values for the equilibrium distance and the binding energy that were later confirmed by experiment [46]. Unfortunately this message was not easy to understand for a non-theoretician. There were also some theoretical papers published, that did not predict a stable Be_2 molecule, but rather a loose van-der-Waals complex — or even a potential curve with both a 'chemical' and a 'van-der-Waals' minimum [48]. The computation of the potential curve of Be_2 is a rather tough problem and it was not sufficient to apply standard methods, rather a much higher degree of sophistication was required than usual. Anyway with a little knowledge of theory it was not too hard to decide which were the relevant papers on this molecule.

The binding energy in Be_2 is unusually small ($D_e \approx 0.1 eV$) for a chemical bond, but larger than for a typical van-der-Waals complex. Especially the bond length ($\sim 4.6 a_0$) is a strong hint that the bond is chemical.

At first glance one would argue that Be_2 is similar to He_2 and displays the typical repulsion between closed shells, now between $2s^2$ rather than $1s^2$. However, we know already that a $2s$-AO behaves sufficiently different from a $1s$-AO, on one hand due to the presence of a core, on the other hand due to the $2s/2p$ near-degeneracy and hence a high polarizability of the $2s$-AO.

It is certainly not possible to explain in a simple qualitative way why Be_2 is chemically bound, since we have seen that to account for this bond requires very sophisticated quantum chemical calculations. However, it must be possible to understand in a simple way why Be_2 does not show the strong 'closed-shell repulsion' that one would expect by analogy with He_2. The answer has certainly to do with the fact that the Be ground state is not a closed-shell state in the same strict sense as He or Ne (which we shall adress shortly at the end of this paragraph). Typical closed-shell states have a large gap between the highest occupied and the lowest unoccupied orbitals. This is not the case for Be, since the $2s \rightarrow 2p$ excitation energy is only 2.7 eV (compared to 21 eV for the $1s \rightarrow 2p$ excitation energy in He or to 16.7 eV for the $2p \rightarrow 3s$ excitation energy in Ne).

SCF calculations, even close to the Hartree-Fock limit don't predict a minimum of the potential curve of Be_2, so the correlation contributions are very important to stabilize Be_2. However, the electron distribution is not strongly changed by electron correlation, so the bond, although is does not show up on SCF level, is essentially prepared at this level. The SCF curve is repulsive, but extremely weakly so and very flat. At the equilibrium distance of the Be_2 molecule ($4.6 a_0$) the overlap integral between the two $2s$-AOs is as large as ~ 0.38, but the SCF repulsion is only 0.02 hartree. At the distance between two He atoms for which the overlap integral has the same value ($R \approx 1.7 a_0$) the repulsion energy is ~ 0.6 hartree. This unusual behaviour of the SCF curve of Be_2 was pointed out to the author by H. Kollmar in 1979, at a time when the whole Be_2 problem was still quite controversial.

The key to an understanding of the bond in Be_2 is to appreciate why the SCF-potential curve is so little repulsive in spite of a large overlap. We have seen that the repulsion in He_2 results from the fact that the antibonding MO is more strongly antibonding than the bonding MO is bonding. Obviously the bonding and antibonding contributions cancel each other much more in Be_2. There must hence be a stabilization mechanism in Be_2 that is not available in He_2. This is mainly the admixture of the $2p\sigma$-AOs to the σ_g and σ_u-MOs, which can loosely be referred to as hybridization. One can admix $2p\sigma$-AOs to the $2s$-AOs in two ways, either such that

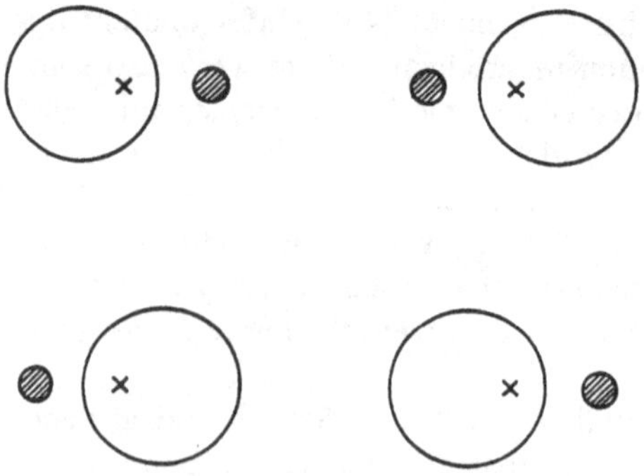

Fig. 11. Deformation of the 2*s*-AOs in the occupied MOs of Be_2

the resulting orbitals are pushed to each other or that they are repelled from each other (see Fig. 11). If one polarizes the 2*s*-AOs in the σ_g-MO in the first way, in the σ_u-MO in the second way, one reaches a stabilization of both MOs, and this is actually what happens.

Another way to look at the bond in Be_2 consists in interpreting it as the result of an avoided crossing between two hypothetical diabatic states, namely one built up from two pure $2s^2$ configurations in ^1S states of the Be atoms and the other from $2s2p$ configurations in ^3P states. The diabatic curve of the latter state is asymptotically higher by $2 \cdot 2.7 = 5.4$ eV, but is able to form a genuine (single or even double) chemical bond with a bond strength comparable in order of magnitude to the promotion energy.

One may argue that the bond in Be_2 is so exceptional that is does not merit to be discussed in a general chapter on the mechanism of the chemical bond. Nevertheless, if one has understood this bond, one has the basis for understanding more interesting features about the chemical bond in general.

One of these features becomes obvious if one compares Be_2 with the iso-valence-electronic Mg_2, where only a van-der-Waals minimum and no chemical bond is present. The reason is that hybridization is much less effective in molecules of elements of the higher rows of the periodic table. This has been discussed elsewhere in detail [49].

A consequence of the stabilization of the $2\sigma_g$ and $2\sigma_u$-MOs in Be_2 ($1\sigma_g$ and $1\sigma_u$ are linear combinations of the K-shell AOs), is that the next σ type MOs $3\sigma_g$ and $3\sigma_u$ are raised in their energy, to the extent that the next available MO (after $2\sigma_u$) is $1\pi_u$ rather than $3\sigma_g$, with the consequence that the bonds in B_2 and C_2 are of π-type [21].

Finally a few words on Ne_2. In an MO-theoretical description one has 4 bonding and 4 antibonding MOs in addition to the two K-shell MOs, all these doubly occupied. Since there is no low-lying unoccupied AO, there is no stabilizing effect like in Be_2 (the mixing of $2p$ to $2s$ is no longer energy-lowering, since the stabilization of $2\sigma_g$ and $2\sigma_u$ is counterbalanced by the destabilization of $3\sigma_g$ and $3\sigma_u$) and the main argument is again that antibonding MOs are more antibonding than bonding ones are bonding. Ne_2 is hence supposed to behave similar to He_2.

8 The Bond in F_2

The next homonuclear molecule with a single σ-bond in the ground state after H_2 and Li_2 is F_2. Like in Li_2 there are occupied cores in F_2. However, the AO involved in bonding in F_2 (if we ignore the certainly present but not very pronounced isovalent hybridization [49]) is of $2p\sigma$-type and orthogonal to the core by symmetry. There is hence no Pauli repulsion and the penetration into the core is prevented by the centrifugal force, so that the only effect of the core is to shield the nuclear charge by ≈ 2 units.

The new feature in F_2 are the fully occupied AOs $2s$, $2p\pi$ on either F atom. Although these don't point into the bond direction, these lone pairs repel each other (as doubly occupied AOs generally do). They hence weaken the bond.

If one builds up F_2 from two atoms one must, of course, in constructing the quasiclassical density, take care of the fact that the two F-atoms (in their ^2P ground states) must be oriented such that the singly occupied $2p_z$-AOs point towards each other. If one starts from spherically averaged F-atoms the unphysical change from these to oriented F-atoms would spuriously appear as a contribution to the change of the density on bond formation.

It is very curious that most experimentalists interested in difference density plots [50] have constructed these in substracting from the experimentally determined molecular electron density a quasiclassical density obtained from spherically averaged atoms. It is certainly in the spirit of the present analysis that a covalent chemical bond is associated with an interference density that shows accumulation of charge in the bond region. It is, however, obvious that a difference density obtained with spherically averaged F atoms has nothing to do with the interference density and does hence not allow conclusions concerning covalent chemical bonds [51].

While the bond in Li_2 is weak because the $2s$-AOs are diffuse and overlap poorly, the bond in F_2 is weak (though much stronger than in Li_2), although the $2p\sigma$-AOs overlap quite satisfactorily (and although these AOs are tightly bound in the atom and have a high kinetic energy), because of the repulsion of the lone pairs.

Ethane $H_3C{-}CH_3$ is, in a way, isoelectronic with F_2. The difference between the two molecules is that the lone pairs in F_2 are replaced by CH bonds. These are localized at larger distance from the bond and from each other. They have hence less effect on the bond and they repel each other much less. The CC-bond in ethane has therefore to be regarded as a more normal single bond than that in F_2 (and in Li_2). In many respects this CC single bond resembles the bond in H_2 [16]. The central bonds in N_2H_4, H_2O_2 lie between C_2H_6 and F_2. In molecules of the higher rows the lone-pair repulsion is less effective such that the bonds in Cl_2, H_2S_2, P_2H_4 are stronger — and hence more genuine single bonds — than in F_2, H_2O_2, N_2H_4 [49].

9 Multiple Bonds, the N_2 Molecule

In molecules with multiple bonds like N_2, in which, in an MO description, 3 more bonding than antibonding MOs are occupied, the constituent atoms approach each other much more closely than in singly bonded molecules. As a consequence, these atoms lose their individuality to a much higher degree than in the case of a single bond. In a typical single bond the interference effect is not only the driving force for bonding, but the interference energy is also the dominant energetic contribution, while the quasiclassical energy as well as the promotion energy are only minor corrections (in the latter respect Li_2 is somewhat untypical). This is now different.

In N_2 the quasiclassical energy is unusually large. This is not too hard to understand. The simplest model for the quasiclassical energy in N_2 is to regard the 1*s* and 2*s* AOs as point charges, which reduce the nuclear charge to $Z_{eff} = 3$, such that

$$\Delta E_{QK} = \frac{9}{R} - 2 \sum_{m=-1}^{1} \langle a_m | \frac{3}{r_b} | a_m \rangle + \sum_{m=-1}^{1} \sum_{m'=-1}^{1} (a_m a_m | b_{m'} b_{m'}) \tag{89}$$

where a_m is a 2*p*-AO on center A with m the azimuthal quantum number. A first approximation consists in taking simple Slater-type orbitals for the a_m and b_m. The analytic expression for (89) is then [52]

$$\Delta E_{QK} = \frac{9}{R} e^{-2\rho} \left\{ 1 + \frac{341}{256} \rho + \frac{93}{128} \rho^2 + \frac{3}{64} \rho^3 - \frac{5}{24} \rho^4 - \frac{1}{20} \rho^5 - \frac{1}{120} \rho^6 - \frac{1}{1260} \rho^7 \right\} \tag{90}$$

with $\rho = \eta \cdot R$. For the equilibrium distance $R = 2.08 a_o$ and with $\eta = 2.05$ according to Slater's rules we obtain

$$\Delta E_{QK} = -0.159 \text{ hartree} \tag{91}$$

This is almost half of the binding energy (—0.364 hartree) in N_2. If one takes care of the 2*s*-AO explicitly i.e. one does not represent it by a point charge, ΔE_{QK} is even lowered more.

Although the estimation that led to (91) is very rough, the concept, on which it is based, is clear. In the neighborhood of the equilibrium distance the sum of the repulsive interactions (core-core and valence-electrons-valence-electrons) is smaller in absolute value than those of the attractive interactions (core-valence electrons). This has already been observed (but to a less extent) in H_2, not in H_2^+ (where there is no electron-electron repulsion).

Not only the quasiclassical contribution to the chemical bond is very large, but also promotion (deformation) of the AOs is very effective. The AOs are under a strong attractive influence of the effective charge of the other core and are hence strongly deformed. This deformation, which is not simply described by a scaling factor (but is more of the angular type), raises the absolute value of the interaction energy considerably. This promotion energy has been estimated to be as high as +1.2 hartree [16] (in an all electron calculation). The deformation of the AOs further reduces the quasiclassical attraction energy to ≈ —0.53 hartree, it also lowers (raises in absolute value) the interference energy, estimated as —0.82 hartree.

The fact that all terms are so large, makes a qualitative description of the bond in N_2 very difficult (the isoelectronic CO and the doubly bonded CC behave, by the way, rather similarly [16]).

Although N_2 is conventionally described by a σ-bond and two π-bonds, only the π-bonds are actually bonding at the equilibrium distance. For the σ-bond one is already in the repulsive region. Anyway a σ-bond between two N-atoms is always much weaker than a π-bond [49].

Bitter et al. [10] have analyzed the contributions of the σ- and π-bonds to the bonding energy separately and also decomposed either contributions into a kinetic- and potential energy part. It has turned out that the kinetic energy of the π-MOs is lowered with respect to the isolated atoms, although the virial theorem requires that the kinetic energy is higher in the molecule. However, the virial theorem need not hold separately for each MO, and in the case of N_2 the σ-MOs take care of the rise in the kinetic energy required by the virial theorem.

Since the decomposition of the binding energy into the various contributions depends sensitively on the chosen basis sets and on other assumptions, one should not give too much importance to the numerical values of the individual contributions. The message to get home is rather that the interpretation of a bond in terms of the constituent atoms becomes the more problematic the stronger the bond is.

One must finally realize that, for N_2, the effects of electron correlation are very important. On SCF-level, corresponding to a MO-description the binding energy is only 5.18 eV, compared to the exact 9.90 eV.

One might think that the analysis of N_2 is representative for multiple bonds in general. This is not entirely so. Already the triple bond in HC≡CH shows a much smaller quasiclassical contribution and much less promotion than N_2 [16]. The main difference between N_2 and C_2H_2 is obviously that in C_2H_2 there is no lone pair. This is a hint that the lone pair — which we have not explicitly mentioned in the analysis just given — plays a major role. This role has been analysed by Bitter et al. [10], but we cannot go into details.

In view of the importance of electron correlation in N_2 an analysis of the chemical bond should explicitly include correlation effects. This has been done by Karafiloglou and Malrieu [20], who analyzed the result of a minimal basis full CI in terms of VB structures. As expected the MO wave function overestimates the weight of ionic structure considerably. On the other hand ionic structures cannot be completely neglected and their actual weight depends sensitively on the accuracy with which one describes intraatomic correlation effects that affect atomic electron affinities. If one is not able to account for these 'dynamic' correlation effects by the chosen computational scheme, one may think of applying semiempirical corrections as suggested by Lievin et al. [19] or Goodgame and Goddard [53]. It is curious that these semiempirical corrections resemble very much the semiempirical adjustment of the electron interaction as suggested long ago by Pariser and Parr [54]. For an analysis of these corrections see Kutzelnigg et al. [55].

Goodgame and Goddard [53] were, by the way, not only able to obtain an almost exact potential curve of N_2 by this procedure, but they even succeeded in accounting for the "hextuple' bond in Cr_2, which has so far not been accessible to rigorous quantum chemical calculations. (For the sake of completeness one should mention that the density functional method seems to work for Cr_2 [56], as it does, by the way qualitatively — with a binding energy too large by a factor ≈ 5 — also for Be_2 [57]. Nobody really understands why this is so). The observation that the quasiclassical energy is of the same order of magnitude as the binding energy does not seem to be limited to N_2 and similar molecules, it appears to hold for all molecules larger than H_2 [69]. It would, nevertheless, be premature to conclude that the chemical bond can be understood, at least approximatively, in terms of a quasiclassical interaction alone. There is obviously no quasiclassical interaction between neutral non-overlapping atoms. Hence a quasiclassical interaction (in non-polar bonds) cannot arise by itself without being accompained by quantum mechanical interference effects (that also depend on overlap). The interference is to a large extent controlled by the Pauli principle. The closed-shell repulsion hence dominates these quantum mechanical effects. Constructive interference (between singly occupied AOs) is necessary to make a system of overlapping charge densities stable.

10 The Ionic Bond

So far we have only considered homonuclear molecules. In these, the binding electrons are shared symmetrically between the two partners. In heteronuclear molecules, a transfer of electronic charge from one partner to the other is possible. This certainly has an effect on the bond. The extreme alternative to the covalent bond discussed so far, is the 'purely' ionic bond which is entirely based on the transfer of an electron from one atom to the other without any appreciable contribution of interference and overlap effects.

In principle an ionic bond is also possible between identical atoms. We come back to this at the end of this section.

The ionic bond is somewhat easier to understand than the covalent bond. It can be formed in a few physical (observable) steps, while the steps chosen to analyze the covalent bond where somewhat artificial.

In the first step an electron (or possibly more than one electron) is transferred from atom A to atom B, at conditions where neither A interacts with B nor A^+ with B^-, i.e. formally at infinite distance. The energy required by this step is the difference of the ionization potential of A and the electron affinity of B (in our nomenclature this is a promotion energy)

$$\Delta E_\infty = IP(A) - EA(B) \tag{92}$$

For all possible pairs of atoms ΔE_∞ is positive, i.e. the transfer of an electron from one atom to the other requires energy. The largest known electron affinity is still smaller than the lowest known ionization potential (this holds for atoms, not necessarily for molecules, an electron transfer between two molecules may in exceptional cases be exoergic).

In a second step we let the ions approach each other. They interact by means of their Coulomb attraction — which is strong and of very long range. Superimposed to the Coulomb interaction (which in our nomenclature is, of course, quasiclassical and the dominating part of the total quasiclassical interaction) is a possible chemical (interference) interaction. For closed shell ions this chemical interaction is the short-range repulsion explained for the example of He_2. So the ions can approach each other until the closed shell repulsion compensates the quasiclassical attraction. The bond is the stronger the closer the cores can approach, i.e.

the smaller the 'ionic radii' are. This is the reason why closed-shell ions are preferred, because these have relatively small ion radii.

The principle of minimal ion radii also explains why doubly or triply charged ions are formed, although the transfer of more than one electron is highly endoergic. But the Coulomb attraction $-Z_A Z_B/R$ is very high as well, especially if a close approach of the multiply charged ions is possible.

Although diatomic molecules with ionic bonds exist, the simplest example being LiH, the real domain of the ionic bond is in solids. In solid salts the charge transfer (92) costs the same energy as in the formation of a diatomic molecule, but the gain in Coulomb energy (Madelung energy) is much larger in the solid, since every positive ion is attracted by several negative ions and vice versa. So if an ionic bond is possible, it will usually not end at the formation of ion-pair type diatomic molecules but at solid salts.

A characteristic of an ionic bond in a diatomic molecule is the long-range attraction $\sim R^{-1}$, while in covalent bonds the attraction decays as $\sim \exp(-\alpha R)$ and is hence of short range.

As mentioned in the introduction to this section there is no reason to exclude the formation of an ionic bond in a homonuclear molecule. The electron transfer described in (92) is also possible for $A = B =$ H or $A = B =$ Li. In either case the formed ions, H^+, H^-, Li^+, Li^- are closed-shell ions (H^+ a somewhat exceptional one without electrons). The electron transfer at $R = \infty$ is certainly possible and the formed ions will certainly attract each other by means of Coulomb's law. What prevents this ionic bond from being realized for H_2 or Li_2, is that the ionic curve ought to cross several (possibly infinitely many) covalent potential curves of the same symmetry species that dissociate into the atoms A and B with either A or B or both in excited states. The non-crossing rule then requires that these crossings are avoided (see Fig. 12) i.e. the ionic curve that starts at $R = \infty$ ends up in an (excited) covalent curve. The partner of this potential curve in this avoided crossing is covalent for large R, at the crossing point it becomes ionic until the next crossing, where it becomes covalent again etc. In plots of the potential curves of ground- and excited states one often sees curves that in some regions (between two avoided

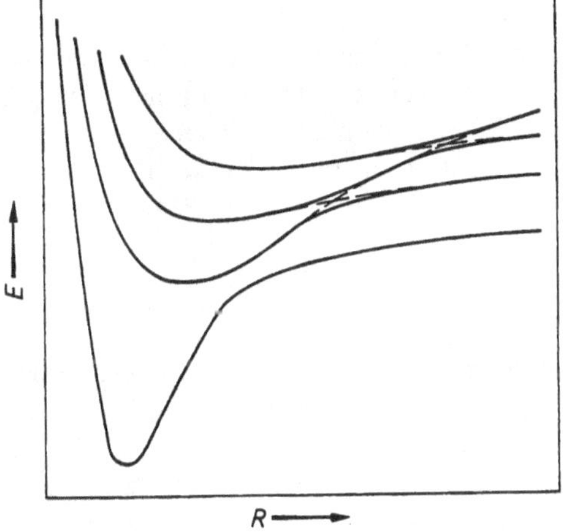

Fig. 12. Illustration of avoided crossing

crossings) go $\sim 1/R$. This is what remains of the ionic bond in these molecules. In Fig. 12 the first excited state is ionic at its equilibrium distance.

If the state A^+B^- lies not very high at $R = \infty$, and the only avoided crossing of the ionic curve is that with the curve of the covalent ground state, such that for small and intermediate distances the curve corresponding to the ionic bond represents the ground state, then an ionic bond is realized.

One should also note that an ionic bond is not necessarily polar and vice versa. For symmetry reasons the wave function that would describe an ionic bond in H_2, is of the form

$$\psi = N\{a(1)\,a(2) \pm b(1)\,b(2)\} \tag{93}$$

which means that with equal probablility both electrons are at atom A or at atom B and the molecule has no dipole moment.

A difference (in addition to those already mentioned) between the ionic and the covalent bond is that the driving force for the former is a lowering of the potential energy, for the latter of the kinetic energy. One further finds covalent one- and two-electron bonds of comparable strength. If one regards LiH as an ionic 'two-electron bond' (one electron of Li and one of H is involved in the bonding), one observes that the corresponding molecule LiH^+ with an ionic 'one-electron' bond is only extremely weakly bound, and this holds generally so [10]. A molecule like LiH^+ consists of Li^+ and H, the Coulomb attraction between oppositely charge ions (which is so important for the stability of the ionic bond) cannot take place. A sharing of valence electron between H and Li is also ineffective due to the different electronegativity. So the only attractive force that remains is induction ($\sim R^{-4}$) which is not a genuinely chemical interaction.

A chemical bond is rarely of purely covalent or purely ionic character. It is, moreover, not easy to decompose a bond into 'covalent' and 'ionic' contributions. Statements found in the literature about the 'percentage of ionic character' of a bond should be taken with great caution. One should mention that the bonds between different partners are usually stronger than those between identical partners. Pauling [58] has studied empirically the bond energy increments

$$E(AB) - \sqrt{E(AA) \cdot E(BB)} \tag{94}$$

and based on these his electronegativity scale. (For an analysis of this definition and its relation to Mulliken's electronegativity [59] see ref. 21). The additional strength of the heteropolar bond is due to electron transfer to the more electronegative atom, hence in a VB language to an increased participation of ionic structures.

11 Conclusions

There is a lot more that can be said about the chemical bond. We have concentrated our interest on the bond in diatomic molecules and we have not considered the important topics of many-center bonds, of bond delocalization and of the conditions for the localizability of bonds. The metallic bond, which is a typical delocalized bond, has hardly been mentioned. We have not discussed the origin of bond angles and the problem of bond saturation. Another term mentioned here only occassionally is hybridization. Some of these problems will be topics of other chapters of this book. They have also been treated elsewhere in detail [21, 22, 49].

We nevertheless claim that no essentially new aspects concerning the 'physical basis of the chemical' bond arise if one addresses the questions just-mentioned. We have seen that the interaction energy between two atoms can be decomposed into mainly three contributions, (a) a quasiclassical one, that is just the electrostatic interaction between the nuclei and undeformed electron clouds of the two atoms, (b) an interference contribution, that represents phase relations between the interacting orbitals, and that takes also care of the Pauli exclusion principle, (c) a rearrangement (promotion) of the atoms. The interference contribution changes the electronic charge density in the regions where the electronic clouds overlap. This interference density gives rise to the reduced resonance integral β defined by Eq. (18), that plays a central role in the theory of the covalent chemical bond. In the examples explicitly considered here, the interference density has little effect on the interaction energy of the valence electrons and a little more effect on the energy of the interaction of the valence electrons with the core, while the dominant contribution is a lowering of the kinetic energy. (As to cases where interference affects the electron interaction see ref. 16). The covalent chemical bond is hence mainly a one-electron effect and can thus, to a large extent, be described, at least qualitatively, in terms of one-electron models (like Hückel theory or extended Hückel theory), a fact which has always been very puzzling. It is still not fully understood what the real physical meaning of the one-electron energies in such models (or e.g. in the foundation [21] of Walsh's rules [60]) is.

The role of the virial theorem has been somewhat disturbing, but one makes no substantial error if one ignores it provisionally. An example of a successful first order theory that purposely does not take the virial theorem into consideration is the free-electron theory of metals.

Another restriction of this chapter is that we have mainly considered ground state potential curves. There is, of course, also bonding in excited states. However the problem with excited states is often that along one potential curve the electronic structure changes drastically due to avoided crossings. In a way, the concept of the chemical bond can only be applied to diabatic rather than to adiabatic potential curves (or surfaces). To say that a certain potential curve represents a chemical bond between two atoms in some given states is only meaningful if at the minimum of this curve the electronic structure of the molecule is directly related to those of the atoms at $R \to \infty$. Avoided crossings spoil this relation.

On the other hand there are bonding situations which are best understood by invoking an avoided-crossing mechanism. This is mainly so in situations where an atom in its ground configuration has no (as Be) or not enough (as C) valence-AOs availbale for bonding. The bond between C and 4 H to CH_4 can, e.g., be rationalized by considering first two diabatic potential surfaces, one built up from the ground state $^3P(s^2p^2)$ of C which is only divalent, and one from the quadrivalent $^5S(sp^3)$ state and finally an avoided crossing between these diabatic states. So in all those cases where the valence state of an atom is very different from the atom in its ground state, an avoided-crossing mechanism may be invoked. This is also the case for most ionic bonds, because by means of avoided crossings they dissociate usually into neutral fragments. One difficulty with the concept of avoided crossings is that a rigorous definition of 'diabatic states' is somewhat problematic.

Some people think that all information concerning the chemical bond is in the electron density and that a theory of the chemical bond should be mainly concerned with an analysis of the electron density. This is at first glance a valid point of view, particularly if one relies on the Hellmann-Feynman theorem according to which all forces in a molecule can be understood electrostatically in terms of the exact molecular density. The problem with this point of view is that one takes the electron density in the molecule for granted. The crucial question is, however, which is the mechanism that causes the electron density in a molecule to be so as it really is. Electron densities are only accessible via solution of the Schrödinger equation, in which the kinetic energy is as important as the potential energy. Our interest has been to understand what causes the electron density to be what it really is. Since we have never used the Hellmann-Feynman theorem, we were never obliged to guarantee that the approximate wave functions that we have considered, do fulfill the Hellmann-Feynman theorem which is not easy to achieve.

We have already admitted that the best understood chemical bonds in H_2^+ or in H_2 are not in all respects representative for the (covalent) chemical bond in general. What changes for other bonds is the quantitative role of the various contributions, but not the fundamental concepts involved in bond formation. The idea often found in recent publications (e.g. ref. 69) that the bond in H_2 is too exceptional to be of any use as a prototype of chemical bonds in general, is certainly not justified.

To close this chapter let us point out that the numerical calculation of molecular wave functions has made enormous progress in the last decades, but that the physical understanding of molecular quantum mechanics is still in many respects an under-developed subject.

Acknowledgement. The author is grateful to W. H. E. Schwarz for permission to use unpublished work (mainly ref. 10) and for valuable comments. He thanks V. Staemmler, S. Koch, and U. Fleischer for critically reading the manuscript.

Appendix: Effective Hamiltonians

Let H be a given Hamiltonian. We want to apply a unitary transformation W to H

$$L = W^+ H W\,; \qquad W^+ W = 1 \tag{A.1}$$

such that the transformed Hamiltonian L is block-diagonal

$$L = L_C + L_O\,; \qquad L_C = H_{eff} \tag{A.2}$$

and that the eigenfunctions of one block are automatically eigenfunctions of the entire L. We even require that the eigenfunctions Φ_μ of L_C are very simple, e.g. linear combinations of Slater determinants constructed from a given small set of spin orbitals.

$$L_C \Phi_\mu = E_\mu \Phi_\mu \tag{A.3}$$

Since L has the same eigenvalues as H, the E_μ in (A.3) are 'exact' eigenvalues, while the relation between the 'model wave function' Φ_μ and the corresponding exact wave function Ψ_μ is easily found. From (A.1) to (A.3) we get

$$L_C \Phi_\mu = L\Phi_\mu = W^+ HW\Phi_\mu = E_\mu \Phi_\mu \tag{A.4}$$

$$HW\Phi_\mu = E_\mu W\Phi_\mu \rightarrow \Psi_\mu = W\Phi_\mu \tag{A.5}$$

The operator W, which transforms the model wave function Φ_μ to the exact wave function Ψ_μ is referred to as 'wave operator', while $L_C = H_{eff}$ is called 'effective Hamiltonian'.

If one is interested in a single eigenstate, say Ψ_0, of H there is usually no serious alternative to solving the Schrödinger equation for this state directly. If one wants to find more than one eigenstate of H it is advantageous to construct first the effective Hamiltonian L_C corresponding to these states, to get then the eigenvalues and the model states from (A.3) and finally (if one wants) the exact eigenstates Ψ_μ from (A.5).

It is often recommended to choose H as a Fock space Hamiltonian, which has eigenstates of arbitrary particle number [61]. In the case of the He atom the Fock space Hamiltonian has eigenstates of He^+ as well as of He. Let us construct the effective Hamiltonian of He from the AOs 1s and 2p! The one-electron model states of the so-called complete model space are then $^2S_g(1s)$ and $^2P_u(2p)$, the two-electron model states are $^1S_g(1s^2)$, $^3P_u(1s2p)$, $^1P_u(1s2p)$, $^3P_g(2p^2)$, $^1D_g(2p^2)$, and $^1S_g(2p^2)$. This effective Hamiltonian is characterized by the matrix elements (values in hartrees; we only give those that we need to describe the two states of the configuration $1s2p$.)

$$\langle 1s|\, L\, |1s\rangle = -2; \quad \langle 2p|\, L\, |2p\rangle = -0.5$$

$$\langle 1s(1)2p(2)|\, L\, |1s(1)2p(2)\rangle = 0.37141 \tag{A.6}$$

$$\langle 1s(1)2p(2)|\, L\, |2p(1)1s(2)\rangle = 0.00478$$

The numerical values have been calculated by an iterative ab-initio method, equivalent to a variational calculation [62]. The desired model wave functions (eigenstates of L) are here determined by symmetry, there is hence no need to diagonalize L. From the matrix elements (A.6) we get

$$E(^3P) = -2.1333\ (-2.1332)$$

$$E(^1P) = -2.1237\ (-2.1239)$$

These values are very close to the exact ones (in parentheses). Note that now, as in the naive explanation of Hund's rule, the difference of the energies of the two states is entirely due to the exchange-type matrix element $\langle 1s(1)2p(2)|\, L\, |2p(1)1s(2)\rangle$.

This example clearly shows that the effective Hamiltonians are often the exact counterparts to simple physical models of the electronic structure of complex systems, a point which has been stressed by Freed [63]. In fact the parameters F_2, F_4 etc. in the theory of atomic states (the central field model) of Slater [64] and Condon [65] are straightforwardly interpreted as matrix elements of an effective Hamiltonian. The same is true for the parameters in Ligand field theory or for those in the π-electron theories of Pariser-Parr-Pople [54, 66] or Hückel [67].

It is obvious that the concept of effective Hamiltonians shoufd be fruitful for a model theory of the chemical bond. This concept has not yet been exploited in a systematic way. Nevertheless, the reduction of Li_2^+ to a one-electron system in the field of a pseudopotential is perfectly in the spirit of the theory of effective Hamiltonians. There have also been applications of effective Hamiltonians to problems of the chemical bond by Malrieu, Durand et al. [20, 68]. If the approach of Goodgame and Goddard [53] can be justified at all, then certainly only by invoking effective Hamiltonians.

12 References

1. see Cycon HL, Froese RG, Kirsch W, Simon B (1987) Schrödinger operators, Springer, Berlin Heidelberg New York; Thirring W, 'Lehrbuch der Mathematischen Physik' (1979) Bd. 3. Quantenmechanik von Atomen und Molekülen, Springer, Wien
2. Kolos W, Wolniewicz L (1964) J. Chem. Phys. 41: 3663, 3674

3. Uang YH, Stwalley WC (1982) J. Chem. Phys. 76: 5069
Feltgen R, Kirst H, Kohler KA, Pauly H, Torello F (1982) J. Chem. Phys. 76: 2360
Cornelius Th, Glöckle W (1986) J. Chem. Phys. 85: 3906
4. Manz J, Meyer R, Pollak E, Römelt J (1982) Chem. Phys. Letters 93: 184; Manz J, Meyer R, Schor HHR (1984) J. Chem. Phys. 80: 1562
5. Primas H, Müller-Herold U (1984) Elementare Quantenchemie, Teubner, Stuttgart
6. Hirschfelder JO, Curtiss CF, Byrd RB (1967) The molecular theory of gases and liquids, Wiley, New York
7. Hafner P, Habitz P, Ishikawa Y, Wechsel-Trakowski E, Schwarz WHE (1981) Chem. Phys. Lett. 80: 311
8. Feinberg MJ, Ruedenberg K, Mehler EL (1970) Adv. Quantum Chem. 5: 27
Feinberg MJ, Ruedenberg K (1971) J. Chem. Phys. 54: 1495, 55: 5804
Ruedenberg K (1975) in: Localization and delocalization in quantum chemistry, vol. I, Reidel, Dordrecht, p 223
9. Layton EM, Ruedenberg K (1964) J. Phys. Chem. 68: 1654 (1964)
Rue RR, Ruedenberg K (1964) J. Phys. Chem. 68: 1676
10. Schwarz WHE, Bitter Th (1976) Ber. Bunsenges. Phys. Chem. 80: 1231
Bitter Th (1982) Thesis, Siegen
Bitter Th, Schwarz WHE, Ruedenberg K, Butscher W, to be published
11. Hellmann H (1937) in: Einführung in die Quantenchemie, Deuticke, Leipzig—Wien,
Feynman RP (1939) Phys. Rev. 36: 340
12. Ruedenberg K (1962) Rev. Mod. Phys. 34: 326
13. Mulliken RS (1949) J. chim. physique 46: 497, 675
14. Hellmann H (1933) Z. Phys. 85: 180
15. Pauling L (1928) Proc. Nat. Acad. Sci. USA 14: 359
London F (1928) Z. Phys. 46: 455
16. Driessler F, Kutzelnigg W (1977) Theoret. Chim. Acta 43: 1307
17. Wilson CW, Goddard III WA (1972) Theoret. Chim. Acta 26: 195
Goddard III WA, Wilson CW (1972) Theoret. Chim. Acta 26: 211
18. Hurley AC, Maslen VM (1961) J. Chem. Phys. 34: 1919
Hurley AC (1976) in: Introduction to electron theory of small molecules, Academic, London
Hurley AC (1982) Int. J. Quant. Chem. 22: 241
19. Lievin J, Breulet J, Clerq P, Metz JY (1982) Theoret. Chim. Acta 61: 513
Lievin J, Metz JY (1985) Theoret. Chim. Acta 67: 391
20. Karafiloglou P, Malrieu JP (1986) J. Chem. Phys. 104: 383
Spiegelmann F, Malrieu JP, Maynau D, Zurru JP, J. chim. phys., to be published
21. Kutzelnigg W (1978) in: Einführung in die Theoretische Chemie, Bd. 2, Die chemische Bindung, Verlag Chemie, Weinheim
22. Kutzelnigg W (1973) Angew. Chemie, 85: 551;
(1973) Angew. Chem. Int. Ed. 12: 46
23. Berlin T (1951) J. Chem. Phys. 19: 208
24. Deb BM (1973) Rev. Mod. Phys. 45: 22
25. Deb BM (ed) (1981) The force concept in chemistry, van Nostrand, New York
26. Bader RFW in ref. 25, and references given there
27. Hirshfeld FL, Rzotkiewicz S (1974) Mol. Phys. 27: 1319,
Spackman MA and Maslen EN (1985) Acta Cryst. A 41: 347
28. Pauling L (1928) Chem. Rev. 5: 173
29. Slater JC (1933) J. Chem. Phys. 1: 687
30. Kauzmann W (1959) Quantum Chemistry, Academic, New York
31. Kutzelnigg W (1980) J. Chem. Phys. 73: 343
32. Finkelstein BN, Horowitz GE (1928) Z. Phys. 48: 118
33. Heine V (1980) Solid State Physics 35: 1
34. Pettifor DG (1973) J. Chem. Phys. 59: 4320
35. Kutzelnigg W, Schwarz WHE (1982) Phys. Rev. A 26: 2361
36. Zener C (1930) Phys. Rev. 36: 51;
Slater JC (1930) Phys. Rev. 36: 57
37. Davidson ER (1965) J. Chem. Phys. 42: 4199

38. Messmer RP, Birss FW (1969) J. Phys. Chem. 73: 2085
39. Katriel J, Pauncz R (1977) Adv. Quant. Chem. 10: 143
40. Shim I, Dahl JP (1978) Theoret. Chim. Acta 48: 165
41. Heitler W, London F (1927) Z. Phys. 44: 455
42. Shull H (1959) J. Chem. Phys. 30: 1405
43. Maeder F, Kutzelnigg W (1979) Chem. Phys. 42: 95
44. Hellmann H (1935) J. Chem. Phys. 3: 61
Gombas P (1967) in: Pseudopotentiale, Springer, Wien
45. Müller W, Jungen M (1976) Chem. Phys. Lett. 40: 199
46. Bondybey VE, English JH (1984) J. Chem. Phys. 80: 568
Bondybey VE (1984) Chem. Phys. Letters 109: 436
47. Liu B, McLean AD (1980) J. Chem. Phys. 72: 3418
Lengsfield BH, McLean AD, Yoshimine H, Liu B (1983) J. Chem. Phys. 79: 1891
Diercksen GF, Kellö U, Sadlej AJ (1985) Chem. Phys. 96: 59
48. Blomberg MRA, Siegbahn PEM, Roos BJ (1980) Int. J. Quant. Chem. 14: 229
Lee YS, Bartlett JR (1984) J. Chem. Phys. 80: 9371
49. Kutzelnigg W (1984) Angew. Chem. 96: 262;
(1984) Ang. Chem. Int. Ed. engl. 3: 272
50. Coppens P, Stevens ED (1977) Adv. Quant. Chem. 10: 1
Dunitz JP, Seiler P (1983) J. Am. Chem. Soc. 105: 7056
51. Schwarz WHE, Valtazanos P, Ruedenberg K (1985) Theoret. Chim. Acta 68: 471;
Schwarz WHE, Mensching L, Valtazanos P, v. Niessen W (1986) Int. J. Quant. Chem. 29: 909;
Mensching L, v. Niessen W, Valtazanos P, Ruedenberg K, Schwarz WHE, to be published
52. Roothaan CCJ (1951) J. Chem. Phys. 19: 445
53. Goodgame MM, Goddard II WA (1985) Phys. Rev. Lett. 54: 661
54. Pariser R, Parr RG (1953) J. Chem. Phys. 21: 466, 767
55. Kutzelnigg W, Del Re G, Berthier G (1971) Topics in Current Chemistry 22: 1
56. Delley B, Freeman AJ, Ellis DE (1983) Phys. Rev. Lett. 50: 488
Bernholc J, Holzwarth NA (1983) Phys. Rev. Lett. 50: 1451
57. Jones RO (1979) J. Chem. Phys. 71: 1300
58. Pauling L (1932) J. Am. Chem. Soc. 54: 3570
Pauling L, Sherman J (1937) J. Am. Chem. Soc. 59: 1450
59. Mulliken RS (1934) J. Chem. Phys. 2: 782; (1935) 3: 573
60. Walsh AD (1966) Ann. Rep. Chem. Soc. 63: 44
61. Kutzelnigg W, Koch S (1983) J. Chem. Phys. 79: 4315
62. Koch S, unpublished
63. Freed KF (1977) in: Segal GE (ed) Modern theoretical chemistry, vol. 7, Plenum, New York
64. Slater JC (1929) Phys. Rev. 34: 1293
65. Condon EU (1930) Phys. Rev. 36: 1121
66. Pople JA (1953) Trans. Farad. Soc. 49: 1375
67. Hückel E (1931) Z. Phys. 70: 204; (1931) 72: 310; (1932) 76: 628
68. Durand Ph (1983) Phys. Rev. A 28: 3184
Malrieu JP, Maynau D (1982) J. Am. Chem. Soc. 104: 3021, 3029
Maynau D, Garcia-Bach MA, Malrieu JP (1986) J. Physique 47: 207
69. Spackman MA and Maslen EN (1986) J. Phys. Chem. 90: 2020

Absolute Electronegativity and Absolute Hardness

R. G. Pearson

Chemistry Department, University of California, Santa Barbara, CA 93106, USA

Two new concepts derived from density functional theory are presented and explained. These are the absolute electronegativity, χ, and the absolute hardness, η. The exact and approximate definitions of these quantities are $\chi = -(\partial E/\partial N)_v \simeq (I + A)/2$, and $\eta = -\frac{1}{2}(\partial \chi/\partial N)_v \simeq (I - A)/2$. While χ must be constant everywhere in a chemical system η need not be. The local hardness, $\tilde{\eta}$, is discussed as well as the local softness, $\tilde{\sigma}$. The global softness is simply $\sigma = 1/\eta$. Using simple MO theory, χ and η are related to orbital energies. The hardness is equal to one half the energy gap between the HOMO and LUMO. The elements of frontier orbital theory are recovered.

Tables of experimental values of χ and η are given for a large number of cations, neutral molecules and radicals. For anions the importance of size and of local softness is discussed. Numerous examples of the practical use of these Tables are given. In many cases quantitative calculations can be made. In other cases only rank ordering is possible. In either case the object is to obtain information about the strength of chemical bonds. It appears that DFT will be increasingly useful in this kind of endeavor.

1 Introduction

In recent years density functional theory (DFT) has proved to be a very valuable tool in considering problems of chemical bonding and chemical reactivity [1]. The well-known theorem of Hohenberg and Kohn [2] led to greatly increased activity in the use of one-electron density functions as a basis for understanding chemical and physical phenomena.

One of the earliest applications of DFT to chemical bonding was the development of the x-alpha scattered wave approach to molecular orbital (MO) theory by Slater and Johnson [3]. Since then there have been great advances in the calculation of molecular properties of complex polyatomic systems, using electron density functions as a basis, rather than the more usual wave functions [4].

Since 1975 R. G. Parr and his coworkers have been particularly active in exploring the use of DFT to solve chemical problems [5]. The present chapter will deal mainly with some of the concepts that have arisen from this work. There will be no attempt to describe density functional theory, and only a limited attempt at the formal derivation of these concepts. Instead the emphasis will be on the concepts themselves, their physical meaning, and, most importantly, their use in the real world of chemical reactions.

It should be noted at the outset that DFT applies strictly only to equilibrium systems. Hence, only the ground states of chemical systems should be considered. In a few cases that follow, there will be use of excited states of atoms, the so-called valence states. These are used as models for the atoms in certain special molecules, for practical reasons.

2 Absolute Electronegativity

Consider a chemical system consisting of several nuclei and N electrons. The nuclei generate a potential v. Holding the nuclei fixed in position, the ground state electron density functional, ϱ, is that which satisfies the variational equation

$$\delta[E[\varrho] - \mu N[\varrho]] = 0 \tag{1}$$

The quantity μ is the Lagrange multiplier which ensures that the integral of ϱ over the volume is equal to N. It is called the electronic chemical potential [6]. From the calculus of variations, it follows that $\mu = (\partial E/\partial N)_v$.

For small changes in N or in v, we have

$$dE = \mu \, dN + \langle \varrho \, dv \rangle \tag{2}$$

The electronic chemical potential in a single molecule plays somewhat the same role that the ordinary chemical potential plays in macroscopic thermodynamics. If the system is in equilibrium, μ is constant everywhere. It measures the escaping tendency of the electrons in the system.

If the system is not in equilibrium, μ will not be constant: electron density will transfer from regions of high μ to those of lower μ, changing ϱ. Finally μ will be constant, ϱ will be the correct electron density for the ground state, and E will have its minimum value, for the given set of nuclear positions.

Suppose we now have two chemical systems, originally isolated from each other. In general they will have different chemical potentials, μ_C° and μ_D^0. If they are now allowed to interact, there will again be a flow of electron density from high μ to low μ, until finally a single value of μ_{CD} prevails. It is often convenient to think of the new chemical potential of each subsystem. Then we have, at equilibrium,

$$\mu_C = \mu_D = \mu_{CD} \tag{3}$$

What is now needed is a way to estimate μ_C^0 and μ_D^0 without the necessity of detailed and difficult calculations. Figure 1 shows a plot of the total electronic energy of a chemical species as a function of the number of electrons, N. The system may be an atom, an ion, a molecule or a radical. The energies are all negative, with zero energy offscale on top. Experimentally we only know points on the curve for integral values of N, from data such as ionization potentials, I, and electron affinities, A. However, it is convenient to consider that a smooth curve connects the points. In a molecule it is natural to think of the individual atoms as having non-integral electron populations.

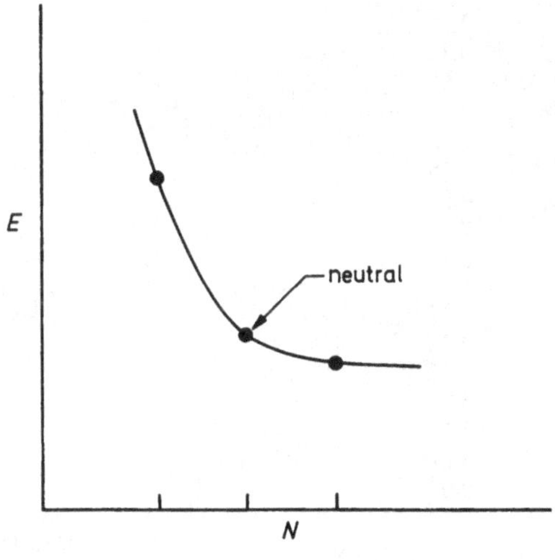

Fig. 1. Plot of electronic energy, *E*, *vs.* number of electrons, *N*, for a chemical system

We do not know the instantaneous slope of the curve in Figure 1. However, if we pick the neutral species (or any other) as our starting point, we do know the mean slope for the change from (N — 1) to N electrons. It is simply equal to —I. Also the mean slope from N to (N + 1) electrons is simply —A. Using the method of finite differences, we can approximate the slope at N as —(I + A)/2. But (I + A)/2 is the Mulliken electronegativity, χ_M [7]. We have

$$-\mu = -\left(\frac{\partial E}{\partial N}\right)_v \simeq \frac{I + A}{2} = \chi_M \tag{4}$$

Table 1. Experimental Parameters for Atoms (eV)
(Reprinted with permission of the copyright owner)

Atom	I[a]	A[b]	χ	η
H	13.60	0.75	7.18	6.43
Li	5.39	0.62	3.01	2.39
Be	9.3	0.4	4.9	4.5
B	8.3	0.28	4.29	4.01
C	11.26	1.27	6.27	5.00
N	14.53	0.07	7.30	7.23
O	13.62	1.46	7.54	6.08
F	17.42	3.40	10.41	7.01
Na	5.14	0.55	2.85	2.30
Mg	7.65	−0.15	3.75	3.90
Al	5.99	0.46	3.23	2.77
Si	8.15	1.39	4.77	3.38
P	10.49	0.74	5.62	4.88
S	10.36	2.08	6.22	4.14
Cl	12.97	3.62	8.30	4.68
K	4.34	0.50	2.42	1.92
Ca	6.1	−1.8	2.2	4.0
Sc	6.54	0.14	3.34	3.20
Ti	6.82	0.08	3.45	3.37
V	6.7	0.5	3.6	3.1
Cr	6.77	0.66	3.72	3.06
Mn	7.44	0	3.72	3.72
Fe	7.87	0.25	4.06	3.81
Co	7.8	0.7	4.3	3.6
Ni	7.64	1.15	4.40	3.25
Cu	7.73	1.23	4.48	3.25
Zn	9.39	−0.49	4.45	4.94
Ga	6.0	0.3	3.2	2.9
Ge	7.9	1.2	4.6	3.4
As	9.8	0.8	5.3	4.5
Se	9.75	2.02	5.89	3.87
Br	11.81	3.36	7.59	4.22
Rb	4.18	0.49	2.34	1.85
Sr	5.7	−1.7	2.0	3.7
Y	6.38	0	3.19	3.19
Zr	6.84	0.43	3.64	3.21
Nb	6.9	1.0	4.0	3.0
Mo	7.10	0.75	3.9	3.1
Ru	7.4	1.5	4.5	3.0
Rh	7.46	1.14	4.30	3.16
Pd	8.34	0.56	4.45	3.89
Ag	7.58	1.30	4.44	3.14
Cd	8.99	−0.33	4.33	4.66
In	5.8	0.3	3.1	2.8
Sn	7.34	1.25	4.30	3.05
Sb	8.64	1.05	4.85	3.80
Te	9.01	1.97	5.49	3.52
I	10.45	3.06	6.76	3.69
Cs	3.89	0.47	2.18	1.71
Ba	5.2	−0.5	2.4	2.9
La	5.6	0.5	3.1	2.6
Hf	6.7	−0.8	3.8	3.0

Table 1 (continued)

Atom	I[a]	A[b]	χ	η
Ta	7.89	0.32	4.11	3.79
W	7.98	0.82	4.40	3.58
Re	7.88	0.15	4.02	3.87
Os	8.7	1.1	4.9	3.8
Ir	9.1	1.6	5.4	3.8
Pt	9.0	2.1	5.6	3.5
Au	9.23	2.31	5.77	3.46
Hg	10.44	−0.63	4.91	5.54
Tl	6.1	0.3	3.2	2.9
Pb	7.42	0.37	3.90	3.53
Bi	8.42	0.95	4.69	3.74

[a] Reference [65]. [b] Hotop H, Lineberger WC (1985) J. Phys. Chem. Ref. Data 14: 731.

Iczkowski and Margrave had earlier proposed that $\chi = -(\partial E/\partial N)_v$ [8]. Because of the fundamental relationship, $\chi_M = -\mu$, it seems reasonable to call χ_M the absolute electronegativity [6]. This name may be unfortunate because χ_M is now rather different from the original definition of electronegativity as given by Pauling [9] — "the power of an atom in a molecule to attract electrons itself." As already mentioned, χ_M applies to the whole system which may be an atom, a molecule, an ion or a radical.

The values of χ for the atoms of the elements can be found, since I and A values are known. These values are roughly linear with the Pauling values, but some deviations exist. Table 1 gives the electronegativity values, χ, for the atoms of the elements. Hydrogen is considerably more electronegative, and oxygen somewhat less electronegative, than on the Pauling scale. These deviations are not alarming, since both the definitions and uses of the scales are quite different.

The chief use of the Pauling, and related scales, is to estimate the polarity of each chemical bond, and the net charge on each atom in a molecule. It does this quite successfully and the results have proved very useful. The absolute scale can be used in a similar way [10], but in addition, it can be used in a unique way to solve for bond polarity. For example, consider a molecule X–Y, consisting of two atoms or radicals held together by a bond. The polarity of the bond determines whether the molecule behaves as X^+, Y^- or X^-, Y^+. The same question can be asked by looking at the reactions

$$X^-_{(g)} + Y^+_{(g)} = X - Y_{(g)} = X^+_{(g)} + Y^-_{(g)} \tag{5}$$

The difference in energy between the products on the right and those on the left is easily found.

$$\Delta E = (I_X - A_Y) - (I_Y - A_X) = 2(\chi_X - \chi_Y) \tag{6}$$

If X has a greater absolute electronegativity than Y, ΔE is positive. This means that X–Y acts as X^-, Y^+. The answer is a thermodynamic one, and involves no assumptions. While the example is in the gas phase, the same result would usually be found in solution. The sum of the solvation energies of X^- and Y^+ would be very nearly equal to the sum for X^+ and Y^-, except when X or Y is the hydrogen atom.

The effect of any environment is usually to stabilize H^+ much more than H^-. For example, the heat of hydration of H^+ is −1125 kJ, and that of H^- is only −452 kJ. This kind of difference has the effect of reducing the apparent electronegativity of hydrogen on the Pauling, and similar scales.

When X and Y are brought together, electrons will flow from Y to X, if $\chi_X > \chi_Y$. This lowers χ_X and raises χ_Y until at equilibrium $\chi_X = \chi_Y = \chi_{XY}$. This is the electronegativity equalization principle, first proposed by Sanderson as an assumption [11]. In density functional theory, it is provable, as long as Eq. (4) is valid. Note that (4) provides an operational definition of the electronic chemical potential. To be consistent with the restriction that v is constant, only vertical values of I and A should be used. For practical reasons, it is often necessary to use adiabatic values.

Instead of bringing two atoms or radicals together, we can bring two molecules into interaction. As already indicated, there will be a flow of electron density from the molecule of low χ, until there is electronegativity equalization. This is extremely useful in looking at the generalized acid-base reaction.

$$C + :D \rightarrow C - D \tag{7}$$

The transfer of electrons from the base, or electron donor, :D, to the acid, or electron acceptor, C, results in the formation of a coordinate, covalent bond. Much of chemistry can be viewed in terms of reaction (7).

3 Absolute Chemical Hardness

To make progress we need to know how the chemical potential changes as the number of electrons in the system (or subsystem) changes. For reasons that will become clear, we define the absolute chemical hardness, η, as follows [12],

$$2\eta = \left(\frac{\partial \mu}{\partial N}\right)_v = \left(\frac{\partial^2 E}{\partial N^2}\right)_v \tag{8}$$

Thus hardness is the resistance of the chemical potential to a change in the number of electrons. Hardness is also the curvature of the plot in Fig. 1.

From the method of finite differences again, we obtain an operational definition of η [12].

$$\eta \simeq \frac{(I - A)}{2} \tag{9}$$

The units of η are the same as those of χ, eV. For any ion, atom, radical or molecule, we can calculate η, if I and A are known. Table 1 has the values for the atoms of the elements. The chemical softness, σ, of a system is also defined as the inverse of the hardness, $\sigma = 1/\eta$.

For isolated reactants, C and D, we can write

$$\mu_C = \mu_C^0 + 2\eta_C \Delta N \tag{10}$$

$$\mu_D = \mu_D^0 - 2\eta_D \Delta N \tag{11}$$

where ΔN is the fractional number of electrons transferred from D to C. Applying electronegativity equalization, or $\mu_C = \mu_D$, we find,

$$\Delta N = \frac{(\chi_C^0 - \chi_D^0)}{2(\eta_C + \eta_D)} \tag{12}$$

Thus the difference in electronegativity drives the process, and the sum of the hardness parameters acts as a resistance.

While (12) seems very reasonable, it is only valid for molecules still quite far from each other. As the molecules approach, there will be additional effects on μ_C and μ_D due to changes in the external potentials, v_C and v_D. For an isolated molecule these potentials are due only to the nuclei, but upon closer approach v_C must include the potential due to the nuclei and electrons of D, and vice versa. These will change as R, the intermolecular distance changes. Also there will be significant changes in ϱ_C and ϱ_D, not only because of electron transfer, but because of reorganization corresponding to covalent bonding.

If semiempirical MO theory is used as a basis, an equation for ΔN can be derived similar to (12), but including ionic bonding and covalency [13]. For neutral atoms, each with one binding electron, the result is

$$\Delta N = \frac{(\chi_C^0 - \chi_D^0)}{2(\eta_C + \eta_D) - 1/R - 2/\beta/(N_C N_D)^{1/2}} \tag{13}$$

Here β is the one electron exchange integral, which has a negative value. Equation (13), and the equation for the energy that can be derived for it, actually estimates bond energies quite well in simple cases [14]. However, it is also important in showing that (12) is incomplete.

Density functional theory also allows corrections to (12) to be estimated. The result for ionic bonding effects has been given by Nalewajski [15]. The more important effect of covalent bonding begins with the definition of the fukui function, f, by Parr and Yang [16]. From equation (2) one can write the Maxwell relation

$$\left[\frac{\delta\mu}{\delta v}\right]_N = \left(\frac{\partial\varrho}{\partial N}\right)_v \equiv f \tag{14}$$

With the "frozen core" assumption, $(\partial\varrho/\partial N)\, v$ becomes simply the normalized density function of the orbital involved. Three different values of f are possible, depending on whether electrons are being accepted, donated, or both by each subsystem.

$$f = \varrho_{HOMO} \text{ donated} \tag{15}$$

$$f = \varrho_{LUMO} \text{ accepted} \tag{16}$$

$$f = 1/2(\varrho_{HOMO} + \varrho_{LUMO}) \text{ both} \tag{17}$$

Equation (14) also yields the differential equation for μ.

$$d\mu = 2\eta \, dN + \langle f \, dv \rangle \tag{18}$$

If we now make the assumption that reaction between molecules occurs at sites in each molecule where the change in chemical potential is greatest we derive the rules of frontier orbtial theory [17]. Reaction in the nucleophile occurs where the HOMO has its largest value, and in the electrophile where the LUMO has its largest value.

This conclusion is most significant. It brings orbitals and covalent bonding into DFT in a natural way. Electron transfer becomes strongly directional, which is not implied in Eq. (12). The relative orientation of reactans C and D is determined by the best overlap of the HOMO and LUMO.

The strongly directional character of the interaction between two molecules is also brought out by further development of (18). Let Eq. (12) be satisfied by an initial transfer of electrons while the molecules are still some distance apart. Then bring them closer, and use (18) to calculate further changes in μ_{CD} [18]. The guiding principle is that $\Delta\mu_C = \Delta\mu_D$.

The result is an additional transfer of electron density given by

$$\Delta N = \frac{\langle(\varrho^{C}_{LUMO} - \varrho^{D}_{HOMO}) \, \Delta v^{n}\rangle}{2(\eta_C + \eta_D - J)} ; \tag{19}$$

$$J = \langle \varrho^{C}_{LUMO} \varrho^{D}_{HOMO} / r_{CD} \rangle ; \tag{20}$$

where J is the Coulomb integral between an electron in the HOMO and one in the LUMO. The changed potential due to moving the nuclei together is Δv^n. Covalent bonding is implicit in the numerator of (19), since it means that the electrons are redistributing themselves to interact more strongly with the nuclei.

It is tempting to return to Eq. (2) and to try to calculate changes in energy from (12) and (19). But there is little point in doing so. At the equilibrium bonding distance, the original HOMO and LUMO of the reactants are no longer meaningful entities. Equation (19) is only valid for the beginning of interaction.

In spite of its shortcomings, Eq. (12) still turns out to be useful. It has the great virtue of trying to predict chemical behavior from only two properties of an isolated system, the ionization potential and the electron affinity. Fortunately, in recent years there have been a great many such values of I and A determined [19]. An important advance has been the measurement of negative electron affinities [20]. These negative values are obviously vertical values.

Hard molecules have large values of (I — A), and soft molecules have small values. Hard molecules resist any change in their number of electrons, whereas soft molecules do not. The factors which determine the value of η are easily discernible. Consider first the atomic case where N valence shell electrons all have the same value of *n* and *l*. The energy is given quite well by a quadratic equation

$$E = aN + bN^2 + E_{core} \tag{21}$$

So that we can use (8) to calculate

$$\chi = -a - 2bN = (I + A)/2 \tag{22}$$

$$\eta = b = (I - A)/2$$

The constant *a* is a combination of a core integral and a valence shell electron-electron repulsion integral. But *b* is exactly one-half the average valence shell electron-electron repulsion integral, as shown by Klopman [14].

If some of the valence shell electrons have different values of *n* and *l*, then E cannot be a continuous function of N [21]. There will be discontinuities at certain values of N and the average values, (I + A)/2 and (I — A)/2, will include these discontinuities. That is, there will be subsantial changes in I or A when *n* or *l* changes for the next electron. These effects will be very large if we leave the valence shell in adding or removing an electron. For molecules the situation is more complicated. However, η will consist of two parts: and averaged interelectronic repulsion term and discontinuities due to going from bonding, or nonbonding to antibonding molecular orbitals.

4 Relation to MO Theory

Molecular orbital theory has proved to be a powerful and versatile tool for both organic and inorganic chemists. It is almost universally used to explain structure and bonding, visible-uv absorption spectra, and detailed mechanisms of chemical reactions. It would be unfortunate if the concepts of absolute electronegativity and hardness conflicted with MO theory, or pretended to supplant it. However they turn out not only to be compatible, but to be complementary in a useful way.

According to MO theory, the ionization potential of a molecule is simply the orbital energy of the HOMO, and the electron affinity is the orbital energy of the LUMO, with changes in sign.

$$I = -\varepsilon_{HOMO} \qquad A = -\varepsilon_{LUMO} \tag{23}$$

These are consequences of Koopmans' theorem. The HOMO is assumed to be doubly occupied. If we make the usual diagram of the molecular orbitals of a chemical system as a function of their energies, we can easily insert χ^0 and η into the same diagram.

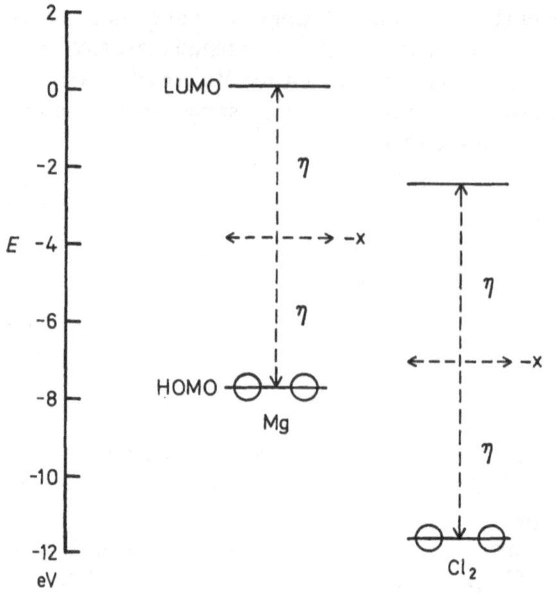

Fig. 2. Orbital energy diagram for Mg and Cl_2, showing χ and η

Two examples, atomic Mg and Cl_2, are shown in Fig. 2. The ionization potential of Mg is 7.65 eV, and the electron affinity -0.15 eV. The χ^0 is 3.75 eV and η is 3.9 eV. The former is shown as a *horizontal dashed line* on the diagram. It is just the average energy of the HOMO and LUMO, with change in sign. Clearly it is a good measure of the ability of the molecules to attract electrons to itself. The *dashed line* is also equal to η, the electronic chemical potential.

The hardness is shown as a *vertical dashed line*. The gap between the HOMO and the LUMO is equal to twice the value of η. This gives a new insight into the meaning of hardness and softness. A hard molecules (or atom, or ion) has a large gap between the HOMO and LUMO, and a soft molecule has a small gap.

For atoms or radicals where the HOMO is only singly occupied (SOMO), the situation is different, since the electron affinity usually refers to adding a second electron to the SOMO. In molecular orbital theory $-\chi$ is now the average energy of the two electrons in the orbital. And $(I - A)$ turns out to be just the average value of the interaction energy of the two electrons.

For Cl_2, I is 11.60 eV and A is 2.4 eV, so that $\chi = 7.0$ and $\eta = 4.6$. Electrons will flow spontaneously from Mg to Cl_2, as they approach each other. The higher value of the electronic potential for Mg is clearly seen. If Eq (12) is used, a value of 0.19 is calculated for ΔN. This is much too low, since the bonding in $MgCl_2$ is very ionic, and ΔN must be between 1.0 and 2.0. Ignoring the ionic bonding underestimates ΔN.

In simple MO theory the HOMO and the LUMO play several important roles. They are the frontier orbitals for chemical reactivity. They also define the lowest excited state, or the lowest energy band in the visible-uv absorption spectrum. $Nh\nu$ for this band should also equal the orbital energy difference. But typically $Nh\nu$ is less than half of (I − A). For example, in Cl_2 the first excited singlet state ($\pi^* \rightarrow \sigma^*$) has a vertical energy 30,000 cm^{-1} above the ground state. This is only

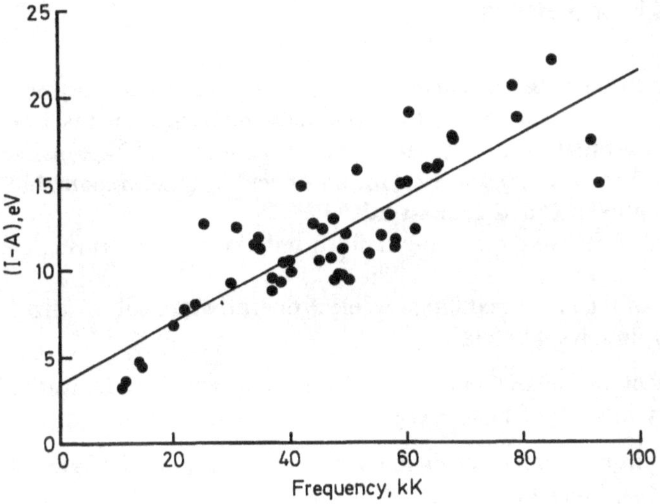

Fig. 3. Plot of (I — A) *vs* υ, the frequency of the first absorption maximum for a number of gas phase molecules

3.7 eV, compared to 9.2 eV for (I — A). For Mg the ($2s \rightarrow 2p$) excited state is 4.3 eV above ground, and (I — A) is 7.8 eV.

The difference exists because of additional electron-electron repulsion in forming Cl_2^- or Mg^-. This difference can readily be calculated, if the approximation of Koopmans' theorem is adopted, namely, no relaxation of the orbitals upon adding, removing or exciting an electron. Then it is found that, for filled shell molecules,

$$(\mathrm{I} - \mathrm{A}) = Nh\upsilon + \mathrm{J}_{\mathrm{if}} - 2\mathrm{K}_{\mathrm{if}}, \qquad (24)$$

where J and K are Coulomb and exchange integrals between an initially occupied MO, φ_i, and one occupied in its place in the excited state, φ_f.

Figure 3 shows a plot of (I — A) for a number of molecules from Table 3, against $Nh\upsilon$ for the same molecules in the gas phase [22]. The value of υ is that of the vertical transition, where the absorption is a maximum. The excited state is the lowest energy one with the same multiplicity as the ground state. Equation (24) is calculated on that assumption.

The slope of the best straight line is 1.45, dimensionless units. The scatter shows that molecules have quite different values of J and K. In addition, orbital relaxation effects are probably quite variable. Still for two or more closely related molecules, a relationship between (I — A) and $Nh\upsilon$ may be usefully assumed. For example, H_2O, H_2S, H_2Se and H_2Te have their first absorption bands at 166, 195, 197 and 200 nm, respectively. Values of (I — A) are 18 eV for H_2O and 12.6 eV for H_2S. From this, reasonable estimates for H_2Se and H_2Te may obviously be made.

5 Hard and Soft Acids and Bases

The terms hard and soft, to describe the chemical properties of molecules or ions, came into chemistry in 1963 [23]. It was applied to Lewis acids and bases, as in reaction (7). It was an attempt to estimate the bond energies for various acid-base pairs. It was already known that there were no universal orders of acid or base strength, but that specific interactions between C and D must exist.

Hard and soft acids and bases were originally defined only in general terms:

soft base — donor atom is of high polarizability, low electronegativity, easily oxidized and assiciated with empty, low-lying orbitals.

hard base — donor atom is of low polarizability, high electronegativity, hard to oxidize and associated with empty orbitals of high energy and hence inaccessible.

soft acids — the acceptor atom is of low positive charge, large size, and has several easily excited outer electrons. Polarizable.

hard acids — acceptor atom is of high positive charge, small size, and does not have easily excited outer electrons. Not polarizable.

Such a qualitative description does not allow for quantification of the property of hardness (or softness). Thus acids and bases were put into one of two boxes, labeled hard and soft, without any rank ordering. It was still possible to summarize a great deal of diverse information by the HSAB principle: hard acids prefer to coordinate to hard bases and soft acids to soft bases [24].

The HSAB Principle is an empirical one based on various kinds of experimental information, such as equilibrium constants, bond energies, rate constants, or even the existence or non-existence of certain compounds. It depends on no theory for its validity, but it was clear from the outset that soft acids and bases relied mainly on covalency for bonding, whereas hard acids and bases relied mainly on ionic bonding. These ideas were theoretically confirmed, and quantified to some extent, by Klopman [25].

To be ultimately useful in correlating with chemical behavior, it is necessary that the new definition of chemical hardness, $\eta = (I - A)/2$, be consistent with the older, empirical information. First it must be shown that the definition fits the general characteristics of hard and soft molecules, and that the HSAB Principle can be deduced from it. Secondly, the hard or soft character of individual molecules or ions must be the same, both empirically and from the definition.

The general correspondence is readily demonstrated. The old definitions of hard and soft stress polarizability as the most important feature. Optical polarizability in quantum theory results from a mixing in of excited state wave functions with the ground state wave function. A small HOMO-LUMO gap automatically means that the whole manifold of excited states can lie closer in energy to the ground state. Small energy differences favor easier state mixing, according to perturbation theory. Soft molecules, with a small energy gap, will be more polarizable than hard molecules, with a large energy gap.

We can extend the qualitative definition of chemical hardness from the above.

Hard molecules resist changes, not only in the number of electrons they contain, but also in the spatial distribution of their electron cloud. A soft molecule has an easily changed electron count and distribution.

The HSAB Principle is already explained in part by Eqs. (12) and (13). When η_C and η_D are both small, ΔN will be large and there will be good bonding, mainly covalent. Large values of η_C and η_D mean little covalent bonding, and a necessary reliance on electrostatic bonding, due to initial charges and dipoles.

More detail can be found by applying the perturbation theory of chemical reactions in the frontier orbital form [26]. Figure 4 shows the MO diagrams of two interacting molecules. The two are of similar electronegativity so that electron transfer can occur in both directions. One molecule is somewhat harder than the other.

Two kinds of interactions are shown:

1. Partial transfer of electrons from the HOMO of each molecules to the LUMO of the other. This occurs by mixing of the orbitals.
2. There is a mixing of the filled MOs of each molecule with its own empty MOs.

The first effect is called delocalization and is the mechanism whereby new bonds are formed between the reactants and old bonds are broken. Other orbitals besides the HOMOs and LUMOs may be involved to a lesser extent.

The second effect is called polarization. It has the effect of lowering the repulsive energy between the two molecules as they approach each other. Polarization is easiest when the energy gap is small for each molecule. A third effect in the interaction of two molecules is not shown in Fig. 4. This is the electrostatic effect, due to charges and dipoles on the reactants.

Considering only the form frontier orbitals, an approximate value for the energy lowering due to delocalization is given by

$$\Delta E = \frac{2\beta_{12}^2}{(A_1 - I_2)} + \frac{2\beta_{21}^2}{(A_2 - I_1)} \tag{25}$$

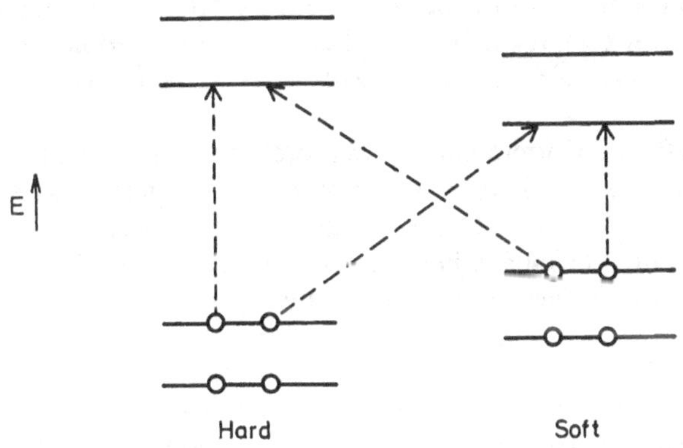

Fig. 4. MO diagram showing interaction of two molecules by electron transfer, and by polarization

The β's are exchange integrals of the perturbation Hamiltonian over the interacting MOs. One β could correspond to σ-bonding, and the other to π-bonding, for example. The stabilization is greatest if A is large (and positive) and I is small, for both molecules.

The second effect is closely related to the optical polarizability discussed earlier. A soft acid and a soft base will interact favorably with each other as a result of mutual polarization. However note that chemical polarizability and optical polarizability are not quite the same, since different excited states may be involved [27].

6 Hardness of Cations

The beginnings of the concept of chemical hardness lay in attempts to understand the ability of different metal ions to form complexes with various ligands [28]. Metal ions were put into two categories, called class (a) and class (b). These later became hard and soft, respectively [23].

Class (a) or (b) behavior was based on empirical results, given by the relative affinities for bases with different donor atoms

(a)	(b)
$F \gg Cl > Br > I$	$F \ll Cl < Br < I$
$O \gg S > Se > Te$	$O \ll S \sim Se \sim Te$
$N \gg P > As > Sb$	$N \ll P > As > Sb$

For ionic reactions, the affinities refer to aqueous solutions.

It is necessary to show that (I — A) indeed does correlate with earlier assignments of hardness and softness for various systems. At the same time (I + A) must correlate with behavior as Lewis acids or bases. Large values of χ characterize Lewis acids, and small values apply to bases.

Table 2 lists the experimental values of I, A, χ, and η for all the important monatomic cations in their common oxidation states. For the heaviest metals some experimental ionization potentials are still not available. It can be seen, that with one exception, the values of η calculated for these ions agree very well with their known chemical hardness.

In addition to the expected effects of ionic charge and size, the influence of *d* electrons is very clear. That the *d* subshell is very important in promoting soft behavior has been emphasized by Ahrland [29]. The role of the *d* electrons in chemical bonding is usually attributed to π-back bonding and/or polarization in the field of the ligands [13]. However ligand filed stabilization is also important and it is interesting to point out that this is also a polarizability effect. Localizing the *d* electrons in certain favored orbitals is a special kind of polarization.

The only cases where the values of η are not a reliable guide to chemical softness occur for Ce^{3+}, and the other trivalent lanthanide ions. These have low values of η compared to La^{3+}, but they are all typical hard Lewis acids. Clearly the *f* electrons

Table 2. Experimental Parameters for Monatomic Cations (eV) (Reprinted with permission of the copyright owner)

Ion	I	A	χ	η
Li^+	75.64	5.39	40.52	35.12
Na^+	47.29	5.14	26.21	21.08
K^+	31.63	4.34	17.99	13.64
Rb^+	27.28	4.18	15.77	11.55
Cs^+	25.1	3.89	14.5	10.6
Cu^+	20.29	7.73	14.01	6.28
Ag^+	21.49	7.58	14.53	6.96
Au^+	20.5	9.23	14.9	5.6
Tl^+	20.43	6.11	13.27	7.16
Co^+	17.06	7.86	12.46	4.60
Rh^+	18.08	7.46	12.77	5.31
Ir^+	16.9[a]	9.1[a]	13.0	3.9
Cl^+	23.81	12.97	18.39	5.42
Br^+	21.8	11.81	16.8	5.0
I^+	19.13	10.45	14.79	4.34
Be^{2+}	153.89	18.21	86.05	67.84
Mg^{2+}	80.14	15.04	47.59	32.55
Ca^{2+}	50.91	11.87	31.39	19.52
Sr^{2+}	43.6	11.03	27.3	16.3
Sc^{2+}	24.76	12.80	18.78	5.98
Ti^{2+}	27.49	13.58	20.54	6.96
V^{2+}	29.31	14.65	21.98	7.33
Cr^{2+}	30.96	16.50	23.73	7.23
Mn^{2+}	33.67	15.64	24.66	9.02
Fe^{2+}	30.65	16.18	23.42	7.24
Co^{2+}	33.50	17.06	25.28	8.22
Ni^{2+}	35.17	18.17	26.67	8.50
Cu^{2+}	36.83	20.29	28.56	8.27
Zn^{2+}	39.72	17.96	28.84	10.88
Ge^{2+}	34.22	15.93	25.08	9.15
Y^{2+}	20.52	12.24	16.38	4.14
Zr^{2+}	22.99	13.13	18.06	4.93
Nb^{2+}	25.04	14.32	19.68	5.36
Mo^{2+}	27.16	16.15	21.60	5.51
Ru^{2+}	28.47	16.76	22.62	5.86
Rh^{2+}	31.06	18.08	24.57	6.49
Pd^{2+}	32.93	19.43	26.18	6.75
Ag^{2+}	34.8	21.5	28.2	6.7
Cd^{2+}	37.48	16.91	27.20	10.29
Sn^{2+}	30.50	14.63	22.57	7.94
Hf^{2+}	23.3	14.9	19.1	4.2
W^{2+}	25.4[a]	16.3[a]	20.9	4.5
Os^{2+}	27.7[a]	16.3[a]	22.3	5.7
Pt^{2+}	35.2[a]	19.2[a]	27.2	8.0
Hg^{2+}	34.2	18.76	26.5	7.7
Pb^{2+}	31.94	15.03	23.49	8.46
B^{3+}	259.37	37.93	148.65	110.72
Al^{3+}	119.99	28.45	74.22	45.77
Ga^{3+}	64.00	30.71	47	17
In^{3+}	54	28.03	41	13
Tl^{3+}	50.7	29.83	40.3	10.4
Sc^{3+}	73.47	24.76	49.11	24.36

Table 2 (continued)

Ion	I	A	χ	η
Y^{3+}	61.8	20.52	41.2	20.6
La^{3+}	49.95	19.18	34.57	15.39
Ce^{3+}	36.76	20.20	28.48	8.28
Lu^{3+}	45.19	20.96	33.08	12.12
Ti^{3+}	43.27	27.49	35.38	7.89
V^{3+}	46.71	29.31	38.01	8.70
Cr^{3+}	49.1	30.96	40.0	9.1
Mn^{3+}	51.2	33.67	42.4	8.8
Fe^{3+}	54.8	30.65	42.73	12.08
Co^{3+}	51.3	33.50	42.4	8.9
Ni^{3+}	54.9	35.17	45.0	9.9
Zr^{3+}	34.34	22.99	28.65	5.68
Nb^{3+}	38.3	25.04	31.7	6.6
Mo^{3+}	46.4	27.16	36.8	9.6
Ru^{3+}	49.9[a]	28.5	39.2	10.7
Rh^{3+}	53.4[a]	31.1	42.2	11.2
Hf^{3+}	33.32	23.3	28.3	5.0
W^{3+}	39.3[a]	25.4[a]	32.4	7.0
Re^{3+}	41.5[a]	25.9[a]	33.7	7.8
Os^{3+}	42.7[a]	27.7[a]	35.2	7.5
Ir^{3+}	45.3[a]	29.5[a]	37.4	7.9
Au^{3+}	54.1[a]	37.4[a]	45.8	8.4

Experimental data from Ref. [65]
[a] Theoretical values. Carlson TA, Nestor CW, Wasserman N, McDowell JD (1970) Atomic Data and Nuclear Data Tables 2: 63

are responsible for the low values of η, since they can be ionized rather easily, so that I is small. However f electrons are chemically quite inert. Because they are well screened, and because of their spatial distribution, covalent bonding, ligand filed effects and chemical polarizability are all small for $4f$ electrons.

The effect of unshared p electrons in the valence shell is similar to that for d electrons. While only the halonium ions are listed, similar low values of η are expected for ions such as RO^+, RS^+, RSe^+, R_2N^+ and R_2P^+. Carbonium ions are expected to be much harder, though it is difficult to measure the second ionization potentials of these ions, to prove this point.

Metal ions have very large values of χ, showing that they are Lewis acids, or electron acceptors. The values increase as the charge on the ion increases. Comparing the electronegativities of these ions with those of the atoms of the elements, shown in Table 1, the values of χ for the atoms are very much less.

Also the ordering may be quite different: For the atoms, χ increases monotonically in going across the fourth Period, from Ca to Zn. For the divalent ions, χ increases in going from Sc^{2+} to Zn^{2+}, but with a bump at Mn^{2+}. Also Ca^{2+} is much more electronegative than any of the transition metal ions.

The correct value of I for these ions is the *third* ionization potential of the atom. This is very much larger for calcium than for the others, and accounts for its large χ value. It may seem absurd to argue that the third ionization potential of calcium has any chemical significance, since it is so large and not in the valence shell. But it is this same value which makes η for Ca^{2+} so large, and correctly classifies it as a hard acid.

7 Hardness of Neutral Molecules

Table 3 gives the values of I, A, χ and η for many of the neutral molecules which are of wide chemical interest, and where the data are available. The negative values of A correspond to vertical electron affinities, i.e., no change in nuclear positions. There are the ones that density functional theory requires, and also the ones that are most appropriate for discussing partial transfer of electron between molecules. While vertical values of the ionization potential would also be preferred, no attempt has been made to correct, or to select, such values of I.

Table 3. Experimental Parameters for Molecules (eV) (Reprinted with permission of the copyright owner)

Molecule	I[a]	A[b]	χ	η
SF_6	15.4	0.5	8.0	7.4
BF_3	15.81	−3.5[k]	6.2	9.7
SO_3	12.7[c]	1.7	7.2	5.5
Cl_2	11.6	2.4	7.0	4.6
H_2	15.4	−2.0[d]	6.7	8.7
SO_2	12.3	1.1	6.7	5.6
N_2	15.58	−2.2	6.70	8.9
Br_2	10.56	2.6	6.6	4.0
O_2	12.2	0.4	6.3	5.9
CO	14.0	−1.8	6.1	7.9
I_2	9.4	2.6	6.0	3.4
BCl_3	11.60	0.33	5.97	5.64
CS	11.71	0.20	5.96	5.23
HNO_3	11.03	0.57	5.80	5.23
CH_3NO_2	11.13	0.45	5.79	5.34
PF_3	12.3	−1.0[e]	5.7	6.7
HCN	13.6	−2.3	5.7	8.0
BBr_3	10.51	0.82	5.67	4.85
PBr_3	9.9	1.6	5.6	4.2
S_2	9.36	1.66	5.51	3.85
$C_6H_5NO_2$	9.9	1.1	5.5	4.4
PCl_3	10.2	0.8	5.5	4.7
N_2O	12.9	−2.2	5.4	7.6
acrylonitrile	10.91	−0.21	5.35	5.56
CS_2	10.08	0.62	5.35	5.56
HI	10.5	0.0[b]	5.3	5.3
CO_2	13.8	−3.8[g]	5.0	8.8
HF	16.0	−6.0[h]	5.0	11.0
CH_3I	9.5	0.2	4.9	4.7
HCl	12.7	−3.3	4.7	8.0

Table 3 (continued)

Molecule	I[a]	A[b]	χ	η
CH_3CN	12.2	−2.8	4.7	7.5
CH_2O	10.9	−0.9	5.0	5.9
HCO_2CH_3	11.0	−1.8[k]	4.6	6.4
CH_3CHO	10.2	−1.2	4.5	5.7
C_2H_4	10.5	−1.8	4.4	6.2
C_5H_5N	9.3	−0.6	4.4	5.0
butadiene	9.1	−0.6	4.3	4.9
H_2S	10.5	−2.1	4.2	6.2
C_2H_2	11.4	−2.6	4.4	7.0
$HCONH_2$	10.3	−2.0	4.2	6.2
styrene	8.47	−0.25	4.11	4.36
CH_3COCH_3	9.7	−1.5	4.1	5.6
PH_3	10.0	−1.9	4.1	6.0
C_6H_6	9.3	−1.2	4.1	5.3
toluene	8.8	−1.1	3.9	5.0
propylene	9.74	−2.0	3.9	5.9
C_6H_5OH	8.5	−1.0	3.8	4.8
C_6H_5SH	8.3	−0.8[k]	3.8	4.6
CH_3Cl	11.2	−3.7	3.8	7.5
p-xylene	8.4	−1.1	3.7	4.8
1,2,5,-trimethylbenzene	8.40	−1.03	3.69	4.72
cyclohexene	8.9	−2.1	3.4	5.5
DMF	9.1	−2.4[k]	3.4	5.8
$C_6H_5NH_2$	7.7	−1.1	3.3	4.4
$CH_3CH{=}C(CH_3)_2$	8.7	−2.2	3.3	5.5
CH_3F	12.5	−6.2[i]	3.2	9.4
H_2O	12.6	−6.4[j]	3.1	9.5
$(CH_3)_3As$	8.7	−2.7	3.0	5.7
$(CH_3)_3P$	8.6	−3.1	2.8	5.9
$(CH_3)_2S$	8.7	−3.3	2.7	6.0
NH_3	10.7	−5.6[k]	2.6	8.2
CH_4	12.7	−7.8[i]	2.5	10.3
$C(CH_3)_4$	10.4	−6.1	2.2	8.3
$(CH_3)_2O$	10.0	−6.0	2.0	8.0
$(CH_3)_3N$	7.8	−4.8	1.5	6.3

[a] Data from Ref. [66], except as indicated
[b] Data from Refs. [20, 67, 68], except as indicated
[c] Lloyd DR et al (1976) Mol. Phys. 31: 1549
[d] Lowe JP (1977) J. Am. Chem. Soc. 99: 5557
[e] Xiao SX, Trogler WC, Ellis DE (1983) J. Am. Chem. Soc. 105: 7033
[f] Spence D, Chupka WA (1982) J. Chem. Phys. 76: 2759
[g] Cadez I et al (1977) J. Phys. B. 10: 3821
[h] Bondeybey V, Pearson PK, Schaefer HF (1972) J. Chem. Phys. 57: 1123
[i] Tossell JA et al (1986) Acc. Chem. Res. 19: 281
[j] Melton CE (1972) J. Chem. Phys. 57: 4218

The molecules are arranged in order of decreasing χ, so that Lewis acids start the list, and Lewis bases are at the bottom. Generally, the ordering is very reasonable, though there are some surprises, samely, SF_6, CO, N_2 and H_2, shown as Lewis acids of high electronegativity. But the order shown must *not* be taken as an order of acid strength, but of inherent tendency to give up or take up electrons. The acid strength depends strongly on two other characteristics: the charge or dipole moment of the acid, and the nature of the orbital which accepts the electron density. The same properties of each base must also be considered.

For example, SF_6 is essentially inert. Being neutral, and non-polar, it must accept electrons to interact with a base, but the accepting orbital is an anti-bonding one, which is physically inaccessible. If filled, it would form unstable products such as BF^+, SF_5^-, where B is the base. Both CO and N_2 are non-polar, and have accepting orbitals of π-type; bonding to σ-donors, such as NH_3, would be weak. Only π-bases of low electronegativity will bind well. H_2 and N_2 are both very hard, which helps account for their low reactivity. But acids such as Cl_2 and I_2 are much softer.

Among the bases, the most significant feature is that molecules where the donor atom is F, O, or N are all very hard. This results from large negative values of the electron affinity. Going to similar molecules where the donor atom in Cl, S or P, there is always a large drop in η. A further, smaller drop occurs for Br, I, Se, and As.

A second feature to note in Table 3 is the softening effect of unsaturation. Compare C_6H_6 and C_2H_4 with CH_4; $C_6H_5NH_2$ with NH_3; CH_3COCH_3 with $(CH_3)_2O$. This effect is consistent with the increased polarizability due to unsaturation.

The values for χ for neutral molecules are all less than those for any cation in Table 2. This indicates, correctly, that the molecules will act as bases, or ligands, towards cations. The coordinate bond strength will depend on χ and η for both the ligand and the cation. No simple equation, such as (12) or (14), can be used to calculate the bond energy, but generalizations can be made.

8 The Hardness of Anions

So far we have not presented a Table of χ and η values for anions. There is an experimental difficulty in this case: there is little likelihood that we will be able to measure electron affinities, to form Br^{2-}, for example [30]. For closed shell anions, such as Br^-, we can be sure that A would be a large negative number, because the added electron is in a higher orbital than the valence shell and because of large interelectronic repulsions.

Since the ionization potentials of anions are small, it would appear that χ for most anions will be a negative number, corresponding to a positive value of μ. This is quite reasonable; the escaping tendency of electrons in Br^- would be indeed large. Also the value of η would be quite large. This is disconcerting, sience we would expect Br^- to be very polarizable, or soft.

This hypothetical hardness of anions is due entirely to their resistance to the addition of electrons, to the large negative A values. But in the theory of chemical bonding we would find almost no cases where states corresponding to Br^{2-} play any role. In short, we have come to the end of the regime where it useful to take average values of the slope and curvature of Fig. 1. The only meaningful slope is that for the loss of electrons, so that the effective value of χ is simply I. In the same way, if it is conceded that an ion such as Al^{3+} will only be an electron acceptor, then the effective χ is equal to A.

There is still a curvature, since for $\Delta N = -1$ we go from Br^- to Br, with a different value of μ. In fact, η would simply be equal to that for the bromine atom. But this seems to ignore an important property of the bromide ion, that it is much more polarizable than the bromine atom.

Clearly the value of η that is needed is one that reflects this property. An important feature for Br^- acting as a base, must be the polarization of its electron cloud, without change in the number of electrons. The best measure of this would be the HOMO-LUMO gap of Fig. 2, in the absence of the extra interelectronic repulsion. For practical reasons, it is difficult to obtain this information from either visible-uv spectra or optical polarizabilities.

If we recall that classically the polarizability of a sphere depends on its volume, there is some reason to believe that I and A values for a radical, χ, may give η values roughly proportional to those for the anions, χ^-. Consider the classical charging energy of a conducting sphere of radium R and charge q, $E = q^2/2R$. Now increase q by one unit, and decrease q by one unit, and calculate the changes in energy, which are I and A.

$$I = \frac{2q + 1}{2R}; \quad A = \frac{2q - 1}{2R}; \quad \chi = \frac{q}{R}; \quad \eta = \frac{1}{2R} \tag{26}$$

The hardness is seen to be independent of the charge, and inversely proportional to the radius. For an atom, of course, R is not constant but increase or decreases with the charge.

This classical result closely matches the actual results for atoms and simple free radicals. In these cases (I — A) is usually due to the mean interelectronic repulsion of two electrons in the same orbital. Larger orbitals mean less repulsion. For closed shell system (I — A) is a combination of mean electron repulsions and the energy gap between the HOMO and LUMO in the absence of such repulsions. But even these energy differences would be a function of atom size. A large atom, such as iodine, has smaller intervals between successive orbitals than a small atom like fluorine. This follows from both experimental and quantum mechanical results.

Table 4. Experimental Values for Radicals (Also Models for Corresponding Anions) (Reprinted with permission of the copyright owner)

Radical	I[a]	A[b]	χ	η
F	17.42	3.40	10.41	7.01
OH	13.17	1.83	7.50	5.67
NH_2	11.40	0.74	6.07	5.33
CH_3	9.82	0.08	4.96	4.87
Cl	13.01	3.62	8.31	4.70
SH	10.41	2.3	6.4	4.1
PH_2	9.83	1.25	5.54	4.29
SiH_3	8.14	1.41[e]	4.78	3.37
Br	11.84	3.36	7.60	4.24

Tafel 4 (continued)

Radical	I[a]	A[b]	χ	η
SeH	9.80[e]	2.2	6.0	3.8
I	10.45	3.06	6.76	3.70
H	13.59	0.74	7.17	6.42
HO_2	11.53	1.19	6.36	5.17
NO_2	>10.1	2.3	>6.2	>3.9
CN	14.02	3.82	8.92	5.10
C_6H_5S	8.63	2.47	5.50	3.08
C_6H_5O	8.85	2.35	5.60	3.25
C_2H_5	8.38	−0.39	4.00	4.39
i-C_3H_7	7.57	−0.48	3.55	4.03
t-C_4H_9	6.93	−0.30	3.31	3.61
C_6H_5	9.20	1.1	5.2	4.1
C_2H_3	8.95	0.74	4.85	4.10
CF_3	9.25	>1.1	>5.18	<4.08
$Mn(CO)_5$	8.44[d]	2[f]	5.2	3.2

[a] Refs. [52 and 66]
[b] Refs. [68 and 69], except as indicated
[c] Interpolated from H_2Se and Se
[d] Bidinosti DR, McIntyre NS (1966) Chem. Comm. 555
[e] Nimlos MR, Ellis GB (1986) J. Am. Chem. Soc. 108: 6522
[f] Beauchamp JL, Stevens AE private communication

Table 4 lists the parameters for most of the radicals where the data are available. The assumption seems to be generally reasonable, since we find hardness orders $F^- > Cl^- > Br^- > I^-$; $OH^- > SH^- > SeH^-$; $NH_2^- > PH_2^-$; $CH_3^- > SiH_3^-$; $F^- > OH^- > NH_2^- > CH_3^-$. However, H^- is definitely out of line, since chemically it is very soft.

To summarize, for anions we cannot evaluate η on the same basis as for cations or neutrals. We can rank order by using the corresponding atoms or radicals as models, or we can make estimates based on size. Large anions are softer than small ones. Unsaturated anions such ar CN^- are soft. Delocalized anions have a large effective size, with special features to be discussed next.

9 Local Hardness and Softness

Table 4 includes some large, delocalized anions, which are soft as expected, but all parts are not equally soft. Unlike the chemical potential, which must be a constant everywhere in the molecule, there is nothing in the concept of hardness which prevents it from having different values in different parts of the molecule. The average value over the molecule then gives the global value, η.

The local hardness, $\tilde{\eta}$, has been defined [31] as

$$\tilde{\eta} = \frac{1}{2N}\left\langle \frac{\partial^2 F}{\partial \varrho\, \partial \varrho} \cdot \varrho \right\rangle \tag{27}$$

Here F is the sum of the electronic kinetic energy and the electron-electron repulsion energy. From (27) one can obtain

$$d\mu = 2\langle \tilde{\eta}\, d\varrho \rangle + \langle f\, dv \rangle\,, \tag{28}$$

which is the "local" counterpart of (18). From (28) it follows that,

$$\tilde{\eta} = \frac{1}{2}\left[\frac{\delta\mu}{\delta\varrho}\right]_V\,, \tag{29}$$

and from (14) and (28),

$$\eta = \langle \tilde{\eta} f \rangle \tag{30}$$

Since there are different values of the Fukui function, f, also there are different values of the local hardness, depending on whether electrons are being added or removed. The local hardness measures the change of μ with changing electron density in each part of the molecule.

Unfortunately, these equations do not allow the local hardness to be easily found even if one knows η, and the HOMO and LUMO orbital functions. A more useful quantity is the local softness, $\tilde{\sigma}$, defined by Parr and Yang [32].

$$\tilde{\sigma} = \sigma f \quad \text{and} \quad \sigma = \langle \tilde{\sigma} \rangle \tag{31}$$

This definition allows σ to be readily calculated. Furthermore, from (8) and (14) it follows that

$$\tilde{\sigma} = \left[\frac{\partial\varrho}{\partial\mu}\right]_V\,. \tag{32}$$

In other words, the softest part of a molecule is that where the electron density can be changed with the least change in the chemical potential. There is now a simple interpretation of frontier orbital theory. Reaction occurs between two molecules so that the HOMO of the donor has maximum overlap with the LUMO of the acceptor. This gives the maximum transfer of charge for a given difference in the chemical potentials. This differs from the classical theory which stresses that good overlap means good covalent bonding.

An interesting analogy exists between the softness of different parts of a molecule and the capacity of condensers combined in parallel. For a given $\Delta\mu$ there will be charge changes, Δq, in different parts of the molecule, which are proportional to $\tilde{\sigma}$ [33]. Huheey had called the reciprocal of $(\partial\chi/\partial q)$ the charge capacity as early as 1971 [34]. The capacity of a spherical condenser is equal to $4\pi e_0 R$, in SI units. Komornicki has suggested that $\eta = (4\pi e_0 R)^{-1}$ for atoms [35]. This is consistent with equation (26).

There is a definite relationship between the local hardness, $\tilde{\eta}$, and the local softness, $\tilde{\sigma}$. But one is not simply the reciprocal of the other [33]. In principle, one can be calculated from the other, but it is not easy. Fortunately, there is another way of estimating local hardness at a given atom in a polyatomic system.

This is based on the distribution of electron density around each atom, leading to some estimated net charge on the atom. The assumption is then made that hardness increases with the net positive charge on the atom. The basis for this is seen in Tables 1 and 2. Increasing the positive charge on an atom does increase the hardness. The underlying reason for the effect of charge is that the electron cloud is drawn closer to the nucleus. This increases the mean interelectron repulsion, and also the kinetic energy.

Unfortunately, there is no way of calculating net charges on an atom in a molecule which is completely unambiguous. Even without any calculations, very simple arguments can often be made about relative values of η. For example, the chemical evidence is strong that CF_3^+ is harder than CH_3^+ (vide infra for the evidence). The explanation is that, because F is more electronegative than H, the net positive charge on C in CF_3^+ is greater than that in CH_3^+. Since the carbon atom is the bonding site in these cations, $\tilde{\eta}$ on carbon is the dominant factor.

10 Effect of Ligands on Cations

Table 2 shows χ and η for bare cations, but of course in chemistry we must deal with metal ions surrounded by various ligands. Some of the examples in Table 3 show that neutralizing the cationic charge by anionic ligands greatly reduces both χ and η. This is an expected result, and a neutral ligand will have a similar, but smaller, effect. Data for this are hard to obtain since we need to know ligand bonding energies to a metal in three different oxidation states. For water as a ligand there is enough information to calculate the following [36]:

$$Fe(H_2O)_6^{2+}(g) = Fe(H_2O)_6^{3+}(g) + e^- \qquad I = 15.6\ \text{eV}\ ; \qquad (33)$$

$$Fe(H_2O)_6^{2+}(g) + e^- = Fe(H_2O)_6^{+}(g) \qquad A = 5.3\ \text{eV}\ . \qquad (34)$$

Thus, χ is 10.4 and η is 5.15, compared to 23.4 and 7.24 for Fe^{2+}.

The rather small lowering of η due to water ligands is probably the minimum effect of any neutral ligand, since water is the hardest ligand of this kind. The basis for this statement is the "symbiotic effect" of Klixbull Jørgensen [37]. Soft ligands make the central atom soft and hard ligands make it hard. One sees this rule operating in the cases of PF_3, PCl_3 and PBr_3, and also BF_3, BCl_3 and BBr_3 (Table 3).

For neutral ligands, the following general argument can be made: hard ligands are usually innocent, and soft ligands are usually noninnocent [38]. An innocent ligand allows an unambiguous assignment of the oxidation state of the central ion.

Noninnocent ligands offer the possibility that they have been oxidized or reduced, and not the central ion. Famous examples are the dithiolate ions, such as mnt^{2-}, $(CN)_2C_2S_2^{2-}$.

What noninnocent ligands do, at the very least, is delocalize changes in charge over the entire complex, rather than concentrate such changes on the metal ion. The effective size of the ion becomes large, and η becomes small, according to Eq. (26). Alternatively, we can say that I is reduced and A is increased, because delocalization lowers the energy for changes in charge. Thus (I — A) becomes smaller for soft ligands, as predicted.

The effect of the ligands on χ and η for metal ions is related to, but not the same as solvent effects. The latter refers to the variation of the properties of $Fe(H_2O)_6^{2+}$, and other complex ions, on being transferred from the gas phase to solutions. The solution values, I′ and A′, can be calculated in many cases [39], but they are not as fundamental as the gas phase values.

11 Applications of the Tables

The chief use of the concept of chemical hardness remains essentially unchanged: to understand a variety of chemical phenomena, which depend ultimately on the strength of bonding between a Lewis acid and a Lewis base. However, we are not restricted to acid-base reactions, since free atoms and radicals may also be considered. Hopefully the availability of numbers for χ and η will greatly extend the usefulness of the concepts.

In principle, Eq. (25) should be the best for calculating total bond energies. Corrections for ionic bonding can be incorporated [25]. The use of (25) however, requires estimates of the exchange integrals, β_{12} and β_{21} as wellas those of various interatomic distances.

Tables 1–4 can be used in two possible ways: as a rank ordering of similar acids or bases to predict relative properties, or as a source of numbers to use in relevant equations such as (12) or (25). Table 2 will be useful only for rank ordering, for two reasons. The first is that we rarely deal with isolated gaseous ions, but with a central metal ion, plus ligands. The second is that it is unlikely that numbers such as the, fourth ionization potential of alumiunum will appear in chemically useful equations. In fact, the semiempirical MO theory of the bonding of Al^{3+} with Cl^- will contain only the first three ionization potentials of Al [40].

As an example of rank ordering, consider the divalent cations of the first transition series. There is a pronounced increase in hardness at Mn^{2+} and Zn^{2+}. The reason for this is clearly the special stability of half-filled and filled *d* shells. But the results correlate nicely with chemical behavior. The sequence of stability constants for high spon complexes in water is almost always the Irving-Williams order $V^{2+} < Cr^{2+} > Mn^{2+} < Fe^{2+} < Ni^{2+} < Cu^{2+} > Zn^{2+}$ [31]. While this order is sometimes attributed to ligand field effects, the drops in stability at Mn^{2+} and Zn^{2+} are greatest for polarizable ligands. In other words, hard Mn^{2+} and Zn^{2+} prefer to remain coordinated to hard H_2O ligands rather than softer ones. The ferric ion shows a similar increased value for η amongst the trivalent ions, consistent with its affinity for oxygen donor ligands.

The numbers in Tables 1 and 3 can be used for both rank ordering and numerical calculations. The same is true for Table 4 for the reactions of the neutral atoms and radicals. However, ions are indirectly included, since the empirical MO theory of bonding between Ag^+ and Br^- is the same as that for Ag and Br. Also, rank ordering of the anions may be done from Table 4, with some reservations.

For many reactions of neutral molecules, or radicals, Eq. (12)

$$\Delta N = \frac{(\chi_C^0 - \chi_D^0)}{2(\eta_C + \eta_D)} \tag{12}$$

may be used for numerical calculations. The assumption is usually that a large value of ΔN is favorable for a reaction. But caution is needed since it is quite possible that two systems of equal electronegativity could react to yield strong bonds, even though ΔN would be zero.

Cases where (12) could be useful are those where extensive electron transfer in one direction is mandated. Since strong ionic bonding would result, we cannot estimate the final bond strengths. But Eq. (12), being an initial driving force, should be related to energy barriers to reaction. A large positive value for ΔN would lower the barrier, and a negative value would mean a contribution to the barrier.

A suitable example is the oxidative-addition reaction of transition metal atoms. Taking iron as the metal,

$$X - Y + Fe = X - Fe - Y \tag{35}$$

Table 5. Electron Transfer in Reactions of X — Y with Ground State Iron Atoms (Reprinted with permission of the copyright owner)

X — Y	ΔN[a]	X — Y	ΔN[a]
F_2	0.26	HF	0.030
Cl_2	0.17	HCl	0.027
Br_2	0.15	H_2S	0.007
I_2	0.12	C_2H_2	0.004
H_2	0.11	CH_3Cl	−0.013
O_2	0.11	CH_3F	−0.035
HI	0.072	H_2O	−0.036
HCN	0.066	CH_4	−0.060
CH_3I	0.048	NH_3	−0.065

[a] Calculated from equation 12.

Table 5 shows the values of ΔN calculated, using Eq. (3), and the data in Tables 1 and 3, for the reaction of an iron atom with a number of common X — Y molecules. The results are in remarkable agreement with expectations. Large positive values of ΔN are calculated for facile, fast reactions and negative values for slow, or unobservable reactions. These statements are based upon experimental results for iron atoms in lowtemperature matrices, [41, 42], and on rate studies of various X — Y molecules with suitable low-valent metal complexes [43].

Naturally, other factors which influence rates of reaction can not be ignored. For example, the rate of addition of HX molecules is some 10^7 times faster than the addition of CH_3X. This refers to the concerted addition of molecular HX to $IrCl(CO) \cdot [P(C_6H_5)_3]_2$ in solvents such as toluene [44]. Clearly steric hindrance is much greater in CH_3X than in HX and accounts for the slower reaction in great part.

In addition to changing the substrate, we can also change the metal atom. Metals of lower electronegativity than iron should be more reactive, and metals of higher χ should be less reactive, up to a point. If χ for the metal becomes comparable to χ for the atoms, or radicals, X and Y, then net electron transfer in one direction is not required, and Eq. (12) is no longer a good criterion.

The experimental results are in good agreement, though not complete very low-temperature matrices Sc, Ti, and V react with H_2O, whereas Fe, Cu, and Ni do not react [42]. At 77 °K Cr reacts, but Mo and W do not [45]. At 77 °K, Ti and Zr react with i-C_4H_{10}, but Fe, Cr, Mn, Ni, Co, Cu and Mo do not react [46]. Both V and W show some reactivity.

In oxidative-addition of alkyl bromides and iodides to metal complexes, the complexes of the third transition series are usually the most reactive [47]. These are examples of the case where the electronegativity of the metal is nearly equal to that of X and Y. Hence, the bonds are not ionic, and net electron transfer in one direction is not required.

The addition of H_2 occurs more readily for the third transition series than for the second. The rates are predicted to be in the order Ni $\sim$ Pt $\gg$ Pd [48]. Similar behavior is expected for oxidative cleavage of hydrocarbons. There is a special reason for strong thermodynamic bonding of H^- and R^-, which will be discussed later.

Equation (12) may also be used to predict trends in bond energies, if certain conditions are met. Electron transfer in one direction should dominate, but ΔN should be small. Also a series of ligands, or of metals, should be compared where the interacting orbitals remain nearly constant, or change in a systemmatic manner. One example is the binding of CO to transition metal atoms in their low spin state [49]. As the values in the Tables show, π-bonding must be predominant since CO is more electronegative than any transition metal atom. The values of ΔN are found to be about 0.10. Good agreement is found between the magnitude of ΔN for different metals and the mean bond energy of the metal carbonyls.

BF_3 has invariably been found to be a weaker Lewis acid than either BCl_3, or BBr_3, which are of similar strength. Taking $(CH_3)_3N$ as the reference base, we can calculate ΔN and compare to estimated heats of reaction [50].

	$-\Delta H$	ΔN
BF_3	79 kJ	0.15
BCl_3	155 kJ	0.19
BBr_3	188 kJ	0.19

The calculated values of ΔN show there is some reluctance to adding electrons to planar BF_3 but the difference is too small to account for the large change in ΔH. The energy needed to pyramidalize the acceptor molecule must also play a role. The two factors are closely related, since strong π-bonding is planar BF_3 leads to both the negative electron affinity, and to a reluctance towards becoming non-planar [51].

Tolman has used Eq. (25), at least in principle, to rationalize the bonding of various olefins to Ni(0) [52]. The data are the equilibrium constants in benzene for the reaction

$$NiL_3 + \text{olefin} = NiL_2(\text{olefin}) + L\ K_{eq}\,, \tag{36}$$

where L is a phosphite ligand.

Table 6. Electron Transfer in Reactions of Olefins with Low Spin Nickel Atoms (Reprinted with permission of the copyright owner)

Olefin	Keq[a]	χ	η	ΔN[b]
maleic anhydride	4×10^{8}	6.3	4.7	0.20
trans-NCCH=CHCN	1.6×10^{8}	6.2[c]	5.6	0.17
CH_2=CHCN	4.0×10^{4}	5.4	5.6	0.12
C_2H_4	250	4.4	6.2	0.053
CH_2=CHF	90	4.2	6.1	0.042
styrene	10	4.1	4.4	0.045
CH_3CH=CH_2	0.5	3.9	5.9	0.024
trans-2-butene	2.7×10^{-3}	3.5	5.6	0.000
cyclohexene	3.5×10^{-4}	3.4	5.5	−0.006
$(CH_3)_2C$=$CHCH_3$	3.0×10^{-4}	3.3	5.5	−0.013
Ni(d^{10})	—	3.5	2.3	—
Pd(d^{10})	—	4.8	3.5	—
Pt(d^{10})	—	5.3	2.9	—

[a] Ref. [52]
[b] Calculated from equation 12
[c] I from Chowdhury S, Kebarle P (1986) J. Am. Chem. Soc. 108: 5453

Table 6 shows values of K_{eq} for various olefins, together with their χ and η values. Assuming that χ and η for NiL_2 are the same as for low spin Ni, the values of ΔN have also been calculated using (12), and are shown in Table 6. These are for the reaction

$$NiL_2 + \text{olefin} = NiL_2(\text{olefin})\,, \tag{37}$$

which is related to (36) by a constant factor.

The calculated ΔN's correlate very well with the equilibrium constants. Large positive values mean strong bonding, with π-bonding from metal to olefin dominating. Negative values mean that σ-bonding to the metal is greater than π-bonding. Clearly σ-bonding is less effective than π-bonding.

Ethylene has more π- than σ-bonding, in agreement with theoretical calculations for the reaction of $Ni(PH_3)_2$ with C_2H_4 [53]. The calculated values of ΔN for the reactions of low-spin Pt and Pd are −0.0495 and −0.0206, compared with Ni,

+0.053. This agrees with the theoretical results for binding energies, Ni > Pt > Pd, especially if π-bonding is better than σ-bonding.

Thermodynamically, the third transition series binds most ligands more strongly than the second [54]. For the ligands H^- and R^-, the binding is also stronger than for the first transition series [55]. The reason for this is clear: these ligands are pure σ-bonders, and Pt has the advantage over Ni, and so on. The advantage must lie largely in the special stability of the *6s* orbital, which in turn is due to relativistic effects [56].

12 Application of Local Properties

The use of local hardness is more qualitative, since we cannot use Eq. (12). However, rank ordering, as suggested for CH_3^+ and CF_3^+ earlier and, in many such cases, an experimental basis for ordering are possible. Eventually, this evidence lies in the values of various bond energies although, theoretically, heterolytic bond energies should be used. Thus, if X is less electronegative than Y,

$$X - Y_{(g)} = X^+_{(g)} + Y^-_{(g)} \qquad -\Delta H = D_0' \tag{35}$$

However, the same conclusions will be reached from the more familiar homolytic bond energies, D_0,

$$X - Y_{(g)} = X_{(g)} + Y_{(g)} \qquad -\Delta H = D_0 \tag{36}$$

Now the strength of the bond in X — Y depends on many factors, such as the electronegativities of X and Y, the sizes of the bonded atoms, and the overlap of the bonding orbitals [57]. How does softness enter in? All that theory has really told us is that if X^+ and Y^- are both soft, or if they are both hard, then they are better suited for bonding than otherwise. This is far from being a basis for estimating bond energies.

It does suggest, however, that if we consider an acid-base exchange reaction,

$$C - D'_{(g)} + C' - D_{(g)} = C - D'_{(g)} + C' - D'_{(g)} \tag{37}$$

then ΔH will be negative, if C′ is softer than C, and D′ is softer than D. An example would be [58]

$$\begin{array}{lccccc} & NaI_{(g)} + & AgF_{(g)} = & NaF_{(g)} + & AgI_{(g)} & \\ D_0, kJ & 305 & 347 & 477 & 255 & \Delta H = -80\ kJ\,. \end{array} \tag{38}$$

While this empirical rule is by no means justified by theory, it does work remarkable well in practice. To be successful, it is desirable to minimize other factors which could change the bond energies. That is, the two acids and the two bases should be of the same charge, of similar size, and use corresponding orbitals for bonding.

The rule may be expressed in the following form

$$hs + sh = hh + ss \qquad \Delta H < 0 \; ; \tag{39}$$

where h and s are translated as the harder of the two acids (bases), and the softer of the two bases (acids). Note that we can use either bond energy data, or heats of formation, to calculate ΔH [59, 60].

In this way can calculate ΔH for the reaction

$$\begin{array}{lccccl} & CF_3I_{(g)} + & CH_3F_{(g)} = & CF_{4(g)} + & CH_3I_{(g)} & \\ D_0 & 230 & 460 & 548 & 234 & \Delta H = -92 \text{ kJ} . \end{array} \tag{40}$$

This result is the empirical basis for saying that CF_3^+ is a harder acid than CH_3^+. To be useful, we must demonstrate consistency. For example,

$$\begin{array}{lccccl} & CF_3H_{(g)} + & CH_3F_{(g)} = & CF_{4(g)} + & CH_{4(g)} & \\ \Delta H_f^0 & -696.8 & -234.1 & -932.1 & -74.8 & \Delta H = -76 \text{ kJ} . \end{array} \tag{41}$$

This result is consistent, provided H^- is a softer base than F^-.

Note that fluoroform may also be regarded as a combination of H^+ and CF_3^-. Normally the values of χ for the radicals C and D determine which forms the Lewis acid and which the base. If the values of χ are similar, either choice can be made. For example,

$$\begin{array}{lccccl} & HCF_{3(g)} + & CH_3F_{(g)} = & CH_3CF_{3(g)} + & HF_{(g)} & \\ D_0 & 443 & 460 & 422 & 568 & \Delta H = -87 \text{ kJ} \end{array} \tag{42}$$

requiring that CF_3^- be softer than F^-.

For polyatomic anions, especially when highly delocalized, it becomes more important to know the local softness, $\tilde{\sigma}$, than the global value, σ. This follows because reaction can occur at different parts of such anions. Equations 15 and 31 show that we need to know the wave function of the anion, or at least of the HOMO. While this may appear formidable, there are two factors which help considerably. One is that wave functions have already been calculated for thousands of molecules and ions, and the results are available in tabulations and bibliographies [61]. The second is that usually only approximate wave functions are needed, such as the results of a Hũckel calculation.

For instance, consider the anion of malonaldehyde, a model for β-diketone anions.

$$\begin{array}{ccccccccc} O & \cdots & CH & \cdots & CH & \cdots & CH & \cdots & O^- \\ 1 & & 2 & & 3 & & 4 & & 5 \end{array}$$

The HOMO is a π-type orbital, for which a Hũckel calculation gives [62]

$$\varphi = 0.435\psi_1 - 0.286\psi_2 - 0.677\psi_3 - 0.2869\psi_4 + 0.435\psi_5 \, . \tag{43}$$

Squaring this to obtain ϱ, we find 0.458 for C_3^2, and 0.184 for C_1^2 and C_5^2. Thus, the central carbon is the soft site, and the oxygen atoms are much harder. Overall, the oxygens are more negative than the carbon. Hard acids such as H^+, Al^{3+} and Mg^{2+} will bind to oxygen, as will most transition metal ions. But Pt^{2+}, Pd^{2+} and Hg^{2+} may bind to carbon, as will Br^+ and HO^+.

Even without calculation, we can often make reasonable predictions. In Table 4 we see that $C_6H_5O^-$ and $C_6H_5S^-$ may not be very different in their values for σ, though the latter is softer. But the HOMO of $C_6H_5O^-$ is known, and has coefficients at the *ortho* and *para* ring carbons which are almost as large as that on oxygen [63]. This delocalization is conventionally shown by resonance structures such as

Since sulfur is less efficient at π-bonding to carbon than oxygen is, we can be fairly sure that $C_6H_5S^-$ has a HOMO which is more concentrated on S. Therefore it is locally very soft at sulfur, whereas $C_6H_5O^-$ is locally much harder at oxygen.

The chemical consequences of this difference are very considerable. First of all, $C_6H_5S^-$ is very much more reactive than $C_6H_5O^-$ in reacting with electrophilic reagents such as alkyl halides where electron transfer is important. Secondly, $C_6H_5S^-$ invariably reacts at the sulfur atom, whereas $C_6H_5O^-$ is just as likely to react at carbon atoms in the ring.

13 Summary

The purpose of this review has been to show, by practical application, that DFT is a powerful tool for elucidating chemical behavior. The new concepts of electronic chemical potential, or absolute electronegativity, absolute global hardness and local hardness (or the related softness properties) can be calculated or estimated for chemical systems. They are equally valid for atoms, ions, molecules or radicals. They are strongly related to chemical behavior, since χ measures the tendency to give up, or take up, electrons and η measures the ease with which this can occur.

Fortunately, the new concepts are completely compatible with MO theory, in its simple form. The frontier orbital theory of chemical reactivity is recovered, but new insights are gained as well. For example, the assumption that reaction occurs at sites where the electron density is large (HOMO), and the hole density is large (LUMO), in conventional theory is explained by arguments of good overlap [64]. In DFT theory these sites are those where the electron density can be changed most easily. These two explanations are consistent and complementary. Density functional theory, of course, offers great hope to chemistry in that the functional which gives the energy may be discovered some day. But even at the present state of development, it appears that chemically useful conclusions can be derived from DFT.

Acknowledgement This work has been supported over the years by grants from the U.S. Department of Energy.

15 References

1. Dahl JP, Avery J (eds) (1984) Loal density approximations in quantum chemistry and solid state physics, Plenum, New York
2. Hohenberg P, Kohn W (1964) Phys. Rev. 136: B864
3. Slater JC, Johnson KH (1972) Phys. Rev. 135: 844
4. For a recent example see Ziegler T, Tschinke V, Ursenbach C (1987) J. Am. Chem. Soc. 109: 4825
5. Parr RG (1983) Am. Rev. Phys. Chem. 34: 631
6. Parr RG, Donnelly RA, Levy M, Palke WE (1978) J. Chem. Phys. 68: 3801
7. Mulliken RS (1934) J. Chem. Phys. 2: 782
8. Iczkowski RP, Margrave JL (1961) J. Am. Chem. Soc. 83: 3547
9. Pauling L (1960) Nature of the Chemical Bond, 2nd ed. Cornell University Press, Ithaca, p 88
10. Mortier WJ, Ghosh SA, Shankar S (1986) J. Am. Chem. Soc. 108: 4315
11. Sanderson WT (1955) Science 121; 207
12. Parr RG, Pearson RG (1983) J. Am. Chem. Soc. 105: 7512
13. Pearson RG (1985) J. Am. Chem. Soc. 107: 6801
14. Klopman G (1964) J. Am. Chem. Soc. 86: 1461, 4550
15. Nalewajski RF (1984) J. Am. Chem. Soc. 106: 944
16. Parr RG, Yang W (1984) J. Am. Chem. Soc. 106: 4049
17. Fukui K (1973) Theory of orientation and stereoselection, Springer Verlag, Berlin Heidelberg New York
18. Berkowitz M (1987) J. Am. Chem. Soc. 109: 4823
19. See references for Tables 1–4
20. Jordan KD, Burrow PD (1987) Chem. Rev. 87: 557
21. Perdew JP, Parr RG, Levy M, Balduz JL (1982) Phys. Rev. Lett. 4: 169
22. Pearson RG (1988) J. Am. Chem. Soc. 110: 2092
23. Pearson RG (1963) J. Am. Chem. Soc. 85: 3533
24. Pearson RG (ed) (1973) Hard and soft acids and bases, Dowden, Hutchinson and Ross, Stroudsburg, Pa
25. Klopman G (1968) J. Am. Chem. Soc. 90: 223
26. Dewar MJS (1952) J. Am. Chem. Soc. 74: 3341. Fukui K, Fujimoto H (1968) Bull. Chem. Soc. Japan 41: 1984
27. Jorgensen CK (1967) Struct. Bonding (Berlin) 3: 106
28. Abrland S, Chatt J, Davies NR (1958) Quant. Rev. (London) 19: 265
29. Ahrland S (1966) Struct. Bonding (Berlin) 1: 207
30. Mark T (1986) Int. J. Mass Spectrum. Ion Proc. 68: 341
31. Berkowitz M, Ghosh SK, Parr RG (1985) J. Am. Chem. Soc. 107: 6811
32. Yang W, Parr RG (1985) Proc. Nat. Acad. Sci. 82: 6273
33. Berkowitz M, Parr RG (1987) J. Chem. Phys. 88: 2554
34. Huheey JE (1971) J. Org. Chem. 36: 204. Politzer P (1987) J. Chem. Phys. 86: 1072
35. Komorowski L (1987) Chem. Phys. 114: 55
36. Basolo F, Pearson RG (1967) Mechanisms of inorganic reactions, 2nd ed. John Wiley, New York, p 60
37. Jorgensen CK (1964) Inorg. Chem. 3: 1201
38. Jorgensen CK (1966) Struct. Bonding (Berlin) 3: 234
39. Pearson RG (1986) J. Am. Chem. Soc. 108: 6109
40. Pearson RG, Gray HB (1963) Inorg. Chem. 2: 358
41. Parker SF, Peden CFH, Barrett PH, Pearson RG (1984) J. Am. Chem. Soc. 106: 1304
42. Kaufman JW (1982) Reactions of metal atoms in low temperature matrices. Thesis, Rice University, Houston
43. Halpern J, Chock PB (1966) J. Am. Chem. Soc. 88: 3511. (1978) Jesson JP 100: 7577 (1974) Collman JP, McLaury MR 96: 3019
44. Walper W, Kelm H (1978) J. Phys. Chemie (N.F.) 113: 207
45. Skell PS, McGlinchey MJ (1975) Angew. Chem. Engl. 14: 195

46. Remick RJ, Asunta TA, Skell PS (1979) J. Am. Chem. Soc. 101: 1320
47. Pearson RG (1987) Adv. Chem. Series No. 215: 233
48. Blomberg MRA, Siegbahn PEM (1983) J. Chem. Phys. 78: 986, 5682
49. Pearson RG (1984) Inorg. Chem. 23: 1675
50. Rothe EW, Mathews BP, Reck GP (1980) Inorg. Chem. 19: 829
51. Tossell JA, Moore JH, Olthoff JK (1986) Int. J. Quantum Chem. 29: 1117
52. Tolman CA (1974) J. Am. Chem. Soc. 96: 2780
53. Kitaura K, Sakaki S, Morokuma K (1981) Inorg. Chem. 20: 2292. (1985) Ziegler T 24: 1547
54. Johnston RD, Basole F, Pearson RG (1973) Inorg. Chem. 10: 247 (1984) Lewis KE, Golden DM, Smith GP J. Am. Chem. Soc. 106: 3905
55. Pearson RG (1985) Chem. Rev. 85: 41 (1986) Norton JR, Sullivan JM J. Am. Chem. Soc. 108: 2257
56. Pitzer KS (1979) Acc. Chem. Res. 12: 271
57. Kutzelnigg W (1984) Angew. Chem. Engl. 23: 272
58. Pearson RG, Mawby RJ (1967) In: Gutman V (ed) Halogen chemistry, Academic, New York, vol 3, p 55
59. Bond energy data from McMillen DF, Golden DM (1982) Annu. Rev. Phys. Chem. 33: 493
60. Heats of formation from Stull DR, Westrum Jr. EF, Sinke GC (1969) Chemical thermodynamics of organic compounds, John Wiley, New York
61. Richards WG, Walker TEH, Hinkley RK (1971) A bibliography of ab initio wave function, Clarendon, Oxford Supplements published in 1974, 1978 and 1981 (1982) Ohno K, Morokuma K Quantum chemistry literature base, Elsevier, Amsterdam. (Supplements published annually in J. Mol. Struct. Theochem) (1965) Streitwieser A, Brauman JI Supplementary tables of molecular orbital calculations, Pergamon, Oxford, Vols I and II
62. Haddon RC (1980) J. Am. Chem. Soc. 102: 1807
63. Pross A, Radom L, Taft RW (1980) J. Org. Chem. 45: 818
64. Klopman G (1974) Chemical reactivity and reaction paths, John Wiley, New York, Chapter 4
65. Moore CE (1970) Ionization Potentials and Ionization Limits, Ref. Data Ser. Nat. Bur. Stand. 34, Washington, D.C
66. Rosenstock HM, Draxl K, Steiner BW, Herron JT (1977) J. Phys. Chem. Ref. Data, 6: Suppl. No
67. Fukuda SK, McIver RT (1985) J. Am. Chem. Soc. 107: 2291 (1985) Grimsrud SA, Caldwell G, Chowdhury S, Kebarle P 107: 4267 (1985) Tossell JA, Moore JH, Giordan JC Inorg. Chem. 24: 1100, J. Am. Chem. Soc. 107: 5600
68. Drzaic PS, Marks J, Brauman JI (1984) Gas phase ion chemistry, Academic, New York, Vol 3, Chapter 4
69. DePuy CH, Bierbaum VM, Damrauer R (1984) J. Am. Chem. Soc. 105: 4051

Valence in Molecular Orbital Theory

K. Jug[1], M. S. Gopinathan[2]
[1] Theoretische Chemie, Universität Hannover, FRG
[2] Indian Institute of Technology, Madras, India

1 Theory

1.1 Introduction

The idea of valence originated from the desire of chemists to measure the extent of covalent bonding in a molecule. In this sense, the valence of an atom in a molecule should count the number of covalent bonds formed by this atom. The valence number or valency of an atom in a molecule can thus be considered as the used covalent bonding capacity of this particular atom. This actual covalent bonding must be distinguished from the potential covalent bonding that a particular type of atom can extend to a number of other atoms. The carbon atom can form four covalent single bonds, so the potential valency of the free atom is 4. It is equal to the number of electrons needed in the bonding to complete an octet shell for carbon forming electron pair bonds in a molecule. This normal valence number is expected in methane CH_4, whereas carbon in CH_3, CH_2 and CH would be described by a valency of 3, 2 and 1, respectively. This represents the number of covalent bonds in these molecular systems (Fig. 1).

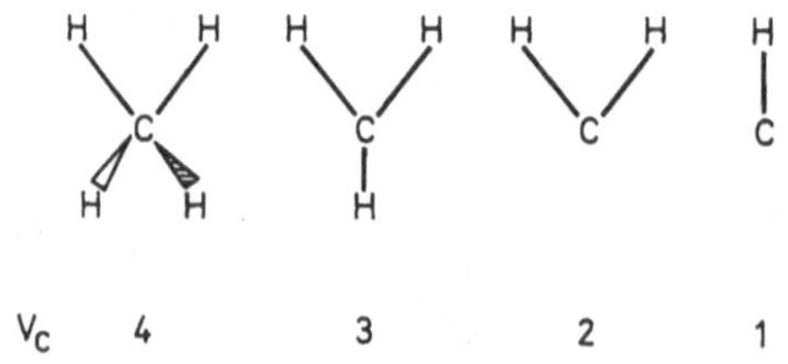

Fig. 1. Expected valence number V_C of carbon in hydrocarbon systems

In a similar way the normal valency for atoms in the first row engaged in bonding would be 3 for N and B, 2 for O and Be, 1 for F and Li. If different types of atoms are bound together, a portion of the potential covalent bonding is converted to ionic bonding due to partial electron transfer from one atom to the other. Here it is expected that the more electronegative atom will receive electronic charge from the less electronegative atom so that its net charge is negative. Of course, in a molecule with many different atoms bound by various bonds, the situation is much more complicated and nothing can be said off-hand about the amount of loss of covalent bonding due to charge transfer. From the foregoing it is evident that the partial charge transfer and the corresponding reduction or increase of covalent bonding will lead to noninteger valence numbers. Open-shell systems also require in general noninteger valence numbers because not all electrons can be paired to form electron pair bonds. There is a further complication in molecular ions where the charge is responsible for ionic bonding or repulsion together with covalent bonding.

A milestone in the study of valence and the first quantitative account of bonding in molecules was Coulson's famous 1939 paper [1] where he defined charge q_μ and bond order $p_{\mu\nu}$ for atomic orbitals (AO's) μ and ν on different atoms in the framework of the Hückel method as:

$$q_\mu = \frac{\partial E}{\partial \alpha_\mu} = \sum_i^{occ} c_{i\mu}^2 \tag{1}$$

$$p_{\mu\nu} = \frac{\partial E}{\partial \beta_{\mu\nu}} = \sum_i^{occ} c_{i\mu}c_{i\nu}$$

from the total Hückel energy E

$$E = \sum_i^{occ} \left(\sum_\mu c_{i\mu}^2 \alpha_\mu + \sum_{\mu,\nu} c_{i\mu}c_{i\nu}\beta_{\mu\nu} \right) \tag{2}$$

In this derivation the atomic orbitals must be considered as orthogonalized (OAO's) and the appropriate way to achieve this is the symmetric orthogonalization by Löwdin [2]. Later Coulson also defined charges and bond orders for nonorthogonal atomic orbitals (AO's) [3] but this procedure is neither straightforward nor unique and is not generally accepted. Rather, alternative definitions of charges and bond orders have been put forward by Mulliken [4] and others which are more widely used. But Coulson can be credited with the first derivation of what was initially called the charge and bond order matrix P and which is now referred to as the first-order density matrix.

$$P_{\mu\nu} = \sum_i^{occ} c_{i\mu}c_{i\nu} \tag{3}$$

When Coulson derived his equations (1) he considered only π electrons with one AO per atom. In general, we need more than one orbital for the description of all electrons and the meaning of the elements of P will be modified. The diagonal elements $P_{\mu\mu}$ are now occupancies of orbitals μ and the charge is the sum of all occupancies on a single atom. The nondiagonal elements $P_{\mu\nu}$ can be subdivided into two classes. If μ and ν are on the same atom they represent intraatomic interactions whereas for μ and ν on different atoms interatomic interactions arise which contribute to the covalent bonding between these two atoms. A generalization of bond order $P_{AB} = P_{\mu\nu}$ in the spirit of Coulson for the case of more than one orbital per atom was proposed by Jug, first with the Mulliken criterion [5] for bonding and antibonding and later modified with a vector projection criterion [6].

1.2 Valence Definitions

To find a proper description for the valency of an atom in a molecule several criteria must be put forward.

1. Invariance under coordinate transformations
2. Measure of electron sharing between atoms
3. Reflection of saturation of covalent bonding
4. Relation to covalent chemical reactivity
5. Deduction from density matrix

These were suggested by us several years ago [7]. To fulfill these conditions we found it appropriate to start from the density matrix P and develop a theory of valence based on an early suggestion by Wiberg [8] which relates charge and covalent bonding in an orbital. A heuristic approach to the definition of a valence number V_A was already contained in papers by Armstrong et al. [9] and Borisova and Semenov [10] but in neither were all the formulas generally valid nor was an in-depth study of the implications from the various definitions made. A review of them can be found in a recent book by Kuznetsov [11]. The main ideas on which the following derivations and applications are based are expressed in the papers by Gopinathan and Jug [7] on the SCF level and by Jug [12] on the CI level. Also a CI extension to natural orbitals in the more general case of interactions between single and double excitations is now available [13].

We start now by subjecting the AO's χ_v to a Löwdin orthogonalization [2] to obtain the OAO's λ_μ:

$$\lambda_\mu = \sum_v (S^{-1/2})_{\mu v}\, \chi_v \tag{4}$$

and we subsequently expand the molecular orbitals ψ_i in this set,

$$\psi_i = \sum_\mu c_{i\mu} \lambda_\mu \tag{5}$$

The elements $P_{\mu v}$ of the density matrix P are then calculated according to (1).

Wiberg [8] has shown that the following relationship holds between the occupancy q_a of orbital a and the squares of the density matrix elements P_{ab} of all other OAO's

$$\sum_{b \neq a} P_{ab}^2 = 2q_a - q_a^2 \tag{6}$$

For the special case of a doubly occupied MO in a diatomic molecule with only one atomic orbital a on an atom A, the left-hand side of (6) provides information about the bonding between atoms A and B and can be used as a valence index. We denote it by V_a and call it the valency of orbital a. Fig. 2 provides an illustration for this case.

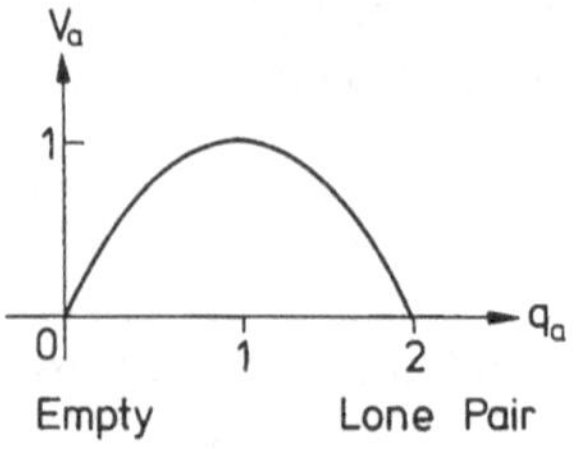

Fig. 2. Valency V_a in dependence of charge q_a for a closed-shell diatomic molecule with a single MO

Here we see that an ideal covalent single bond is formed for equal sharing of the total charge between the atoms ($q_a = 1$). If orbital a is empty ($q_a = 0$) or contains a lone pair ($q_a = 2$) the valency V_a is zero and the situation is non-bonding.

Since the density matrix elements referring to two centers are a measure of covalent bonding, the valency V_a extended from an atomic orbital a to other centers can be generalized as:

$$V_a = \sum_{B} \sum_{b \neq a} P_{ab}^2 \tag{7}$$

The valency V_a of an atom A is the sum of all atomic orbital contributions

$$V_A = \sum_{a}^{A} V_a \tag{8}$$

To derive a more general expression from (6) we now rewrite (8) for a closed shell case [7]

$$V_A = \sum_{B \neq A} \sum_{b}^{B} \sum_{a}^{A} P_{ab}^2 = \sum_{a} (2q_a - q_a^2) - \sum_{a'}^{A} P_{aa'}^2 \tag{9}$$

From (9) it is easy to derive the following inequality:

$$V_A \leqq \sum_{a} (2q_a - q_a^2) \tag{10}$$

The equal sign holds for the case where a transformation to natural hybrid orbitals (NHO's) eliminates all nonvanishing intraatomic nondiagonal density elements $P_{aa'}$ [14]. We conclude from (10) that for any element in the first row the atomic valency V_a cannot exceed 4 if only 2*s* and 2*p* orbitals are considered as valence orbitals in the basis set. A special case is carbon for which tetravalence is the maximum valence. The search for pentavalent carbon as it has been suggested by some experiments [15] is therefore useless. At best we can find pentacoordinated carbon where the valence multiplicity of the average bond is less than that of a single bond.

P	A	B	C
A	P_{AA}	P_{AB}	P_{AC}
B	P_{BA}	P_{BB}	P_{BC}
C	P_{CA}	P_{CB}	P_{CC}

Fig. 3. General density matrix in atomic block form

The valence properties of atoms in molecules will now be presented in a systematic way. Starting from the block form of the density matrix (Fig. 3). We define the bond valence number V_{AB} between two atoms as the sum of squares of all orbital contributions:

$$V_{AB} = \sum_{\mu, \nu} P_{\mu_A \nu_B}^2 \tag{11}$$

The atomic valence number V_A is then the sum of all bond valence contributions:

$$V_A = \sum_{B \neq A} V_{AB} \tag{12}$$

The number of covalent bonds in the molecule can then be defined by the molecular bond number M:

$$M = 1/2 \sum_A V_A \tag{13}$$

At this point we can say that four of the five criteria set forth in the beginning are fulfilled by these definitions. Eq. (7) is found to be invariant under coordinate transformation [7], which is required by the first criterion. Eq. (11) is a measure of electron sharing and therefore fulfills the second criterion. Inequality (10) limits the covalent bonding of an atom in line with the third criterion. Finally all these formulas are derived from the density matrix according to the fifth criterion. The fourth criterion will be discussed in the section 1.4.

To complete the discussion of these definitions several more points should be mentioned. It can be shown that the relevant formulas (7) to (13) are invariant under transformation to natural hybrid orbitals (NHO's). For natural bond orbitals (NBO's) [5] the sum over $P^2_{\mu_A \nu_B}$ in (11) has to be replaced by a sum over the squares of the positive eigenvalues p_μ [12] corresponding to the NBO's. This was demonstrated on the SCF level [7] and CI level [12, 13]. The general formulas were applied in semiempirical methods and minimal basis set ab initio calculations. This will be the topic of the following sections. A more general assessment for ab initio calculations will be given in the section on the basis set dependence.

It is now appropriate to point out that definitions of bond valence number V_{AB}, atomic valence numbers V_A and bond numbers M can also be given on the basis of nonorthogonal orbitals χ. Giambiagi et al. [16] used the matrix $P^\chi S$ to replace P^λ in the formulas. Later Mayer [17] independently arrived at the same formulas, but extended the scope and applications. Since the relation between P^λ and P^χ is

$$P^\lambda = S^{1/2} P^\chi S^{1/2}, \tag{14}$$

the results would be the same for orthogonal and nonorthogonal basis if P^χ and S commute. This is generally not the case. There are serious problems with Mayer's definition in extended basis sets [18] and we shall not present further details of it. Moreover Mayer's approach yields valence numbers for carbon close to 4 both for CH_4 and CH_3 which should not be the case. It is clear, however, that a theory on nonorthogonal density matrix elements can be pursued both in semiempirical and ab initio methods.

1.3 Molecular Orbital Valence

A different approach which leads to the concept of valence numbers as a sum of MO contributions can be pursued in the following way [12]. We use the CI expansion for the total wavefunction Ψ:

$$\Psi = \sum_I A_I \Psi_I \tag{15}$$

These Ψ_I are Slater determinants which are composed of MO's ψ_i. The MO's are in turn expanded in OAO's λ [11]

$$\psi_i = \sum_\varrho c_{i\varrho}\lambda_\varrho \tag{16}$$

The density operator can then be defined as

$$\Gamma = |\Psi\rangle\langle\Psi| . \tag{17}$$

We select now the diatomic portions of the density operator to describe the bonding between atoms A and B

$$\Gamma^{AB} = |\Psi^A\rangle\langle\Psi^B| , \tag{18}$$

and reduce the MO's to diatomic contributions of these two atoms

$$\psi_i^{AB} = \psi_i^A + \psi_i^B . \tag{19}$$

From Γ^{AB} the corresponding first-order reduced density operator P^{AB} can be derived [19],

$$P^{AB} = \sum_{I,J} A_I A_J \sum_{i,j} \varrho_{IJij} |\psi_{iI}^A\rangle\langle\psi_{jJ}^B| \tag{20}$$

$$\text{with} \quad \varrho_{IJij} = \begin{cases} (-1)^{s-t} & \text{for} \quad \{I - i_s\} = \{J - j_t\} \\ 0 & \text{otherwise} \end{cases}$$

Eq. (20) can be rewritten in diagonal form in terms of natural orbitals (NO's) ω_i

$$\omega_i = \sum_\varrho c'_{i\varrho}\lambda_\varrho ; \tag{21}$$

and their occupation numbers n'_i

$$P^{AB} = \sum_I A_I^2 \sum_i |\omega_{iI}^A\rangle\langle\omega_{iI}^B| = \sum_i n'_i |\omega_i^A\rangle\langle\omega_i^B| \tag{22}$$

We now define the bond valency V_{AB} as the expectation value of P^{AB} over reduced natural orbitals ω_i^{AB}

$$V_{AB} = \sum_i n'_i \langle\omega_i^{AB}| P^{AB} |\omega_i^{AB}\rangle = \sum_i V_i^{AB} \tag{23}$$

where V_i^{AB} is the orbital contribution of NO ω_i to the bond valency V_{AB}. Substitution of (21) and (22) into (23) yields,

$$\begin{aligned} V_{AB} &= \sum_{i,j} n'_i n'_j \langle\omega_i^A | \omega_j^A\rangle\langle\omega_j^B | \omega_i^B\rangle \\ &= \sum_{i,j} n'_i n'_j \sum_\varrho c'^A_{i\varrho} c'^A_{j\varrho} \sum_\sigma c'^B_{j\sigma} c'^B_{i\sigma} \\ &= \sum_{\varrho,\sigma} \sum_i n'_i c'^A_{i\varrho} c'^B_{i\sigma} \sum_j n'_j c'^A_j c'^B_j \\ &= \sum_{\varrho,\sigma} (P^{AB}_{\varrho\sigma})^2 \end{aligned} \tag{24}$$

The result of (24) is equivalent to (11). Definition (23) provides a convenient way of describing the total bonding as an exclusive sum of partially occupied natural orbitals. Alternatively, one could express V_{AB} as a sum of MO contributions [13],

$$V_{AB} = \sum_{i} n_i \langle \psi_i^{AB} | P^{AB} | \psi_i^{AB} \rangle + \sum_{i,j} n_{ij} \langle \psi_i^{AB} | P^{AB} | \psi_j^{AB} \rangle \; ; \qquad (25)$$

where n_i is the MO occupation number and n_{ij} the coupling element between two MO's [11]

$$n_{ij} = \sum_{\substack{I,J \\ I \neq J}} A_I A_J (-1)^{s-t} \delta_{i i_{sI}} \delta_{j_t J j} \qquad (26)$$

1.4 Free Valence

The original use of the valence concept as developed by Coulson focussed on chemical reactivity of radical reactions. As a prototype for discussion the butadiene molecule was used [20]. Under the assumption of a fixed σ core with unit σ bond between the carbon a π electron calculation with the Huckel method was performed to obtain the π electron bond orders between pairs of carbon atoms. Coulson then argued that the sum of bond orders from adjacent atoms to the reference atom would be a measure of how much this atom is engaged in covalent bonding. Interactions between nonadjacent atoms are neglected.

This procedure resulted in distinctly different atomic valence numbers for the inner and outer carbon atoms of butadiene. Since the outer atoms had smaller valence numbers, Coulson argued that they had more free valence than the inner atoms and would therefore be the sites for radical attack.

Free valence F_A in the sense of Coulson was then defined as the difference between a maximum atomic valence number $V_{A\,max}$ and the actual valence number V_A of an atom A in a molecule

$$F_A = V_{A\,max} - V_A \, . \qquad (27)$$

Comparisons were made between free valence and calculated polarization energies [21] in a series of unsaturated hydrocarbons.

Later Kooyman and Forenhorst [22] correlated the position of maximum free valence in hydrocarbons and the center of attack for CH_3 and CCl_3 radical addition and found a linear relationship between free valence and the logarithm of the reaction rate.

Despite the success of these tests for unsaturated hydrocarbons, the absolute atomic valence numbers from Coulson's definition are unphysical. In a model calculation on trimethylenemethane $C(CH_2)_3$ the valence number of the inner carbon atom would be 4.71 [23]. This was considered as the maximum and the amount of free valence was measured in comparison with this value. It can be easily demonstrated that planar tetramethylenemethane $(CCH_2)_4$ would result in an even larger valence number for the inner carbon [24] and therefore increase the difficulty with these numbers. It is clear that the number of 4 is exceeded because the bond orders are not additive as we have pointed out [12]. In trimethylenemethane, for instance, the same central π orbital is used to describe the π bonding to all three other atoms. Summing up the square of π orbital density matrix elements in this particular case results in a π valency V_C of 1 instead of $\sqrt{3}$. Physically this means that the maximum amount of exclusive sharing that a single atomic orbital can extend to other atoms is 1.

We therefore suggested that free valence of an atom in a molecule is the difference between the atomic reference value $V_{A\,normal}$ for this type of atom and the actual valency V_A of this atom in a molecule.

$$F_A = V_{A\,normal} - V_A \tag{28}$$

Since all contributions from other atoms are counted towards the valence number V_A, it usually includes also some small antibonding contributions.

However, one could also follow Coulson's argument that for reactivity only those bonds are counted which are the nearest neighbor bonds of an atom. This free valence F'_A would be defined as

$$F'_A = V_{A\,normal} - V'_A \tag{29}$$

$$\text{with} \quad V'_A = \sum_{B \neq A}^{\text{bonds}} V_{AB}$$

The trends of this result are the same as Coulson's but normalized to physical standards. So Coulson's predictions can be considered as a special case of (29).

If we consider the attack of a radical on a π electron system, it is only this subset of electrons related to the π orbitals which should be taken into account in valence calculations. So if we can separate the occupied orbitals in subsets of different symmetry, only the subset directly involved in the reaction would be taken to measure reactivity.

1.5 Subvalence and Hypervalence

It is apparent to chemists that the normal valency $V_{A\,normal}$ for atoms in the first row engaged in bonding is 4 for carbon, 3 for nitrogen and boron, 2 for oxygen and beryllium and 1 for fluorine and lithium. If we take this as a standard we can compare the actual valency V_A of atom A in a molecule with this reference value. We term an atom A with $V_A < V_A^{normal}$ as subvalent and with $V_A > V_A^{normal}$ as hypervalent.

However, since there is a continuous scale of deviations from the reference value, these terms become meaningful only if a certain interval e.g., ± 0.1 about this reference value is exceeded. One could further differentiate between moderately sub- and hypervalent compounds with $\pm(0.1$ to $0.5)$ and highly sub- and hypervalent compounds with deviations larger than 0.5.

In this context it is possible to define a radical as a subvalent compound with an atom whose valency is one unit less than normal and a diradical as a subvalent compound with one atom with valency by two units less than normal or two atoms by one unit less than normal [25] whereas all other atoms have valencies around normal. Of course, the latter definition has to be modified if the symmetry of the molecule prescribes a distribution of the deficient bonding over more than two atoms. This definition agrees quite well with Berson's [26] who called a molecule a diradical if the total number of bonds is one less than allowed by the standard rules of valence.

Typical radicals would be CH_3 and NH_2 with expected values $V_C = 3$ and $V_N = 2$. A typical diradical would be CH_2 with only two electrons engaged in the bonding or trimethylenemethane $C(CH_2)_3$ where an average valency $V_C = 3.33$ is expected for the outer carbons.

A zwitterion is a neutral molecule with a substantial charge separation where two, usually not adjacent atoms, appear with a positive and negative unit charge. If such a positive charge appears on Li, Be, B and C, its valency should be decreased whereas it will be increased for N, O and F. Alternatively, a negative charge will create an increase of valency for Li, Be and B and a decrease for C, N, O and F. In general, a zwitterion will appear as a molecule with a pair of sub- and hypervalent atoms.

In the second row the question of hypervalence comes up in sulfur and phosphorus compounds. From inequality (10) it is clear that the valency in such compounds cannot exceed 4 in a minimal basis set of *s* and *p* orbitals. It is, therefore, necessary to study the effect of *d* orbitals in second-row compounds.

1.6 Basis Set Dependence

The concept of valency as introduced in Sect. 1.2 and 1.3 is immediately applicable if only minimal basis sets are employed. In semiempirical all-valence electron methods only valence orbitals e.g., 2*s* and 2*p* orbitals in the first row, are included for the description of bonding. In ab initio methods minimal basis sets comprise also 1*s* orbitals. Minimal basis sets appeal to the chemist because they are usually needed even in the description of the atom. Exceptions are Li and Be where only 2*s* orbitals are occupied in the atom. If only a 2*s* orbital is used for the description of bonding in molecules, it is clear from (10) that the atomic valence number cannot exceed 1. More generally, it can be shown that the valence number of an atom cannot exceed 1 if there is only one doubly occupied MO in a molecule. A realistic case is Li_2 where the inclusion of 2*p* orbitals cannot lead to an increase of V_A above 1. Here the hybrid on Li is, at the same time, a natural atomic hybrid (NHO) [27]. If more than one MO is occupied, the transformation to NHO's will, in general, create occupancies different from zero for more than one NHO. It can now be demonstrated [27] that an increase of valence number in particular for Li atoms occurs if there is a tendency towards equal occupancy of several NHO's. This usually happens in compounds with several lithium atoms if the ionic bonding is not pronounced in the basis set.

The same situation arises in the second row if 3*d* orbitals are included in the basis set in addition to 3*s* and 3*p* orbitals. So hypervalence in sulfur and phosphorus compounds with valencies larger than 4 is linked to the inclusion of 3*d* orbitals. This special group of compounds includes molecules with pentavalent and higher valent atoms. We shall call the increase in valency of an atom due to 3*d* orbitals extended valence. Three situations can be distinguished: a) the influence of *d* orbitals is small, b) the influence of *d* orbitals is larger, but the valency of the atom does not exceed 4, c) the influence of *d* orbitals is large and the atomic valency exceeds 4.

In ab initio calculations with split-shell bases like double zeta or 4-31G and polarization functions like 6-31G*, the basis functions are not physically interpretable, because they do not represents atomic orbitals. But they serve to lower the total energy by creating more sophisticated electron distributions in space. Since it can be shown [7] that these additional basis functions tend to increase the valence numbers, we suggest that a contraction procedure be used which reduces the degrees of freedom to a function space of dimension 5 in the first row and to 9 in the second row. Details will be described elsewhere [27].

2 Molecular Structure

2.1 Multiplicity of Bonding in the First Row

To understand how much the concept of valence developed in the first section reflects chemical knowledge, the theory must be applied to a representative variety of compounds. For first-row atoms the semiempirical MO method SINDO1 [28] seemed most appropriate for a uniform comparison. Its accuracy for geometries

Table 1. Subvalence, normal valence and hypervalence in the first row

Subvalence			Normal Valence		Hypervalence	
Atom	Molecule	Valency	Molecule	Valency	Molecule	Valency
C	CH_3	2.97	CH_4	4.00		
	$^3CH_2$	1.91	C_2H_4	3.99		
	CH	1.00	C_2H_2	3.97		
N	NH_2	1.99	N_2	3.00	NH_4^+	3.64
	^3NH	0.99	NH_3	2.97	HNC	3.54
O	BO_2	1.57	CH_3OH	2.02	O_3 (inner)	2.81
	$^3O_2$	1.50	H_2O	1.96	CO	2.57
F	HF	0.96	OF	1.01	BF	1.27
			F_2	1.00	CF	1.22
B	NH_2	1.93	B_2H_4	3.07	B_2H_6	3.41
	BN	1.87	BH_3	2.92	HBO	3.36
Be	BeB	1.06	BeF_2	1.95	BeC_2	2.21
	BeH	0.92	BeH_2	1.87		
Li	LiH	0.92	Li_2	1.00	CLi_4	1.40
	LiF	0.81			OLi_4	1.35

and binding energies was tested and found reliable [29]. Table 1 contains atomic valence numbers of structures optimized with SINDO1. We have classified the compounds as subvalent, normal valent and hypervalent according to the criteria outlined in Sec . 1.5. It is apparent from this table that the expectations from Fig. 1 are met for atomic valencies of carbon atoms in the sequence CH_4, CH_3, $^3CH_2$ and CH. No compounds with hypervalent carbon atoms were found as expected from wave functions with minimal basis sets. Whereas for nitrogen compounds the valency of 3 is frequently exceeded, common oxygen compounds exceeded the valency of 2, but not 3. The largest value for such compounds is 2.81 in O_3. However, in principle, oxygen valencies can also approach the value of 4. One such case is OF_4^{2+} where the oxygen valency of 3.30 exceeds 3 substantially. The case of O_2 is also interesting where single occupancy of two degenerate orbitals of this open-shell triplet state prevents the expected bond valence of 2. We relate this decrease of valency to an increase of reactivity. Another important feature of valence analysis is the occurrence of hypervalent lithium atoms. This can be explained by the *p* orbital participation in the bonding. Distribution of the valence electrons of lithium atoms in CLi_4 among several hybrids on Li can increase the atomic valency above 1. An unresolved question is the absolute polarity in lithium bonds. Since net charge and

Table 2. Bond valencies V_{AB} and atomic valencies V_A and V_B of molecules with first-row atoms

Molecule	Bond	Bond valency	Atomic valency	
	AB	V_{AB}	V_A	V_B
C_3H_6	CC	0.99	3.94	3.94
C_2H_6		1.04	4.00	4.00
C_6H_6		1.44	3.98	3.98
C_2H_4		2.04	3.99	3.99
C_2H_2		2.99	3.97	3.97
CH_3NH_2	CN	1.03	3.98	3.00
H_2CNOH		1.98	3.97	3.00
HNC		2.65	2.71	3.54
HCN		2.98	3.96	3.00
CH_3OH	CO	1.02	3.95	2.02
CO_2		1.87	3.74	2.13
H_2CO		2.05	3.91	2.13
CO		2.57	2.57	2.57
H_2NOH	NO	1.02	2.99	2.01
NO_3		1.01	3.72	1.19
		1.36	3.72	1.64
NO_2		1.52	3.04	1.78
HNO		2.05	2.99	2.11
NO^+		2.85	2.85	2.85
N_2H_4	NN	1.03	3.00	3.00
N_2H_2		2.05	3.00	3.00
N_2		3.00	3.00	3.00
H_2O_2	OO	1.02	2.01	2.01
O_3		1.41	1.76	2.81
O_2		1.50	1.50	1.50

valency are related, more extensive studies with extended basis are necessary to see whether the absolute magnitude of lithium valencies in LiH, LiF and LiC bonds can be predicted with minimal basis sets. To complement the picture, the multiplicity of bonding in CC, CN, CO, NO, NN and OO bonds is presented in terms of bond valencies V_{AB} and atomic valencies V_A and V_B in Table 2.

The values are in the expected range. Whereas carbon and nitrogen can easily form single, double and triple bonds, oxygen forms usually only single or double bonds with the exception of CO which comes close to a triple bond.

2.2 Multiplicity of Bonding in the Second Row

In the second row with atoms Na to Cl molecular structure may be influenced by the presense of close lying 3*d* orbitals which are vacant in the atom. The situation is here similar to that of Li and Be in the first row where the vacant 2*p* orbitals are partially occupied in molecules with these atoms. Even so we expect that the occupation numbers of 3*d* orbitals in the second row are not substantial, in the sense that they are not valence orbitals of the same weight as 3*s* and 3*p* orbitals. The SINDO1 semiempirical MO method was recently parametrized [30] under inclusion of 3*d* orbitals. Tests for sulfur [31], phosphorus [32] and chlorine [33] compounds with this method showed that 3*d* orbital occupancies are similar to those from ab initio calculations. We therefore took the SINDO1 method to generate wave functions which could be analyzed with respect to atomic and bond valencies. The results for a variety of molecules with second-row elements are presented in Table 3.

Table 3. Bond valencies V_{AB} and atomic valencies V_A and V_B of molecules with second-row atoms

Molecule	Bond	Bond valency	Atomic valency	
	AB	V_{AB}	V_A	V_B
Si_2H_6	SiSi	1.02	3.84	3.84
Si_2H_2		1.44	2.43	2.43
$^3Si_2$		1.86	1.86	1.86
Si_2H_4		1.98	3.87	3.87
SiH_3CH_3	SiC	1.02	3.85	3.80
SiC_2		1.87	2.42	3.87
SiC		2.75	2.75	2.75
SiO_2	SiO	2.15	4.31	2.30
SiO		2.42	2.42	2.42
SiF_4	SiF	0.95	3.79	1.01
SiF_2		1.08	2.16	1.11
SiF		1.12	1.12	1.12
P_4	PP	0.97	2.92	2.92
P_2F_4		0.98	2.96	2.96
P_2F_2		1.95	2.99	2.99
P_2		3.02	3.02	3.02
P_4O_6	PO	0.97	3.26	2.18
$P(OH)_3$		1.10	3.53	2.08

Table 3 (continued)

Molecule	Bond	Bond valency	Atomic valency	
	AB	V_{AB}	V_A	V_B
PO_2		1.74	3.52	1.93
H_3PO		2.12	4.39	2.27
PO		2.19	2.19	2.19
PF_5	PF	0.91, 0.97	4.74	1.06, 0.99
PF_3		1.04	3.11	1.10
PF_2		1.05	2.10	1.09
PF		1.06	1.06	1.06
$^3S_2$	SS	1.66	1.66	1.66
SO_2	SO	1.49	2.98	1.86
SO		1.60	1.60	1.60
SO_3		1.67	5.02	2.08
SF_6	SF	0.85	5.06	1.01
SF_4		0.83, 0.99	3.66	0.95, 1.10
SF_2		1.05	2.09	1.08
CCl_4	ClC	0.95	3.80	1.17
C_2Cl_6		0.96	3.84	1.17
C_2Cl_4		6.98	3.88	1.18
C_6Cl_6		1.00	3.91	1.17
CCl_2		1.20	2.41	1.33
$^3Al_2$	AlAl	1.11	1.11	1.11
AlO_2H	AlO	1.13, 2.14	3.37	2.06, 2.20
AlOH		1.24	1.37	2.10
Al_2O		1.28	1.49	2.57
AlO		1.40	1.40	1.40
$AlCl_2$	AlCl	1.01	2.03	1.05
$AlCl_3$		1.06	3.18	1.16
AlCl		1.24	1.24	1.24
$Mg(OH)_2$	MgO	0.87	1.77	1.86
MgS	MgS	2.23	2.23	2.23
Na_2	NaNa	1.00	1.00	1.00
Na_2O	NaO	0.88	0.96	1.79
NaO		0.91	0.91	0.91
NaOH		1.15	1.28	2.30

Table 4. Atomic valencies of hypervalent compounds in the second row without (A) and with (B) *d* orbitals

Molecule	Atom	Valency		Molecule	Atom	Valency	
		A	B			A	B
ClF_3	Cl	1.97	3.04	PO_2	P	2.82	3.52
ClF_4		(1.87)	3.67	PCl_5		3.77	4.34
ClF_5		2.83	4.97	H_3PO		3.91	4.39
ClF_6		2.12	6.35	PF_5		3.49	4.40
SF_4	S	2.93	3.66	H_3PO_4		3.62	4.76
SO_2		2.99	3.85	SiF_6^{2-}	Si	2.80	4.03
SO_3		3.67	5.02	SiO_3^{2-}		3.64	5.02
SF_6		3.82	5.06				
H_2SO_4		3.56	5.24				

The influence of bonding of the 3*d* orbitals can be seen from Table 4. Here a number of hypervalent compounds with Si, P, S and Cl atoms is presented. Valence numbers of these second row atoms were calculated without and with 3*d* orbitals. We call the difference between the two numbers extended valence. It is clear that in all cases the influence of *d* orbitals is substantial reaching a maximum for Cl_2O_7. These additional orbitals provide new degrees of freedom for orbital occupation in second row elements. They provide a channel to reduce the polarity of the otherwise rather polar compounds. For example, in Cl_2O_7 the net charges on oxygen indicate ionic bonding, whereas the introduction of 3*d* orbitals enhances the covalent bonding in these compounds. In some compounds the minimal basis set limit of 4 is substantially exceeded. These compounds can be called hypervalent in the usual way.

2.3 Valence versus Coordination Numbers

To understand that coordination number and valence number do not necessarily coincide three sets of compounds will be compared:
a) tetracoordinated nitrogen compounds, b) tetracoordinated carbon compounds, c) pentacoordinated carbon and boron compounds.

Atomic valencies and net charges of nitrogen compounds are shown in Fig. 4.

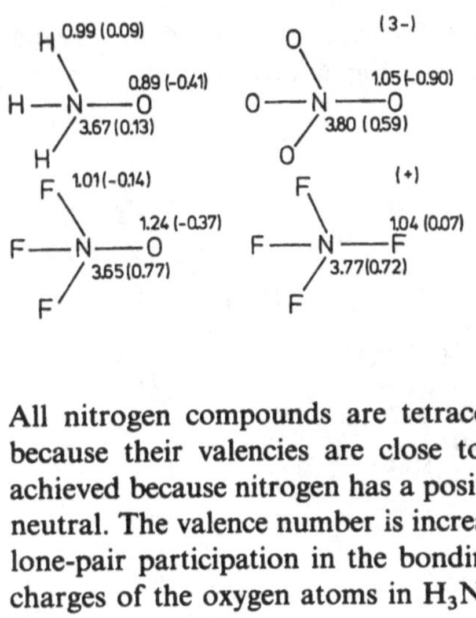

Fig. 4. Atomic valence numbers (net charges) of tetracoordinated nitrogen compounds

All nitrogen compounds are tetracoordinated and can also be called tetravalent because their valencies are close to 4. However, these valence numbers are not achieved because nitrogen has a positive unit charge. In H_3NO the N atom is almost neutral. The valence number is increased above the standard valency of 3 because of lone-pair participation in the bonding to oxygen. If we compare valencies and net charges of the oxygen atoms in H_3NO and F_3NO we see that equal charges do not

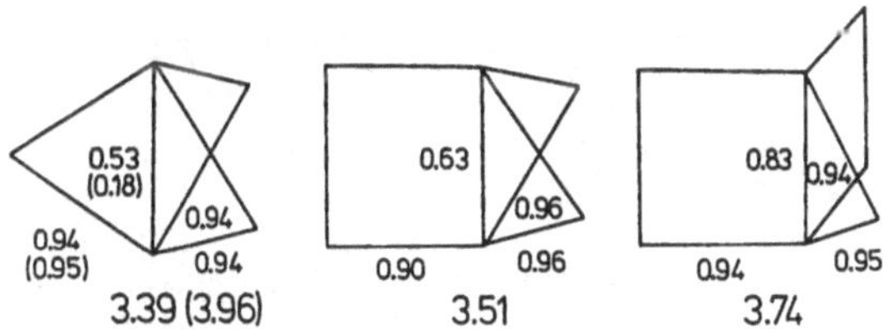

Fig. 5. Atomic and bond valencies of propellanes (hydrated propellanes)

imply equal valencies. The oxygen atom of H_3NO has a substantially lower valency and should be more reactive to radical attack than the one in F_3NO.

In the second set of compounds we compare the valence numbers of the carbon bridge atoms in propellanes (Fig. 5). All bridge carbon atoms can be considered as tetracoordinated if we draw a valence line between two bridge atoms. This would be in line with the fact that the distances between two bridge atoms are close to single bonds distances. This is predicted both by SINDO1 and ab initio calculations [34]. Wiberg [35] found that the breaking of the C—C bridge bond required 65 kcal/mol in (1.1.1) propellane, 30 kcal/mol in (2.1.1) propellane and 5 kcal/mol in (2.2.1) propellane. But there was not enough density between the C atoms to support a normal single bond in (1.1.1) propellane [36]. The calculated valence numbers can help to resolve this discrepancy. The bond valence increases from 0.53 to 0.83 from (1.1.1) to (2.2.1) propellane. So the bridge bond becomes gradually more like a single bond. The bridgehead atoms reflect this trend with valencies of 3.39, 3.51 and 3.74 in this sequence. Actually we must consider the C—C bridge bond in (1.1.1) propellane as the weakest, but the strain in the system keeps the bridge bond length close to a single bond also in this system. Although all three compounds have formally tetracoordinated bridgehead atoms, the valencies range between 3 and 4. The reduced valence is connected with radical activity at the bridgehead atoms. Hydrogen atom addition at these two centers reduced the bond valence of the central bond in (1.1.1) propellane to 0.18 and increases the atomic valency to the normal value of 3.96 (Fig. 5).

The third set of compounds are closoboranes and closocarboranes. Here we find tetra- and pentacoordinated boron and carbon atoms. Fig. 6 shows that the bonding

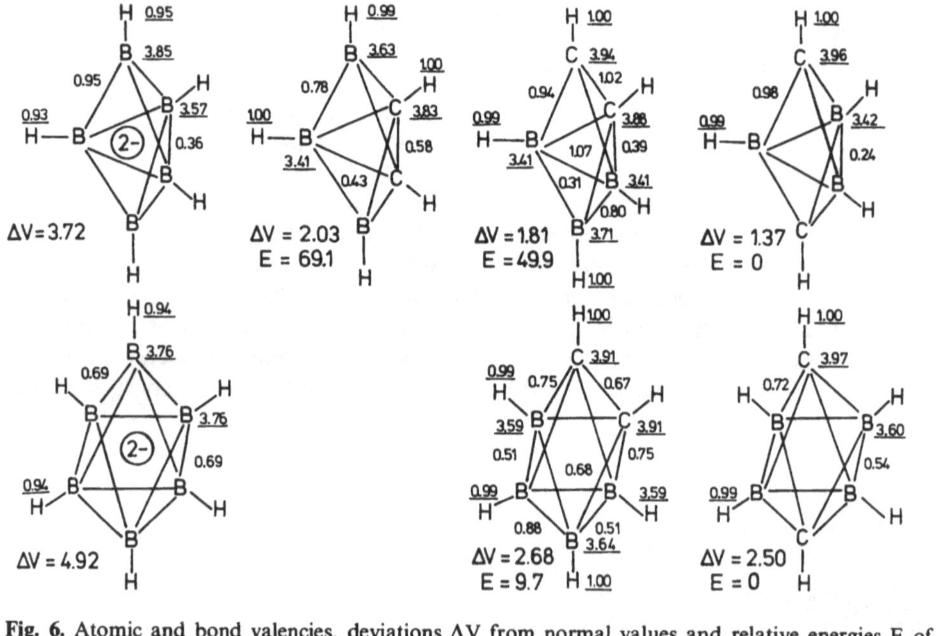

Fig. 6. Atomic and bond valencies, deviations ΔV from normal values and relative energies E of closoboranes and closocarboranes

in these compounds is unequally distributed between the bonds. All boron atoms have valencies between 3 and 4, but the valencies of carbon atoms cannot exceed 4 either in a minimal basis set.

2.4 Bond Ionicity

An analysis of pair bonding in heteropolar bonds shows that the AO coefficients of MO's are no longer equal for equivalent orbitals on the two atoms of the considered pair of atoms. This polarity is expressed in atomic dipole moments and charge transfer moments. For the example of Fig. 2 this means that the charge distribution is no longer equal ($q = 1$) for the two atoms but moves to the left or right. This results in a reduction of covalent bonding expressed by the bond valency. Structures of molecules are influenced by the tendency of the atoms to diminish or enhance ionic character in the bonding. Hydrogen atoms have a tendency to enhance covalent bonding, whereas lithium atoms enhance ionic bonding. The water molecule is bent rather than linear and ammonia is pyramidal rather than planar because in the more stable geometry the ionicity of the bonds is decreased [37]. The stabilization is accompanied by an increase in the occupation number of the 2*s* orbital. This result is in line with Epiotis' MOVB theory [38, 39] which explains stabilization in these compounds by a dominance of intrafragmental relaxation from $2p \rightarrow 2s$ over interfragmental relaxation between the fragments. The same situation arises for acetylene where the bridged form is much less stable than the linear form because of decreased ionicity of the latter. The situation is reversed for lithium compounds which are quite ionic. Here we find linear Li_2O, planar Li_3N and bridged C_2Li_2 because these forms enhance ionicity. But valence numbers show only small changes for these compounds compared to the hydrogen compounds.

2.5 Strained Molecules

Atomic valence in strained ring systems is invariably lower than that in unstrained compounds [40]. A quantitative evaluation of strain energy from valence is possible by decomposing atomic valence into bond valencies using Eq. (12) and considering the reduction in bond valences in strained compounds compared to unstrained ones. In cyclopropane, for instance, the C—C bond valence is 0.995 compared to the unstrained value of 1.000. This bond valency of 0.995 in cyclopropane can be further decomposed into its σ component, V_σ along the C—C bond with a value of 0.856 and a π component, V_π perpendicular to the C—C bond with a value of 0.139. If the unstrained bond is taken as $V_\sigma = 1.00$ and $V_\pi = 0.0$, we have a loss in σ component of 0.144 and a gain of π valency of 0.139. If the bond energy of an unstrained C—C σ bond is taken as 85.4 kcal/mol and that of the π bond as 42.7 kcal/mol, the above numbers lead to a net loss of 6.4 kcal per C—C bond due to valency reduction. This amounts to a total strain energy of 19.2 kcal/mol for the three C—C bonds in cyclopropane. The CH bond also has a reduced valence of 0.986 compared to 1.00 in normal alkanes, and this leads to a loss of 1.4 kcal/mol per CH bond or 8.4 kcal/mol for the molecule, when CH bond energy is taken as 99.4 kcal/mol. Thus we have a

Table 5. Strain energy (kcal/mol) estimated from valency

Molecule	Strain energy	
	from valency	exp.[a]
Cyclopropane	27.6	27.6
Cyclopropene	41.6	52.0
Cyclobutane	24.0	26.9
Cyclopentane	9.0	4.6
Bicyclo(1,1,0)butane	62.3	65.6
(1,1,1)propellane	97.5	98.0
Tetrahedrane	102.2	105.0
Spiropentane	68.0	63.5
Prismane	119.4	114.0
Cubane	176.0	166.0
Oxirane	26.5	27.2
Aziridine	29.2	30.2
Thiirane	16.9	19.8
Cyclotrisilane	25.5	36.1
Cyclotetrasilane	13.6	13.7
Cyclopentasilane	5.5	6.2
Tetrasilatetrahedrane	81.2	88.9
Tetraphosphirane	30.6	36.0

[a] see [42] and [45] for sources of experimental and other estimates of strain

total strain energy of 27.6 kcal/mol for cyclopropane which is identical with the experimental value [41]. The strain energy estimated from valency in this manner is listed for some typical molecules in Table 5. There is generally a very close quantitative agreement with experimental values for hydrocarbons as well as heteroatom ring systems.

The valency formulation of strain energy enables discussion of the contribution of various bonds to strain energy. Generally, CH bond contributes about 30% to the strain energy. Nonbonded interactions, as estimated from valency, are found to be negligibly small. As would be expected, the bond between the two bridgehead atoms in (1.1.1) propellane has the largest strain energy of 20.7 kcal/mol. The C—C bond strain in tetrahedrane and cubane are 14 and 12 kcal/mol, respectively. The CH strain is also large being around 4.0 kcal/mol in both these systems compared to 1.4 in cyclopropane.

Another interesting observation is that the ring additivity rule, namely that the strain energy of polycyclic hydrocarbons is the sum of the strain energies of the monocyclic components, holds fairly well [24]. The additivity rule works because the strain of the bridging bond in bicyclic systems turns out invariably to be twice that of the C—C bond in the monocyclic compound. Similarly, it is three times this quantity for the bridging bond in tricyclic compounds.

2.6 Photoelectron Spectra

The molecular bond number M of Eq. (13) which may be alternatively called molecular valency, can also be expressed as a sum over occupied molecular orbitals as

$$M = \sum_{i}^{occ} V_i , \tag{30}$$

where V_i is the valency of the ith MO and is given for a closed shell molecule by [43]:

$$V_i = 1/2 \sum_{A} \sum_{B \neq A} \sum_{a}^{A} \sum_{b}^{B} n_i n_j c_{ia} c_{ib} c_{ja} c_{jb} \tag{31}$$

This definition of molecular orbital valency is easily extended to open shell cases also. V_i as defined here turns out to be a good measure of the bonding nature of the MO, being close to zero for core, nonbonding and antibonding MO's and close to unity for strong bonding MO's. That the MO valency is a good index of the strength of bonding can be seen from Table 6 where the valencies of the MO's in some simple molecules are given and compared with predictions of bonding nature from photoelectron spectra.

Table 6. Molecular orbital valency

Molecule	Orbital	MO Valency	PES[a]
N_2	$1\sigma_g$	0.004	
	$1\sigma_u$	0.001	
	$2\sigma_g$	0.777	
	$2\sigma_u$	0.027	n
	$1\pi_u$	0.971	s
	$3\sigma_g$	0.184	w
CO	1σ	0.001	
	2σ	0.001	
	3σ	0.695	
	4σ	0.140	w
	1π	0.812	s
	5σ	0.112	n
H_2O	$1a_1$	0.000	
	$2a_1$	0.614	
	$1b_2$	0.997	s
	$3a_1$	0.360	b
	$1b_1$	0.000	n

[a] n = nonbonding; s = strongly bonding; w = weakly bonding, Photoelectron spectral (PES) information for N_2 and H_2O from Turner DW, Baker C, Baker AP, Brundle RC (1970) Molecular photoelectron spectroscopy, Wiley, London and for CO from Carlson TA (1975) Photoelectron and auger spectroscopy, Plenum, New York

2.7 Correlation Diagrams

The result of the last section that valency of the MO is a measure of its bonding power opens up avenues for various applications for this quantity. If the valencies of occupied MO's in a molecule are plotted as a function of bond angle, the resulting correlation diagrams have a close resemblance to the well known Walsh diagrams [44]. The MO valency correlation diagrams have been shown to satisfy the general criteria for successful Walsh-type correlation diagrams. Universal MO valency correlation diagrams for symmetry such as AH_2, AH_3, HAB and AB_2 have been constructed [45] and are remarkably similar to the original qualitative Walsh diagrams. The valency correlation diagram for AB_2 species, for instance, is given in Fig. 7 along with the corresponding Walsh diagram.

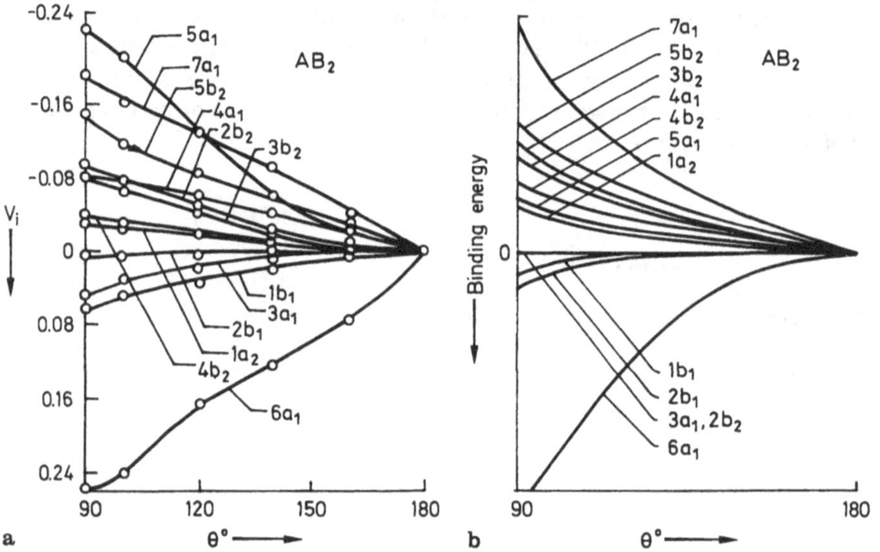

Fig. 7. Correlation diagrams for AB_2 systems: *a* MO valency diagram, b Walsh diagram

Moreover, the valency correlation diagram for a given molecule is quantitative in the sense that the diagram predicts the equilibrium bond angle exactly, unlike the original Walsh diagram or its variants. This finding has been utilized [45] to construct correlation diagrams for various excited and ionized states of molecules in a simple way using just its ground state wave function. Such diagrams successfully predict the geometry of these states. For instance, the predicted bond angles by the valency method for the $^2B_1$, $^2A_1$ and $^2B_2$ states of H_2O^+ are 100°, 180° and 50° which may be compared with ab initio results of 111°, 180° and 55°. More examples of such predictions may be found in [45].

3 Reactivity

3.1 Rotation Barriers

Rotation barriers fall mainly into three categories: a) rotations about single bonds, b) rotations about double bonds and c) rotations during reactions about bonds which are in between single and double bonds. In this subsection we shall discuss only the first two categories. Relations between rotation barriers and intramolecular valence changes were presented on the basis of SINDO1 calculations for H_2O_2, N_2H_4, C_2H_6, B_2H_4, B_2F_4, N_2H_2 and C_2H_4 [46]. For rotations about single bonds changes in bond numbers M during the rotation are negligible except for B_2H_4. Here ΔM between the orthogonal stable form and the planar form is 0.13. For this compound increased stabilization is achieved in the orthogonal form by increased π bonding due to hyperconjugation. For rotations about double bonds the changes in bond number are much larger. For the *trans* versus the orthogonal arrangement of N_2H_2 $\Delta M = 0.88$ and for the planar versus the orthogonal arrangement in C_2H_4 $\Delta M = 0.83$. The barriers for such compounds must be calculated with CI. The magnitude of ΔM reflects that a bond is broken during the rotation. From these data one can see that a correspondence between the magnitude of the rotation barrier and the change of bond number exists.

3.2 Radicals, Diradicals and Zwitterions

In the definition of the chemist a radical is a compound with an atomic center which has a single nonbonding electron. Such centers are particularly reactive in addition reactions. In Table 1 the valence numbers reflect that CH_3, NH_2 or BeH are radicals with such radical centers. In the case of NO the valencies for both atoms are 2.12. The oxygen atom is engaged in more bonding than in normal compounds with valencies of about 2. This atom is therefore unreactive. However, the nitrogen atom has 0.88 free valence and this reflects that N is the radical center. Whereas the discussion of radical activity is usually quite straightforward, more care has to be taken in the case of diradicals. Diradicals are species with usually two radical centers. But symmetry may prescribe that the two lone electrons which are not engaged in the bonding are distributed among more than two atoms. Such a case is trimethylenemethane $C(CH_2)_3$ with D_{3h} symmetry for the triplet ground state. This system is a case where Salem's singlet-triplet degeneracy criterion for diradicals [47] does not hold. Döhnert and Koutecky [48] have instead used natural orbital occupation numbers and suggested single occupancy of degenerate orbitals as an alternative criterion. This allows a definition of diradicals without singlet-triplet degeneracy. At variance with these two global criteria, the valence method provides a local criterion which allows us to determine the diradical centers. This is particularly important in large molecules or in thermal reactions.

A similar problem arises with idea of a zwitterion. A neutral molecule can be called a zwitterion if it contains two ionic centers of opposite polarity. This shift of charge is accompanied by a change in the covalent bonding of the system. If a hydrogen atom of hydroxylamine H_2NOH migrates from oxygen to nitrogen, the zwitterionic isomer ammoniaoxide H_3NO originates. Whereas the normal compound has valencies $V_N = 2.99$ and $V_C = 2.01$, the zwitterionic compound shows increase of valency of nitrogen to $V_N = 3.67$ and a decrease of the valency of oxygen to $V_0 = 0.85$. These compounds also differ in their dipole moments with 0.56 Debye for the former and 4.80 Debye for the latter compound. By analogy, we discovered the zwitterion H_4CO of methylalcohole H_3COH [25]. However, in this case the valence of carbon cannot increase because it is already close to 4 in the normal compound.

A combined search for diradicals and zwitterions arises when the geometry of a system leads to close lying ground and first excited singlet states which can be characterized as diradical and zwitterionic. Such situations arise in twisted ethylene and in tetramethylene.

In the first case a 4×4 CI results in a lower lying pair of singlet and triplet diradicals and a higher lying pair of singlet zwitterions. The diradical pair is characterized by bond numbers of 4.94, whereas the zwitterions have bond numbers of 5.30 and 5.19. So the zwitterionic states show more covalent bonding than the diradical states. It is now important that this statement also holds in the case of polar bonds for perpendicular H_2BNH_2, $H_2BCH_2^-$ and $H_2CNH_2^+$. So this criterion can be used to distinguish diradical and zwitterionic states. The dipole moment cannot be used in general for such a purpose. Whereas the dipole moments of the diradical states of polarized perpendicular ethylene [25] are much smaller with values of 1.07 Debye than for the zwitterionic states with 2.21 and 4.36 Debye, the values of 4.76 Debye for the diradical states of aminoborane are in between the values of 1.60 and 7.92 Debye for the zwitterionic states.

In the case of tetramethylenes there is a specific interest in the distinction between diradical and zwitterionic states. In a lucid discussion of the polymerization mechanism of substituted tetramethylenes, Hall [49] has advanced the idea that diradical tetramethylenes foster copolymerization whereas zwitterionic tetramethylenes can achieve only homopolymerization. In the former case a chain polymer with alternating donor and acceptor groups is generated whereas in the latter donor and acceptor group fragments polymerize separately or the tetramethylene dimerizes. Although most of the experiments reported were done in solution, a comparison in terms of bond numbers led to very good agreement in the trend. Table 7 shows ten substituted tetramethylenes of which the first five were predicted to initiate diradical mechanisms whereas the last five were causing zwitterionic mechanisms [49]. Optimization of the ground state planar tetramethylenes and vertical excitation to the first excited singlet resulted in energy differences $E_z - E_d$ between zwitterionic and diradical states. For diradical ground states this difference is positive and decreases the more the substituents approach the zwitterionic ground states where this difference is negative. Energetically only the last three compounds should lead to zwitterionic mechanisms. Cases 6 and 7 are borderline cases in Hall's diagram. The differences in bond numbers $M_z - M_d$ show a parallel trend. For diradical ground states bond numbers are very similar for the two states, but the zwitterionic ground states are substantially more covalent than the diradical excited states. LUMO-HOMO energy differences can also be used on a scale. Within this scheme it is possible to

Table 7. Total energy E (Hartree) and bond number M changes between zwitterionic (z) and diradical (d) states and LUMO-HOMO energy differences (Hartree) of ground states S_0

Molecule Donor	Acceptor	$E_z - E_d$	$M_z - M_d$	$E_{LUMO} - E_{HOMO}$
H	H	0.38	0	0.007
H	CN	0.34	0	0.009
4F	CN	0.32	0.09	0.006
OCH_3	4F	0.27	0.01	0.013
C_6H_5	CN	0.22	0.01	0.018
OCH_3	2CN	0.19	0.11	0.034
OCH_3	NO_2	0.09	0.23	0.090
2 CH_3, NH_2	2 CN	−0.05	0.71	0.173
2 OCH_3	$COOCH_3$	−0.06	0.69	0.187
2 CH_3, NH_2	$COCH_3$	−0.06	0.73	0.176

predict polymerization mechanism of tetramethylenes from bond number differences if the line is drawn from experimental test cases [61].

3.3 Acid-Base Reactions

If we assume that a reaction tends to products which have normal covalent bonds with atomic valencies around their normal values, rather than to subvalent or hypervalent atoms, it is tempting to investigate the relative stabilities of reactants and products in acid-base reactions. In reactions of the type $H_3O^+ + OH^- \rightarrow 2\,H_2O$, the deviation of actual atomic valence numbers from normal atomic valence numbers is small for the right side whereas it is large for the left side. Energetically the products are more stable. Similar results can be found for other reactions where a cation and an anion combine to two neutral species [46]. A typical case is $NH_4^+ + OH^- \rightarrow NH_3 + H_2O$. The situation is different when both reactants and products contain the same number of neutral and ionic species. Here we would expect that the sums of absolute deviations from normal valencies are similar for reactants and products and that the energy differences between reactants and products are correspondingly small. An example is $NH_3 + H_3O^+ \rightarrow NH_4^+ + H_2O$ where this assumption holds. More comprehensive investigations [46] show that stability is indeed accompanied by a trend towards normal atomic valencies and standard covalent bonds.

4 Excited States

4.1 Vertical and Adiabatic Excitations

The use of valence numbers and bond numbers in excited states is faced with a particular problem. Whereas the contributions from molecular orbitals add up to the typical numbers for ground states, the replacement of one occupied MO by a virtual

MO is usually accompanied by an increase of the number of nodes. As a consequence, the atomic valencies are reduced on the average and this reduction is expressed in the bond number. Most curiously, the amount of reduction is in the order of one bond unit in most cases and reflects an overall decrease in bonding of excited states. Let us take a SINDO1 wave function [28, 29] for 2-methylfuran generated by a 17×17 CI by excitation from the two highest occupied MO's to the four lowest unoccupied MO's. The bond number for the ground state is 14.02. For vertical excitation of the three lowest excited singlet states the bond numbers are 12.69, 12.85 and 12.72. It is now essential to realize that these numbers do not change significantly if these excited states are optimized. For the equilibrium geometries the corresponding numbers are 12.74, 12.80 and 12.82. The effect of excitation (vertical) is larger than the effect of optimization (adiabatic) by an order of magnitude. Whereas reduction can be used in ground state to characterize diradicals and zwitterions, this is no longer true for excited states. In excited states the reduction is usually spread over many centers instead over two for diradicals.

4.2 Triplets

Excitation to a triplet state from a closed-shell singlet is accompanied by a change of nodal properties of the molecular orbitals involved. If we consider the first excited triplet of ethylene, it is characterized by a $\pi\,\pi^*$ excitation. One electron is moved from a bonding π orbital to an antibonding π^* orbital. This results in a net loss of one bond. This change is reflected in the bond number M. Whereas $M = 5.99$ in the ground state, $M = 4.98$ in the first triplet in its optimized perpendicular geometry. Now it is essential to realize that rotation about a single bond does not change the bond number significantly. This was already discussed in Sect. 3.1. Therefore the bond number of the triplets is also close to 5 for the planar arrangement. Its value of 4.87 is somewhat less because the perpendicular arrangement is stabilized by hyperconjugation.

Another interesting case is the first excited triplet state of benzene. It has a planar structure with two bonds shortened and four bonds lengthened compared to the ground state. If we look at the atomic valencies, two atoms in *para* position have valence numbers of 3.27 whereas the other four valence numbers are 3.80 [50]. We could call this state a diradical, but it is not degenerate with a singlet state. The bond number is again reduced by 1.08 compared with the ground state.

Whereas for the above two cases the triplet is an excited state, there are also cases of triplet ground states: Examples are CH_2, $C(CH_2)_3$ or O_2. In all these cases the bond number is reduced compared to a closed-shell singlet. For symmetry reasons O_2 is an exception with a reduction from 2 to 1.5. In a different way CH_2 is also an exception with a reduction from 2 to 1.92, because the singlet CH_2 can also only form two bonds. It may be called a diradical with the two lone electrons on one center. In trimethylenemethane the central atom has an atomic valence of less than 4 in contrast to Coulson's definition.

We have searched a large variety of compounds and found that the bond number is always reduced compared to the ground state value. So this number can no longer serve in a simple way to define the radical or ionic character of systems even in triplet states. Considerations of the location of radical or charged centers have to supplement the valence considerations.

4.3 Aromaticity and Antiaromaticity

Aromaticity can be defined in various ways. The main ideas are a) stabilization by resonance energies, b) ring current by delocalization, c) reactivity under addition reactions [51]. We have developed a method related to the second criterion and suggested that the weakest bond order in a ring can serve as an index of aromaticity [52]. It is now suggestive to replace the bond order between two atoms by the bond valence and search for the same regularities that were found with the bond-order criterion. In particular, it was found previously that systems, which can be classified as aromatic in the ground state, are only moderately aromatic or nonaromatic in excited states. On the other hand, antiaromatic ground state systems show nonaromaticity or aromaticity in excited states.

These trends are also shown with the bond-valence criterion. Only the reference values change. In terms of bond valencies, the standard values are 1.04 for the single bond of ethane and 2.04 for the double bond of ethylene. The standard aromatic compound benzene has a bond valence of 1.44. Substituted benzenes like toluene, aniline, phenole of fluoroaniline are characterized by aromatic indices between 1.37 and 1.40. For excited singlet and triplet states of these compounds this index is between 0.99 and 1.09 except for S_1 of fluoroaniline with an index of 1.22. Only the latter state can be called moderately aromatic, the rest is nonaromatic. A typical change from antiaromaticity to aromaticity is in cyclobutadiene where the ground state index is 0.94, whereas for the quadratic form of the excited singlet and triplet it is 1.23. In the case of cyclooctatraene the nonaromatic ground state with an aromatic index of 1.02 changes to moderate aromaticity with an index of 1.28 in the first excited singlet and triplet state. Another case is the triplet ground state of $C_5H_5^+$ with an index of 1.26 which is aromatic, whereas the singlet has an index of 0.95 and must be classified as antiaromatic. It is not quite clear at this point whether the bond order criterion or the bond valence criterion is more suitable for an aromaticity scale.

5 Reactions

5.1 Thermal Woodward-Hoffmann Reactions

Woodward and Hoffmann [53] have classified various classes of reactions as allowed or forbidden by means of correlation diagrams. In these diagrams special attention was given to the HOMO and LUMO of reactants and products. All molecular orbitals involved in the interpretation of the reaction were given symmetry labels related to symmetry operations which are considered as essential for a particular reaction pathway. Reactant and product orbitals of the same symmetry were linked together under observance of the noncrossing rule i.e., the rule that orbitals of the same symmetry cannot cross. If the diagram shows no crossing of HOMO and LUMO, the reaction is called allowed whereas a crossing indicates a forbidden reaction. This crossing of HOMO and LUMO must involve a degeneracy or quasi-degeneracy of these orbitals. We have made this quasi-degeneracy visible in correlation diagrams calculated with SINDO1 wave functions in a multiconfigurational variation of moments approach [54]. As a consequence of this orbital degeneracy the electrons are distributed among these two orbitals and reduction in bonding results. This is the typical situation of a diradical. In contrast, in an allowed pathway no diradical is encountered. Therefore a forbidden pathway must pass through a transition structure with two diradical centers which have a reduced valence by approximately one unit for each compared to the normal value. No substantial valence reduction is to be expected in an allowed pathway. We illustrate this in Fig. 8 for the allowed conrotatory and the forbidden disrotatory pathways of the cyclobutene → butadiene rearrangement. Valence numbers for the transition structures are determined from SINDO1 calculations. Similar situations will be encountered in other thermal Woodward-Hoffmann reactions as we have shown [55].

Ponec [56] has recently discussed topological aspects of chemical reactivity in terms of reorganization of electron density in allowed and forbidden reactions. By incorporating the idea of valence numbers into a model with a reaction coordinate, φ transforms the reactant bonding function continuously into the product bonding function. In a different way he also demonstrates that valency changes are enormous during the forbidden process and small during the allowed process [57].

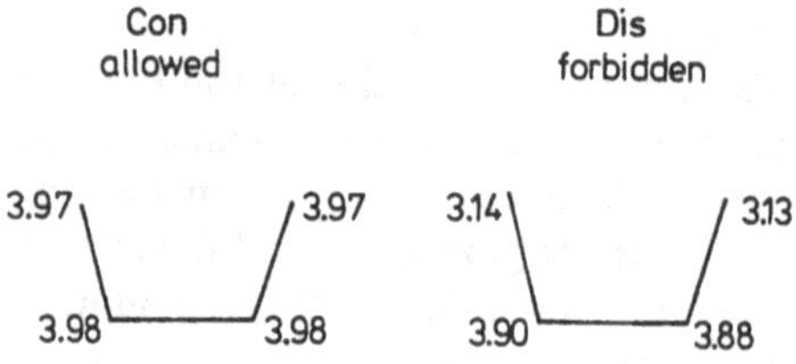

Fig. 8. Atomic valencies for transition structures of cyclobutene → butadiene rearrangement

5.2 Diradical Reactions

The Woodward-Hoffmann rules are applicable only for certain classes of reactions under the assumption that the reaction is concerted i.e., that no intermediates are encountered along the reaction pathway. The diradicals of forbidden pathways are no such intermediates and should really be related to conical intersections of potential surfaces. However, there are cases in which diradicals can play a major role in a thermal reaction mechanism. Such a case is the involvement of the 2,3-dimethylene-1,4-cyclohexadiyl in some rearrangement reactions. Roth and collaborates [58] have investigated the role of this diradical quite extensively over many years in a series of experiments. We could confirm the central role of this diradical by SINDO1 calculations [59]. From a valence analysis of the essential stationary points on the potential surface it became apparent that 9 out of 12 transition structures have substantial diradical character. If is possible with this method to locate the diradical centers which can be different in different regions of the potential surface.

5.3 Photochemical Reactions

Excited state reactions involve the replacement of occupied orbitals of the ground state by unoccupied or virtual orbitals. These latter orbitals are characterized by an increase of the number of nodes compared to lower lying orbitals in order to maintain orthogonality between all orbitals. This leads to partial reduction in bonding in the excited states. To define a diradical on an excited state surface, we therefore require that reduction in atomic valence is concentrated on two centers. Most important for a photochemical reaction are usually the lowest singlet surface S_1 and the lowest triplet surface T_1. In the study of a variety of photochemical reactions it was found that triplet diradical intermediates govern the photoisomerization and photofragmentation of cyclopentanone [60]. The mechanistic scheme of this reaction is presented in Fig. 9. It is clear from this analysis that the changes in bonding of a great variety of molecular systems can be followed during a photoreaction with a valence analysis of the wavefunction.

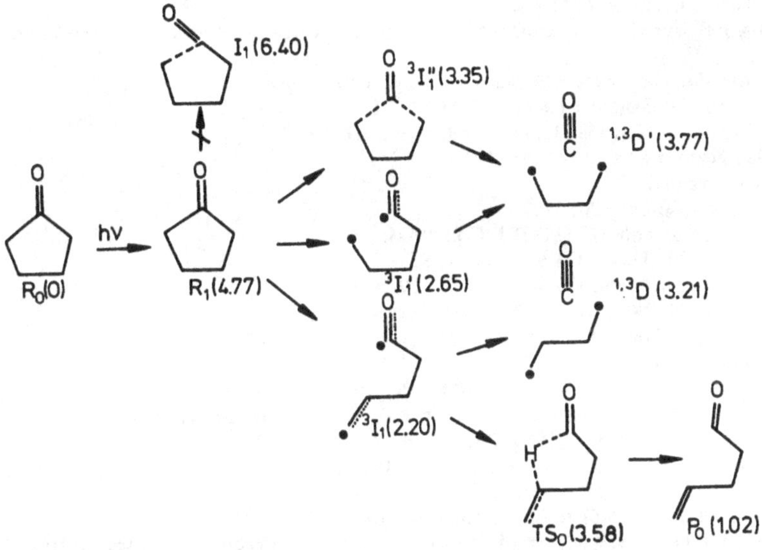

Fig. 9. Mechanistic scheme and relative energies (eV) for the photoreaction of cyclopentanone

Acknowledgement. A major portion of this work was done when K. J. was a DAAD Guest Professor at IIT Madras. He thanks the Department of Chemistry of this institution for its hospitality.

6 References

1. Coulson CA (1939) Proc. Roy. Soc. A169: 413
2. Löwdin PO (1950) J. Chem. Phys. 18: 365
3. Chirgwin BH, Coulson CA (1950) Proc. Roy. Soc. A201: 196
4. Mulliken RS (1955) J. Chem. Phys. 23: 1841
5. Jug K (1977) J. Am. Chem. Soc. 99: 7800
6. Jug K (1979) Theor. Chim. Acta 51: 331
7. Gopinathan MS, Jug K (1983) Theor. Chim. Acta 63: 497
8. Wiberg K (1968) Tetrahedron 24: 1093
9. Armstrong DR, Perkins PG, Stewart JJP (1973) J. Chem. Soc. Dalton 838: 2273
10. Borisova NP, Semenov SG (1973) Trans. Leningrad Univ. 16: 119
11. Kuznetsov VI (1980) Theory of Valency in Progress, Mir Publishers, Moscow
12. Jug K (1984) J. Comput. Chem. 5: 555
13. Jug K In: Maruani J (ed) Topics in molecular organization and engineering, Vol. 3. p. 149ff. Reidel: Dordrecht
14. McWeeny R (1960) Rev. Mod. Phys. 23: 335
15. Martin JČ (1979) J. Am. Chem. Soc. 101: 5057
16. Giambiagi M, de Giambiagi MS, Grempel DR, Heynman CD (1975) J. Chim. Phys. 72: 15
17. Mayer I, (1983) Chem. Phys. Lett. 97: 270
18. Baker J, (1985) Theor. Chim. Acta 68: 221
19. Davidson ER (1976) Reduced density matrices in quantum chemistry, Academic Press, New York, p 31
20. Coulson CA (1961) Valence, Second Ed., Oxford Univ. Press, p 271
21. Burkitt FH, Coulson CA, Longuet-Higgins HC (1951) Trans. Faraday Soc. 47: 553
22. Kooyman EC, Forenhorst E (1953) Trans. Faraday Soc. 49: 58
23. Moffit WE (1949) Trans. Faraday Soc. 45: 373
24. Jug K (unpublished results)
25. Jug K (1985) Tetrahedron Lett. 26: 1437
27. Jug K, Fasold E, Gopinathan MS (1989) J. Comput. Chem. 10
28. Nanda DN, Jug K (1980) Theor. Chim. Acta 57: 95
29. Jug K, Nanda DN (1980) Theor. Chim. Acta 57: 107: 131
30. Jug K, Iffert R, Schulz J (1987) Int. J. Quantum Chem. 32: 265
31. Jug K, Iffert R (1987) J. Comput. Chem. 8: 1004
32. Jug K, Schulz J (1988) J. Comput. Chem. 9: 40
33. Jug K, Schulz J (1987) J. Comput. Chem. 8: 1040
34. Michl J, Radziszewski GJ, Downing JW, Wiberg KB, Walker FH, Miller RD, Kovacic P, Jawdosiak M, Bonacic-Koutecky V (1983) Pure Appl. Chem. 55: 315
35. Wiberg KB, Walker FH (1982) J. Am. Chem. Soc. 104: 5239
36. Wiberg K (1983) J. Am. Chem. Soc. 105: 1227
37. Jug K, Epiotis ND, Buss S (1986) J. Am. Chem. Soc. 108: 3640
38. Epiotis ND (1982) Unified valence bond theory of electronic structure, Lecture Notes in Chemistry, Vol 29, Springer, Berlin Heidelberg New York
39. Epiotis ND (1983) Unified valence bond theory of electronic structure, applications, Lecture Notes in Chemistry, Vol 34, Springer, Berlin Heidelberg New York
40. Siddarth P, Gopinathan MS (1986) J. Mol. Struct. (Theochem) 148: 101
41. Schleyer PvR, Williams JE, Blanchard KR (1970) J. Am. Chem. Soc. 92: 2377
42. Siddarth P (1988) Ph. D. Thesis, Indian Institute of Technology, Madras
43. Gopinathan MS, Siddarth P, Ravimohan C (1986) Theor. Chim. Acta 70: 303
44. Walsh AD (1953) J. Chem. Soc. 2260
45. Siddarth P, Gopinathan MS (1988) J. Am. Chem. Soc. 110: 96
46. Jug K, Buss S (1985) J. Comput. Chem. 6: 507
47. Salem L, Rowland C (1972) Angew. Chem. Int. Ed. Engl. 11: 92
48. Döhnert D, Koutecky K (1980) J. Am. Chem. Soc. 102: 1789
49. Hall HK Jr (1983) Angew. Chem. Int. Ed. Engl. 22: 440
50. Malar EJP, Jug K (1984) J. Phys. Chem. 88: 3509

51. Bergmann ED, Agranat I (1971) In: Bergmann ED, Pullman B (eds) Aromaticity, pseudo-aromaticity antiaromaticity; The Jerusalem Symposia on Quantum Chemistry and Biochemistry III, Israel Academy of Sciences and Humanities, Jerusalem p 9
52. Jug K (1983) J. Org. Chem. 48: 1344
53. Woodward RB, Hoffmann R (1971) The conservation of orbital symmetry, Verlag Chemie, Weinheim
54. Müller-Remmers PL, Jug K (1985) Int. J. Quantum Chem. 28: 691
55. Jug K, Gopinathan MS (1985) Theor. Chim. Acta 68: 343
56. Ponec R (1985) Coll. Czech. Chem. Commun. 50: 1121
57. Ponec R (1987) Coll. Czech. Chem. Commun. 52: 2603
58. Roth WR, Erker G (1973) Angew. Chem. 85: 510, 512; Roth WR, Biermann M, Erker G, Jelich K, Gerhartz W, Görner H (1980) Chem. Ber. 113: 586
59. Jug K, Iffert R, (1982) Theor. Chim. Acta 62: 183
60. Müller-Remmers PL, Mishra PC, Jug K (1984) J. Am. Chem. Soc. 106: 2538
61. Jug K (1988) J. Am. Chem. Soc. 109: 3534

The Magnetic Description of Conjugated Hydrocarbons

J.-P. Malrieu

Laboratoire de Physique Quantique, U.A. 505 du CNRS, Université Paul Sabatier, 118, route de Narbonne, 31062 TOULOUSE CEDEX, FRANCE

A nonempirical approach has been recently proposed for treating the π system of conjugated hydrocarbons — it is an Heisenberg Hamiltonian i.e. a magnetic model, with two operators only. This is a very paradoxical approach since the magnetic Hamiltonians are usually proposed for the weak delocalization limit (strong correlation), while the π systems of aromatic hydrocarbons are considered as the most-metallic molecular frames. The amplitude of the two basic operators may be extracted from ab initio calculations with the ethylene molecule as functions of bond distance and twist angle. The model has proved to give very reliable results for conformations, energy, torsional barriers, properties of neutral excited states of a wide series of molecules (alternant or not, closed shell or open shell . . .). The present review presents, in a very simple way:

— the definition and construction of an Heisenberg Hamiltonian
— the extraction of the basic information
— the numerical performances of the model

However, its essential theme is the new qualitative description of the electronic assembly of the molecule which emerges from this model and one derives simple rules, to predict, without any calculation, properties regarding spin multiplicity, spin densities, bond lengths and isomerisation energies. The limits of the model are discussed as well as possible extensions to other half-filled band problems.

1 Introduction

Any chemist necessarily has an implicit or explicit view of the electronic assembly occupying the molecular building. His qualitative ways of thinking about the molecular world rest on some kind of sociology or ethnology.

Two descriptions usually prevail in the chemical literature, and they are both based on the notion of electron pairs: the electrons are supposed to be coupled (except for a few unpaired electrons in free radicals, diradicals or excited states). This couple of electrons, — for which, for sixty years, it has been known that the two electrons are of opposite spins —, is the cornerstone of these descriptions, and in a rather clear metaphor where spin replaces sex, the electron pair is described as a "faithful couple" [1].

The oldest description is the Lewis picture [2]; it assigns one electron-pair per chemical bond, i.e. in a definite region of space between two bonded atoms, (for bonds) or in a certain section of space around a single atom (for core orbitals and lone pairs). In that description the electron pairs are spread in small boxes of the molecular space (the multiple bond complexifies the picture, but banana bonds fit into the basic scheme). Nothing is said about the energies of these electrons, which are almost equivalent for weakly polar molecules. The picture coincides with the prequantum chemistry models, and does not lead to any direct prediction regarding electronic ionization and excitations, which are delocalized processes.

The alternative description in terms of delocalized Molecular Orbitals (MO) became the new generally-accepted description, especially after the success of the stereoselectivity rules of concerted reactions as interpreted by Woodward and Hoffmann [3]. In this picture each electron pair occupies a level, a flat spread from one side to the other of the molecular space; the flats are distinguished according to their height (i.e. their energy) and their internal architecture (i.e. the number of nodes or walls). One main advantage of this approach is its ability to give a zeroth-order description of ionization and electronic excitation. In its simplest version (the Hückel approximation) this picture rests upon a diagonalization of a topological matrix, which represents an essentially computational step (i.e. a black box, despite the possible use of theorems issued from the graph theory) [4].

The equivalence between these two pictures, although insufficiently revealed in text-books, is well-established from the theory of electronic localizability, through unitary transforms [5]. The set of delocalized MOs in equivalent to a set of localized MOs. These localized MOs are bond-MO or lone pairs MO in saturated systems, while the localizability is poor for conjugated systems, for which the delocalized schemes had received the most convincing applications. The ab initio methods as precise computational tools have reinforced the success of the delocalized description, since their first step, the SCF approximation, provides such a set of delocalized MOs.

The VB approach [6] also had a tendency to build local electron pairs of opposite spins (i.e. two electron bond singlets), but considering all linearly-independant ways of partitioning the electronic assembly into bicentric ($\alpha - \beta$) electron pairs. This formulation did not lead to facile computational tools [7] nor to qualitative rules to help us understand collective or delocalized effects.

The present paper is intended to present a really orthogonal description of the electronic assembly of the molecule: it uses very local basic stones, the *single* electron in a single *Atomic* Orbital (AO), without any prejudice in favor of electronic coupling into *bond-pairs*. As will be shown in Sect. 4 this model is essentially an approximate treatment of the VB method, the space in which it works is the ensemble of neutral VB structures and the energies and interactions of the neutral VB determinants are "effective" energies and interactions, i.e. dressed by the effect of ionic VB structures and non-valence determinants. This effective Hamiltonian acts on spins since the space part of all the neutral VB determinants is unique for a half-filled band problem. It is therefore a spin-Hamiltonian, or a magnetic Hamiltonian; it may take the form of the well-known Heisenberg Hamiltonian, [8] first proposed in solid state physics as a phenomenological Hamiltonian, later on rationalized by Anderson [9].

The possible usefulness of such a magnetic description in chemistry has been demonstrated by the works of our group (following a few preliminary suggestions [10–11]) which essentially concern the conjugated systems [12–18]. The present review describes the principle of the method and the mechanism of the Hamiltonian-matrix building. It shows how that method allows the deduction of *qualitative rules*, without any numerical computation, directly from the molecular graph, regarding

— the spin multiplicity of the ground state (especially for diradicals),
— the spin distribution in free radicals and triplet states,
— structural properties concerning bond lengths,
— isomerization energies.

Later on, this review article describes the quantitative use of the Heisenberg Hamiltonian, especially in its non-empirical version, and its usefulness for the prediction of ground state equilibrium geometries, rotational barriers, the position of vertical (neutral) excited states, the geometry relaxation in (neutral) excited states, their adiabatic energy and some photochemical processes.

Besides the ease of use of this new computational scheme, which enables us to treat on an equal footing typical closed-shell problems (for the well-behaved molecules in their ground state) and open-shell problems (regarding free radicals and excited states), the main interest of this magnetic description rests in its qualitative description of the electronic population: although some electron pairs of opposite spins may have a tendency to concentrate in short (i.e. strong) double bonds, the basic picture is dominated by the collective spin ordering: the electrons tend to occupy the molecular space (i.e. the various atomic centres) in such a way that an electron of α spin is surrounded by electrons of β spins as far as possible, and vice-versa. The priviledged VB components are then spin-waves [15] (more precisely called Neel states by Solid State physicists).

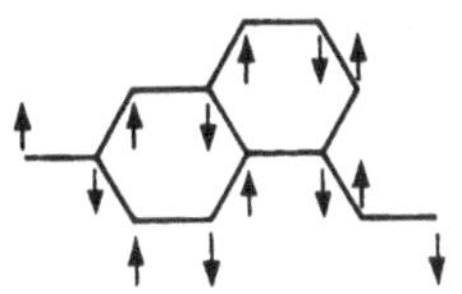

This ordered, most probable distribution is not the only one, there may exist two such interfering (or resonating) ordered distributions; but less ordered distributions, obtained from the preceding ones by spin exchange between adjacent atoms also occur, with a smaller probability and play an energetic role too. The electronic assembly is governed by two features
— the reluctance to have parallel spins on adjacent atoms, i.e. the preference for spin alternations on all bonds (or the largest number of bonds) creating a tendency to a collective order,
— the possible spin exchanges between adjacent atoms, damping this collective order through some kind of noise.

We let the reader translate in terms of a sexual metaphor this new description, orthogonal to the usual faithful electron-couples. In our opinion the magnetic picture might be taught rather extensively in a qualitative manner while the non-empirical computational version should be used more extensively than it has been, especially in photochemistry. The last section of the present review, while suggesting possible improvements and generalizations, clearly underlines some fundamental limitations of that type of approach, limitations which one should not ignore.

2 Neutral vs Ionic States in the VB Theory

2.1 Ethylene

Let us consider the simplest conjugated hydrocarbon, the ethylene molecule. The ground state wavefunction may be approximated as a single determinant describing four CH bonds, one CC σ bond and one π bond.

One may write it as

$$\Phi_0 = \mathscr{A}(\Sigma \cdot \pi\bar{\pi})$$

where $\mathscr{A}$ is an antisymmetrizer and Σ is the product of σ MOs; the π MO is equally spread on the p_π AOs of the atoms 1 and 2 (or linear combinations of p_π AO on these atoms)

$$\pi = a(p_1 + p_2) .$$

The π^2 distribution gives an equal mixture of neutral ($p_1\bar{p}_2$ and $p_2\bar{p}_1$) and ionic ($p_1\bar{p}_1$ and $p_2\bar{p}_2$) VB components (cf. H_2)

$$\pi^2 = \pi\bar{\pi} = a^2(p_1\bar{p}_2 + p_2\bar{p}_1 + p_1\bar{p}_1 + p_2\bar{p}_2) .$$

But the valence antibonding π^* MO

$$\pi^* = b(p_1 - p_2)$$

may be used to define a doubly excited determinant Φ_D

$$\Phi_D = \mathscr{A}(\Sigma \cdot \pi^*\bar{\pi}^*)$$

of ^1A_g symmetry, as Φ_0; the two determinants interact

$$\langle \Phi_D | H | \Phi_0 \rangle = \langle \pi^*\pi^* | 1/r_{12} | \pi\pi \rangle = K_{\pi\pi^*}$$

through an exchange integral, which is positive, and the exact wavefunction may be written as

$$\psi = \Phi_0 - \frac{K_{\pi\pi^*}}{2\,\Delta E_{\pi\pi^*}} \Phi_D + \text{other contributions}$$

where $\Delta E_{\pi\pi^*}$ is the $\pi\pi^*$ excitation energy.

Therefore

$$\psi \simeq \mathscr{A}(\Sigma \cdot (\pi\bar{\pi} - \lambda\pi^*\bar{\pi}^*))$$
$$= \mathscr{A}(\Sigma \cdot [(a^2 + \lambda b^2)(p_1\bar{p}_2 + p_2\bar{p}_1) + (a^2 - \lambda b^2)(p_1\bar{p}_1 + p_2\bar{p}_2)]$$

where $\lambda = K^*_{\pi\pi}/\Delta E^*_{\pi\pi}$ is positive.

Thus in the correlated wavefunction the ground state is essentially neutral, having larger components on the neutral VB situations than on the ionic ones.

The lowest excited state of the ethylene molecule is the $\pi\pi^*$ triplet state ^3B_u, which may be represented as

$$^3\Phi^*_{\pi\pi} = \mathscr{A}(\Sigma \cdot {}^3(\pi\pi^*))$$

where $^3(\pi\pi^*)$ may be either the $\pi\pi^*$ ($S_z = 1$), the $\overline{\pi\pi^*}$ ($S_z = -1$) or the $(\pi\bar{\pi}^* - \pi^*\bar{\pi})/\sqrt{2}$ ($S_z = 0$) products. In the three cases it is straightforward to see the VB content of the π parts. For sake of simplicity we shall assume that the two AO p_1 and p_2 are orthogonal ($a = b = 1/\sqrt{2}$) or consider their $S^{-1/2}$ transforms. Then

$$\pi\pi^* = p_1p_2$$
$$\overline{\pi\pi^*} = \overline{p_1p_2}$$
$$(\pi\overline{\pi^*} - \pi^*\bar{\pi})/\sqrt{2} = (p_1\overline{p_2} - p_2\overline{p_1})/\sqrt{2}$$

i.e. the triplet state is a purely neutral VB state.

In contrast the lowest valence singlet state, the $\pi\pi^*$ singlet state (of ^1B_u symmetry) is purely ionic

$$\Phi^*_{\pi\pi} = \mathscr{A}(\Sigma \cdot {}^1(\pi\pi^*))$$

with $^1(\pi\pi^*) = (\pi\overline{\pi^*} + \pi^*\bar{\pi})/\sqrt{2} = (p_1\overline{p_1} - p_2\overline{p_2})/\sqrt{2}$

Due to its ionic character it is strongly correlated, coupled with $\sigma\sigma^*$ valence excitations and with other (dynamical) correlation corrections. The correct description of the ionic excited state is much more difficult than that of the ground and lowest triplet states, which are neutral in character, and the theoretical reproduction of the experimental valence excitation at 8.2 eV has been a ten-year fight for quantum chemists. From the valence π and π^* MOs one might also produce a ^1A_g statn orthogonal to the ground state

$$\psi^* = \mathscr{A}(\Sigma \cdot (\pi^*\overline{\pi^*} + \lambda\pi\bar{\pi}))$$

$$\langle\psi^* \mid \psi\rangle = 0$$

which is essentially ionic, having $(b^2 + \lambda a^2)$ components on the ionic VB determinants and $(b^2 - \lambda a^2)$ components on the neutral VB determinants.

One may therefore classify the eigenstates according to their leading VB character; the two lowest states are essentially neutral. When the conjugation (i.e. the overlap between the two p_π AOs, or the amplitude $|\langle p_1| F |p_2\rangle|$ of the extradiagonal Fock operator) descreases, the neutral character of these wavefunction increases. Actually

$$\Delta E_{\pi\pi^*} \simeq \langle\pi^*| F |\pi^*\rangle - \langle\pi| F |\pi\rangle = \varepsilon_{\pi^*} - \varepsilon_\pi = -2\langle p_1| F |p_2\rangle$$

while $K_{\pi\pi^*} \simeq ((p_1p_1, p_1p_1) - (p_1p_1, p_2p_2))/2$

where $(p_1p_1, p_1p_1) = \langle p_1p_1| 1/r_{12} |p_1p_1\rangle$

$(p_1p_1, p_2p_2) = \langle p_1p_2| 1/r_{12} |p_1p_2\rangle$

When the ethylene molecule is twisted around the C—C bond by angle θ between the two CH_2 groups, the $\langle p_1| F |p_2\rangle$ hopping integral decreases as $\cos\theta$ and when $\theta \to 90°$ the two lowest states tend to become nearly degenerate and purely neutral. Similarly for planar structures the longer the C—C bond the more neutral is the ground state.

2.2 Larger Conjugated Hydrocarbons

The difference between neutral and ionic (in the sense of VB theory) eigenstates of a neutral molecule is a basic feature of chemistry [19], hidden by the MO language. It is relevant to any conjugated system. Going to butadiene for instance one may show that the ^1A_g ground state, the two lowest $\pi\pi^*$ triplet states (of ^3B_u and ^3A_g symmetries) are neutral, while the lowest ^1B_u $\pi\pi^*$ singlet excited state is ionic (and even purely ionic). These statements are true for all linear polyenes:

- the ground state and the two lowest triplet states are neutral
- the lowest $\pi\pi^*$ singlet excited of ^1B_u symmetry obtained by a dipolarly-allowed transition from the ^1A_g ground state is ionic
- there exists an excited singlet of ^1A_g symmetry which is neutral in the sense of VB theory, and which lies quite low in energy. Of course that state being of the same symmetry as the ground state, the transition from the ground state is

dipolarly-forbidden, and the transition is not easy to observe, it is frequently called the "hidden" ^1A_g state.

There has been great controversy [20], which is not yet settled, regarding the energy ordering of the ionic ^1B_u and neutral ^1A_g excited states. It seems now well accepted that for linear polyenes of more than three conjugated double bonds, the ^1A_g state is *below* the ^1B_u state; for smaller polyenes (buadiene and hexatriene) the two states are at least quite close in energy.

While experimentalists had accessible information regarding the ^1B_u state, theoreticians were in great trouble (and are still in great trouble) in calculating the ^1B_u and ^1A_g excited states on an equal footing. The ^1B_u state is essentially obtained from the (homo (φ_n) → lumo (φ^*_{n+1})) single excitation as zeroth-order description, while the ^1A_g state is more difficult to describe in the MO language, since it is a mixture of three configurations, namely the $\varphi_n \rightarrow \varphi^*_{n+2}$ and $\varphi_{n-1} \rightarrow \varphi^*_{n+1}$ single excitations and the $(\varphi_n \rightarrow \varphi^*_{n+1})^2$ double excitation. Nevertheless, it is much easier to position quantitatively the ^1A_g state than the ^1B_u state, precisely because of the ionic VB nature of the second state. Its mixing with diffuse states is greater because one of the atoms necessarily bears two π electrons in any of its VB components; the angular correlation (mixing with excited configurations involving d AO) is larger for the same reason, because $(p^2 \rightarrow d^2)$ excitations are crucial on an atom bearing two electrons in the same p AO, and the (σ/π) correlation effects are much greater, since the σ core must react to the instantaneous field created by the ionic π instantaneous distribution. For all these reasons, as it was for ethylene, the ^1B_u state is difficult to describe, it requires large basis sets and extensive CI [21]. Most standard type MO-CI calculations overestimate the energy of the ^1B_u state by 1 eV while the lowest ^3B_u state is quite easy to obtain, due to its neutral VB character and so is the ^1A_g hidden excited state.

The same comments are true for aromatic hydrocarbons. For benzene for instance, the ground state, the $^3B_{1u}$, $^3E_{1u}$ and $^1B_{2u}$ are neutral while the dipolarly-allowed excited singlet $^1E_{1u}$ is ionic as are the $^3B_{2u}$ and $^1B_{1u}$ states and it has been shown that the (σ/π) correlation effects are 1 eV larger in these states than in the former ones [22].

The difference between neutral and ionic states explains clearly a paradox of the semi-empirical π theories [23], which were so famous during the 1960's, namely the impossibility of having a common set of parameters to treat the ground state properties and the lowest triplet state energies on one hand and the transition energies to the dipolarly-allowed excited singlets on the other. For the same value of the β_{pq} monoelectronic hopping integral between adjacent atoms, one had to use larger bielectronic integrals to treat the neutral eigenstates than to obtain the dipolarly allowed electronic transition energies, which involve ionic states. This is due to the fact that in ionic states non-valence excited configurations, involving diffuse MOs, polarization AOs or $\sigma \rightarrow \sigma^*$ excitations play a much larger role; their effect is simulated by a reduction of the bielectronic γ_{pp} integrals, but if one reduces these integrals in order to have correct neutral → ionic transition energies, the delocalization effects become too large in the neutral eigenstates, the ground state delocalization energy is too large, the lowest triplet states are too low in energy . . . The same contradiction takes place in all valence electron semi-empirical schemes such as the

CNDO or INDO parametrized methods [24], and to obtain correct allowed transition energies, modified parametrizations have been proposed (cf. CNDO-S) [25], which are not satisfactory for ground state energy variations . . .

All these comments based on the history of quantum chemistry prove the importance of VB theory in providing clarifying concepts even for computation. The distinction between neutral and ionic states, although neglected by the dominant MO-CI "doxa", is of primary importance.

3 Magnetic Hamiltonian for π Systems; Its Definition, and How It Works

3.1 Ethylene

Let us move back first to ethylene, and restrict the two neutral eigenstates, the ground ^1A_g state and the ^3B_u state, to their valence neutral components; considering a common Σ core, which will not be formulated, and calling a and b the orthogonal(ized) p_z AOs on atoms A and B, one gets

$$\psi_N(^1A_g) = \mathscr{A}(\Sigma \cdot (a\bar{b} - \bar{a}b))/\sqrt{2}$$
$$\psi_N(^3B_u) = \mathscr{A}(\Sigma \cdot (a\bar{b} + \bar{a}b))/\sqrt{2}$$

where $\mathscr{A}$ is an antisymmetrizer.

Let us consider an effective Hamiltonian which would have $\psi_N(^1A_g)$ and $\psi_N(^3B_u)$ as eigenstates with exact energies

$$H^{eff}\psi_N(^1A_g) = E(^1A_g)\,\psi_N(^1A_g)$$
$$H^{eff}\psi(^3B_u) = E(^3B_u)\,\psi_N(^3B_u)$$

This Hamiltonian is such that

$$\langle|\Sigma \cdot a\bar{b}|\,H^{eff}\,|\Sigma \cdot a\bar{b}|\rangle = \langle|\Sigma \cdot \bar{a}b|\,H^{eff}\,|\Sigma \cdot \bar{a}b|\rangle$$
$$= [E(^1A_g) + E(^3B_u)]/2 = R'$$
$$\langle|\Sigma \cdot a\bar{b}|\,H^{eff}\,|\Sigma \cdot \bar{a}b|\rangle = \langle|\Sigma \cdot \bar{a}b|\,H^{eff}\,|\Sigma \cdot a\bar{b}|\rangle$$
$$= -[E(^1A_g) - E(^3B_u)]/2 = g$$

i.e. it is a two-by-two matrix for $S_z = 0$ VB components

$\lvert\Sigma \cdot a\bar{b}\rvert$	$\lvert\Sigma \cdot \bar{a}b\rvert$
R′	g
g	R′

where g is a positive quantity.

Since the (^3B_u) state also has $S_z = 1$ ($|\Sigma \cdot ab|$) and $S_z = -1$ ($|\Sigma \cdot \overline{ab}|$) components of energy $E(^3B_u)$, one may define

$$\langle|\Sigma \cdot ab|\, H^{eff}\, |\Sigma \cdot ab|\rangle = \langle|\Sigma \cdot \overline{ab}|\, H^{eff}\, |\Sigma \cdot \overline{ab}|\rangle = E(^3B_u) = R$$

One may rewrite H^{eff} as a 4×4 matrix, noticing that $R' = R - g$

$\|\Sigma \cdot a\bar{b}\|$	$\|\Sigma \cdot \bar{a}b\|$	$\|\Sigma \cdot ab\|$	$\|\Sigma \cdot \overline{ab}\|$
R − g	g	0	0
g	R − g	0	0
0	0	R	0
0	0	0	R

The diagonalization of that matrix gives of course the singlet ground state energy (and the neutral VB component of the ground state) and the three-fold degenerate lowest $\pi\pi^*$ triplet state energy. This effective Hamiltonian acts only on neutral VB componets of the π system in the field of a non-specified σ core and produces exact eigenenergies. Notice that all the neutral VB components have a common space component, they only differ by their spin distribution. Thus this effective Hamiltonian is a spin Hamiltonian, or a magnetic Hamiltonian, and it will be shown later that it is identical to a Heisenberg Hamiltonian.

3.2 Transfer to Other Conjugated Molecules

One may attempt to transfer the above-defined Hamiltonian to larger systems, involving more than two conjugated carbon atoms. One may actually formulate the following rules:

i) the Hamiltonian is *defined in the space of π neutral VB states*, i.e. on all the determinants having, besides a common implicit σ core, one electron per atom, with all possible spin distributions. For a four-centre problem (butadiene for instance), the space component of all the determinants will be $|\Sigma \cdot abcd|$. Since H and S_z commute (i.e. H does not mix determinants of different S_z values), one may consider independantly the $S_z = 0, 1, 2 \ldots$ values for even numbers (2n) of atoms, or $S_z = 1/2, 3/2 \ldots$ for radicals with 2n + 1 atoms. The $S_z = 0$ (or 1/2) case concerns the electron distributions where one has n (or n + 1) α spins and n β spins. This subspace has dimensions of $\binom{2n}{n}$ or $\left(\binom{2n+1}{n}\right)$ and it will generate all the possible neutral states of the problem (Singlet (or doublet), triplet (or quadruplet) etc ... up to the state of the upmost multiplicity (a 2n + 1 (or 2n + 2—)-uplet which is unique).

For butadiene one has $\binom{4}{2} = 6$ neutral determinants, namely $|a\bar{b}c\bar{d}|$, $|abcd|$, $|a\overline{bc}d|$, $|\bar{a}bc\bar{d}|$, $|ab\overline{cd}|$, $|\overline{ab}cd|$. As a convention we shall write all the determinants with the same order of AOs, of increasing label a < b < c < d, whatever the spin they bear.

ii) The *diagonal matrix element* relative to one of these determinants is simply given by a sum of bond contributions between adjacent atoms

$$\langle \Phi_I | H^{eff} | \Phi_I \rangle = \sum_{\text{bonds } ij} \left(R_{ij} - g_{ij} \left(\frac{1}{2} - 2S_i S_j \right) \right) \tag{1}$$

where S_i is the spin momentum on atom i (1/2 if it bears an α spin, —1/2 if it bears a β spin) and R is a scalar. This expression simply means that one has an R_{ij} "repulsion" on *all* bonds, and an attractive term $-g_{ij}$ on the bonds which bear an $(\alpha - \beta)$ *electron pair*.

iii) The *off-diagonal terms* of the matrix concern the couples of determinants which only differ by one spin-exchange between bonded atoms i, j, and the amplitude of the coupling is then g_{ij}

$$\langle \Phi_K | H^{eff} | \Phi_L \rangle = g_{ij}$$

if $\Phi_K = ex_{(ij)} \Phi_L$
where $ex_{(ij)}$ is a spin-exchange operator between two-electrons of opposite spins on the ij bond. In a more rigourous way, using creation a_i^+ and annibilation a_j operators

$$\begin{aligned} &\text{if} \quad \Phi_K = a_i^+ a_{\bar{j}}^+ a_j a_{\bar{i}} \Phi_L \quad \text{or} \quad \Phi_K = a_{\bar{i}}^+ a_j^+ a_{\bar{j}} a_i \Phi_L \\ &\langle \Phi_K | H^{eff} | \Phi_L \rangle = g_{ij} \end{aligned} \tag{2}$$

The hamiltonian may be rewritten more concisely in an operatorial form. Using second-quantization notation it is possible to write it

$$H^{eff} = \sum_{\substack{ij \\ \text{bonded}}} R_{ij} - g_{ij} [(a_i^+ a_{\bar{j}}^+ - a_{\bar{i}}^+ a_j^+)(a_j a_{\bar{i}} - a_{\bar{j}} a_i)]$$

or in a bracket notation with two-particle projectors

$$H^{eff} = \sum_{\substack{ij \\ \text{bonded}}} R_{ij} - g_{ij} |i\bar{j} - \bar{i}j\rangle \langle i\bar{j} - i\bar{j}| \tag{3}$$

One may also notice that the diagonal terms (Eq. (1)) may be written as

$$\langle \Phi_I | H^{eff} | \Phi_I \rangle = \sum_{\substack{ij \\ \text{bonded}}} \left[R_{ij} - \frac{1}{2} g_{ij} + 2g_{ij} \langle \Phi_I | S_{zi} \cdot S_{zj} | \Phi_I \rangle \right]$$

and that

$$\langle \Phi_I | H^{eff} | \Phi_J \rangle = \sum_{\substack{ij \\ \text{bonded}}} g_{ij} \langle \Phi_I | S_i^+ S_j^- + S_i^- S_j^+ | \Phi_J \rangle$$

where S^+ and S^- are the step up and step down spin operators

$$s^+\beta = \alpha \qquad S^+\alpha = 0$$

$$S^-\alpha = \beta \qquad S^-\beta = 0$$

So that finally, noticing that

$$\vec{S}_i \cdot \vec{S}_j = S_{zi}S_{zj} + (S_i^+S_j^- + S_i^-S_j^+)/2$$

it becomes possible to write the effective Hamiltonian as

$$H^{eff} = \sum_{\substack{ij \\ \text{bonded}}} \left[\left(R_{ij} - \frac{g_{ij}}{2}\right) + 2g_{ij}\, \vec{S}_i \cdot \vec{S}_j\right] \qquad (4)$$

in which one recognizes the classical expression of an Heisenberg Hamiltonian. Notice that while R_{ij} is the energy of relative to the triplet organization on the ij bond, $R_{ij} - g_{ij}/2$ is the mean energy of all possible spin states on that bond (three triplet states of energy R_{ij}, one singlet state of energy $R_{ij} - 2g_{ij}$).

As examples one may write the $S_z = 0$ Hamiltonian matrices for butadiene A—B—C—D (upper-half matrix)

$\|\bar{a}b\bar{c}d\|$	$\|a\bar{b}c\bar{d}\|$	$\|a\bar{b}\bar{c}d\|$	$\|\bar{a}bc\bar{d}\|$	$\|ab\bar{c}\bar{d}\|$	$\|\bar{a}\bar{b}cd\|$
$\Sigma R - g_{ab} - g_{bc} - g_{cd}$	0	g_{ab}	g_{cd}	0	g_{bc}
	$\Sigma R - g_{ab} - g_{bc} - g_{cd}$	g_{cd}	g_{ab}	g	0
		$\Sigma R - g_{ab} - g_{cd}$	0	0	0
			$\Sigma R - g_{ab} - g_{cd}$	0	0
				$\Sigma R - g_{bc}$	0
					$\Sigma R - g_{bc}$

where $\Sigma R = R_{ab} + R_{bc} + R_{cd}$

For equal bond-lengths (equal values of R and g on all bonds) the eigenstates and eigenergies are, by decreasing order of energy,

$$^5\Sigma_g^+ = 3R$$

$$^3\Sigma_u^+ = 3R - (2 - \sqrt{2})\, g$$

$$^1\Sigma_g^+ = 3R - (3 - \sqrt{3})\, g$$

$$^3\Sigma_g^+ = 3R - 2g$$
$$^3\Sigma_u^+ = 3R - (2 + \sqrt{2})\, g$$
$$^1\Sigma_g^+ = 3R - (3 + \sqrt{3})\, g$$

which is the ground state. The energy ordering is in agreement with experiment, at least for the four lowest states. Notice that the dipolarly-allowed ^1B_u singlet excited state does not appear in the partial spectrum given by the Magnetic Hamiltonian.

4 The Hierarchy of Spin-Distributions and Its Qualitative Implications

4.1 The Most-Spin Alternant VB Structures

From the preceding rules defining the matrix elements of an Heisenberg Hamiltonian, a series of qualitative implications may be deduced, which may help to rationalize a certain number of chemical features.

One may notice first that the diagonal matrix elements of the Heisenberg Hamiltonian are not equal; the larger the number n_I^a of bonds presenting a spin-alternation in a spin-distribution Φ_I, the lower the corresponding diagonal matrix element $\langle\Phi_I| H |\Phi_I\rangle$. As a second remark, one may notice that a spin exchange is possible on each spin-alternated bond, and thus a determinant Φ_I having n_I^a spin-alternated bonds will interact with n_I^a other determinants Φ_J. It is well known that (in general) the lowest eigenvector has its largest components on the basis vectors which have the lowest diagonal energy and/or the largest number of extradiagonal matrix elements. One may thus expect that *the ground-state vector of the Heisenberg Hamiltonian will have its largest coefficients on the determinants presenting a spin-alternation on the maximum number of bonds.*

Now one may distinguish several types of conjugated hydrocarbons, namely

— the "*alternant*" *hydrocarbons* for which it is possible to have a spin-alternation on *all* bonds, i.e. for which one may surround each α spin by β spin neighbors and vice versa. For instance for 4 atoms, cyclobutadiene, butadiene and trimethylene methane are alternant.

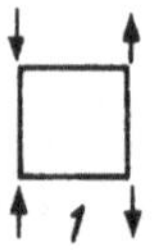
1 $S_z = 0$

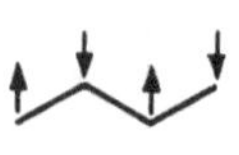
2 $S_z = 0$

3 $S_z = 1$

— the "*non-alternant*" *hydrocarbons,* for which it is impossible to have a spin alternation on all bonds. It is clear that the hydrocarbons (and molecular graphs) presenting *odd-membered cycles* cannot receive a fully-alternated spin distribution.

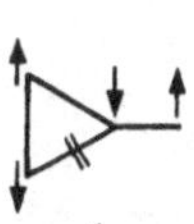
4

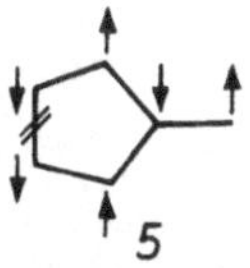
5

Also notice the following differences

— the alternant hydrocarbons with an *odd* number of centers (free radicals) may only receive one fully-alternated spin distribution with a positive value of S_z (i.e. with more α spins than β spins). In general $n_\alpha = n_\beta + 1$ i.e. $S_z = 1/2$, but it may happen that $n_\alpha = n_\beta + 3$ i.e. $S_z = 3/2$, or $n_\alpha = n_\beta + 5$ i.e. $S_z = 5/2$

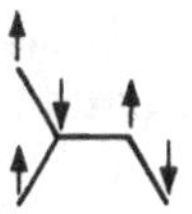
6
$S_z = 1/2$

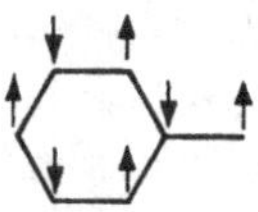
7
$S_z = 1/2$

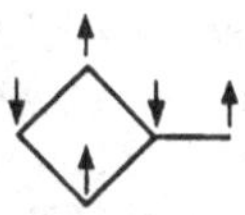
8
$S_z = 1/2$

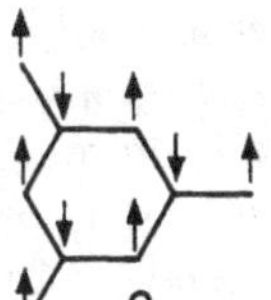
9
$S_z = 3/2$

— the alternant hydrocarbons with an even-number of conjugated carbons may in general receive *two* fully-alternated spin distributions, with $n_\alpha = n_\beta$ (and therefore $S_z = 0$)

and
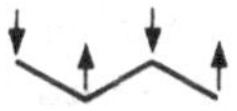
2

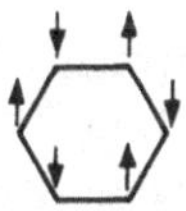
and
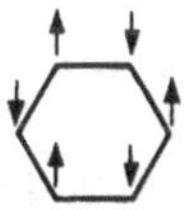
10

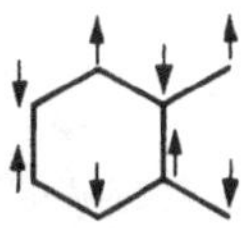
and
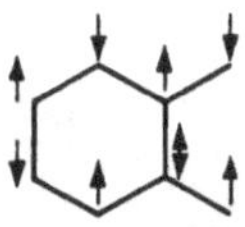
11

but in some cases the fully-alternated spin distribution is obtained for $n_\alpha = n_\beta + 2$ (and then $S_z = 1$). Such is the metaxylilene

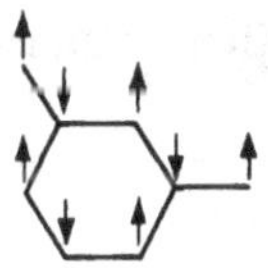
12

— In the non-alternant hydrocarbons there are several determinants with a maximum number of spin-alternations, (lower than the number of bonds), for instance six distributions in fulvene, *5*, *5′* and *5″* (each of them being two fold degenerate, by permutations of α and β spins).

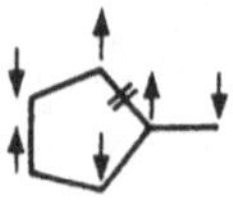
5′

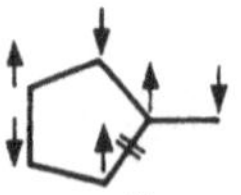
5″

From these remarks, and due to the larger stability of the fully-alternated structures, it is easy to infer the following rules:

1. *Prefered multiplicity of molecules*

— *The lowest eigenstate of a molecule will have the multiplicity associated with the* S_z *value of the most spin-alternated spin distributions,* i.e. it will be a singlet if $n_\alpha = n_\beta$ in the fully alternated (or most alternated) spin distribution, a triplet if $n_\alpha = n_\beta + 2$, a doublet if $n_\alpha = n_\beta + 1$ and a quadruplet if $n_\alpha = n_\beta + 3$ (and so on).

This rule, formulated by Ovchinnikov [10], rationnalizes the prefered multiplicity of diradicals or triradicals in an elegant manner. It makes it easy to understand why ortho and paraxylilenes (*11* and *13*) are Singlet, why metaxylilene *12* has a Triplet

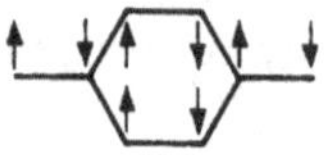
13

ground state, why the 1-3-5 trimethylene benzene will be a Quadruplet, or the difference between 2 — 3 (*14*) and 2 — 4 (*15*) dimethylene butadiene, the former being Singlet, the second one being Triplet.

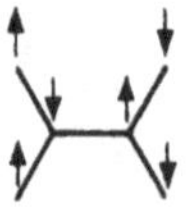
14

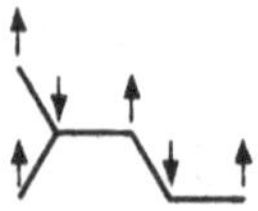
15

2. *Spin distribution in alternant radicals*

The most alternated spin-distributions having the largest coefficients in the ground state wavefunction one may expect that for free radicals, triplet or quartet alternant

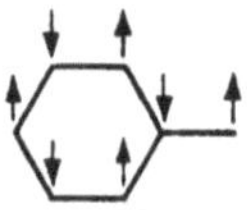
16

polyradicals, *the dominant spin will have larger amplitudes on the atoms which bear an α spin in the most spin-alternated distribution.* Actually in that case the most alternant spin distribution is unique. One may expect that in benzyl (*16*) the atoms having major (positive) spin densities will be the outer atoms (.68 e^- exp.) and the ortho and meta atoms of the cycle which experimentally bear .26 and .21 e^- respectively.

3. *Spin distribution in Triplet excited states*

Considering the excited Triplet states of an alternant hydrocarbon, one may focus on its $S_z = 1$ components in order to establish which are the most-spin alternated VB components, and deduce some information regarding the spin distribution. Of course all bonds cannot be spin-alternant now, but some VB components have a lower energy; for instance in linear polyenes the parallel spins will tend to occupy the ground state short bonds (i.e. 1–2, 3–4, etc . . .) rather than the ground state long bonds (2–3, 4–5, etc . . .). The major spin distribution will appear on the even-numbered atoms.

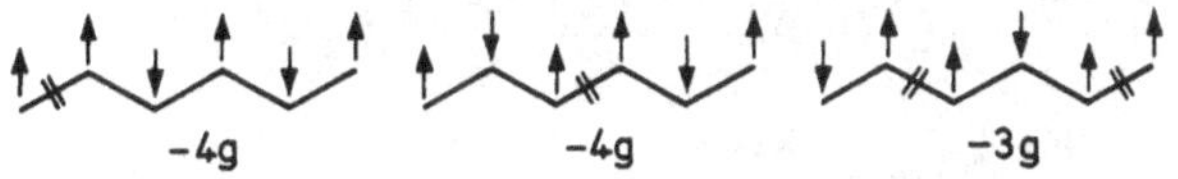

In naphthalene the triplet excited state will have major components on situations which only spin-frustrate two bonds and

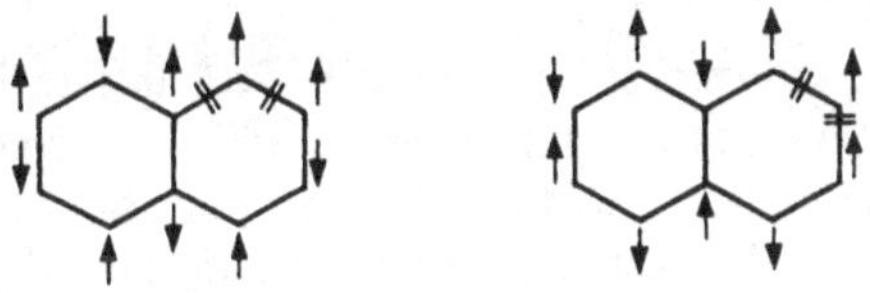

therefore induce induce a spin-density on atoms 1 and 2, as is well-known.

4.2 Hierarchy of Less Spin-Ordered Structures

The preceding conclusions are based on the consideration of the most-alternated spin-distribution Φ_0 (and Φ'_0). Then one may consider the spin distributions (i.e. VB

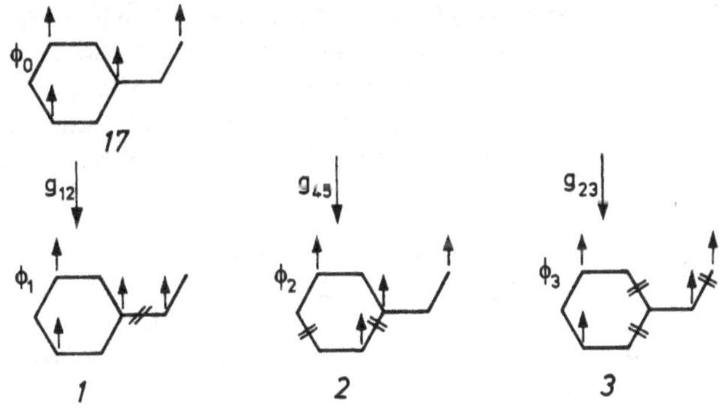

neutral components) which have lost only one, two or three spin-alternations with respect to the preceding ones; they are usually coupled by a factor g to the most-alternated structure Φ_0, i.e. are obtained from Φ_0 by a spin exchange on a spin-alternated bond. For instance in styrene *17*, starting from Φ_0, one spin exchange on the extracyclic terminal (double) bond (1–2) will lead to a determinant Φ_1 which has lost 1 spin-alternation only, while exchanging the spins on the extracyclic single bond (2–3) will lead to a loss of 3 spin-alternations (Φ_3). Exchanging the spins between ortho-meta atoms 4 and 5 of the cycle will lead to a determinant Φ_2 which has lost two spin-alternations. In view of their intrinsic energies the situations Φ_1 should appear in the ground state eigenvector ψ with a probability larger than Φ_2, which will in turn have a larger probability than Φ_3

$$|\langle\psi|\Phi_0| > |\langle\psi|\Phi_1\rangle| > |\langle\psi|\Phi_2\rangle| > |\langle\psi|\Phi_3\rangle|$$

This may easily be understood by a perturbative approach. If Φ_0 was taken as zeroth order wavefunction one would have

$$\psi = \Phi_0 + \frac{\langle\Phi_0|H|\Phi_1\rangle}{E_0 - E_1}\Phi_1 + \frac{\langle\Phi_0|H|\Phi_2\rangle}{E_0 - E_2}\Phi_2 + \frac{\langle\Phi_0|H|\Phi_3\rangle}{E_0 - E_3}\Phi_3$$

$$= \Phi_0 + \frac{g_{12}}{g_{23}}\Phi_1 + \frac{g_{45}}{g_{34} + g_{56}}\Phi_2 + \frac{g_{23}}{g_{12} + g_{34} + g_{38}}\Phi_3$$

Therefore *the probability of a spin exchange on a bond presenting a spin alternation in* Φ_0 *will be larger for the bonds having only one adjacent bond than for those having two or more adjacent bonds.*

In some cases the second most-alternated spin distribution is not immediately coupled with the most alternated one, but it will appear with a large coefficient in ψ. For instance in benzyl *16*, the second most-alternated situation keeps the spin alternation in the benzenic cycle and destroys the spin alternation on the extracyclic bond (loss of 1 g). Notice that in the two most-alternated spin distributions *16* and *16'*,

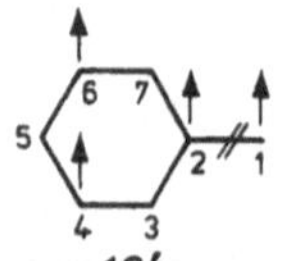

16'

which appear with coefficients .575 and .245 respectively in the wavefunction, the extracyclic carbon bears an α spin, and actually the spin density has a large probability of occurence in that atom (.68 e⁻ experimentally) against .26 and .21 on the ortho and para carbon atoms of the cycle. The fact that the spin-density on the ortho atom (3 or 7) is larger than that on the para atom (5) is certainly related to the larger probability of permuting spins on the 4–5 or 6–5 bonds from *16* than permuting spins on the 2–7 or 2–3 bonds;

from *16* exchange on 4–5 or 6–5 → loss of 2 g
exchange on 2–3 or 2–7 → loss of 3 g

This remark would also explain why the negative spin density is larger on atom 2 (—.090) than on atom 6 (—.067).

4.3 Structural Consequences

So far we have concentrated on magnetic properties (spin multiplicity and spin-distributions), which is of course a natural target of a magnetic Hamiltonian. But this Hamiltonian and the qualitative considerations developed above are able to lead us to some qualitative consequences regarding geometries, even before performing any calculation.

Hypothesis: Let us assume the following correlation, which will be demonstrated later on: *the more probable the spin alternation on one bond in the most-stable VB-neutral structures, the shorter the bond.*

In non-alternant hydrocarbons, for the most-spin alternated VB stuctures, some bonds bear parallel spins, and one may expect that they will have longer bonds. This is actually the case for instance on the following cyclic hydrocarbons

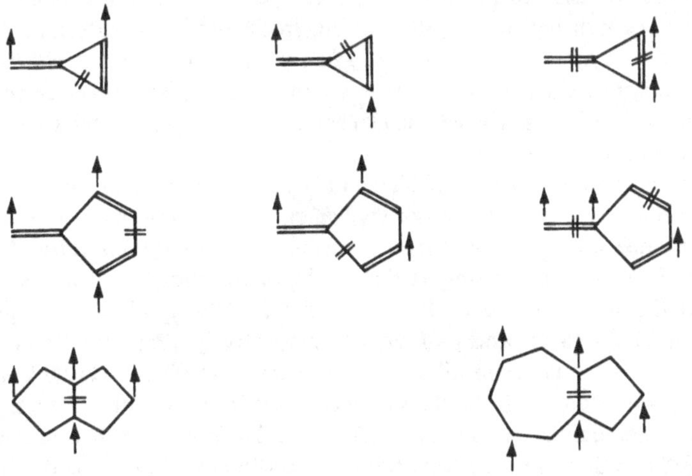

This rule rationalizes easily some tendencies to bond alternations or to very long bonds as occur in fused bicyclic non-alternant hydrocarbons such as azulene for which the bond common to the two cycles has a 1.50 Å length (0.10 Å more than the other bonds).

In *alternant hydrocarbons*, the most spin-alternated structures Φ_0 exhibit spin alternation on all-bonds, and the discrimination between short and long bonds cannot originate from these structures; it may be understood by the facility of spin exchanges on the various bonds according to the following *rule: the larger the number of bond spin-alternations lost by a spin reversal on bond ij in Φ_0, the longer the bond.* For instance in butadiene, the external bond will be shorter than the internal one, since a spin-exchange on bond 1–2 only loses one spin-alternation (on bond 2–3) while a spin-exchange on bond 2–3 will lead to a loss of two spin alternations

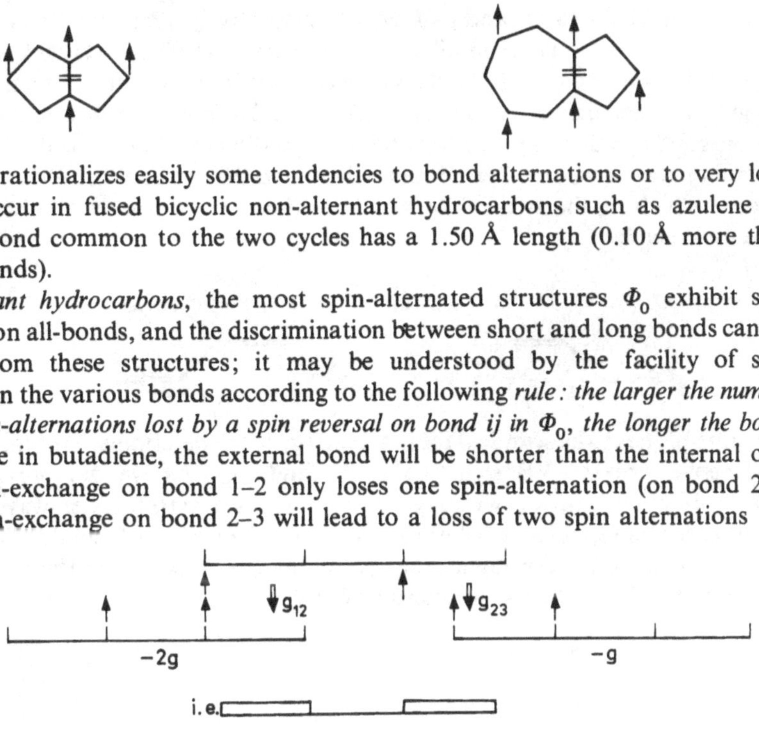

The external bonds of polyenes will always be the shortest ones. Actually the bond-length alternation of internal bonds, which is correctly reproduced by numerical calculations using the Heisenberg Hamiltonian, cannot be interpreted easily from these simple considerations alone.

From such qualitative views one may predict that the longest bond in naphtalene will be the bond common to the two cycles since a spin exchange on that bond will cost 4 spin-alternations.

4.4 Energetic Qualitative Predictions

Such simple counting of bond alternations in the most stable neutral VB components may sometimes be used to predict the π energy ordering between isomers. One may first consider the number of spin alternations in the most spin-alternant VB structure and conclude that *for the same number of bonds, alternant isomers will be more stable than non alternant ones* (this is not true when the comparison concerns a molecule with four-membered rings for which Heisenberg Hamiltonians miss some important four-body effects (vide infra)).

When one compares isomers of alternant hydrocarbons with the same number of bonds, one may expect that *the graphs which support two spin alternant VB determinants will have a lower energy than those supporting only one fully-alternated spin distribution.* For instance among xylilenes, the ortho and para isomers will be more stable than the meta one (actually PPP full CI with equal bond lengths leads to energy −18.34, −18.25 and −17.25 eV respectively [26]; topologic Heisenberg Hamiltonian [13] (equal g on all bonds), predict −11.03 g, −11.00 g and −9.92 g respectively, while an ab initio Heisenberg Hamiltonian with geometry relaxation [15] gives a π delocalization energy of 275, 274, 254 kcal/mole respectively).

More subtle predictions may be obtained regarding the π energy ordering of isomers presenting the same number of fully-alternated VB structures. One may ask for instance which of the two isomers of hexatriene (the linear or the branched one) is the most stable

The principle has to be searched for in the energy of the determinants obtained from the most spin-alternated VB structure(s) Φ_0 (and Φ_0'). As we did previously when looking at spin densities, we may consider Φ_0 as a zeroth-order wavefunction of zeroth-order energy $-n_b$ g where n_b is the total number of bonds. Then we may perturb this wavefunction by all the spin exchange processes on the various bonds. The second order perturbation energy will be

$$\varepsilon^2 = \sum_{\substack{i \\ \text{all bonds}}} \varepsilon_i^2$$

where ε_i^2 is a second order correction brought by the spin permutation on the bond i concerning the atoms i_1 and i_2.

$$\varepsilon_i^2 = \frac{g_{i_1 i_2}^2}{\Delta g}$$

where Δg is the number of spin alternations lost by the spin exchanges on atoms i_1 and i_2. Let us call n_i^v the number of bonds adjacents to the bond i

$$\Delta g = n_i^v g \quad \text{thus} \quad \varepsilon_i^2 = \frac{g}{n_i^v}$$

The energy brought by a spin exchange on a bond is inversely proportional to the number of adjacent bonds. Therefore for equal bond lengths

$$\varepsilon^2 = g \sum_i \frac{1}{n_i^v}$$

(notice that for $n_i^v = 1$ (i.e. for terminal bonds) the perturbation expansion from Φ_0 would diverge immediately, but for our qualitative considerations, this does not matter).

Then moving back to hexatrienes it is easy to see that for the linear one

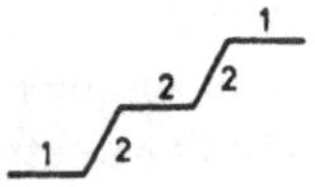

$$\varepsilon^2 = -(2/1 + 3/2)g$$

for the branched one

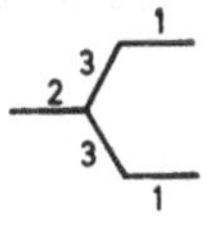

$$\varepsilon^2 = -(2/1 + 1/2 + 2/3)g$$

which is smaller than the energy of the linear one. Similarly one may compare the energies of the ortho and paraxylilene

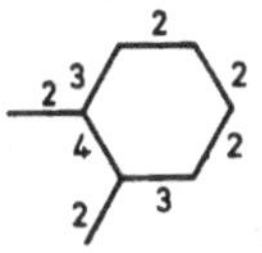

$$\varepsilon_0^2 = -(5/2 + 2/3 + 1/4)g$$

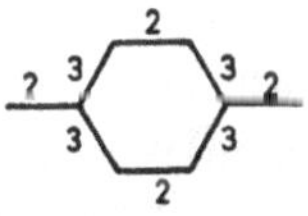

$$\varepsilon_p^2 = -(4/2 + 4/3)g$$

$$\varepsilon_0^2 - \varepsilon_p^2 = -(1/2 + 1/4 - 2/3)g = -0.09g$$

and actually the π energy of the ortho-one is lower by .1 eV in PPP full CI calculation [26] (by 1 kcal/mol only in the ab initio Heisenberg Hamiltonian with geometry relaxation) [15].

5 Heisenberg Hamiltonians as Quantitative Tools for Chemical Predictions

In this section the lowest roots of the Heisenberg Hamiltonian are supposed to be obtained numerically.

5.1 Semiempirical Level

The first importance of Heisenberg Hamiltons is their ability to simulate the full CI results of a π semiempirical Hamiltonian such as a Pariser Parr Pople (PPP) or a Hubbard (simplified PPP) Hamiltonian. Assuming equal bond lengths Dohnert and Koutecky [26] calculated the two lowest exact (i.e. full CI) eigenvalues of a PPP Hamiltonian for a series of hydrocarbons between four and ten atoms, i.e. the exact Singlet Triplet separation. The Singlet-Triplet ordering was correctly reproduced for all the compounds from a simple topological (equal values of R and g) Heisenberg Hamiltonian [13]. The agreement was even qualitative with a linear relation between the PPP and Heisenberg eigenergies (or S — T energy differences) as far as four-membered rings were disregarded in the series (for reasons which will be discussed in Sect. 6). The size of the Heisenberg Hamiltonian matrix, even without taking benefit of symmetry is of course much smaller than that of the full CI (even when working in a symmetry-adapted MO-CI scheme). But of course the Heisenberg Hamiltonian can only provide the neutral eigenstates.

Also of importance for such a crude model, i.e. a tight-binding topological Heisenberg Hamiltonian, one may quote the good linear correlation observed between the calculated probabilities to find a Singlet spin ordering between the bonded atoms i and j and the experimental r_{ij} bond-length [14]. If ψ_m is an eigenvector of the Heisenberg Hamiltonian, the probability of finding a Singlet spin-ordering on the bond ij is given by

$$\mathscr{P}^{s,m}_{ij} = \frac{1}{2} \langle \psi_m | i\bar{j} - \bar{i}j \rangle \langle i\bar{j} - \bar{i}j | \psi_m \rangle$$

Ref. 14 suggested a linear correlation which was numerically satisfied for a series of 6 molecules (18 different bonds)

$$r_{ij} = k\mathscr{P}^{s,0}_{ij} + r_0$$

where a least square fit gives

$$k = -.225\ \text{Å}\,, \qquad r_0 = 1.57\ \text{Å}$$

One may notice at that stage that 1.57 Å is actually the optimal bond length of the planar $\pi\pi^*$ Triplet ethylene (vide infra and ref. 15) for which $\mathscr{P}^{S}_{ij} = 0$, and that $1.57 - 0.23 = 1.34$ Å is the equilibrium bond length of the ethylene ground state Singlet for which $\mathscr{P}^{S}_{ij} = 1$. The linear fit observed for a series of hydrocarbons is therefore in perfect agreement with the properties of the basic subunit, the ethylenic double bond.

One may explain that relation by noting that the energy of a state is directly related to the Singlet bond indexes. From Eq. (3) it is clear that

$$E_m = \langle\psi_m| H |\psi_m\rangle = \sum_{\substack{ij \\ \text{bonded}}} (R_{ij} - \langle\psi_m| i\bar{j} - \bar{i}j\rangle g_{ij} \langle i\bar{j} - \bar{i}j |\psi_m\rangle)$$

$$= \sum_{\substack{ij \\ \text{bonded}}} R_{ij} - 2\mathscr{P}^{s,m}_{ij} g_{ij}$$

Then if R_{ij} and g_{ij} are linear functions of the distance r_{ij}

$$R_{ij} = R^0 + \frac{dR}{dr} r_{ij}$$

$$g_{ij} = g_0 + \frac{dg}{dr} r_{ij}$$

and if $\mathscr{P}^{s,m}_{ij}$ only varies weakly with r_{ij}

$$\frac{dE_m}{dr_{ij}} = \frac{dR}{dr} - 2\mathscr{P}^{s,m}_{ij} \frac{dg}{dr}$$

which is the desired linear relation.

We shall not devote more place to that semiempirical version, since the use of an ab-initio version does not cost more computation time.

5.2 An Ab-Initio Heisenberg Hamiltonian

This version represents an approximate quantum chemical method, since the exact Hamiltonian is replaced by an approximate Hamiltonian, assumed to work in a special configuration-space and keeping a well-defined form (with two-body interactions between adjacent atoms), but is not a semiempirical method since it does not involve any parameter fitted on experiment. Its interactions are extracted from accurate ab initio MO-CI calculations on ethylene performed in a double zeta + polarization basis set. Moving back to Eq. (3) and if the lowest singlet and triplet states energies are known as functions of the CC bond length, the torsion angle θ between the two CH_2 planes, but also eventually of CH bond lengths r', the bending angles φ and pyramidalization angle α of the carbon atoms

$$E(S^t_g) = f(r, \theta, r', \varphi, \alpha \ldots)$$

$$E(T^t_g) = f'(r, \theta, r', \varphi, \alpha \ldots)$$

The Heisenberg Hamiltonian might permit geometry optimizations for any neutral state of a conjugated system, and the research of transition states in chemical processes. In practice the R and g parameters have been defined as functions of r and θ only and are given in ref. 16, (in its Appendix the summations should be read

$$g = \sum_{i=0}^{5} -a_i x^i \quad \text{and} \quad R = \sum_{i=0}^{5} a_i' x^i)$$

5.2.1 Ground State Properties

The calculated *geometries* of small-sized molecules (n $\leqq$ 10) agree quite well with experimental determinations. Figure 1 takes the example of four molecules with six conjugated carbon atoms. Notice the bond alternation in hexatriene (with a somewhat longer internal double bond than the external one), a feature which also appears in larger polyenes (the most internal double and simple bonds in $C_{12}H_{14}$ have a 1.36 Å and 1.44 Å length [27]. The agreement for benzene is perfect, despite the neglect of important six-body effects (see Sect. 4). It is as good for non-alternant hydrocarbons such as or fulvene. The deviation from experiment for naphtalene and azulene is less than .02 Å.

Of course the procedure may be applied to free radicals or unstable molecules for which experiments are lacking. Here comparison must be done with the best ab-initio calculations, and they actually agree within the same deviation [16]. The Jahn-Teller distortions of, say, cyclopentadienyl or trimethylene methane are perfectly reproduced. An interesting picture of odd linear polyenes emerges from the calculations, namely the appearance of a *soliton* on the center of the molecule, with opposite bond-alternations on the two terminal regions.

The out-of-equilibrium geometries and energies may be calculated as well, for instance the *rotational barriers* around all single and double bonds, with full geometry relaxation at the transition state since the parameters have been extracted for all values of the torsion angle of ethylene. It has been shown that the rotational barrier around double bonds is lower for internal double bonds, and decreases when the chain length increases [16]. It has been estimated as 35 kcal/mole for a 10-carbon chain internal bond, and this agrees with the sensitization of the retinal isomerization. The rotational barrier around single bonds increases slightly when the chain length increases.

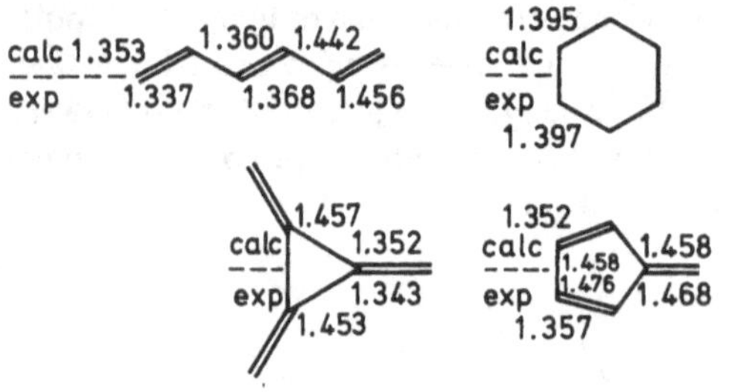

Fig. 1. Comparison of calculated and experimental bond lengths (in Ångstrom) of four conjugated molecules with six carbon atoms: linear hexatriene, benzene, trimethylenecyclopropane and fulvene. The calculated values have been obtained from the ab-initio Heisenberg Hamiltonian given in ref. 16.

It has been shown that the transition state for cis-trans isomerization of even linear polyenes, i.e. for the rotation around a double bond, defines a *soliton pair* on both sides of the rotated double bond. From these remarks the energy to create a neutral soliton pair in a large even polyene has been estimated as 10 kcal/mole.

5.2.2 Excited States Properties [17]

From the geometry optimization of the ground state, the spectrum of the Heisenberg Hamiltonian provides an evaluation of the *vertical transition energies* from the ground state to the neutral excited states, i.e. the ^3B_u, ^3A_g and ^1A_g excited states in linear polyenes [17].

Table 1. Spectroscopic properties of linear polyenes (see references to other theoretical estimates and experimental values in [17])

	Vertical absorption (eV)				Adiabatic transition (eV)	
	Heisenberg		MO—CI	exp	Heisenberg	exp
Butadiene	^3B_u	3.03	3.24, 3.35, 3.45	3.22	2.45	2.58
	^3A_g	5.28	4.95, 5.04, 5.08	3.04		
	^1A_g	7.23	7.06, 6.77	7.07	6.0	
Hexatriene	^3B_u	2.44	2.71	2.6	1.89	
	^3B_g	4.25	4.33	4.2		
	^1A_g	5.72	5.87	~5.	4.75	4.5
Octatetraene	^3B_u	2.38		2.2	1.57	
	^1A_g	4.08			3.91	3.54
Infinite chain	^3B_u	.87 eV			.43	
	^1A_g	1.3 eV			.47	

Table 1 permits a comparison between the predictions of the Heisenberg Hamiltonian, the experimental evidences and the best ab-initio MO-CI calculations for the lowest members of the linear polyene series. The agreement is quite satisfactory. The extrapolated values for an infinite chain are given in the same Table.

Of course one advantage of the Heisenberg Hamiltonian is its ability to perform geometry optimizations for excited states, i.e. to calculate adiabatic absorption energies from the ground state or an excited state, as well as vertical and adiabatic emission energies (fluorescence and phosphorescence).

The most fascinating conclusion concerns the *relaxed geometry* of the linear polyene in its lowest triplet state, which appears to represent a neutral soliton pair. The first soliton (unpaired electron) being located at the first fourth of the chain length, while the second one is located at the third fourth. The asymptotic *adiabatic transition* from the ground state to the lowest triplet is .43 eV (infinite chain length) (and .47 eV toward the ^1A_g hidden state).

The calculations are as easy on odd linear polyenes, for which experimental evidence is lacking. However the study of such systems is very relevant for the analysis of the mechanism of *cis-trans photoisomerizations*. Twisting a polyene around a double bond defines a transition state where the π system is cut into two orthogonal delocalized free radicals, i.e. two polyenes with odd numbers of conjugated carbons. The mechanisms of cis-trans isomerizations of butadiene and hexatriene in their lowest triplet (^3B_u) and lowest neutral excited singlet (^1A_g) states have been studied in great detail. For large enough polyenes, the method predicts that the singlet cis-trans isomerization must proceed through the neutral ^1A_g state; it also predicts that cis-trans isomerization from vertical excitations would become impossible around external double bonds for large polyenes, while it will remain possible around the internal double-bonds.

The photochemical applications of Heisenberg Hamiltonians are only at their beginning. The photochemical isomerizations of conjugated imines have been studied as well [28–29]. The cis-trans isomerization around the external bond of styrene ($C_6H_5-CH=CH_2$) has been reinterpreted [30] as occuring through neutral states rather ionics states (i.e. through $[C_6H_5-CH]\,CH_2$ rather than through $[C_6H_5-CH]^+$ CH_2^-). The mechanisms of the corresponding photoreactions of stilbene and di-phenyl polyenes should be studied using the same procedure.

6 Theoretical Foundations, Possible Extensions and Intrinsic Limits of Magnetic Hamiltonians

After this brief review of its conceptual and practical effiency [31], the Magnetic approach must be analysed critically, examining first its theoretical foundation since in the above presentation the transfer from ethylene to larger systems was performed arbitrarily.

6.1 Theoretical Foundations

The Heisenberg Hamiltonian may be legitimated as an effective Hamiltonian spanned by the neutral VB determinants and obtained from the exact Hamiltonian through a second-order expansion of the Quasi Degenerate Perturbation Theory. The Heisenberg Hamiltonian was historically introduced as phenomenological, but it was rationalized later on by Anderson [9] as an effective Hamiltonian in the rigourous meaning of that term.

If one considers an N-dimensional finite subspace of n-electronic functions Φ_I, which are not eigenfunction of H, they may be chosen as spanning what will be called the model space S_0. If they are orthonormal

$$\langle \Phi_I \mid \Phi_J \rangle = \delta_{ij}$$

$$P_0 = \sum_{I=1}^{N} |\Phi_I\rangle \langle \Phi_I|$$

and the effective Hamiltonian defined within S_0 is such [32] that

$$H^{eff}\,|P_0\psi_m\rangle = \varepsilon_m|P_0\psi_m\rangle$$

with $H|\psi_m\rangle = \varepsilon_m|\psi_m\rangle$
i.e. its N roots are eigenvalues of the exact Hamiltonian, and its N eigenvectors are projections onto S_0 of the associated exact eigenvectors.

It is expected that the relevant ψ_m are the eigenstates of H having the largest components on S_0. In our problem S_0 will be the space spanned by the neutral VB components, and the ψ_m are supposed to be the *neutral eigenstates* defined in Section 3.

The Quasi Degenerate Perturbation Theory [33] is an order-by-order expansion of H^{eff} from P_0 and H. The second order expression of the matrix elements is very simple

$$\langle\Phi_I|\,H^{eff(2)}\,|\Phi_I\rangle = \langle\Phi_I|\,H\,|\Phi_J\rangle + \sum_{\alpha\notin S_0}\frac{\langle\Phi_I|\,H\,|\Phi_\alpha\rangle\,\langle\Phi_\alpha|\,H\,|\Phi_J\rangle}{E_J - E_\alpha}$$

where, for instance,

$$E_J = \langle\Phi_J|\,H\,|\Phi_J\rangle$$
$$E_\alpha = \langle\Phi_\alpha|\,H\,|\Phi_\alpha\rangle$$

The determinants Φ_α may be valence ionic VB determinants, or non-purely valence determinants. Let us consider for simplicity the valence determinants only and a Hubbard [34] or PPP [23] Hamiltonian (i.e. a neglect of differential overlap between the AO for bielectronic integrals)

$$H = \sum_p \alpha_p a_p^+ a_p + \sum_{\substack{p\neq q\\ \text{bonded}}} \beta_{pq} a_p^+ a_q + \sum_{p,q} \gamma_{pq} a_p^+ a_q^+ a_p a_q$$

Then the neutral determinants do not interact among themselves at the first order $\langle\Phi_I|\,H\,|\Phi_J\rangle = 0$ since they differ by at least two orbitals and may only interact through bicentric exchange integrals K_{pq}, which are neglected if they only differ by a spin exchange on a bond. The neutral determinants are coupled to some of the ionic determinants where the positive and negative charges are on adjacent atoms:

$$<\Phi_I|H|\Phi_\alpha> = \beta_{pq} \quad \text{if} \quad \Phi_\alpha = a_p^+ a_q \Phi_I$$

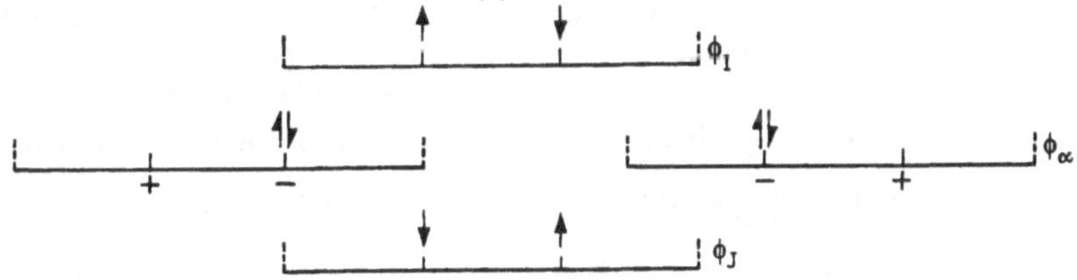

Such electron jumps are only possible on the bonds of Φ_I bearing a spin-alternation. Then

$$\langle \Phi_I | H^{eff\,(2)} | \Phi_I \rangle = \langle \Phi_I | H | \Phi_I \rangle - \sum_{\substack{p \neq q \\ \text{spin alternant} \\ \text{bonds in } \Phi_I}} \frac{2\beta_{pq}^2}{\Delta E_{pq}}$$

where $\Delta E_{pq} = \gamma_{pp} - \gamma_{pq}$ is the energy difference between the neutral determinants and the singly ionic VB determinants with a + — or — + zwitterion on the chemical bond pq.

Writing

$$g_{pq} = 2\beta_{pq}^2/\Delta E_{pq} \quad \text{and} \quad \langle \Phi_I | H | \Phi_I \rangle = \sum_{\substack{pq \\ \text{bonded}}} R_{pq}$$

one sees that $\langle \Phi_I | H^{eff\,(2)} | \Phi_I \rangle = \sum_{pq} [R_{pq} - g_{pq} n_{pq}^I]$ where n_{pq}^I is one if the bond pq is spin-alternant in Φ_I and zero otherwise. One recognizes here the diagonal term of an Heisenberg Hamiltonian.

Extradiagonal terms appear at the second-order only between determinants Φ_I and Φ_J which are coupled to the same singly ionic determinants Φ_α by electron jumps. This is possible, as pictured above, if Φ_I and Φ_J only differ by a spin exchange on one bond (pq for instance). Then

$$\langle \Phi_I | H^{eff\,(2)} | \Phi_J \rangle = + \frac{2\beta_{pq}^2}{\Delta E_{pq}}$$

if $\Phi_I = a_{\bar{p}}^+ a_q^+ a_{\bar{q}} a_p \Phi_J$ or $\Phi_I = a_p^+ a_{\bar{q}}^+ a_q a_{\bar{p}} \Phi_J$.

Note that

$$\langle \Phi_I | H^{eff\,(2)} | \Phi_J \rangle = g_{pq}$$

and also note the change of sign with respect to the diagonal contribution, the origin of which appears clearly in the above scheme.

Then the 2nd order effective Hamiltonian takes the form of an Heisenberg Hamiltonian. Of course this is only an approximate effective Hamiltonian, the exact one being obtained through a summation of *higher orders*. This expansion has been explained in detail elsewhere [12], and compared with fully converged results [14], and one may summarize the main results as follows:

— The third-order corrections are zero

— The fourth-order correction would introduce some dependance on the bicentric exchange to the environment and 1–3 effective exchanges between second neighbours, but these are, in general, minor corrections. A specific fourth-order correction concerns the four-membered rings. It acts especially strongly on the fully spin-alternated distributions, coupling  and  ; this interaction is a operator since it acts on four spins and four centers simultaneously. Its amplitude is large;

to the fourth-order, this correction has a $(40\beta^4/\Delta E^3)$ amplitude. After infinite summation one finds that the exact amplitude is about 60% of the amplitude of the effective exchange g between adjacent atoms. This coupling destabilizes the singlet organization of the cycle and stabilizes the triplet organization. This is the magnetic explanation of the *antiaromatic* character of four-membered rings.

One may easily understand its origin through fourth-order movements of electrons along the square, such as

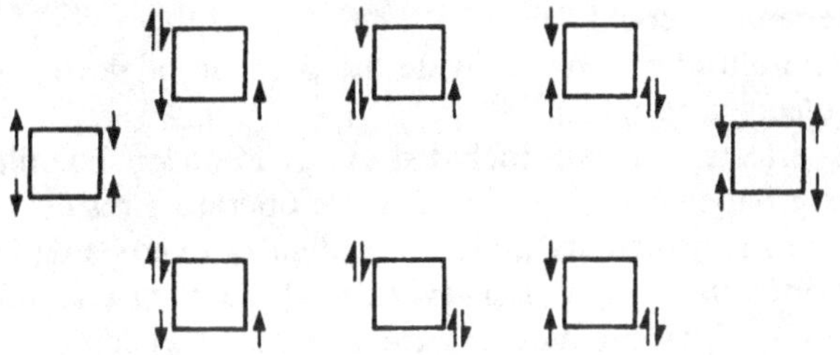

The just-mentioned large coefficient 40 results from the large number of possible independant processes. Notice that fourth-order electronic processes represent a *ring current*, since the electron jumps occur in the same clockwise (or anticlockwise) circular movement. There is a direct connection between collective effects occuring in rings and the ring currents. Since these four-body effects destabilize the ground state singlet state, i.e. explain the antiaromaticity of four-membered rings, the connection between ring currents and antiaromaticity appears clear.

— Sixth-order contributions are important in six-membered rings only; the largest operator couples the fully spin-alternant neutral VB distributions

In practice the amplitude of that coupling is about 40% of that of the effective exchange g. This six-body operator lowers the energy of the lowest singlet and destabilizes the lowest triplet, i.e. is responsible for the *aromatic* stabilization. The same remark concerning the connexion with ring currents would be valid for six-membered rings.

6.2 Limits of Heisenberg Hamiltonians

Several lessons may be learned from these analyses:

— The Heisenberg Hamiltonians with effective exchanges between nearest neighbours are only approximations of some exact effective Hamiltonians. The two-body terms are not *strictly* transferable, some many-body effective operators should be introduced in principle.

— The most important many-body effects concern the four-membered rings. Calculations with classical Heisenberg Hamiltonians (two-body exchanges between adjacent atoms) are reliable for systems which do not involve squares; their application to chemical systems involving four-membered rings must be regarded cautiously; they are only valid when the coordination number of the atoms is sufficiently large (i.e. they are more reliable for  or for a square or cubic lattice than for cyclobutadiene).

– The six-body terms which one should introduce in principle in the treatment of aromatic molecules apparently do not play an important role.

One must of course remember that

i) the Heisenberg Hamiltonians only provide the position of *neutral* states, i.e. a limited part of the molecular spectrum.

ii) some *collective* effects are not included in Heisenberg Hamiltonians. As apparent from its perturbative building, the effective operators result from local or regional travels of electrons, which are limited in extent or exponentially decreasing with the number of bonds they imply. However the Heisenberg Hamiltonian being N-electronic, i.e. defining different matrices for different topologies or molecular graphs, is able to reflect collective effects. The collective effects not included in simple Heisenberg Hamiltonians are those which involve circulations of electrons along circuits, leading for instance to the four-body and six-body operators occuring on four- or six-membered rings. These operators are responsible for the 4n/4n + 2 antiaromatic/aromatic character, and therefore of the 4n/4n + 2 rule for the *stereospecificity* of electrocyclic concerted reactions. If the cyclic operators are not included in the Hamiltonian, the stereospecificity will not be reproduced.

Despite these limits, we believe that the Heisenberg Hamiltonians are a fantastic tool, both for the direct and elegant picture of the electronic assembly which they provide and for the quantitative and predictive power of their ab-initio versions. The method may be extended in various directions. It may be applied to *weakly polar conjugated systems*. In imines for instance, i.e. in C=N— double bonds, the N atom also brings one electron in one AO and R and g parameters may be obtained for the CN bond; they may be used, with apparent success, to treat larger imines, i.e. conjugated molecules with CC and CN bonds [28–29] in both their ground state and lowest $\pi\pi^*$ excited states. The same process might be applied to C=O bonds for the treatment of conjugated aldehydes and ketones [35]. Of course the $n \rightarrow \pi^*$ excited states are not provided by that model, which works with a constant number of π electrons. One may similarly extract information on the $H_2C{-}NH{-}CH_2$ and $H_2C{-}CH{-}NH_2$ systems to define effective spin interactions between conjugated carbons separated by an amine group (i.e. a π lone part) or adjacent to an amine group [36]. As far as the polarity of the conjugated system remains small enough, Heisenberg Hamiltonians may be used for the treatment of the neutral eigenstates. Notice that in the case of imines the π electron of the Nitrogen is included in the half-filled band, and the N-atom appears explicitely, while in the case of amines the π electrons of the Nitrogen are kept outside the half-filled band and the Nitrogen atom is implicit; the effect of the CN distance will then appear through the distance of R_{cc} and g_{cc} to the CN distance. The exclusion of the amine nitrogen atom π electrons is an intrinsic consequence of the Heisenberg Hamiltonian structure: the Heisenberg Hamiltonian only concern half-filled bands i.e. systems where the number of electrons is equal to the number of valence AO.

Other *half-filled bands*, with several electrons per atom might be considered and treated in a similar manner. The carbons of a triple bond bring two π electrons in two orthogonal π AO, and Heisenberg Hamiltonians have been proposed for the treatment of *polyines* (i.e. $H(C{\equiv}C)_nH$). Their parameters might certainly be applied with success for the treatment of cumulenes which mix carbon atoms involved in one and two double bonds (i.e. bring one or two π electrons in one or two AO). The Nitrogen, Phosphorus or Arsenic atoms have an s^2p^3 ground state configuration; the p bond is half-filled in homogeneous clusters of these atoms (or bulk) and an attempt to extract a Heisenberg Hamiltonian from the diatoms N_2, P_2 or As_2 is in progress [37]; it will be used to study the properties of medium-size clusters and the various phases generated from a cubic lattice by Peierls distorsion.

Of course one may also generalize the strategy leading to a Heisenberg Hamiltonian for the case of half-filled bands, applying it to other situations than the ideal half-filled. The method always consists in the building of an effective Hamiltonian spanned by the most neutral VB configurations. Effective Hamiltonians have been derived for the positive ions of conjugated systems [38], the model space being defined by the one-hole VB configurations. Other systems may be treated along the same lines [39], which are based on a general remark concerning the hierarchy of VB configurations [40].

7 Bibliography

1. Salem L (1978) J. Chem. Educ. 55: 344
2. Lewis GN (1916) J. Amer. Chem. Soc. 38: 762
3. Woodward RB, Hoffmann B (1965) J. Amer. Chem. Soc. 87: 395; (1965) 87: 2511
4. see for instance Trinajstic N (1977) Chemistry of conjugated molecules. Springer, Berlin Heidelberg New York
5. a. Foster JM, Boys SF (1960) Rev. Mod. Phys. 34: 300
   b. Edminston C, Ruedenberg K (1963) Mod. Phys. 35: 457
   c. Millié P, Levy B, Berthier C (1975) in: Chalvet O, Daudel R, Diner S, Malrieu JP (eds) Localization and delocalization in quantum chemistry, vol. 1. Reidel, Dordrecht, p 59
6. a. Heitler W, London F (1927) Z. Physik 44: 455
   b. Rumer G (1932) Gottinger Nach. 337
   c. Pauling L, Wheland GW (1933) J. Chem. Phys. 1: 362
7. Gerratt J (1987) Chem. in Britain 327
8. Herring C (1986) in: Rado CT, Suhl H (eds) Magnetism. Academic, New York, vol II B
9. a. Anderson PW (1959) Phys. Rev. 115: 2
   b. Anderson PW (1963) in: Rado CT, Suhl H (eds) Magnetism. Academic, New York, vol I p 25
10. a. Ovchinnikov AO (1978) Theor. Chim. Acta 47: 297
    b. Bulaewski LN (1967) JETP 24: 154 Translation of (1966) JETP 51: 230
11. a. Klein DJ, Garcia-Bach MA (1979) Phys. Rev. B 19: 877
    b. Garcia-Bach MA, Klein DJ (1977) Int. J. Quant. Chem. 12: 273
    c. Klein DJ (1982) J. Chem. Phys. 77: 3098
    d. Klein DJ, Nelin CJ, Alexander S, Matsen F (1982) J. Chem. Phys. 77: 3101
12. Malrieu JP, Maynau D (1982) J. Amer. Chem. Soc. 104: 3021
13. Maynau D, Malrieu JP (1982) J. Amer. Chem. Soc. 104: 3029
14. Maynau D, Durand P, Daudey JP, Malrieu JP (1983) Phys. Rev. A, 28: 3193
15. Maynau D, Said M, Malrieu JP (1983) J. Amer. Chem. Soc. 105: 5244
16. Said M, Maynau D, Malrieu JP, Garcia-Bach MA (1984) J. Amer. Chem. Soc. 106: 571

17. Said M, Maynau D, Malrieu JP (1984) J. Amer. Chem. Soc. 106: 580
18. Kuwajima S (1982) J. Chem. Phys. 77: 1930
19. a. Cizek J, Paldus J, Hubac I (1974) Int. J. Quant. Chem. 8: 951
    b. Paldus J, Cizek J, Hubac I (1974) Int. J. Quant. Chem. 85: 923
20. for a review see Malrieu JP (1984) Pure and Appl. Chem. 56: 1241
21. a. Buenker RJ, Shih S, Peyerimhoff SD (1976) Chem. Phys. Lett. 44: 385
    b. Hosteny RP, Dunning TM, Gilman RR, Pipano A, Shavitt I (1975) J. Chem. Phys. 62: 4764
    c. Nascimento MAC, Goddard WA (1980) Chem. Phys. 53: 251
22. Rancurel P, Huron B, Malrieu JP, Berthier G, Praud L (1976) J. of Molec. Spectr. 60: 267
23. a. Pariser R, Parr RG (1963) J. Chem. Phys. 21: 466, 767
    b. Pople JA (1953) Trans. Far. Soc. 49: 1375
24. a. Pople JA, Segal GA (1965) J. Chem. Phys. 43: 5136
    b. Pople JA, Segal GA (1966) J. Chem. Phys. 44: 3289
25. a. Del Bene J, Jaffé HH (1968) J. Chem. Phys. 48: 1807
    b. Del Bene J, Jaffé HH (1968) J. Chem. Phys. 49: 1221
26. Dohnert D, Koutecky J (1980) J. Amer. Chem. Soc. 102: 1789
27. Sanchez-Marin J, Maynau D, Malrieu JP (1987) Int. J. Quant. Chem. 31: 903
28. Sanchez-Marin J, Malrieu JP (1985) J. Phys. Chem. 89: 975
29. Sanchez-Marin J, Malrieu JP (1985) J. Amer. Chem. Soc. 107: 1985
30. Said M, Malrieu JP (1983) Chem. Phys. Lett. 102: 312
31. for further applications see
    a. Mujica W, Correia N, Goscinski O (1985) Phys. Rev. B 32: 4175, 4186
    b. Cioslowski P (submitted for publication)
32. Bloch C (1960) Nucl. Phys. 20: 321
33. a. Van Vleck JH (1929) Phys. Rev. 33: 467
    b. des Cloizeaux J (1960) Nucl. Phys. 20: 321
    c. Brandow BH (1967) Rev. Mod. Phys. 39: 771
    d. Shavitt I, Redmon I (1980) J. Chem. Phys. 73: 5711
34. Hubbard J (1963) Proc. Roy. Soc. London, A 276: 283
35. Sanchez-Marin J (to be published)
36. de Loth P, Maynau D, Malrieu JP (1987) Phys. Rev. B 36: 3365
37. Pelegatti A, Trinquier G, Malrieu JP (work in progress)
38. Gadéa X, Maynau D, Malrieu JP (1984) Int. J. Quant. Chem. 26: 1
39. see for instance for a 1/6th filled band,
    a. Pelegatti A, Marinelli F, Maynau D, Malrieu JP (1987) J. de Phys. 48: 29
    b. Pelegatti A, Marinelli F, Roche M, Maynau D; Malrieu JP (1987) J. Phys. C 20: 5145
40. Durand P, Malrieu JP (1987) Adv. Chem. Phys. 57: 321

Directional Properties of Covalent Bonding in Molecules

Zvonimir B. Maksić

Theoretical Chemistry Group, The "Rudjer Bošković" Institute, Bijenička 54, 41001 Zagreb, Yugoslavia and Faculty of Natural Sciences and Mathematics, University of Zagreb, Marulićev trg 19, 41000 Zagreb, Croatia, Yugoslavia

Extensive evidence is presented showing that hybridization is one of the most important models of covalent bonding in molecules. Hybridization is physically well founded because it represents a local density distortion as a response to a descent in symmetry of an atom in a chemical environment. It is nicely reflected in electron isopycnic maps and in the shapes of Fermi holes. Hybridization provides an essential link between the quantum theory and chemical phenomenology giving a simple rationalization of empirically established relations between various observables. Hybridization is a phenomenon, but it is hidden because it can not be directly measured. Since it is intimately related to a large number of measured entities, we would like to suggest a term pseudo- or quasi-observable for hybridization.

1 Introduction

Since van't Hoff and Le Bel's brilliant anticipation of the tetrahedral carbon atom (1874), it has been known that molecules have a three-dimensional structure. Their predecessor L. Pasteur (1848) was the first to realize that asymmetric atomic configurations are the very cause of molecular optical activity. The herculean efforts of L. Pauling in the first half of this century showed conclusively that geometric structure is a central and most fruitful theme in chemistry [1]. Indeed, knowledge of structural parameters enables a deep insight into the electronic structure and properties of molecules. Whereas these great scientists had to use ingenuity and imagination as one of the most important tools in their research, determination of molecular and crystal structures by modern X-ray, electron and neutron diffraction techniques and spectroscopic methods is almost a routine nowadays. Contemporary quantum chemistry substantially contributes to the performance of experimental methods [2, 3]. It is therefore common knowledge that molecules (apart from some exceptional cases) possess relatively rigid geometric structures. Spatial arrangement of chemical bonds gives rise to beautiful and aesthetically pleasing symmetrical structures [4] which provide a key for understanding many essential molecular features [5]. One of the first tasks of the quantum theory of molecules is to interpret molecular geometries in quantitative and qualitative terms. Whilst the former aspect is well covered by modern computational quantum chemistry [2], the latter is a subject matter of numerous discussions. The best model for rationalization of directional properties is the concept of hybridization put forward by Pauling [6] as early as 1928. It is the aim of this chapter to provide arguments for this conjecture.

We would like to give some evidence that hybridization is a genuine physical process being a response of an atom in its chemical environment to perturbations exerted by nearest neighbors. Hybrid atomic orbitals (HAOs) reflect not only the local symmetry defined by directly bonded atoms, but also an increase in overlapping due to their pronounced polarization in the bonding direction. They contribute to the increase of electron density in the region between linked atoms thus diminishing repulsion of the nuclei. At the same time they minimize repulsion between nonbonded atoms. HAOs provide the best basis for perfect-pairing approximation and localized molecular orbitals (LMOs), being an essential ingredient of the valence bond (VB) method(s). Some attention will be paid to the meaning of hybridization within the molecular orbital (MO) theory and various recipes of its extraction will be discussed. It is also argued that HAOs are the most suitable basis functions for semiempirical theories of chemical bonding. Further, it will be shown that HAOs coupled with the maximum overlap criterion offer a transparent and intuitively appealing model of covalent bonding which found wide application in organic chemistry. A vast variety of molecular properties can be correlated with hybridization indices and/or the corresponding overlap integrals thus offering their qualitative and sometimes a semiquantitative interpretation. A number of relations between observables can be rationalized by the hybridization concept. Hybridization can not be directly measured. However, since it explains relationships between observables which are totally unrelated otherwise, playing in this connection the role of a unifying principle, we conclude that hybridization represents a "pseudo-observable" par excellence.

It should be mentioned that the literature on hybridization is very abundant and that no attempt has been made to cover all papers because of space limitations. Hence, this is necessarily an "ice-berg spike" coverage. Fortunately, there are several extensive review articles [5, 7–11] and books available [12]. In particular, we would like to draw the reader's attention to the recent special of the Journal of Molecular Structure (Theochem) dedicated to the "Six Decades of the Hybridization Model" [13].

2 The Hybridization Model and its Early Development

Extensive knowledge of X-ray structural data and magnetic properties of substances as well as a deep understanding of quantum mechanics enabled L. Pauling to design a hybridization model of chemical bonding. In his landmark paper in 1931 [6] he built one of the most important corner stones of the quantum theory of molecules. Generalizing and articulating findings of Heitler, London and others [14–17] ~~he~~ formulated the first three rules of covalent bonding:

(1) The electron-pair bond is formed through the interaction of an unpaired electron on each of two atoms.
(2) The spins of the electrons are opposed when the bond is formed, so that they cannot contribute to the paramagnetic susceptibility of the substance.
(3) Two electrons which form a shared pair cannot take part in forming additional pairs.

Pauling's analysis revealed that the bond is stronger if the concentration of orbitals in the internuclear region is greater and he was the first who discussed the principle of maximum overlapping. However, instead of calculating overlap integrals, he defined the bond strengths of the orbital which is equal to the maximum value of the angular part of the orbital function normalized to 4π over the surface of the sphere. Pauling put forward the following three additional postulates:

(4) The main resonance terms for a single-pair bond are those involving only one eigenfunction from each atom.
(5) Of two eigenfunctions with the same dependence on r, the one with the larger value in the bond direction will give rise to the stronger bond, and for a given eigenfunction the bond will tend to be formed in the direction with the largest value of the eigenfunction.
(6) Of two eigenfunctions with the same dependence on ϑ and φ, the one with smaller mean value of r that is, the one corresponding to the lower energy level for the atom, will give rise to the stronger bond.

Eigenfunction here means atomic or hybrid orbital. Rule (4) states something which is generally accepted now: bonding stabilization is given predominantly by resonance integrals. The last but one statement formulates the maximum overlap criterion, but in a less accurate form (vide infra). What is actually maximized is in fact the amplitude of atomic one-electron wavefunctions. Finally, rule (6) reflects a finding that, e.g., bond energies decrease in a series H_2O, H_2S, H_2Se and H_2Te inspite of the fact that the bond orbitals have similar composition (apart from the principal quantum number n).

Since the l-quantization does not hold anymore in perturbed atoms embedded in the intramolecular field, one can freely mix s, p, d, f :.. AOs. Assuming hybrids' orthonormality and employing maximum bond strength criterion equivalent tetrahedral sp^3 hybrids are easily constructed:

$$\begin{aligned}
\chi_1 &= \frac{1}{2}[(\mathrm{n}s) + (\mathrm{n}p_x) + (\mathrm{n}p_y) + (\mathrm{n}p_z)] \\
\chi_2 &= \frac{1}{2}[(\mathrm{n}s) + (\mathrm{n}p_x) - (\mathrm{n}p_y) - (\mathrm{n}p_z)] \\
\chi_3 &= \frac{1}{2}[(\mathrm{n}s) - (\mathrm{n}p_x) + (\mathrm{n}p_y) - (\mathrm{n}p_z)] \\
\chi_4 &= \frac{1}{2}[(\mathrm{n}s) - (\mathrm{n}p_x) - (\mathrm{n}p_y) + (\mathrm{n}p_z)]
\end{aligned} \tag{1}$$

Pauling observed that a degree of hybridization is strongly dependent of the (ns) — (np) energy gap. Hence, hybridization is most effective for the first row atoms. A tacit assumption that HAOs should be orthogonal is plausible because they are populated with electrons with parallel spins according to Hund's rule. In an asymptotic case where both hybrids are identical ($\langle\chi_1|\chi_2\rangle = 1$) electrons possess antiparallel spins and their valencies are mutually saturated due to the Pauli principle. This is apparently unfavorable. However, some overlapping of the hybrids placed on the same nucleus actually takes place — a point which will be discussed later.

Interestingly, Pauling explained double and triple bonds by two tetrahedra sharing a common edge and side, respectively, giving to an old idea of van't Hoff and Le Bel a quantum mechanical interpretation. Description of multiple bonds in terms of Pauling's bent bonds is highly apprecitated in modern VB theory (vide infra).

Considering atoms possessing *d*-orbitals Pauling constructed hybrid orbitals directed to vertices of a square, octahedron and tetrahedron again.

Trigonal sp^2 hybridization was discusses too, but it was found that its bonding strength is somewhat lower than that of sp^3 hybrids. The corresponding bonding strengths of sp^3, sp^2 and sp^1 hybrids are 2.00, 1.99 and 1.93, respectively. They are higher than strength of pure s (1.00) and p (1.73) orbitals given in the parentheses. However, the relative values for sp^3, sp^2 and sp^1 schemes are not correct since the C—H bond is shorter and stronger in acetylene than in ethylene, ethane being on the other end in this series. Trends in properties of C—H bonds described by sp^3, sp^2 and sp^1 hybrids are nicely reproduced by overlap integrals, however. In spite of this "Schönheitsfehler", Pauling's paper [6] is a milestone in the development of the theory of chemical bond. His hybrid orbitals bridge the gap between the quantum theory and traditional chemical concepts[1].

[1] It should be pointed out that similar ideas were indenpendently developed by Slater (Slater JC (1931) Phys. Rev. 37: 481).

In 1940 Kimball demonstrated how $s^m p^n d^p$ hybrid bond orbitals can be derived for molecules belonging to point group symmetries by utilizing a group-theoretic approach [18]. Racah used hybrids in elucidating the structure of the $Mo(CN)_8^{4-}$ complex [19], whereas Kuhn discussed some quadricovalent complexes of the transition metal ions in terms of HAOs [20]. Duffey and co-workers [21—23] used the criterion of maximum amplitude in studying a number of large molecules. Kimball's group theoretic approach was extended to encompass $s^m p^n d^p f^r$ hybrids by Shirmazan and Dyatkina [24] and by Eisenstein [25]. An important contribution to the field was a paper of McClure [26] who showed that it is possible to construct nine spd hybrid orbitals with favourable bonding characteristics. Additionally, McClure found that a restriction of equivalence $R_{ns}(r) = R_{np}(r) = R_{nd}(r)$ of radial parts of AOs can be removed but little is gained in physical insight by that refinement, whereas mathematical intricacies are increased to a great extent. We shall present more applications of the hybridization model in subsequent sections.

3 The Optimization of Hybridization Parameters

Kimball's [18] group theoretic approach is applicable only in highly symmetrical cases. If a molecule has a small number of symmetry operations (or none), other criteria are needed in order to determine the composition of hybrids. It is worth mentioning that even in symmetric moieties, group theory alone cannot offer a unique hybridization scheme. For example, one easily finds out that in tetrahedral T_d symmetry possible modes of hybridization are: sp^3, sd^3, sf^3, fd^3 f^4 or any combination thereof [5]. The most general hybrid of the tetrahedral symmetry takes the form:

$$\chi = a(s) + b[(p_x) + (p_y) + (p_z)] + c[(d_{xz}) + (d_{yz}) + (d_{xy})] + d[(f_{x3}) + (f_{y3}) + (f_{z3})] \qquad (2)$$

Optimal hybridization depends on a specific system in question and the method applied. Apart from the maximum amplitude criterion [6], there are many other ways of constructing local hybrid AOs. We shall briefly consider the ab initio type of calculation first.

3.1 Ab Initio Optimization of Hybrid Parameters

Canonical Hartree-Fock molecular orbitals (CMOs) have some computational advantages. They belong to particular irreducible representations of the molecular point symmetry groups and have a high interpretational value in molecular spectroscopy (via Koopman's theorem etc.). Gimarc's qualitative MO theory yields an insight into the numerous molecular properties [27]. Woodward-Hoffmann's rules of symmetry conservation in concerted chemical reactions is a great aid in understanding molecular reactivity [28]. CMOs are, however, delocalized over the entire molecule, whilst the monumental edifice of phenomenological chemistry rests on the concept of localized electron pair bonds (apart from some mobile π-electrons). An enormous body of experimental data can be encompassed by bond-additivity rules and sometimes by relatively small deviations which in turn can be traced down to special effects. Indeed one can identify characteristic bond distances, bond angles, bond energies, force constants, stretching frequencies, indirect spin-spin coupling constants etc. The conceptual basis of bond additivities is provided by transferable electron pairs. Extensive discussion of coupled electron pairs as building blocks in molecules was given by Surjan [29, 30]. Relative constancy of the properties of characteristic electron pairs imbedded in a potential region of a specific pair of nuclei gives rise to characteristic atomic groupings or fragments which can be clearly recognized in molecules. Furthermore, formation of localized electron pairs upon chemical bonding does not destroy atoms on the whole, but rather perturbs them which yields a conceptual basis for the model of modified or distorted atoms in a molecular environment [31, 32]. Thus it seems that electrons behave sometimes as if they were localized whereas they exhibit delocalized nature in other occasions, just like the ancient Roman deity Janus who had two different faces. How can one reconcile this apparent contradiction? The matter of fact is that CMOs are determined only up to a unitary transformation of MOs, because the total wave function has a form of a determinant due to the antisymmetry requirement. This degree of freedom can be used to obtain localized molecular orbitals (LMOs) which correspond to "inert" inner-shell electron pairs, unshared lone pairs in the valence shell, and coupled bond electron pairs, as observed by Lennard-Jones and co-workers [33, 34]. To be more specific let's denote the many-electron closed shell wave function by Ψ

$$\Psi = \|\psi_1(1)\,\bar{\psi}_1(2)\,\ldots\,\psi_n(2n-1)\,\bar{\psi}_n(2n)\| \tag{3}$$

where a bar denotes β spin and ψ_i are real molecular orbitals for the sake of simplicity. Any set φ_i obtained by orthogonal transformation

$$\varphi_i = \sum_k O_{ki}\psi_k \tag{4}$$

will do equally well because Ψ is not changed:

$$\begin{aligned} &\|\varphi_1(1)\,\bar{\varphi}_1(2)\,\ldots\,\varphi_n(2n-1)\,\bar{\varphi}_n(2n)\| \\ &= \|\psi_1(1)\,\bar{\psi}_2(2)\,\ldots\,\psi_n(2n-1)\,\bar{\psi}_n(2n)\| \end{aligned}$$

The total Hartree-Fock energy of a molecule is given by

$$E = \langle\Psi|\hat{H}|\Psi\rangle = H + C - X \tag{5}$$

where H is the one-electron contribution:

$$H = 2\sum_k \langle\psi_k|\hat{h}|\psi_k\rangle \tag{6}$$

the term C represents Coulomb repulsions:

$$C = 2\sum_k\sum_m \langle\psi_k(1)\,\psi_m(2)|\,r_{12}^{-1}\,|\psi_k(1)\,\psi_m(2)\rangle \tag{7}$$

and finally the exchange part is denoted by X

$$X = \sum_k \sum_m \langle\psi_k(1)\,\psi_m(2)|\,r_{12}^{-1}\,|\psi_m(1)\,\psi_k(2)\rangle \tag{8}$$

It can be easily shown that each component of the energy H, C and X is separately invariant to orthogonal transformations according to Eq. (4). Consequently, there are many quite different recipes leading to localized MOs. It was pointed out by Lennard-Jones and Pople that LMO probably maximize the sum of the orbital self-repulsions minimizing thereby Coulomb repulsions between different LMOs and the "non-classical" off-diagonal exchange terms [35]. This is indeed the case because the diagonal term

$$\sum_k \langle\psi_k(1)\,\psi_k(2)|\,r_{12}^{-1}\,|\psi_k(1)\,\psi_k(2)\rangle = \max \tag{9}$$

appears both in C and X terms. Equation (9) is exactly the criterion of Edmiston and Ruedenberg (ER) [36] for constructing LMOs. Another widely adopted method is that of Foster and Boys (FB) [37] who advocate minimizing the square of the relative distances in LMOs:

$$\sum_i \langle\varphi_i(1)\,\varphi_i(2)|\,r_{12}^{2}\,|\varphi_i(1)\,\varphi_i(2)\rangle \tag{10}$$

which is equivalent to maximazing the sum of squares of the centroids of charge of localized orbitals

$$\sum_i \langle\varphi_i|\,r\,|\varphi_i\rangle^2 = \max\,. \tag{11}$$

Both criteria although widely different yield similar LMOs. Arbitrariness in the choice of Eq. (4) is a subject matter of criticizm of LMOs in general. This is not justified. Actually, this freedom and flexibility should be employed in obtaining LMOs offering the best description of local bond properties. As far as (ER) and (FB) local orbitals are concerned they give results in accordance with traditional chemical concepts. Their density plots reveal that lone-pairs do have hybrid-like shape and that bond LMOs are superpositions of HAOs. Furthermore, localized MOs can be resolved

Table 1. Localized molecular orbitals in CH_4 and C_2H_6

CH_4	$1s_H$	$1s_{H2}$	$1s_{H3}$	$1s_{H4}$	$1s_C$	$2s_C$	$2p_{xC}$	$2p_{zC}$
iC	−0.02	−0.02	−0.02	−0.02	1.00	−0.04	0	0
bCH	0.52	−0.09	−0.09	−0.09	0.03	0.34	0.48	−0.17

C_2H_6	$1s_{H1}$	$1s_{H2}$	$1s_{H3}$	$1s_{C1}$	$1s_{C2}$	$2s_{C1}$	$2p_{xC1}$	$2p_{zC1}$
iC	−0.02	−0.02	−0.02	1.00	0	−0.04	0	0
bCC′	−0.07	−0.07	−0.07	0.03	0.03	0.34	0	0.47
bCH	0.54	−0.09	−0.09	0.03	−0.01	0.33	0.47	−0.17

	$1s_{H4}$	$1s_{H5}$	$1s_{H6}$	$2s_{C2}$	$2p_{xC2}$	$2p_{zC2}$
iC	0.01	0.01	0.01	−0.01	0	−0.02

in local hybrid AOs and small delocalized "tails". This type of hybrids will be used and compared with hybridization obtained by less rigorous criteria in Sect. 3.2. As an illustration we give here localized orbitals in CH_4 and C_2H_6 as calculated by the ER method [38] (Table 1). Inner- core and bond orbitals are denoted by i and b, respectively. One observes that coefficients corresponding to AOs involved in direct bonding are larger by at least an order of magnitude than those representing "tails". Furthermore, CH_4 and C_2H_6 serve as illuminating examples of one of the most important notions in chemistry. This is transferability of properties of localized fragments which states that similar moieties have similar features. In this case CH bond orbital has practically the same composition in both molecules. Similarity of fragment or bond orbitals yields a basis for understanding the fragment or bond additivity schemes of various molecular properties. As far as hybridization is concerned, it can be defined by the s-character:

$$(s\,\%) = \frac{c^2(2s)}{c^2(2s) + c^2(2p_x) + c^2(2p_y) + c^2(2p_z)} \tag{12}$$

which in turn is given by the squares of coefficients of one-center AOs taking part in a particular LMO, see Eq. (12). S-characters (in %) of CH bonds in CH_4 and C_2H_6 assume values 30.9 and 30.1, respectively, being very close indeed. On the other hand, s(CC) in C_2H_6 is 34.1%. One notices several characteristic features: s-characters differ from canonical hybridization values (25%, 33.3%, 50%) and a sum of single-center s-contents may exceed 100%. The latter is a consequence of the fact that the *s*-orbital has lower energy and higher electron population than 2p AO. It should be pointed out that the results presented in Table 1 refer to an ab initio minimal STO basis set. Generally speaking, different methods give different hybridizations. Hence, hybridization has not an absolute meaning, but if an appropriate definition is adopted and consistently used within the same theoretical framework, then a fairly consistent picture emerges offering a transparent model of covalent bonding.

It is noteworthy that the ER method yields rabbit ears for lone pairs in H_2O and banana bonds in the multiple bonding situation. Since the ER recipe minimizes Coulomb and exchange energies between different bonds the corresponding LMOs closely approach the classical description being at the same time within the quantum framework.

The ER localization procedure was an inspiration for some related but distinctly different schemes. One of them is the localized atomic orbital method (LAO) designed by Aufderheide and Chung-Phillips [39–42]. They maximize function L_A for each atom A:

$$L_A = \sum_{\mu}^{A} \sum_{\lambda}^{occ} \sum_{\sigma}^{occ} \frac{1}{4} P_{\mu\sigma} P_{\lambda\mu} \langle \chi_\mu(1)\, \chi_\lambda(2) |\, r_{12}^{-1}\, | \chi_\mu(1)\, \chi_\sigma(2) \rangle \tag{13}$$

where $P_{\mu\sigma}$ denotes the conventional bond order matrix. The basis set orbitals χ which yield maximal value of L_A are local hybrid orbitals. They posses desirable directional properties and can be combined to yield localized MOs suitable for simple interpretation of intramolecular bonded and nonbonded interactions thus providing chemically and conceptually useful information. Since orthogonal transformations are employed in producing hybrid AOs the latter are orthogonal too. Let us consider the H_2O molecule in some more detail [42]. Apart from the oxygen 1*s* inner-core lone pair, there are two OH bond hybrids and two equivalent rabbit ears lone pairs HAO obtained by the localization procedure (13). STO-3G ab initio calculation gives:

$$\chi(b)_{\pm} = 0.396(2s) + 0.586(2p_x) \pm 0.707(2p_y) \quad (14a)$$
$$\chi(l)_{\pm} = 0.586(2s) - 0.396(2p_x) \pm 0.707(2p_z) \quad (14b)$$

Lone pair hybrids $\chi(l)_{\pm}$ have a higher s-content ($sp^{1.92}$) than bond hybrids $\chi(b)_{\pm}$ ($sp^{5.39}$) compatible with the lower 2*s*-level. The angle between $\chi(l)_+$ and $\chi(l)_-$ lone pairs is correspondingly 121.5° whereas bond hybrids close up an angle of 100.7°.

LAOs obtained by the maximization of L_A in Eq. (13) may be subjected to some preconceived requirements. For example, one can constrain localization to yield $\chi(l)_\sigma + \Pi$ lone pairs meaning that they are nonequivalent. Then the hybrid AOs take the form:

$$\chi(b)_{\pm} = 0.516(2s) + 0.483(2p_x) \pm 0.707(2p_y) \quad (15a)$$
$$\chi(l)_{\sigma} = 0.683(2s) - 0.730(2p_x) \quad (15b)$$

A very high s-content of the $\chi(l)_\sigma$ lone pair gives rise to $sp^{1.14}$ hybridization. A considerable increase in the s-character is found in OH bond hybrids, too ($sp^{2.75}$). Concomitantly, the angle between $\chi(b)_{\pm}$ hybrids is larger assuming a value of 111.3°. The question arises which picture is more reliable: rabbit ears or $\sigma + \pi$ lone pairs? Both correspond to the same total wave function and the same total energy, but we are inclined to believe that rabbit ears provide a better description of lone pairs in H_2O for a simple reason: the interhybrid $\chi(b)_{\pm}$ angle of 100.7° is in good accordance with the more sophisticated bond path SCF calculations of Wiberg (101°) [43]. This example illustrates rather nicely that the non-unique nature of the localization procedure should be better scrutinized and employed for extracting conceptually the best physical pictures of covalent bonding.

One can start ab initio calculations with HAOs at the very outset. The first use of the hybrid orbitals in nonempirical calculations was made in the early thirties. Dickinson [44a] and Rosen [44b] employed polarized AOs of the type ($1s_H + 2p_H$) for the calculation of the ground states of H_2^+ and H_2, respectively. The SCF MO based on HAOs were performed on hydrocarbons by Hoyland [45] and on small heteroatomic molecules by Petke and Whitten [46]. The simulated ab initio MO scheme (SAMO) heavily relies on the transferability of the Hartree Fock (HF) matrix elements from smaller fragment molecules to larger molecular and polymer systems [47]. A use of hybrid orbitals is not a necessary prerequisite for the SAMO formalism, but they are beneficial for several reasons. One does not have to worry about the orientation of coordinate systems in molecules studied by a technique of transferable fragments. In the HAO basis one can clearly distinguish relevant matrix elements which are magnified and the rest which can be neglected in an approximate treatment. Lest but not least, hybrids can assume different screening constants in different spatial directions thus being more flexible in describing anisotropy of the atomic environment than a pure AO basis set (vide infra).

The self-consistent group function method employs the antisymmetrized product of pair functions describing the inner-core, bond and lone pairs of electrons. This approach can transcend the limitations imposed by the HF method because correlation can be admitted within each electronic group and can be further refined by allowing the admixture of local excited functions [48]. Applications gave hybridization parameters in several diatomics and small hydrocarbons [49]. A slight shortcomming of the performed calculations is the assumed orthogonality of the hybrid AOs.

Hybridization is the best way of preserving the idea of coupled electron pairs. Hence, HAOs offer the most natural basis set for VB method which experiences a renaissance nowadays. The classical VB method has a serious drawback that ionic structures are very important even in highly nonpolar hydrocarbons. They apparently compensate for inadequate covalent perfect-pairing scheme [50] (vide infra). The modern spin coupled VB theory (SCVB) is still based on the simple orbital picture with one orbital per each electron but includes all possible modes of spin coupling [51–53]. The orthogonality requirements of the single-center orbitals are omitted enabling a full variational freedom. The SCVB method is much more flexible than the original VB and yields energies which are considerably lower than those obtained by the HF calculations in the same basis set. This is a consequence of the fact that the SCVB function is highly correlated. It is important to strongly point out that hybrid orbitals are not a result of some preconceptions but an outcome of the calculations. Hence, the SCVB function is not invariant to arbitrary linear transformations of the orbitals and the form of the total wave function is unique. It was shown that a relaxation of the orthogonality constraint is more important than including all the modes of spin coupling. The perfect-pairing spin function dominates with a weight of more than 90%. It is interesting to mention that for large nuclear separations the spin coupled orbitals take the form of localized pure AOs. At the equilibrium geometries it is usually energetically more favorable for the atomic functions to form hybrids which resemble conventional HAOs. Additionally, hybrids tend to distort as the atoms approach one another, the distortion being greatest in the regions where formation of new bonds takes place. For example. SCVB calculations on CH_4 [53] gave eight spin-coupled orbitals for valence electrons which fall into two groups. Four of the orbitals are localized on hydrogens being composed essentially of $1s_H$ AOs slightly distorted toward the C atom. The other group of four degenerate orbitals is localized on the carbon. They consist of approximately sp^3 slightly distorted hybrids pointing toward H atoms. Hybrids have appreciable mutual overlapping of 0.48. Eeach of them overlaps strongly (0.73) with only one local hydrogen orbital. Needless to say, the orbitals conform to the T_d symmetry. Energetic properties are worth mentioning. The perfect-pairing spin-coupled wave function yields energy of (in hartrees) —40.26259 is superior to Hartree-Fock SCF energy of —40.20086. Full-valence VB gives energy of —40.27772. As the most important corrollary follows that hybridization is a consequence of the full variational calculation and not of an a priori postulate.

Generalized Valence Bond (GVB) method enjoys a significant popularity. Essentially, the GVB wavefunction can be viewed as a closed shell HF function:

$$\hat{A}(\psi_1\alpha\psi_1\beta\psi_2\alpha\psi_2\beta\ \ldots) \tag{16}$$

where $\hat{A}$ represents a customary antisymmetrizer, in which some doubly occupied singlet pairs are replaced by GVB pairs

$$\psi_i\alpha\psi_i\beta \rightarrow (\psi_{ia}\psi_{ib} + \psi_{ib}\psi_{ia})\,\alpha\beta \tag{17}$$

where the first position in products implies electron 1 and the second electron 2. The paired GVB orbitals ψ_{ia} and ψ_{ib} strongly overlap as a rule. This is a technical difficulty but conceptual strength and quantitative merit of the method. A large number of molecules were treated by the GVB procedure and the corresponding hybrid AOs are available. The results are covered by review articles and consequently need not be repeated here [54, 55]. We shall, however, discuss briefly the results of Messmer et al. [56, 57] because they bear some importance in representation of double and triple bonds. By using GVB perfect-pairing functions they have shown that a bent bond picture describes multiple bonds in CO_2, C_2H_4 and C_2H_2 better than the more common σ and π-orbitals, in line with Pauling's early conjecture. Inclusion of correlation between pairs of coupled electrons makes the bent bond picture even more accurate [57]. In difluoroacethylene, GVB with correlation yields a triple bent bond between carbons and modified tetrahedral hybrids for fluorine lone pairs. Contour plots of GVB orbitals clearly reveal left-right correlation in bond hybrids and in-out correlation in lone pairs [57].

An important conceptual and computational step forward in the VB theory was discussed by McVeeny recently [58]. He considered introduction of Coulson-Fischer [59] orbitals into the VB procedure in studying the H_2O molecule:

$$\bar{\chi}_1 = \chi_1 + \lambda 1s_{H1}, \qquad \overline{1s}_{H1} = 1s_{H1} + \mu\chi_1 \tag{18}$$

where χ_1 and $1s_{H1}$ are original oxygen bond hybrid and hydrogen AO, respectively. It appears that $\bar{\chi}_1$ and $\overline{1s}_{H1}$ orbitals exhibit enhanced overlapping. McWeeny has shown that full-CI limit may be quite closely approached by using a small number of structures among which the perfect-pairing term is predominant of course. The heavily non-orthogonal orbitals of Eq. (18) designated "semi-localized" by Eyring [60] are obviously advantageous. They increase not only mutual overlap but take also into account the ionic character to some extent.

We stress once again that a choice of basis functions is extremely important in quantum theory of molecules.

Although molecular spectroscopy was a traditional playground of molecular orbital theories and, in particular, photoelectron spectroscopy (PES) served as the experimental confirmation of the adequacy of the MO viewpoint of chemical bonding, it was recently shown that the VB method can do equally well [61].

Finally, we note in passing that VB-type wave functions are a better starting point in perturbation treatments than HF functions, because they usually involve a good portion of correlation already [62].

To summarize, the VB method possesses not only interpretive and conceptual advantages, but also yields superior numerical results which is by no means accidental. It is a happy marriage of accuracy and visuality. The localized orbital picture retains its validity even in archetypical delocalized molecules like benzene [63], pyridine and the related six membered rings [64]. It appears that aromatic stability arises more from the mode of spin coupling than from orbital delocalization.

The hybrid orbitals discussed so far were used as basis functions in various ab initio treatments (SCF, VB) or *ex post facto* by constructing localized orbitals (MOs, AOs) employing some localization recipes. Another way of extracting hybridization parameters from molecular wave functions computed in basis of pure (free atom) AOs is by making use of the first order density matrix.

Let us first consider the early prescription of McWeeny and Del Re [65] in constructing optimum hybrid orbitals. They assume that AOs are symmetrically orthogonalized in the least-squares sense [66]. The localized bond formed by coupled hybrid orbitals χ_r and χ_s takes a form $\cos\theta_{rs}\chi_r + \sin\theta_{rs}\chi_s$, where θ_{rs} is the ionic-character parameter. The corresponding charge and bond order matrix elements are: $P_{rr} = 1 + \cos 2\theta_{rs}$, $P_{ss} = 1 - \cos 2\theta_{rs}$ and $P_{rs} = \sin 2\theta_{rs}$. The matrix $\tilde{P}$ will schematically read:

$$\tilde{P} = \begin{array}{c} \\ r \\ s \\ t \end{array}\begin{pmatrix} r & s & t \\ P_{rr} & P_{rs} & \\ P_{sr} & P_{ss} & \\ & & P_{tt} \end{pmatrix} \tag{19}$$

where $P_{tt} = 2$ describes the lone pair corresponding to two coinciding hybrids. If $\theta_{rs} = 45°$ one has a purely covalent bond where $P_{rr} = P_{ss} = P_{rs} = P_{sr} = 1$. McWeeny-Del Re method implies a unitary transformation of initial pure AOs and subsequent SCF calculations. In general, the hybrids are neither completely localized nor are the bonds 100% covalent. However, HAOs can be constrained to satisfy one or both of these conditions leading to an increase in total energy of a molecule. In this way effects of ionicity and delocalization may be estimated. Our main concern is here the form of the matrix $\tilde{P}$ (19) which inspired Foster and Weinhold to design the so called natural hybrid orbitals [67]. A partitioned form of $\tilde{P}$ where blocks $\tilde{P}_{AA}$ and $\tilde{P}_{AB}$ are associated with the atomic center A or a pair of centers A and B, respectively, reads:

$$\tilde{P} = \begin{pmatrix} \tilde{P}_{AA} & \tilde{P}_{AB} & \cdots & \tilde{P}_{AD} & \cdots \\ \vdots & \vdots & & \vdots & \\ \tilde{P}_{BA} & \tilde{P}_{BB} & \cdots & \tilde{P}_{BD} & \cdots \\ \vdots & \vdots & & \vdots & \\ \tilde{P}_{DA} & \tilde{P}_{DB} & \cdots & \tilde{P}_{DD} & \cdots \\ \vdots & \vdots & & \vdots & \end{pmatrix} \tag{20}$$

Assumming that AOs do mutually overlap and that the actual SCF (or some other) calculation is executed and the matrix $\tilde{P}$ (20) is known, Foster and Weinhold diagonalize atomic blocks

$$\tilde{P}_{AA}\tilde{h}_i^{(A)} = n_i^{(A)}\tilde{S}_{AA}\tilde{h}_i^{(A)} \tag{21}$$

where $h_i^{(A)}$ is the eigen-hybrid given as a column vector. The eigenvalues $n^{(A)}$ are expected to have values close to 1 or 2 corresponding to bond or lone pair hybrids, respectively. This is in full analogy with the values of diagonal elements in the matrix (19). Similarly, if one diagonalizes extended submatrix associated with bonded atoms A and B

$$\tilde{P}^{(AB)} = \begin{pmatrix} \tilde{P}_{AA} & \tilde{P}_{AB} \\ \tilde{P}_{BA} & \tilde{P}_{BB} \end{pmatrix} \tag{22}$$

one obtains

$$\tilde{P}^{(AB)}\tilde{h}_i^{(AB)} = n_i^{(AB)}\tilde{S}^{(AB)}\tilde{h}_i^{(AB)} \tag{23}$$

$\tilde{h}_i^{(AB)}$ describes the A—B bond orbital and $n_i^{(AB)}$ is expected to have a value close to 2. In order to perform the calculation a minimal threshold occupancy of an electron "pair" has to be adopted. It is taken to be 1.90, but usual bond occupancies $n_i^{(AB)}$ assume values between 1.98 and 2.00. Finally, the hybrids are symmetrically orthogonalized.

Instead of diagonalizing the density matrix $\tilde{P}$, a straightforward calculation of hybrids' s- and p-characters is possible [68, 69]. The idea was based on an early observation of Wiberg [70] that the square of the bond order in CH bonds is related to their s-characters. The bond charge can be expressed with a full appreciation of overlapping as

$$W^{AB} = 2\sum_{\mu<\nu}^{A}\sum^{B} P_{\mu\nu}S_{\mu\nu} \tag{24}$$

A part of the $2s_A$ orbital involved in the A—B covalent bond is given by

$$W_{s(AB)} = \sum_{\nu}^{B} P_{2s_A\nu}S_{2s_A\nu} \tag{25}$$

The bonding charge of all four $2s$, $2p_\nu$ (ν = x, y, z) AOs of the atom A taking part in the A—B bond is $(1/2)W^{AB}$. Then the s-character of the χ_{AB} hybrid is the properly normalized $W_{A(AB)}$ quantity:

$$W_S^{AB} = W_{S(AB)}/(1/2)W^{AB} \tag{26}$$

An analogous formula holds for approximate methods employing neglect of the diatomic overlap simplifications [68].

Equations (21), (23), and (26) wll be used in comparing hybridization obtained by more accurate (ab initio) method with those calculated by the maximum overlap criterion.

3.2 Semiempirical Estimates of Hybridization Indices

Simple methods for the description of the σ-framework of molecules based on hybrid AOs were deviced by Sandorfy [71] and Fukui et al. [72]. This approach was reexamined and refined by Pople and Santry [73]. The matrix elements of the Hartree-Fock hamiltonian between overlapping hybrids have been considered as disposable parameters. Approximate ab initio schemes employing HAOs lend themselves to the semiempirical development. It was shown by McWeeny, Cook, and co-workers [74–77] that various NDDO integral approximations can be at least partially justified if proper basis sets are employed. This implies hybrid orbitals which exhibit a desired localization. In addition, they are orthogonalized usually in a symmetric fashion[2]. It appears that integrals abandoned by NDDO approximation are small indeed if this type of HAOs is used. It is rather unfortunate that subsequent development of semiempirical methodology neglected hybrids relying on a very large number of adjustable parameters instead on the physical picture. Unitary (i.e. hybridization) invariance of the final wave function is the central dogma in the CNDO and related schemes, which should be considered as a very unfortunate circumstance. It has been pointed out several times already that the basic problem in any approximate method is the choice of the most appropriate set which fits the adopted theoretical framework the best (and vice versa). Hence, freedom in the choice of unitary transformations of the basis functions should be utilized so that the integral approximation scheme introduces the least possible error. This problem was discussed in depth by Cook [77].

The semiempirical PCILO method is one of the rare procedures which is built upon the concept of localized MOs and hybrid orbitals [78–83]. This is not the only difference with standard semiempirical methods. Most of them are variational, whereas PCILO is a perturbation treatment. Current semiempirical schemes are based on the independent particle approximation whilst PCILO works in a basis of n-electron wave function at the configurational level. Concomitant with basic assumptions, PCILO is effective in treating well localized systems. Its performance is however negatively affected by the adopted CNDO integral approximation.

The hybrid orbitals proved useful in approximate treatment of clusters and surfaces [84–86]. They are an invaluable tool for the simple description of any system involving a large number of subunits, e.g. myriads of atoms. Hence, HAOs are pivotal in discussing both qualitative and computational aspects of the theory of covalent solids [87–89].

3.2.1 Maximum Overlap Method(s)

The first general method for constructing hybrid orbitals was designed by Murrell [90] who considered molecular systems of the type $M(X_1 \ldots X_k)$, where M represents the central atom and X_i are ligands. Let us denote desired hybrid orbitals by $\chi_1 \ldots \chi_n$ $(n \geqq k)$ whereas each ligand has one orbital θ_i $(i = 1 \ldots k)$ at disposal for bonding with the central atom. A set of pure AOs of the latter is given by φ_i $(i = 1 \ldots n)$. The optimal HAOs are determined by a matrix $\tilde{A}$:

$$\chi_i = \sum_{j=1}^{n} A_{ij}\varphi_j \tag{27}$$

If the matrices $\tilde{S}$ and $\tilde{R}$ are defined by overlaps $S_{ij} = \langle\theta_i|\chi_j\rangle$ and $R_{ij} = \langle\theta_i|\varphi_j\rangle$, respectively, then

$$\tilde{S} = \tilde{A}\tilde{R} \tag{28}$$

---

[2] Schmidt orthogonalization procedure is required in delineating inner core and valence shell orbitals.

The unknown matrix $\tilde{A}$ is obtained by the maximum overlap condition

$$\sum_{i=1}^{k} \langle \theta_i | \chi_i \rangle = \max . \tag{29}$$

The first step of the procedure is construction of the intermediate orbitals of the central atom $\eta_i(i = 1 \dots n)$ $\eta_i = \sum_{l=1}^{n} B_{ij}\varphi_j$, which involve $n - k$ members satisfying orthogonality conditions with ligand AOs:

$$\langle \eta_i | \theta_j \rangle = 0 \qquad (i = n - k + 1 \dots n; \quad j = 1 \dots k) \tag{30}$$

These $n - k$ orbitals do not contribute to the overall bonding and may be used for accommodating lone pairs for instance. If the overlap matrix between η_i and θ_j is denoted by $\tilde{P}$ (i.e. $P_{ij} = \langle \eta_i | \theta_j \rangle$) then the hybridization matrix $\tilde{A}$ is given by:

$$\tilde{A} = \tilde{P}^{-1}(\tilde{P}\tilde{P}^t)^{1/2}\,\tilde{B} \tag{31}$$

where a superscript t denotes a transposed matrix. It should be pointed out that the last $n - k$ rows of the $\tilde{P}$ matrix filled out by zeros are deleted and consequently $\tilde{P}$ is a square $k \times k$ matrix. The square root of the matrix $\tilde{P}\tilde{P}^t$ is calculated by the diagonalization technique

$$\tilde{U}^t(\tilde{P}\tilde{P}^t)\,\tilde{U} = \tilde{d}(p_1 \dots p_k) \tag{32}$$

where $\tilde{U}$ is an orthogonal matrix, $\tilde{d}$ is a diagonal matrix and p_i $(i = 1 \dots k)$ are the eigenvalues. Then $(\tilde{P}\tilde{P}^t)^{1/2}$ is given by

$$(\tilde{P}\tilde{P}^t)^{1/2} = \tilde{U}\tilde{d}(p_1^{1/2} \dots p_k^{1/2})\,\tilde{U}^t \tag{33}$$

Signs of the square roots $p_i^{1/2}$ should be chosen so that the extremum in Eq. (29) is satisfied.

A more elegant matrix form was found by Golebiewski [91]. He showed that the matrix $\tilde{A}$ may be calculated by a matrix equation:

$$\tilde{A} = \tilde{U}^t d(s_1^{-1/2} \dots s_k^{-1/2})\,\tilde{U} \cdot \tilde{S} \tag{34}$$

where the overlap matrix is here given in the basis of pure AOs: $S_{ij} = \langle \theta_i | \varphi_i \rangle$. The eigenvalues of $\tilde{S}\tilde{S}^t$ are denoted by s_i $(i = 1 \dots k)$. The maximum overlapping is expressed in terms of $s_i^{1/2}$: $\sum_{i=1}^{k} \langle \theta_i | \chi_i \rangle = \sum_{i=1}^{k} s_i^{1/2}$.

Various generalization of the Murrell-Golebiewski method were subsequently made. Volkov and Dyatkina [93] considered σ- and π-bonding between a central atom and ligands whereas simultaneous optimization of ligand HAOs was discussed by several researchers [93, 94].

Maximum overlap molecular orbitals (MOMO) received a detailed description [95]. They are eigenvectors of the metric matrix formed by overlap integrals of the nonorthogonal basis set. Let us denote a basis set of real AOs φ_i (i = 1 ... n) which are normalized and linearly dependent. The desired MOMOs are of the form

$$\psi = \sum_{j=1}^{n} c_j \varphi_j \tag{35}$$

They can be determined by the requirement that a sum S_m of the squares of the projections of ψ (35) on the basis set functions is at maximum

$$S_m = \sum_{i}^{n} \langle \varphi_i | \psi \rangle^2 = \max \tag{36}$$

By substituting Eq. (35) into Eq. (36) and introducting the normalization condition of ψ into the extremum condition by the Lagrange multiplier method (λ), the problem is reduced to finding the absolute extremum of the expression S'_m:

$$S'_m = \sum_{j,k}^{n} c_j c_k (\tilde{S}^2)_{jk} - \lambda \sum_{j,k}^{n} c_j c_k S_{jk} \tag{37}$$

Derivatives over coefficients ($\partial S'_m/\partial c_i$) must vanish thus yielding a set of linear homogeneous equations:

$$\sum_{k}^{n} c_k [(\tilde{S}^2)_{jk} - \lambda S_{jk}] = 0 \tag{38}$$

which has a nontrivial solution if

$$|\tilde{S} - \lambda \tilde{I}| = 0 \tag{39}$$

following from the nonsingularity of the $\tilde{S}$ matrix. Hence, one has to find the eigenvalues of the overlap matrix.

Since the overlap matrix reflects the symmetry of a molecule, the MOMOs contain a certain degree of similarity with true wave functions. Lykos and Schmeising have shown that they are equivalent with Hückel MOs in π-systems. The eigenvalues of the overlap matrix are identical with Hückel MO energies if a proper scale and choice of the origin is made.

Lykos and Schmeising's method was thoroughly examined and discussed by Bartlett and Öhrn [96]. Their analysis indicates that MOMO charge densities and dipole moments are reliable for predominantly covalent molecules where the intramolecular charge transfer is not highly pronounced. Then the overlap and Hartree-Fock matrices $\tilde{S}$ and $\tilde{F}$ nearly commute and consequently their eigenvectors are similar. Interestingly, constrained MOMOs are significantly better than other constrained nonempirical and semiempirical SCF wave functions [97]. The Lykos-Schmeising method is easily implemented even in the largest systems unlike the scheme of Murrell-Golebiewski. It should be pointed ouf that the Lykos-Schmeising method does not employ hybrid orbitals but it could be and should be generalized, in this sense.

Del Re [98–100] discussed a way of obtaining hybrid AOs within the framework of the effective one-electron hamiltonian where the wavefunctions are expressed in a form of linear combinations of atomic orbitals (LCAO). This approach is sometimes called the tight binding approximation and the two center matrix elements are usually proportional to the overlap integrals. The tight binding matrix equation

$$\tilde{H}\tilde{C} = \tilde{S}\tilde{C}\tilde{E} \tag{40}$$

is solved by a unitary transformation of the pure AOs so that the overlap matrix $\tilde{S}$ has the pseudo-diagonal 2×2 form as close as possible. The transformed basis set orbitals are hybrids.

The Del Re method assumes that maximum localization is possible and consequently it works well in systems which satisfy this condition. In this connection it is interesting to mention that there is a relation between Löwdin's canonical orthogonalization [101] and the maximum overlap.

The energy weighted maximum overlap (EWMO) method [103, 104] is worthy of mention. It is based on the idea that a molecule is a collection of essentially undisturbed atoms. The bonding interactions arise from overlapping of atomic electron distributions. The one-electron energies of the latter are explicitly taken into account. The EWMO method was successfully applied to the calculation of a number of molecular properties particularly of radical species [104]. We mention in passing that the EWMO scheme is easily extended and adapted to HAOs basis sets.

Finally, we shall briefly discuss Pauling's latest results obtained by using the bond strength criterion [105, 106]. Strictly speaking, this approach is not semiempirical because it does not employ any adjustable parameter. Nevertheless, it is considered in this section in view of its highly approximate and qualitative nature. The maximum strength M_s of two equivalent and orthonormal hybrids is determined by their mutual angle ϑ:

$$M_s(\vartheta)_{sp^3} = (0.5 + 1.5w)^{1/2} + (1.5 - 1.5w)^{1/2} \tag{41a}$$

$$M_s(\vartheta)_{sp^3d^5} = (3 - 6w + 7.5w^2)^{1/2} + (1.5 + 6w - 7.5w^2)^{1/2} \tag{41b}$$

$$M_s(\vartheta)_{sp^3d^5f^7} = (3 + 15w - 45w^2 + 35w^3)^{1/2} + \\ + (5 - 15w + 45w^2 - 35w^3)^{1/2} \tag{41c}$$

where $w = \cos^2(\vartheta/2)$.

Equations (41a–c) give optimal angles for sp^3, sp^3d^5 and $sp^3d^5f^7$ hybridization schemes, respectively. Maximum strength for sp^3 hybrids is achieved at $\vartheta = 109.47°$ as expected. Enneacovalence sp^3d^5 would have optimal angles at $\vartheta = 73.2°$ and 134.6°, whereas strong $sp^3d^5f^7$ hybrids would be pointed at angles $\vartheta = 54.9°$, 100.4° and 145.4° relative to the first (reference) HAO. It is interesting to mention that optimal angles coincide with nodal cones of the reference hybrid. In other words, the symmetry axis of the second hybrid possessing maximum bonding power lies on the nodal cone of the reference hybrid. The strength of Pauling's model is its perfect generality and a lack of emprirical parameters. It predicts structural properties which depend solely on the characteristics of the central atom and the results are independent of some specific properties of ligands. Thus the optimal angles cited above should hold on average in a large number of various molecules. The weakness of the model lies in the fact that the influence of the ligands is neglected which leads to a loss of finer details of molecular structure. Hence, a price is paid for the simplicity of the model which is reflected in its crudeness. Pauling applied this model for prediction and interpretation of structural properties of inorganic compounds, in particular transition metal complexes [107–110] employing the *spd* and *spdf* basis set. The fundamental role of enneacovalence in determining structures of transition metal compounds was found. Interestingly, two optimal polyhedra for enneacovalent transition metal atoms were detected. They have a form of a trigonal prism with equatorial caps on the three rectangular faces and a tetragonal antiprism with a polar cap on one of the two bases. These polyhedra seem to play the same crucial role in transition metal chemistry as the tetrahedron does in organic chemistry. A large number of transition metal structures was analyzed along these lines and reasonable agreement with available experimental data was obtained. More experimental work is required to assess these remarkable predictions.

3.2.2 The Iterative Maximum Overlap (IMO) Method

The IMO method [111–113] was designed for the calculation of bond lengths and angles in molecules characterized by localized bonds, i.e. systems which are well described by a predominating single VB structure. Since the details of the method are available elsewhere we provide here its compact sketch as applied to hydrocarbons. A basis set of local hybrid orbitals

$$\chi_i = a_i(2s) + (1 - a_i)^{1/2}(2p)_i \tag{42}$$

is varied until the sum of bond energies

$$E_b = \sum_{CC} E_b(CC) + \sum_{CH} E_b(CH) \tag{43}$$

is maximized. A plausible assumption that the bond energies are linear functions of the corresponding overlap integrals $E_{AB} = k_{AB}S_{AB} + l_{AB}$ is made, where k_{AB} and l_{AB} are empirically adjusted parameters. The hybridization indices a_i are subject to orthonormality conditions

$$a_i a_j + (1 - a_i^2)^{1/2}(1 - a_j^2)^{1/2} \cos \vartheta_{ij} = \delta_{ij} \qquad (i, j = 1 \ldots 4) \tag{44}$$

where ϑ_{ij} is the angle between the axially symmetric hybrids ψ_i and ψ_j sharing the same carbon atom. During the calculations, the HAOs are allowed to follow the directions of the straight lines passing through the directly bonded nuclei except in small rings where bent bonds necessarily appear. In other words, it is assumed that perfect orbital following holds for CH and acyclic CC bonds. To be quite precise, bond angles are determined by the interhybrid angles ϑ_{ij} of the optimal hybrids with the notable exception of small cyclic molecules. Orthogonality and perfect orbital following of hybrids are not completely justified suppositions, but they work very well in most cases. They considerably add to the simplicity of the method.

Finally, interatomic overlap integrals are forced to follow the predetermined linear bond overlap — bond distance correlations

$$d(C{-}C) = -1.166\, S(C{-}C) + 2.298\,\text{Å} \quad (45a)$$
$$d(C{=}C) = -0.677\, S(C{=}C) + 2.087\,\text{Å} \quad (45b)$$
$$d(C{-}H) = -0.869\, S(C{-}H) + 1.726\,\text{Å} \quad (45c)$$

This is the salient feature of the IMO procedure. In the first iterative step the initial bond distances are selected by an educated guess and the independent hybridization parameters a_i are varied until the maximum of E_b according to Eq. (43) is achieved. Then the next set of bond distances is deduced via Eqs. (45a–c) and the hybridization indices are reoptimized. The whole iterative procedure is repeated until a consistency between the input and output bond distances is established. Hence, the IMO scheme is a constrained weighted maximum overlap method. Finally, the effect of mobile π-electrons in weakly delocalized systems is considered as a correction imposed on the σ-skeleton. It is estimated by making use of MOMOs of Lykos-Schmeising — Eq. (39) as described elsewhere [113]. The IMO method has a remarkable predictive power providing useful information on the size and shape of hydrocarbons. It is capable of rationalizing a number of other properties discussed in the next section.

4 Description of Molecular Properties by the Hybridization Model

Hybrid orbitals are natural constituents of local bond orbitals. Consequently, they offer a simple interpretation of essentially local properties which characterize pairs of bonded atoms and global properties (vide infra) which follow bond additive schemes. The main body of this section will present IMO results for the simple reason that they are most abundant. We shall discuss local molecular properties first.

4.1 Local Molecular Properties

4.1.1 Molecular Shape and Size

A deep connection between geometry and matter was expressed in the sentence "Ubi materia, ibi geometria" by the famous astronomer Johannes Keppler. It means that wherever there is matter, there is geometry too thus reflecting the fact that geometry is something fundamental and universal. Quantum theory has certain difficulties in the rigorous definition of molecular geometries. We shall adopt here the concept of molecular structure which follows from the Born-Oppenheimer or clamped nuclei approximation and the concomitant idea of a potential energy surface as a determinant of the nuclear motion. This is a pragmatic point of view, but it enables an understanding of the behaviour of molecules and provides a substantial help in rationalizing and classifying a plethora of experimental data.

Strictly speaking, molecular shape and size are global properties. However, in what follows, we shall mostly consider, bond distances and angles which imply geometric relations between two or three nearest neighbor atoms. Hence, they belong to a class of properties localized in nature thus being closely related to hybridization. Concomitant with previous discussions this is indeed the case. Namely, in spite of the fact that hybrids to not always coincide with straight lines passing through the neighboring nuclei, due to steric effects (vide infra), they give some useful information about bond angles and regularities in their changes. The key relation is given by the orthogonality requirement in Eq. (44) $\cos \vartheta_{ij} = -1/(n_i n_j)^{1/2}$, where $n_i = (1 - a_i^2)/a_i^2$. Furthermore, it follows that for the C_{2v} local symmetry case there is a relation:

$$\cos \vartheta_{34} = (\cos \vartheta_{12} + 1)/(3 \cos \vartheta_{12} - 1) \tag{46}$$

ϑ_{34} denoting the angle between the second pair of the 2 × 2 equivalent hybrids placed at the same nucleus. It appears that an increase in one angle is accompanied by a decrease in the other, which is empirically established fact known as the Thorpe-Ingold effect [114]. Therefore, a functional relationship between the structurally independent parameters of Eq. (46) is explained through the hybridization concept. It explains nicely the opening up of the bond angle in CH_2 groups attached to small carbocycles

Table 2. Comparison of the angular structural parameters as obtained by the maximum overlap criterion with the experimental data for some bicyclic systems (in degrees)

Molecule	Angle	Calcd.	Exp.
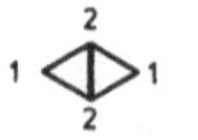	$C_1C_2C_2C_1$	131	126
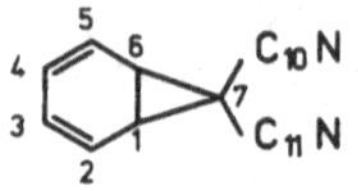	$C_1C_2C_3$	118.6	121.8
	$C_1C_2C_8$	117.2	117.5
	$C_2C_3C_4$	122.3	119.2
	C_7C_1H	111.0	110.2
	$C_{10}C_7C_{11}$	115.2	114.9
	$C_2C_1C_6C_7$	107.1	108.1
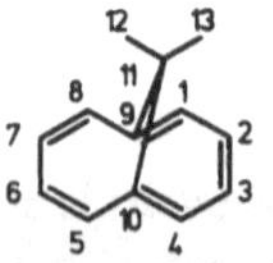	$C_9C_1C_2$	121	124
	$C_1C_2C_3$	125.2	123
	$C_{12}C_{11}C_{13}$	111.6	109
	$C_1C_9C_{10}C_5$	143.4	136
	$C_1C_7C_4$	92	93.2 ± 1.5
	$C_2C_1C_4C_5$	111.5	113.0 ± 1.5
	$C_1C_7C_4$	90	92.0 ± 0.8
	$C_2C_1C_4C_5$	112.5	115.0 ± 0.8

[111, 115]. The same feature exhibits the SiH_2 group in silacarbocycles [116]. This is easily understood by a transfer of the s-content from the ring hybrids to *exo*-bonds. Strained bonds of the small rings prefer, namely, high p-character. This rehybridization is a consequence of a tendency of the molecule to reduce the angular strain. The opposite takes place in ethylene. A multiple bond is always stronger in competing for the s-character leaving more *p*-orbitals for the CH_2 moiety. Concomitantly, the HCH angle decreases from the ideal 120° value to 117.5° in agreement with experiment. It should be kept in mind that Eq. (46) is realized in practice only approximately due to neglect of higher AOs in hybrids, because of disregarding of steric effects etc. It holds rather accurately in the cases mentioned above, however. In addition to bond angles, the IMO method gives relatively good estimates of dihedral angles in polycyclic systems like in bicyclo(1.1.0)butane [117], norcaradiene [118], 9,10-methano(10)-annulene [119], norbornane systems [120] or puckering in larger rings as in cyclooctatetraene [113]. They are shown in Table 2.

Before considering bond distances the problem of bent bonds should be scrutinized. It follows straightforwardly from the orthogonality condition of Eq. (44) that two equivalent hybrids close up an angle which is always larger than 90°. Consequently, in cyclopropane C—C bonds should be bent and maximum electron density shifted off the center of the C—C straight line. In order to circumvent the problem of bond bending Mårtensson and Öhrn [121] suggested the use of complex hybrid AOs:

$$h_k = a_{k1}(2s) + f(r)\left[a_{k2}\frac{x}{r} + a_{k3}\frac{y}{r} + a_{k4}\frac{z}{r}\right] \qquad (47)$$

where a_{ki} (i = 1 ... 4) are the elements of the k-th row of the unitary matrix $\tilde{A}$. Since the phase of the complex hybrid is arbitrary, they were chosen so that the first column of the matrix $\tilde{A}$ was real. The direction of the hybrid is defined as the direction of the line connecting the carbon nucleus and the center of the hybrid density. The direction cosines of the hybrid center are:

$$\gamma_{k\mu} = \frac{\langle h_k|\,\mu\,|h_k\rangle}{\left|\sum_{\nu=x,y,z}\langle h_k|\,\nu\,|h_k\rangle^2\right|^{1/2}} \qquad (48)$$

where the Greek indices denote the coordinates x, y and z. It can be shown that the adopted phase convention gives direction cosines by the formula:

$$\gamma_{k\mu} = \frac{\mathrm{Re}\,\{a_{kn(\mu)}\}}{\left|\sum_{m=2}^{4}(\mathrm{Re}\,\{a_{km}\})^2\right|^{1/2}} \qquad (49)$$

where n(x) = 2, n(y) = 3 and n(z) = 4. It appears that the direction cosines are completely determined by the real parts of *p*-orbitals. The complex hybrids are more flexible than the real ones since the unitary matrix $\tilde{A}$ has twelve real parameters. Eight of them are defined by the direction cosines of the 4 nonequivalent hybrids. Hence, there are four parameters left free. Let us consider the complex hybrids in cyclopropane which have C_{2v} local symmetry implying that we have 2 independent hybridization parameters. Taking the coordinate system as in Ref. [121] the matrix $\tilde{A}$ takes the form:

$$\tilde{A} = \begin{pmatrix} \mathrm{Re}\,\{a_{11}\} & a_{12} & a_{13} & 0 \\ \mathrm{Re}\,\{a_{11}\} & a_{12} & -a_{13} & 0 \\ \mathrm{Re}\,\{a_{31}\} & a_{32} & 0 & a_{34} \\ \mathrm{Re}\,\{a_{31}\} & a_{32} & 0 & a_{34} \end{pmatrix} \tag{50}$$

where the first two rows and the last two rows represent C—H and C—C hybrids, respectively. Taking into account the direction cosines and fixing independent parameters it is possible to construct hybrids which lie along the C—C bonds. Postulating that all elements of the first column are equal 1/2 and that $\mathrm{Im}\{a_{13}\} = 0$ Mårtensson and Öhrn obtained that all hybrids are of the sp^3 type [121]. However, assuming that the bonding strength is proportional to the absolute value of overlap integrals we have shown that these complex hybrids have poor energetic performance [122]. In the first place, the matrix $\tilde{A}$ can be simplified because the interhybrid angle between CH HAOs is larger than 90°. Therefore, one can safely assume that $\mathrm{Im}\{a_{12}\} = \mathrm{Im}\{a_{13}\} = 0$ and retain complex coefficients only for C—C bonds. Employing Clementi AOs [123] and standard weighting factors for S(C—C) and S(C—H) overlaps [124] one finds that the maximum overlap method yields the following best hybrids:

$$\begin{pmatrix} h_1 \\ h_2 \\ h_3 \\ h_4 \end{pmatrix} = \begin{pmatrix} 0.542 & 0.455 & 0.707 & 0 \\ 0.542 & 0.455 & -0.707 & 0 \\ 0.455 & -0.542 & 0 & 0.707 \\ 0.455 & -0.542 & 0 & 0.707 \end{pmatrix} \begin{pmatrix} 2s \\ 2p_x \\ 2p_y \\ 2p_z \end{pmatrix} \tag{51}$$

Hence, the optimum hybrids are real and the resulting HCH angle is 114° in fine agreement with the experimental value of 113.6° [125]. Furthermore, bending of the C—C hybrids is 22.5° in line with experimental X-ray measurements in *cis*-1,2,3-tricyanocyclopropane [126] and earlier theoretical results [127–129, 124].

It should be pointed out that the maximum overlap calculations of hybrid orbitals in bicyclo(1.1.0)butane have led to the discovery of the twisted bond [117]. The hybrids describing C—C side bonds are not only bent relative to the straight line linking bonded C atoms, but they also pass by each other. In other words they do not lie in the same plane.

It is a common knowledge nowadays that bent bonds in small rings are a well established experimental and theoretical fact. Recent investigations have shown that bond bending is a general feature but deviation angles in OH_2, NH_3 etc. are rather small [130, 131]. Steric effects introduce bending of a couple of degrees which is reflected e.g. in tilted CH_3 groups. One can conclude that bent bonds are more a rule than an exception.

Let us consider the ab initio results of Wiberg and Murcko [130] in some detail. According to Bader [132] bond path is defined as a ridge of maximum electron density between a pair of bonded atoms. The 6–31 G* basis set gives the equilibrium bond

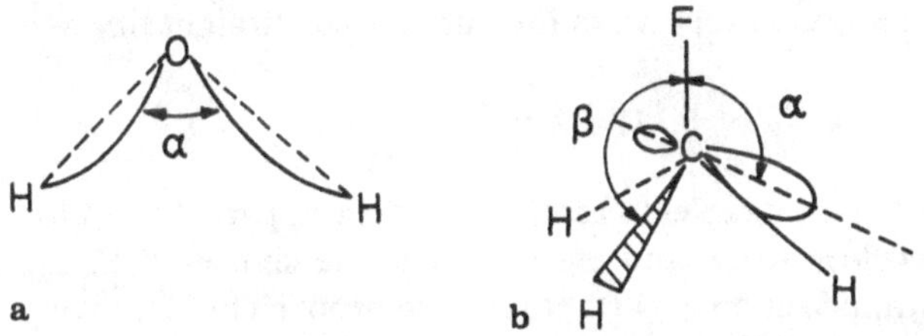

Fig. 1a, b. Bent bonds in H_2O (**a**) and CH_3F (**b**) molecules

angle in H_2O of 105.4° in good accordance with the experimental value. A qualitative guess of hybridization in H_2O could be made in two distinct and conceptually widely different ways. One might assume the canonical sp^3 hybridization and argue that the HOH angle is reduced by a strong repulsion between the lone pairs and O—H bond pair electrons. This would be in line with the valence shell electron pair repulsion (VSEPR) model [133]. According to the opposite point of view, OH bonds are formed by almost pure *p*-orbitals because two lone pairs of oxygen prefer high s-character. The observed bond angle is a consequence of the repulsion of two hydrogens [1]. The ab initio 6–31 G* calculation is in favour of the second picture. The bond path angle is found to be 101° meaning that hybrids are bent inside the H—O—H triangle ($\delta \sim -2.5°$) as shown on Fig. 1a. The estimated OH s-character is 16%. This is in fine agreement with STO-3 G calculation of hybridization which gave 15.6% by the LAO procedure (14a).

Bent bonds are helpful in resolving situations which seemingly violate the Walsh-Bent [134] rule stating that a more electronegative substituent absorbs more of the p-character of the central atom. For example, in CH_3F molecules 6–31 G** basis set gives 109.2° for the HCF angle in excellent agreement with experiment. This finding apparently contradicts the Walsh-Bent rule since a highly electronegative F atom causes a substantial shift of p-character to the carbon hybrid belonging to a CF bond. The increased s-character in CH bonds would enlarge HCH and diminish HCF angles. This is indeed the case if bond paths are considered. Then HCF and HCH have angles of 106.7° and 112.1° [131], respectively (Fig. 1b) thus being smaller and larger than the tetrahedral angle as required by the rule. Appearance of bent CH bonds gives support to the hybridization concept indicating that the VSEPR model is insufficient. The repulsion between the fluorine lone pairs and a diffuse $1s_A$ electron distribution outweighs Coulomb H ... F attraction and H ... H repulsion. The Pauli repulsion between lone pairs and CH hybrids is present too but the requirement "more p-character of the CF bond for stronger bonding" prevails. This is a rather nice example illustrating that bonding effects are more important than various nonbonding interactions, which is plausible because molecules are stable species.

It appears that one has to distinguish interatomic bond angle defined by straight lines passing through the nuclei in question and bond path angle defined by bent bonds. Analogously, it is conceptually useful to distinguish interatomic distance and bond length as suggested by us some time ago [135, 136]. The latter coincides with bond path length. An approximate estimate of the bond path length is provided by a parabola which approaches nuclei in such a way that bent hybrid AOs are its tangents. Then the relation between the bond length (l) and interatomic distance (d) reads:

$$l(CC)_{bent} = [1/K(\delta)]\, d(CC) \tag{52}$$

where δ is the angle of bending measured relative to the internuclear straight line and

$$[1/K(\delta)] = \{\mathrm{tg}\,\delta(1 + \mathrm{tg}^2\,\delta)^{1/2} + \ln[\mathrm{tg}\,\delta + (1 + \mathrm{tg}^2\,\delta)^{1/2}]\}/2\,\mathrm{tg}\,\delta \quad (53)$$

The concept of the bond path length enables a resolution of an apparent paradox that interatomic distances in cyclopropane and cyclopropene are shorter than the corresponding values in the unstrained counterparts ethane and propylene. Employing IMO results for δ and d(CC), one can easily estimate the bond path length by making use of Eqs. (52) and (53). It follows that the CC bond length e.g. in cyclopropane is 1.56 Å meaning that it is ~0.024 Å longer than the straight CC length in ethane. Hence, the highly strained CC bond is stretched relative to the "normal" CC bond in accordance with chemical intuition [136]. It should be pointed out that the Coulson-Moffitt VB calculation also gives the hybrid deviation angle δ = 22° [128], but more

Table 3. Comparison of some IMO bond distances with the corresponding experimental data and/or ab initio results (in Å)

Molecule	Bond	El. diffr.	X-ray (or ab initio)	IMO
	C—C	1.510[a]	(1.513)[b]	1.516
	C—H	1.089	(1.08 ass.)	1.088
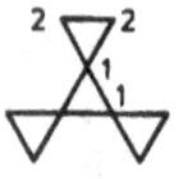	C_1—C_1	—	1.465[c]	1.468
	C_1—C_2	—	1.500	1.494
	C_2—C_2	—	1.496	1.520
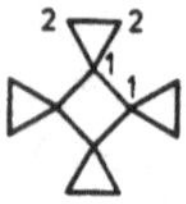	C_1—C_1	1.534[d]	1.525[d]; 1.477[c]	1.516
	C_1—C_2	1.491	1.502; 1.518	1.503
	C_2—C_2	1.528	1.521; 1.505	1.519
	C—H	1.103	1.05	1.088
	C—C	—	1.485[e] (1.482)[f]	1.491
	C—H	—	— (1.054)	1.065
	C—C	1.575[g]	1.55[h] (1.570)[i]	1.552
	C—H	1.100	1.0—1.1 (1.081)	1.089
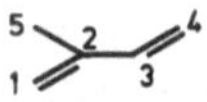	C=C	1.340[j]	(1.322)[k]	1.339
	C_2—C_3	1.463	(1.470)	1.467
	C_2—C_5	1.512	(1.509)	1.512
	C_3—H	1.076	(1.070)	1.081
	C_5—H	1.110	(1.083)	1.100
	C=C	1.349[j]	(1.344)[k]	1.339
	C_2—C_3	1.491	(1.477)	1.466
	C_2—C_5	1.511	(1.505)	1.512
	C_4—H	1.093	(1.073)	1.081
	C_5—H	1.123	(1.085)	1.100

[a] Ref. [139]; [b] Ref. [140]; [c] Ref. [141]; [d] Ref. [142]; [e] Ref. [143]; [f] Ref. [144]; [g] Ref. [145]; [h] Ref. [146]; [i] Ref. [147]; [j] Ref. [148]; [k] Ref. [149]

Table 4. Theoretical and experimental estimates of bond distances (in Å) in some strained hydrocarbons involving an exo-double bond*

Molecule	Bond	IMO	STO-3G	3-21G	4-21G	Exptl.
$H_2C=CH_2$	C=C	1.338	1.306	1.316	1.313	1.337
	C—H	1.085	1.073	1.079	1.082	1.086
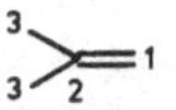	C(1)=C(2)	1.336	1.312	1.318	1.317	1.330
	C(2)=C(3)	1.508	1.530	1.520	1.516	1.508
	C(1)—H	1.082	1.080	1.074	1.073	1.072
	C(3)—H	1.100	1.086	1.084	1.085	1.095
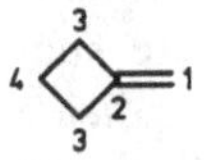	C(1)=C(2)	1.321	1.306	1.309	1.309	1.331
	C(2)—C(3)	1.522	1.531	1.537	1.536	1.517
	C(3)—C(4)	1.556	1.556	1.557	1.573	1.565
	C(1)—H	1.081	1.088	1.089	1.074	1.104
	C(3)—H	1.094	1.086	1.086	1.081	1.104
	C(4)—H	1.094	1.081	1.080	1.079	1.104
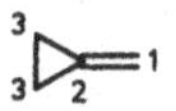	C(1)=C(2)	1.316	1.298	1.301	1.300	1.332
	C(2)—C(3)	1.475	1.474	1.472	1.476	1.457
	C(3)—C(3)	1.522	1.521	1.545	1.546	1.542
	C(1)—H	1.081	1.083	1.074	1.073	1.088
	C(3)—H	1.089	1.083	1.073	1.073	1.090
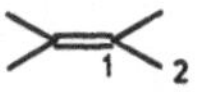	C(1)=C(1)	1.330	1.321	1.324	1.325	1.351
	C(1)—C(2)	1.512	1.531	1.523	1.522	1.511
	C(2)—H	1.100	1.086	1.084	1.086	1.110
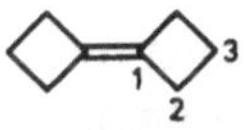	C(1)=C(1)	1.311	1.306	1.303	1.304	1.338
	C(1)—C(2)	1.521	1.531	1.535	1.536	1.518
	C(2)—C(3)	1.556	1.557	1.575	1.574	1.571
	C(2)—H	1.093	1.089	1.082	1.081	1.106
	C(3)—H	1.093	1.087	1.080	1.079	1.106
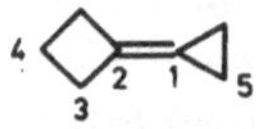	C(1)=C(2)	1.305	1.297	1.293	1.298	
	C(2)—C(3)	1.472	1.533	1.537	1.524	
	C(3)—C(4)	1.525	1.556	1.557	1.560	
	C(1)—C(5)	1.519	1.474	1.475	1.467	
	C(5)—C(5)	1.556	1.520	1.546	1.533	
	C(3)—H	1.094	1.089	1.089	1.082	
	C(4)—H	1.094	1.087	1.082	1.080	
	C(5)—H	1.089	1.083	1.082	1.074	
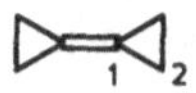	C(1)=C(1)	1.298	1.287	1.284	1.286	1.304[a]; 1.314[b]
	C(1)—C(2)	1.475	1.476	1.475	1.478	1.442; 1.468
	C(2)—C(2)	1.522	1.521	1.545	1.544	1.504; 1.554
	C(2)—H	1.089	1.083	1.080	1.073	1.099

* References to experimental papers are to be found in [151] if not stated here.

[a] X-ray data from Ref. [154]; [b] ED data from Ref. [154]

recent SCF calculations [137, 138] indicate that bond bending is actually much less pronounced.[3]

Keeping in mind that interatomic bond distances are not identical in general with bond path lengths, we shall present some characteristic results which illustrate the performance of the IMO procedure in predicting geometries of hydrocarbons (Table 3). It should be mentioned that cyclopropane was used in a calibration procedure and that a particular empirical constant was used in Eq. (45a) to reproduce the CC distance of 1.51 Å. However, the rest of the results is obtained without any additional adjustments and the accuracy of the predicted geometries is remarkable. Perusal of the data reveals several interesting features. The bond distances in the central carbocycle of [*n*]-rotanes are appreciably shorter than in the parent molecules. This finding was confirmed by later X-ray and ED measurements [142]. Similar results were subsequently obtained by the consistent force-field method [150]. Our estimates indicate that the X-ray data for [4]-rotane of Conia et al. [141] are in error. The IMO results for elusive tetrahedrane are in fine accordance with ab initio computations [144] and X-ray data for its derivative [143]. A single serious discrepancy is found in cubane, where IMO falls short in estimating the C—C distance by 0.02 Å relative to the ED result [145] and rather accurate DZ + d calculations [147]. The substituted 1,3-butadienes serve the purpose of illustrating a performance of the IMO scheme in estimating the geometric structure of molecules involving weakly delocalized π-electrons. It is noteworthy that the tub structure of cyclooctatetraene is a consequence of relieving the angular strain. The unfavourable number of π-electrons seems to be of less importance.

The IMO procedure may be conveniently used for pilot calculations and as a forerunner of more accurate (and expensive) ab initio methods. A nice example is offered by the question of bond distances of double bonds emanating from the small rings. IMO calculations indicates a substantial rehybridization at the carbon junction atom with a transfer of s-character to the exo-bond(s). Concomitantly, their bond distances were considerably shortened relatively to the free ethylene C=C distance. Available experimental data were not unambiguous but subsequent ab initio calculations carried out at STO-3 G, 3-21 G 4-21 G levels confirmed that the C=C bond attached to three and four-membered rings are indeed shrunk [151, 152]. Some characteristic results are displayed in Table 4. Survey of the presented experimental C=C bond distances along the series *2–4* possessing the $(C)_2C{=}CH_2$ structural unit does not exhibit a monotonic variation. On the contrary, one could conclude that C=C distances are approximately constant if experimental errors are taken into account. This conjecture would be in harmony with some theoretical arguments put forward by Allen et al. [153]. However, IMO and ab initio methods indicate a smooth decrease of the C=C distance. The same holds for the series *5–8* involving a central $(C)_2C{=}C(C)_2$ moiety. Some more recent experimental measurements of *8* lend credibility to our theoretical findings [154, 155]. The first study makes use of X-ray and ED-techniques and both predict shortening of the central double bond [154]. However, other CC distances estimated by these two methods exhibit significant discrepancies which widely exceed the stated limits of experimental errors. This example shows that there

[3] Discrepancy between SCF bond bendings and X-ray results is probably a consequence of the fact that the former are not corrected for thermal motion.

is nothing sacrosanct about experimental results and that they should be constantly tested against other experimental techniques and theoretical methods. In this connection it is useful to mention the X-ray data of Schenk [155] which for C(1)=C(1), C(1)—C(2) and C(2)—C(2) read 1.305, 1.467 and 1.539 Å, respectively. They show that the adjacent CC bond is shorter whereas the distal bond is considerably longer than in the parent cyclopropane molecule in line with IMO, STO-3G, 3-21G and 4-21G results. Hence, we can safely conclude that X-ray data for *8* reported by Traetteberg et al. [154] are in error.

To summarize, theoretical results and some reliable experimental data show that a double bond attached to strained ring(s) undergoes a substantial compression caused by rehybridization. It is noteworthy that the 4-21G basis set gives a C=C bond distance in *8* in fine agreement with ED data [154] if the empirical "offset" correction of 0.025 Å, which holds for all carbon double bonds, is taken into account. Similar structural variations were found in small ring hydrocarbons involving allylic double bonds [156].

It follows as an important corollary that significant structural changes may be expected whenever appreciable rehybridization takes place. This is obviously the case, e.g., when lone pairs are protonated [157]. We give here some characteristic results for small base molecules and their conjugated acids as computed by the SCF 3-21G procedure and Møller-Plesset perturbation calculations to the second order (MP2) to allow, at least partly, for electron correlation. The estimated bond distances and stretching force constants are presented in Table 5. The results can be interpreted in terms of the hybridization model. It is a common knowledge that lone pairs have a high s-content. Consequently, protonation of a lone pair leads to redistribution of a large portion of the s-character to neighbouring bonds. Their bond distances are compressed and the force constants are increased as a rule. This

Table 5. Bond distances (in Å) and stretching force constants (in mdyn $Å^{-1}$) in some small bases and their conjugated acids as obtained by SCF/3-21G and MP2/3-21G procedures*

Molecule	Bond distance	3-21G	MP2/3-21G	Force const.
N≡N	d(NN)	1.083	1.148	28.2
$[N{\equiv}NH]^+$	d(NN)	1.076	1.138	29.3
$[NH{\equiv}NH]^{2+}$	d(NN)	1.063	1.120	32.4
HC≡N	d(CN)	1.137	1.186	23.9
C≡NH	d(CN)	1.160	1.195	20.5
$[HC{=}NH]^+$	d(CN)	1.122	1.158	26.0
HN=NH	d(NN)	1.239	1.322	12.4
$[HN{=}NH_2]^+$	d(NN)	1.224	1.271	13.1
$H_2C{=}NH$	d(CN)	1.256	1.296	(9.8) 13.3
$[H_2C{=}NH_2]^+$	d(CN)	1.268	1.292	(10.9) 12.7
C≡O	d(CO)	1.129	1.171	21.7
$[HC{\equiv}O]^+$	d(CO)	1.100	1.148	26.1
$[C{\equiv}OH]^+$	d(CO)	1.166	1.193	17.1
O=NH	d(NO)	1.216	1.329	12.2
$[O{=}NH_2]^+$	d(NO)	1.189	1.279	13.5

* Force constants are calculated within the SCF/3-21G formalism. Exception is made for values given in parentheses, where MP2/3-21G procedure was employed

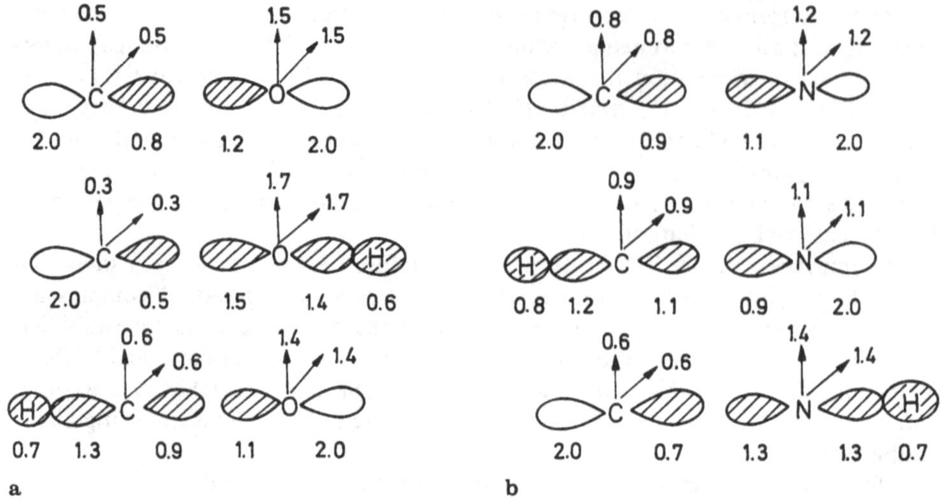

Fig. 2 a, b. MNDO bond and lone pair orbital populations in: (**a**) CO molecule and its protonated forms; (**b**) CN system and its conjugated acids HCN and HNC

is borne out by the present results. Double protonation of N_2 enhances the rehybridization effect as expected. There are, however, exceptions. They occur when the protonated atom is linked to a highly electropositive atom or functional group. Carbon monoxide serves as an excellent illustrative example. Protonation at the carbon site $[HC{\equiv}O]^+$ causes shortening and strengthening of the CO bond. On the contrary, protonation at oxygen has exactly the opposite effect. How can one rationalize this difference in behavior? A simple qualitative explanation of the dichotomy is given by the Mulliken population analysis. Electron populations of the chemically adapted local hybrid AOs (in σ—π picture) as obtained by the MNDO scheme are presented in Fig. 2(a). One observes that protonation at oxygen site increases population of the oxygen's π-orbitals to a high value of 1.7 while π(C) AOs contain only 0.3 electrons. This is unfavorable because it is close to the case where both oxygen π-orbitals are occupied by lone pairs, whereas π-AOs of the carbon are empty. A large disparity in populations of the bonded orbitals is obviously not advantageous for optimal covalent bonding. The same holds for the HAOs forming the carbon-oxygen σ bond. Their populations are 0.5 and 1.5 respectively. The opposite effect takes place when a proton attacks the carbon end of the CO molecule. Then, a charge flow toward carbon yields a more even distribution of electrons over the bonded AOs and concomitantly a more efficient covalent bonding. One can say that protonation at oxygen increases the difference in electronegativity between two fragments, whereas protonation of the carbon atom contributes to electronegativity equalization. It is noteworthy that an enhanced lone pair character of π-AOs in $[C{\equiv}OH]^+$ is accompanied by a decrease in π-bond order relative to the parent molecule CO. The corresponding values are 1.5 and 1.7, respectively. It appears that rehybridization is overshadowed by the unfavorable charge redistribution which acts in the opposite sense and prevails. In $[HC{\equiv}O]^+$ both mechanisms act synergetically in the same direction. It is therefore important to keep in mind that hybridization gives only a part of the bonding picture.

Table 6. Bond distances (in Å) in some characteristic small molecular anions and the corresponding neutral species obtained by SCF formalism employing various basis sets*

Molecule	Bond distances	TZ	TZ + P	TZ + 2P + diffuse	Near HF
$[HC{\equiv}C]^-$	d(CC)	1.233	1.226	1.223	1.223
	d(CH)	1.055	1.056	1.057	1.058
$HC{\equiv}CH$	d(CC)	1.187	1.183	1.179	1.180
	d(CH)	1.051	1.056	1.054	1.054
$[C{\equiv}N]^-$	d(CN)	1.166	1.154	1.151	1.152
$H{-}C{\equiv}N$	d(CN)	1.134	1.126	1.123	1.124
	d(CH)	1.053	1.058	1.057	1.057
$[HO]^-$	d(OH)	0.968	0.948	0.943	0.944
H_2O	d(OH)	0.951	0.944	0.941	0.941
$[H_3C]^-$	d(CH)	—	—	—	1.094
CH_4	d(CH)	—	—	—	1.082
$[H_2N]^-$	d(NH)	—	—	—	1.012
NH_3	d(NH)	—	—	—	0.999

* Taken from Ref. [158]

If the bonded atoms exhibit a smaller difference in electronegativity, the effect of charge redistribution is less pronounced. An example — $[CN]^-$ and the protonated species HCN and HNC — is shown in Fig. 2(b).

Additional evidence which conclusively shows the importance of the rehybridization and its effect on structural properties is provided by large basis set calculations on molecular anions and corresponding neutral species [158] (Table 6). Compression of the central bond is appreciable in all cases a notable exception being $[HO]^-$ perhaps because hybridization in oxygen is small at any rate.

Structural changes upon protonation and rehybridization are not surprising. It was shown a long time ago that the bonding radius is directly dependent on the hybrid's composition [159]. Dewar and Schmeising [160] have quite convincingly argued that the characteristic CC and CH bond distances could be classified by the canonical hybridization schemes (sp^n — sp^m, where n and m are integers), thus implying an intimate relation between the hybridization indices and interatomic distances. This relationship can be used to retrieve hybridization from X-ray data as evidenced by extensive work of Allen [161]. Last but not least, hybrid orbitals produced by the IMO procedure are flexible enough to rationalize a great deal of stereochemistry of hydrocarbons. Its success, this should be kept in mind, is based on the implicit relation between hybridization and bond distances as examplified by the empirical formulas of Eqs. (45a–c).

4.1.2 Bond Energies, Stretching Frequencies and Force Constants

The relation between directional properties of covalent bonds and the total molecular energy is something which is intuitively clear. If the idea of the covalent bond formed by localized pairs of electrons is essentially correct, then it is plausible to assume that their spatial distribution will minimize their repulsion. The shape of a hybrid and its direction in space depends on its s, p, etc. composition just like its bonding power and overlapping ability [1]. This conjecture is supported by analyses of numerous semiempirical methods and energy decomposition techniques. Hence, hybrids' composition and their bonding strength yield additional and, in our opinion, an even more important link between the energy and directional features. Accordingly, an intimate relation between hybridization and bond energies may be anticipated. This point was discussed in a series of papers most notably by Mulliken [162]. We found a very good correlation between bond overlap integrals and instantaneous bond dissociation energies (IBDE) in hydrocarbons [118]. The former were defined as the energies necessary for a bond rupture, when the formed fragment radicals remain unrelaxed. The correlation

$$E(C{-}C)_{IBD} = [372.3\,S(C{-}C) - 143.7]\ \text{kcal/mole} \tag{54}$$

yields good IBDEs as evidenced by the data presented in Table 7.

McKean et al. [164] found a very good correlation between the "isolated" CH bond stretching frequencies $\bar{\nu}_{CH}$ and bond dissociation energies $D^{\circ}_{298}(CH)$. The linear least squares fit gives a relation between these dissociation energies and S(CH) overlaps:

$$D^{\circ}_{298}(CH) = [617.8\,S(C{-}H) - 347.5]\ \text{kcal/mole} \tag{55}$$

The estimated IMO bond energies are presented in Table 8 and compared with experimental data. The difference Δ indicates a discrepancy which in toluene is untolerably high. Since $D^{\circ}_{298}(\bar{\nu}_{CH})$ and $D^{\circ}_{298}(S_{CH})$ values are compatible to a large extent, one concludes that the experimental bond energy of the methyl CH bond energy in toluene is in error. It seems also that experimental CH dissociation energies in cyclopropane and cyclobutane are too low.

Another two entities which are intuitively related to bond strengths are stretching frequencies and force constants. They should be proportional to s-characters of

Table 7. Experimental and calculated instantaneous bond dissociation energies (IBDE) for some molecules involving characteristic C—C single bonds*

Molecule	Bond hybridization	S(C—C)	IBDE	
			Correlated	Exptl.
$H_3C{-}CH_3$	$sp^{3.20} - sp^{3.20}$	0.648	97.4	98 ± 3
$C_6H_5{-}CH_3$	$sp^{2.23} - sp^{3.15}$	0.670	105.8	105 ± 4
$C_6H_5{-}C_6H_5$	$sp^{2.19} - sp^{2.19}$	0.702	117.5	116 ± 3
$H_3C{-}C{\equiv}CH$	$sp^{1.18} - sp^{3.01}$	0.718	123.7	126
$H_2C{=}CH(C_2H)$	$sp^{1.20} - sp^{2.23}$	0.737	130.7	130
$HC{\equiv}C{-}C{\equiv}CH$	$sp^{1.15} - sp^{1.15}$	0.789	150.0	150

* In kcal/mole. Experimental data are taken from Ref. [163]

Table 8. Comparison between the calculated dissociation energies of CH bonds with experimental data (in kcal/mole)

Molecule	S(C—H)	D°_{298}(corr.)	$D^{\circ}_{298}(\bar{\nu}_{CH})$	Exptl.	Δ
H—C≡C—H	0.769	127.3	132.9	128	0.7
C_2H_4	0.740	109.8	109.1	108	− 1.8
C_6H_6	0.741	110.4	109.5	110.5	0.1
$H_2C{=}C{=}CH_2$	0.739	109.3	108.1	—	—
cyclopropane	0.733	105.2	108.7	100.4	− 4.8
cyclobutane	0.729	102.6	100.1	96.8	− 5.8
H_3C^*—C≡C—H	0.730	103.2	100.3	—	—
C_6H_5—C^*H_3	0.725	101.1	99.1	88.3	−12.8
C_2H_6	0.720	97.0	99.6	98	1.0
CH_4	0.722	98.8	103.2	104	5.2
$(CH_3)_4C$	0.723	99.3	98.2	99.3	0

hybrid AOs involved in covalent bonds. The least squares fitting gave the following relation for the isolated $\bar{\nu}$ stretching frequencies of C—H bonds [165]

$$\bar{\nu}_{CH} = 20.8(s\,\%)_{CH} + 2411.6\ \mathrm{cm}^{-1} \tag{56}$$

The results presented in Table 9 reveal a good qualitative accordance with experimental frequencies. If cyclopropane, cyclobutane and methane are excluded from the correlation, a standard deviation of 10 cm^{-1} is obtained. Reasons for these discrepancies are unclear.

A similar correlation may be obtained for symmetrical and asymmetrical CH stretchings by using experimental data for a wider set of molecules. Although the agreement with measured values is more qualitative than quantitative, these correlations proved useful in asignment of infrared spectra of norbornane and norbornene [166].

Table 9. Experimental and calculated CH stretching frequencies (in cm^{-1}) in some simple hydrocarbons

Molecule	$(s\,\%)_{CH}$	$\bar{\nu}_{corr.}$	$\bar{\nu}_{exptl.}$	Δ
CH_3—C≡C^*—H	44.6	3341	3334	− 7
C_2H_2	43.5	3320	3336	16
C_2H_4	31.5	3070	3060	−10
C_2H_6	31.5	3068	3065	− 3
$H_2C{=}C{=}CH_2$	31.2	3062	3049	−13
cyclopropane	28.6	3008	3056	48[a]
cyclobutane	27.4	2983	2956	−27[a]
H_3C^*—C≡C—H	25.9	2951	2958	7
C_6H_5—C^*H_3	25.5	2943	2953	10
C_2H_6	25.4	2941	2950	9
CH_4	25.0	2932	2992	59[a]
$(CH_3)_4C$	25.0	2933	2934	1

[a] Values excluded from the correlation

Table 10. Comparison of F(C=C) stretching force constants calculated by the 4-31 G basis set and correlated values obtained by the IMO hybridization parameters (in mdyn/Å)

Molecule	Hybridization parameters	F(C=C) force constant	
	$a^2(C_1C_2)\ a^2(C_2C_1)$	Ab initio	IMO
$(CH_3)_2C=C(CH_3)_2$	0.1459	8.86	8.9
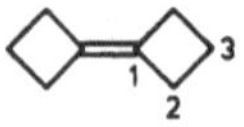	0.1722	9.28	9.4
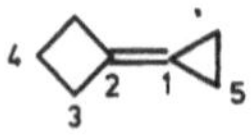	0.1839	9.57	9.6
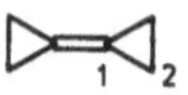	0.1954	9.94	9.9
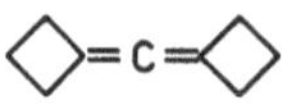	0.2075	—	10.1
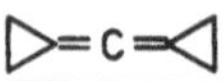	0.2212	—	10.4

Whereas a simple relation between $\bar{\nu}$(stretch.) and bonding parameters may be blurred by mass effect etc., a more direct insight into hybridization indices is expected by force constants. A linear relationship between a product of hybridization s-contents and force constants might be anticipated. As an outcome of our studies of structural properties of double bonds attached to strained rings we found the correlation [152, 167].

Table 11. Correlations of hybridization parameters with C—H and C—C stretching force constants (in mdyn/Å) in some characteristic compounds

C—H bonds* Molecule	$(s\%)_{CH}$	$K(C{-}H)_{exptl.}$	IMO	Δ	Ab initio
$H_3C{-}CH_3$	25.4	4.767[a]	3.9	0.9	5.59[b]
$H_2C{=}CH_2$	31.5	5.129	4.9	0.2	5.91
$HC{\equiv}CH$	43.5	6.096	6.8	−0.7	6.98

C—C bonds** Molecule	$a^2(c_1c_2)\ a^2(c_2c_1)$	$K(C{-}C)_{exptl.}$	IMO	Δ
$H_3C{-}CH_3$	0.057	4.45[c]	1.9	2.6
$H_3C{-}CH{=}CH_2$	0.078	4.69[d]	3.1	1.6
$H_2C{=}CH{-}CH{=}CH_2$	0.114	5.10[d]	5.1	0
$H_3C{-}C{\equiv}CH$	0.096	5.50[e]	4.1	−1.4
$HC{\equiv}C{-}C{\equiv}CH$	0.216	8.79[f]	10.7	−1.9

[a] Ref. [164]; [b] Ref. [168]; [c] Ref. [169]; [d] Ref. [170]; [e] Ref. [171]; [f] Ref. [172].
* Correlation reads: $K(C{-}H) = 0.162(s\%)_{CH} - 0.2$.
** Least squares fit yields: $K(C{-}C) = 55.2a^2(C_1C_2)\ a^2(C_2C_1) - 1.2$

$$F(C{=}C) = 19.8\, a^2(c_1c_2)\, a^2(c_2c_1) + 6.0 \qquad \text{(in mdyn/Å)} \tag{57}$$

where F(C=C) are calculated at the SCF 4-31 G level of accuracy and empirical 4-21 G offset values are used. Results obtained by correlating s-characters can be favourably compared with ab initio data (Table 10). Agreement is almost quantitative. It should be kept in mind that relations like Eq. (57) hold only in families of closely related molecules. In other words, it is taken for granted that many other things remain equal or at least approximately constant. This contention is substantiate by the correlation of C—H and C—C force constants over a large range of hybridization parameters (Table 11). Trend of changes is well reproduced by the IMO K(C—H) and K(C—C) constants but in quantitative sense much is left to be desired. Variation of hybridization indices is obviously important but the model is too simple for quantification of this property. Ab initio results are not accurate either (Table 11), but they are more consistent and empirical offset corrections may put force constants in good agreement with experiment.

Thermodynamic Acidity of Hydrocarbons

Acids are defined as species which show a tendency to lose a proton. Nonconjugated hydrocarbons are weak acids whose strength increases as the s-character of CH hybrids becomes larger. This was rationalized by the lower orbital energy of 2*s* AO which is more favorable for accommodating the negative charge of the carbanion concerned. Furthermore, hybrids with high s-content are more electronegative, meaning that the hydrogen atom carries more positive effective charge. Maximum overlap calculations [173] gave the following relation:

$$pK = 83.1 - 1.33(s\,\%)_{CH} \tag{58}$$

which yields the thermodynamic acidity of localized hydrocarbons with good accuracy (Table 12).

Table 12. Comparison between experimental and theoretical estimates of acidities in hydrocarbons

Molecule	C—H s-character	pk(exp.)*	pk(corr.)
Acetylene	43.6%	25	25.1
Propene(2-)	31.2%	40	41.6
Ethylene	31.6%	42	41.1
Cyclopropane	29.4%	44	44.0
Cyclobutane	27.6%	47	46.4
Methane	25.0%	47	50.0
Cyclopentane	26.6%	48	47.7
Ethane	25.4%	48	49.3
Propane(2-)	25.9%	49	48.7
Cyclohexane	25.8%	49	48.8

* Experimental data are taken from the Ref. [174]. Their accuracy is ±2

4.1.3 Indirect Spin-Spin Coupling Constants

Indirect spin-spin coupling constants across one bond are very sensitive probes of the electron density in the immediate vicinity of the nuclei. Spin coupling of the latter takes place via the coupled electron pair. Hence, the interpretation in terms of LMOs seems to be an obvious choice. In covalent bonds involving carbon and hydrogen atoms, the Fermi contact (FC) term determines the dominant mode of interaction. Recent theoretical studies have shown that other (spin-orbit) terms might be sometimes of importance [175, 176] even in hydrocarbons if multiple bonds are present. But in general, the FC term prevails. Since the FC mechanism depends on the extent of electron penetration into the nuclei, it is proportional to the product of s-characters of hybrids in question. Early experimental work on CH couplings [177] and CC constants [178] confirmed this expectation. Maximum overlap calculations proved useful in providing simple interpretation of spin-spin couplings of the directly bonded nuclei [179]. Our extensive studies [180] have led to semiempirical relations:

$$J(C—H) = 10.79(s\%)_{CH}/(1 + S^2_{CH}) - 54.9\ \text{Hz} \tag{59}$$

and

$$J(C_1—C_2) = 0.10205(s\%)_1(s\%)_2/(1 + S^2_{CC}) - 8.2\ \text{Hz} \tag{60}$$

Based on the model of perfectly localized HAOs they lent support to the earlier work of Muller-Pritchard [177] and Frei-Bernstein [178]. Overlap integrals in Eqs. (59) and (60) appear because of normalization of the VB type of coupling of HAOs. The role of hybridization in determining CH and CC coupling constants was questioned by Gil and Geraldes [181], but subsequently, more serious theoretical analyses [182–184] have convincingly shown that their criticism was unjustified.

We have shown that mutatis mutandis the relations similar to the types in Eqs. (59) and (60) hold for Si—H [185] and C—Si [186] directly bonded atoms.

Geminal $^2J(H—H)$ coupling constants depend on the hybridization of the central atom [187]. On the other hand, geminal $^2J(C—C)$ constants are determined by hybridization of all three atoms concerned [188] but there are exceptions too [187].

4.1.4 Atomic Multipoles and Hybridization

Dissection of the electron charge distribution in molecules into atomic or alternatively to atomic and bond terms proved extremely useful. Intermolecular Coulomb interactions are best described by microscopic atomic multipoles [189–192]. The latter are a consequence of the anisotropic electron density distribution of atoms in a chemical environment. It is intuitively clear that all multipoles higher than monopole should explicitly depend on hybridization. We note in passing that monopoles indirectly depend on hybridization since an increased s-character in a bond hybrid increases its electronegativity. Hence, the carbon atoms e.g. in acetylene should be more negative then those in ethylene or ethane. Notwithstanding the general importance of atomic multipoles, little is done in elucidating their connection with hybridization dipole moments being an exception. Atomic dipole due to hybrid χ_A is

$$\langle \chi_A | r_z | \chi_A \rangle = 2a(1 - a^2)^{1/2} \langle 2s | r_z | 2p_z \rangle \tag{61}$$

where the z coordinate coincides with the symmetry axis of the hybrid and the origin is placed at the nucleus. It is tacitly assumed that only $2s$ and $2p_z$ AOs contribute to a formation of a hybrid. The polarization integral equals $\langle 2s| r_z |2p_z\rangle = 160(\xi_{2s}\xi_{2p})^{5/2}/[\sqrt{3}(\xi_{2s} + \xi_{2p})^6]$ and consequently the atomic dipole in Eq. (61) may assume appreciable values particularly if lone pairs are concerned. Functional dependence of bond dipoles on hybridization in C—H and C—C bonds was discussed by a number of researchers [193–196]. It appears that rehybridization is an important ingredient in interpreting infrared intensities [196–199].

4.1.5 Miscellaneous Molecular Properties

The list of molecular properties considered above is by no means complete. We shall mention *en passant* several additional features without elaborating explicit formulas because of space limitations.

Compton profiles which depend on the momentum distribution in a molecule [200], exhibit in well-localized systems a characteristic additivity which can be rationalized in terms of localized MOs and their building blocks — HAOs [201, 202]. It can be shown that the average momentum in, e.g., the CH bond increases with the p-content of the corresponding hybrid.

The isotropic hyperfine constant a_H of a proton bonded to a trigonal carbon atom which is a part of a planar radical, is a function of hybridization. The so called McConnell's "constant" Q_{CH} depends quadratically on the s-character [203]. Unfortunately, the empirical correlation is based on the wrong experimental hyperfine constants a_H of the cyclopropenyl radical as pointed out by Ingold et al. [204]. Nevertheless, we feel that a proper a_H value for this elusive compound when available will alter only the empirical weighting factors, but not the final conclusions. The HAOs seem to be useful in discussing anisotropic coupling between a carbon $2p$ AO and a β-proton [205].

The electric field gradients depend, inter alia, on hybridization [206] as well as their asymmetry parameters [207]. It is noteworthy that bending force constants can be estimated with the aid of HAOs [208, 209]. Hybrid orbitals provide a natural choice of a basis set in various molecular force field methods [210] but care must be exercised regarding the incomplete orbital following of the nuclear motions [211, 212].

4.2 Global Molecular Properties

Classification of molecular properties into global and local ones is neither unambiguous nor is it used here in a very consistent way. In the first place, there are, strictly speaking, no completely local properties. For example, the C—H distance in CH_2F_2, local in nature, is affected by fluorine substitution. According to the Walsh-Bent rule, the s-character is shifted to C—H bonds making them shorter and stronger relative to the parent molecule. Hence, the C—H bond is influenced by the specific characteristics of atom(s) on the other side of the molecule. Its local character is still preserved, however, in a sense that the altered C—H distance is relatively close to the C_{sp^3}—H value.

Further, there is no sharp borderline between global and local properties. For instance, diamagnetic shielding of a nucleus σ_N^d is a microscopic property of a partcular nucleus and consequently it is a local label. However, in its estimate one has to take into account the electron densities of all atoms in a molecule meaning that σ_N^d is a collective property. Electron density contours give an overall picture of the charge distribution in a molecule, but their detailed examination yields information on local features. In particular, Fermi holes resemble local electron distributions (hybrids!) but they are considered here for convenience just as IR-intensities and Compton profiles were mentioned in a previous section. Keeping in mind that separating of properties into local and global is somewhat arbitrary, we shall discuss the latter in some detail.

4.2.1 Isopycnic Maps

Electron isopycnic (meaning that of the same density) maps reflect the changes in charge distribution upon the formation of chemical bonds [213, 214]. Particularly informative are for this purpose the deformation density contours defined as the difference between the total molecular electron density and the superposition of densities of neutral atoms placed at the equilibrium position. The latter yield the so called promolecule charge distribution which precedes reorganization due to bonding effects. They have to be produced by accurate ab initio calculations meaning that deformation isopycnic maps are strictly speaking not observable. Indeed, their definition is not unique but they give, nevertheless, a nice visual picture of changes caused by chemical bonding. For our purposes it is useful to observe that changes are rather small thus supporting the idea of perturbed and modified atoms in molecules. We shall mention some of their salient features. Atoms which possess moderate electronegativities exhibit when bonded humps of electron density in the regions between bonded nuclei. Concomitantly, a depletion of the charge density in the peripheral parts of the molecule takes place. Further, buildups of the bond densities illustrate

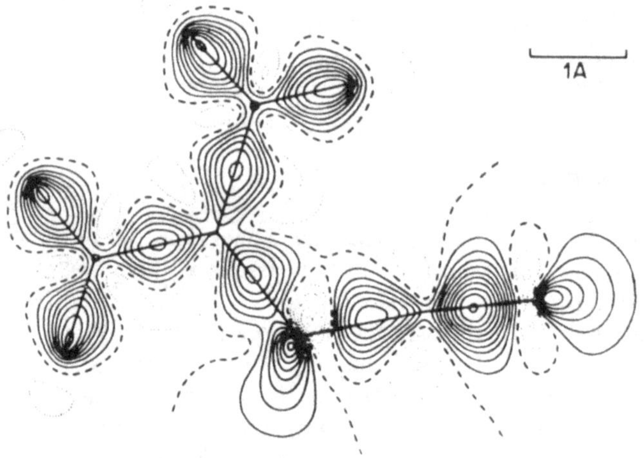

Fig. 3. Isopycnic deformation contours in 2-cyanoguanidine (Courtesy of the Israel Journal of Chemistry)

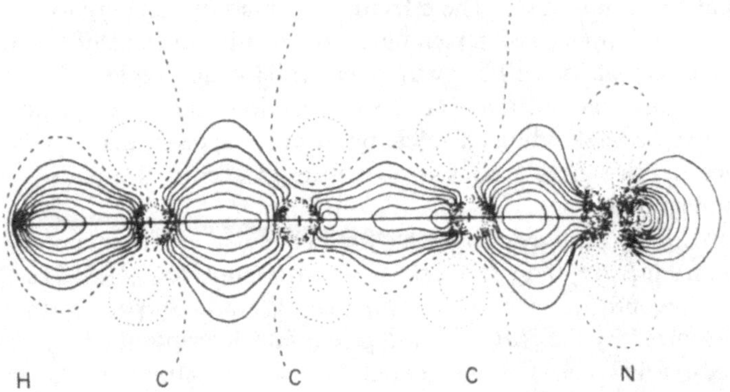

Fig. 4. Isopycnic deformation contours in H—C≡C—C≡N (Courtesy of the Israel Journal of Chemistry)

a descent of symmetry of an atom in the chemical environement. We shall illustrate this assertion by the X-ray deformation density map of 2-cyanoguanidine $(NH_2)_2C{=}N{-}C{\equiv}N$ in the mean molecular plane [215] (Fig. 3). One can easily recognize the σ-sp^2 moiety of the central carbon atom. The nitrogen in the NH_2 group reveals the approximate sp^2 hybridization. Significant distortion of the $sp^2(D_{3h})$ local symmetry is observed at the nitrogen of the C=N—C group where the lone pair in the simplest description assumes hybridization somewhere between sp^2 and sp^3. The hybrid orbital occupied by a lone pair in C≡N group can be roughly characterized by the sp^1 composition. Diffuse humps found along the bond reflect the π-character or better to say bent bond feature of multiple bonding. Similar deformation isopycnic contours of the CN group are found in H—C≡C—C≡N (Fig. 4) [216]. The C—H and C—C single bonds can be approximately described by the sp^1 hybrids whereas the C≡C triple bond exhibits a bent bond character as expected. These two examples provide pictorial evidence that HAOs are well adapted for a description of atomic charge anisotropy and reinforcement of densities in the covalent bonding regions.

It should be stressed that for electronegative atoms possessing several valence lone pairs, a promolecule consisting of spherical neutral atoms provides a poor reference state. Then the deformation maps may show a depletion of charge in bonding domains as in the cases of O—O and F—F bonds thus being in conflict with intuition. It is easy to show that a spherically averaged atomic electron density is not suitable as a gauge of changes caused by positive interference of (singly occupied) bond AOs. For example, the average valence population of the F atom is $(2s)^2(2p_x)^{1.67}(2p_y)^{1.67}(p_z)^{1.67}$ which yields appreciable electron density along the bond (z-)axis. A high density of the lone-pair-like $2p$ AOs obviously is not appropriate for chemical bonding. On the other hand, common sense implies that the F—F bond is of a σ-type and that interacting atoms reorient their singly occupied $2p_z$ orbitals. They become axially directed to achieve maximum overlapping and the electrons involved form a covalent bond by perfect pairing. Hence, a prepared F atom has a configuration $(s)^2(2p_x)^2(2p_y)^2(2p_z)^1$ which yields a better reference point for bonding superposition $[(2p_z)_{F1} + (2p_z)_{F2}]/[2(1 + S_{pp\sigma})]^{1/2}$ density which indeed shows in this case a relative accumulation [217] at the bond center. The effect is increased by some hybridization which enhances bond overlapping and screening of the positive nuclei shifting at the same time electron density of lone pairs away from the bonding region. This brief analysis shows once again that intra-atomic reorganisation and hybridization are extremly important in understanding chemical bonding. It appears that a judicious choice of a reference density gives a consistent picture in accordance with general knowledge and chemical inuition.

Isopycnic maps in highly strained rings call for particular attention because of unusual features exhibited by these remarkable compounds [126, 213, 218–220]. Their presentation is beyond the scope of this chapter. We note in passing that the local hybrid AOs produced by the IMO method give a fair description of the charge distribution in three-, four- and five-membered carbocycles although the bond-bending seems to be exaggerated [221].

Laplacian of the electronic charge density reveals humps and troughs in accordance with the Lewis idea of localized electron pairs [222–228]. This feature is an intrinsic property of the $-\nabla^2$ operator [229] thus giving a picture of density reorganization from the first principles being free of a more or less arbitrary choice of promolecule charge distribution. It gives shell-structure of atoms [222, 223] and energy density distribution. Maxima of $-\nabla^2\varrho$ correspond to humps of electron density thus indicating regions which are electrostatically favorable. They contribute the most to the local total energy. The portion of the outer valence shell where $-\nabla^2\varrho > 0$ exhibits local maxima, corresponds to electron pairs [230]. For example, in the molecules CH_4, NH_3 and H_2O one finds four valence charge maxima. Hence, Laplacian $-\nabla^2\varrho$ and hybrids are closely related. One should observe that four maxima in H_2O indicate the rabbit ears picture concomitant with our earlier discussion. Two maxima corresponding to lone pairs subtend a larger angle in full accord with the hybridization model which predicts higher s-character on the basis of a lower 2*s* orbital energy.

Taking into account that the total electron density ϱ exhibits sharp spikes only at the near vicinity of nuclei monotonically decaying as the distance from the nucleus decreases, one can say that Laplacian $-\nabla^2\varrho$ gives a very useful tool for studying density redistribution upon molecular formation providing detailed information on local charge concentrations and its depletion.

Another useful concept is given by the idea of a Fermi hole [231–239], which is a double-ganger of each electron. According to the Pauli exclusion principle, electrons of the same spin tend to avoid each other as much as possible. Hence, there is a domain around each electron where the probability of finding the second electron of a like spin is diminished. The reader is referred to chapter 8 of this volume where definition(s) and properties of Fermi holes are elaborated in detail [238]. A Fermi hole is determined by the first- and second-order reduced density matrices meaning that it can be calculated from the first principles. It appears that the Fermi hole assumes the shape of a local orbital which dominates a particular region of the three-dimensional molecular volume. For instance, if a probe electron is confined within the local core AO, then the Fermi hole will take the shape of this inner-shell orbital. On the other hand, if a probe electron is placed in a domain between two bonded atoms, it strikingly resembles local hybrid or bond AOs. The Fermi hole changes very little as the probe electron moves inside its (central) region. Overlapping of Fermi holes is generally small. A very interesting interpretation of the Fermi hole is provided by a notion that it represents a probability distribution of a probe electron itself. The Fermi hole (or orbital) is not constrained to look like anything and yet it assumes the shape of a LMO or even a hybrid. This is remarkable and shows again that hybrid AOs appear without any preconceptions. In our opinion, this is a very strong indication that hybrid-like distributions of density take place in the SCF-MO theory too, in spite of the fact that the total wave function, first- and second-order reduced density matrices are invariant to unitary transformations of the basis set.

Finally, analyses of Fermi holes in molecules show that charge distribution differs markedly from that of a free-electron gas as evidenced by the shifts of the probe electron into various molecular domains. It is possible to construct localized MOs by the Fermi hole formalism [237–239].

4.2.2 Molecular Heats of Formation and Strain Energies

The standard heat of formation $\Delta_f H$ of a compound $A_k B_l$ is defined as the change in heat content for the synthesis of this molecule from its elements:

$$\frac{k}{m} A_m + \frac{l}{n} B_n \rightarrow A_k B_l \tag{62}$$

where the process is carried out isothermally at room temperature (25 °C) and the elements A_m and B_n as well as the final product $A_k B_l$ are assumed to be in their standard thermodynamic states [240]. Since bond energies are linearly related to overlap integrals, Eqs. (54) and (55), we tried to fit the experimental heats of formation by the expression

$$\Delta_f H = k_{CC}^{\sigma} \sum_{CC} S_{CC}^{\sigma(b)} + k_{CC}^{\pi} \sum_{CC} S_{CC}^{\pi(b)} + k_{CH} \sum_{CH} S_{CH} + n_{CC} l_{CC} + n_{CH} l_{CH} \tag{63}$$

where n_{CC} and n_{CH} are the numbers of CC and CH bonds in a molecule. Additionally, $S_{CC}^{\sigma(b)}$ and $S_{CC}^{\pi(b)}$ denote the σ- and π-components of overlapping in a bent bond. For axially symmetric C—C bonds, $S_{CC}^{\pi(b)}$ vanishes. The highly strained cyclopropyl ring requires a separate empirical factor k_{CC}^{π}. Other details of parametrization are described elsewhere [241]. It should be mentioned that Eq. (63) tacitly assumes some sort of additivity of zero point energies. This is fully justified as Schulman and Disch have subsequently shown [242]. The zero-point vibration energy is indeed a linear function of numbers of carbon and hydrogen atoms in hydrocarbons.

The correlated maximum overlap and experimental heats of formation are compared in Table 13. The general trends are well reproduced the standard deviation being 3 kcal/mole. For example, insertion of a CH_2 group in the aliphatic chain decreases $\Delta_f H$ by roughly 5 kcal/mole in pretty good accordance with observations. The same holds for substitution of the CH_3 group in the ethylene moiety which leads to a decrease in $\Delta_f H$ by approximately 8 kcal/mole. Taking into account the utmost simplicity of the method and relatively small number of employed empirical parameters a degree of accordance is surprising. It should be recalled that ab initio calculations with DZ basis sets have to be empirically adjusted, too [242]. Ab initio results are more consistent of course.

Table 13. Comparison between correlated maximum overlap and experimental heats of formation for some characteristic hydrocarbons (in kcal/mole)

Molecule	$\Delta_f H$(corr.)	$\Delta_f H$(exptl.)	Δ
methane	−16.6	−17.9	1.3
ethane	−20.7	−20.24	−0.5
propane	−26.8	−24.82	−2.0
isobutane	−30.3	−32.15	1.9
neopentane	−35.2	−39.67	4.5
ethylene	13.2	12.50	0.7
propene	5.8	4.88	0.9
2-methylpropene	− 1.9	− 4.04	2.1
2-butene (*trans*)	− 1.8	− 2.67	0.9
trimethylethylene	− 9.6	—	—
tetramethylethylene	−15.7	−16.68	1.0
1,3-butadiene	31.8	38.77	−7.0
n-butane	−30.6	−30.15	−0.5
n-pentane	−35.5	−35.0	−0.5
methylcyclobutane	1.7	− 0.60	2.3
cyclobutene	34.4	37.50	−3.1
1,2-dimethylcyclobutene	17.7	19.80	−2.1
cyclopentane	−23.4	−18.46	−4.9
cyclohexane	−27.6	−29.43	1.8
methylcyclopentane	−28.6	−25.50	−3.1
1,1-dimethylcyclopentane	−33.9	−33.05	−0.8
cyclopentadiene	36.2	32.24	4.0
1,4-cyclohexadiene	24.6	26.30	−1.7
1,3-cyclohexadiene	24.0	26.00	−2.0
1,3,5-cycloheptatriene	48.2	44.50	3.7
cyclooctatetraene	69.8	71.30	−1.5
norbornane	−9.0	−12.42	3.4
norbornene	24.7	24.70	0
norbornadiene	60.5	59.70	0.8
1,3-dimethylenecyclobutane	56.5	53.30	3.2
bicyclo(2.2.2)octane	−27.5	−24.09	−3.4
cubane	148.8	148.70	0.1
adamantane	32.8	32.94	−0.1
bicyclo(1.1.1)pentane	78.6	72.81*	5.8
cyclopropane	14.6	12.74	1.9
spiropentane	39.4	44.23	−4.8

* The force field calculation of Allinger NL, Tribble MT, Miller MA., Wertz DH (1971) J. Am. Chem. Soc. 93: 1637

The estimated heats of formation by hybrids can be used in a predictive fashion to give additional information on molecular energetics. For instance, one can obtain a good idea of the heat of hydrogenation of olefins by employing data presented in Table 13. The resulting $\Delta_h H$ values are in good agreement with measurements in view of the crudeness of the model (Table 14).

A good performance of the maximum overlap hybridization model can be traced to the fact that a large portion of nonbonded repulsions can be absorbed in effective bond energies [243]. This conjecture is supported by the energy partitioning of the

Table 14. Comparison between the correlated maximum overlap and experimental heats of hydrogenation for some characteristic olefins

Molecule	$\Delta_h H$(corr.)	$\Delta_h H$(exptl.)	Δ
ethylene	33.9	32.82	1.1
cyclobutene	32.3	31.1	1.2
propene	32.6	30.12	1.5
2-methylpropene	28.4	28.39	0
2-butene (*trans*)	28.8	27.62	1.2
1,3-butadiene	62.4	57.07	5.3
1,3-cyclohexadiene	51.6	55.37	−3.8
1,3-dimethylenecyclobutane	55.5	60 ± 1	−4.5
norbornene	33.7	31.1	−0.6
norbornadiene	69.5	68.1	1.4

approximate SCF CNDO/2 method which has shown that the most important contribution to the molecular binding energy comes from the resonance term $E^r_{AB} = \sum_{\mu}^{A}\sum_{\nu}^{B} P_{\mu\nu}S_{\mu\nu}\beta_{\mu\nu}$, where A and B are two directly bonded neighboring atoms and $\beta_{\mu\nu}$ denotes an appropriate empirical weighting factor [244]. The contribution of the nonbonding interactions to $\Delta_f H$ is small. Essentially the same results were obtained by energy decomposition performed by employing more refined semi-empirical schemes [245].

A very useful concept is provided by the strain destabilization energy (SDE) which measures a degree of structural deformation and/or nonbonded contacts. Angular SDE or Baeyer strain dominates in small ring compounds. Since SDE is related to molecular reactivity, it continuously attracts considerable attention of researchers. To begin with, strain energy can not be unequivocally defined and, consequently, it is a somewhat loose term. In principle, one could obtain SDE of a molecule in question by comparing the $\Delta_f H$ value with that of the strain-free analog. The problem is that there are no unique strain-free reference molecules [246]. The same applies to preconceived strainless bond or group increments [247, 248]. A more appealing SDE scale is given by a set of gedanken homodesmotic reactions [249]. The latter are defined as reactions which preserve the number of primary, secondary, tertiary and quaternary C atoms. Additionally, the number of CC bonds classified by the coordination numbers of participating C atoms remains the same in reactants and products. What is changed in products is the hybridization state of the C atoms involved [250]. For example:

$$\text{cyclobutane} + 4\,CH_3{-}CH_3 \rightarrow 4\,CH_3{-}CH_2{-}CH_3 \quad (64a)$$

and

$$\text{methylenecyclobutane} + 4\,CH_3{-}CH_3 \rightarrow (CH_3)_2C{=}CH_2 + 3\,CH_3{-}CH_2{-}CH_3 \quad (64b)$$

The strain energy is then obtained as a negative $\Delta_f H$ change in these reactions regardless of the method employed in producing molecular heats of formation. The IMO method yields SDE values (in kcal/mole) in cyclobutane and methylenecyclobutane

Table 15. Comparison between maximum overlap strain energies and experimental data for some cyclic and polycyclic alkanes and alkenes (in kcal/mole)

Molecule	E_s(corr.)	E_s(exptl.)	Δ
cyclobutane	24.6	26.9	−2.3
methylcyclobutane	24.9	27.0	−2.1
cyclopentane	4.5	7.2	−2.7
cyclohexane	5.2	1.4	3.8
cyclobutene	29.6	30.6	−1.0
1,2-dimethylcyclobutene	29.6	29.6	0
cyclopentadiene	6.9	2.9	4.0
norbornane	17.1	17.6	−0.5
norbornene	20.7	27.2	−6.5
norbornadiene	35.0	34.7	0.3
1,4-cyclohexadiene	1.6	2.2	−0.6
1,3,5-cycloheptatriene	0.8	−2.0	2.8
cyclooctatetraene	0.7	2.5	−1.8
1,3-dimethylenecyclobutane	28.9	30.4	−1.5
1,3-cyclohexadiene	1.3	1.9	−0.6
bicyclo(2.2.2)octane	7.7	11.0	−3.3
cubane	167.4	166.0	1.4
adamantane	11.9	6.5	5.4
methylcyclopentane	6.4	7.87	−1.5
1,1-dimethylcyclopentane	7.2	7.23	0
bicyclo(1.1.1)pentane	105.2	92.7*	12.5*
cyclopropane	31.0	28.0	3.0
spiropentane	62.8	65.1	−2.3
bicyclopropyl	62.5	55.8	6.7
bicyclo(1.1.0)butane	65.5	66.5	−1.0
1,3-dimethylbicyclo(1.1.1)butane	65.9	70.0	−4.1

* The force field calculation is referenced at the bottom of Table 14

of 24.6 (24.8) and 28.8 (27.9), respectively [250]. Excellent agreement with experimental estimates given in parentheses is somewhat fortuitous, but the homodesmotic reactions of Eqs. (64a and b) illustrate the point that the IMO method can yield useful qualitative information about strain energies.

Another possibility is offered by correlating "experimental" strain energies with overlap integrals leading to a formula of the type of Eq. (63):

$$SDE = k^{\sigma(b)}_{C-C} \sum S^{\sigma(b)}_{C-C} + k^{\pi(b)}_{C-C} \sum S^{\pi(b)}_{C-C} + l_{CC} n_{CC} \tag{65}$$

By using the data of Schleyer et al. [247] one obtaines the linear least squares fit parameters $k^{\sigma(b)}_{C-C} = -18.9$, $k^{\pi(b)}_{C-C} = 1110.1$ and $l_{CC} = 13.2$ (in kcal/mole) [251]. The estimated maximum overlap SDE values are favourably compared with empirical information (Table 15). The additive l_{CC} constant serves to cancel out the stabilizing effect of σ-overlapping of bent HAOs. This is most easily seen in ethane. If one substitutes $n_{CC} = 1$, $k^{\sigma(b)}_{CC} = -18.9$, $S^{\pi(b)}_{CC} = 0$ and $l_{CC} = 13.2$ (kcal/mole) and takes into account that overlap $S_{CC} \sim 0.7$, it turns out that angular strain is zero as it should be. A very large $k^{\pi(b)}_{C-C}$ constant indicates that bond bending plays a crucial role in determining angular SDE values. Since the correlation coefficient is 0.996, the probability that the calculated strain energies are uncorrelated is less than 0.01%. Equation (65), although highly approximate, enables simple rationalization of the additivity of angular SDE values in polycyclic systems.

It should be stressed, however, that the strain destabilization in small rings is a complex phenomenon. Dewar [252] and Cremer et al. [253, 254] argued that actual strain in cyclopropane is much higher than 28 kcal/mole, but destabilization is diminished by the aromatic stabilization of 6 local ring hybrid AOs. Taking the bond angle into account (Baeyer strain), bond eclipsing (Pitzer strain) stretching effect of C—C bonds and increase in CH bond energy due to higher s-character, Cremer and Gauss estimated by ab initio calculation that conjugative interaction of CC orbitals stabilizes the cyclopropyl ring by 16.4 kcal/mole. On the other hand, Wiberg et al. [255] have conclusively shown that intramolecular charge transfer from hydrogen toward carbon atoms in cyclopropane, estimated by Bader's atoms in molecule theory, accounts for destabilization in this molecule relative to $3CH_2$ standard groups. A charge drift of 0.048e from each hydrogen to the neighboring carbon leads to an increase of 15.5 kcal/mole but simultaneously to a decrease in stability of a hydrogen by 12.6 kcal/mole relative to a standard value. A total of 28.8 kcal/mole of SDE is in good accord with "measured" strain. Both approaches are probably not contradictory at all because increased carbon density is unfavorably distributed around the 3-membered ring due to bending. It is quite possible that angular destabilization is alleviated by conjugation. Hence, the resulting strain is a consequence of interplay of several effects. As far as the approximate relation of Eq. (65) is concerned, all contributions other than bond bending are relatively successfully absorbed in adjustable parameters.

4.2.3 Diamagnetic Properties in Hydrocarbons

It is common knowledge that diamagnetic shielding $\sigma^{d}_{av.}$ and diamagnetic susceptibility $\chi^{d}_{av.}$ depend only on atomic identity cards (i.e. atomic numbers) and the equilibrium geometries [31, 32]. According to Ramsey [256], diamagnetic shielding of the nucleus A is related to the electrostatic potential exerted at its site by:

$$\sigma^{d}_{av.}(A) = (e^2/3mc^2)\langle 0|\, 1/r_A\, |0\rangle \tag{66}$$

It can be expressed in a simple form which corresponds to the independent atom (IAM) model:

$$\sigma^{d}_{av.}(A) = \sigma^{d}_{av.}(FA) + (e^2/3mc^2) \sum_{B}{}' Z_B/R_{AB} \tag{67}$$

where $\sigma^{d}_{av.}(FA)$ refers to the free atom shielding. Adopting a $\sigma^{d}_{av.}(FA)$ value for carbon of 260.7 ppm as calculated by Malli and Froese [257] and IMO calculated geometries, one obtains quite reasonable estimates of carbon nucleus shieldings in representative hydrocarbons. They can be favorably compared with ab initio data [9] (Table 16).

Similarly, we have shown that a diagonal element of the Langevin's diamagnetic tensor may be expressed with the IAM model with good accuracy [31, 32]:

$$\chi^{d}_{av.} = (Ne^2/4mc^2) \sum_{A} (b_A^2 + c_A^2)\, Z_A + \sum_{p} n_p k_p \tag{68}$$

Table 16. Comparison between the IMO diamagnetic shieldings of carbon atoms in some representative hydrocarbons and the corresponding ab initio DZ results (in ppm)

Molcule	IMO	Ab initio (DZ)
methane	294.8	296.7
ethane	335.7	337.2
ethylene	328.9	330.1
propylene		
C(1)	357.1	355.5
C(2)	371.1	369.5
C(3)	356.7	355.8
cyclopropane	368.7	370.7
cyclopropene		
C(1)	362.3	360.4
C(2)	363.6	360.4

where a, b, and c are inertial coordinates of the nucleus A and fundamental constants have their usual meaning. The second term is isotropic and relatively small. It arises due to spatial extension of AOs. The empirical constant k_p is the same for all atoms

Table 17. Comparison of diamagnetic contributions to magnetic susceptibility in some hydrocarbons as calculated by the IMO method with corresponding experimental results and ab initio values (in 10^{-6} cm^3/mole)*

Molecule	Axis	IMO	Exptl. or ab initio
methane	aa	− 29.0	− 27.8[a]
ethane	aa	− 54.7	− 52.6[a]
	bb	−103.9	−103.0
	cc	−103.9	−103.0
ethylene	aa	− 38.4	− 39.6[a]
	bb	− 72.5	− 72.8
	cc	− 87.2	− 84.1
cyclopropane	aa	−113.5	−112.4[a]
	bb	−113.5	−112.4
	cc	−149.3	−149.3
cyclopropene	aa	− 82.5	− 82.3 ± 0.3
	bb	− 94.9	−100.4 ± 0.3
	cc	−130.7	−133.3 ± 0.3
methylene- -cyclopropane	aa	−123.9	−115.9
	bb	−233.8	−240.6
	cc	−285.9	−284.3
cyclobutene	aa	−155.8	−154.1
	bb	−159.4	−160.8
	cc	−244.8	−242.8

* Inertial coordinate systems are used.
[a] Ab initio results

belonging to a particular period of the Mendeleev system. Correspondingly, n_p is their number in a molecule. The other two diagonal elements χ^d_{bb} and χ^d_{cc} are obtained by cyclic permutation of coordinates. The calculated IMO values are close to experimental or ab initio data (Table 17).

4.2.4 Photoelectron Spectroscopy (PES) and Hybridization

Simple MO schemes are very effective in interpreting PES spectra [258]. Hybrid AOs seem to be redundant in this respect. This is indeed the case in highly symmetric environments. Consider, for example, the CH_4 molecule. Its molecular orbitals can be constructed consecutively by forming first the localized two-center bond orbitals:

$$\lambda_i = \frac{1}{\sqrt{2}}(\chi_i + H_i) \tag{69}$$

where χ_i are sp^3 hybrids (1) and H_i denotes $1s_{H_i}$ hydrogen function. It is assumed that the C—H bond is purely covalent and the overlap is neglected in the normalization constant. The delocalized MOs of A_1 and T_2 symmetry species are readily produced by using projection operators [5]:

$$A_1: \qquad \Psi_1 = \frac{1}{2\sqrt{2}}\sum_{i=1}^{4}(\chi_i + H_i) \tag{70}$$

yielding

$$\Psi_1 = \frac{1}{\sqrt{2}}\left[(ns) + \frac{1}{2}(H_1 + H_2 + H_3 + H_4)\right] \tag{71}$$

Similarly:

$$\Psi_2 = \frac{1}{\sqrt{2}}\left[(np_x) + \frac{1}{2}(H_1 - H_2 + H_3 - H_4)\right] \tag{72a}$$

$$T_2: \qquad \Psi_3 = \frac{1}{\sqrt{2}}\left[(np_y) + \frac{1}{2}(H_1 - H_2 - H_3 + H_4)\right] \tag{72b}$$

$$\Psi_4 = \frac{1}{\sqrt{2}}\left[(np_z) + \frac{1}{2}(H_1 + H_2 - H_3 - H_4)\right] \tag{72c}$$

Hence, hybrids disappeared being transformed to the conventional (ns) and (np) AOs. The set of formulas in Eqs. (71–72) is a source of a widespread misconception that hybrids do not "exist". In fact, it shows only that s and p AOs are a more sensible choice for the MO scheme in T_d symmetry. Both basis sets are completely equivalent since they are related by an orthogonal transformation. On the other hand, HAOs may be more favorable in less symmetric moieties as discussed in detail by Heilbronner

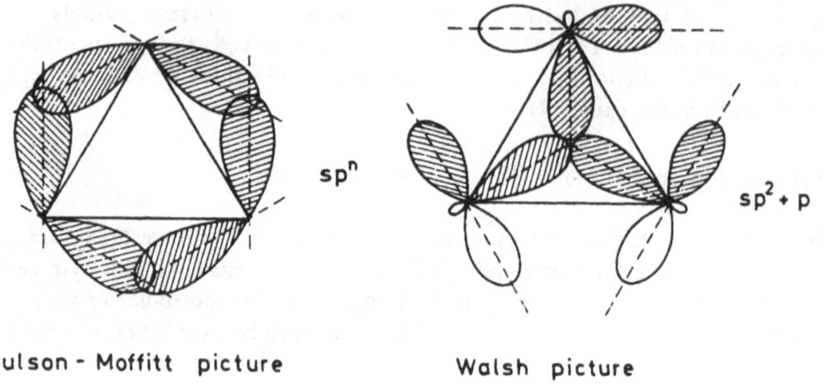

Fig. 5. Description of the cyclopropane ring by local bent bond hybrids and Walsh orbitals

et al. [259]. They have shown that Förster-Coulson-Moffitt [127, 128] bent bond (BB) hybrid AOs (Fig. 5a) are well adapted to a highly strained cyclopropyl structural unit being superior to the e.g. widely used Walsh [260] (Fig. 5b) basis set consisting of a pure tangential *p*-orbital and radial sp^2 hybrid. Both basis sets BB and W are again related by orthogonal transformation but it appears that the bent bond picture leads to a final result in a smooth and more direct way. Hence the BB set is a much better starting point and offers a simpler interpretation of the cyclopropane PE spectrum [259]. The Walsh basis set can achieve the same result only after heavy CI mixing. In the BB picture one can safely skip the CI step thus illustrating a general feature that a more appropriate basis set in a physical sense is computationally more economical. The reader is referred to the original papers for details [259]. Pictorial argument based on unfavorable overlapping of tangential p_t AOs (Fig. 5b) should actually suffice for dissmisal of the W-model.

A similar analysis can be performed for bicyclo(1.1.0) butane [261]. The Walsh scheme representing the generalization of the cyclopropyl case is not compatible with the geometric structure of the molecule. A refined W-model employing sp^1 hybrids at the bridgehead carbons accompanied by a pair of p_r radial and p_t tangential AOs is more suitable and will be utilized in the following discussion. The terminal

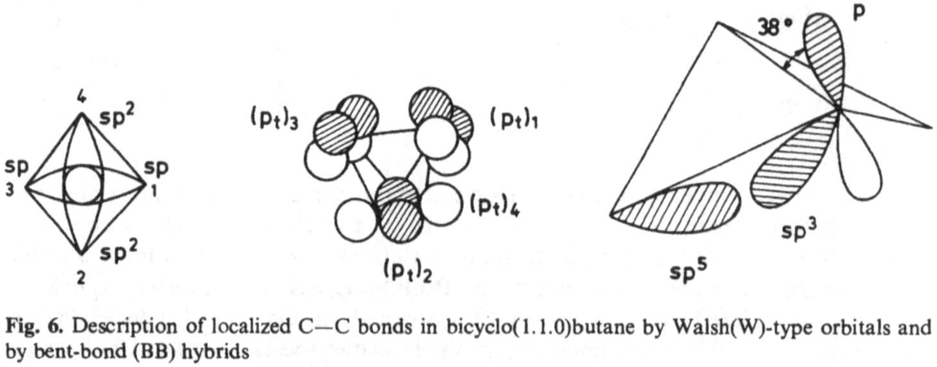

Fig. 6. Description of localized C—C bonds in bicyclo(1.1.0)butane by Walsh(W)-type orbitals and by bent-bond (BB) hybrids

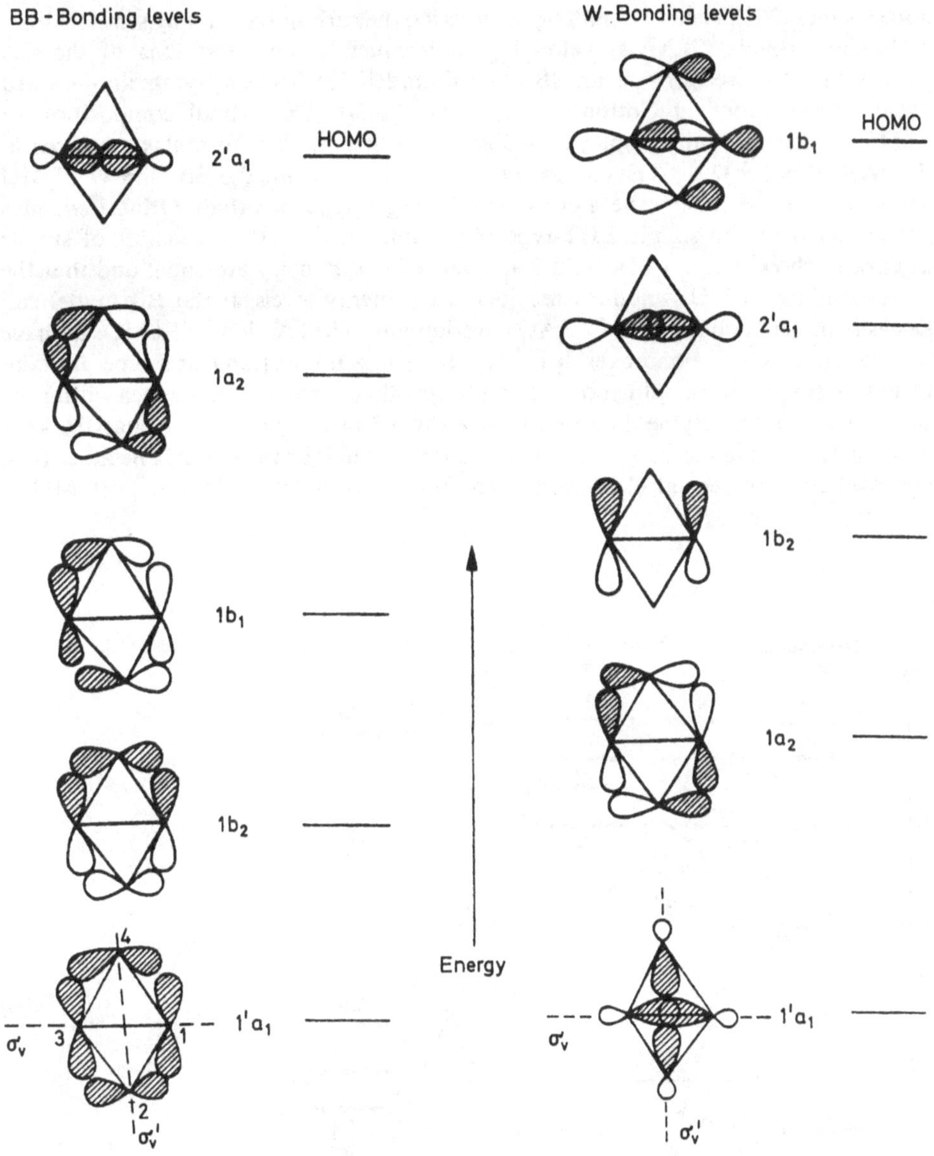

Fig. 7. Schematic ordering of delocalized orbitals describing C—C bonds in bicyclo(1.1.0)butane obtained by using bent bond (BB) and Walsh (W) type of basis sets

carbons are described by $p_t + sp^2$ basis as in cyclopropane (Fig. 6a). The bent bond picture will employ sp^5—sp^3 hybrids for envelope C—C bonds (Fig. 6b) and p AOs inclined by 38° to the central bond. This is concomitant with ab initio results [262]. In order to make a fair comparison between BB and W basis sets, we shall slightly modify the latter. This is achieved by using $sp^{0.5}$ hybrids at the bridgehead instead of sp^1 HAOs. Under these condition BB and W bases correspond to the same hybridization schemes [261]. Next, we shall try to describe the carbon skeleton by semilocalized molecular orbitals (SLMOs) belonging to irreducible representations of the C_{2v} point-symmetry group by using BB and W models (or SALCs: symmetry-adapted linear combinations in Cotton's terminology [263]). The orbital composition of SLMOs is schematically displayed in Fig. 7 and their explicit formulas are given in the original paper [261]. The energy scale is not the same for the BB- and W-SLMO levels because the latter have a considerably larger span of values. Orbital energies are obtained by the simple EHT-type of parametrization. It is a matter of simple algebra to show that $1'a_1$, $1a_2$ and $2'a_1$ levels of A symmetry are equal and that the corresponding SLMO functions are identical. Energy levels in the BB model can be easily understood in terms of HAOs overlapping. The four lowest SLMOs all have four favorable C—C bond overlaps. They differ in a number and/or a type of nodal planes. A simple overlapping criterion will show that negative interference within the same ring is less favorable than that between two 3-membered rings. Hence $1b_2$ level is below $1b_1$ level in the BB picture. The highest orbital is obviously the one describing the most strained central C—C bond. The final most stable SLMO and HOMO in

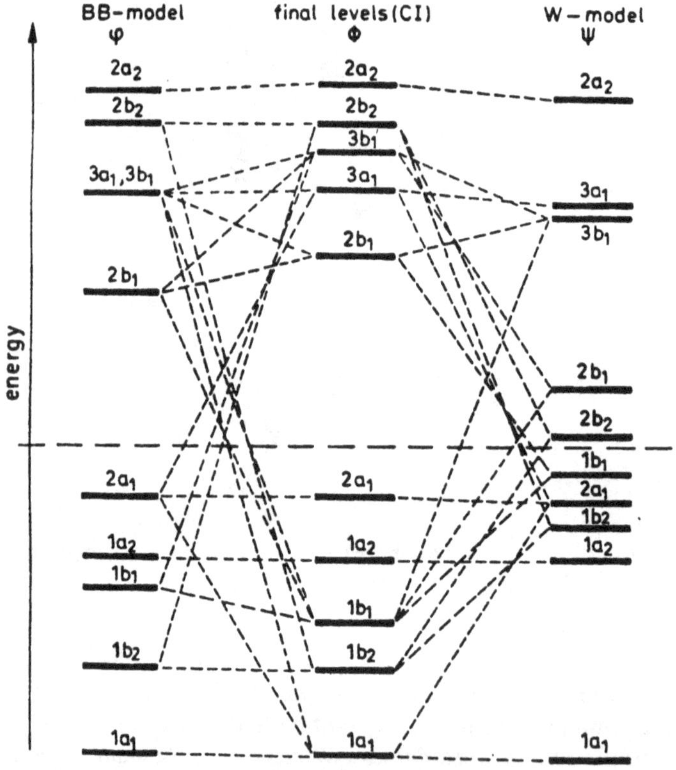

Fig. 8. Spacing of CC levels in bicyclo(1.1.0)butane obtained by BB and W basis sets. Final levels after CI are the same

BB basis set is obtained by slight mixing of $1'a_1$ and $2'a_1$ levels which are of the same symmetry. Principal contributions are of course given by SLMOs presented in Fig. 7 in view of the large energy difference of $1'a_1$ and $2'a_1$ levels. It appears that HOMO in BB and W sets belong to different symmetry species. The BB model gives a correct answer. The final C—C energy levels are obtained by mixing occupied and virtual levels (Fig. 8). One observes that a heavy CI mixing is necessary for W set whereas the BB model slightly lowers only its $1b_1$ level. The final result is again identical by using both basis sets. They differ, however, considerably in their performance. The point is that BB HAOs give a qualitative one-electron eigenvalues and eigenvectors at once and the CI effect may be neglected. They have a simple and transparent composition. Their spacing is easily interpreted in terms of intrabond and interbond overlapping. This is not the case for W-SLMOs. The levels of the b_1 and b_2 symmetries are by far too unstable. This is a consequence of the shift of s-character to the virtual (unoccupied) levels belonging to the same symmetry species. Hence, excited levels are too low. A closer crutiny of the W-SLMOs shows that they suffer from two major drawbacks. The first is the notoriously poor overlapping ability. The second is their decreased flexibility relative to the BB HAOs basis set. Consider terminal atoms C_2 and C_4 in Fig. 7. Local hybrids can assume the same sign as in $1'a_1$ and $1b_2$ levels or the opposite signs like in $1b_1$ and $1a_2$ levels. On the other hand, the tangential p_t orbital is antisymmetric by its nature. Consequently, it must disappear in W-levels of $1'a_1$, $1b_2$ and $2'a_1$ species which are symmetrical relative to a reflection in the plane passing through terminal carbon atoms. It follows that HAOs are physically more acceptable and computationally more convenient. *Ceterum censeo*, we stress once again that a shrewd choice of the basis set can save a lot of effort. The significance of the analysis presented above lies in the fact that hybrid AOs are highly transferable between similar moieties and that symmetry reasoning together with the nodal plane (or overlapping) criterion gives enlightening insight into the ordering of the one-electron levels.

Finally, it should be mentioned that HAOs are useful also in some more refined semiempirical MO schemes [264] than the simple EHT approach adopted above.

5 Maximum Overlap vs Other Semiempirical and Ab Initio Estimates of Hybridization Indices

The iterative maximum overlap procedure is one of the simplest semiempirical methods designed for approximate description of the electronic structure of molecules. It is based on the idea that redistribution of the s- and p-character between bonds emanating from atoms has a strong influence on stereochemical and other properties of molecules. This pictorial and intuitively appealing model seems to work very well as evidenced by numerous correlations of hybridization parameters and/or overlap integrals with various observables. Its reproduces trends of changes in large families of molecules rather successfully, as conclusively shown in the previous section. The model is undoubtedly useful when used within the limits of its applicability. It has passed numerous tests against experimental data and has appreciable predictive power. On the other hand, the model is so simple that comparison with more rigorous theoretical results is highly desirable.

Here we shall briefly discuss hybridization parameters obtained by some common semiempirical and ab initio molecular orbital methods in relation to maximum overlap results. The general characteristics of the former procedures is given by the fact that they do not employ HAO basis sets. In addition, they are all invariant to hybridization mixing of AOs which should be kept in mind, because the results will nevertheless show polarizations of local atomic densities very similar to those predicted by the hybridization model. For this purpose we selected widely different semiempirical methods: EHT, SCC-MO, and CNDO/2. The first two employ the model hamiltonian approach fully appreciating orbital overlapping. The SCC-MO method alias IEHT represents an improved EHT scheme where the effect of electron repulsion is implicitly taken into account by the self-consistent charge technique [265]. Extensive calculations of one-electron molecular properties have shown that the resulting charge distribution is quite reliable [266, 267]. In contrast, the CNDO/2 method [268] makes use of the zero-differential overlapping approximation and all integrals are forced to be invariant to hybridization. EHT, SCC-MO, and CNDO/2 hybridization indices were retrieved from first order density matrix according to Eq. (26) [68, 69]. Finally, hybridization may be estimated empirically by using Muller-Pritchard [177] relation for C—H bonds $(s\%)_{CH} = 0.5\,J(C\text{—}H)$, where J denotes the spin-spin coupling constant. Hybridization in C—H bonds for a large family of hydrocarbons is presented in Table 18. Compounds *1*—*7* are depicted in Fig. 9. Perusal of the displayed data shows

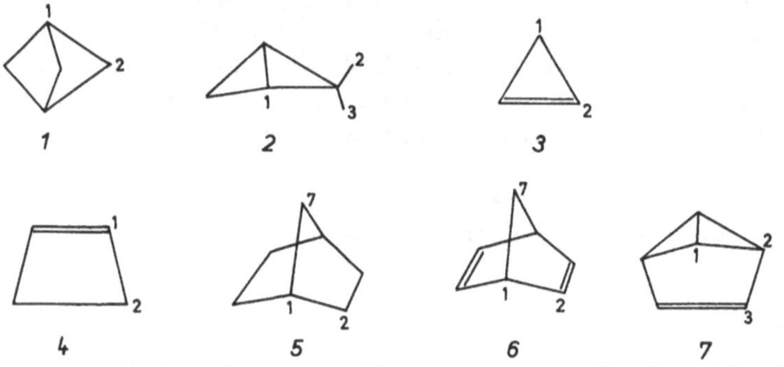

Fig. 9. Some characteristic strained molecules exhibiting large deviation from canonical hybridization states

that the resulting hybridization parameters are similar to a large extent in spite of the difference in the methods of their deduction. This compatibility shows that hybridization is an essential ingredient of covalent bonding. Each theoretical scheme defines its own scale of hybridization, but the emerging picture is fairly consistent within the same framework. The degree of similarity between maximum overlap (MO) hybridization and other theoretical results offers additional support to the simple overlap criterion. Particularly close to the MO hybridization results are those from the CNDO/2 method. It seems, however, that the CNDO/2 method performs somewhat worse in highly strained systems. In this connection, it is noteworthy that the introduction of the hybridization concept into the CNDO/2 scheme may remedy some of its intrinsic inconsistencies. For instance, the standard CNDO/2 method failed to give a

Table 18. Comparison of s-character of C—H hybrids estimated by different semiempirical methods (in %)

Molecule	Bond	EHT	SCC[a]	CNDO/2	MO[b]	Emp.[c]
methane	C—H	27.5	27.8	25.0	25.0	25.0
ethane	C—H	29.3	25.6	24.6	25.4	25.0
cyclobutane	C—H	32.2	—	26.1	27.4	26.8
ethylene	C—H	34.5	32.5	30.8	31.5	31.3
benzene	C—H	34.8	32.5	28.8	31.5	31.8
cubane	C—H	—	35.6	30.6	32.1	32.0
cyclopropane	C—H	34.9	33.4	30.3	28.6	32.2
acetylene	C—H	45.0	42.2	44.5	44.5	49.7
prismane	C—H	38.3	38.8	34.8	34.1	—
1	C_1—H	38.3	38.9	34.2	32.1	32.8
	C_2—H	31.8	32.0	27.5	28.9	28.8
2	C_1—H	40.2	41.0	35.0	38.6	40.6
	C_2—H	32.4	32.8	32.1	33.0	30.4
	C_3—H	33.0	33.3	38.6	33.0	34.0
3	C_1—H	33.7	34.1	32.8	30.0	—
	C_2—H	41.8	42.6	38.7	39.1	44.0
4	C_1—H	36.3	37.0	33.3	35.4	34.0
	C_2—H	31.0	31.3	26.5	27.9	28.0
5	C_1—H	31.9	32.1	25.1	28.7	27.8
	C_2—H_{exo}	29.4	29.6	24.2	25.9	26.0
	C_7—H	30.6	30.8	25.3	27.4	26.0
6	C_1—H	31.6	31.8	25.5	28.3	29.2
	C_2—H	35.5	36.1	31.6	33.4	34.5
	C_7—H	30.7	30.9	24.9	27.4	27.1
7	C_1—H	38.8	39.6	33.1	35.8	—
	C_2—H	36.9	37.1	35.3	34.0	—
	C_3—H	33.6	34.1	32.0	33.3	—

* Taken from Ref. [69]. Molecules *1*–*7* are shown in Fig. 9

[a] SCC-MO is self-constistent charge alias IEHT method

[b] Maximum overlap results

[c] Empirical s(%)-characters are obtained by Muller-Pritchard [177] formula $s(\%)_{CH}^{emp} = 0.2\,J(^{13}C{-}H)$

minimum for a variation of the HCH angle in cyclopropane in a range between 102° and 122° [264]. Meyer imposed a constraint

$$\cos \vartheta_{HCH} = -P^2_{2s(CH)}/(1 - P^2_{2s(CH)}) \tag{73}$$

which can be easily derived from Eqs. (24–26). For this purpose we shall write the hybrid orbital as:

$$\chi_{AB} = \sqrt{W_{2s(AB)}}\,(2s)_A + \sqrt{W_{2p(AB)}}\,(2p)_{AB} \tag{74}$$

where $(2p)_{AB}$ is the $2p$ orbital oriented along the A—B bond and $W_{2p(AB)} = W_{2p_x(AB)} + W_{2p_y(AB)} + W_{2p_z(AB)}$. If two equivalent and orthogonal hybrids describing C—H bonds are designated by χ_{AH_1} and χ_{AH_2} one obtains:

$$\cos \vartheta_{H_1AH_2} = -W_{2s(AH_1)}/W_{2p(AH_1)} \tag{75}$$

Table 19. Comparison of IMO hybridization s-characters with MINDO/3 and MNDO results for some strained hydrocarbons* involving a double bond

Molecule	Bond	IMO $s_A - s_B$	MINDO/3 $s_A - s_B$	MNDO $s_A - s_B$
1	C=C	36.8–36.8	40.3–40.3	35.6–35.6
	C—H	31.5	27.8	27.8
2	C(1)=C(2)	36.8–37.9	40.2–36.2	36.0–35.3
	C(2)=C(3)	31.1–23.7	31.2–30.7	30.5–24.5
	C(1)—H	31.5	27.5	30.7
	C(2)—H	25.4	27.2	24.6
3	C(1)=C(2)	37.3–41.7	39.6–39.6	35.2–38.5
	C(2)—C(3)	29.2–23.3	29.0–26.3	28.3–23.9
	C(3)—C(4)	22.3–22.7	26.7–26.4	22.1–22.1
	C(1)—H	31.3	27.8	30.9
	C(3)—H	27.2	22.2	26.5
	C(4)—H	27.3	22.3	26.3
4	C(1)=C(2)	37.3–44.2	38.6–45.8	34.9–44.5
	C(2)—C(3)	27.8–22.1	26.8–23.1	26.2–20.0
	C(3)—C(3)	20.8–20.8	23.7–23.7	19.4–19.4
	C(1)—H	31.3	28.7	31.2
	C(3)—H	28.6	26.7	29.5
5	C(1)=C(1)	38.2–28.2	36.4–36.4	36.1–36.1
	C(1)—C(2)	31.0–24.0	30.6–30.7	30.5–24.6
	C(2)—H	25.3	22.2	24.5
6	C(1)=C(1)	41.5–41.5	39.0–39.0	39.4–39.4
	C(1)—C(2)	29.2–23.2	29.0–26.3	27.9–21.8
	C(2)—C(3)	22.3–22.7	26.8–26.3	22.4–22.0
	C(2)—H	27.2	22.2	26.5
	C(3)—H	27.3	22.3	26.4
7	C(1)=C(2)	44.1–41.7	45.2–38.2	44.1–39.1
	C(2)—C(3)	29.2–23.7	29.7–26.2	29.1–22.4
	C(3)–C(4)	22.3–22.8	26.6–26.2	22.1–22.1
	C(1)—C(5)	27.9–22.5	26.8–23.1	27.0–20.0
	C(5)–C(5)	20.5–20.5	23.7–23.7	19.3–19.3
	C(3)—H	27.0	22.2	26.5
	C(4)—H	27.2	22.3	26.4
	C(5)—H	28.5	26.7	29.5
8	C(1)=C(1)	44.2–44.2	44.1–44.1	43.7–43.7
	C(1)—C(2)	27.9–22.1	27.6–23.1	27.3–19.9
	C(2)—C(2)	20.8–20.8	23.8–23.8	19.4–19.4
	C(2)—H	28.5	26.7	29.5

* Molecules denoted by *1–8* correspond to compounds presented in Table 4 keeping the same order

Taking into account that in the CNDO/2 method one abandons overlap integrals, W entities should be replaced by appropriate $P_{\mu\nu}$ bond order matrix elements. By assuming that normalization relation $P^2_{2s(CH)} + P^2_{2p(CH)} = 1$ holds one straightforwardly obtains Eq. (73). Employing the latter constraint Meyer [264] obtained a resonable HCH angle in cyclopropane and the s-characters $(s\%)_{CH} = 28.7\%$ and $(s\%)_{CC} = 21.3\%$ which can be favorably compared with the maximum overlap values of 28.6% and 21.4%, respectively. Empirical indices are in good agreement with theoretical estimates. Some discrepancies may be ascribed to the fact that J(C—H) is a more complicated function of the s-character than a simple proportionality (vide supra). Some calculations indicate that MNDO results are much closer to the IMO hybridization indices than the less sophisticated MINDO/3 method [151] (Table 19). Since the IMO method is based on the perfectly localized hybrids, a degree of agreement with semi-empirical methods (Tables 18 and 19) is remarkable indeed. It should be mentioned that the sum of s-characters in molecular orbital semiempirical methods exceeds 100% which is a slight imperfection in definition. The point is that the 2*s* orbital has a population which is higher than 1 because of its lower orbital energy. For example, in CH_4, the SCC-MO method gives 1.169 and 0.955 e for 2*s* and 2*p* AOs, respectively.

Finally, we shall test IMO hybridization against some ab initio results of Newton et al. [269, 270] obtained via the molecular orbital localization (ER) procedure.

Table 20. Comparison of maximum overlap hybridization parameters with the corresponding ab initio values and INDO natural hybridization indices

Molecule	Bond	IMO	INDO[a]	Ab initio[b]
CH_4	C—H	25.0	25.0	27.4
C_2H_6	C—H	25.4	24.2	26.9
	C—C	23.9	27.4	34.0
C_2H_4	C—H	31.5	29.7	35.1
	$(C—C)_\sigma$	36.9	40.8	36.5
C_2H_2	C—H	44.4	44.8	49.8
	$(C—C)_\sigma$	55.6	55.2	49.8
cyclopropane	C—H	28.6	28.1	32.9
	C—C	21.4	22.2	20.8
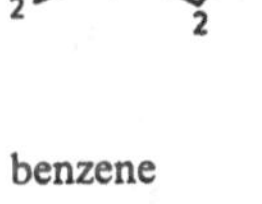	C_1—H	28.7	24.3	26.0[c]
	C_2—H	25.9	22.7	24.3
	C_3—H	27.4	23.2	25.6
	C_1—C_2	24.2–24.1	26.2–27.1	27.3–26.1
	C_2—C_2	24.1–24.1	27.5–27.5	27.5–27.5
	C_1 C_3	22.9–22.6	23.5–26.8	24.4–26.3
benzene	C—H	31.5	—	34.8
	C—C	34.3	—	36.6

[a] Natural hybridization [67]
[b] Ab initio LMO results [269, 270]
[c] Ref. [271]

Hybridization indices were extracted from LMOs by using Eq. (12). These hybrids are not orthogonal but the mutual overlap for geminal hybrids never exceeds 0.15. Hence they are comparable to the IMO local orbitals. A good accordance is in general obtained with IMO hybridization (Table 20), a striking exception being $(s\,\%)_{CC} = 34\,\%$ in ethane, where ab initio results deviates considerably from the standard values which are all close to 25%. We feel that ab initio calculations are not adequate here because a difference between $(C{-}C)_\sigma$ bonds in ethane and ethylene is far too small. Another failure is found in acetylene where ab initio calculation does not distinguish hybrids describing C—H and $(C{-}C)_\sigma$ bonds. Apart from these discrepancies a good qualitative agreement is found. The same holds for natural hybrids produced at the INDO level [67] (Table 20). It should be pointed out that ab initio hybridization is not very sensitive on the quality of the basis set [269].

To summarize, hybridization is insensitive to the method applied since highly diverse procedures yield qualitatively similar indices. An important corollary, which follows from the fact that hybridization is hidden but present even in SCF wave functions formally invariant to all orthogonal transformations, indicates that the Copernicus' reversal in thinking is necessary. One has to conclude that pure AO basis sets are acceptable in SCF theory of chemical bonding only because they are equivalent to implicitly present hybrid AOs which in turn serve the purpose well.

6 Discussion of the General Features of the Hybridization Model

In this section, hybridization of the carbon atom will be mainly considered. A general conclusion can be drawn that the hybridization index is an important internal parameter which determines a vast number of the properties of covalent bonds. It changes continuously as the local symmetry and local geometry are changed. Hence, noninteger hybridization parameters are a rule whereas the canonical hybridization ratios n = 3, 2 and 1 are exceptions. The latter occur extremely rarely. For example, sp^3 hybridization takes place in tetrahedral molecules and diamond, sp^2 in graphite, sp^1 in, e.g., the central bond of a cumulene $H_2C{=}C{=}C{=}C{=}C{=}CH_2$ etc. Needless to say, sp^2 and sp^1 hybridization schemes hold only in the $\sigma{-}\pi$ approximation. Particularly pronounced departures from the integer canonical values are found in strained rings. They favour hybrids with a very high p-orbital participation in order to reduce angular strain. Since this is not completely possible, bent bonds necessarily appear. Bond bending is a general feature but in angular undistorted fragments it is very small. On the other hand, it is highly pronounced in three-membered rings particularly if they are fused as, e.g., in the central bond of bicyclo(1.1.0)butane. Incidentally, envelope bonds in these molecules are twisted [117]. It is interesting to observe that hybrids forming the same bond tend to achieve the same bending angles [272]. This insures optimal overlapping as illustrated by the following brief analysis. Let's denote deviation angles of hybrid orbitals χ_A and χ_B by δ_A and δ_B. Maximum penetration of one electron density distribution into the other is obtained when $\delta_A = \delta_B = 0$. Since this condition is not always met, we shall suppose that $\delta_A \neq 0$ and $\delta_B \neq 0$.

Assuming that hybrids have a form as given by Eq. (42) their overlap integral reads:

$$\begin{aligned}\langle\chi_A|\chi_B\rangle &= a_A a_B\langle(2s_A)|(2s_B)\rangle + a_A(1-a_B^2)^{1/2}\cos\delta_B\langle(2s_A)|(2p_B)_\sigma\rangle + \\ &+ a_B(1-a_A^2)^{1/2}\cos\delta_A\langle(2s_B)|(2p_A)_\sigma\rangle + \\ &+ (1-a_A^2)^{1/2}(1-a_B^2)^{1/2}[\cos\delta_A\cos\delta_B\langle(2p_A)_\sigma|(2p_B)_\sigma\rangle + \\ &+ \sin\delta_A\sin\delta_B\langle(2p_A)_\pi|(2p_B)_\pi\rangle] \end{aligned} \quad (76)$$

where $(2p_\sigma)$ and $(2p_\pi)$ are p-orbitals oriented along and perpendicular to the bond, respectively. Suppose that the hybrids' composition is fixed, but they can freely orient relatively to the straight line passing through the nuclei A and B. Denoting $\delta_A = \delta_0 + \Delta$ and $\delta_B = \delta_0 - \Delta$ and using shorthand notation $\langle(2s_A)|(2s_B)\rangle = S_s$, $\langle(2s_A)|(2p_B)_\sigma\rangle = S_{sp}$, $\langle(2p_A)_\sigma|(2p_B)_\sigma\rangle = S_p^\sigma$, $\langle(2p_A)_\pi|(2p_B)_\pi\rangle = S_p^\pi$ and $\langle\chi_A|\chi_B\rangle = S_{AB}$ one obtains:

$$\begin{aligned} S_{AB} &= a_A a_B S_s + [a_A(1-a_B^2)^{1/2}\cos(\delta_0-\Delta) + \\ &+ a_B(1-a_A^2)^{1/2}\cos(\delta_0+\Delta)]\cdot S_{sp} + \\ &+ (1-a_A^2)^{1/2}(1-a_B^2)^{1/2}[\cos(\delta_0+\Delta)\cos(\delta_0-\Delta)S_p^\sigma + \\ &+ \sin(\delta_0+\Delta)\sin(\delta_0-\Delta)S_p^\pi] \end{aligned} \quad (77)$$

The maximum of the overlapping S_{AB} is obtained by requiring $(\partial S_{AB}|\delta\Delta) = 0$. Employing a Taylor expansion for the sine and cosine functions and terminating after the linear terms in the small quantity Δ, one gets

$$\Delta_m = \sin\delta_0[n_A^{-1/2} - n_B^{-1/2}]\,S_{sp}/2[\sin^2\delta_0 P_p^\sigma + \cos^2\delta_0 P_p^\pi] \quad (78)$$

where Δ_m is the optimal difference Δ and $n = (1-a^2)/a^2$. It appears that Δ_m vanishes if $\sin\delta_0 = 0$ implying that bending should be zero if optimal optimal overlapping is desired. Analogously, one can show that two co-axial hybrids overlap maximally if $n_A = n_B$ [250]. It is obvious, therefore, that two hybrids will assume hybridization parameters $n_A = n_B$ and deviation angles $\delta_A = \delta_B$, if they are left to adjust completely free without any (geometrical) constraints. Namely, $n_A = n_B$ implies that Δ_m vanishes (Eq. (78)). Therefore, the hybrids describing a particular bond and their deviation angles are as close as possible. This is not feasible if an sp^3 carbon is linked to an sp^2 or sp^1 carbon or in fused systems where small rings are annelated with larger cycles as, e.g., in benzocyclobutenes [273]. But a tendency to equalize hybridization parameters and/or bending angles is obvious. Two remarks are in place here. Bending equalization explains more or less even distribution of the angular strain over bonds. Secondly, orbital electronegativity of a hybrid is a function of its s-character. To be more specific, electronegativity ε of the carbon HAO is:

$$\varepsilon_c = 0.0324(s\,\%) + 1.67 \quad (79)$$

where electronegativities of the canonical hybridization states [274] were used in the process of derivation of Eq. (79). It follows that a part of the orbital electronegativity equalization is a consequence of the maximum overlap requirement. The rest is insured by the intrabond electron charge transfer.

Table 21. Centroids of "classical" hybrids and of LMOs describing C—H bonds in ethane, ethylene and acetylene

Property	AOs				
	$2s$	$2p$	sp^3	sp^2	sp^1
Centroid[a]	0	0	0.415	0.452	0.479
Size	1.677	1.827	1.610	1.560	1.501
Total size[b]	—	—	6.440	6.507	6.656
C—H LMOs			C_2H_6	C_2H_4	C_2H_2
Centroid	—	—	0.733	0.726	0.688
Size	—	—	1.582	1.537	1.473
Bond critical point[c]	—	—	0.659	0.661	0.680

* All data are calculated at the 4-31 G level of accuracy [275] if not otherwise stated
[a] Distance between the heavy nucleus and the center of gravity of the electronic charge (in Å)
[b] Total size of one ($2s$) and three ($2p$) AOs is 7.158. A total size of an sp^n center is given by a sum of sizes of (n + 1) sp^n hybrids plus a size of (3 — n) ($2p$) AOs
[c] 6-31 G** results of Wiberg [276]

Another point of great importance is the high degree of transferability of the hybrid orbitals between similar structural groupings. This feature allows one to understand the properties of very large molecular systems in terms of smaller fragments. The significance of this finding is impossible to overestimate because it enables reduction of a problem of immense complexity to several elementary problems of low dimension. If a set of characteristic hybridizations describing typical building blocks is available, approximate charge distribution of a complex molecule is possible without any calculation. Finally, hybridization is a general concept necessary for classification of atoms, atomic groupings, and chemical bonds in molecules.

The fact that hybrid AOs can serve as local wave functions of the zeroth order can be illustrated by the calculated centers of charge of sp^3, sp^2 and sp^1 HAOs which can be favourably compared with the corresponding bond critical points of C—H bonds in C_2H_6, C_2H_4 and C_2H_2 (Table 21). It means that HAOs are well prepared for the formation of chemical bonds and that polarization by mixing local AOs brings the electron density close to the bond critical points and centroids of localized molecular orbitals. In contrast, centroids of pure *s*, *p* etc. AOs coincide with the respective nucleus. A shift of the charge density toward the center of the chemical bond by hybridization may be further refined by introduction of semilocalized Coulson-Fischer [59] orbitals.

A comparison of the sizes of AOs against HAOs is of equal interest. The size of atomic or localized molecular orbitals is related to their second moments [275]. The calculated total size of one ($2s$) and three ($2p$) carbon AOs is 7.16 at the 4-31 G level (Table 21). This is appreciably higher than the sizes of sp^3 (6.44), sp^2 (6.51) and sp^1 (6.66) carbon centers, conclusively showing that hybrid AOs are more localized than free-atom AOs. Furthermore, the sizes of single HAOs are comparable to the spatial extentions of C—H LMOs (Table 21).

The intimate relation between hybridization and various molecular properties were thoroughly discussed in previous sections. As a final comment we shall mention that a long standing problem of the hindered rotation in ethane was finally resolved by concluding that it arises due to overlap (i.e. exchange) repulsion of vicinal hybrids [277–279]. Direction properties of moderately strong hydrogen bonds seem to be closely related with conventional lone pair(s) positions, at least in gas-phase dimers [280]. Relevance of hybridization in this respect was discussed by Lin [281]. Finally, hybridization has been found essential in understanding Van der Waals forces between helium metastable system He* ($2\,^1S$) and He* ($2\,^3S$) and rare gas atoms (Ne, Ar, Kr, Xe) [282]. It accounted for the main features of the potential energy curves derived from scattering measurements.

There are a number of empirical relationships between various observables as we shall see below. Some of them are intuitively expected as, e.g., relations between bond dissociation energies, bond distances and stretching frequencies. What it surprising, however, is the accuracy of these empirical correlations which is usually high. How could this be explained? Let us consider in some more detail McKean's relation between C—H bond distance and isolated (C—H) stretching frequencies [164]:

$$d(C{-}H) = 1.402 - 0.0001035\,\bar{\nu}(C{-}H) \quad (Å) \tag{80}$$

yielding an accuracy of ± 0.0005 Å! Since both entities are related to s-characters of hybrids calculated by the maximum overlap method one obtains by excluding theoretical parameter the expression:

$$d(C{-}H) = 1.412 - 0.000106\,\bar{\nu}(C{-}H) \quad (Å) \tag{81}$$

The degree of compatibility between Eqs. (80) and (81) is remarkable. Although such good quantitative agreement can not be expected to hold generally, a good qualitative interpretation of empirical relations interconnecting measured observables is found as a rule. We shall present some more examples illustrating the point. McKean [164] gives

$$D^{\circ}_{298}(C{-}H) = 0.0861\,\bar{\nu}(C{-}H) - 154.6 \text{ (kcal)} \tag{82}$$

which compares with the relation deduced by the hybridization model

$$D^{\circ}_{298}(C{-}H) = 0.0741\,\bar{\nu}(C{-}H) - 117.8 \text{ (kcal)} \tag{83}$$

Employing maximum overlap results one can derive approximate relations like the one describing a well known observation that shorter bonds are stronger

$$D^{\circ}_{298}(C{-}H) = 863.2 - 694.6\,d(C{-}H) \tag{84}$$

Hybridization is the underlying principle which can rationalize empirical relations between properties which are not a priori obvious. For example, there is a relationship between J(C—H) couplings and d(C—H) distances [177] or between J(C—H) and thermodynamic acidity of the corresponding C—H bonds. By using acidity data for hydrocarbons one obtains an empirical linear relation utilizing the least squares technique:

$$pk = -0.1874\,J(C{-}H) + 72.0 \tag{85}$$

whereas maximum overlap method yields:

$$pk = -0.2128\,J(C{-}H) + 75.9 \tag{86}$$

which is in satisfactory agreement with Eq. (85). Finally, let us mention that various relations between structural features (bond angles) and molecular properties (e.g. spin-spin coupling constants [283]) are rooted in hybridization phenomenon. One concludes that a large number of correlations of molecular properties with s- or p-characters of hybrids and mutual relations thereof rationalized by hybridization can not be accidental. Hence, hybridization has a touch of reality.

Hybridization is undoubtedly effective in hydrocarbons. A pertinent question arises, however, whether it is of the same importance in other atoms too? The answer is not so simple because several conditions should be satisfied in order to have an efficient hybridization. The first is provided by Maccoll's theorem [284] which states that *s*, *p*, *d* etc. AOs of the same center will mix only if their overlap integrals with the neighbouring atoms' AOs are comparable and appreciable in magnitude. A necessary requirement is that AOs placed at the common nucleus are of a similar size. This condition seems to be of general importance. Namely, 2*s* and 2*p* AOs will effectively mix forming hybrids if their local overlap $\langle (2s)_A | |(2p)_A| \rangle \geqq 0.63$ [38], where $|(2p)_A|$ denotes absolute values of the 2*p* function. Only in this case will hybrids be more localized than free 2*s* and 2*p* AOs implying that the latter should be confined within approximately the same volume. This argument is intuitively clear because 2*s* and 2*p* AOs need to have a sizeable interpenetration space where they are enhanced by positive interference and diminished by a negative interference. Therefore, one concludes that roughly equal spatial extension of *s* and *p* etc. AOs is a necessary prerequisite for hybridization to be operative.

Another important condition is that promotion energy gap $E_s - E_p$ is not highly pronounced, because promotion of electron(s) is too costly otherwise. Everything said for *s* and *p* AOs holds for orbitals of higher angular momentum too. Since a large difference in orbital energies is directly related to a difference in their spatial extension, because AOs of lower energy are more compact, it appears that both effects act parallel against hybridization. They are both responsible for a fact that hybridization is more important in first row atoms than in atoms belonging to higher rows.

(Non)-orthogonality of hybrids deserves special scrutiny. It is a consequence of unequal population of *s* and *p* AOs that hybrid orbitals are non-orthogonal. This feature was discussed in depth by Kutzelnigg [10] and Verwoerd [285]. It appears that orthogonality of hybrids is tolerable for the first row atoms but it fails badly for heavier atoms. According to Kutzelnigg's analysis, hybrids of the carbon atom approach the orthogonality conditions best. However, this conclusion should be taken cum grano salis because it is obvious that orthogonality breaks down at least in highly strained small rings.

A distinct advantage of hybrid AOs is the possibility of optimatization of their nonlinear parameters for each direction separately. Contraction or extension of atomic orbitals is an integral part of the bonding process in molecules leading to a sustantial stabilization. Magnusson advocates optimization of screening constants of AOs for each molecular orbital [286]. Hybrids enable a choice of best nonlinear parameters for each particular localized MO which seems to be a more natural approach. Additional radial distortion caused by charge reorganization upon formation of chemical bonds may be imposed by mixing in a small portion of the bonding partner HAO in the spirit of Coulson and Fischer [59].

Finally, one should strongly point out that symmetrically orthogonalized hybrid

AOs offer a convenient basis for a semiempirical development of the theory of chemical bonding. It is easy to see that many integrals will then become negligibly small. Let us denote two symmetrically orthogonal hybrids by $\bar{\chi}_A$ and $\bar{\chi}_B$. Then the following expression holds:

$$\bar{\chi}_A\bar{\chi}_B = \left[\chi_A\chi_B - \frac{1}{2} S_{AB}(\chi_A^2 + \chi_B^2)\right] \Big/ (1 - S_{AB}^2) \tag{87}$$

Since the Mulliken approximation

$$\chi_A\chi_B \simeq \frac{1}{2} S_{AB}(\chi_A^2 + \chi_B^2) \tag{88}$$

works very well in most cases, it follows that the expression in square brackets in Eq. (87) is close to zero meaning that a mixed density $\bar{\chi}_A\bar{\chi}_B$ can be neglected in the first approximation. We feel confident that the next generation of semiempirical methods will be based on suitably orthogonalized HAOs.

As a final comment let us focus attention on the maximum overlap criterion. Its performance is very good as evidenced by numerous IMO results which are in fine accordance with experiment. This is not quite unexpected because more elaborated methods such as EHT or IEHT employ overlapping in describing stabilization interactions. For example, resonance integrals are approximated by Wolfsberg and Helmholtz as [287]:

$$\beta_{\mu\nu} = \frac{1}{2} KS_{\mu\nu}(H_{\mu\mu} + H_{\nu\nu}) \tag{89a}$$

or by Cusachs [288]:

$$\beta_{\mu\nu} = \frac{1}{2}(K - |S_{\mu\nu}|)\, S_{\mu\nu}(H_{\mu\mu} + H_{\nu\nu}) \tag{89b}$$

The larger overlap integrals lead to lower resonance terms and increased bonding interactions. The same conclusion holds in the angular overlap method [289]. The maximum overlap criterion may be extended to involve a power series in the overlap integral [290] or could be modified to take into account overlap population density [291]. An interesting possibility not yet explored in practice is offered by the generalization of the maximum overlap molecular orbital (MOMO) method as suggested by Gilbert and Lykos [292].

7 Epilogue

We commence the plaidoyer by stating a general fallacy concerning hybridization as mentioned by Kutzelnigg [10]: "Hybridisierung stammt aus der Fühzeit der Theoretischen Chemie und ist bei vielen Chemikern nach wie vor sehr beliebt, obwohl er in der modernen Quantenchemie eigentlich keinen Platz hat und daher bei Theoretikern als problematisch, wenn nicht gar unseriös gilt." Indeed, many theoreticians believe that the model of hybridization is obsolete. It is considered as an ad hoc assumption and possesses at best some heuristic value. Since the Hartree-Fock wave function in the MO-LCAO approximation is invariant to transformations of atomic basis set — hybridization "does not exist". We have convincingly shown that a radical Copernicus reversal in thinking is necessary. Hybrid AOs appear without any preconceptions as a result of spin paired valence bond approach. They give shape to Fermi holes and can be recognized in electron isopycnic deformation maps. Hybridization is physically well founded because it represents a local distortion in density as a response to a descent in symmetry of atoms in molecular environments. It is a reaction to local perturbation exerted by the nearest neighbours. Since it can be correlated with a large number of molecular properties, it provides an important link between the quantum theory and chemical phenomenology. Perhaps more importantly, it gives a simple rationalization of various empirically established relationships between observables playing a role of the underlying principle. We conclude that hybridization has a grain of reality as much as the one-electron model can possess it. Hybridization is a phenomenon, but since it can not be measured directly it is not an observable. In order to distinguish hybridization from some other artificial theoretical contructs, we suggested that it should be termed a pseudo-observable. Interestingly, hybridization is excommunicated in molecular orbital theories at the SCF level as a heresy and yet it appears in the corresponding first order density matrices or in equivalent LMOs. This is fortunate because LCAO approximation formed by pure AOs and Slater determinants would not be acceptable in chemical theory otherwise.

Hybridization is one of the most fundamental facets of chemical bonding just like electronegativity equalization accompanied by intramolecular charge transfer. It is an essential ingredient of the modified atom in molecule (MAM) model [31, 32] and a precious jewel in the treasury of the interpretative quantum chemistry.

Acknowledgement:

A part of this work has been done at the Organisch-chemisches Institut der Universität Heidelberg and I would like to thank Alexander von Humboldt-Stiftung for financial support and Professor R. Gleiter for hospitality.

8 References

1. Pauling L (1960) The nature of the chemical bond and the structure of molecules and crystals, 3rd eds, Cornell University Press, Ithaca, N.Y.
2. Boggs JE (1989) In: Maksić ZB (ed) Theoretical models of chemical bonding, vol 1, Springer, Berlin Heidelberg New York
3. Schäfer L, Siam K, Ewbank JD, Caminati W, Fantoni AC (1987) In: Maksić ZB (ed) Modelling of structure and properties of molecules, Ellis Horwood, Chichester, p 79
4. Hargittai I, Hargittai M (1987) Symmetry through the eyes of a chemist, VCH, N.Y.
5. Maksić ZB (1986) Comp. & Maths. with Appls. 12B: 697
6. Pauling L (1928) Proc. Natl. Acad. Sci. USA 14: 359; (1931) J. Am. Chem. Soc. 53: 1367
7. Bingel WA, Lüttke W (1981) Angew. Chem. 93: 944
8. Herman ZS (1983) Int. J. Quantum Chem. 23: 921

9. Maksić ZB (1983) Pure Appl. Chem. 55: 307
10. Kutzelnigg W (1984) Angew. Chem. 96: 262
11. Cao Y (1988) J. Mol. Structure (Theochem) 169: 1
12. Verkade JG (1986) A pictorial approach to molecular bonding, Springer, Berlin Heidelberg New York
13. Maksić ZB, Orville-Thomas WJ (eds) (1988) Six decades of the hybridization model, J. Mol. Structure (Theochem) 169: 1–568
14. Heitler W, London F (1927) Z. Phys. 44: 455
15. Gentile G (1930) Z. Phys. 63: 795
16. Slater JC (1927) Phys. Rev. 32: 349
17. Delbrück M (1930) Ann. Phys. 5: 36
18. Kimball G (1940) J. Chem. Phys. 8: 188
19. Racah G (1943) J. Chem. Phys. 11: 214
20. Kuhn H (1948) J. Chem. Phys. 16: 727
21. Duffey GH (1950) J. Chem. Phys. 18: 943
22. Macek JH, Duffey GH (1961) J. Chem. Phys. 34: 288
23. Canon JR, Duffey GH (1961) J. Chem. Phys. 35: 1657
24. Shirmazan MG, Dyatkina ME (1953) Zh. Fiz. Khim. 27: 491
25. Eisenstein JC (1956) J. Chem. Phys. 25: 142
26. McClure V (1970) Ph.D. Dissertation, University of California, San Diego
27. Gimarc BM (1979) Molecular structure and bonding. The qualitative molecular orbital approach, Academic, New York
28. Woodward RB, Hoffmann R (1970) The conservation of orbital symmetry, Academic, New York
29. Surjan PR (1984) Croat. Chem. Acta 57: 811
30. Surjan PR (1990) in chapter 6 of this book
31. Maksić ZB, Eckert-Maksić M, Rupnik K (1984) Croat. Chem. Acta 57: 1295
32. Maksić ZB (1988) J. Mol. Structure (Theorem) 170: 39
33. Lennard-Jones JE (1949) Proc. Roy. Soc. (London) A198: 14
34. Hall GG, Lennard-Jones JE (1950) Proc. Roy. Soc. (London) A202: 155
35. Lennard-Jones JE, Pople JA (1951) Proc. Roy. Soc. (London) A210: 190
36. Edmiston C, Ruedenberg K (1963) Rev. Mod. Phys. 35: 457
37. Foster JM, Boys SF (1960) Rev. Mod. Phys. 32: 300; Boys SF (1966) In: Löwdin PO (ed) Quantum theory of atoms, molecules and solid state, Academic, New York, p 253
38. Edmiston C, Ruedenberg K (1966) In: Löwdin PO (ed) Quantum theory of atoms, molecules and solid state, Academic, New York, p 263
39. Aufderheide KH (1980) J. Chem. Phys. 73: 1777
40. Aufderheide KH, Chung-Phillips A (1980) J. Chem. Phys. 73: 1789; (1982) J. Chem. Phys. 76: 1885
41. Aufderheide KH (1984) Croat. Chem. Acta 57: 811
42. Chung-Phillips A (1988) J. Mol. Structure (Theochem) 169: 211
43. Wiberg KB (1990) In: Maksić ZB (ed) Theoretical Models of chemical bonding, vol 1, Springer, Berlin Heidelberg New York
44 (a). Dickinson BN (1933) J. Chem. Phys. 1: 317; (b). Rosen N (1931) Phys. Rev. 38: 2099
45. Hoyland JR (1968) J. Am. Chem. Soc. 90: 2227; (1969) J. Chem. Phys. 50: 473
46. Petke JD, Whitten JL (1969) J. Chem. Phys. 51: 3166
47. Eilers JE, Whitman DR (1973) J. Am. Chem. Soc. 95: 2067; Duke BJ, O'Leary B (1983) J. Chem. Phys. 79: 3424; (1983) Theoret. Chim. Acta 62: 223
48. Klessinger M, McWeeny R (1965) J. Chem. Phys. 42: 3343; Klessinger M (1965) J. Chem. Phys. S 43: 117
49. Klessinger M (1967) J. Chem. Phys. 46: 3261; (1968) Chem. Phys. Lett. 2: 562; (1968) Symp. Farad. Soc. No. 2: 73; (1969) Chem. Phys. Lett. 4. 144
50. Raimondi M, Campion W, Karplus M (1977) Mol. Phys. 34: 1483
51. Cooper DL, Gerratt J, Raimondi M (1987) Adv. Chem. Phys. 59: 319
52. Gerratt J (1971) Adv. At. Mol. Phys. 7: 141
53. Penotti F, Gerratt J, Copper DL, Raimondi M (1988) J. Mol. Structure (Theochem) 169: 421 and the references cited therein
54. Goddard III WA, Dunning Jr TH, Hunt WJ, Hay PJ (1973) Acc. Chem. Res. 6: 368; Bobrowicz

FW, Goddard III WA (1977) In: Schaefer III HF (ed) Methods of electronic structure theory, Plenum, New York
55. Goddard III WA, Harding LB (1978) Ann. Rev. Phys. Chem. 29: 363; Goddard III WA (1985) Science 227: 917
56. Messmer RP, Schultz PA, Tatar RC, Freund HJ (1986) Chem. Phys. Lett. 126: 176; Schultz PA, Messmer RP (1987) Phys. Rev. Lett. 58: 2416
57. Messmer RP, Schultz PA (1986) Phys. Rev. Lett. 57: 2653
58. McWeeny R, Jorge FE (1988) J. Mol. Structure (Theochem) 169: 459
59. Coulson CA, Fischer I (1949) Philos. Mag. 40: 306
60. Mueller CR, Eyring H (1951) J. Chem. Phys. 19: 1495
61. Messmer RP (1988) J. Mol. Structure (Theochem) 169: 137
62. Kirtman B, Palke WE, Chipman DM (1980) Isr. J. Chem. 19: 82
63. Cooper DL, Gerratt J, Raimondi M (1986) Nature 323: 699
64. Cooper DL, Wright SC, Gerratt J, Raimondi M (1989) J. Chem. Soc. Perkin Trans II: 255
65. McWeeny R, Del Re G (1968) Theoret. Chim. Acta 10: 13
66. Löwdin PO (1950) J. Chem. Phys. 18: 365; Carlson BC, Keller JM (1957) 105: 102
67. Foster JP, Weinhold F (1980) J. Am. Chem. Soc. 102: 7211
68. Trindle C, Sinanoglu O (1969) J. Am. Chem. Soc. 91: 853
69. Maksić ZB, Randić M (1973) J. Am. Chem. Soc. 95: 6522
70. Wiberg KB (1968) Tetrahedron 24: 1083
71. Sandorfy C (1955) Can. J. Chem. 33: 1337
72. Fukui K, Kato H, Yonezawa T (1960) Bull. Chem. Soc. (Japan) 33: 1197
73. Pople JA, Santry DP (1964) Mol. Phys. 7: 269
74. Cook DB, Hollis PC, McWeeny R (1967) Mol. Phys. 13: 553
75. Cook DB, McWeeny R (1968) Chem. Phys. Lett. 1: 588
76. Cook DB, Palmieri P (1969) Mol. Phys. 17: 271; Cook DB (1975) Theoret. Chim. Acta 40: 297
77. Cook DB (1978) Structures and approximations for electrons in molecules, Ellis Horwood, Chichester
78. Diner S, Malrieu JP, Claverie P, Jordan F (1968) Chem. Phys. Lett. 2: 319
79. Diner S, Malrieu JP, Claverie P (1969) Theoret. Chim. Acta 13: 1
80. Malrieu JP, Claverie P, Diner S (1969) Theoret. Chim. Acta 13: 18
81. Diner S, Malrieu JP, Jordan F, Gilbert M (1969) Theoret. Chim. Acta 15: 100
82. Jordan F, Gilbert M, Malrieu JP, Pincelli U (1969) Theoret. Chim. Acta 15: 211
83. Malrieu JP (1977) In: Segal GA (ed) Modern theoretical chemistry, vol 7, Plenum, New York
84. Sauer J (1989) Chem. Rev. 89: 199 and the references cited therein.
85. Barone V, Del Re G (1983) J. Mol. Structure (Theochem) 94: 173
86. Bauschlicher Jr CW, Bagus PS, Nelin CJ (1983) Chem. Phys. Lett. 98: 439
87. Levin AA (1977) Solid state quantum chemistry, McGraw-Hill, New York
88. Julg A (1978) Lecture Notes in Chemistry, No. 9, Springer, Berlin Heidelberg New York
89. Harrison WA (1980) Electronic structure and the properties of solids, Freeman, San Francisco
90. Murrell J (1960) J. Chem. Phys. 32: 767
91. Golebiewski A (1961) Trans. Farad. Soc. 57: 1849
92. Volkov VM, Dyatkina ME (1963) Zh. Strukt. Khim. 4: 610; Levin AA (1963) Zh. Strukt. Khim. 4: 739
93. Hubač I, Laurinc V, Kvasnička V (1972) Chem. Phys. Lett. 13: 357; Boča R, Pelikan P, Valko L, Miertuš S (1975) Chem. Phys. 11: 229
94. Miertuš S, Boča R, Pelikan P, Valko L (1976) Chem. Phys. 11: 237; Boča R, Pelikan P, Valko L (1978) J. Mol. Struct. 50: 161
95. Lykos PG, Schmeising HN (1961) J. Chem. Phys. 35: 288
96. Bartlett RJ, Öhrn Y (1971) Theoret. Chim. Acta 21: 215
97. Whitehead MA, Zeiss GD (1975) In: Saunders VR, Brown J (eds) Quantum chemistry. The state of the art, Science Research Council, Oxford, p 181
98. Del Re G (1963) Theoret. Chim. Acta 1: 188
99. Veillard A, Del Re G (1964) Theoret. Chim. Acta 2: 313
100. Del Re G, Esposito U, Carpentieri M (1966) Theoret. Chim. Acta 4: 36
101. Löwdin PO (1956) Adv. Phys. 5: 1
102. l'Haya Y, Morikawa T (1980) Adv. Quant. Chem. 12: 43

103. Lindenberg J, Öhrn Y (1973) Propagators in quantum chemistry, Academic Press, London
104. Spanget-Larsen J (1973) J. Electron Spectrosc. Relat. Phenom. 2: 33, 3: 369; Byberg J, Spanget-Larsen J (1973) Chem. Phys. Lett. 23: 247; Pedersen JA, Spanget-Larsen J. (1975) Chem. Phys. Lett. 35: 41; Spanget-Larsen J, Pedersen JA (1975) J. Magn. Res. 18: 383
105. Pauling L (1975) Proc. Natl. Acad. Sci. USA 72: 4200; (1976) 73: 274
106. Pauling L, Herman ZS, Kamb BJ (1982) Proc. Natl. Acad. Sci. USA 79: 1361
107. Pauling L (1978) Proc. Natl. Acad. Sci. USA 75: 569; (1984) 81: 1918
108. Pauling L (1978) Acta Crystallogr. B 34: 746
109. Pauling L (1979) In: Ovchinnikov YA, Kolosov MN (eds) Frontiers in bioorganic chemistry and molecular biology, Elsevier, Amsterdam
110. Pauling L, Herman ZS (1984) J. Chem. Ed. 61: 582
111. Kovačević K, Maksić ZB (1974) J. Org. Chem. 39: 539
112. Vujisić Lj, Maksić ZB (1976) J. Mol. Struct. 33: 49
113. Maksić ZB, Rubčić A (1976) J. Am. Chem. Soc. 99: 4233
114. Testa B (1979) Principles of organic stereochemistry, Dekker, New York
115. Maksić ZB, Kovačević K, Moguš A (1980) Theoret. Chim. Acta 55: 127
116. Eckert-Maksić M, Kovačević K, Maksić ZB (1979) J. Organomet. Chem. 168: 295
117. Maksić ZB, Klasinc L, Randić M (1966) Theoret. Chim. Acta 4: 273
118. Vujisić Lj, Maksić ZB (1971) J. Mol. Struct. 7: 431
119. Maksić ZB, Kovačević K, Vampola M (1981) Z. Naturforsch. 36a: 1196
120. Eckert-Maksić M, Maksić ZB (1974) J. Mol. Struct. 22: 445
121. Mårtensson O, Öhrn Y (1967) Theoret. Chim. Acta 9: 133
122. Maksić ZB, Vujisić Lj (1969) Theoret. Chim. Acta 14: 396
123. Clementi E (1965) Tables of atomic functions: a supplement to IBM J. Res. Develop. 9: 2
124. Klasinc L, Maksić ZB, Randić M (1966) J. Chem. Soc. 755
125. Almeningen A, Bastiansen O, Skancke PN (1961) Acta Chem. Scand. 15: 711
126. Hartman A, Hirshfeld FL (1966) Acta Cryst. 20: 80
127. Förster Th (1939) Z. Phys. Chem. B 43: 58
128. Coulson CA, Moffitt WE (1949) Phil. Mag. 40: 1
129. Randić M, Maksić ZB (1965) Theoret. Chim. Acta 3: 59
130. Eisenstein M, Hirshfeld FL (1979) Chem. Phys. 38: 1
131. Wiberg KB, Murcko MA (1988) J. Mol. Structure (Theochem) 169: 355
132. Runtz G, Bader RFW, Messer RR (1977) Can. J. Chem. 55: 3040
133. Gillespie RJ (1972) Molecular Geometry, Van Nostrand, Princeton, N.Y.
134. Walsh AD (1947) Disc. Farad. Soc. 2: 18; Bent HA (1961) Chem. Rev. 61: 275
135. Maksić ZB, Eckert-Maksić M (1970) Croat. Chem. Acta 42: 433
136. Eckert-Maksić M, Maksić ZB (1982) J. Mol. Structure (Theochem) 86: 325
137. Cremer D, Kraka E (1985) J. Am. Chem. Soc. 107: 3800, 3811
138. Cremer D (1987) In: Maksić ZB (ed) Modelling of structure and properties of molecules, Ellis Horwood, Chichester, p. 125
139. Bastiansen O, Fritsch FN, Hedberg K (1964) Acta Cryst. A 17: 538
140. Durmaz S, Kollmar H (1980) J. Am. Chem. Soc. 102: 6942
141. Prangé T, Pascard C, de Meijere A, Behrens U, Barnier JP, Conia JM (1980) Nouv. J. Chim. 4: 321
142. Almeningen A, Bastiansen O, Cyvin BN, Cyvin S, Fernholt L, Romming C (1984) Acta Chem. Scand. A 38: 31
143. Irngartinger H, Goldmann A, John R, Nixdorf M, Rodenwald H, Maier G, Malsch KD, Emrich R (1984) X-ray data on tetra-*tert*.-butyltetrahedrane. Angew. Chem. Int. Ed. Engl. 23: 993
144. Schulman JM, Venanzi TJ (1974) J. Am. Chem. Soc. 96: 4739
145. Almeningen A, Jonvik T, Martin HD, Urbanek Th (1985) J. Mol. Structure (Theochem) 128: 239
146. Fleischer EB (1964) J. Am. Chem. Soc. 86: 3889
147. Almlof J, Jonvik T (1982) Chem. Phys. Lett. 92: 267
148. Traetteberg M, Paulen G, Cyvin SJ, Panchenko YuN, Mochalov VI (1984) J. Mol. Structure (Theochem) 116: 141
149. Kavane-Saebo K, Saebo S, Boggs JE (1984) J. Mol. Structure (Theochem) 106: 259
150. Rasmussen K, Tossi C (1985) J. Mol. Structure (Theochem) 121: 233

151. Eckert-Maksić M, Maksić ZB, Skancke A, Skancke PN (1987) J. Phys. Chem. 91: 2786
152. Eckert-Maksić M, Maksić ZB, Skancke A, Skancke PN (1987) In: Maksić ZB (ed) Modelling of structure and properties of molecules. Ellis Horwood, Chichester
153. Deakyne CA, Allen LC, Laurie VW (1977) J. Am. Chem. Soc. 99: 1343
154. Traetteberg M, Simon A, de Meijere A (1984) J. Mol. Structure (Theochem) 118: 333
155. Schenk H (1984) Presented at the XIII Congress of the International Union of Crystallography, Hamburg
156. Eckert-Maksić M, Maksić ZB, Skancke A, Skancke PN (1988) J. Mol. Structure (Theochem) 164: 25
157. Maksić ZB, Eckert-Maksić M, Skancke PN, Skancke A (1988) J. Mol. Structure (Theochem) 169: 447
158. Lee TJ, Schaefer III, HF (1985) J. Chem. Phys. 83: 1784
159. Coulson CA (1948) Victor Henry Memorial Volume, Desoer, Liege
160. Dewar MJS, Schmeising HN (1959) Tetrahedron 5: 166
161. Allen FH (1987) In: Maksić ZB (ed) Modelling of structure and properties of molecules, Ellis Horwood, Chichester, p. 51 and the references cited therein
162. Mulliken RS (1950) J. Am. Chem. Soc. 72: 4493; (1952) J. Chem. Phys. 56: 295; (1955) J. Chem. Phys. 23: 1841
163. Hay JM (1970) J. Chem. Soc. B 45
164. McKean DC, Duncan JL, Batt L (1973) Spectrochim. Acta A29: 1037
165. Maksić ZB, Meić Z, Kovačević K (1973) Z. Naturforsch. 28a: 797
166. Maksić ZB, Meić Z, Randić M (1972) J. Mol. Structure 12: 482
167. Skancke A, Skancke PN, Eckert-Maksić M, Maksić ZB (1987) J. Mol. Structure (Theochem) 150: 259
168. Pulay P, Meyer W (1974) Mol. Phys. 27: 473
169. Duncan JL (1964) Spectrochim. Acta 20A: 1197
170. Kamienska-Trela K (1980) Spectrochim. Acta 36A: 239
171. Duncan JL, McKean DC, Nivellini GD (1976) J. Mol. Structure 32: 255
172. Phibbs MK, Sipos PA (1972) Can. J. Spectr. 17: 93
173. Maksić ZB, Eckert-Maksić M (1969) Tetrahedron 25: 5113
174. Kosower EM (1968) Physical organic chemistry, Wiley, New York
175. Lee WS, Schulman JM (1979) J. Am. Chem. Soc. 101: 3182; (1980) 202: 5184
176. Laaksonen A, Kowalewski J, Saunders VR (1983) Chem. Phys. 80: 221; Overill RE, Saunders VR (1984) Chem. Phys. Lett. 106: 197
177. Muller N, Pritchard DE (1959) J. Chem. Phys. 31: 768, 1471
178. Frei K, Bernstein HJ (1963) J. Chem. Phys. 38: 1216
179. Goldstein JH, Hobgood RT Jr (1964) J. Chem. Phys. 40: 3592
180. Maksić ZB (1971) Int. J. Quant. Chem. S 5: 301; Maksić ZB, Eckert-Maksić M, Randić M (1971) Theoret. Chim. Acta 22: 70
181. Gil VMS, Geraldes CFGC (1974) In: Axenrod T, Webb GA (eds) Nuclear magnetic resonance of nuclei other than protons, Wiley-Interscience, New York
182. Van Alsenoy C, Figeys HP, Geerlings P (1980) Theoret. Chim. Acta 55: 87
183. Schulman JM, Newton MD (1974) J. Am. Chem. Soc. 96: 6295
184. Jansen HB, Meeuvis A, Pyykkö P (1979) Chem. Phys. 38: 173
185. Kovačević K, Krmpotić K, Maksić ZB (1977) Inorg. Chem. 16: 1421
186. Kovačević K, Maksić ZB (1973) J. Mol. Structure 17: 203
187. Klessinger M, Bolte P (1988) J. Mol. Structure (Theochem) 169: 119
188. Klessinger M, Barfield M (1987) In: Maksić ZB (ed) Modelling of structure and properties of molecules, Ellis Horwood, Chichester, p 269
189. Scrocco E, Tomasi J (1973) Top. Curr. Chem. 42: 95; (1978) Adv. Quant. Chem. 11: 115
190. Bentley J (1981) In: Politzer P, Truhlar DG (eds) Chemical applications of atomic and molecular electrostatic potentials, Plenum, New York
191. Stone AJ (1981) Chem. Phys. Lett. 83: 233
192. Stone AJ (1990) In: Maksić ZB (ed) Theoretical models of chemical bonding, vol 3, Springer, Berlin Heidelberg New York, and the references cited therein
193. Petro AJ (1958) J. Am. Chem. Soc. 80: 4230
194. Coulson CA (1942) Trans. Farad. Soc. 38: 433

195. Pritchard RH, Kern CW (1968) J. Am. Chem. Soc. 91: 1631
196. Wiberg KB, Wendoloski JJ (1978) J. Am. Chem. Soc. 100: 723
197. Wiberg KB (1979) J. Am. Chem. Soc. 101: 1718
198. Wiberg KB, Wendoloski JJ (1981) J. Comp. Chem. 2: 53
199. Jalsowsky G, Orville-Thomas WJ (1971) J. Chem. Soc. Farad. Trans. 2, 67: 1894; Galabov B, Orville-Thomas WJ (1972) J. Chem. Soc. Farad. Trans. 2, 68: 1778
200. Epstein IR (1973) Acc. Chem. Res. 6: 145 and the references cited therein
201. Coulson CA (1941) Proc. Cambridge Philos. Soc. 37: 55; (1973) Mol. Phys. 26: 507
202. Rozendal A, Baerends EJ (1984) Chem. Phys. 87: 263
203. Maksić ZB, Kovačević K (1980) Chem. Phys. Lett. 73: 592
204. Sutcliffe R, Lindsay DA, Griller D, Walton JC, Ingold KU (1982) J. Am. Chem. Soc. 104: 4674
205. Geoffroy M, Lucken EAC (1981) Mol. Phys. 43: 887
206. Lucken EAC (1969) Nuclear quadrupole coupling constants, Academic, London
207. Vega S (1977) Isr. J. Chem. 16: 213
208. Jordan PCH, Longuet-Higgins HC (1962) Mol. Phys. 5: 121
209. Offenhartz P, O'Donnell (1973) Chem. Phys. Lett. 21: 388
210. Mills IM (1963) Spectrochim. Acta 19: 1585
211. Linnett JW, Wheatly PJ (1949) Trans. Farad. Soc. 45: 33; Cohan NV, Coulson CA (1956) Trans. Farad. Soc. 52: 1163; Coulson CA, Stephen MJ (1957) Trans. Farad. Soc. 53: 272
212. Chipman DM, Palke WE, Kirtman B (1980) J. Am. Chem. Soc. 102: 3377; Palke WE, Kirtman B (1983) J. Mol. Structure (Theochem) 104: 207
213. Coppens P, Stevens ED (1977) Adv. Quant. Chem. 10: 1; Coppens P (1977) Angew. Chem. 89: 33
214. Smith VH Jr, Price PF, Absar I (1977) Isr. J. Chem. 16: 187
215. Hirshfeld FL (1977) Isr. J. Chem. 16: 198
216. Hirshfeld FL (1977) Theoret. Chim. Acta 44: 129
217. Kunze KL, Hall MB (1986) J. Am. Chem. Soc. 108: 5122
218. Hall MB (1982) In: Coppens P, Hall MB (eds) Electron distribution and the chemical bond, Plenum, New York, p 205; Irngartinger H (1982) *ibid.*, p 361
219. Fritchie CJ Jr (1966) Acta Cryst. 20: 80
220. Newton MD (1977) In: Schaefer HF III (ed) Applications of electronic structure theory, Plenum Press, New York, p 223
221. Vujisić Lj, Vučković DLj, Maksić ZB (1984) J. Mol. Structure (Theochem) 106: 323
222. Bader RFW, Beddall PM (1972) J. Chem. Phys. 56: 3320
223. Bader RFW, Essen H (1984) J. Chem. Phys. 80: 1943
224. Bader RFW, Nguyen-Dang TT (1981) Adv. Quant. Chem. 14: 63
225. Cremer D, Kraka E (1984) Croat. Chem. Acta 57: 1259
226. Bader RFW, MacDougall PJ, Lau CDH (1984) J. Am. Chem. Soc. 106: 1594
227. Cremer D, Gauss J, Cremer E (1988) J. Mol. Structure (Theochem) 169: 531
228. Kraka E, Cremer D (1990) In: Maksić ZB (ed) Theoretical models of chemical bonding, vol 2, Springer, Berlin Heidelberg New York
229. Morse P, Feshbach H (1953) Methods of theoretical physics, Part 1, McGraw-Hill, New York
230. Bader RFW, Gillespie RJ, MacDougall PJ (1988) J. Am. Chem. Soc. 110: 7329
231. Wigner EP, Seitz F (1933) Phys. Rev. 43: 804; (1934) Phys. Rev. 46: 509
232. Slater JC (1951) Phys. Rev. 81: 385
233. Maslen VW (1956) Proc. Phys. Soc. A69: 734
234. Boyd RJ, Coulson CA (1974) J. Phys. B7: 1805
235. Cooper IL, Pounder CNM (1980) Int. J. Quant. Chem. 17: 759
236. Bader RFW, Stephens ME (1975) J. Am. Chem. Soc., 97: 7391; (1974) Chem. Phys. Letters 26: 445
237. Luken WL, Beratan DN (1982) Theoret. Chim. Acta 61: 265; Luken WL (1984) Croat. Chem. Acta 57: 1283
238. Luken WL (1990) In: Maksić ZB (ed) Theoretical models of chemical bonding, vol 2, Springer, Berlin Heidelberg New York
239. Luken WL, Culberson JC (1984) Theoret. Chim. Acta 66: 279
240. Skinner HA, Pilcher G (1963) Quart. Rev. 17: 264

241. Kovačević K, Eckert-Maksić M, Maksić ZB (1974) Croat. Chem. Acta 46: 249
242. Wiberg KB (1984) J. Comp. Chem. 5: 197
243. Dewar MJS, Pettit R (1952) J. Am. Chem. Soc. 74: 3357
244. Fischer H, Kollmar H (1970) Theoret. Chim. Acta 16: 163
245. Eckert-Maksić M, Bischof P, Maksić ZB (1985) Croat. Chem. Acta 58: 407; (1986) J. Mol. Structure (Theochem) 139: 179 and the references cited therein
246. Nelander B, Sunner S (1966) J. Chem. Phys. 44: 2476
247. Engler EM, Andose JD, Schleyer PvR (1973) J. Am. Chem. Soc. 95: 8005
248. Van Vechten D, Liebman JF (1981) Isr. J. Chem. 21: 105
249. George P, Trachtman M, Bock CW, Brett AM (1976) Tetrahedron 32: 317
250. Eckert-Maksić M, Maksić ZB (1982) J. Mol. Structure (Theochem) 86: 325; (1983) 91: 295
251. Maksić ZB, Kovačević K, Eckert-Maksić M (1975) Tetrahedron Lett. 101
252. Dewar MJS (1984) J. Am. Chem. Soc. 106: 669
253. Cremer D, Kraka E (1985) J. Am. Chem. Soc. 107: 3800, 3811
254. Cremer D, Gauss J (1986) J. Am. Chem. Soc. 108: 7467
255. Wiberg KB, Bader RFW, Lau CDH (1987) J. Am. Chem. Soc. 109: 1002
256. Ramsey NF (1961) Am. Sci. 49: 509
257. Malli G, Froese C (1967) Int. J. Quant. Chem. 15: 95
258 Gleiter R (1979) Top. Curr. Chem. 86: 197 and the references cited therein
259. Honegger E, Heilbronner E, Schmelzer A (1982) Nouv. J. Chim. 6: 519; Honegger E, Heilbronner E, Schmelzer A, Jian-Qi W (1982) Isr. J. Chem. 22: 3
260. Walsh AD (1947) Nature 159: 167; (1949) Trans. Farad. Soc. 45: 179
261. Eckert-Maksić M, Maksić ZB, Gleiter R (1984) Theoret. Chim. Acta 66: 193
262. Newton MD, Schulman JM (1972) J. Am. Chem. Soc. 94: 767
263. Cotton FA (1971) Chemical applications of group theory, 2nd edn, Wiley-Interscience, New York
264. Meyer AY (1971) Theoret. Chim. Acta 22: 271; (1977) Pasternak R, Meyer AY (1971) Theoret. Chim. Acta 43: 287
265. Van der Voorn PC, Drago RS (1966) J. Am. Chem. Soc. 88: 3255; White WO, Drago RS (1970) J. Chem. Phys. 52: 4717
266. Graovac A, Maksić ZB, Rupnik K, Veseli A (1977) Croat. Chem. Acta 49: 695 and the references therein
267. Maksić ZB, Rupnik K (1986) J. Mol. Structure 141: 309 and the references cited therein
268. Pople JA, Santry DP, Segal GA (1965) J. Chem. Phys. 43: S 129; Pople JA, Segal GA (1966) J. Chem. Phys. 44: 3289
269. Newton MD, Switkes E, Lipscomb WN (1970) J. Chem. Phys. 53: 2645
270. Newton MD, Switkes E (1971) J. Chem. Phys. 54: 3179
271. Lehn JM, Wipff G (1974) Theoret. Chim. Acta 33: 43
272. Maksić ZB, Kovačević K, Moguš A (1981) J. Mol. Structure (Theochem) 85: 9
273. Eckert-Maksić M, Hodoščck M, Kovaček K, Mitić D, Maksić ŻB, Poljanec K (1990) J. Mol. Structure in print
274. McWeeny R (1979) Coulson's valence, 3rd edn. Oxford University Press, Oxford, and the references given therein
275. Leroy G, Peeters D (1988) J. Mol. Structure (Theochem) 169: 257 and the references cited therein
276. Wiberg KB (1989) In: Maksić ZB (ed) Theoretical models of chemical bonding, vol 1, Chapt. 1, Springer, Berlin Heidelberg New York
277. Sovers OJ, Kern CW, Pitzer RM, Karplus M (1968) J. Chem. Phys. 49: 2592
278. Pitzer RM (1983) Acc. Chem. Res. 16: 207
279. Magnasco V, Musso GF (1987) In: Maksić ZB (ed.) Modelling of structure and properties of molecules, Ellis Horwood, Chichester
280. Millen DJ (1982) Croat. Chem. Acta 55: 133
281. Lin SH (1970) In: Eyring H, Henderson D, Jost W (eds) Physical chemistry, vol 5, Academic, New York
282. Siska PE (1979) J. Chem. Phys. 71: 8942
283. Foote, CS (1963) Tetrahedron Lett.: 579; Mislow, K. (1964) Tetahedron Lett.: 1415
284. Maccoll A (1950) Trans. Farad. Soc. 46: 369
285. Verwoerd WS (1979) Chem. Phys. 44: 151

286. Magnusson E (1986) Chem. Phys. Lett. 131: 224; (1988) Aust. J. Chem. 41: 827 and the references cited therein
287. Wolfsberg M, Helmholtz L (1952) J. Chem. Phys. 20: 837
288. Cusachs LC (1965) J. Chem. Phys. 43: 157
289. Schäffer CE, Jørgensen CK (1965) Mol. Phys. 9: 401; Schäffer CE (1970) Pure Appl. Chem. 24: 361
290. Burdett JK (1979) Inorg. Chem. 18: 1024
291. Manne R (1968) Int. J. Quant. Chem. 2: 69
292. Gilbert TL, Lykos PG (1961) J. Chem. Phys. 34: 2199
293. Maksić ZB (1989) In: Maruani J (ed) Molecules in physics, chemistry and biology, vol 3, Kluwer Academic, Dordrecht

The Two-Electron Bond as a Molecular Building Block

P. R. Surján
CHINOIN Research Centre, H-1325 Budapest, P.O.B. 110, Hungary, and Central Research Institute of Chemistry of the Hungarian Academy of Science, H-1525 Budapest P.O. Box 17

Good models contribute to a better understanding of the structure of matter and offer a possibility of developing effective computational schemes. Rigorous quantum chemical approaches usually work within a well defined framework which is often very sophisticated but run into serious difficulties in treating extended systems or when interpretation of computational results in chemical terms is desired. It is also common knowledge that quantum theory still needs to explain some basic facts of empirical chemistry; it is very difficult to derive phenomenological rules and concepts from first principles. Many of these features are connected to the existence of local (two centre, two-electron) chemical bonds in polyatomic molecules, reflecting the basically pairwise nature of interatomic interactions in molecular systems. Although we cannot derive the existence of such chemical bonds a priori, it is possible to determine, using quantum mechanics, the best possible wavefunction for each assumed two-electron bond and to construct the many-electron wave function as an antisymmetrized product of these bond wave functions. The resulting wave function, the *antisymmetrized product of strictly localized geminals*, can be considered as a correlated zeroth order reference state for a many-body perturbation theory approaching the exact solution of the Schrödinger equation in the given basis set. This formalism becomes especially convenient in the second quantized formulation in which the chemical bonds are represented by composite-particle creation operators which obey Bose-like commutation rules. The internal structure of these quasi Bose-particles is affected by local characteristics as well as certain electrostatic interbond interactions which do not give rise to delocalization. Although this zeroth order picture can be in principle refined by perturbation theory ad infinitum, it should not to be considered as a quantitative approach. Rather it is a tool for exploring how far the electronic structure can be described in terms of local two-electron bonds, thus being conceptually important.

1 Introduction

The notion of the chemical bond originates from phenomenological chemistry. Modern theoretical chemistry still has to explain the majority of empirical properties which follow from this important concept. It is possible, for example, to predict quantitatively binding energies on a theoretical basis, but the question of the observed universality and transferability of local bond properties, as well as the basically bielectronic nature of the chemical bond in polyatomic molecules remains to be answered and interpreted.

The situation is somewhat controversial [1–9] because some scientists want to eliminate the bond concept as superflous. They claim that this concept is redundant since everything can be computed without it. On the other hand, some researchers tend to reduce the whole of quantum chemistry to two-centre bonding [10–12] introducing sometimes oversimplified calculational schemes. Accordingly, significance of the bond concept and its proper use deserve some more comment. This point will be addressed in the next Section. In Sect. 3 and 4, we shall briefly review the theory of localized molecular orbitals which offers a pictorial tool permitting a representation of local bonds at the SCF level. This formalism allows also for a development of a priori theories which approach the Hartree-Fock solution in terms of local bond wave functions. In Sect. 5, an extension of this theory to the level of correlated wave functions will be presented. An application of this latter scheme is given in Sect. 6.

2 Importance and Reliability of the Bond Concept

Some authors argue that the "chemical bond" is 'not measurable' and consequently should be dropped from the modern vocabulary of quantum chemistry. This point of view is unjustified because of the following reasons:

(i) The existence of local bonds is **empirically** well established. It is a concept which permits a systematization of otherwise hardly understandable measured data. For example, a carbonyl group is easily identified in an NMR spectrum by its characteristic value. Empirical bond distances for a given bond show a surprising transferability in different molecules. The relativley small observed deviations can always be explained in terms of changes in bond-bond interactions or intra-bond differences. The empirical facts should give impetus to finding a proper microscopic model which provides a proper explanation.

(ii) It is well known in theoretical physics that the concept of 'observables' or 'measurable quantities' is a rather problematic one. These concepts are often inherently coupled to a specific model when they refer to a microscopic system, since any measurement of a microscopic system is, by necessity, always indirect and statistical. This point was clearly emphasized in the monograph by Del Re et al. [13]. Hence, if a picture of a localized chemical bond disappears in a particular theoretical framework, in spite of the abundant empirical evidence, it is a fault of the model and not of the concept.

(iii) Although most quantum chemical methods are based on quantum mechanical theories of isolated systems, one has to realize that — especially in the microscopic world — it is hard to find a system left to itself. Consequently, if one argues that the energy or dipole moment of a chemical bond are not measurable, one should also take into account that the energy or the dipole moment of an individual molecule are not measurable for the very same reason.

(iv) One should keep in mind that in discussing bonds in polyatomic molecules, we have a theory at a different level (or 'hierarchy') of intricacy as compared to that of a simple quantum mechanical calculation on a few electrons. It is well known in physics that one should turn to a different level of theory if a different organization of the matter is studied. Although quantum theory is, in principle, valid for any (nonrelativistic) system of any size, an aircraft is never designed by solving the Schrödinger equation for $\sim 10^{30}$ atoms but one uses the specific laws of hydro- and aerodynamics or Hooke's law instead. Concepts such as 'turbulency', 'elasticity', 'Young-modulus', etc. do not correspond to any microscopic observable but they have their well-defined meaning at a level of theory describing a higher organization of matter. In chemistry, one often uses seemingly strange concepts such as that of atomic charge, electronegativity or the chemical bond, but if they offer any help in systematizing experimental data [14], they should be considered as useful concepts and theory should provide their firm basis.

(v) Finally, it should be mentioned that in different branches of theoretical physics (solid state physics for example) a phenomenon is claimed to be understood only if it has been explained in simple terms, preferable by analytical models. Understanding is definitely not provided by a brute force calculation yielding a single number. The relevance of this attitude for quantum chemistry was emphasized recently by Durand and Malrieu [15] and Maksić [16]; see also the discussion in Ref. [17]. Certainly, the problem of the two-electron bond being much simpler than that of the whole molecule, hence the concept of the local bond offers some conceptual advantages.

Briefly, adopting general patterns of theoretical physics, one should accept the concept of local bonds in a polyatomic molecule, and try to develop theories which account for this property. In principle, there are three different kinds of method of attacking this type of problem:

(i) *A posteriori* analysis. This approach tries to *extract* information from standard quantum chemical calculations of the properties of chemical bonds.

(ii) *A priori deductive* treatment. In this method one attempts to derive the empirically observed features of the chemical bond from the first principles of quantum mechanics of a many-body system.

(iii) *A priori inductive* treatment. This is a constructive approach in which the wave functions of individual chemical bonds are calculated, and general tendencies are guessed on the basis of these calculations. It has also the advantage that approximate many-electron wave functions can be constructed in terms of local two-electron bond wave functions.

Most efforts in the past have been dedicated to the a posteriori analysis mentioned in point (i). The theory of localized molecular orbitals [18–23] falls in this category which will be briefly discussed in the next Section. Much less is known about a deductive theory of localized bonds quoted in point (ii) above. Mayer's chemical Hamiltonian approach [24, 25] can be considered as an initial step in this direction.

A central theme of the present article is related to the inductive treatment belonging to point (iii), which has been extensively studied, too. It is worth mentioning that the discussion above is not necessarily restricted to localized bonds only, but can be easily generalized to encompass larger fragments (functional groups, etc.) in molecules. In principle, the localized bonds and larger units can be theoretically treated on an equal footing. Some of the most important fragment theories are: McWeeny's group function formalism [26], various "Atoms-in-molecules" (AIM) [27–30] or "Molecules-in-molecules" (MIM) [31] methods, the PCILO idea [32–35], perturbation theories for constructing localized molecular orbitals [36–40], and the related geminal-type models [41–44]. We may also mention Mehler's nonorthogonal group function method [45–46], Kirtman's density matrix treatment [47–48], the fragment SCF methods [49–53] or various other group function techniques [54–57]. In spite of these considerable efforts, the problem is not completely solved, and a widely accepted scheme is still not available. The state of the art of the perturbative construction of localized orbitals and geminals will be discussed in Sect. 3–5, whereas some illustrative examples will be presented in Sect. 6.

Before a detailed elaboration of the theory we would like to give some physical hints which are helpful for understanding the formalism. The most important clue is the concept of a **"building block"**, which plays a key role in the (statistical) theory of matter.

The building block is a subunit, or a 'particle', which is small enough to possess a finite (preferably small) number of internal degrees of freedom, but it is appropriately chosen so that the interactions between subunits are significantly smaller compared to forces responsible for their formation. The interaction between properly chosen building blocks should be so small that various physical properties are (nearly) additive in terms of the block contributions. The additivity of extensive thermodynamic quantities can be considered as a criterion in selecting the smallest possible building blocks of the matter.

In many cases not much choice is left. In the thermodynamics of the gas phase, for example, the closed-shell atoms or molecules represent the natural building blocks of a sample; they are often considered as point-like. These subunits are really weakly interacting, and most physical properties on the macroscopic scale can be approximated as sums of particle contributions.

This simple picture cannot be further maintained in chemistry. Molecules can no longer be considered as indivisible particles. Their internal structure is fairly complicated, leading to inapplicability of the standard quantum chemical approach for large molecules. Hence, one has to find a smaller building block. The atoms from which the molecules are formed can be considered as building blocks, since the binding energies are small fractions of the molecular total energy. However, atomic fragments are not good starting points for further calculations, and inter-atomic interactions are not small enough to justify a perturbational treatment [29, 58]. This indicates that another building block is to be searched for when actual calculations are desired. One has to realize that in chemistry one does not have necessarily to go to the level of electrons and nuclei, and an electron cloud moving in the field of nuclei is not the only possible picture of molecules. Hence, it appears that in the theory of molecules at an intermediate level of sophistication, the *two-electron bonds represent the most natural building blocks*.

This choice seems to be straightforward because (i) bonds have rather small number of internal degrees of freedom as compared to the whole molecule; and (ii) global molecular properties (energy, dipole moment, etc.) can be well described in terms of bond contributions. It should be emphasized that this statement does not imply that bond contributions are strictly transferable. In fact, the extent of transferability is usually smaller than that of additivity. Why are properties of local chemical bonds nearly transferable is an interesting question which requires more subtle studies. We only mention in passing that this is a very important point which should be eluciated because otherwise a gap between empirical and theoretical chemistry will remain, leaving a no-man's land in the structure of this science.

The *limitations* of the corresponding two-electron model are obvious. Some systems show collective phenomena (cf. aromaticity), and cannot be described in terms of local subunits. Here a larger building block should be considered (the 6 π-electrons for benzene, for example) [59, 60]. Another problem arises when excited state properties are treated, because they often exhibit collective properties. However, a localized treatment can sometimes be useful here, too [61–65].

In spite of these limitations, a rather wide class of molecules can be viewed as a system of weakly interacting two-electron bonds. This simple picture yields a transparent interpretation of a number of physical and chemical properties. In what follows we shall discuss some theoretical aspects of two-electron chemical bonds.

3 Localized Orbitals

First, we will discuss the representation of bonds and the possibility of constructing molecular wave functions in terms of bond wave functions at the Hartree-Fock level. For the sake of better understanding, it is worthwhile to overview usual localization schemes. This will be done very briefly, since a lot of good papers and reviews of the subject are available (see for example, Refs. [1, 18–23, 66–67]). Since the literature on the theory and application of localized orbitals is quite abundant, we cannot cite all relevant papers, but some of the most important contributions will be mentioned.

3.1 A Posteriori Localization

The term '*a posteriori*' means that one has first to determine the standard self-consistent solution of the canonical Hartree-Fock (HF) equations within a given basis set, and a *localization transformation* of the molecular orbitals (MOs) is carried out afterwards. This transformation is based on the determinantal nature of the HF wave function: owing to the mathematical properties of determinants, the total many-electron wave function is invariant (up to an irrelevant phase factor) against any unitary transformation among the occupied MOs.

The invariance of a many-electron wave function (and also that of any expectation values) against mixing the occupied orbitals shows that one-electron orbitals forming the many-electron determinant, have no physical significance. In most cases one uses the so called *canonical* MOs $\{\psi_i\}_{i=1}^n$, which are eigenfunctions of the Fockian $\hat{F}$:

$$\hat{F}\psi_i = \varepsilon_i\psi_i \qquad (i = 1, 2, \ldots, n) \tag{1}$$

The canonical orbitals being eigenfunctions of the Fockian, have the special property that they are symmetry adapted, i.e. they transform according to the irreducible representations of the molecular point group symmetry. Consequently, they are usually spread over the whole molecule[1].

Equation (1) is a sufficient but **not necessary** condition for the solution of the Hartree-Fock problem. As a matter of fact, the HF problem is equivalent [70, 71] to the *Brillouin theorem* which states:

$$\langle \Psi_0 | \hat{H} | \Psi_{i \to k^*} \rangle = 0 \tag{2}$$

where $\hat{H}$ is the Born-Oppenheimer electronic Hamiltonian, Ψ_0 is the ground state Hartree-Fock wave function, while $\Psi_{i \to k^*}$ is an excited determinant resulting from Ψ_0 by substituting an occupied orbital ψ_i with a virtual one ψ_{k^*}. Equation (2) is the global or integral form of the Brillouin theorem, which may be converted into a local from by expanding the determinants in Eq. (2) and applying Slater's rules:

$$\langle \psi_i | \hat{F} | \psi_{k^*} \rangle = 0 \,. \tag{3}$$

This condition is equivalent to the Hartree-Fock problem. It expresses that the Fockian must be *block diagonal*: its matrix elements between occupied and virtual orbitals are required to be zero. The Fockian need not be *diagonal*: this holds only for the canonical orbitals. In the general case, we demand only that the occupied and virtual subspace do not interact[2]. It is immediately seen that a unitary transformation within the occupied space does not affect the validity of Eq. (3). The same holds true for transformations within the virtual space, too. These transformations permit us to localize the occupied and the virtual orbitals, respectively.

Earliest attempts to get localized orbitals were based on symmetry considerations by searching *equivalent orbitals*[3] [72–74]. In more complicated cases, symmetry

[1] If the this is not the case, i.e. if the canonical orbitals turn out to be localized for some reason, some attention is required because it may reflect a special physical meaning (cf. Anderson-localization [68], see also [69]).

[2] This can also be formulated as both the occupied and virtual subspaces are *invariant subspaces* of the Fockian.

[3] The equivalent orbitals do not correspond to irreducible representations, but they are transformed into each other by the symmetry operations of the point group serving as a basis of a reducible representation.

arguments alone are not sufficient to determine a localization transformation, sometimes also problems are encountered with finding equivalent orbitals corresponding to the local site symmetries [75]. Therefore, more automatized methods have been developed to find localized orbitals. These methods can be either

(i) *internal*, if only certain inherent properties of the orbitals are required, or
(ii) *external*, if one specifies certain fragments onto which the searched orbitals are required to be localized.

All internal criteria are based on the following scheme:

(1) We define a functional of the orbitals, denoted by M, which is written as a sum of orbital and orbital pair contributions:

$$M = \sum_{i,j} M_{ij} = \sum_{i} M_i + \sum_{i \neq j} M_{ij} \tag{4}$$

Of course, the summation extends only to the occupied MOs if they are to be localized. The functional M is constructed in such a way that it is invariant against a unitary transformation of the orbitals. The two terms at the right-hand-side of Eq. (4), however, are not invariant separately: both the orbital and orbital pair contributions may change.

(2) We determine the unitary transformation which maximizes the orbital contribution ΣM_{ii}, while, due to the invariance of M, minimizes the pair contributions ΣM_{ij} $(i \neq j)$ at the same time, or vice versa. By a proper choice for M it is possible to achieve that the transformed orbitals are localized.

(3) The extremal value of the orbital contribution, ΣM_{ii}, is usually not searched for directly. Instead, two-by-two rotations are employed:

$$\left.\begin{aligned} \psi_i' &= \cos\alpha\,\psi_i + \sin\alpha\,\psi_k \\ \psi_k' &= -\sin\alpha\,\psi_i + \cos\alpha\,\psi_k \end{aligned}\right\} \tag{5}$$

The angle of this rotation, α, is determined so that the involved part of the orbital contributions is stationary:

$$\frac{\partial}{\partial\alpha}[M_{ii} + M_{kk}] = 0\,. \tag{6}$$

(4) The above 2 by 2 rotation is performed for all pairs (i, k), and the whole procedure is repeated iteratively until convergence is achieved.

Various internal a posteriori localization criteria differ from each other in the actual choice for the functional M. The most important criteria are the following ones:

(a) *Edmiston-Ruedenberg or energetic criterion* [18, 19, 66]. This maximizes the self-repulsion of the orbitals:

$$M = \sum_{i,k} (ii|kk) \tag{7}$$

that is

$$\sum_i (\mathrm{ii}|\mathrm{ii}) = maximal \tag{8}$$

Here the usual notation for the two-centre repulsion integral in the (11 | 22) convention is applied.

(b) The *Foster-Boys or spatial criterion* [20, 23, 67] maximizes the distances between the centroids of orbitals:

$$M = \sum_{i,k} |\mathbf{r}_i - \mathbf{r}_k|^2 \tag{9}$$

that is

$$\sum_{i \neq k} |\mathbf{r}_i - \mathbf{r}_k|^2 = max \tag{10}$$

(c) The *von Niessen or Density-criterion* [21, 22] makes use of the operator of the electron density $\delta(r - r_i)$:

$$M = \sum_{i,k} \int \langle\psi_i|\, \delta(\mathbf{r} - \mathbf{r}_i)\, |\psi_i\rangle \langle\psi_k|\, \delta(\mathbf{r} - \mathbf{r}_k)\, |\psi_k\rangle\, \mathbf{d}^3\mathbf{r}$$

$$= \sum_{i,k} \int \varrho_i(\mathbf{r})\, \varrho_k(\mathbf{r})\, d^3\mathbf{r} \tag{11}$$

with the extremum condition

$$\sum_i \int |\langle\psi_i|\, \delta(\mathbf{r} - \mathbf{r}_i)\, |\psi_i\rangle|^2\, d^3\mathbf{r} = \sum_i \int \varrho_i^2(\mathbf{r})\, d^3\mathbf{r} = max \tag{12}$$

where ϱ_i is the contribution of orbital ψ_i to the electrón density.

Application of the above three localization criteria yields very similar, though not exactly the same, localized orbitals in most molecules [22, 76]. Most frequently the criteria (a) and (b) are used. Although it is not widely used, a very simple and effective method for localizing molecular orbitals was proposed by Perkins and Stewart [77]. It is based simply on the localization of the expansion coefficient matrix.

The automatized procedures discussed above are called internal criteria because they use only the properties of the localized orbitals themselves. The *external* criteria (vide supra) are usually based on a partitioning of the basis set. Since in quantum chemistry one uses most frequently LCAO-type basis sets, each basis function is uniquely assigned to one of the atoms[4]. This permits one to introduce different partitionings of charge densities.

The most often used external localization criterion is that of Magnasco and Perico [78, 79]. It is based on maximizing the local Mulliken populations

$$\sum_i \sum_{\mu,\nu \in \Gamma_i} C_{i\mu} C_{i\nu} S_{\mu\nu} = max \tag{13}$$

[4] This assignation is problematic, however, if using very diffuse or off-centred basis sets.

where $C_{i\mu}$ is the coefficient of the i-th molecular orbital on the μ-th basis function, S is the overlap matrix, while Γ_i is the subspace of the basis fragments assigned to the i-th MO. The selection of these subspaces is a matter of intuition — this makes this criterion 'external'. For more recent developments on this localization method, including its extension to the virtual space, the reader is referred to Refs. [80–81].

There are several other external localization criteria as well, which usually utilize the properties of the density matrix [82–84]. They are only marginally important and will not be discussed here in detail (see also Refs. [85–87]).

As emphasized in Sect. 2, a posteriori localization criteria have mainly interpretative value, though localized orbitals can sometimes be used to simplify configuration interaction calculations [88–90]. In fact, by performing the localization of the MOs by one of the above criteria on a large number of molecules, the following picture emerges [91–95].

The transformed orbitals are mainly localized on a small part of the molecule, possessing only relatively small components on other fragments. These latter small components are called *tails*. The central part of the localized MOs corresponds usually to classical chemical concepts (two-centre bonds, lone pairs, inner-shells). The central parts describing similar fragments in different molecules are similar, too. Briefly, the central parts of localized MOs are *transferable* to a good approximation. This transferability does not hold for the tails.

According to this picture, the idealized chemical bonds are represented by the central parts of localized MOs, while the tails correspond to the weak *delocalization interaction* between chemical bonds.

The above ideas are valid only for the so called "well localizable" systems. There are, however, molecules where the delocalization is significant. As mentioned above, these (sub)systems have to be treated separately.

Characterization of the degree of localizability for different systems is therefore important. This is possible by certain localization indices [96–98] which measure the relative weight of the central part in some manner. In regular polymers the localizability can also be characterized unambigously on the basis of the SCF LCAO density matrix D. The elements of this matrix in non-metallic regular polymers depend exponentially on the distance between the corresponding atomic orbitals:

$$D_{\mu\nu}(n) \sim e^{-\xi n} \tag{14}$$

where the unit cells the atomic orbitals μ and ν belong to are n-th neighbours and the exponent ξ is constant for a given μ, ν pair. This formula can be analytically derived in simple Hückel type models [99–102] but holds also true at the ab initio level to a very good approximation [103], see also Fig. 1. The elements of the

▶

Fig. 1 a–c. Exponential decay of density matrix elements $D_{\mu\nu}$ for some polymers between AOs μ and ν separated by n unit cells. (**a**): linear chain of H_2 molecules, STO-4G basis; (**b**): equidistant chain of H atoms, charge-density wave solution, STO-4G basis; (**c**): polyene $(CH)_x$ chain, alternating geometry, STO-3G basis. In the latter figure the basis orbitals are numbered as follows: $1s_a$, $2s_a$, π_a, $2p_{ya}$, $2p_{za}$, $1s_b$, $2s_b$, ... etc. The results are taken from Ref. [103]

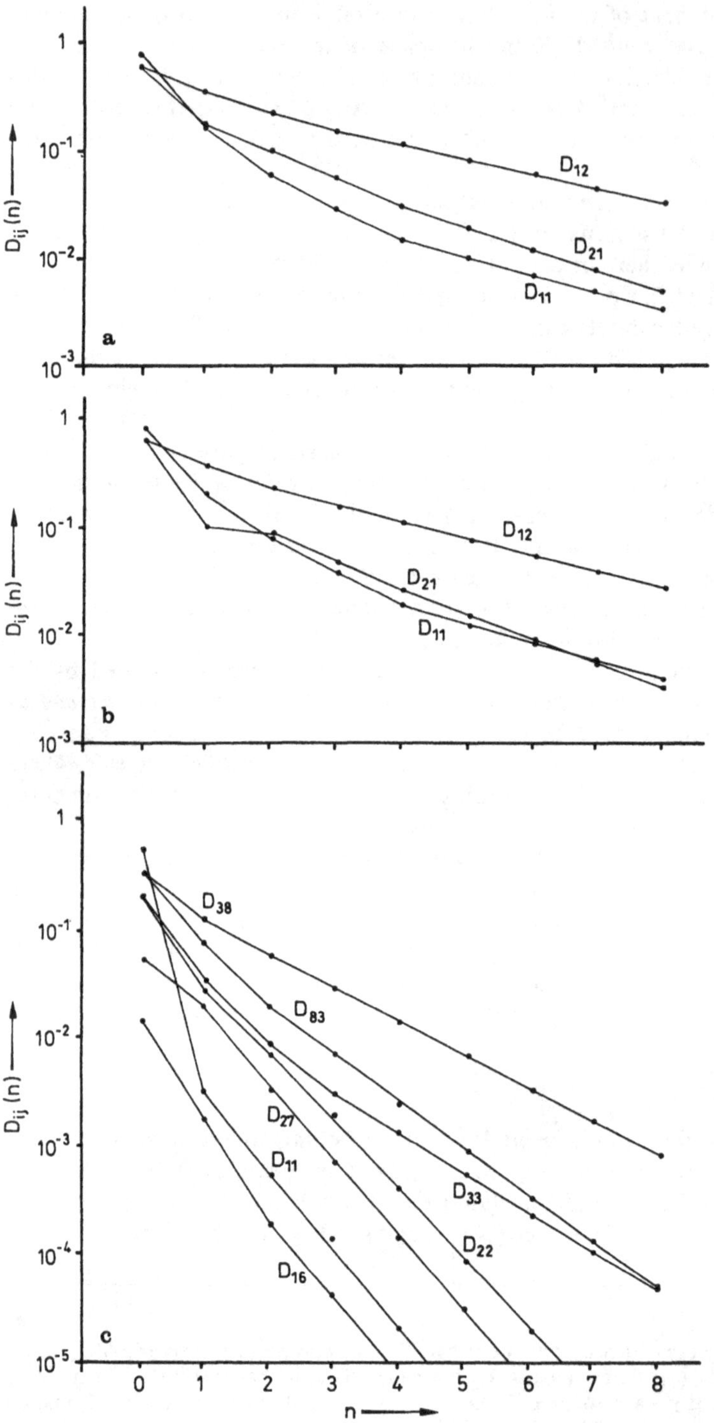

1
10⁻¹
10⁻²
10⁻³
Dij (n)
D12
D21
D11
a
b
D38
D83
D27
D11
D33
D22
D16
c
10⁻⁴
10⁻⁵
0
1
2
3
4
5
6
7
8
n

density matrix D between far-lying atoms can be related to the localizability [103, 104]. The $D_{\mu\nu}$ element is calculated in the closed shell case as

$$D_{\mu\nu} = 2 \sum_{i}^{occ} C_{i\mu} C_{i\nu} \tag{15}$$

where $C_{i\mu}$-s are the expansion coefficients of the MOs (real orbitals are assumed). If the orbitals are localized in the given basis set, the exponential decay of $D_{\mu\nu}$ assumes the similar decrease of the orbital coefficients. Consequently, in regular polymers the extent of localizability depends on the exponent of the asymptotic decay for the density matrix: The greater the exponent, the better is the localizability of the system. Since the matrix D is invariant to a unitary transformation of the orbitals, this condition allows us to characterize the localizability in these systems unambiguously [103].

3.2 A Priori Localization

Besides the a posteriori schemes discussed above, there are various procedures which yield directly localized orbitals without solving first the canonical Hartree-Fock equations (1).

The mathematical background of this problem is as follows. Since Eq. (1) yields canonical orbitals, the Hartree-Fock equations for localized orbitals can be found by considering the general system of equations, Eq. (3). It is easy to see that the latter are satisfied if

$$\hat{F}\psi_i = \sum_{k}^{occ} \varepsilon_{ik}\psi_i \tag{16}$$

where the summation extends over the occupied orbitals, and the elements of the ε matrix are defined as

$$\varepsilon_{ik} = \langle\psi_i|\, \hat{F}\, |\psi_k\rangle\,. \tag{17}$$

Equation (16) just states that the occupied subspace is an invariant subspace of the Fockian $\hat{F}$, since a transformed occupied function $\hat{F}\psi_i$ can be completely expanded within this space.[5]

Equation (16) is sometimes referred to as the general Hartree-Fock equation. It is valid even for non-orthogonal orbitals.

Unfortunately, this equation does not determine the orbitals unequivocally, only the subspace spanned by them. To fix the MOs ψ_i, one has to fix the ε_{ik} matrix elements in some manner. The diagonal choice ($\varepsilon_{ik} = \varepsilon_i\sigma_{ik}$) results in canonical orbitals, the localized ones should correspond to some other choice.

The elements of the matrix ε cannot be fixed arbitrarily, if the resulting orbitals are required to be orthogonal. Besides the basic symmetry of this matrix ($\varepsilon_{ik} = \varepsilon_{ki}$), we have to conserve any quantities which are invariant to a unitary transformation, such as the trace for example:

$$\mathrm{Tr}(\varepsilon) = \sum_{k} \varepsilon_{kk} \tag{18}$$

---

[5] An equivalent way of deriving Eq. (16) is to introduce a unitary transformation into Eq. (1) explicitly.

The quantity 2Tr(ε) should be equal to the sum of ionization energies in light of the Koopmans theorem [105].

Fixing of the ε_{ik} matrix elements is done according to the localization criterion in question. This, following Gilbert [106], can be done as follows:

Instead of solving the general Hartree-Fock equations, Eq. (16), one determines the solution of the pseudo-eigenvalue equation

$$[\hat{F} + \hat{G}]\,\psi_k = \varepsilon_k \psi_k \tag{19}$$

where $\hat{G}$ is a properly chosen **localization potential.** The role of the localization potential is just to ensure that the orbitals ψ_k are localized.

A simple localization potential was introduced by Edmiston and Ruedenberg [18], in connection with their a posteriori criterion discussed above. They defined the matrix elements of the localization potential as

$$G_{ik} = [(ii \mid ik) - (kk \mid ki)]^2 - F_{ik} \qquad (i \neq k) \tag{20a}$$

$$G_{ii} = 0\,. \tag{20b}$$

It is possible to show that the resulting orbitals fulfil the Edmiston-Ruedenberg localization criterion [18]. One can also see that the localization potential of Eq. (20) does not bring out the MO ψ_k from the occupied subspace, thus the solutions for Eq. (19) are the solutions for the general Hartree-Fock equation, Eq. (16), too. In practice, it is rather complicated to solve examples of Eq. (19) with the localization potential of Eq. (20); it is certainly a simpler task to find the standard canonical orbitals and then localize them by successive two-by-two rotations.

Similar a priori localization schemes were introduced by Gilbert [106–108], Adams [109–113], Peters [114–116], and Wilhitte and Whitten [117]. The latter scheme was recently applied by Tseng and Whitehead [118].

These direct localization schemes, though they are certainly very important from the theoretical point of view, did not help too much in practical applications. We have to stress, however, the importance of the localization scheme of Daudey [119]. This scheme is based on a first-order perturbation theory utilizing the Brillouin theorem [120–122]. The essential feature of this method is that one starts from an intuitively selected approximate set of localized orbitals $\{\psi_i^{(0)}\}$, which do not obey exactly the local Brillouin theorem of Eq. (3). To improve the approximation, one perturbes the initial localized orbitals by the Ansatz

$$\psi_i^{(1)} = \psi_i^{(0)} + \sum_{j*} \eta_{ij*} \psi_{j*}^{(0)} \tag{21}$$

where η_{ij*} is a small coefficient. The summation label runs over the virtual orbitals.

Substituting Eq. (21) into the local Brillouin theorem Eq. (3), up to the first order in η-s one gets:

$$\eta_{ij*} = -\frac{F_{ij*}}{F_{j*j*} - F_{ii}} \tag{22}$$

where F now denotes the Fockian constructed from the approximate orbitals $\{\psi_i^{(0)}\}$. The orbitals can be subsequently orthogonalized and, constructing a new Fockian from the perturbed orbitals, one can improve them further by the linear expression having the form of Eq. (21). This can be done iteratively until convergence is achieved. The converged orbitals will obey the local Brillouin theorem of Eq. (3), and will therefore be exact Hartree-Fock orbitals. If the original orbitals $\{\psi_i^{(0)}\}$, used as the initial guess, were localized, the converged orbitals will be localized, too, though they will not obey any of the standard criteria discussed previously.

In the next section we shall outline a theory for approximating and perturbing localized orbitals by a scheme which is, in a certain sense, a generalization of this first-order perturbation approach. First of all, we shall introduce the concept of strictly localized molecular orbitals.

4 Strictly Localized Orbitals and Their Perturbation

4.1 The Zeroth Order

As discussed above, exact Hartree-Fock localized molecular orbitals (LMOs) consist of a central part and small delocalized tails. The approximate orbitals which result from exact LMOs by cutting the tails of the latter are often referred to as *strictly localized molecular orbitals*, SLMOs. They are approximate[6] since they do not exactly obey Eq. (3) or Eq. (16). In well-localizable systems, the tails being small, SLMOs can, however, give a good approximation. The above definition of SLMOs is not unique; it depends on how do we define the 'central part' and the 'tails', and that on what type of LMOs we are considering. In the present section we shall treat a method for direct, a priori determination of SLMOs, without referring to the exact Hartree-Fock orbitals. This method makes it possible to represent local two-electron bonds, as well as to consider certain types of interbond interactions.

To being with, we have to investigate first the connection between basis orbitals and chemical bonds. This is possible by means of a partitioning of the underlying basis set.

Let us consider a basis $\{\chi_\mu\}_{\mu=1}^{N}$ in which the localized orbitals are expanded. Following the LCAO principle, each basis function χ_μ is assigned to one of the atoms (see, however, footnote 4). The assignment of the orbital χ_μ to the atom A can be expressed by the symbol $\mu \in A$.

The LCAO formalism can be applied to studying the properties of atoms in molecules in a natural manner. Some examples are (i) the population analysis as developed by Mulliken [123] which permits one to define e.g. atomic charges; (ii) the binding energy analysis by Clementi [124] in which the electronic energy of the molecule is partitioned into atomic contributions and two-, three- and four-center

[6] However exact SLMOs may exist in some special systems, such as first-row hydrides in a minimal basis set for example [170]. These are non-orthogonal, however.

terms; (iii) and the chemical Hamiltonian approach by Mayer [24–25] which allows us to represent atoms in molecules and yields an energy partitioning scheme where only atomic and pair contributions are present.

These methods do not give us the possibility to find the mathematical representation of two-center chemical bonds. This latter requirement makes it necessary to assign basis functions to bonds. An important tool to derive such basis sets is *hybridization*. This involves a unitary transformation[7] of the basis orbitals which mixes only orbitals assigned to the same atom. Usually, the hybridization does not affect the inner shell orbitals, while the valence shell AOs are transformed into orbitals which are directed towards two-centre bonds or lone pairs.

The hybridization transformation is not unique. Actual choice of hybrids can only be done intuitively. The simplest hybrids are those of Pauling [125, 126] ('standard hybrids'). There have been several other procedures proposed for finding hybrid orbitals. These can be classified into two groups. One group consists of a posteriori procedures, which determine the hybridization parameters by knowing the exact Hartree-Fock orbitals or the corresponding density matrix [83, 84, 86, 87, 127–131]. From the practical point of view, the second group is more important. This consists of procedures which yield hybridized basis sets directly, without prior solution of the Hartfree-Fock problem [132–140].

In this article we do not want to review the details of the above cited hybridization procedures — these could be the topic of a separate paper. As concerns the theory outlined below, the actual choice of hybrids is a question of secondary importance. The only important point is that, as a result of a proper hybridization transformation, each basis function χ_μ can be uniquely assigned to a particular chemical bond, inner shell or lone pair. This assignation is denoted by $\mu \in \mathrm{i}$, where i stands for the corresponding two-electron subunit.

We note that the hybridization transformation is not the only possibility of connecting basis orbitals to chemical bonds. There are also so called 'bond functions' used in quantum chemistry, which are not centered on any of the atoms, but are placed between two neighboring ones to represent bonds, or in the area of a lone pair to improve the description of the latter [141–145]. In most cases, bond functions are used to substitute polarization functions, and they augment a basis set consisting of inner-shell and valence-shell orbitals, preferably of at least split-shell quality. Bond functions improve the basis set as a consequence of the better description of the electronic structure of chemical bonds, and they can be uniquely assigned to the bond in question. Note also that to describe the valence shell, basis sets consisting merely of off-centred bond functions have also been proposed and applied [146–153].

Once each basis function has been assigned to one of the two-electron bonds[8], the list of integrals (matrices) can be partitioned as shown by Fig. 2. The large intrabond matrix elements are grouped into the diagonal blocks, while the interbond integrals appear in the off-diagonal ones. Since the latter are expected to be smaller, we

[7] In general, one can consider nonsingular linear transformations as well, which do not preserve the orthogonality of the atomic basis functions. In an ab initio model, especially in extended basis sets, there is no reason to restrict ourselves to unitary hybridization transformations.

[8] Here and further on, for the sake of brevity, the term 'bond' will denote any two-electron fragment (two-center bond, lone pair, inner shell) of the molecule.

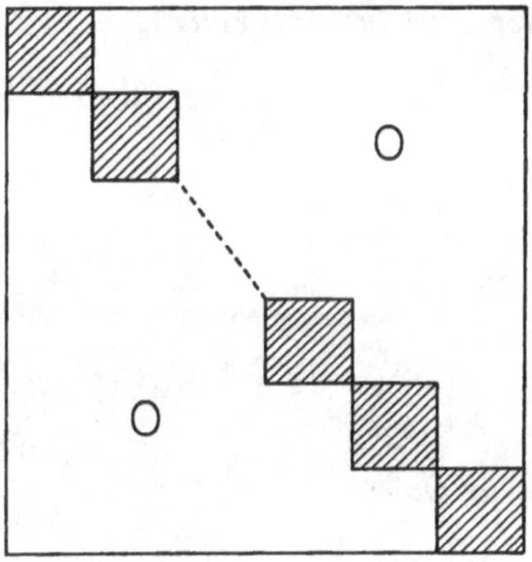

Fig. 2. Partitioning of interaction matrices in a hybridized basis set. The diagonal blocks contain all intrabond matrix elements, while the offdiagonal ones are neglected at the zeroth order

may neglect them at a zero order description, or, more precisely, we may consider them as first-order perturbations.

The above partitioning of the basis set permits one to define the strictly localized molecular orbitals (SLMOS) in a natural manner. Dealing with two-electron subunits, there is only one occupied SLMO in each bond. This can be expanded within the subspace of basis orbitals assigned to the bond in question:

$$\psi_i^{(0)} = \sum_{\mu \in i} C_{i\mu} \chi_\mu \tag{23}$$

where the $\mu \in i$ symbol expresses the strict localization on the bond i. The best (optimised) SLMOs are obtained by determining the expansion coefficients $C_{i\mu}$ variationally yielding to an energy minimum for the many-electron determinantal wave function built up by the set of SLMOs. The corresponding formalism is in a close relationship with the group function methods [26, 50–52, 154–156]. In fact, the present theory is a special case of the latter where the fragments are not necessarily identified with two-centre subunits.

The SLMOs are often written in a more special form as compared to Eq. (23). If one restricts the expansion to a minimal basis set, an SLMO possesses only two orbitals, and it has only a single variational parameter which is called the bond polarity [157]. The bond polarity is often determined empirically [158–160], or the appropriate value can be estimated a posteriori from the Hartree-Fock solution [83, 86, 87, 127, 161–165]. Determination of bond polarities has been outlined for the case of special model Hamiltonians (CNDO, INDO, etc.) by several authors. In what follows, we review the determination of the expansion coefficients $C_{i\mu}$ in Eq. (23) as generally as possible, that is, the following discussion will be valid for the case of an arbitrary basis set and at the ab initio level. Overlap effects will also be properly handled.

The variational optimization of the coefficients $C_{i\mu}$ in Eq. (23) can be done either by a direct optimization technique [166] or by solving a Hartree-Fock-Roothaan (HFR) type equation for each bond. This latter approach will be considered here in some detail. To derive the corresponding equations, we have to investigate first the Fockian whose matrix elements are defined as [167]

$$F_{\mu\nu} = H_{\mu\nu} + \sum_{\lambda\sigma} D_{\lambda\sigma}\left[(\mu\nu\,|\,\lambda\sigma) - \frac{1}{2}(\mu\sigma\,|\,\lambda\nu)\right] \tag{24}$$

where D is the density matrix as defined by Eq. (15),

$$H_{\mu\nu} = \langle\chi_\mu| - \frac{1}{2}\Delta - \sum_A Z_A/r_A\,|\chi_\nu\rangle \tag{25}$$

is the matrix element of the one-electron Hamiltonian, while

$$(\mu\nu\,|\,\lambda\sigma) = \iint \chi_\mu(1)\,\chi_\nu(1)\,r_{12}^{-1}\chi_\lambda(2)\,\chi_\sigma(2)\,dV_1\,dV_2 \tag{26}$$

is a two-electron repulsion integral in the (11 | 22) convention.

We will first investigate under what conditions the general Fockian of Eq. (24) can be factorized into intrabond blocks, permitting us to separate the full HFR equations into local equations for each bond. To achieve this, we have to utilize the partitioning of matrices as shown in Fig. 2. First of all, we neglect the interbond elements of the overlap matrix S and the one-electron Hamiltonian H. Let us assume further that the initial density matrix D has a similar blockdiagonal structure:

$$S_{\mu\nu} = 0 \tag{27a}$$

$$H_{\mu\nu} = 0 \qquad \text{for} \qquad \mu\in i\,, \qquad \nu\in k\,, \qquad i\neq k \tag{27b}$$

$$D_{\mu\nu} = 0\,. \tag{27c}$$

Now the *inter*bond elements of the Fockian (which have to be eliminated) reduce to

$$F_{\mu\nu} = \sum_j \sum_{\lambda,\sigma\in j} D_{\lambda\sigma}\left[(\mu\nu\,|\,\lambda\sigma) - \frac{1}{2}(\mu\sigma\,|\,\lambda\nu)\right] \qquad \mu\in i,\ \nu\in k,\ i\neq k\,, \tag{28}$$

where $H_{\mu\nu}$ is absent due to Eq. (27b), while the summation restriction follows from Eq. (27c).

If this expression is not zero, the Fockian is not block diagonal. This means that, even if we start with a block-diagonal initial density, the interbond elements of the Fockian will yield to interbond components (tails) of the MOs, so the strict localization as expressed by Eq. (23) cannot be maintained. This involves also that at the next SCF iteration the density matrix D will not be block diagonal. In other words, the block-diagonal nature of matrix D is not self-consistent.

This problem can only be solved by introducing a partitioning also for the hyper-matrix containing the two-electron integrals [44]. Similarly to Eqs. (27), in calculating

SLMOs one has to neglect any integral containing a differential overlap between basis functions centred on two different bonds:

$$(\mu\nu \mid \chi\sigma) = 0 \quad \text{if} \quad \begin{cases} \mu \in i\,, & \nu \in k\,, \quad i \neq k\,, \quad \text{or} \\ \chi \in i\,, & \sigma \in k\,, \quad i \neq k\,. \end{cases} \tag{29}$$

Just these integrals give contributions in Eq. (28), thus they yield to delocalization. Substituting the conditions of Eq. (29) into Eq. (28), the latter expression vanishes, thus the Fock matrix F becomes block diagonal. The neglect of interbond differential overlap, together with the partitioning of the one-electron matrices according to Eq. (27), makes the block-diagonal nature of the density matrix self-consistent.

Partitioning of the two-electron integrals as described by Eq. (29) resembles to the usual 'zero differential overlap' (ZDO) condition widely applied in semiempirical quantum chemistry [168]. The essential difference is that in Eq. (29) the integrals in which the differential overlap of two functions of the **same bond** are present, are not neglected. So, the differential overlap of two hybrids directed to each other is not neglected even at the zero order. Though the corresponding overlap is certainly rather large, it would be neglected even by the most sophisticated NDO scheme, the NDDO method [168]. In searching for optimum SLMOs, one only neglects the really small interbond matrix elements and differential overlaps.

Utilizing now the block-diagonal character of each relevant matrix, the local Roothaan equation for the bond i will be written as

$$\sum_{\nu \in i} F^{i}_{\mu\nu} C_{i\mu} = \varepsilon_i \sum_{\nu \in i} S_{\mu\nu} C_{i\nu} (\mu \in i) \tag{30}$$

with the local Fockian F^i which is defined as

$$F^{i}_{\mu\nu} = H_{\mu\nu} + \sum_{\lambda,\sigma \in i} D^{i}_{\lambda\sigma} \left[(\mu\nu \mid \lambda\sigma) - \frac{1}{2} (\mu\sigma \mid \lambda\nu) \right]$$

$$+ \sum_{j(\neq i)} \sum_{\lambda,\sigma \in j} D^{j}_{\lambda\sigma} (\mu\nu \mid \lambda\sigma) \qquad (\mu, \nu \in i) \tag{31}$$

where the latter sum describes the Coulomb repulsion between the electrons of bonds i and j. Due to this interaction, the local HFR equations for different bonds have to be solved iteratively.

We can see from Eq. (31) that the restriction of strict localization does not involve the model of non-interacting bonds. In fact, the elements of the local Fockian are affected by the interbond Coulombic interactions (the interbond nuclear attraction terms being contained in $H_{\mu\nu}$). This means that the internal structure of local bonds is affected by interbond electrostatic interactions which yields the well known *inductive effects*. Note that, as a consequence of condition (29), no exchange integrals contribute to the inductive interaction in Eq. (31). A similar approximation was used in the "mutually consistent field" method [169] which was developed to study intermolecular interactions.

The SLMOs resulting as eigenvectors of the local Fockians of Eq. (31) are not orthogonal to each other, since they are expanded within mutually exclusive sets of overlapping basis functions [cf. Eq. (23)]. Explicit orthogonalization of the SLMOs yields to the so called **orthogonality tails** which do not correspond to any physical interaction. Although the interbond overlap effects are neglected in deriving the equations for the SLMOs[9], Eqs. (30–31), they must not be neglected when calculating the total electronic energy. The neglect of the overlap between SLMOs affects the normalization of the determinantal wave function introducing an enormous error in energy. For this reason, the energy of an SLMO wave function has to be calculated either by explicitly orthogonalizing the SLMOs, or by the general energy expression [171] valid in the non-orthogonal case.

Naturally, the Coulombic interaction of bonds accounted for by Eq. (31) is not the only interbond interaction which should be considered in an exact theory. This is reflected by the fact that the energy of the best optimized SLMOs is higher than the exact HF energy in the same basis set[10]. The interactions absent from the SLMO model lead to the tails of LMOs, and they are called delocalization interactions [38–39].

The question whether the tails (or delocalization in the above sense) are important from the practical point of view has been investigated for a long time [36–40, 44, 50–60,

Table 1. Some energies and energy differences as computed by the SLMO wave function at the ab initio level. For comparison, test data showing full SCF reference calculations are also shown. Total energies are given in atomic units, energy differences are in kJ/mol

system	basis set	quantity	SLMO result	full SCF result	Ref.
H_6 chain	STO-6G	⟨F⟩[a]			[39]
equidistant geom.			− 3.5510	− 3.7155	
alternating geom.			− 3.5528	− 3.6028	
CH_3CH_3	STO-3G	E_{tot}			[182]
eclipsed			− 78.2371	− 78.3010	
staggered			− 78.2407	− 78.3057	
		barrier	9.3	12.5	
BH_3NH_3	STO-3G	E_{tot}			[182]
eclipsed			− 81.5155	− 81.5860	
staggered			− 81.5183	− 81.5904	
		barrier	7.4	11.6	
CH_3NH_2	STO-3G	E_{tot}			[40]
eclipsed			− 93.9407	− 94.0262	
staggered			− 93.9435	− 94.0297	
		barrier	7.2	9.2	
CH_3OH	STO-3G	E_{tot}			
eclipsed			−113.4762	−113.5426	
staggered			−113.4781	−113.5449	
		barrier	4.9	6.2	
LiH	(*5s, 1p/5s*)	E_{tot}	− 7.9698	− 7.9710	[195]

[a] expectation value of the converged Fockian

[9] The overlap of SLMOs is neglected also in calculating the local density matrices by the formula $D^i_{\mu\nu} = C_{i\mu}C_{i\nu}$.

[10] The only exception is the special case when exact SLMOs exist in the molecule [170], cf. footnote (6).

86, 87, 172–174]. Based on these investigations, the following general conclusions can be drawn.

The SLMOs provide sufficiently good approximation for well-localizable systems if only a rough estimation of the electronic structure is necessary. For example, delocalization effects may be neglected in estimating net atomic charges or electrostatic potentials [175–178]. If, however, the charge distribution of the molecule results from significant long-range charge transfer (cf. zwitterions), it cannot be calculated without considering delocalization effects. A rough estimation of barriers to internal rotations around single bonds is also possible the the SLMO approach [179–184]. Here the role of interbond overlap in the energy expression is imperative which is necessary to describe the Pauli repulsion [182–186].[11] Tails become important, however, if fine details of the electronic structure are to be accounted for, or if one needs better quantitative estimates. To illustrate the power and limitations of the SLMO model, some numerical results are collected in Table 1. It is seen that the total energies obtained by SLMOs usually account for more than 99% of the full SCF result, while the SLMO estimation of barriers to internal rotations around single bonds in ethane-like molecules is reliable up to 60 to 80% of the exact result.

4.2 Delocalization Corrections

The zero-order fully localized description of the electronic structure by means of SLMOs discussed in the previous section can be improved towards the exact Hartree-Fock solution in the given basis set. There are basically two ways of achieving this: (a) the SLMOs can be used as an initial guess in a variational calculation [188–189], or (b) they can be considered as a zero-order description and can be improved by perturbation theory (PT) [119, 36–40]. Here we review a general approach of type (b) developed in our laboratory [38, 39] — considering delocalization effects by perturbing SLMOs. We have to cite here also some other related approaches [32–35, 190–198] which are either too special or complicated; or also consider correlational effects [32–35], and mostly give low-order energy corrections [40, 182]. We discuss a PT for interbond delocalization, and derive practically useful expressions for the tails of LMOs. First of all, we have to fix the form of the LMOs to be considered.

The occupied and virtual SLMOs (ψ_i^0 and $\{\psi_{i\alpha}^0\}$) together span the same space as the original basis orbitals $\{\chi_\mu\}$ do (here i is the label for the bonds and α is a quantum number for the virtual SLMOs of the bond i). Consequently, the **exact** HFR orbitals can be expanded in terms of SLMOs the latter serving as a basis set:

$$\psi_i = \psi_i^0 + \sum_k \eta_{ik}\psi_k^0 + \sum_k \sum_\alpha{}' \eta_{i,k\alpha}\psi_{k\alpha}^0 \tag{32}$$

[11] In orthogonal models such as CNDO, INDO, etc. the SLMOs cannot account for barriers [37, 179, 187] which appear only as a consequence of delocalization.

where we applied the so called intermediate normalization which, if the basis $\{\psi_k^0\}$ were orthonormal, would be defined by $\langle\psi_i \mid \psi_i^0\rangle = 1$. The η_{ik} is the expansion coefficient of the exact MO ψ_i over the occupied SLMOs of various bonds while $\eta_{i,k\alpha}$ is that over virtuals with quantum number α on the bond k. Expansion (32) is fully general in the sense that it is valid for an MO of any type (canonical or localized). It is clear, however, that if the expansion coefficients η_{ik} and $\eta_{ik\alpha}$ are small, ψ_i will be well localized in the bond i, since it contains the SLMO ψ_i^0 with a coefficient of 1. In turn, if the set of SLMOs is a good approximation to an exact set of localized orbitals, the η coefficients in Eq. (32) are expected to be small. In this case, these coefficients can be considered as the **tails** of the LMO, and the aim of the PT being discussed is the calculation of these η — s.

As it was emphasized as early as 1954 [191], in treating delocalization corrections, the admixture of occupied orbitals does not necessarily have to be considered. Recently, it has been formally proved [201] that there exists a set of **exact nonorthogonal** LMOs for which all coefficients η_{ik} mixing the occupied orbitals are zero, that is

$$\psi = \psi^0 + \sum_{k*}^{(\mathrm{virt})} \eta_{ik*}\psi_{k*}^0 \tag{33}$$

where the label $\{k\alpha\}$ has been combined into $\{k^*\}$ the asterisk referring to the virtual nature of the orbitals. It can be shown [201] that this form of the LMOs determines the orbitals uniquely, provided that the SLMO approximation is not singularly wrong[12].

In the perturbation theory for delocalization corrections, we cannot proceed in the usual manner, since the localized orbitals are not eigenvectors of the Fockian. Instead of expanding an eigenvalue equation, the expansion of the local Brillouin theorem Eq. (3) appears to be more appropriate. Unfortunately this equation is valid only for orthogonal orbitals, while the LMOs with the form of Eq. (33) we were searching for are not orthogonal. The generalization of Eq. (3) to the non-orthogonal case can be done as follows.

We start from the general Hartree-Fock equation (16) which is valid for any set of orbitals, even for non-orthogonal cases. To eliminate the ε coefficients, which make the application of this equation rather unpleasant, we define a set of virtual bra-orbitals $\langle\tilde{\psi}_{k^*}|$ which are biorthogonal to the occupied ones:

$$\langle\tilde{\psi}_{k^*} \mid \psi_i\rangle = 0\,. \tag{34}$$

Multiplying the general Hartree-Fock equations by this set from the left and using Eq. (34) we get

$$\langle\tilde{\psi}_{k^*}| \hat{F} |\psi_i\rangle = 0 \tag{35}$$

[12] The $\psi_i^0 \approx \psi_i$ approximation is singularly wrong [201] if the weight of ψ_i^0 in the exact LMO is zero. In this case the intermediate normalization breaks down. Although such a bad situation is not expected to appear for interbond interactions generally, a thorough analysis shows that it can be responsible for the breakdown of certain perturbation theories [202–206] for intermolecular interactions [207].

which can be as the **biorthogonal local Brillouin theorem** [39]. Nonorthogonal Hartree-Fock orbitals span the exact subspace if, and only if, they satisfy Eq. (35). The utilization of biorthogonal orbitals in deriving the above equation is analogous to the method of moments [208–210]. One possible choice of the biorthogonal virtuals is

$$\tilde{\psi}_{k*} = \tilde{\psi}^{0}_{k*} - \sum_{l*} \eta_{kl*} \tilde{\psi}^{0}_{l*} \tag{36}$$

where the **biorthogonal virtual SLMOs** $\tilde{\psi}^{0*}_{k}$ are defined as the reciprocal orbitals with respect to the direct-space virtual SLMOs:

$$\langle \tilde{\psi}^{0}_{k*} \mid \psi^{0}_{i*} \rangle = 0 \tag{37}$$

which is satisfied by

$$\tilde{\psi}^{0}_{k*} = \sum_{i} S^{-1}_{k*i} \psi^{0}_{i} \tag{38}$$

where S^{-1} is the inverse of the overlap matrix of SLMOs. Such biorthogonal formulation is widely applied in field theories (cf. co- and contravariant coordinates), in crystallography (cf. reciprocal lattice vectors), etc. In quantum chemistry, it has proved to be an efficient tool for dealing with the basis set overlap problem [39–43, 202–206, 211–213]. By means of this formalism it is possible to avoid the problem of calculating matrix elements between nonorthogonal determinants which is a rather complicated task [171, 214–216].

Now we are in a position to derive the perturbation corrections for the tails η_{ik*}. For the sake of simplicity, and to obtain more transparent, easily interpretable formulae for the delocalization corrections, we assume an effective one-electron Hamiltonian F to be known. (The generalization to the explicit consideration of SCF effects is obvious [44]; the corresponding first-order theory is called the 'linearized SCF approximation' [36, 44] which has been successfully applied to study conformational properties of large molecules at the semiempirical level [37]).

Let us expand the tail coefficients in the perturbation series

$$\eta_{ik*} = \sum_{\mu=1}^{\infty} \lambda^{\mu} \eta^{(\mu)}_{ik*} \tag{39}$$

where λ is a formal perturbation parameter. The effective one-electron Hamiltonian F is partitioned as

$$F = F^{(0)} + \lambda F^{(1)} \tag{40}$$

where $F^{(0)}$ contains the intrabond matrix elements and it has a blockdiagonal structure (cf. Fig. 2), while the interbond matrix elements are collected into $F^{(1)}$.

Substituting these expansions into the biorthogonal local Brillouin theorem,

Eq. (35) and collecting all terms of a given order separately, we get the following results:

$$\eta_{ik*}^{(1)} = -\frac{\tilde{F}_{ik*}}{\Delta E_{ik*}} \tag{41 a}$$

$$\eta_{ik*}^{(2)} = \left(\sum_{k*}^{occ} \eta_{lk*}^{(1)} \tilde{F}_{li} - \sum_{j*}^{virt} \eta_{ij*}^{(1)} \tilde{F}_{k*j*}\right) \Big/ \Delta E_{ik*} \tag{41 b}$$

and for $n \geqq 3$:

$$\eta_{ik*}^{(n)} = \frac{\sum_{l} \eta_{lk*}^{(n-1)} \tilde{F}_{li} - \sum_{j*} \eta_{ij*}^{(n-1)} \tilde{F}_{k*j*} + \sum_{\mu=1}^{n-2} \sum_{j*,l} \eta_{ij*}^{(\mu)} \eta_{lk*}^{(n-\mu-1)} \tilde{F}_{lj*}}{\Delta E_{ik*}} \tag{41c}$$

where we introduced the notations

$$\tilde{F}_{ik} = \langle \tilde{\psi}_i | F^{(1)} | \psi_k \rangle \tag{42}$$

and

$$\Delta E_{ik*} = F_{k*k*}^{(0)} - F_{ii}^{(0)} . \tag{43}$$

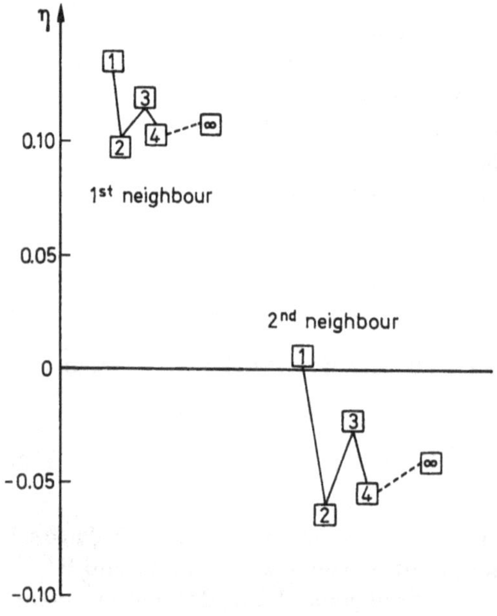

Fig. 3. Convergence of the PT series for the tails of LMOs in a H_6 chain with weakly alternating geometry ($r_< = 93.73$ pm, $r_> = 101.60$ pm) in STO-6G basis set. The figures in the circles specify the order of the PT. The infinite order results have been obtained by *a posteriori* localization [201]

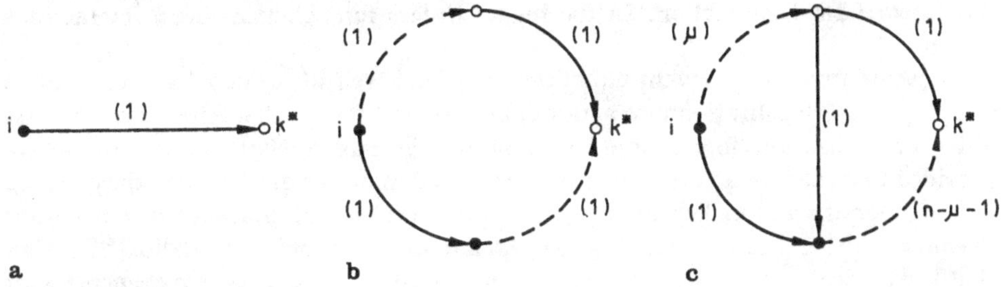

Fig. 4a–c. Diagrammatic representation of the recursive tail formulae (41) giving the PT expression for the delocalization correction η^*_{ik} at various orders. **(a)**: first order diagram showing throught space delocalization; **(b)**: second order graph illustrating through bond interaction; **(c)**: n-th order diagram exhibiting through-many-bond delocalization for n > 2. The vertices of the graphs represent zero-order SLMOs while the edges correspond to interaction matrix elements (solid line) or tails from previous orders (dashed line). The numbers in parentheses indicate the order of the given line. For other rules, see text

The recursive formulae (41) give the values of the delocalization corrections at any order n. The convergence is illustrated in Fig. 3 for a chain of hydrogen atoms. More detailed numerical results are reported in Ref. [39].

Fig. 4 shows a diagrammatic representation of Eqs. (41) in order to visualize the results obtained. The specific features of the problem justify the application of a special diagrammatic technique suited for representing **tail formulas** instead of energy corrections. Accordingly, the graphs in Fig. 4 are constructed in the following manner:

(i) the SLMOs serve as the vertices of the graph; occupied SLMOs are denoted by full circles, virtual ones by empty circles.
(ii) each diagram represents all terms contributing to the tail coefficient $\eta^{(n)}_{ik^*}$ at the given order; every possible path along the oriented lines from the initial vertex i· to the final vertex ∘k* represents a definite type of terms in the corresponding tail formula;
(iii) full lines mean the interaction matrix element $\tilde{F}_{ik^*}$, while dashed lines represent tails generated already at the lower orders;
(iv) the sum of the orders along a path must be equal the order of the diagram n (the interaction matrix element is considered as a first-order quantity);
(v) except for the initial vertex i· and the final vertex ∘k* all vertex involve a summation over the corresponding index. Similarly, one must perform a summation over the possible values of the label μ for the path containing both dashed lines of order μ and $n - 1 - \mu$ in Fig. 4 (c);
(vi) each sum is to be divided by the corresponding excitation energy, ΔE_{ik^*}, and multiplied by $(-1)^q$ q being the number of full vertices i· along the path in question.

The above diagrams are useful because they can be applied as mnemotechnic rules as well as they make the interpretation of the results easier. The first order graph illustrates a direct (**through space**) interaction between an occupied and a virtual SLMO. At the second order one finds that the interaction is propagated *via* a third intervening bond. Both the occupied and the virtual SLMOs may transmit this **through bond interaction.** At the higher orders **through-many-bond** interactions appear.

As seems from the relevant equations (41), and well illustrated by Fig. 3, these three types of delocalization interactions have rather different characteristics. The first order delocalization interaction resulting in $\eta_{ik^*}^{(1)}$ has a short radius of action provided that the interaction matrix element $\tilde{F}_{ik^*}$ diminishes quickly with the distance between bonds i and k. Typically, the first order tails are accurate for first neighbor ('geminal') delocalization (see Fig. 3). At the second order, this radius of action is at least twice as large, since the intervening bonds may have matrix elements with both bonds i and k. Figure 3 shows, for instance, that a good estimation for the second neighbor tail is obtained at the second order of the PT. The higher order tails determine the long range asymptotics of the LMOs as well as that of the density matrix discussed in Sect. 3.

Tails of LMOs at various orders of PT can also be classified according to their sensitivity to the relative orientation of bonds. Clearly, the direct first order through space interaction may be rather sensitive to the relative orientation of the two bonds involved. The opposite holds for the second order tails: as illustrated by Fig. 5, no change is expected for a second order ('vicinal') tail $\eta_{ij^*}^{(2)}$ if the interaction is propagated by the bond k as the axis of the rotation, this being clearly the dominating path in most cases (cf. ethane-like molecules). Consequently, first order theories are expected to give nearly accurate results for conformational energy differences. This, in fact, has been found to be true in ab initio as well as semiempirical bond orbital approaches [36–40, 60, 98, 179–183, 217–224].

The tail formulae in the form of Eq. (41) reflect the above statements somewhat implicity, since the interaction matrix $\tilde{F}$ in terms of which these formulae are written in nonlocal as a consequence of the biorthogonal formulation [cf. Eq. (42)]. This problem can be eliminated by expanding the inverse of the overlap matrix in a Taylor series [39].

In summarizing, the following model can be constructed to represent two electron bonds in well-localizable molecules at the Hartree-Fock level. Molecules are viewed as systems of weakly interacting bonds. Each bond corresponds to a strictly localized MO expanded in mutually exclusive, though overlapping, subspaces of (hybridized) basis functions. The internal structure of bonds is affected by inductive interbond interactions. The antisymmetrized product (determinant) of SLMOs is considered as a zero-order wave function. The corresponding zero-order energy is evaluated as an expectation value of the full Hamiltonian with the zero-order determinant. Due to the overlap between basis orbitals, the SLMOs corresponding to different bonds also overlap with each other. Interbond delocalization can be taken into account by perturbation theory which has been elaborated for any order resulting in recursive tail formulae. The latter can be visualized by a special diagrammatic representation.

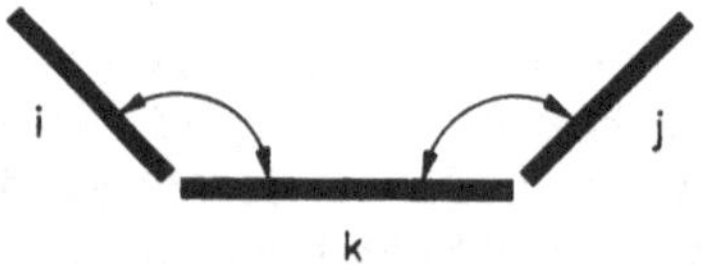

Fig. 5. Through space interaction between vicinal bonds (i, k) and (k, j) results in delocalization interaction between (i, j) through the bond k. The latter is insensitive to the relative orientation of bonds i and j

The above theory is in a close relationship with the PCILO method [32–35]. Apart from some minor differences, the main deviations from PCILO are that

(i) we consider only PT corrections to LMOs, while the energy is calculated as the expectation value of the Hamiltonian;
(ii) we did not consider electron correlation effects which are treated in PCILO with delocalization corrections on an equal footing. In the next section we outline a different strategy of quantum chemical calculations including also correlation effects.

5 Strictly Localized Geminals

5.1 On the Strategy of Quantum Chemical Calculations

Accurate evaluation of the electronic energy of a many-electron system in most quantum chemical approaches consists of two main issues. First, one determines an approximate energy on the basis of the independent particle (Hartree-Fock) model. Second, this energy is gradually improved taking into account electron correlation effects by some (mostly variational or perturbational) method.

This procedure is visualized by path (1) in Fig. 6, and can be called the "usual path". The HF wave function is considered as a zero-order reference state. It is computed accurately in the given basis set, while the correlation energy is guessed usually by some truncated calculation. This is illustrated by the dashed line of path (1) in Fig. 6.

Quite another approach is followed by the PCILO scheme [32–35], see path (2) in Fig. 6. The zero-order reference state is represented by the SLMO approximation discussed in Sect. 4, then perturbation theory is used to account for both delocalization and correlation effects. This approach can be justified because the SLMO level is very easy to reach, then PCILO takes care of the above two difficulties simultaneously. Also, we can say that there is not too much reason to compute E_{HF} exactly while E_{corr} is only approximately calculated. PCILO gives us a balanced account of Hartree-Fock and correlation effects in a molecule.

A conceptual problem with the PCILO method is, however, that the effects handled by PT are by no means small. Although the SLMO energy is often quite close to the

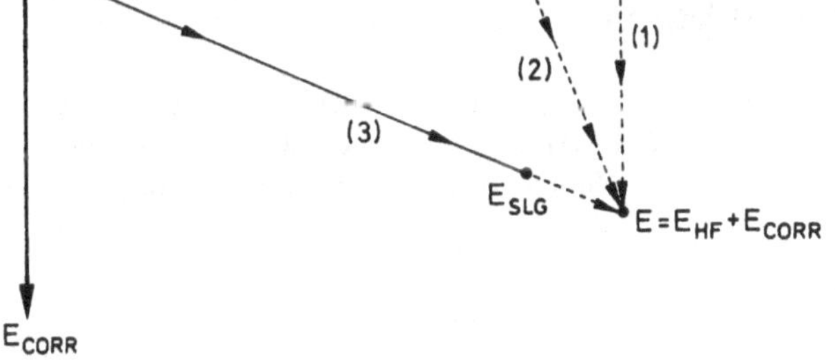

Fig. 6. Different "paths" of quantum chemical methods approaching the exact energy. (*1*): usual path; (*2*): PCILO path; (*3*): proposed SLG + PT path

E_{HF} in a given basis, the correlation energy can be classified to intra- and interbond contributions the former being much larger in most cases. The large intrabond ("left-right") correlation neglected at the zero order may cause unfavorable convergence properties of the PCILO expansion [225]. {This objection should, of course, hold also for the usual path (1)}.

Intrabond correlation effects nevertheless need to be calculated by PT: they can be estimated at the zero order. The corresponding strategy is illustrated by path (3) of Fig. 6. According to this approach, one first solves the quantum mechanical problem of each two electron bond within the corresponding subspace of basis orbitals leading to the best possible local wave function representing the bonds of the molecule. The antisymmetrized product of these correlated local bond wave functions is considered as the zero-order reference state, which is expected to be quite close to the exact wave function. The aim of PT is then merely to account for interbond interactions. Since the two-electron functions (geminals) for each bond are strictly localized, they can be called as **strictly localized geminals** (SLG). The corresponding zero-order many-electron wave function is the antisymmetrized product of strictly localized geminals (AP-SLG). The SLG method is similar to various electron-pair approaches [226–229] and to some valence bond (VB) methods [230–232] which also follow, more-or-less, the ideas of path (3) in Fig. 6. The recent extended geminal method by Roeggen [233] should be mentioned. The essential difference between this method and the SLG + PT approach is that Roeggens's zero order is not strictly localized being obtained from the standard antisymmetrized product of strongly orthogonal geminals (APSG) [234, 235] approximation. Also, the strict localization in an overlapping basis means that the SLG-s will not be orthogonal (neither strongly nor weakly) to each other. Though the standard APSG wave function is not fully localized, the corresponding geminals can be well associated with local bonds [233, 236, 237]. Accordingly, the relation between the SLG and the APSG wave functions is analogous to that between the SLMO and the HF wave functions. As to the relation between the SLG wave function and those obtained by traditional VB schemes, the main difference is that the ionic and covalent VB structures are not treated separately in the SLG method, their weight being optimized in a geminal.

The three paths indicated in Fig. 6 approaching the exact solution within a given basis set correspond to **different partitionings** of the Hamiltonian. The usual path (1) is the most elaborated one, but we feel that path (3) is the most useful one in many cases, since it works with an easy-to-obtain and rather accurate zero-order reference state, possessing a balanced account of SCF and correlation effects. In the next section we shall specify the formalism for obtaining the SLG wave function.

5.2 Antisymmetrized Product of Strictly Localized Geminals

We assume the existence of a local subset of basis orbitals assigned to each chemical bond of the molecule. The most general local two-electron wave function for the bond i, expanded in the corresponding subset, can be written in terms of spinorbitals as

$$\psi_i(1,2) = \sum_{\substack{\mu,\nu \in i \\ (\mu<\nu)}} C^i_{\mu\nu} a^+_\mu a^+_\nu \,|\mathrm{vac}\rangle \tag{44}$$

where $C^i_{\mu\nu}$ is the expansion coefficient to be determined, a^+_μ is a creation operator creating an electron on the basis orbital χ_μ, and $|vac\rangle$ is the vacuum state. This formula is analogous to Eq. (23) giving the form of SLMOs. The restriction $\mu, \nu \in i$ expresses the **strictly localized** character of the two-electron function or **geminal** ψ_i.

Application of the second quantized formalism is not necessary at this point, but as it will turn out later, this approach has some advantage in deriving and interpreting the results. The partitioning of the Hamiltonian is especially straightforward in the second quantized representation.

Eqation (44) is quite general: no restrictions of spin, etc. are yet introduced. If the geminals are singlet coupled and are described by an antisymmetric spin part, the spatial coefficient matrix is symmetric. Diagonalization of this symmetric matrix leads to the so called "natural geminals" [237, 240–242] which are expanded by a simple summation, in contrast to the general form of Eq. (44) which exhibit a double summation. We shall keep this general form, however.

In order to derive the local equations for determining the optimum values of the expansion coefficients $C^i_{\mu\nu}$, we have to analyze the structure of the Hamiltonian.

Partitioning of the Hamiltonian in terms of bonds is possible only through partitioning of the basis orbitals. This requires to use an effective Hamiltonian projected into the space of finite basis, which can be most conveniently achieved in the second quantized formalism. Due to the overlap of the basis functions we have to apply a somewhat special formalism: the second quantization for non-orthogonal orbitals which has been elaborated previously by several authors [243–247].

The Born-Oppenheimer electronic Hamiltonian can be written in a non-orthogonal basis set using the second quantized representation as [24, 25, 43, 243]:

$$H = \sum_{\mu\nu} h_{\tilde{\mu}\nu} a^+_\mu a_\nu + \frac{1}{2} \sum_{\mu\nu\lambda\sigma} [\tilde{\mu}\tilde{\nu}|\lambda\sigma]\, a^+_\mu a^+_\nu a_\sigma a_\lambda \qquad (45)$$

where a_ν is the true annihilation operator obeying the fermion anticommutation rule

$$[a^+_\mu a_\nu]_+ = \delta_{\mu\nu} \qquad (46)$$

and the modified set of integrals is defined by

$$h_{\tilde{\mu}\nu} = \sum_\lambda S^{-1}_{\mu\lambda} h_{\lambda\nu} \qquad (47)$$

and

$$[\tilde{\mu}\tilde{\nu}|\lambda\sigma] = \sum_{\tau,\varrho} S^{-1}_{\mu\tau} S^{-1}_{\nu\varrho} [\tau\varrho|\lambda\sigma] \qquad (48)$$

with S^{-1} being the inverse of the basis overlap matrix. Since the basis functions are overlapping, the corresponding creation and annihilation operators are **not** the adjoints of each other, but the following relation holds [24, 25, 244]:

$$a_\nu = \sum_\lambda S^{-1}_{\nu\lambda} (a^+_\lambda)^\dagger \qquad (49)$$

where the dagger (†) indicates the adjoint. We say that the annihilation operators are defined in the **reciprocal space.** Similarly, all bra-indices of the integral list of the Hamiltonian are transformed into the reciprocal space.

Several other forms of the second quantized Hamiltonian could have been written down for an overlapping basis. The above form resembles greatly the usual Hamiltonian written in an orthogonal matrix, and it will serve as a convenient basis of the forthcoming deduction.

Since each basis function χ_μ is assigned to a particular bond, any summation over basis orbitals can be formally substituted by a summation over bonds i and over orbitals centred on the bond i:

$$\sum_{\mu} \equiv \sum_{i} \sum_{\mu \in i} \tag{50}$$

By means of this identity, the Hamiltonian of Eq. (45) becomes

$$H = \sum_{i} H^{i} + \sum_{ij}' H^{ij} + \sum_{ijk}' H^{ijk} + \sum_{ijkl}' H^{ijkl} \tag{51}$$

where the primes on the summations indicate the restriction i ≠ j, etc. The operator H^i is a one-bond Hamiltonian in which the summation labels run only within the i-th subspace:

$$H^{i} = \sum_{\mu\nu \in i} h_{\tilde{\mu}\nu} a_{\mu}^{+} a_{\nu} + \frac{1}{2} \sum_{\mu\nu\lambda\sigma \in i} [\tilde{\mu}\tilde{\nu} \,|\, \lambda\sigma] \; a_{\mu}^{+} a_{\nu}^{+} a_{\sigma} a_{\lambda} \tag{52}$$

This operator is considered as the Hamiltonian for the bond i. Apart from some overlap effects which are reflected by the biorthogonal integral list, it describes an isolated bond; we shall introduce below effective bond Hamiltonians which describe the electronic structure of localized bonds in the field of each other. The next three terms of the total Hamiltonian in Eq. (51) describe interbond interactions explicitly, with H^{ij} defining the pairwise interaction, while H^{ijk} and H^{ijkl} accounting for the simultaneous interaction of three and four bonds, respectively. The expressions of these interaction Hamiltonians can be obtained easily by the partitioning of the summation labels in Eq. (51) according to Eq. (50). The pairwise interaction operator H^{ij} after simple algebraic manipulations can be given in the following compact form:

$$\begin{aligned} H^{ij} = {} & \sum_{\mu \in i} \sum_{\nu \in j} h_{\tilde{\mu}\nu} a_{\mu}^{+} a_{\nu} + \sum_{\mu\nu\lambda \in i} \sum_{\sigma \in j} [\tilde{\mu}\tilde{\nu} \,|\, \lambda\sigma] \; a_{\mu}^{+} a_{\nu}^{+} a_{\sigma} a_{\lambda} \\ & + \sum_{\mu\lambda\sigma \in i} \sum_{\nu \in j} [\tilde{\mu}\tilde{\nu} \,|\, \lambda\sigma] \; a_{\mu}^{+} a_{\nu}^{+} a_{\sigma} a_{\lambda} \\ & + \sum_{\mu\lambda \in i} \sum_{\nu\sigma \in j} \{[\tilde{\mu}\tilde{\nu} \,|\, \lambda\sigma] - [\tilde{\mu}\tilde{\nu} \,|\, \sigma\lambda]\} \; a_{\mu}^{+} a_{\nu}^{+} a_{\sigma} a_{\lambda} \\ & + \frac{1}{2} \sum_{\mu\nu \in i} \sum_{\lambda\sigma \in j} [\tilde{\mu}\tilde{\nu} \,|\, \lambda\sigma] \; a_{\mu}^{+} a_{\nu}^{+} a_{\sigma} a_{\lambda} \end{aligned} \tag{53}$$

Different terms of H^{ij} have simple physical meanings. The operator string in the first term, $a_\mu^+ a_\nu$, annihilates an electron in the bond j and creates one on the bond i. Accordingly, it describes a j → i electron delocalization (charge transfer). The next two terms have a similar meaning: they describe delocalization as a consequence of electron interaction. The fourth term has a different significance: its operator string, $a_\mu^+ a_\nu^+ a_\sigma a_\lambda$ conserves the number of electron on both bonds i and j, thus it describes the interbond Coulomb and exchange repulsion as well as dispersion. Finally, the last term describes a simultaneous transfer of two electrons which is expected to be of smaller importance as compared to the previous effects.

Expressions for the three- and four-body operators H^{ijk} and H^{ijkl} are available [44], but we do not put them down here since they do not enter the working formulae for a low-order perturbation treatment.

Let us now study the formalistic aspects of the SLG theory. This will be useful in simplifying the following development, as well as introducing a special language characteristic of this field. First of all, we introduce the notations

$$\psi_i^+ = \sum_{\substack{\mu,\nu \in i \\ (\mu<\nu)}} C^i_{\mu\nu} a_\mu^+ a_\nu^+ \tag{54}$$

and[13]

$$\psi_i^- = \sum_{\substack{\mu,\nu \in i \\ (\mu<\nu)}} C^i_{\mu\nu} a_\nu a_\mu \tag{55}$$

Using Eq. (54), the strictly localized geminal of Eq. (44) can be written simply as

$$\psi_i(1,2) = \psi_i^+ \mid \text{vac}\rangle \tag{56}$$

Operators $\psi_i^+ (\psi_i^-)$ are formally create (annihilate) the wave function of the bond i on the vacuum state. (The annihilation property of ψ_i^- will be proved below.) One can say that they create (annihilate) a **composite particle** identified with a two-electron bond. In this language, each local chemical bond of a môlecule is viewed as a **quasiparticle** whose internal structure is defined by the expansion coefficients $C^i_{\mu\nu}$. The mathematical step passing from the elementary fermion operators a_μ^+, a_ν to the composite particle operators ψ_i^+, ψ_i^- is a special **quasiparticle transformation.** Similar transformations have been used by several authors in a somewhat different context [248–254].

To understand the nature of the quasiparticles introduced above, we have to study the algebraic properties of the corresponding composite particle creation operators. Evaluating the commutators directly, we find

$$\psi_i^+ \psi_i^+ - \psi_i^+ \psi_i^+ \equiv [\psi_i^+, \psi_i^+] = 0 \tag{57a}$$

$$\psi_i^- \psi_i^- - \psi_i^- \psi_i^- \equiv [\psi_i^-, \psi_i^-] = 0 \tag{57b}$$

---

[13] For ψ^- the Longuet-Higgins notation [240] is used, the superscript "—" indicating that this mathematical object is an annihilation operator, to be distinguished from the wave function $\psi_i(1, 2)$.

and

$$\psi_i^- \psi_i^+ - \psi_i^+ \psi_i^- \equiv [\psi_i^-, \psi_i^+] = \hat{Q}_i \delta_{ik} \tag{57c}$$

where operator $\hat{Q}_i$ is defined as

$$\hat{Q}_i = 1 + \sum_{\mu\nu\lambda \in i} C^i_{\nu\lambda} C^i_{\lambda\mu} a^+_\mu a_\nu \tag{58}$$

provided that the convention

$$C^i_{\mu\nu} = -C^i_{\nu\mu} \tag{59}$$

is used for the (so far undefined) elements ($\mu > \nu$) of the coefficient matrix. In Eq. (58), 1 symbolizes the unity operator[14].

The first thing we observe is that in Eq. (57) we are dealing with *commutators* rather than anticommutators. This is quite natural since, according to Eqs. (54–55), the quasiparticles are composed of two electrons, thus any transposition of ψ_i^+ – s involves a simultaneous transposition of two fermions so no sign change may occur. Equations (57) are **Bose-like** commutation rules, and the quasiparticles introduced above are composite Bose-particles. There is an important deviation in Eq. (57c) from the usual Bose commutation rules, the appearance of operator $\hat{Q}_i$ which would be the unity operator in the usual Bose algebra. The deviation of $\hat{Q}_i$ from the unity operator (the second term in Eq. (58)) is a consequence of the composite nature of the quasiparticles.

The presence of operator $\hat{Q}_i$ in the commutator equations results in some complications. The situation is not too bad, however, because the commutator of Eq. (57c) is diagonal in the indices i and k. This is a consequence of the strict localization of the geminals, that is, of the expansion of ψ_i^+ and ψ_k^+ on mutually exclusive subspaces of fermion operators $\{a_\mu^+\}$ for $i \neq k$. Also, operator $\hat{Q}_i$ has the important property that acting on the vacuum state it acts as the unity operator:

$$\hat{Q}_i \,|\,\mathrm{vac}\rangle = |\mathrm{vac}\rangle \tag{60}$$

which also results in considerable simplifications of the formalism. First of all, ψ_i^- can be considered as a true annihilation operator with respect to ψ_i^+:

$$\psi_i^- \psi_i^+ \,|\,\mathrm{vac}\rangle = (\psi_i^+ \psi_i^- + \hat{Q}_i)\,|\mathrm{vac}\rangle = |\mathrm{vac}\rangle$$

This result is very important because it permits us to use the standard calculus of creation-annihilation operators and the strange operator $\hat{Q}_i$ will not appear in any further formulas.

---

[14] It may be of interest to note that operator $\hat{Q}_i$ can easily be expressed in terms of the first order density matrix P^i (vide infra Eqs. (68, 69)) as $\hat{Q}_i = 1 - \sum_{\mu,\nu \in i} P^i_{\mu\nu} a^+_\mu a_\nu$.

The many-electron wave function in the SLG approximation can be written as

$$|\Psi\rangle = \psi_1^+ \psi_2^+ \dots \psi_n^+ \ |\, \text{vac}\rangle \tag{61}$$

where we have n two-electron bonds in the molecule (closed shell systems are treated throughout for the sake of simplicity). This wave function is symmetric in terms of the quasiparticles ψ_i^+, but it is properly antisymmetric in terms of electrons. Accordingly, the wave function $|\Psi\rangle$ of Eq. (61) can be defined as the antisymmetric product of strictly localized geminals, AP-SLG.

It is clear from the definitions of ψ_i^+ and ψ_i^- that they are not the adjoints of each other, reflecting the nonorthogonality of the one-electron basis set expressed by Eq. (49). Therefore, evaluation of the usual expectation values as e.g. $\langle\Psi|\, H\, |\Psi\rangle$, is very complicated. However, we may define the counterpart of the AP-SLG wave function $|\Psi\rangle$ in the reciprocal space as

$$\langle\tilde{\Psi}| = \langle \text{vac}\ |\, \psi_n^- \dots \psi_2^- \psi_1^- \ , \tag{62}$$

and define our energy functional by the following generalized Rayleigh-quotient:

$$E = \frac{\langle\tilde{\Psi}|\, H\, |\Psi\rangle}{\langle\tilde{\Psi}\ |\ \Psi\rangle} \tag{63}$$

Evaluation of matrix elements of type of Eq. (63) can be done exactly in the same manner as in the orthogonal case. All formal rules (Wick's theorem, etc.) remain valid. For example, application of the commutation rules in Eq. (57) gives

$$\langle\tilde{\Psi}\ |\ \Psi\rangle = \langle \text{vac}|\ \psi_n^- \dots \psi_2^- \psi_1^- \psi_1^+ \psi_2^+ \dots \psi_n^+\ |\text{vac}\rangle = 1 \tag{64}$$

This means that the wave function $|\Psi\rangle$, though not normalized, is **'binormalized'** to unity. Similarly, with Eq. (63) one does not evaluate the energy as the usual expectation value, but one uses a different energy functional. However, if the wave function $|\Psi\rangle$ is a true eigenvalue of the Hamiltonian, that is, the Schrödinger equation

$$H\ |\ \Psi\rangle = \mathscr{E}\ |\ \Psi\rangle \tag{65}$$

is fulfilled, we find:

$$E = \frac{\langle\tilde{\Psi}|\ H\ |\Psi\rangle}{\langle\tilde{\Psi}\ |\ \Psi\rangle} = \mathscr{E}\,\frac{\langle\tilde{\Psi}\ |\ \Psi\rangle}{\langle\tilde{\Psi}\ |\ \Psi\rangle} = \mathscr{E} \tag{66}$$

That is, the energy functional (63) has the value of the exact energy in this limiting case. Using approximate wave functions which do not obey Eq. (65), one does not get the usual expectation value, and the use of Eq. (63) corresponds to the method of moments rather than to the variational method. Further on, when for the sake of brevity we write 'expectation value', this will be understood as the generalized matrix element of Eq. (63).

Let us see now how these expectation values can be evaluated in practice. We study first the matrix element of the one-bond Hamiltonian H^i:

$$\langle\tilde{\Psi}|\, H^i \,|\Psi\rangle = \langle \mathrm{vac}|\, \psi_i^- H^i \psi_i^+ \,|\mathrm{vac}\rangle =$$

$$= \sum_{\mu\nu\in i} h_{\tilde{\mu}\nu} P^i_{\nu\mu} + \frac{1}{2} \sum_{\mu\nu\lambda\sigma\in i} [\tilde{\mu}\tilde{\nu}\,|\,\lambda\sigma]\, \Gamma^i_{\lambda\sigma\mu\nu} \tag{67}$$

where P^i and Γ^i are the analogues of the first- and second-order density matrices for the bond i. They are defined as the (generalized) expectation values of corresponding operator strings:

$$P^i_{\nu\mu} = \langle \mathrm{vac}|\, \psi_i^- a_\mu^+ a_\nu \psi_i^+ \,|\mathrm{vac}\rangle \tag{68a}$$

$$\Gamma^i_{\lambda\sigma\mu\nu} = \langle \mathrm{vac}|\, \psi_i^- a_\mu^+ a_\nu^+ a_\sigma a_\lambda \psi_i^+ \,|\mathrm{vac}\rangle \tag{68b}$$

Explicit evaluation of these matrix elements can be done in a straightforward manner by the substitution of the form of SLGs, Eqs. (54–55), and by means of the usual fermion anticommutation rules. We get:

$$P^i_{\nu\mu} = \sum_{\lambda\in i} C^i_{\mu\lambda} C^i_{\nu\lambda} \tag{69a}$$

$$\Gamma^i_{\lambda\sigma\mu\nu} = C^i_{\mu\lambda} C^i_{\nu\sigma} \tag{69b}$$

(Here again the antisymmetry condition of Eq. (59) for the coefficients $C^i_{\mu\nu}$ is utilized.)

Let us turn now to the problem of defining the effective bond Hamiltonians $H^{i(\mathrm{eff})}$. This can be done on the basis of the group function theory [26, 45, 46] or various fragment SCF approaches [49–52, 60], or analogously to the methods describing a molecule within the fields of others proposed e.g. by Otto and Ladik [169] and introduced independently by Weinstein et al. [255]. The effective Hamiltonians will be used to put down the local Schrödinger equations for each bond, which determine the expansion coefficients of the SLGs:

$$H^{i(\mathrm{eff})} \psi_i^+ \,|\mathrm{vac}\rangle = E_i \psi_i^+ \,|\mathrm{vac}\rangle \qquad i = (1, 2, \ldots n) \tag{70}$$

Considering the interbond interaction terms H^{ij}, etc., see Eqs. (51–53), we find that the fourth term of Eq. (53), which describes interbond Coulomb interactions, has an average value. This can be incorporated into an effective one-body core in a symmetrized form as

$$h^{i(\mathrm{eff})}_{\mu\nu} = h^i_{\mu\nu} + \frac{1}{2}(G_{\tilde{\mu}\nu} + G_{\tilde{\nu}\mu}) \tag{71}$$

where

$$G_{\tilde{\mu}\nu} = \sum_{(j\neq i)} \sum_{\lambda\sigma\in j} P^j_{\lambda\sigma}([\tilde{\mu}\tilde{\lambda}|\nu\sigma] - [\tilde{\mu}\tilde{\lambda}|\sigma\nu]) \qquad (\mu, \nu \in i) \tag{72}$$

These formulae describe a pairwise interaction between bonds. This interaction is electrostatic in nature, containing pure coulomb and exchange effects as seems from the two-electron integrals present in Eq. (72). Such type of interbond interactions do not lead to electron delocalization between different bonds, but they give rise to the so called **inductive effect** determining the overall charge distribution of most molecules.

At this zeroth order description we neglect the biorthogonal transformation in the list of one- and two-electron integrals in the Hamiltonian of Eq. (52). They represent essentially interbond overlap effects, and will give first-order contributions to the total energy (vide infra). As to the handling of intrabond overlap effects, this is done by appropriate local orthogonalization for convenience. Thus, the final form of the effective Hamiltonian of the bond i becomes:

$$H^{i(eff)} = \sum_{\mu\nu\in i} h^{i(eff)}_{\mu\nu} a^+_\mu a^+_\nu + \sum_{\mu\nu\lambda\sigma\in i} [\mu\nu|\lambda\sigma]\, a^+_\mu a^+_\nu a_\sigma a_\lambda \tag{73}$$

The wave function and the effective energy of the bond i are evaluated by the local Schrödinger equation (70) which results a "full CI" solution in the corresponding local basis set. The total electronic energy of the system is not simply the sum of effective bond energies, but the latter has to be corrected for a double counting of interbond electron repulsion incorporated into the effective core. Similarly, the difference between the biorthogonal and the direct two-electron integral list will enter the energy formula as a first-order correction. It follows than that the final energy expression for the AP-SLG wave function takes the form[15]:

$$\begin{aligned} E_{SLG} = \sum_{i=1}^{n} E_i - \frac{1}{2} \sum_{i\neq j} \sum_{\mu\nu\in i} \sum_{\lambda\sigma\in j} P^i_{\mu\nu} P^j_{\lambda\sigma} ([\mu\lambda|\nu\sigma] - [\mu\lambda|\sigma\nu]) \\ + \sum_{i=1}^{n} \sum_{\mu\nu\in i} (h_{\tilde{\mu}\nu} - h_{\mu\nu})\, P^i_{\nu\mu} \\ + \sum_{i=1}^{n} \sum_{\mu\nu\lambda\sigma\in i} ([\tilde{\mu}\tilde{\nu}|\lambda\sigma] - [\mu\nu|\lambda\sigma])\, \Gamma^i_{\lambda\sigma\mu\nu} \end{aligned} \tag{74}$$

with the density matrices defined by Eqs. (69).

Due to the presence of the first-order density matrix P^i in the effective core matrix elements, the local Schrödinger equations have to be solved iteratively. According to the numerical experience, this iterative procedure converges very fast.

The energy as defined by Eq. (74) gives only a crude approximation of the exact electronic energy as a consequence of the strict localization of the geminals. Still, this zeroth order method can be useful in some cases, because it is extremely cheap and simple and represents a fairly correlated description taking into account the so called "left-right" intrabond correlation. As a consequence, it yields qualitatively correct potential curves for the fission of single bonds [43, 239]. Treatment of dissociation of multiple bonds is possible in a similar framework [256]. Further,

---

15 Note that the result for the energy expression is different from that given in the original paper [43], since there an unfortunate symmetrization of the two-electron integrals was introduced.

Table 2. Zeroth order total energies (a.u.) obtained by the SLG wave function in STO-3G basis set. For comparison, SCF results are also indicated

molecule	SCF	SLG
H_2	− 1.1167	− 1.1373 (full CI)
CH_4	− 39.7265	− 39.7302
H_2O	− 74.9632	− 75.0069
CH_3F	−137.1686	−137.1321

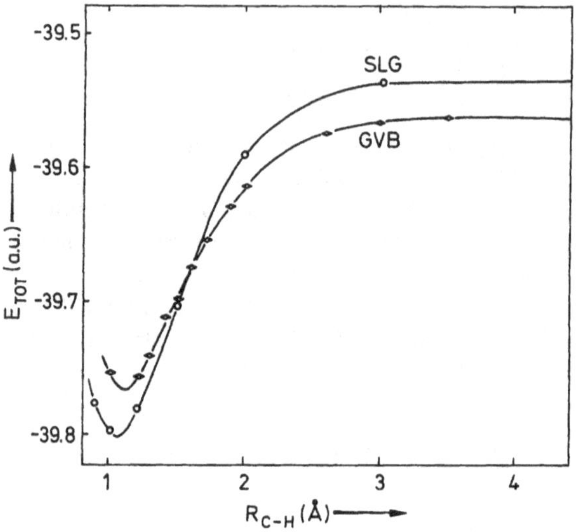

Fig. 7. Potential curve of the $CH_4 = CH_3 + H$ dissociation as obtained by the GVB and SLG approximations in STO-3G basis sets

assuming proper hybridization, this description provides us the best possible local representation of chemical bonds, thus the shortcomings of the SLG wave function reflect the limitations of the localized bond models in the given basis set.

Table 3. Bond energies E_i, polarities λ_i, ionic I_i^2 and covalent C_i^2 characters for CH and OH bonds in various chemical environment as computed by the SLG method in STO-3G basis. Energies are given in atomic units, distances in pm. (Data taken from Ref. [239].)

BOND	Molecule/environment		$-E_i$	λ_i	I_i^2	C_i^2
CH	$H-^R-CH_3$	R = 105	1.993	0.081	0.417	0.583
		R = 200	1.966	0.046	0.415	0.585
R_{CH} = 1.0936		R = 500	1.968	0.042	0.415	0.585
	FCH_3	R_{FC} = 140	2.190	0.192	0.419	0.581
	F_2CH_2	R_{FC} = 140	2.190	0.192	0.419	0.581
	F_3CH	R_{FC} = 140	2.305	0.259	0.421	0.579
OH	$H-^R-OH$	R = 90	2.615	0.355	0.431	0.569
R_{OH} = 96		R = 400	2.573	0.322	0.426	0.574
	CH_3-^R-OH	R = 140	2.587	0.332	0.431	0.569
		R = 400	2.576	0.322	0.426	0.574

To illustrate the accuracy of the SLG method, Table 2 and Fig. 7 are included to show the total energies of some simple molecules and the potential curve of the $CH_4 \rightarrow CH_3 + H$ dissociation.

The SLG wave function can also be used to study the transferability of various bond properties. It is obvious that transferability is a stronger condition than localizability; even apparently local properties may be non-transferable as a consequence of effective interactions. In Table 3 we present bond energies E^i and some wave function characteristics for CH and OH bonds based on a recent investigations using the SLG method [239]. The bond polarities (λ_i), ionic (I_i^2) and covalent (C_i^2) characters are defined by expanding the SLGs in a minimal basis set as

$$\psi_i^+ = \frac{I_i}{2}\left[\sqrt{1+\lambda_i}\, a_{1\alpha}^+ a_{1\beta}^+ + \sqrt{1-\lambda_i}\, a_{2\alpha}^+ a_{2\beta}^+\right] + \frac{C_i}{\sqrt{2}}\left[a_{1\alpha}^+ a_{2\beta}^+ + a_{2\alpha}^+ a_{1\beta}^+\right]$$

where the subscripts 1 and 2 label the two hybrids of bond i. It is apparent from the table that the energies are very sensitive to the chemical environment; this is quite natural because E_i contains the inductive interaction energy with other bonds. On the other hand, the ionic and covalent characters (note that these would be equal for a one-determinantal wave function) are practically constant for one bond. For instance, the covalent character of the CH bond is 0.58 while that of the OH bond is 0.57 for all cases. This indicates that the local electron correlation is rather insensitive to the chemical environment, even in the case when a neighbouring bond is dissociating thus its electron correlation is significantly altered.

The bond polarity parameters, however, show quite a different behaviour. They are very sensitive to the environment, thus completely nontransferable. Because of inductive effects, the CH bond polarity increases to 0.08, 0.13, 0.19 and 0.26 as 0, 1, 2 or 3 fluorine substituents are present on methane, respectively. Concerning the inductive role or the neighbouring CH bonds, we observe a 100% change in the polarity for methane when one of the H atoms in the neighborhood is dissociating. This means that the small bond polarity of methane is mainly determined by the interbond dipole-dipole inductive interaction rather than simply the electronegativity difference between carbon and hydrogen. Note that, according to the present analysis by Reed and Weinhold [257], bond dipoles have nothing to do with the bond polarity λ_i.

The shortcomings of the SLG approximation can be corrected by perturbation theory. This will be the topics of the next section, while in Sect. 6. an informative qualitative application of the zero-order scheme will be presented.

5.3 Perturbation of Geminals

Improvement of the SLG wave function is possible by means of a perturbative approach. Perturbation of geminals was discussed previously by Kapuy [41, 42] and recently by Roeggen [241]. A related valence-bond method followed by perturbation theory was developed by Kirtman and Cole [258]. The present problem is considerably simpler because the geminals are kept in the strictly localized form. On the other hand, in contrast to the APSG-type theories, the SLG-s are not orthogonal which requires a special formalism. Here, as previously, we shall use the biorthogonal formulation to overcome this non-orthogonality problem and to deal with the formally non-hermitian operators [43, 205, 239, 243, 259, 260].

Second quantization provides a powerful tool for developing a many-body theory suitable for deriving interbond delocalization and correlation effects. The first question concerns the partitioning of the Hamiltonian to a zeroth order part and perturbation. Using a straightforward generalization of the Moller-Plesset partitioning [261], the zeroth order Hamiltonian is chosen as the sum of the effective intrabond Hamiltonians:

$$H^0 = \sum_{i=1}^{n} H^{i(eff)} \tag{75}$$

Where $H^{i(eff)}$ is defined by Eqs. (71) and (73). The perturbation operator W is defined, as usual, by the difference between the exact Hamiltonian of Eq. (45) and $H°$:

$$W = H - H° \tag{76}$$

If the zeroth order solution of the problem provides a good approximation, one expects that it will be sufficient to consider low-order PT contributions. In this case, not all the terms of the perturbation operator W will contribute to the energy formula. For example, only two-bond terms will enter the second-order energy expression.

A somewhat unusual feature of the partitioning defined by Eqs. (75–76) is that the zeroth order Hamiltonian $H°$ is not hermitian. This non-hermiticity is rather formal: we have a properly symmetric list of integrals in the effective bond Hamiltonians, but the creation and annihilation operators are not the adjoints of each other, reflecting the overlap between strictly localized wave functions. The non-hermiticity has also the consequence that various eigenstates of $H°$ do not form an orthonormal set. This requires a careful rederivation of the well-known results of perturbation theory [43, 202–205].

The perturbed Schrödinger equation is written as

$$[H^0 + W]\, \Psi_K = E_K \Psi_K \tag{77}$$

where neither H^0 nor W are hermitian, but, of course, the total Hamiltonian is. The index K labels the various (ground and excited) many-electron states of the system. The corresponding wave functions can be expanded in terms of the zeroth order approximations as

$$\Psi_K = \sum_{KL} C_{KL} \Psi_L^0 \tag{78}$$

The coefficients C_{KL} as well as the energies E_K are expanded in the PT series in the usual manner

$$C_{KL} = \delta_{KL} + \sum_{\mu=1}^{\infty} C_{KL}^{(\mu)} \tag{79a}$$

$$E_K = \sum_{\mu=0}^{\infty} E_K^{(\mu)} \tag{79b}$$

These expansions can be substituted into the Schrödinger equation (77). In the convetional PT the resulting formulae are multiplied by the zeroth order functions from the left, providing expressions for $C^{(\mu)}$-s and $E^{(\mu)}$-s. In the present problem the zeroth order wave functions neither form an orthonormal set, nor they represent eigenfunctions from the left, being obtained from a formally non-Hermitian eigenvalue problem. We may use, however, the reciprocal-space functions as bra-vectors $\langle\tilde{\Psi}^0_K|$ which are biorthogonal to the direct-space ket-functions $|\Psi^0_L\rangle$:

$$\langle\tilde{\Psi}^0_K \mid \Psi^0_L\rangle = \delta_{KL}\,.$$

Multiplying with the biorthogonal functions from the left, one obtains the generalization of the usual Rayleigh-Schrödinger PT expressions:

$$E^{(1)}_K = \langle\tilde{\Psi}^0_K|\, W\, |\Psi^0_K\rangle \tag{80}$$

$$E^{(2)}_K = -\sum_{L(\neq K)} \frac{\langle\tilde{\Psi}^0_K|\, W\, |\Psi^0_L\rangle\,\langle\tilde{\Psi}^0_L|\, W\, |\Psi^0_K\rangle}{E^0_L - E^0_K} \tag{81}$$

That is, we recovered the familiar results with the exception that all bra-functions are taken from the reciprocal space. Consequently, the matrix

$$W_{\tilde{L}K} = \langle\tilde{\Psi}^0_L|\, W\, |\Psi^0_K\rangle$$

is not symmetric. The same result holds for the higher orders, too. Evaluation of these matrix elements is extremely simple as compared to the difficulties which one would encounter without utilizing biorthogonality, i.e. using merely the overlapping direct space wave functions. If the perturbation series of Eqs. (79) converge, one obtains the exact solutions E_K and Ψ_K in the given basis set.

As noted before, this approach is similar to the method of moments [208–211]. One argument against the method of moments in connection with configuration interaction [243] or VB [211] type approaches is that the resulting energies do not serve as upper bounds to the true energy. This objection is, however, not relevant to a perturbational approach since the latter does not give upper bounds even in the traditional case. At the expense of working with non-Hermitian operators, the above formalism permits one to incorporate some of the important overlap effects already at the zeroth order of PT. This offers the promise that a rapidly convergent perturbation expansion will be obtained.

Application of the PT formulae presented in Eqs. (80–81) requires the construction of direct- and reciprocal space excited wave functions which are biorthogonal to each other. For the ground state, the corresponding wave functions are defined in Eqs. (61) and (62). The excited states can be distinguished according to the number of electrons excited, and according to the number of bonds involved in the transition. A **local excitation** on the bond i is described by

$$|\Psi^0_K\rangle_{pol} = \psi_i^{+q}\psi_i^{-}\, |\Psi^0_0\rangle \tag{82}$$

where $|\Psi^0_0\rangle$ is the zeroth order AP-SLG ground state wave function, and the label

K refers to the o → q excitation on the bond i. A **charge transfer** or **delocalization** excitation from the bond i to the bond j can be given as:

$$|\Psi_K^0\rangle_{del} = \psi_j^{+(3)q}\psi_i^{+(1)p}\psi_j^-\psi_i^- \, |\Psi_0^0\rangle \tag{83}$$

where we are left with a single electron on the bond i in state p and a three-electron system on the bond j in state q..Interbond **dispersion** effects can be accounted for by considering simultaneous local excitations on two different bonds:

$$|\Psi_K^0\rangle_{disp} = \psi_j^{+q}\psi_i^{+p}\psi_j^-\psi_i^- \, |\Psi_0^0\rangle \tag{84}$$

The biorthogonal bra-function counterparts of the last three formulas are readily seen to be

$$\langle\Psi_K^0|_{pol} = \langle\Psi_0^0|\, \psi_i^+\psi_i^{-q} \tag{82a}$$

$$\langle\Psi_K^0|_{del} = \langle\Psi_0^0|\, \psi_i^+\psi_j^+\psi_i^{-(1)p}\psi_j^{-(0)q} \tag{83a}$$

$$\langle\Psi_K^0|_{disp} = \langle\Psi_0^0|\, \psi_i^+\psi_j^+\psi_i^{-p}\psi_j^{-q} \tag{84a}$$

The one- and three-electron states $\psi^{+(1)p}$ and $\psi^{+(3)q}$ appearing in Eqs. (83) are expanded analogously to Eq. (44):

$$\psi^{+(1)q} = \sum_{\mu\in i} C_\mu^{iq} a_\mu^+ \tag{85}$$

and

$$\psi^{+(3)p} = \sum_{\mu\nu\lambda\in i} C_{\mu\nu\lambda}^{ip} a_\mu^+ a_\nu^+ a_\lambda^+ \tag{86}$$

both of them are Fermion operators, describing a system of an odd number of electrons. The coefficients C_μ^{iq} and $C_{\mu\nu\lambda}^{ip}$ can be obtained from solving local one- and three-electron Schrödinger equations for each bond.

Substituting the above representation of various excited states into the general PT results, and evaluating the matrix elements by the usual rules of the second quantized algebra, we get the relevant PT corrections to the energy. The first order results have already been included in the SLG energy formula given in Eq. (74). At the second order, we get corrections due to (i) the overlap effects neglected at the zeroth order; (ii) delocalization between different bonds; and (iii) interbond dispersion. In what follows we report the final second order PT corrections for the energy in terms of spatial orbitals.

(i) The overlap correction is given by

$$\Delta E_{ov} = -\sum_i \sum_q^{\text{singlets}} \frac{S_{\tilde{O}K}^i S_{\tilde{K}O}^i}{E_i^q - E_i^o} \tag{87}$$

where i runs over all bonds, q runs over all singlet excited state of bond i, and the corresponding many-electron excited state is labeled by K [cf. Eqs. (82, 82a)]. The energy E_i^q is the eigenvalue of the effective Hamiltonian of the bond i corresponding to state q (cf. Moller-Plesset partitioning). The elements of the matrix S^i result from the neglect of the biorthogonal transformation at the zeroth order:

$$S^i_{\tilde{O}K} = \sum_{m,\,n \in i} (h_{\tilde{m}n} - h_{mn})\, P^{iq}_{nm} + \frac{1}{2} \sum_{mnrs \in i} ([\tilde{m}\tilde{n}\,|\,rs] - [mn\,|\,rs])\, C^{io}_{mn} C^{iq}_{rs} \quad (88)$$

where m, n, r and s label the spatial basis orbitals and P^{iq} is the spatial first-order transition density matrix of bond i which is defined as

$$P^{iq}_{nm} = \sum_{\sigma} P^{iq}_{n\sigma,\,m\sigma} \equiv \sum_{\sigma} P^{iq}_{\mu\nu} = \sum_{\sigma} \langle vac|\, \breve{\psi}_i^- a^+_\mu a_\nu \psi_i^{+q}\, |vac\rangle = \sum_{\lambda \in i} \sum_{\sigma} C^{io}_{\mu\lambda} C^{iq}_{\nu\lambda} \quad (89)$$

Here σ is the spin label (either α or β), $\mu = \{m\sigma\}$ is a spinorbital, and the SLG expansion coefficients $C^{iq}_{\mu\nu}$ were defined previously. The $S_{\tilde{K}O}$ matrix elements appearing also in Eq. (87) can be given analogously to Eq. (88).

(ii) Delocalization corrections result from charge-transfer excitations of type (83). The corresponding energy correction reads:

$$\Delta E_{deloc} = -\sum_{ij} \sum_{pq}^{doublets} \frac{W^{del}_{\tilde{O}K} W^{del}_{\tilde{K}O}}{E^p_i + E^q_j - E^o_i - E^o_j}$$

where p runs over all doublet states of the bond i while q runs over those of j. Again, K labels the corresponding excited state of the molecule given by Eqs. (84) and (84a). The matrix elements of the delocalization operator can be expressed as

$$\begin{aligned} W^{del}_{\tilde{O}K} = {} & 2 \sum_{mn \in i} \sum_{rst \in j} h^{eff}_{\tilde{m}r} c^{ip}_{n} c^{iq}_{rts} c^{io}_{mn} c^{jo}_{ts} \\ & - 2 \sum_{hek \in i} \sum_{rts \in j} [\tilde{h}\tilde{e}\,|\,kr]\, c^{ip}_{k} c^{io}_{he} c^{jq}_{rts} c^{jo}_{ts} \\ & + 2 \sum_{hkrs \in j} \sum_{ef \in i} [\tilde{h}\tilde{e}\,|\,kr]\, c^{ip}_{f} c^{i}_{ef} c^{j}_{hs} (c^{jq}_{krs} - c^{jq}_{rsk}) \end{aligned}$$

where the two-bond effective core $h^{eff}_{\tilde{m}r}$, $m \in i$, $r \in j$, is defined as

$$h^{eff}_{\tilde{m}r} = h_{\tilde{m}r} + \sum_{k(\neq i,\,j)} \sum_{l,\,s \in k} P^k_{ls}([\tilde{m}\tilde{l}\,|\,rs] - [\tilde{m}\tilde{l}\,|\,sr]) \qquad (m \in i,\ r \in j) \qquad (92)$$

which takes into account the electrostatic (Coulomb + exchange) interaction of bond k with the charge distribution (mr) giving rise to delocalization from bond i to j. Although the two-electron integrals in Eq. (92) are three-center integrals in terms of bonds, they contain only one interbond differential overlap if the exchange part is neglected. Similarly, the last term of the pairwise interaction Hamiltonian of Eq. (53) contains two interbond differential overlap, thus it was dropped from the delocalization correction of Eq. (91).

(iii) Finally, we present the result for the interbond dispersion correction which takes the form

$$\Delta E^{disp} = -\sum_{\substack{ij\\(i<j)}} \sum_{pq}^{M_S=0} \frac{\left| \sum_{hk\in j} \sum_{er\in i} \left([\tilde{h}\tilde{e}|rk] - \frac{1}{2}[\tilde{h}\tilde{e}|rk] \right) P_{hk}^{jp} P_{er}^{iq} \right|^2}{E_i^p + E_j^p - E_i^o - E_j^o}$$

$$-\frac{1}{4}\sum_{j<i} \sum_{p,q}^{M_S^{p,q}=\pm 1;\, M_S^p+M_S^q=0} \frac{\left| \sum_{hk\in j} \sum_{er\in i} [\tilde{h}\tilde{e}|rk] P_{hk}^{jp} P_{er}^{iq} \right|^2}{E_i^q + E_j^p - E_i^o - E_j^o}$$

with obvious notations for the S_z quantum numbers M_S.

The above equations are the working formulae for the perturbation calculation of interbond interaction in the present scheme. Although they may appear to be complicated, they are computable with a rather small effort. The working integral list is obtained by a local four-index hybridization transformation and a global two-index transformation to the biorthogonal set. Only one- and two-bond integrals are required with the exception of the two-bond effective core in which certain types of three-bond integrals appear. Anyway, only a small portion of the whole integral list should be transformed and manipulated. On the other hand, since the zeroth order results are already fairly correlated, the second order PT contributions are expected to give rather accurate results. Numerical checking of these second order formulas has not been done yet, however.

6 An Application of Strictly Localized Geminals: Retinal Schiff-Base

Even though the computational aspects of the localized bond model treated in Sect. 5 are not yet completely elaborated, the zeroth order wave function which is very easy to compute may be useful in some cases. The essential feature of the SLG model is that, as mentioned previously, it offers the best possible representation of the local two-electron bonds in the given basis set. Consequently, the limitations of the SLG approach indicate the limitations of the localized picture of the molecule in question. This fact will be utilized in the application discussed below [262].

Several quantum chemical investigations have been done on the retinal molecule and its various substituted forms and analogs [262–265]. These studies may be relevant from the biological point of view since this molecule plays an important role in vision. The basic structure of the molecule is shown in Fig. 8 where the standard numbering of carbons is also indicated which will be used to identify the atoms in the following figures. The substituent X can be oxygen (retinal), =NH group (retinal Schiff base), or $=NH_2^+$ group (protonated retinal Schiff base).

Recently, we have performed a detailed ab initio study [262] on the ground state properties of various retinal analogs by considering only a limited number of double

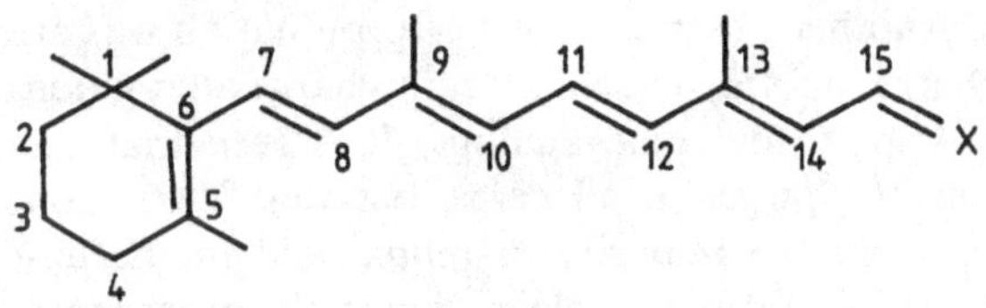

Fig. 8. Standard numbering of atoms in the retinal molecule

bonds (that is, a shorter chain) and cutting also the ring of the molecule. In other words, retinal is modeled by its chromophore in the spectroscopical sense which is a part of a polyene chain. In the course of these investigations, a remarkable difference has been noticed between the protonated and unprotonated forms. The unprotonated Schiff base analogs have been found to be very similar to retinal in most respect. First of all, optimizing the geometry of several retinal analog [262] it turned out that the $=NH_2^+$ terminal group causes a remarkable increase in conjugation along the chain which is reflected by the equalization of the bond lengths

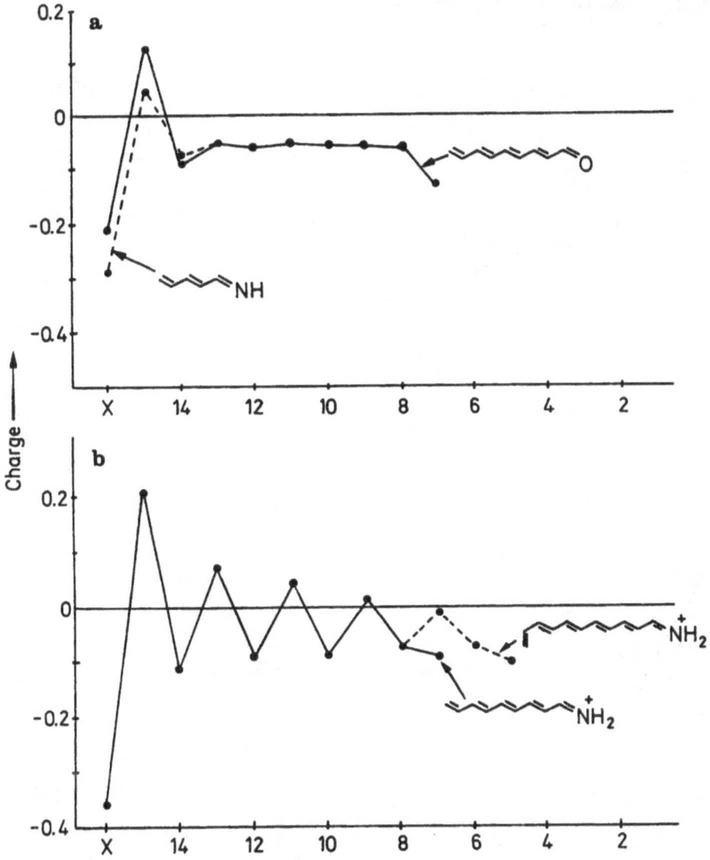

Fig. 9a and b. Oscillation of Mulliken atomic charges in retinal analogs. (**a**): retinal and unprotonated Schiff base analogs; (**b**): protonated Schiff base analogs

for the single and double CC bonds. Another observed effect of protonation was an electrostatic one: the extra positive charge on NH_2^+ causes a strong charge alternation along the chain. This is shown by Fig. 9. for some analogs. It is seen that the heteroatom introduces an alternation of charges in all cases, but this fluctuation diminishes rapidly with the distance from the impurity in retinal and in the unprotonated Schiff base. It extends however over the whole system in the protonated form exhibiting also a significantly increased amplitude. The fact that heteroatoms introduce a charge alternation along a chain has been well known for a long time; it can therefore be understood even by simple topological Hückel models [266].

Now the question arises whether this enhancement of the charge alternation upon protonation is a consequence of the observed increase in conjugation reflected by the equalization of bond lengths, or it should be attributed to some other reason. The SLG model can be applied to decide this question, since it inherently neglects any conjugation effects as a consequence of the strict localization. In Fig. 10 we present net atomic charges as calculated by the delocalized SCF model which accounts for conjugation in comparison with those obtained by the strictly localized model. Although the SCF and SLG charges are not exactly the same, it is clearly seen that they show the same pattern demonstrating that the protonation causes much stronger

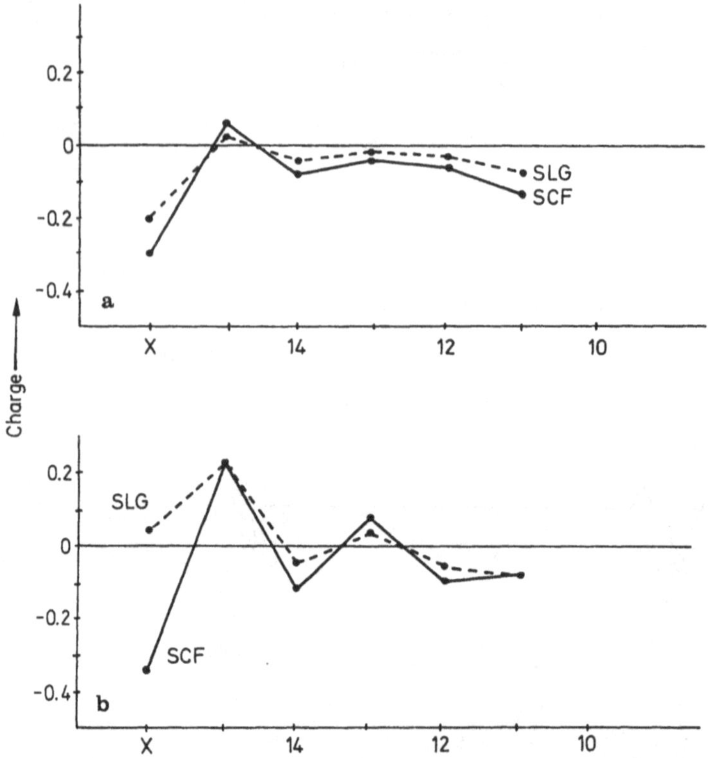

Fig. 10a and b. Comparison of Mulliken charges in retinal-like molecules as obtained from the SLG and SCF methods. For (**a**) and (**b**) see Fig. 9

charge oscillation. This means that conjugation effects do not play a direct role in this phenomenon. The only interbond interaction which is included even in the simple SLG wave function is the inductive effect, thus the increased charge alternation in the protonated Schiff base should be attributed to the induction of local bond dipoles which is not sensitive to the changes in conjugation or geometry.

It was possible to draw the above conclusion only on the basis of an inherently approximate scheme, because we wanted just to investigate whether a local approach can explain the general finding obtained by a more sophisticated delocalized approach.

Of course, the correlated nature of the SLG wave function was not utilized in the above consideration. There is, however, another problem found in retinal analogs where we can benefit from the account of electron correlation in the SLG method. This problem concerns the separation of charges introduced by twisting the chain around a double C=C bond as observed at the SCF level [262]. In contrast to this finding the SLG charges for the twisted retinal do not show this "sudden polarization" effect [262]. The reason is clear: the one-configuration RHF method is not suitable for describing the electronic structure of the twisted conformation which is of open-shell character. The RHF method will always place two electrons on one orbital, while the double bond is cut upon twisting. This problem is analogous to the improper dissociation of the RHF wave function which often describes an ionized structure at large distances. The SLG wave function is properly correlated for each bond and can describe the fission of the π-bond upon twisting the retinal thus such artefact effects do not appear.

The above examples indicate that simple local models can contribute to a better understanding of the electronic structure. The effective computer realization of the SLG + PT approach is not yet completed[16], and a lot of work should be done until this method can be routinely applied in electronic structure calculations. It is hoped however that it will offer a useful alternative augmenting delocalized computational schemes.

7 Conclusions

In this article we have discussed a theory for the representation of local two-electron chemical bonds in a many-electron system both at the SCF and correlated levels. Emphasis was laid on the approaches developed in our laboratory [10, 12, 36–39, 43–44, 49, 63, 86–87, 103, 164, 170, 177, 179, 184, 205–206, 238, 239], though we are aware that several other viewpoints are possible [1–9, 32–35, 106–118, 125–126, 160–161, 221–224, 270–282]. We have tried to refer to the related theories throughout

[16] By the time this review article was written, the SLG program [268] had been attached to the MONSTERGAUSS ab initio program system [269] by means of which the numerical calculations reported here were performed. The SLG program at present can handle only atoms with $Z < 10$ in minimal basis set, work to extend it is in progress.

the text, but the literature of this field is so extensive that we had to cut the number of references.

The conclusions one can draw from the presented results and discussions are summarized as follows.

Molecules can be divided into families according to the extent of the localizability of their electronic structure. Easily localizable are those systems which can be characterised properly by their classical Lewis-structure. Their overall electronic structure can be understood as a system of two-electron subunits (bonds, lone pairs, inner shell electron pairs). These subunits are localized in different regions of space, but interact with each other. Interbond interactions are much weaker than interatomic ones, which justifies saying that easily localizable molecules are composed of two-electron bonds as building blocks.

If the system under consideration also contains aromatic rings or similar inherently delocalized fragments showing collective phenomena, these fragments should be considered as a whole while the two-electron model is considered valid only for the rest of the molecule. The situation is more complicated for certain radicals and excited systems, where the localized picture may not be adequate at all, though electron excitations can also be classified according to their localized character and a local treatment may be advantageous in some cases [61–65, 283].

The above ideas have become well known in the past few decades and several effective computational and interpretative schemes have been developed and applied on these grounds [284–290]. The best known achievement is that of the PCILO method [32–35, 289–294]. The basic idea of PCILO is to start from a fully localized SCF reference state and approach the exact solution by means of perturbation theory.

In our laboratory, we developed basically two types of theories bracketing PCILO in a certain sense. Both are elaborated at the ab initio level. The first approach, reviewed in Sect. 4, starts from the same SLMO zero-order reference state, but only SCF effects are considered for the PT treatment. The aim of this PT is to derive perturbation expressions for the tails of LMOs. Total molecular energies are calculated in this approach as expectation values of the determinants built up by the perturbed LMOs.

The second approach, reviewed in Sect. 5, is more complicated and more powerful than PCILO. It is based on a correlated, although fully localized, reference state. The corresponding zeroth order functions are called strictly localized geminals (SLG) which form the many-electron wave function

$$|\Psi\rangle = \psi_1^+ \psi_2^+ \dots \psi_n^+ \,|\mathrm{vac}\rangle$$

where each composite particle creation operator ψ_i^+ refers to a particular two-electron bond. In a certain sense, this wave function offers us the best possible representation of local two-electron bonds in a many-electron system.

The creation operators ψ_i^+ follow a Bose algebra which suggests a pictorial model of the molecular electronic structure. Each chemical bond is considered as a quasi bose-particle composed of two electrons. The internal structure of these quasiparticles is determined by the local characteristics of the corresponding chemical bond as well as inductive interbond interactions. Quasiparticle transformations of this type have been applied by several authors [248–254, 295]; the connections with Valdemoro's approach [296–301] must be mentioned explicitly. Some formal relations to the theory of superconductivity [302] are also interesting to realize. Actually, geminal wave functions of this type can be interpreted as number-projected versions of the BCS wave function [256, 295]. However, the non-elementary nature of the quasi-particles expressed by their commutation rules of Eqs. (57) is essential[17].

Interactions between bonds is taken into account by an effective potential which does not influence localization and by perturbation theory for delocalization, polarization and dispersion corrections. The corresponding theory is similar to some extent to the PT for intermolecular interactions [202–204].

The limitations of this localized approach are obvious. Except for some systems where a localized treatment is not adequate at all, the above theories are based on the assumption that the wave functions of different chemical fragments are expanded in disjunct subspaces of basis orbitals. This assumption makes the mathematical formalism feasible and the computational strategy effective. Conceptually, however, it reflects a limitation of such theories which cannot be generalized to a level where numerical solution of the Schrödinger equation are sought for [303, 304] without any basis set expansion. In this case, density-matrix-based approaches may help [47, 48, 240], though the latter also utilize some basis expansion in practical calculations in most cases.

Another disadvantage of basis set partitioning is that this approach limits the size of the local basis set. Namely, if we want to assign each basis orbital to a particular chemical bond or fragment, we cannot use nearly complete sets where an orbital of a fragment could be expanded in terms of orbitals of another fragment to a good approximation. In very large basis sets partitioning is possible for example by Bader's approach [305] which has, however, only interpretative power since it cannot be used in a priori localized calculations. Practical quantum chemical calculations in usual basis sets do not reach this "overcompletness" limit, and partitioning is possible by means of hybridization techniques. It has to be mentioned, however, that although hybridization has been extensively studied and applied [83–86, 125–139, 306–309], determination of proper hybrids in extended basis sets is still an unsolved problem. A split-shell bond-orbital calculation was recently published by Musso and Magnasco [310].

Within all these difficulties and limitations in mind, one may hope that ab initio local approaches such as the SLG + PT treatment discussed in Sect. 5 will represent a useful alternative to more standard delocalized methods. In some applications, the evaluation of higher order terms of PT might be necessary, but if the many-body PT with the localized and correlated reference state converges, it approaches the same final result as other methods do. Computationally, the localized treatment is superior to standard techniques, and it represents practically the only possibility of performing sophisticated quantum chemical calculations for extended but not strictly periodic systems. The significance of this perspective is evident in the field of molecules of biological interest.

---

[17] It may be useful to investigate under what conditions the composite nature of the quasiparticles could be neglected. This study should concern the structure and role of operator $\hat{Q}$ in Eq. (58).

Acknowledgements. A vast majority of the results summarized in this article was achieved in cooperation with several colleagues. Among others, the inspiration, continuous interest, and a great deal of joint work and effort from Professors G. Náray-Szabó, I. Mayer, and R. A. Poirier are gratefully acknowledged. The author is indebted to Professors Z. B. Maksiç and G. Naray-Szabó for reading the manuscript; their critical remarks have significantly improved the presentation. This work was partly supported by the grant OTKA 64/1987.

8 References

1. Chalvet O, Daudel R, Diner S, Malrieu JP (1976) Localization and Delocalization in Quantum Chemistry, Vols I–II, Reidel, Dordrecht
2. Nelander B, Del Re G (1970) J. Chem. Phys. 52: 5225
3. Allen TL, Shull H (1961) J. Chem. Phys. 35: 1644
4. Del Re G (1973) Int. J. Quantum. Chem. S7: 193
5. Julg A (1984) Int. Quantum Chem. 26: 790
6. Del Re G (1976) in: Chalvet O, Daudel R, Diner S, Malrieu JP (eds) Localization and delocalization in quantum chemistry Vol. II, Reidel, Dordrecht, p 153
7. Hoffman DK, Ruedenberg K, Verkade JG (1977) Structure and Bonding 33: 57
8. Kirtman B, Palke WE, Chipman DM (1980) Isr. J. Chem. 19: 82
9. Berthier G (1981) Int. J. Quantum. Chem. 19: 985
10. Náray-Szabó G (1976) Acta Physica Hungarica 40: 261
11. Del Re G (1972) Gazzetta Chimica Italiana 102: 929
12. Surján PR (1982) Chem. Phys. Letters 92: 483
13. Del Re G, Berthier G, Serre J (1980) Electronic states of molecules and atom clusters, Springer, Berlin Heidelberg New York
14. Siegbahn K, Nardling C, Johansson G, Hedman J, Heden PF, Hamrin K, Gelins U, Bergmark T, Werme LO, Manne R, Baer Y (1969) ESCA applied to free molecules, North-Holland, Amsterdam
    Gussoni M, Castiglioni C, Zerbi G (1985) Chem. Phys. Letters 117: 263
15. Durand Ph, Malrieu J-P (1987) Advances in Chemical Physics, 67: 321
16. Maksić ZB (1984) Prologue to: Conceptual approach in quantum chemistry. Models and applications, Croatica Chemica Acta 57(5)
17. Evangelisti S, Daudey JP, Malrieu JP (1987) Phys. Rev. A35: 4930
    Roeggen I, Nilssen EW (1987) J. Chem. Phys. 86: 2869
18. Edmiston C, Ruedenberg K (1963) Rev. Mod. Phys. 35: 457
19. Edmiston C, Ruedenberg K (1965) J. Chem. Phys. 43: S97
20. Foster JM, Boys SF (1960) Rev. Mod. Phys. 32: 300
21. Niessen W von (1790) J. Chem. Phys. 56: 4290
22. Niessen W von (1972) Theor. Chim. Acta 27: 9
23. Boys SF (1960) Rev. Mod. Phys. 32: 296
24. Mayer I (1983) Int. J. Quantum Chem. 23: 341
25. Mayer I (1983) Chem. Phys. Letters 97: 270
26. McWeeny R (1959) Proc. Roy. Soc. London A253: 242
27. Moffitt W (1951) Proc. Roy. Soc. London A210: 245
    Arai T (1960) Rev. Mod. Phys. 32: 370
28. Maksić ZB (1988) THEOCHEM 47: 39
29. Adams WH, Clayton MM (1986) Int. J. Quantum Chem. S19: 333
30. Colle R, Fortunelli A, Salvetti O (1984) J. Chem. Phys. 80: 2654
31. Niessen W von (1973) Theor. Chim. Acta 31: 111
32. Diner S, Malrieu JP, Clavery P (1969) Theoret. Chimical Acta 13: 1
33. Malrieu JP, Claverie P, Diner S (1969) Theoret. Chim. Acta 13: 18
34. Kvasnička V (1974) Theoret. Chim. Acta 34: 61

35. Malrieu J-P (1977) in: Segal GA (ed) Semiempirical methods in electronic structure calculation, Plenum, New York
36. Surján PR, Mayer I (1981) Theoret. Chim. Acta 59: 603
37. Surján PR, Náray-Szabó G, Mayer I (1982) Int. J. Quantum. Chem. 22: 929
38. Surján PR, Mayer I, Kertész M (1982) J. Chem. Phys. 77: 2454
39. Mayer I, Surján PR (1984) J. Chem. Phys. 80: 5649
40. Musso GF, Figari G, Magnasco V (1983) J. Chem. Soc. Faraday Trans. 2 79: 1283
41. Kapuy E (1969) Chem. Phys. Letters 3: 43
42. Kapuy E (1966) Theoret. Chim. Acta 6: 281
43. Surján PR, Mayer I, Lukovits I (1985) Phys. Rev. A32: 749
44. Surján PR (1984) Croatica Chemica Acta 57: 833
45. Mehler EL (1977) J. Chem. Phys. 67: 2728
46. Mehler EL (1981) J. Chem. Phys. 74: 6298
47. Kirtman B (1982) J. Phys. Chem. 86: 1059
48. Kirtman B, de Melo C (1981) J. Chem. Phys. 75: 4592
49. Náray-Szabó G, Surján PR (1983) Chem. Phys. Letters 96: 499
50. Náray-Szabó G (1984) Croatica Chemica Acta 57: 901
51. Klessinger M, McWeeny R (1965) J. Chem. Phys. 42: 3343
52. Klessinger M (1983) Int. J. Quantum. Chem. 23: 535
53. Schule J, Klessinger M (1987) Theoret. Chim. Acta 72: 35
54. Leasure SC, Martin TP, Bálint-Kürti GG (1984) J. Chem. Phys. 80: 1186
55. Adams WH (1974) Phys. Rev. Letters 32: 1093
56. Adams WH (1975) Int. J. Quantum Chem. S9: 367
57. Huzinaga S, Cantu AA (1971) J. Chem. Phys. 55: 5543
58. Hurley AC (1955) Proc. Roy. Soc. London A68: 149
59. Boca R (1982) Theoret. Chim. Acta 61: 179
60. Náray-Szabó G, Surján P, Kiss AI (1985) THEOCHEM 123: 85
61. Langlet J (1972) Theoret. Chim. Acta 27: 223
62. Langlet J, Malrieu JP (1976) in: Chalvet O, Daudel R, Diner S, Malrieu JP (eds) Localization and delocalization in quantum chemistry, Vol. II, Reidel, Dordrecht, p 15
63. Ángyán J, Surján PR (1983) Theoret. Chim. Acta 63: 43
64. Langlet J, Malrieu JP (1973) Theoret. Chim. Acta 30: 59
65. Bouman TD, Voigt B, Hansen AE (1979) J. Am. Chem. Soc. 101: 550
66. Edmiston C, Ruedenberg K (1966) in: Löwdin PO (ed) Quantum theory of atoms, molecules and the solid state, Academic, New York, p 263
67. Boys SF (1966) in: Löwdin PO (ed) Quantum theory of atoms, molecules and the solid state, Academic, New York, p 253
68. Anderson PhW (1958) Phys. Rev. 109: 1492
69. Kürti J, Kuzmany H (1987) in: Kuzmany H, Mehring M, Roth S (eds) Proc. Int. Winterschool, Kirchberg, Tirol, Springer Series in Solid State Sci 76., Springer, Berlin Heidelberg NewYork
70. Mayer I (1971) Acta Physica Hung. 30: 373
71. Mayer I (1974) Acta Physica Hung. 36: 11
72. Coulson CA (1942) Trans. Faraday Soc. 38: 433
73. Lennard-Jones JE (1949) Proc. Roy. Soc. London A198: 1, 14
74. Hall GG, Lennard-Jones JE (1950) Proc. Roy. Soc. London A202: 155
75. Kapuy E (1981) Acta Physica Hungarica 51: 21
76. Burke LA, Leroy G, Daudel R, Stephens ME (1978) Chem. Phys. Letters 57: 15
77. Perkins PG, Stewart JJP (1982) J. Chem. Soc. Faraday Trans. 2 78: 285
78. Magnasco V, Perico A (1967) J. Chem. Phys. 47: 971
79. Magnasco V, Perico A (1968) J. Chem. Phys. 48: 800
80. Bohm MC (1981) Theoret. Chim. Acta 59: 609
81. Rajzmann M, Brenier B, Purcell KF (1987) Theoret. Chim. Acta 72: 13
82. McWeeny R (1960) Rev. Mod. Phys. 32: 355
83. Jug K (1978) J. Am. Chem. Soc. 100: 6581
84. Foster JP, Weinhold F (1980) J. Am. Chem. Soc. 102: 7211
85. Caldwell D, Redington P, Eyring H (1979) Proc. Natl. Acad. Sci. USA 76: 3042

86. Mayer I, Révész M (1980) Kém. Közl. 54: 205
87. Surján PR, Révész M, Mayer I (1981) J. Chem. Soc. Faraday 2 77: 1129
88. Kutzelnigg W (1976) in: Chalvet O, Daudel R, Diner S, Malrieu JP (eds) Localization and delocalization in quantum chemistry, Vol I, Reidel, Dordrecht, p 143
89. Kapuy E, Csépes Z, Kozmutza C (1984) Croatica Chemica Acta 57: 855
90. Kapuy E, Bartha F, Kozmutza C, Bogar F (1988) THEOCHEM 50: 237
91. Daudel R, Kapuy E, Kozmutza C, Goddard JD, Csizmadia IG (1976) Theoret. Chim. Acta 44: 197
92. Haddon RC, Williams GRJ (1976) Chem. Phys. Letters 42: 453
93. Leonard JM, Luken WL (1982) Theoret. Chim. Acta 62: 107
94. Luken WL, Beratan DN (1982) Theoret. Chim. Acta 61: 265
95. Fleischer U, Schindler M, Kutzelnigg W (1987) J. Chem. Phys. 86: 6357
96. Bonaccorsi R, Ghio G, Scrocco E, Tomasi J (1980) Isr. J. Chem. 19: 109
97. Náray-Szabó G, Peterson MR (1981) THEOCHEM 85: 249
98. Náray-Szabó G, Surján PR, Ángyán JG (1987) Applied quantum chemistry, Akademiai Kiado-Reidel, Budapest-Dordrecht
99. Salem L (1961) Proc. Camb. Phil. Soc. 57: 353
100. Ukrainskii II (1975) Theoret. Chim. Acta 38: 139
101. Monkhorst J, Kertész M (1981) Phys. Rev. B24: 3015
102. Piela L, Andre JM, Fripiat JG, Delhalle J (1981) Chem. Phys. Letters 71: 143
103. Surján PR, Kertész M, Koller J, Karpfen A (1983) Phys. Rev. B27: 7583
104. England W, Gordon MS (1972) J. Am. Chem. Soc. 94: 4818
105. Koopmans T (1933) Physica 1: 104
106. Gilbert TL (1964) in: Löwdin PO, Pullman B (eds) Molecular orbitals in chemistry, physics and biology, Academic, New York, p 405
107. Gilbert TL (1972) Phys. Rev. A6: 580
108. Gilbert TL (1974) J. Chem. Phys. 60: 3835
109. Adams WH (1961) J. Chem. Phys. 34: 89
110. Adams WH (1962) J. Chem. Phys. 37: 2009
111. Adams WH (1965) J. Chem. Phys. 42: 4030
112. Adams WH (1971) Chem. Phys. Letters 11: 71
113. Adams WH (1971) Chem. Phys. Letters 12: 295
114. Peters D (1967) J. Chem. Phys. 46: 4427
115. Peters D (1969) J. Chem. Phys. 51: 1559
116. Peters D (1969) J. Chem. Phys. 51: 1566
117. Wilhite DL, Whitten JC (1973) J. Chem. Phys. 58: 948
118. Tseng TJ, Whitehead MA (1982) J. Comp. Chem. 3: 552
119. Daudey JP (1974) Chem. Phys. Letters 24: 574
120. Lefebvre R (1955) Comt. Rend. Acad. Sci. Paris 240: 1097
121. Lefebvre R (1957) J. Chim. Phys. 54: 168
122. Lefebvre R, Moser CM (1956) J. Chim. Phys. 53: 393
123. Mulliken RS (1955) J. Chem. Phys. 23: 2343
124. Clementi E (1967) J. Chem. Phys. 46: 3842
125. Pauling L (1931) J. Am. Chem. Soc. 53: 1367
126. Pauling L (1960) The nature of the chemical bond, 3rd edn, Cornell Univ. Press. Ithaca, New York
127. McWeeny R, Del Re G (1968) Theoret. Chim. Acta 10: 13
128. Polák R (1970) Int. J. Quantum Chem. 4: 271
129. Polák R (1972) Int. J. Quant. Chem. 6: 1077
130. Reed AE, Weinhold F (1983) J. Chem. Phys. 78: 4066
131. Rives AB, Weinhold F (1980) Int. J. Quantum. Chem. S14: 201
132. Maksić ZB (1986) Comp. Math. Appl. 12B: 697
133. Del ReG (1963) Theoret. Chim. Acta 1: 188
134. Randic M, Maksić ZB (1965) Theoret. Chim. Acta 3: 59
135. Randic M, Maksić ZB (1972) Chem. Rev. 72: 43
136. Maksić ZB, Rubcic A (1977) J. Am. Chem. Soc. 99: 4233
137. Boca R, Pelikan P, Valko I, Miertus S (1975) Chem. Phys. 11: 229

138. Bingel WA, Lttke WA (1981) Angew. Chem. Int. Ed. Engl. 20: 899
139. Bálint I, Bán M (1983) Computers and Chem. 7: 199
140. Barone V, Del Re G, Lami A, Abbate G (1983) THEOCHEM 105: 191
141. Preuss H (1956) Z. Naturforsch. 11: 823
142. Shipman LL, Christoffersen RE (1972) Chem. Phys. Letters 15: 469
143. Pyan PB, Todd HD (1977) J. Chem. Phys. 67: 4787
144. Neisius D, Verhangen G (1979) Chem. Phys. Letters 66: 358
145. Neisius D, Verhangen G (1981) Chem. Phys. Letters 78: 147
146. Frost AA (1967) J. Chem. Phys. 47: 3707
147. Frost AA (1967) J. Chem. Phys. 47: 3714
148. Frost AA (1968) J. Phys. Chem. 72: 1289
149. Frost AA, Rouse RA (1968) J. Am. Chem. Soc. 90: 1965
150. Frost AA, Rouse RA, Vescelius L (1968) Int. J. Quantum. Chem S2: 43
151. Christoffersen RE (1972) Adv. Quant. Chem. 6: 333
152. Christoffersen RE, Spangler D, Hall GG, Maggiora GM (1973) J. Am. Chem. Soc. 95: 8526
153. Gáspár RJr, Gáspár R (1979) Int. J. Quantum Chem. 16: 57
154. Meerman CM, van der Hart WJ (1975) Theoret. Chim. Acta 37: 319
155. Franchini PF, Vergani G (1969) Theoret. Chim. Acta 13: 46
156. Parr RG, Ellison FO, Lykos PG (1956) J. Chem. Phys. 24: 1106
157. Huggins ML (1953) J. Am. Chem. Soc. 75: 4123
158. Pople JA, Gordon M (1967) J. Am. Chem. Soc. 89: 4253
159. Gordy W (1950) J. Chem. Phys. 19: 792
160. Tyrrell J, Weinstock RB, Weinhold F (1981) Int. J. Quantum Chem. 19: 781
161. Polák R (1969) Theoret. Chim. Acta 14: 163
162. Polák R (1973) Coll. Czech. Chem. Commun. 38: 1450
163. Polák R (1975) Coll. Czech. Chem. Commun. 40: 1975
164. Náray-Szabó G, Grofcsik A, Kosa K, Kubinyi M, Martin A (1981) J. Comp. Chem. 2: 58
165. Zahradnik R, Polák R (1980) Elements of quantum chemistry, Plenum, New York
166. Smits GF, Altona C (1985) Theor. Chim. Acta 67: 461
167. Roothaan CCJ (1951) Rev. Mod. Phys. 23: 69
168. Pople JA, Beveridge DL (1970) Approximate molecular orbital theory, McGraw Hill, New York
169. Otto P, Ladik J (1977) Chem. Phys. 19: 209
170. Surján PR (1988) J. Mol. Struct (THEOCHEM) 169: 95
171. Löwdin PO (1955) Phys. Rev. 97: 1474
172. Gordon MS, England W (1972) J. Am. Chem. Soc. 94: 5168
173. Gordon MS, England W (1973) J. Am. Chem. Soc. 95: 1753
174. Musso GF, Magnasco V (1982) J. Chem. Soc. Faraday 2 78: 1609
175. Del Re (1958) J. Chem. Soc.: 4031
176. Holder JM, Peters D (1979) J. Mol. Struct 56: 221
177. Náray-Szabó G (1979) Int. J. Quantum. Chem. 16: 265
178. Barone V, Douady J, Ellinger Y, Subra R, Del ReG (1979) J. Chem. Soc. Faraday 2 75: 1597
179. Surján PR (1983) Croatica Chemica Acta 56: 289
180. Hoyland JR (1968) J. Am. Chem. Soc. 90: 2227
181. Hoyland JR (1969) J. Chem. Phys. 50: 473
182. Musso GF, Figari G, Magnasco V (1983) J. Chem. Soc. Faraday 2 79: 931
183. Musso GF, Magnasco V (1984) Mol. Phys. 53: 615
184. Surján PR (1985) Chem. Phys. Letters 117: 386
185. Veillard A (1976) in: Pullman B (ed) Quantum mechanics of molecular conformations, Wiley, New York
186. Sovers OJ, Kern CW, Pitzer RM, Karplus J (1968) Chem. Phys. 49: 2592
187. England W, Gordon MS (1972) J. Am. Chem. Soc. 94: 4818
188. Lether JH (1971) J. Chem. Phys. 54: 3215
189. Lether JH, Absar I, Van Wazer JR (1972) Int. J. Quantum. Chem. 86: 451
190. Pople JA, Santry DP (1963) Mol. Phys. 7: 269
191. Dewar MJS, Pettit R (1954) J. Chem. Soc. 1954: 1625
192. Lochman R (1977) Int. J. Quantum. Chem. 12: 795

193. Lochman R (1977) Int. J. Quant. Chem. 12: 841
194. Lochman R (1977) Int. J. Quant. Chem. 12: 851
195. Stoll H, Wagenblast G, Preuss H (1987) Theor. Chim. Acta 49: 67
196. Douady J, Barone V, Ellinger Y, Subra R (1980) Int. J. Quantum. Chem. 17: 211
197. Boca R, Pelikan P (1980) Int. J. Quantum, Chem. 18: 1361
198. Magnasco V, Musso GF (1981) Chem. Phys. Letters 84: 575
199. Magnasco V, Musso GF (1982) Chem. Phys. Letters 87: 447
200. Linderberg J, Ratner MA (1981) J. Am. Chem. Soc. 103: 3265
201. Mayer I (1982) Chem. Phys. Letters 89: 390
202. Gouyet JF (1973) Chem. Phys. 59: 4637
203. Kochanski E, Gouyet JF (1975) Mol. Phys. 29: 693
204. Kochanski E, Gouyet JF (1975) Theor. Chim. Acta 39: 329
205. Surján PR, Mayer I, Lukovits I (1985) Chem. Phys. Letters 19: 538
206. Surján PR, Poirier RA (1986) Chem. Phys. Letters 128: 358
207. Mayer I, Surján PR to be published
208. Hegyi MG, Mezei M, Szondi T (1971) Theoret. Chimica Acta 15: 273
209. Hegyi MG, Mezei M, Szondi T (1971) Theoret. Chim. Acta 15: 283
210. Preuss H (1958) Z. Naturforsch. 13a: 439
211. Norbeck JM, McWeeny R (1975) Chem. Phys. Letters 34: 206
212. Lathouwers L (1976) in: Calais JL, Goscinski O, Linderberg J, Öhrn Y (eds) Quantum science, Plenum, New York
213. Stoll H, Wagenblast G, Preuss H (1980) Theoret. Chim. Acta 57: 169
214. Raimondi M, Campion W, Karplus M (1977) Mol. Phys. 34: 1483
215. Löwdin PO, (1950) J. Chem. Phys. 18: 365
216. Hayes IC, Stone AJ (1984) Mol. Phys. 53: 69
217. Brunck TK, Weinhold F (1979) J. Am. Chem. Soc. 101: 1700
218. Santry DP (1977) Theor. Chim. Acta 42: 67
219. Ladik J (1976) in: Calais JL, Goscinski O, Linderberg J, Öhrn Y (eds) Quantum science, Plenum, New York, p 539
220. Boca R (1979) Coll. Czech. Chem. Commun. 44: 3041
221. Musso GF, Magnasco V (1974) Chem. Phys. Letters 111: 468
222. Musso GF, Magnasco V (1984) Chem. Phys. Letters 107: 585
223. Magnasco V, Rui M, Musso GF, Figari G (1985) Chem. Phys. Letters 115: 395
224. Musso GF, Rui M, Magnasco V (1986) Chem. Phys. Letters 126: 335
225. Cullen JM, Zerner MC (1982) Int. J. Quantum. Chem. 22: 497
226. Parks JM, Parr RG (1958) J. Chem. Phys. 28: 335
227. Kapuy E (1969) Acta Physica Hungarica 27: 179
228. McWeeny R, Steiner E (1965) Adv. Quant. Chem. 2: 93
229. Meyer W (1976) J. Chem. Phys. 64: 2902
230. Coulson CA (1961) Valence, 2nd edn, Oxford Univ. Press, London
231. Kuwajima S (1981) J. Chem. Phys. 74: 6342
232. van Lenthe JH, Bálint-Kürti GG (1983) J. Chem. Phys. 78: 5699
233. Roeggen I (1987) Int. J. Quantum. Chem. 31: 951
234. Ahlrichs R, Kutzelnigg (1968) J. Chem. Phys. 48: 1819
235. Hurley AC, Lennard-Jones JE, Pople JA (1953) Proc. Roy. Soc. A220: 446
236. Hurley AC (1976) Electron correlation in small molecules, Academic, New York
237. Luken WL (1983) J. Chem. Phys. 78: 5729
238. Surján PR (1984) Phys. Rev. A30: 43
239. Poirier RA, Surján PR (1987) J. Comput. Chem. 8: 436
240. Kirtman B (1983) J. Chem. Phys. 79: 835
241. Roeggen I (1983) J. Chem. Phys. 79: 5520
242. Kutzelnigg W (1964) J. Chem. Phys. 40: 3640
243. Payne PhW (1982) J. Chem. Phys. 77: 5630
244. Longuet-Higgins HC (1966) in: Löwdin PO (ed), Quantum theory of atoms, molecules and the solid state, Academic, New York, p 105
245. Löwdin PO (unpublished)
246. Moshinsky M, Seligman TH (1971) Ann. Phys. (NY) 66: 311

247. Ng B, Newman DJ (1985) J. Chem. Phys. 83: 1758
248. Bratos S, Durand Ph (1963) J. Chem. Phys. 43: 2670
249. Girardeau M (1963) J. Math. Phys. 4: 1096
250. Girardeau M (1971) Phys. Rev. Lett. 27: 822
251. Girardeau M (1980) Int. J. Quantum Chem. 17: 25
252. Takahashi M, Fukutome (1983) Int. J. Quant. Chem. 24: 603
253. Kvasnička V (1982) Czech. J. Phys. B32: 947
254. Kvasnička V (1984) Croatica Chem. Acta 57: 1643
255. Weinstein H, Eilers JE, Chang Sh-Y (1977) Chem. Phys. Letters 51: 534
256. England WB, Silver DM, Steinborn EO (1984) J. Chem. Phys. 81: 4546
257. Reed AE, Weinhold F (1985) J. Chem. Phys. 83: 1736
258. Kirtman B, Cole SJ (1978) J. Chem. Phys. 69: 5055
259. Löwdin PO (1983) J. Math. Phys. 24: 70
260. Löwdin PO (1983) J. Math. Phys. 24: 88
261. Moller C, Plesset MS (1934) Phys. Rev. 46: 618
262. Poirier RA, Yadav A, Surján PR (1987) Canadian J. Chem. 65: 892
263. Salem L, Bruckman P (1975) Nature 258: 526
264. Warshel A, Barboy N (1982) J. Am. Chem. Soc. 104: 1469
265. Tavan P, Schulten K, Oesterhelt D (1985) Biophys. J. 47: 415
266. Gutman I (1979) Theor. Chim. Acta 50: 287
267. Daudel R (1974) Quantum theory of the chemical bond, Reidel, Dordrecht
268. Surján PR, Lukovits I, Mayer I (1987) Program BOSE, Budapest
269. Peterson M, Poirier RA (1986) Program MONSTERGAUSS, Univ. Toronto & Memorial Univ. Newfoundland
270. Sándorfy C (1955) Can. J. Chem. 33: 1337
271. Ruedenberg K (1962) Rev. Mod. Phys. 34: 326
272. Kapuy E, Kozmutza C, Daudel R (1980) Theoret. Chim. Acta 56: 259
273. Kapuy E, Kozmutza C, Daudel R (1980) Chem. Phys. Letters 72: 275
274. Zivkovic T, Maksić ZB (1976) Croatica Chemica Acta 48: 301
275. Sundberg K, Bicerano J, Lipscomb WN (1979) J. Chem. Phys. 71: 1515
276. Corcoran CT, Weinhold F (1980) J. Chem. Phys. 72: 2866
277. Maksić ZB (1981) Z. Naturforsch. 36a: 373
278. Kirtman B, Palke WE (1984) Croatica Chemica Acta 57: 1247
279. Maksić ZB, Eckert-Maksić M, Rupnik K (1984) Croatica Chemica Acta 57: 1295
279. Maksić ZB (1982) God. Jugosl. cent. kristalogr. ZAGREB 17: 89
280. Leonard JM, Luken WL (1984) Int. J. Quantum Chem. 25: 355
281. Dmitrieva IK, Zenevich VA, Plindov GI (1986) Chem. Phys. Letters 123: 541
283. Denis A, Langlet J, Malrieu JP (1973) Theor. Chim. Acta 29: 117
284. Eckert-Maksić M, Kovacevic K, Maksić ZB (1979) J. Organomet. Chem. 168: 295
285. Maksić ZB, Kovacevic K, Vampola M (1981) Z. Naturforsch. 36a: 1196
286. Eckert-Maksiç M, Maksiç ZB (1982) THEOCHEM 86: 325
287. Maksić ZB (1983) THEOCHEM 91: 295
288. Kovacevic K, Maksić ZB, Mogus-Milankovic A (1984) Croat. Chem. Acta 57: 187
289. Gerhards J, Ha T-K, Perlia X (1982) Helv. Chim. Acta 65: 105
290. Lochmann R, Weller Th (1984) Int. J. Quantum. Chem. 25: 475
291. Lochmann R, Weller Th (1984) Int. J. Quantum Chem. 25: 1061
292. Lochmann R, Weller Th, Kohler HI (1984) Int. J. Quantum Chem. 25: 1055
293. Daudey JP, Langlet J, Malrieu JP (1974) J. Am. Chem. Soc. 96: 3393
294. Otto P, Ladik J (1982) Int. J. Quantum Chem. 22: 169
295. England WB (1982) J. Phys. Chem. 86: 1204
296. Valdemoro C (1978) Phys. Rev. A 18: 2443
297. Valdemoro C, Rubio J (1980) An. Fis. 76: 26
298. Valdemoro C (1981) Proc. 5-th seminar on computation methods in quantum chemistry, Gröningen, Max Planck Inst. für Physik und Astrophysik
299. Valdemoro C (1985) Phys. Rev. A31: 2114
300. Valdemoro C (1985) Phys. Rev. A31: 2123
301. Valdemoro C (1986) Phys. Rev. A33: 1525

302. Bardeen J, Cooper LW, Schrieffer JR (1957) Phys. Rev. 108: 1175
303. Defranceschi M, Suard M, Berthier G (1984) Int. J. Quantum. Chem. 25: 863
304. Sundholm D, Pyykkö P, Laaksonen L, Sadlej AJ (1984) Chem. Phys. Letters 112: 1
305. Bader FW, Nguyen-Dang TT (1981) Adv. Quantum. Chem. 14: 63
306. Maksić ZB, Kovacevic K, Modus A (1981) THEOCHEM 85: 9
307. Magnusson E (1984) J. Am. Chem. Soc. 106: 1185
308. Liu F, Zhan ChG (1987) Int. J. Quant. Chem. 32: 1
309. Zhan ChG, Liu F, Hu ZM (1987) Int. J. Quantum Chem. 32: 13
310. Musso GF, Magnasco V (1987) THEOCHEM 37: 161

Interpretation of Molecular Behaviour by Localized Molecular Orbitals (LMOs)

C. Edmiston
Department of Chemistry, University of Wyoming, Laramie, Wyoming 82071/USA

Localized molecular orbitals (LMOs) as SCF MOs are described in various systems. Brief descriptions of their methods of calculation are given. These LMOs are used to understand: 1) electron-pair correlations using pair-natural-orbitals (PNOs) and pseudonatural-orbitals (PSNOs) for LMO electrons, 2) the symmetries of occupied and unoccupied "spectroscopic" (canonical) SCF MOs (CMOs) and the associated frontier-MO ideas and Walsh diagrams, 3) the approximate transferability of "chemical bonds", 4) "resonance energies" from unlocalizable (inherently delocalized) electrons, via "truncated LMOs" (TLMOs), and 5) other total molecular properties. The emphasis is on *conceptual* understanding of LMOs and molecular properties. Other reviews of LMOs are cited and briefly described.

1 General Ideas

For the general chemist, we will sketch the conceptual and mathematical basis of self-consistent-field (SCF) localized molecular orbitals (LMOs) of closed-shell molecules and atoms with N electrons [1, 2]. We will use the Ne, CH_4, and H_2O ground states to illustrate the concepts. For systems like these, which have no "inherently delocalized" electrons, the optimum LMOs correspond closely to the chemist's conception of "bond", "lone-pair", and "inner-shell" electron pairs. Furthermore, these LMOs (ψ_I) are fairly transferable between molecules for which chemical intuition indicates similar bonds, lone-pairs, and inner-shells, e.g., for certain homologous series of molecules and similar "functional groups". There are several different methods for calculation of these LMOs which we will describe, and all give similar LMOs. Each method has its own advantages and each can be used to study and refine chemical intuition through the study of related molecules. In the LCAO SCF LMO approach, the LMOs, ψ_I, are linearly expanded in terms of (NB) *basis functions* (atomic orbitals: AOs), which are usually "contracted (grouped) gaussians" (χ_u): $\Psi_I = \sum_u^{NB} c_{Iu}\chi_u, \left(\chi_u = \sum_m^{NG} K_{um}G_m\right)^*$. Basis functions optimized for a given type of LMO in a small molecule can be usefully transferred to a large molecule involving the same general type of LMO. This is true also when electron-pair correlation energies, e_{IJ}, are considered.

The LMOs offer a simplified picture of and approach to electron correlation, via the "pair natural orbitals" (PNOs) of electron pairs of the LMOs [3]. Electron-pair correlations tend to be most "uncoupled" when considered for LMO SCF electron pairs. The PNOs are the natural orbitals [4a] of an "independent electron pair (IEP)" [4b] moving in the Hartree-Fock potential of the other N-2 electrons. Pseudo-natural orbitals (PSNOs: $\psi^*_{I,K}$) are the PNOs of the "doubly occupied" SCF LMO electron pairs, after energy-weighted orthogonalization. These PSNOs are a good basis for describing *all* electron correlations, including those involving unpaired and inherently delocalized electrons.

Many conjugated systems and "activated complexes" can be treated as "resonance hybrids" of valence-bond theory structures, with each structure involving fairly localized LMOs (and corresponding PNOs). For example, for the square-planar singlet cyclobutadiene "activated complex", the LMOs of two equally-weighted structures are highly localized, even for the pi-electrons, and correspond closely to the two structural formulas:

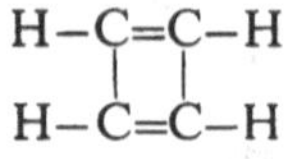

and

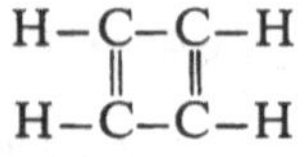

* The "primitive gaussian", G_m, is usually of the form $x^n y^l z^k e^{-a_m r^2}$ (cartesian gaussian) or simply $e^{-a_m r^2}$ (lobe-type gaussian).

Similar ideas apply for the "resonance" of the two B_2H_6 classical structures:

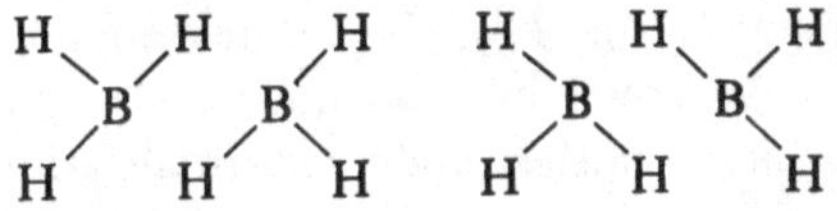

(See Section Below)

Linear combinations of LMOs, to form symmetry bases for irreducible representations of the point group of a molecule, give insight into the symmetries of the delocalized (canonical: spectroscopic) MOs (CMOs). This is particularly useful for "activated complexes", but is also useful for understanding the shapes of molecules from perturbation theory and the integral Hellmann-Feynman theorem (IHFT).

LMOs allow us to examine certain simplified models of molecular geometries, such as the "valence shell electron pair repulsion" (VSEPR) model, and the hybridization model.

1.1 Closed-shell States

The electron density, $\varrho(1)$, and several other properties of closed-shell states of atoms and molecules are given well by the "restricted Hartree-Fock" (HF or SCF) wavefunction, Ψ_{HF}, which is a Slater determinant of variationally optimized orthonormal spin-orbitals, φ_is $\left(\text{the electronic density is } \varrho = \sum_i^N |\varphi_i|^2\right)$. Here $\varphi_i = \psi_k$ (α or β) where ψ_k is a variationally optimized space orbital". For example, for the Ne atom ground state, $\Psi_{HF} = |1s\alpha 1s\beta 2s\alpha 2s\beta 2p_x\alpha 2p_x\beta 2p_y\alpha 2p_y\beta 2p_z\alpha 2p_z\beta|$ represents this normalized Slater determinant. The α and β are the usual "spin functions", which are simultaneously eigenfunctions of S^2 and S_z for a single electron. For a closed-shell state it is a self-consistent assumption for each 'space orbital", ψ_k, to be "doubly occupied". The φ_is are eigenfunctions of the "Fock equation",

$$\hat{F}(1)\,\varphi_i(1) = \varepsilon_i\varphi_i(1) \tag{1}$$

where $\hat{F}(1) = \hat{h}^\circ(1) + \sum_j^N (\hat{J}_j - \hat{K}_j)$, $\hat{h}^\circ(1) = (-1/2)\,\hat{\Delta}(1) + \sum_n^{NN} (-Z_n/r_{n1})$ is the "bare nuclei" hamiltonian, and $\hat{J}_j - \hat{K}_j = \int \varphi_j(2)\,((1 - \hat{P}_{12})/r_{12})\,\varphi_j(2)\,d\tau_2$ is the combined coulombic and exchange potentials of $\varphi_j(2)$ acting upon $\varphi_i(1)$. This can be shown by the variation of the orthonormal φ_is in the total variational energy expression:

$$E = OE + ER + NR = \sum_i^N h_i^\circ + \sum_i^N \sum_{j>i}^N (J_{ij} - K_{ij}) + NR \tag{2}$$

where $h_i^\circ = \int \varphi_i(1)\,\hat{h}^\circ(1)\,\varphi_i(1)\,d\tau_1 = \langle \varphi_i(1)|\,\hat{h}^\circ(1)\,|\varphi_i(1)\rangle = \langle I|\,\hat{h}^\circ\,|I\rangle$ and $J_{ij} - K_{ij} = \int\int \varphi_i(1)\,\varphi_j(2)\,[(1 - \hat{P}_{12})/r_{12}]\,\varphi_i(1)\,\varphi_j(2)\,d\tau_1\,d\tau_2 = \langle \varphi_i\varphi_j\,||\,\varphi_i\varphi_j\rangle = \langle IJ\,|\,IJ\rangle - \langle IJ\,|\,JI\rangle = (II\,|\,JJ) - (IJ\,|\,IJ)$. Here NR is the nuclear-nuclear repulsion, OE is the one-electron energy, and ER is the electron-electron repulsion energy.

Since first being pointed out by Fock (1c), it has been known that the φ_is (and therefore the ψ_is) can be subjected to a unitary transformation (orthogonal transformation, $\hat{O}$, for real φ_is) without changing the Ψ_{HF}, since a determinant formed by linear combinations of the columns (or rows) of a previous determinant leaves it unchanged, except by a constant which vanishes upon renormalization. Thus any $\varphi'_k = \sum_i^N O_{ki}\varphi_i$, which are also orthonormal, can be used in Ψ_{HF}. However these φ'_ks are solutions of $\hat{F}(1)\,\varphi'_k(1) = \sum_j^N \lambda_{kj}\varphi'_j(1)$, where the Hermitian "Lagrangian-multiplier matrix", $\hat{\lambda}$, of the λ_{ij}s, is given by $\hat{\lambda} = \hat{O}^t\hat{\varepsilon}\hat{O}$ with $\hat{\varepsilon}$ being the diagonal matrix of the ε_is. (It is easily shown that $\hat{F}(1)$ is invariant to the unitary transformation, $\hat{O}$, of the φ_is to the φ'_is).

Now the unitary (or orthogonal) transformation, $\hat{O}$, can be chosen so that the φ'_is are "localized", and, where symmetry allows, "equivalent". Thus for the Ne atom ground state

$$\Psi_{HF} = |1s\alpha 1s\beta t_1\alpha t_1\beta t_2\alpha t_2\beta t_3\alpha t_3\beta t_4\alpha t_4\beta|$$

where t_i is an "sp^3" hybrid atomic orbital (HAO) of the 2s and 2p SCF ψ_is. In Table 1 we show this $\hat{O}$ matrix, where each element is to be multiplied by (1/2) for normalization. The t_i are "equivalent AOs" because symmetry operations take them into each other. For the CH_4 ground state Ψ_{HF}, a similar situation holds:

$$\begin{aligned}\Psi &= |1a_1\alpha 1a_1\beta 2a_1\alpha 2a_1\beta 1t_x\alpha 1t_x\beta 1t_y\alpha 1t_y\beta 1t_z\alpha 1t_z\beta| \\ &= |1a_1\alpha 1a_1\beta T_1\alpha T_1\beta T_2\alpha T_2\beta T_3\alpha T_3\beta T_4\alpha T_4\beta|\end{aligned}$$

Here the $2a_1$ and $1t_x$, etc., are the delocalized (canonical: spectroscopic) molecular orbitals (CMOs), and T_i is a localized (C—H bond) MO (a LMO) "pointing toward" the ith H atom. In Fig. 1 we show the orientation of the axes x, y, and z, relative to the positions of the H atoms (at the alternate corners of the cube). Exactly the same $\hat{O}$ transformation gives the T_is as gave the t_is for Ne: Table 1, where $1t_x$ replaces $2p_x$, etc, and $2a_1$ replaces 2s.

In Table 2 we give the occupied CMOs of CH_4 approximately in terms of a minimal basis of AOs: 1s, 2s, $2p_x$, $2p_y$, $2p_z$ on the C atom, and $1s(H_1) = h_1$, etc., on the H atoms. N is the normalization constant of the CMO. Let us call the matrix

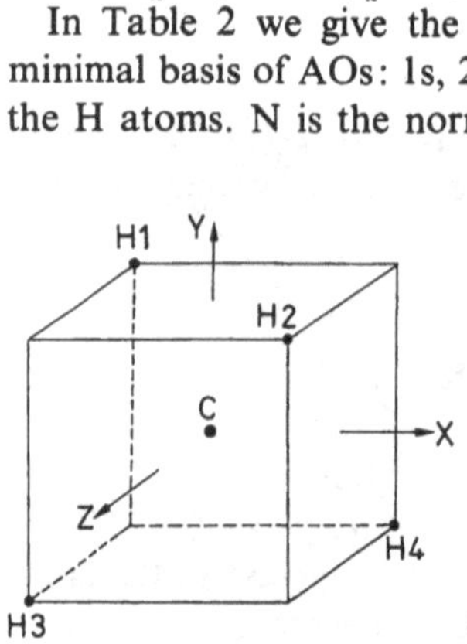

Fig. 1. CH_4 coordinate system

Table 1.2 $\hat{O}$ Matrix

	2s	$2p_x$	$2p_y$	$2p_z$
t_1	1	−1	1	−1
t_2	1	1	1	1
t_3	1	−1	−1	1
t_4	1	1	−1	−1

Table 2. $\hat{C}_{CMO}$

	1s	2s	$2p_x$	$2p_y$	$2p_z$	h_1	h_2	h_3	h_4
$1a_1$	k'	m'				−L	−L	−L	−L
$2a_1$	n'	N'				K	K	K	K
$1t_x$			N			−Nc	Nc	−Nc	Nc
$1t_y$				N		Nc	Nc	−Nc	−Nc
$1t_z$					N	−Nc	Nc	Nc	−Nc

k' = .9915, m' = .0530, L = .0106, n' = −.1917, N' = .6413, K = .1700, N = .6024, Nc = .2800.

of these "linear combination of atomic orbitals" (LCAO) coefficients $\hat{C}_{CMO}$, so that $\hat{C}_{LMO} = \hat{O}\hat{C}_{CMO}$. We give $\hat{C}_{LMO}$ in Table 3, both in terms of the original AOs and in terms of HAOs (sp^3 hybrids on the C atom): t_1, t_2, t_3 and t_4. By means of Table 3 one can choose either set of AOs on the C atom, but the sp^3 HAOs show clearly the "chemist's intuition". We can define $t_i' = Nt_i + \sum_{j\neq i}^{4} (q)\, t_j$ as a new hybrid (HAO) in terms of which the qs would vanish.

Note that the T_iC—H bond LMOs have small negative contributions (P ≅ −.0264) from H-atom 1s AOs for H-atoms other than the one toward which it "points": is directed. These are necessary to maintain orthogonality of the T_is, and are often called

Table 3. $\hat{C}_{LMO}$ (LMOs are "energy localized" LMOs)

	1s	2s	$2p_x$	$2p_y$	$2p_z$	h_1	h_2	h_3	h_4	t_1	t_2	t_3	t_4
ic = $1\tilde{a}_1$	k	m	0	0	0	−L'	−L'	−L'	−L'	θ	θ	θ	θ
T_1	n	r'	−r	r	−r	N_C	P	P	P	N	q	q	q
T_2	n	r'	r	r	r	P	N_C	P	P	q	N	q	q
T_3	n	r'	−r	−r	r	P	P	N_C	P	q	q	N	q
T_4	n	r'	r	−r	−r	P	P	P	N_C	q	q	q	N

k = 1.0097, m = −.0489, L' = .0131, n = −.0393, r' = .2692, r = .3009, N_C = .5010, P = −.0264, N = .5860, q = −.0159, θ = −.0244.

Each atomic orbital is an optimized linear combination of four primitive Gaussians.

"LMO tails" or "nonprinciple components". The t_i' and h_i components of T_i are called the "principal components" or "HAO components" of the LMO.

Even before LCAO MO SCF calculations had been carried out on small molecules such as CH_4, Coulson, Lennard-Jones, Hurley, Hall, Pople, etc. [2] had worked out the above ideas of "equivalent LMOs". They [2b] had even pointed out that presumably the equivalent MOs would nearly minimize |EX| for the nonclassical off-diagonal "exchange energy", $EX = -\sum_{i}^{n}\sum_{j>i}^{n} 2K_{ij}$, $n = N/2$, of the Hartree-Fock (SCF) energy. However, they did not give any method for obtaining LMOs when symmetry did not uniquely determine "equivalent LMOs". In 1963 an iterative method was proposed [1a] and applied for obtaining those LMOs which do minimize |EX|, using successive binary (2×2) orthogonal-rotations of the ψ_is until the optimum ψ_Is are obtained: $\psi_I = \psi_i(\cos\theta) + \psi_j(\sin\theta)$, $\psi_J = \psi_j(\cos\theta) - \psi_i(\sin\theta)$. This was shown to nearly always give well-defined "inner-shell", "lone-pair", and "bond-orbital" LMOs, and that, furthermore, these "energy localized" LMOs were fairly transferable between molecules with similar types of bonds, lone-pairs and inner-shells. In fact, to first-order in the perturbation theory of the LMOs, these are the *most transferable* LMOs, since they minimize all interactions with each other, *both* coulombic and exchange. $COD = \sum_{i}^{n}\sum_{j>i}^{n} 4J_{ij}$, which represents the "coulombic off-diagonal" interactions between the space MOs, is minimized when |EX| is minimized, because $D = \sum_{i}^{n} J_{ii} = \sum_{i}^{n} K_{ii}$ is maximized. The total electronic repulsion energy, $ER = C - TSR$, and $C = 2D + COD$ are of course not changed since Ψ_{HF} is not changed. $TSR = -EX + D$ is appropriately called the "total self-repulsion" of the electrons, but has also been called the negative of the "total exchange energy", TEX: $TSR = -TEX = -(EX - D)$.

The (2×2) rotation method is analogous to Jacobi diagonalization, and occurred to the present author upon reading a paper by Nesbet [1d] which stated that it would be good to be able to calculate those LMOs which minimize $|EX| = \sum_{I}^{n}\sum_{J\neq I}^{n} \langle IJ \mid JI\rangle$ and therefore maximize $D = \sum_{I}^{n} \langle II \mid II\rangle$. (Jacobi diagonalization reduces off-diagonal elements of a real-symmetric matrix gradually to zero, via (2×2) orthogonal rotations of the matrix basis, while *not* changing the trace). Calculations show that:

$D_{IJ} = \langle II \mid II\rangle + \langle JJ \mid JJ\rangle = \langle ii \mid ii\rangle + \langle jj \mid jj\rangle + A_{ij} - A_{ij}\cos 4\theta + B_{ij}\sin 4\theta$ where $A_{ij} = (ij \mid ij) - (1/4)(ii - jj \mid ii - jj)$ and $B_{ij} = (ii - jj \mid ij)$. Setting $(dD_{ij}/d\theta) = 0$, we see that $\tan 4\theta = -B_{ij}/A_{ij}$. This is a maximum for D_{ij} when $(d^2D_{ij}/d^2\theta) = 16(A_{ij}\cos 4\theta - B_{ij}\sin 4\theta) < 0$.*

* Alternately, suppose the $\psi_I = I$ which minimize the |EX| are slightly "mixed" from i and j. Then $I = K(i + Tj)$ and $J = K(j - Ti)$, where $T = \tan\theta$ and $K^2 = 1/(1 + T^2)$, for any given I, J pair. Then, since $IJ = K^2[(ij)(1 - T^2) - T(ii - jj)]$, we have $(IJ \mid IJ) = K^4[(1 - T^2)^2 (ij \mid ij) - 2T(1 - T^2)(ij \mid ii - jj) + T^2(ii - jj \mid ii - j)]$, so, neglecting terms in T^2, T^3, etc., we have $(IJ \mid IJ) = (ij \mid ij) - 2T(ij \mid ii - jj)$. Thus we see that *when* the ψ_is are optimum (LMOs), $(ij \mid ii - jj) = (ij \mid ii) - (ij \mid jj) = 0$. (This is also true when they are *most delocalized*; $D = \sum_{I}^{n} (II \mid II)$ is a *minimum*.) If into

For *each* (2×2) rotation, one must transform all (ij | kl)s to those involving (IJ | xy), (II | xy) and (JJ | xy), for all occupied x and y MOs. For N occupied MOs, this is essentially 3/[N(N + 1)/2] of a full integral transformation (in computer time). However, *if* we are close to convergence, we can multiply the N(N — 1)/2 calculated small rotations, $\hat{O}_{ij}$, *before* transforming any integrals. Then we can transform all integrals, using symmetrically orthogonalized φ_i''s from $\hat{O} = \Pi_{i,j>i}\hat{O}_{ij}$. This is then essentially a "steepest ascent" approach for the function $D = \sum_i^N \langle ii \mid ii \rangle$.

If $\varphi = \theta - \theta_{max}$ is very small (near convergence), then since $\tan \varphi = \varphi + a(\varphi^2) + \ldots \cong \varphi$, $4T = 4 \tan \varphi \cong \tan (4\varphi)$. W. England [1e] made this the basis of a first-order scheme for energy localization, finding it to work even when starting with the CMOs, provided a symmetric orthogonalization of all MOs are made after each set of N(N — 1)/2 pair-rotations.

The "orthogonal Hartree model (OH)" of a system [1 a] assumes that each electron interacts with the other electrons *only* through the repulsion of their "orbital probability charge clouds", $|\psi_i|^2$: in an average way, and via the "Pauli-principle constraint" that the φ_is be orthogonal: $\langle \varphi_i \mid \varphi_j \rangle = \delta_{ij}$. This corresponds to a wavefunction, Ψ_{OH}, which is a simple product of orthonormal spin-orbitals φ_i: $\Psi_{OH} = \varphi_1(1)\,\varphi_2(2) \ldots \varphi_N(N)$. In this model the instantaneous motions of the electrons are in no way correlated, either by "Fermi holes" or by "coulombic correlations holes" (see below). This very simple conceptual model should give the gross features for the electron density, $\varrho(1) \cong \sum_i^N \varphi_i^2(1)$, of the system, and an energy, E_{OH}, in error by only a few percent at most. Calculations on molecules with this model are slightly more difficult than Hartree-Fock (HF) ones, when each φ_i is represented as a linear combination of NB basis functions: $\varphi_i = \sum_n^{NB} c_{in}\chi_n$. The same electron repulsion integrals, $\langle \chi_m(1)\,\chi_n(2)|\, 1/r_{12}\, |\chi_k(1)\,\chi_l(2)\rangle$, must be calculated in either case and there is a different, and nonhermitian "Hartree operator", $F_i^H(OH)$, for each $\varphi_i(OH)$. Therefore this model has never been used to any extent. By the Pauli antisymmetry-principle, the proper wavefunction must be antisymmetric in the exchange of electrons. The E_{HF} differs from E_{OH} only in the subtraction of the K_{ij} ($i \neq j$) "exchange integrals", for the same set of φ_is. The K_{ij}s (and K_{IJ}s for space SCF MOs) arise from the "Fermi holes" which keep two electrons of the same spin from ever being at the same point in physical space, due to the Pauli principle.

Sometimes these "Fermi holes" are extremely strange in shape. For the $\Psi_{HF} = |1s\alpha 1s\beta 2s\alpha 2s\beta|$ of the Be atom ground state, if one electron of a given spin (α or β) is at a given "r" (radial distance) from the nucleus, the other electron of this same spin will not be found anywhere on the sphere of radius "r", because the 1s

(IJ | II — JJ) = 0 we substitute again, we get $0 = (1 - T^2)^2$ (ii — jj | ij) — $T(1 - T^2)$ (ii — jj | ii — jj) + $4T(1 - T^2)$ (ij | ij) — $4T^2$(ij | ii — jj) which gives $\tan (4\theta) = -B_{ij}/A_{ij}$, which gives the *optimum* $\theta(\theta_m)$, $0 < \theta_m < \pi/2$, for the given (2×2) rotation. By comparing this $D(\theta_m)$ with the "unrotated" D, we can see that unless $D(\theta_m) - D = \Delta D \geqq 0$, we must replace θ_{m_2} with $\theta_m + (1/4)\,\pi$, since $\Delta D = 2[(ij \mid ij) - (IJ \mid IJ)] = A_{ij} + (A_{ij}^2 + B_{ij}^2)^{1/2} \cos 4(\theta - \theta_{max})$ [1 a].

and 2s depend only on "r". For an electron in the valence-shell of the ground state Ne atom, an electron at the radial maximum of that shell ($4\pi r^2 \varrho$ maximum) will have a Fermi-hole which is roughly the same shape as an sp^3 hybrid AO for that shell. Thus for LMOs, when the electron is near the maximum of $|\varphi_i'|^2$, its Fermi-hole is nearly identical with that LMO density. This is the basic reason why the K_{ij} integrals are so small between different φ_i's (LMOs). The K_{ij} can also be considered as the self-repulsion of the orbital product $\varphi_i(1)\,\varphi_j(1)$, which product is quite small everywhere for well localized LMOs. Nevertheless, even for the most localized LMOs of a molecule, the K_{ij} ($i \neq j$) can never be neglected, for they usually constitute about 1% to 5% of the total electron repulsion energy, ER. They are in general a bigger effect on chemical binding energies and energy surfaces than is the "coulombic correlation energy", E^c, left out of the SCF model [5a]. However their effects on the gross forms of the φ_i's is minor, and it is expected that the φ_i(OH) of the orthogonal-Hartree model would be quite similar to the LMOs of Hartree-Fock theory.

The Pauli antisymmetry (exclusion) principle demands *orthogonal* MOs (CMOs or LMOs) only if we wish to use the comparitively *simple* and visualizable energy expressions and Fock operator [Eqs. (*1*) and (*2*)]. However such orthogonality is no constraint on the Ψ_{HF}, since any optimized nonorthogonal SCF MOs could be orthogonalized with no change in the overall Ψ_{HF} determinant. Once the optimum orthonormal φ_is (or φ_i's) have been calculated, they can be replaced by linear combinations which are *non*orthogonal, although this is usually not done due to the much greater complexity of the energy expression.

To summarize, the electronic structure of the ground state of any closed shell molecule *can be visualized* as resulting in the following way, with energy effects considered in descending order of their importance and confining the electrons to orthonormal MOs to keep energy expressions and concepts comparatively simple. Each electron's orbital, φ_i', takes on its optimum form while the electron moves in the "classical-coulombic-like" field of the probability-charge-clouds, $|\varphi_j'(2)|^2$, of the other N-1 electrons *and* the nuclear attraction potentials. (This is the orthogonal-Hartree model, which would probably give the electron density and many other properties quite well, but is computationally more difficult than the HF model.) In the HF wavefunction, the electrons each become "surrounded" by modified and fluctuating "Fermi holes", which make *zero* the probability of finding any other electron of like spin at the same point in physical space. This introduces the K_{ij} ($i \neq j$) in the HF energy expression, and slightly changes the MOs as well, changing them to the LMOs of the HF wavefunction.

This Hartree-Fock Ψ_{HF} leaves out the "coulombic correlation hole" which roughly spherically surrounds each electron. This last effect is usually "built into" Ψ using configuration-interaction: CI. When LMOs are used in Ψ_{HF} this CI is comparatively simple, as we will now describe.

1.2 Coulombic Correlation Effects Between Electrons in LMOs

Within an error of a few percent the correlation energy, E^c, is given by the sum of "independent electron pair approximation (IEPA)" pair-correlation energies, $\tilde{e}_{ij}$ [4b]. An $\tilde{e}_{ij}$ is the improvement in energy, below E_{HF}, when only the electrons in spin-orbitals "i" and "j" are allowed to correlate: we allow only excited configurations in which these spin orbitals are replaced by excited (unoccupied) spin-orbitals in the CI expansion. Even for a rather large basis set of $M = NB - N/2$ excited space orbitals, this leads to a secular equation of size $(M + 1)(M + 2)/2$, which is quite easily handled provided one has been able to afford the transformation of the two-electron integrals from the elemental basis to the NB SCF orbitals (or similar orthonormal functions). A faster method for this transformation of integrals, which goes as NB^5, was worked out in 1969, and is much used today [5b]. Previously known methods went as NB^6, and so were much less efficient for large NB [5c]. Once the $(M + 1)(M + 2)/2$ pair-excitation CI-coefficients, C_{ij}^{kl}, are determined, one can diagonalize the first-order density matrix (or the matrix of the c_{ij}^{kl}s, with off-diagonal elements multiplied by $(1/\sqrt{2})$), to obtain the pair-natural-orbitals (PNOs) for the (i, j) electron pair [3].

When we apply this approach to a "doubly occupied" SCF MO electron-pair, in space orbital Ψ'_i, the "electron pair function" becomes $\Lambda_{ii}(1, 2) = N\left[\Psi'_i(1)\Psi'_i(2) + \sum_{k>n}^{M} c_k \Psi_k(1)\Psi_k(2)\right] S(1, 2)$, where the singlet spin function $S(1, 2) = (1/\sqrt{2}) \times (\alpha(1)\beta(2) - \beta(2)\alpha(1))$. It is found that usually Ψ'_i is very similar to the SCF Ψ'_i, since usually "single excitations" from the SCF ground state configuration make very small contributions, due to Brilloin's theorem.*

Because of the minimax properties of eigenvalues, the c_ks decrease in magnitude at optimum rate, so that only a *few* PNO configurations can account for a very high percentage of the correct $\tilde{e}_{ii}$, provided a "nearly complete" elemental basis is used. This basis can be approximately optimized economically for LMO $\tilde{e}_{ii}$s, since we need consider only a *large* basis in the region of the SCF LMO: on one or two atoms usually. For small and intermediate sized molecules, one can optimize the exponents *and* contraction coefficients of the "excited PNO basis". This is important since the time-consuming NB^5 integral transformation requires keeping NB as small as possible.

The natural orbitals for the ground state H_2 electrons have been discussed by Davidson and Jones [6], and their rapid convergence demonstrated. About half the correlation is of the "left-right type" (lrc), well described by the Weinbaum wavefunction [7] at $R = R_0$ and becoming the only type of correlation as $R \to \infty$. This lrc corresponds to one electron of the pair being on one atom as the other is instantaneously on the other atom. It approaches $\sim 20\%$ of E^c of He as $R \to 0$. The two electrons also "angularly correlate" (ac) with one tending to be on the opposide side of the internuclear axis from the other. This ac is $\sim 35\%$ of E^c for $R = R_0$ but becomes about 40% of E^c as $R \to 0$. Finally "in-out correlation" (io) increases to about 40% of E^c as $R \to 0$, but is only about 15% of E^c at $R = R_0$. For this io correlation one electron tends to be near the nuclei (or internuclear axis) while the other tends to be far away. This is reasonably typical for the correlation in any "single bond" LMO, except that strong "exclusion effects" often decrease some types of correlation, as the excited PNOs (and PSNOs) must remain orthogonal to

* This is not true for H_2 at large internuclear distances, for example, where the first natural orbital, $1\sigma_g^N$, becomes quite different from the SCF MO, due to "coupling" of the single excitations through the very strong $1\sigma_u^2$ double excitation.

any occupied SCF MOs. For example, the $2s^2$ electron pair of the Ne atom has little angular correlation due to the occupation of the 2p AOs it most would use for this. Also for *long* bond distances, as in the F_2 molecule at R_0 or H_2 at large R, "left-right" correlation (lrc) becomes much larger. When lrc becomes dominat, due to the near degeneracy of the σ_g and σ_u^* LMOs, the other two types of correlation rapidly approach zero, because the two electrons are efficiently kept from each other by the lrc.

In order to obtain reasonably accurate SCF energies, polarization basis functions must be included, e.g., 2p-type basis on H atoms and 3d-type basis on first-row atoms. For very high accuracy work, even higher angular momentum bases must be included. These basis functions *can be* chosen (optimized) to serve efficiently for both occupied SCF MO polarization, and for excited PNOs, although some compromise is involved. A good approach is to calculate SCF LMOs using only a "medium sized" double-zeta unpolarized basis: (8s, 4p → 4s, 2p) on first row atoms, and (4s → 2s) on H atoms.*

This can be followed with a PNO calculation, for a given LMO, with polarization and other basis introduced on the atoms heavily involved in the LMO. At this point the polarization basis can be optimized, even with exponent optimization. (One can use W. Meyer's method [8] for avoiding the two-electron transformation in these optimization steps.) After carrying out this PNO calculation for all doubly-occupied SCF LMO electron pairs, one can then proceed to the final CI all-system calculation in the following efficient way. The excited PNOs are Schmidt-orthogonalized to the "occupied" SCF LMO, which usually changes them very little, depending on the magnitude of "single and triple excitation" CI effects. The PNOs from different SCF LMOs are *not* orthogonal to each other, although they are orthogonal (now) to all ground-state SCF LMOs. They can be "energy-weighted orthogonalized" (EWO) to each other. This is a process intermediate between symmetric (Lowdin) orthogonalization and Schmidt orthogonalization.⁺ In this way we can efficiently obtain a set of NB orthonormal PNOs (PSNOs) with each localized "doubly-occupied" SCF LMO electron-pair having several important excited PSNOs for its correlation.

The PSNOs strongly associated with (localized over) two neighboring SCF LMOs are also quite efficient at "interorbital" (inter-pair) correlations. We can illustrate this with the simple example of two strongly interacting He atoms a and b (in their ground states). Let "z" be the internuclear axis. There are corresponding $2p_z$-like, $2p_x$-like and $2p_y$-like PSNOs "on" the two atoms, with each having small "tails"

---

* This basis has eight s-type primitive gaussians on first-row atoms contracted (grouped) to four s-type groups, four p-type gaussians contracted to two p-type groups, and on H atoms four s-type gaussians are contracted to two s-type groups. Usually the gaussians of lowest exponent are made groups in themselves.

⁺ If φ_{ik} and φ_{jl} are two excited PNOs associated strongly with the SCF LMOs ψ_i and ψ_j respectively, then we set $\varphi'_{ik} = N(\varphi_{ik} - S(a/2)\,\varphi_{jl})$ and $\varphi'_{jl} = N(\varphi_{jl} - S(b/2)\,\varphi_{ik})$, where $a^2 + b^2 = 2$. N is the normalization constant, and $S = \langle \varphi_{ik} \mid \varphi_{jl} \rangle$; a/b is taken as the ratio of $(c_{ii}^{kk}/c_{jj}^{ll})^2$. We do this for all pairs of nonorthogonal PNOs, using the NB — (N/2) PNOs with the largest c_{ii}^{kk}s. We then iterate this process to convergence: to exact orthogonality. (If all the PNOs used should happen to have the same c_{ii}^{kk}, then this would be symmetric orthogonalization, by iteration.) Note that if a given PNO is very important (large c_{ii}^{kk}) then another PNO of much smaller importance (small c_{jj}^{ll}) is essentially Schmidt-orthogonalized to that important PNO.

on the other atom. It turns out that pair-excitations of the type $2p_z^a(1)\ 2p_z^b(2)$, $2p_x^a 2p_x^b$, and $2p_y^a 2p_y^b$ are very efficient at describing the van der Waals dispersion-force interatom-correlation effects, even for close range (repulsive range) for the two He atoms. This is understandable in the following way. When an electron of the 1s (He) SCF MO instantaneously "feels" the repulsion of another electron, it tends to shift its probability density somewhat to the other side of the atom (angular correlation), and also inward if the other electron is far from the nucleus (in-out correlation). This is true in an isolated He atom, as is well known. When only a single interatom ($\alpha\beta$) pair-correlation is being considered (uncoupled correlations), this takes place in much the same way. Of course in the final CI wavefunction these correlations become "coupled" and interfere with each other to some extent, but this is a secondary effect responsible for only a small percentage of E^c. (It is however an important reason why the IEPA overestimates the E^c.) When the probability of the electron is thereby polarized (away from the other electron), it is done using PSNOs which are *least* increasing of the energy of the electron, whether the correlation is intra-atomic or inter-atomic in nature. Thus nearly the same PSNOs are used in both cases.

Because LMOs do not much penetrate, the $\tilde{e}_{ij}$ between them are not so large as are those $\tilde{e}_{ii}$ for the pair of electrons in a LMO. Nevertheless, because there are so many such interorbital $\tilde{e}_{ij}$s ($i \neq j$), they can add up to more than the $\tilde{e}_{ii}$s in a system like the Ne ground state or the CH_4 ground state [8]. Using the same basis and configurations, this approach gives results very close to those of the SOGVB and SOGVB-CI method [9].

1.3 Other Intrinsic (Internal) and External (Extrinsic) Methods of SCF MO Localization to LMOs

We will here briefly review several methods by which LMOs have been obtained even for systems which have no symmetry, so that "equivalent MOs" could not be obtained. These methods usually give LMOs which are roughly equivalent MOs, if the system nearly has the necessary symmetry, and usually the LMOs will go over smoothly into equivalent MOs as symmetry is approached.

In 1960 Boys [10a] proposed the transformation of the occupied CMOs to LMOs which, it was hoped, would be highly "chemically invariant" and which were approximated as "exclusive MOs". These exclusive-MOs were defined to be those φ_i's which maximized the *product* of the squares of the distances between the "centers of charge" of the LMOs. This causes no "rotation" between two MOs of the same symmetry, when their centers-of-charge coincide. Following the work of Edmiston and Ruedenberg, Boys [10b] redefined the "exclusive MOs" as those which *minimize* $I = \sum_i^n (\varphi_i^2(1)|\, f(r_{12})\, |\varphi_i^2(2))$ where $f(r_{12}) = r_{12}^2$. (For the Edmiston-Ruedenberg localization method, $f(r_{12}) = (1/r_{12})$, and $I = D$ is *maximized.*) This type of localization has been called an "intrinsic localization" criterion by Ruedenberg [1b], since it in no way depends on the LCAO basis, and could be applied in principle even to numerical SCF MOs. Boys also adopted the "pair-wise rotation" of φ_is, proposed by Edmiston and Ruedenberg [1a]. Boys' second criterion is equivalent to the maximation of the *sum* of the squares of the distances between the LMO centers-of-charge. This method has been used to a fair extent over the years, and usually does not give LMOs differing greatly from those of other methods.

Since one need only compute and transform the "dipole integrals", $\langle\varphi_i(1)|\ \vec{r}_1\ |\varphi_j(1)\rangle$, for larger molecules this method is faster than the Edmiston-Ruedenberg one, which requires transformation of electron-repulsion (2-electron) integrals. Lipscomb and coworkers [11 a] have made a detailed comparison of these two types of LMOs in a number of boron hydrid compounds. Von Niessen [11 b] has proposed the "intrinsic" criterion where $f(r_{12}) = \delta(\vec{r}_2 - \vec{r}_1)$, so that one maximizes the sum of the self-overlaps $\langle\varphi_i^2(1)\ |\ \varphi_i^2(1)\rangle$ of the LMO *densities*. Because the self-overlap of the total charge density with itself does not change, this is equivalent to minimizing the charge-overlaps of different LMOs with each other. This brings in new types of integrals which must be evaluated between the basis functions.

There have also been proposed other localization schemes, particularly of the "extrinsic criterion" type: they depend on the nature of the LCAO basis set used. That of Magnasco and Perico [12] has been most used.

A Mulliken population analysis can be carried out for each separate real orthonormal MO, φ_i, with the results then being summed over occupied MOs to give the total $\chi_u\chi_v$ population components: $\varphi_i = \sum_u^{NB} C_{iu}\chi_u$, $2 = 2\ \Sigma_u\Sigma_v C_{iu}C_{iv}S_{uv}$, where $S_{uv} = \int\chi_u\chi_v\,d\tau$. (Here both summations are, of course, over all the NB basis functions.) Then $N = 2\ \Sigma_u\Sigma_v p_{uv}S_{uv}$ where $p_{uv} = \sum_i^n C_{iu}C_{iv}$, $n = N/2$. Here p_{uv} is the bond-order, for χ_u and χ_v. In the localization scheme of Magnasco and Perico, one assigns by chemical intuition *certain* basis AOs, χ'_{iv}, to be "associated" with a given LMO, φ_i, in the MO population analysis. In a "balanced basis set", as defined by Mulliken [13], one can usually identify those AO bases which principally contribute to the inner-shell LMO of atom A: $\Gamma(i_A)$. The remaining AOs on that atom are called the "valence AO basis": $\Gamma(V_A)$. If we expect a certain bond between two adjacent atoms A and B, then we expect any LMO, φ_i, associated with that bond to have population with heavy contribution from $P_i = 2\left(\sum_u\sum_v c_{iu}c_{iv}s_{uv}\right)$ where here the sums run only over certain AOs of $\Gamma(V_A)$ *and* $\Gamma(V_B)$: $\Gamma_i(V_AV_B)$. We expect any inner-shell φ_i on atom A, to have nearly all (population) contributions from P_i, where $u \subset \Gamma(i_A)$ and $v \subset \Gamma(i_A)$. We expect, similarly, any lone-pair φ_i on atom A to have heavy contribution from P_i where $u \subset \Gamma(V_A)$ and $v \subset \Gamma(V_A)$. They then determine LMOs by the maximization of $\sum_i^n P_i$, using the (2×2) MO-pair "rotations" method introduced by Edmiston and Ruedenberg [1 a] for their method. P_i is called the *local* total-population, and will be nearly 2 (electrons) if the given LMO, φ_i, is nearly completely described (expandable) in terms of the "local" AOs *chosen* for it.

Recently Rajzmann, Brenier and Purcell [14] have proposed an "external" localization method which they maintain is applicable separately to both occupied *and* unoccupied MOs. Here they *minimize* the sum of the "remote populations", $\sum_i^n P_i^r$, where $p_i^r = 2\sum_{u\not\subset\Gamma_i}\ \sum_{v\not\subset\Gamma_i}[c_{iu}c_{iv}s_{uv}]$. (Note that $2 = (P_i + P_i^r + 2\,P_i^c)$ where $P_i^c = 2\sum_{u\subset\Gamma_i}\ \sum_{v\not\subset\Gamma_i}[c_{iu}c_{iv}s_{uv}]$). The unoccupied localized MOs (ULMOs) should be quite useful for describing localized electron correlations with a smaller CI.

Pipek and Mezey [53] have proposed an intrinsic method called "population localization", which minimizes D where $D^{-1} = N^{-1} \sum_{i}^{N} d_i^{-1}$, $d_i^{-1} = \sum_{A}^{n} (Q_A^i)^2$, where Q_A^i is the gross Mulliken population of MO "i" from atom "A". D is a measure of the mean extension (delocalization) of the MOs. $Q_A^i = \sum_{n}^{NB} \sum_{m}^{NB} P_{nm}^i S_{nm}$, $P_{nm}^i = C_{in} C_{im}$, where the atomic-orbital basis function χ_n or χ_m (or both) are centered on atom A. d_i measures the "number of atoms where the MO is delocalized". For example, suppose that a LMO was essentially confined to two atoms, and described a homopolar covalent bond between them. $Q^i = 1/2$ for both atoms, and $d_i = 2$. If the MO existed equally on three atoms, as for the MO of $H_3(D_{3h})$, then $Q^i = 1/3$ and $d_i = 3$. If the LMO was a "lone pair" LMO essentially confined to one atom, then $Q_A^i = 1$ for that atom, and zero for the other atoms. Then $d_i = 1$, As with the ER energy-localized LMOs, these LMOs often keep sigma and pi LMOs separate, which the Boys method does not. This method requires only the AO overlap integrals, similarly to the Magnasco and Perico method, and it is similarly very fast computationally.

1.4 Methods for Faster Calculation of Energy-localized LMOs

Leonhard and Luken [15] have shown that convergence to near-LMOs can be obtained with much less computation if *all* W_{ijkl} (ij | kl) less than a small "editing threshold" (10^{-2} to 10^{-4}, say) are neglected. Here W_{ijkl} is basically the "bond-order product", $P_{ijkl} = 4p_{ij}p_{kl} - p_{ik}p_{jl} - p_{il}p_{jk}$, except that W_{ijkl} changes the negative signs to positive ones and uses absolute values for the terms. Note that the total electron repulsion energy (ER) is $ER = \frac{1}{2} \sum_i \sum_j \sum_k \sum_l p_{ijkl}$ (ij | kl), so that this will eliminate any insignificant components of ER in the transformation to LMOs. Note also that $2p_{ij}S_{ij}$ is the (i, j) AO-product "population component" of the electron density, ϱ, so that P_{ijkl} (ij, kl) can be thought of as AO-product population repulsions. Therefore they call their method "population weighted editing of integrals" in the transformations. Thus as we perform the binary (2 × 2) orbital-pair rotations, the MO integrals (II | IJ), (JJ | IJ), (II | JJ), (II | II) and (JJ | JJ) are all calculated *essentially* correctly. Since the LMOs usually essentially involve only the AO basis of one or two atoms, as we converge toward the LMOs very many terms fall below this "editing threshold". For the early iterations (2 × 2 rotations), *exact* transformations are not important anyway. This approach is in the same spirit which allowed the CNDO, INDO and other ZDO (zero-differential-overlap) methods to easily carry out the transformations, since only a relatively small number of nonzero (ij | kl) integrals are involved [16].

Stoll *et al.* [17a] have derived equations for obtaining the energy-optimum SCF LMOs, *without tails* (minor components). These are a correction of earlier nonorthogonal LMO equations given by Payne [17b]. It is difficult to judge the physical significance of these LMOs which have tails only to the extent demanded by orthogonality. They certainly will not describe any "inherent delocalization" such as one has in the π-electrons of benzene, where the two-centered Kekule-structure π-LMOs are much more delocalized onto the other four C-atoms (particularly the neighboring two), than just that delocalization demanded by orthogonality [17c].

1.5 Review of Recent Calculations of LMOs

Many workers have calculated energy-localized LMOs and investigated their properties. Daudel *et al.* have given contour plots and first and second moments, in order to understand dipole moments of molecules; ten-electron hydrides are examined here [18]. They have modeled these moments with spheres and ellipsoids, to show their most essential features, both for lone pairs and bond pairs. They are based on *ab initio* double-zeta level SCF MO wavefunctions. They found chemically understandable and uniform changes in the LMOs along an isoelectronic series, e.g., from N^{-3} to H_4N^+, or C^{-4} to Ne.

LMOs have been used [19] to study the inherent delocalization energy associated with the dimerization of BH_3, *within* the SCF MO model. By placing two distorted BH_3 molecules and their LMOs with the geometry of B_2H_6 "half molecules", symmetrically orthogonalizing the two sets of BH_3 LMOs at that geometry, we found a single determinant energy ~628 kj/mole higher than the SCF MO energy of B_2H_6 in the same basis. Thus within the SCF MO model (no correlation) the "resonance energy" of the two classical structures of B_2H_6 would be a similar value. The major difference between the two wavefunctions (before and after delocalization) is of course that a B—H LMO "spreads itself" over three atoms: B—H—B is a "three centered LMO", and this takes place symmetrically for both the B—H LMOs directed towards the other "half molecule". PNOs for the three-centered B—H—B bond LMO were also calculated and reported, and found to be analogous to those found earlier for the H_3 activated complex (linear) [20]. These PNOs only involved s and p-type basis functions on the B atoms, and s-type basis functions on the H atoms, so only "left-right type correlation" was described, which was about 50 kj/mole. This indicated that electron correlations were not the major effect leading to dimerization. Some pair-correlation energies were also very roughly estimated by assuming them to be very rougly proportional to the exchange integrals between the LMOs. This does not seem to be generally the case.

2 Correspondence Between LMO and CMO Descriptions of Molecules and Their Properties

2.1 VSEPR Model Examined with LMOs

Some properties are best understood in terms of CMOs (canonical:spectroscopic:delocalized) MOs, and some in terms of LMOs. For closed shell ground states, one description is really not more fundamental than another, and any property which can be understood in terms of one set of MOs can also be understood in principle in terms of the other, although it may not be so convenient. Thus, the VSEPR model [21] (Valence Shell Electron Pair Repulsion Model) seems readily interpreted via the LMOs corresponding to lone-pairs and bond-pairs of electrons. This interpretation [5a] allowed it to be shown that "Pauli repulsions" (without Coulombic repulsions) of electrons can not keep the equilibrium structure of H_2O from being linear of NH_3 from being planar, even with no change in the SCF LMO wavefunctions: if the electron repulsion (ER) energy is deleted. This shows that equilibrium geometries of even simple molecules are determined by a complex mixture of many effects. Four protons on a sphere of radius R_0(C—H) are more "stiffly" tetrahedral than is the CH_4 molecule itself [22]. Therefore, addition of electrons to the nuclear framework makes CH_4 *less* rigidly tetrahedral, on the whole. Therefore, it seems strange from this point of view to credit the CH_4 structure to sp^3 AO hybrids on the C atom, or to the VSEPR model "forces" between electron pairs.

2.2 Earlier Reviews of LMOs

Over the years since 1963, there have been a number of very good reviews of LMOs. England et al. [23] gave contour plots of very many pi-electron LMOs for conjugated ring systems. England et al. [24] gave contour plots for many LMOs of small molecules and π-systems, including those from earlier work [1a].

A good review by Weinstein et al. [25] covered the early developments. Caldwell and Eyring [26] have considered the spectroscopic implications of LMOs, as well as reviewed other aspects.

Kapuy et al. [27] have investigated the transferability of the first and second electric moments of energy-localized LMOs which correspond in the ten and eighteen-electron systems CH_4, CH_3CH_3, CH_3NH_2, CH_3OH, NH_2NH_2,NH_2OH, H_2O and H_2O_2. The "first moments" are given by $\bar{u} = \langle\psi_i|\, u\, |\psi_i\rangle$, u = x, y, or z, which are relative to one of the two nuclei of a bond LMO or the one nucleus of a lone-pair LMO, ψ_i. They give the location of the center-of-charge of the LMO. They find that bond-pair LMOs have first moments transferable with a standard deviation, SD, of about 1.0%, and for lone-pair LMOs about 1.5–2.0%. The second moments are $q_u^2 = \langle\psi_i|\, u^2\, |\psi_i\rangle$, u = x, y, or z, where x, y and z are relative to the center-of-charge of the LMO, and diagonalize the tensor $\tilde{Q}$. The q_x^2 have SD = 2.9—4.6%, and for the q_y^2 and q_z^2 the SDs are smaller, for CH, NH and OH bond LMOs. The lone pair LMOs have SD = 10% for the O atom q_x^2 and the SDs for q_y^2 and q_z^2 are 5.4% and 1.6% respectively. For the N atom they are respectively SD(q_x^2) = 6.9%, SD(q_y^2) = 3.5% and SD(q_z^2) = 1.7%. $q_x > q_y \geqq q_z$ in all cases, so the "x axis" is roughly the "principle axis" of the LMO, to the extent it has one. As expected q_x^2 (bond LMO) $> q_x^2$ (lone pair LMO) in all cases, due to the presence of nuclei at both "ends" of the bond LMO. However the opposite is true for q_y^2 and q_z^2, which is *probably* because $\langle\psi_i|\, r^2\, |\psi_i\rangle = q_x^2 + q_y^2 + q_z^2$ has some tendency to remain fairly constant for energetic reasons: so that J_{ii} and $\bar{T}_i$ do not become too large, in acting against the attractive-potential terms of $\langle\psi_i|\, F\, |\psi_i\rangle$. For example *imagine* HF formed from Ne by the "pulling out" of a proton from the Ne nucleus, to its position in HF. Now the sp^3 HAO (LMO) along the axis of which the proton moves would be expected to be most affected in *shape*, being "drawn out" somewhat with the proton. We could also imagine HF formed by bringing a proton from infinity toward an

F^- ion, along the axis of the bond LMO, b(H — F). Again this LMO from F^- would be most affected in *shape*. The lone-pair LMOs (L_F) are intermediate between the sp^3 L_{Ne} and sp^3 L_{F^-}, we have found [1b], although having a small "tail" (component) on the H atom. However the "net" shielded nuclear attraction on the b(H — F) has not greatly changed from Ne, so there is *roughly* the same balance of attractive and "repulsive" forces (including the "kinetic energy pressure", $\bar{T}$, much considered in Ruedenberg's analysis of chemical bonding [28]).

In earlier work, Kapuy, Kozmutza, Daudel and Stephens [29a] considered the transferability of $\bar{T}_i$s, J_{ii}s and J_{ij}s (for similarly neighboring LMOs). The SD($\bar{T}$, iC) = .01%, SD($\bar{T}$, CH) = 1.5%, SD($\bar{T}$, iN) = .01%, SD($\bar{T}$, NH) = 1.26%, SD($\bar{T}$, N) = 1.7%, SD($\bar{T}$, Oi) = .1%, SD($\bar{T}$, OH) = 1.7%, SD($\bar{T}$, O) = 1.1%, indicating the great transferability, although not perfect of course. Similar results were obtained for the SDs of J_{ii}s, (Table 3b of reference 21a) and the $J_{ij} - K_{ij}$ SDs. The SDs of the $\bar{T}_i$s are particularly interesting, because of the implications for the transferability of bond-energies, since $\bar{T} = -E$, and $\bar{T} = \sum_i^N \bar{T}_i$, for completely accurate SCF LMO wavefunctions. This work should be repeated in the better basis sets possible today, where the virial theorem will be better obeyed. (We are presently doing this in our group.) The above work was carried out with a quite large basis of s and p-type contracted gaussians: (13s 7p/4s) contracted to [4s 2p/2s]. However, no polarization functions (3d on heavy atoms and 2p on H atoms) were included, so this basis set is basically of "double zeta" quality. They chose this basis set because it was small enough to handle and earlier work [29b] on H_2O had shown that a minimal basis did not give very accurate moments for LMO charge distributions. However this work also showed that polarization functions affected these moments less than 5%, in H_2O.

England, Gordon, and Ruedenberg [30] have carried out an exhaustive analysis of the transferability of energy-localized LMOs in acyclic hydrocarbons, including a careful analysis of when "nonbond fragments" (tails) of LMOs can be transferred. This is done within the INDO SCF method, but should be a rough indication of what to expect when more accurate SCF LMOs are considered. They consider geminal (neighbor bond), vicinal (next nearest bond), and third and fourth neighbor "fragments". They also give tables from which one can *predict* both these contributions and "bond fragments" (principle components) of LMOs, given a certain nuclear framework with "standard geometry" (standard bond lengths for the type of bond and standard bond angles). Using the tables they can quite accurately predict the LCAO LMO coefficients of a rigorous SCF treatment, without any energy computations. The elemental bases are sp^3, sp^2, and sp hybrids on the C atoms and $1s_H$ on the H atoms, with the hybrids being chosen by "chemical intuition", for a molecule to be treated. All components (including tails) of a LMO are sensitive to the "environment" of the given LMO bond, out to third, or fourth neighbors, both because of orthogonality constraints and because of different potentials in the Fock operator, for different environments for a LMO. They also showed that *ad hoc* homopolar two-center LMOs, projected onto the SCF orbital space and then symmetrically orthogonalized, (called "starting LMOs" (SMOs)) were fairly close to the correct energy localized LMOs. Much of the above will become more involved when heteroatoms are considered, which induce very large polarizations of bonds.

2.3 Predicting Symmetries of Occupied and Unoccupied CMOs from LMOs and the Use of the LMOs in Frontier MO Theories

By considering the bonding and antibonding LMOs which one can *conceptually* construct from the minimal AO basis, one can then go on to ascertain the symmetries of the occupied and low-lying unoccupied CMOs. We illustrate this for the H_2O (HOH′) molecule. The $1s_O$ is taken as the iO (oxygen atom inner-shell LMO). Two of the $\sim sp^3$ HAOs of the O-atom become approximately the lone-pair LMOs, LO and L′O. The other two $\sim sp^3$ HAOs, each pointing toward one of the H nuclei, are combined in both bonding and antibonding combinations with the H-atom $1s_H$ orbitals, to give b(O—H), b(O—H′) and b*(O—H), b*(O—H′). These seven LMOs are not strictly orthogonal, but could be symmetrically orthogonalized if this were not just a conceptual construction. We can now take linear combinations of these LMOs which are symmetry MOs ($N = 1\sqrt{2}$.) Thus, $3a_1 = N(b(O—H) + b(O—H'))$, $1b_2 = N(b(O—H) - b(O—H'))$, $1a_1 \cong iO \cong 1s_O$, $2a_1 = N(LO + L'O)$, $1b_1 = N(LO - L'O)$, $4a_1^* = N(b^*(O—H) + b^*(O—H'))$ and $2b_2^* = N(b^*(O—H) - b^*(O—H'))$. For linear (or near linear) H_2O, $4a_1^* \cong k(2s_O) - cN(1s_H + 1s_{H'})$, with $c > k$, and $2a_1 \cong c'(2s_O) + k'N(1s_H + 1s_{H'})$, with $c' > k'$. Also, $3a_1 \cong 1\pi_z \cong 2p_z$, $1b_1 = 1\pi_x \cong 2p_x$, $1b_2 = \bar{c}(2p_y) + \bar{k}N(1s_H - 1s_{H'})$, with $\bar{c} > \bar{k}$, and $2b_2^* = \bar{c}'(2p_y) - \bar{k}'N(1s_H - 1s_{H'})$, with $\bar{c}' < \bar{k}'$. (Here c, k, c′, k′, $\bar{c}$, k, $\bar{c}'$, and $\bar{k}'$ are all *positive* coefficients. The z axis is the principle axis of the bent H_2O and the y axis is the principle axis of the linear molecule, so the x axis is perpendicular to the plane of the bent H_2O.) Now since we know that $\varepsilon(1s_O) \ll \varepsilon(2s_O) < \varepsilon(2p_O) \cong \varepsilon(1s_H)$, we can reason that $\varepsilon(1a_1) \ll \varepsilon(2a_1) < \varepsilon(1b_2) < \varepsilon(3a_1) \cong \varepsilon(1b_1) < \varepsilon(4a_1^*) < \varepsilon(2b_2^*)$ for both linear and bent H_2O. For linear H_2O, $3a_1 = 1\pi_z \cong 2p_z(O)$ and $1b_1 = 1\pi_x \cong 2p_x(O)$ are degenerate, but $\varepsilon(3a_1)$ decreases strongly as H_2O bends from linear and $\varepsilon(1b_1)$ stays constant, in Walsh diagrams [31]. This strong decrease in $\varepsilon(3a_1)$ is usually taken to be the basic cause of the "spontaneous bending" of linear H_2O.

We have used the minimal basis of seven AOs to conceptually construct seven approximate LMOs, and using an orthogonal transformation of these LMOs we have constructed seven "symmetry (delocalized) MOs". Linear combinations of these latter, of like symmetry, give the CMOs: delocalized (spectroscopic) SCF MOs. The CMOs are the orthogonal basis in which the Fock matrix is diagonal, with the ε_i along its diagonal. The true SCF MOs contain polarization-AOs to a small extent, e.g., 3d-type AOs on the O atom and 2p-type AOs on the H atoms, but we need not consider that here. Any higher-energy unoccupied SCF MOs must be largely composed of these "polarization AOs" and other higher-energy AOs, and so have MO energies still higher than the seven we have considered. They are here the "Rydberg MOs". Since H_2O has ten electrons, we have shown, conceptually, that $4a_1^*$ is the LUMO (lowest unoccupied MO) and $3a_1 = 1\pi_z$ and $1b_1 = 1\pi_x$ are the degenerate HOMO (highest occupied MO) for linear H_2O. Since both $3a_1$ and $4a_1^*$ have the same symmetry (A_1), they can "mix" in the bent H_2O to help form the new $3a_1$ of the bent molecule.

Since the early work of Walsh [31], it has been generally believed that the *changes* in $3a_1$ as the molecule bends (H_2O or NH_3) is the primary "driving influence" to the energy lowering, ΔE. Thus BeH_2 and BH_3 are respectively linear and planar, in the equilibrium configuration, since the $3a_1$ is unoccupied. The occupation of the $3a_1$ MO is indeed the important factor (in accord with Walsh's *rules*) for the spontaneous bending. In fact H_2O^{+2} and NH_3^{+2} become linear and planar respectively, when we remove the $3a_1$ electrons keeping the other SCF MOs unmodified. However this is not primarily due to *changes* in the $3a_1$, and certainly not much due to HOMO ($3a_1$) — LUMO ($4a_1^*$) mixing, as the frontier-MO ideas would indicate [32]. The Integral Hellmann-Feynman Theorem (IHFT) [33] *and* second-order perturbation theory [34] have been used to show this [35]. The exact IHFT gives: $\Delta E = \langle\psi^0|\,\Delta\hat{H}\,|\psi\rangle/\langle\psi^0\,|\,\psi\rangle = \langle\psi^0|\,\Delta\hat{NA}\,|\psi^0\rangle + \langle\psi^0|\,\Delta\hat{NA}\,|\Delta\tilde{\psi}\rangle + \Delta NR$, $\Delta\hat{H} = \Delta\hat{NA} + \Delta\hat{NR}$, $\Delta\tilde{\psi} = (\tilde{\psi}\,|\,\eta) - \psi^0$, $\eta = \langle\psi^0\,|\,\psi\rangle$. Letting ψ and ψ^0 be the SCF MO wavefunctions, we have: $\Delta E \cong \sum_i^n 2\,\Delta E_i + \Delta NR$, where $\Delta E_i = \langle\varphi_i^0|\,\Delta\hat{NA}\,|\varphi_i^0\rangle + \langle\varphi_i^0|\,\Delta\hat{NA}\,|\Delta\tilde{\varphi}_i\rangle, \Delta\tilde{\varphi}_i = (\varphi_i\,|\,\eta_i) - \varphi_i^0, \eta_i = \langle\varphi_i^0\,|\,\varphi_i\rangle$. This relates to the perturbation theory when we take $\Delta\hat{H} \cong S\hat{H}' + \lambda\hat{H}''$, $\lambda = S^2/2$, $\Delta E \cong SE' + \lambda E''$, and $\psi \cong \psi^0 + S\psi'$. $\Delta H = \Delta NA + \Delta NR$ since electron-repulsions (ER) and kinetic-energies (T) operators cancel. For linear H_2O, or planar NH_3, $\langle\psi^0|\,\hat{H}'\,|\psi^0\rangle = 0$ and all $\langle\varphi_i^0|\,\hat{H}'\,|\varphi_i^0\rangle = 0$, so $\lambda H''$ essentially represents $\Delta\hat{H}$, and $\lambda NA''$ represents ΔNA in the $\langle\varphi_i^0|\,\Delta NA\,|\varphi_i^0\rangle$ terms. We have shown that although $\langle\psi^0|\,\Delta NA\,|\psi^0\rangle$ is very positive, and $\langle\psi^0|\,\Delta NA\,|\Delta\tilde{\psi}\rangle$ is very negative, the term $\langle 3a_1|\,\Delta NA\,|4a_1^*\rangle^2/\Delta E_{01}$ is a trivial part of $\langle 3a_1|\,\Delta NA\,|\Delta 3\tilde{a}_1\rangle$ which is only a fraction of ΔE, *and positive*. The very negative $\langle 3a_1|\,\Delta NA\,|3a_1\rangle$ is the essential "bending component" in ΔE, associated with $3a_1$. At any angle of bending up to $120^0 = {<}HOH$, if we bend by 5^0 more, say, keeping $3a_1 = 3a_1^0$ "frozen", we find that $2\langle 3a_1^0|\,\Delta NA\,|3a_1^0\rangle$ is more negative than is ΔE, so this is not just the key factor in the *initial* bending at HOH $= 180^0$. *Changes* in the $3a_1$ are not how its electron density primarily attracts the protons, at any angle. However, for the electron density *as a whole*, ϱ, the protons move (from ${<}HOH = 180^0$) because of *changes* in ϱ, by the Hellmann-Feynman theorem. The same ideas apply to NH_3.

Advocates of the frontier MO ideas maintain that the allowedness of reactions (nuclear rearrangements) is mainly controlled by the allowed mixing of HOMO and UMO, or, at the very least, by mixing of the two or three highest OMOs with the lower UMOs. By the variation theorem the mixing of *any* UMO will necessarily lower the energy: *forced* orthogonality of any of the OMOs of the new geometry to any UMO of the old geometry will result in a higher *total* energy in the new geometry. We find that if *all five* OMOs of ${<}HOH = 170^0$ are held orthogonal to the two lowest UMOs ($3\sigma_g^* = 4a_1^*$ and $2\sigma_u^* = 2b_2^*$) of linear H_2O then $\Delta E(180^0 \rightarrow 170^0) = -1.36$ mH compared to $\Delta E_{SCF} = -2.02$ mH, so H_2O still spontaneously bends. This result is typical for other angles of bending as well. At ${<}HOH = 170^0$, the removal of this constraint changes the ε_is by $\Delta\varepsilon(1a_1) = +.10$ mH, $\Delta\varepsilon(2a_1) = -.38$ mH, $\Delta\varepsilon(1b_2) = -.01$ mH, $\Delta\varepsilon(3a_1) = -.05$ mH, and $\Delta\varepsilon(1b_1) = -.07$ mH. ($2\sigma_g$, $3\sigma_g^*$) mixing is a major effect prevented by the constraint, in agreement with the IHFT and perturbation theory results. During the geometry change (bending), each OMO must remain orthogonal to all OMOs of *lower* $\varepsilon_i = \langle\varphi_i|\,\hat{F}\,|\varphi_i\rangle$. Because lower a_1 OMOs of bent H_2O mix in the LUMO of linear H_2O, this can cause HOMO-LUMO mixing which tends to *raise* ε(HOMO) and $E(3a_1)$ because $\langle 1\pi\,|\,\Delta NA\,|\Delta 1\tilde{\pi}_z\rangle$ is *positive*. The HOMO can *not* change freely to minimize its own individual orbital energy, ε_i. This important effect has been ignored by many workers with the frontier MO ideas. The frontier MO ideas can not be saved in this case by an orthogonal transformation of $2a_1$ and $3a_1$ since such transforms are not symmetry OMOs for the linear H_2O. These constrained SCF calculations are being extended to other systems. All variational calculations, including these constrained ones, have an energy error varying as the square of the error in the wavefunction. Thus they have a reliability not always shared by other approaches.

3 Transferability of LMOs and Associated Bonds

3.1 LMO Analyses of Empirical Bond-Energy Additivity Schemes: An Important Use of LMOs

For systems with well-localized "bonds", a number of empirical methods have been developed to estimate the total molecular binding energy, BE, as a sum of "bond energies", $E_b(AB)$: $BE = \sum_b E_b(AB)$, where the sum is over all the traditional "chemical bonds" drawn by the chemist in the structural formula. These methods range from extensions of Pauling's original electronegativity (EN) method [36], particularly by Sanderson [37], to that of Fliszar [38]. The success of these methods for many compounds, even with their *ad hoc* assumptions, leads one to believe that we should be able to analyze accurate wavefunctions and energies (using LMOs) into "bond energy change components", e.g. $\Delta\tilde{E}_b(AB)$, which reflect the perturbation of some "standard bond energy", $E_b^s(AB)$, by its *immediate* environment. Such an analysis would not only give better theoretical support to the empirical methods, but would also probably point to possible improvements as well as inherent limitations. For example, we can take the C—H bond LMO of CH_4, truncated of its "tails" (minor components from atoms other than these two) as the "standard C—H LMO" (b_{C-H}), SLMO ($b_{C=O}$) from H_2CO, SLMO ($b_{C=C}$) from $H_2C{=}CH_2$, SLMO(b_{O-H}) from H_2O, etc. By truncating away H-atom components in defining the SLMOs, we keep with the simplest compounds for the SLMOs and also the H atom is roughly midway in electronegativity. This allows great accuracy for the SLMOs and also allows most visualization of their corresponding "parent LMO". Calculations on CH_4, C_2H_6, C_2H_4, and HCCH show that the major part of the "tails" of the LMOs actually arise from subsequent symmetric orthogonalization of the SLMOs: the truncated LMOs (TLMOs) [17c]. When we transfer a SLMO to some larger molecule containing that *type* of LMO, and symmetrically orthogonalize the SLMOs (to OSLMOs) we have a reasonable guess for the LCAO MO SCF wavefunction in the SCF procedure. Here both the small "standard molecule" and the larger molecule would be at "standard geometries" for the given types of bonds. As the OSLMOs (guess SCF MOs) "relax" to the correct SCF LMOs of the large molecule, they change somewhat due to the following causes, which can not be *uniquely* separated. Because each OSLMO is surrounded by other OSLMOs and other nuclei, than was the case in the small SLMO molecule, it will "polarize" and "delocalize onto other atoms" differently. Thus, the final LMOs are not identical in different molecules. Even if we hold the SLMOs "frozen" in the new environment of the large molecule, their interactions energy ΔE(SLMO) will be different, except for those "transferred" together on a standard geometry nuclear framework, e.g., two b_{C-H} transferred from CH_4 to the same C atom of C_2H_6.

The empirical theories can often deal with conjugated systems like benzene simply by assuming an empirical bond energy for the "conjugated-bond" between two C atoms [37a]. However, when this same value is used for C_4H_4 (cyclobutadiene), it gives much too much stability for that molecule, since it then predicts that ΔE_4 for the reaction 2(HCCH) → C_4H_4 is about (2/3) of the $\Delta E_6 = -143$ kJ/mole for 3(HCCH) → C_6H_6. In fact if the same C—C and C—H bond lengths are used for C_4H_4 as are true for C_6H_6, then these empirical theories seem not able to avoid this prediction [37b]. However, at its optimum bond lengths, C_4H_4 should be even more stable. It is well known from both experiment and approximate MO-CI calculations that C_4H_4 is not nearly this stable, and is almost impossible to prepare. The simple Huckel MO theory immediately explains this difference in stability, as is well known. The π-type LMOs of C_4H_4 are almost exactly confined to the two C atoms, whereas the "Kekule structure" π-type LMOs of C_6H_6 are highly delocalized onto the other four C atoms. England and Ruedenberg [23] have carefully shown why the delocalization energy (the ID energy) is usually negative and roughly proportional to the number of π-type electrons. (However it is actually somewhat *positive* for C_4H_4.) They also display electron density contour graphs of LMOs for π-type electrons in various conjugated systems, including C_4H_4 and C_6H_6.

Analyses by Ruedenberg and coworkers and Goddard and coworkers [28] show why electron delocalization leads to lower energy, in nearly all cases. It is much the same explanation as for the chemical bonding energy of H_2^+ or H_2. As the orthogonal MO wavefunction "spreads out" over more atoms, its average kinetic energy $\bar{T}_i$, decreases as does that for the "particle in the box" as the box becomes larger. This transfers some charge into the interbond region from near the nuclei, raising somewhat the potential energy, V. These are the essential effects at large R (internuclear distance), as seen from virial theorem graphs of $\bar{E}$, $\bar{T}$ and $\bar{V}$ [39]. However, at smaller R (near the equilibrium R_0), the AOs of the LCAO MOs "contract" toward the nuclei, greatly increasing the $\bar{T}$ and decreasing the $\bar{V}$, so that at R_0 we have $\Delta\bar{T} = -\Delta\bar{E} = -1/2\,\Delta\bar{V}$. However the "spreading out" (delocalization) of the MO, "relieving the kinetic energy pressure", allows the contraction effect without violation of the uncertainty principle. So it can be *thought of* as the critical initial effect giving *binding* in H_2^+ or H_2, or delocalization stabilization in π-electron conjugated systems or other "conjugated" or "hyperconjugated" systems. The "orbital charge cloud" spreads out for $\varrho = N^2(1s_A^2 + 1s_B^2 + 2\,1s_A\,1s_B)$ compared to $\varrho^0 = (1/2)\,(1s_A^2 + 1s_B^2)$, for H_2^+ or H_2, simply because the *superposition* of $1s_A$ and $1s_B$, by constructive interference, leads to the "overlap density", $N^2 2(1s_A 1s_B)$, which is transferred from the (1/2) $1s_A^2$ and (1/2) $1s_B^2$ densities.

Many wavefunctions analyzed today do not obey the virial theorem very well because of lack of optimization of the exponents. Overall "scaling" of the wavefunctions is not a proper answer either because this mainly affects the very large $\bar{V}$ and $\bar{T}$ of the inner-shell electrons, whereas the real shortcomings of the wavefunction are mainly with the valence electrons (due to nonoptimization of exponents and small basis). Thus, these wavefunctions are somewhat analogous to some of the early ones for H_2^+ and H_2 which did not optimize the 1s exponent, and so showed chemical binding mainly due to a *lowering* of $\bar{T}$, with $\bar{V}$ often increasing. (The famous original Heitler-London wavefunction [40] for H_2 and Pauling wavefunction [41] for H_2^+ are examples). Things are additionally complicated in systems with more than one occupied MO, in that the virial theorem is not obeyed for *individual* MOs, as they are not acted upon by a hamiltonian with only inverse-square-law forces, as is the total wavefunction. Each SCF MO "sees" the *average* repulsion from the other MO charge clouds, which is not "Coulomb's law" (point charges) in nature. Thus, without properly scaling *all* the orbital exponents, so the virial theorem is satisfied by the total $\bar{E}$, $\bar{T}$, and $\bar{V}$, we really have not much idea of how the $\bar{T}_i$ and $\bar{V}_i$ of individual SCF LMOs (φ_is) change as binding takes place. However scaling of all exponents in the molecule, at every configuration, is very time consuming, and is almost never done.

An important use of LMOs is in an attempt to justify and refine various semiempirical "bond energy" methods, such as that of Pauling's electronegativity (EN) approach and the molecular mechanics (MM) approaches.

The molecular mechanics (MM) approach is basically an "atomballs (masses) and bond-springs" model of molecules with *well localized* electron-pair bonds (often with SCF MO corrections for any inherently delocalized electrons). It has been much used in organic chemistry. Force constants in a valence-force-field type approach are assigned to each type of localized bond-stretch and interbond-angle-bend, including often cubic terms of the potentials as well. Barriers for "internal rotation" about single bonds and/or "nonbonded interactions" are also usually built into the potential. The computer calculates a relative minimum-energy configuration, starting with some reasonable guess. While the method has worked quite well in many cases, sometimes it fails badly for molecules with unusual geometries, large strain energies, or nearly dissociated bonds. It seems not reliable for activated complexes and reaction intermediates, where it could be most useful to chemists. This is largely because even well-localized bonding electron-pairs are not quite so independent as this model would indicate, e.g., they constantly induce polarizations in each other both electrostatically and through orthogonality constraints. Also many MM approaches count the so called "nonbonded interactions" in more than one way, or sometimes not at all. Theoretically, the conceptually seemingly simple MM model is not on a good foundation, but that foundation should be created to the extent possible. Surely such a theoretical basis lies with the energy-localized LMOs, since they are least sensitive to changes in each other, to first order in perturbation theory.

W. von Niessen [42] has discussed bond energies defined in terms of energy-localized LMOs. Arbitrary partitions of the nuclear charges between the LMOs is made: each lone-pair (including "innershells") is given +2 of its own nucleus, and each bond-LMO is given +1 from each of the nuclei it "binds". $E_{SCF} = INT + EXT$ where $INT = \sum_i^n B_{ii}$.

$B_{ii} = 2h_i^0 + J_{ii} + (1/2) \sum_{\lambda \neq \mu} Z_{i\lambda} Z_{j\mu}/R_{\lambda\mu}$ and $EXT = \sum_{i \neq j}^n B_{ij}$, $\lambda \neq \mu$, where $B_{ij} = 2J_{ij} - K_{ij} + \sum_{\lambda \neq \mu} Z_{i\lambda} Z_{j\mu}/R_{\lambda\mu}$, ($n = N/2$). This total energy can be arbitrarily written as a simple sum of "bond energies", B_i, if (as with electron-repulsion partitioning among the ($h_i^0 + \varepsilon_i$) s) we partition the B_{ij}s ($i \neq j$) equally among the two LMOs, i and j. These B_is are each many Hartrees, much larger than conventional bond energies, so von Niessen uses a "scaling procedure" by shifting all B_is according to the C—C bond of CH_3—CH_3 being 347 kJ/mole (from Cottrell [43]). These scaled B_is are in rough agreement with those from semiempirical tables.

For both the SCF MO and "separated pair approximation" (strongly orthogonal geminals) $\bar{T} = \sum_i^n \bar{T}_i'$, $n = N/2$, where $\bar{T}_i'$ is the average kinetic energy associated with a given doubly occupied LMO or "separated pair geminal" [44]. (Here we consider only closed shells, but any unpaired electrons can be treated using the $\bar{T}_i$ of their SCF MO, even in the "separated pairs" wavefunction.) Since we imagine the usual two-electron chemical bond as a superposition of two HAO principal components of a LMO (or "geminal" pair function), we could write $-\Delta E = +\Delta\bar{T} = \sum_i^n (\bar{T}_i' - (\bar{T}_a + \bar{T}_b)_i)$, where $\bar{T}_a$ and $\bar{T}_b$ are the kinetic energies of the two electrons in the separated atoms A and B. As an example, for CH_4 we have $-\Delta E(C{-}H) = \bar{T}'(C{-}H) - (\bar{T}(C(sp^3)) + \bar{T}(1s_H)) - (1/4)\,\Delta\bar{T}_p$, where $\Delta\bar{T}_p = -\Delta\bar{E}_p$ and $\Delta\bar{E}_p$ is the promotion energy for the C atom from the ground (^3P) state to the $(sp^3)^4\ {}^5\bar{S}$ "valence state". $\Delta E = 4\,\Delta E(C{-}H)$ is the total molecular binding energy. $\bar{T}(C(sp^3)) = (1/4)\,(\bar{T}(^5s) - \bar{T}'(1s_C^2))$, where $\bar{T}'(1s_C^2)$ is the $\bar{T}$ for that geminal. For extensions of this approach to wavefunctions accounting for all the electron pair correlations, we can reasonably divide the individually small "inter-space-LMO" correlation energies, ε_{ij} ($i \neq j$), equally between space-LMOs (or geminals) "i" and "j". Ideas such as these were proposed long ago by Allen and Shull [44] (1961), but have yet to find application on wavefunctions with enough virial theorem accuracy to evaluate them. Allen and Shull proposed the ideas as a way to hopefully understand the great transferability of bond energies, especially for C—C and C—H bonds in saturated hydrocarbons.

3.2 Review of Transferability of LMOs and Fock Matrix Elements Between Similar Molecules

There have been a number of good reviews on localized molecular orbitals (LMOs) [24, 25, 26] and chemical bonds in recent years, as well as some [45a] on the transferability of Fock-matrix elements, $F_{ij} = \langle\chi_i|\,\hat{F}\,|\chi_j\rangle$, for both LMOs, φ_i, and their hybrid atomic orbital HAO principal-components, χ_i. Reference [45a] reviews several methods for transferring F_{ij} elements and/or LMOs, from small "pattern molecules", PMs, to a large "target molecule", TM. Here the PMs contain the same general type of LMOs as the TM, up to nearest-neighbor atoms, or better, next-nearest-neighbor atoms. If one transfers F_{ij} elements for HAOs of corresponding LMOs, one has ambiguity as to which (or which average) of the F_{ij}s to transfer. Thus for the $\langle b_C|\,\hat{F}\,|b_C'\rangle$ element between two sp^3 hybrids on the same C atom of CH_4 or C_2H_6, from which pattern molecule should we take this F_{ij} for use in the TM C_3H_8? Deplus et al. [45b] take an average for a range of small molecules, ignoring interactions for hybrids on one atom and third-neighbor or more remote.

Also they take all F_{ij}s smaller than 7×10^{-3} as zero. This leads to a *small* list of "standard" F_{ij}s between "standard HAOs". Their results, compared to rigorous LCAO MO SCF results on the TM, are poorer than those of the SAMO (Simulated *Ab Initio* MO) method [45a].

The SAMO method uses PMs which are "chemically-like" the TM over large regions, while being not more than about half as large. Since the number of electron repulsion integrals to calculate and transform increases as N^4 and N^5, for N basis functions, this makes for a great reduction in calculation time, even if about four to ten PMs must be used for every TM. PMs are chosen to "overlap" each other when "superposed" on the TM, conceptually, and it is hoped that "termination" (edge) effects of the PMs does not cause too great error in the transferred F_{ij}s. The results are very good and indicate that for molecules without "inherently *highly* delocalized electrons" (as in some conjugated systems), "environmental effects" on F_{ij}s "die out" very rapidly with separation (distance through space). The one-electron matrix elements, $\langle\chi_i|\,\hat{h}^0\,|\chi_j\rangle = h^0_{ij}$, can be quickly calculated rigorously (only N^2 of them). Then the SCF MOs, φ_i, can be determined by one diagonalization of the F matrix, followed by calculation of $E = \sum_i^N (h_i^0 + \varepsilon_i)$ + NR where $\varepsilon_i = \langle\varphi_i|\,\hat{F}\,|\varphi_i\rangle$, $h_i^0 = \langle\varphi_i|\,\hat{h}^0\,|\varphi_i\rangle$, and NR is the nuclear-nuclear repulsion $\left(NR = \sum_I^{NN} \sum_{I>J}^{NN} Z_I Z_J / R_{IJ}\right)$.

The SAMO and related methods have accuracies better than most semiempirical methods (CNDO, INDO, Huckel, etc.), and greater than the NEMO (Nonempirical MO) method, all of which use Wolfsberg-Helmholtz type approximations for F_{ij}: $F_{ij} = K_{ij}S_{ij}(F_{ii} + F_{jj})/2$, where $S_{ij} = \int \chi_i\chi_j\, dr$ and K_{ij} is more-or-less a constant. (In the NEMO method, the $F_{ij} = T_{ij} + U_{ij}$, and the kinetic energy component, T_{ij}, is calculated correctly as are also the S_{ij}s. The U_{ii}s are taken from small "model compounds", not so large as PMs.) The main drawback of the SAMO method, compared to the rigorous LCAO MO SCF one, is that its energy is not variational, and so does badly when comparing isomers which do not differ much in energy, e.g., comparing the chair and boat forms of cyclohexane, the experimental ΔE is about 29 kJ/mole, but SAMO gives about 167 kJ/mole. *Ab initio* results give about 9 kJ/mole. Also the SAMO method seems restricted to a minimal basis of HAOs, and furthermore does not include any of the electronic correlation effects.

Degand et al. [45c] have considered the transferability of *both* LMOs and the F_{ij}s between LMOs of PMs obtained by the localization method of Magnasco and Perico [12], for various small hydrocarbons: CH_4, C_2H_6, C_3H_8 n-C_4H_{10} and iso-C_4H_{10}. Here again we have more than one choice for which F_{ij} to transfer, and even for which LMOs, so a scheme of average over all PMs is used to make some "standard" values to transfer to the TM. They consider various levels of accuracy for the parameters, depending on the correspondence of the two environments: neglect of neighbors, inclusion of nearest neighbors and finally inclusion of next-nearest neighbors as well. This method is comparable in accuracy to the SAMO method, but involves fewer parameters. Only orbital energies ε_i are calculated, since one-electron matrix elements were not calculated, although they could be without great effort. The ε_is are of course obtained by the diagonalization of the F_{ij} matrix for the LMOs, after the latter are (in principle) orthogonalized in the TM, and the F matrix accordingly transformed to the orthogonalized LMOs.

There seems to be little doubt that as larger and larger molecules can be treated with the LCAO MO SCF approach, to serve as PMs for still larger TMs, we will soon be able to obtain quite accurate SCF MO results even for very large molecules including polymers. This might not be true for electrons so inherently delocalized that they lead to "collective effects" throughout the molecule: if they can not be essentially localized into LMOs extending over no more than 3 to 5 atoms, with appreciable LCAO coefficients. Some metals, for example, might be expected to have LMOs with appreciable LCAO coefficients on 13 atoms: on a central atom and its neighbors in "closest packing" crystals. Even these "worst cases" hopefully can eventually be dealt with in a few years using approximate representations of the environment of the "13-atom cluster", within the metal. Although some methods have been developed for treating "exchange integrals", K_{ij}, between delocalized (Bloch type) orbitals of solids, often using transformation to momentum-space representations, it still appears computationally easier to use LMOs for high accuracy for the exchange interaction, both within the SCF MO model and its CI extensions.

More importantly perhaps, results using the pair-CI approach of the IEPA approach for LMOs indicates the "pair natural orbitals" (PNOs) for LMO pairs will be highly transferable between chemically similar situations. These can be used as nonorthogonal sets of PNOs, as in early work by Meyer on H_2O and CH_4 [8], or energy-weighted orthogonalized into one rapidly converging CI basis set (PSNO), as we have done. These localized excited PNOs can be well represented (expanded) in terms of basis functions on only one or two atoms, in most cases, prior to orthogonalization, which then introduces small "tails" on neighboring atoms. The "exclusion effect" on the electron-pair correlation is also "built in" through this Schmidt orthogonalization to the occupied LMOs. However, pair correlations are not very sensitive to the details of the LMO electron density and the "LMO and PNO tails", as has been shown by numerous PNO calculations. Therefore one can use a quite large basis set on the one or two principle atoms of a LMO (or three atoms of two LMOs) to construct *approximate* PNOs for an intraorbital (or interorbital) pair correlation. The PNOs are held strictly orthogonal to the one (or two) LMOs concerned, and *approximately* orthogonal to neighboring LMOs, via simplified (smaller basis) descriptions of the latter. Here a combination of one or two gaussians and some point charges can simulate a distant LMO. In this way the PNOs "feel" the correct "orthogonality constraint" and Fock operator field, and so do not differ much from those obtained from a rigorous treatment using the full basis set of the large molecule. However, the size of the calculations is greatly reduced, and PNOs can often be transferred (like LMOs) from related PMs. The elemental basis of the PNOs, which augments of the SCF MO basis, can be optimized both for exponents and contraction coefficients (particularly the latter). If then the most important PNOs from all the doubly-occupied LMOs are retained, and "weighted orthogonalized", the "full basis" transformation of the integrals can be handled, and we can have a rigorous variational PNO-CI of *all* electrons simultaneously (all "double excitations" and unlinked clusters, for example). For large molecules this is still beyond most computer capabilities, but with the increasing speed and storage of those machines, it may not be too far in the future.

4 Recent Uses of LMOs

4.1 Understanding of NMR Coupling Constants using LMOs

LMOs give much insight into NMR coupling constants, either through their use with an unrestricted-SCF wavefunction, Ψ_{UHF}, or equivalently by considering the CI with triplet-spin configurations, brought in by the "hyperfine contact" hamiltonian: $HS = K \sum_k \sum_n \delta(\vec{r}_{kn})\, \vec{I}_n \cdot \vec{S}_k$. Conceptually, the $s = +1/2$ electron spin-density of a LMO tends to *slightly* polarize toward the $I_z = (+$ spin) nucleus, of the LMO bond. Because of the correlation of the two LMO electrons (mainly left-right type), the $s = -1/2$ electron spin-density is slightly polarized toward the other nucleus, and thereby transmits the spin-spin coupling ($J_{nn'}$) of the two nuclei *through* the LMO electrons bonding them. The Ψ_{UHF} allows the different LMO space orbitals ($\psi_{I\alpha}$ and $\psi_{I\beta}$) for the paired LMO electrons. The spin-polarized LMOs will spin-polarize other neighboring LMOs, giving a transmission of the spin-spin coupling through several bonds. Since the direct "through space" spin-spin coupling is eliminated by molecular "tumbling", this is the important coupling. Caldwell and Eyring give the mathematical details of this treatment (p. 106 of Ref. [26]), using LMOs and second-order perturbation theory. It should be noted that the spin-polarization of the space SCF MOs is very much less here than arises from the unbalanced "exchange potential" in systems with unpaired electrons, e.g., the N-atom ^4S ground state, which gives rise to the hyperfine effects in ESR spectra due to coupling of nuclear and electron spins. However the slight polarization is *transmitted* from LMO to neighboring LMO by the resulting unbalanced "exchange potential", which is a very strong "coupling" between the spin-polarized LMOs.

4.2 Brief Review of the Use of LMOs in Complex Molecules

Energy localized LMOs have been used in the following treatments of chemically complex molecules and reactions. Minato and Yamabe [46] have studied gas-phase reactions of F^- with FCH_3CH_3 in both E2 and S_N2 reactions, using "configuration analysis" and expanding the wavefunctions in terms of LMOs. The mechanism of the retro-Diels-Alder reaction in the 4-vinylcyclohexene cation radical was studied by Pancir and Turecek, using LMOs [47]. Gleiter et al. studied several ring compounds involving SiR_2 and —C=C— groups using MINDO/3 and LOMOs. They studied "through-bond" and through-space" interactions between the LMOs [48]. In their study of the electronic structures of the phosphoryl and thiophosphoryl bonds, Schmidt and Gordon [49] determined energy-localized LMOs for the phosphine oxides. A study [50] of oxygen lone pairs in charge-transfer interactions with a proton in H_2O, CH_3OH, HOOH, furan, and acetaldehyde was made by Fujimoto et al. They also studied the similar interactions with the N-atom lone pair in NH_3, CH_3NH_3, $(CH_3)_3N$, H_2NNH_2, $HONH_2$, pyridine, pyrazine, pyridazine, and pyrimidine. They expand the localized MO most involved in the interaction (the "interacting orbital) in terms of the lone-pair LMO of the molecule before the protonation, and the

H-atom's 1s. They show that the contribution of the lone pair LMO is about 80%, whereas the contribution of the delocalized HOMO is often much lower (about 60% in H_2O, for example). (However, in furan the situation seems reversed.) They also give the extent of localization on the heavy atom (O or N), and this is typically about 90% or more. These calculations used the STO 3-G basis, and so are preliminary in nature.

4.3 Understanding Circular Dichroism (CD) and Optical Rotatory Dispersion (RD) Spectra in Terms of LMO Components of CMOs

Fukui, Imamura, and associates have analyzed the optical rotational strength of optically active L-5-methypyrrolid-2-one (pyrrolidone), using a "configuration analysis" based on energy-localized LMOs [51]. They studied the n → π* and π → π* transitions rotational strength. They were able to find simultaneously the contributions from the chromophore (—HN—C′=O), the substituent ($—CH_3$) and the remainder of the molecule (shown in Fig. 1 of Ref. [51]). By studying the intramolecular interactions between the molecular fragments (represented by the LMOs) in terms of excited local and charge transfer configurations (of a CI), the origin and effective path giving the rotational strength were found. The wavefunction analyzed was a MINDO/3 version, with CI. They studied how the symmetric (nonchiral) chromophore is coupled with its unsymmetric environment to give the rotations to the transitions. They included 42 single excitations from the seven upper OMOs to the six lowest UMOs. The CMOs are represented as linear combinations of the LMOs of the fragments, for both occupied and unoccupied MOs. The observed transitions are of course between CMOs in the simplest picture. The reduced rotational strength for the transition $\varphi_g \rightarrow \varphi$ (CMOs), $[R_{ge}] = -[7313 \mid E - E_g)] \cdot \langle\varphi_g|\,\nabla\,|\varphi_e\rangle\,\langle\varphi_e|\, rX\nabla\,|\varphi_g\rangle$ is represented in terms of the LMOs. For both the gradient and magnetic dipole integrals the expansion in LMOs leads to intrafragment excitations (A → A*, for fragment A) and interfragment charge-transfer excitations (A → B*). This second-order perturbation approach is then generalized to the CI wavefunctions of ground (Ψ_0) and excited (Ψ_1) states. Let n_O, $n_{O'}$, C=O(π), and C=O(σ) be the LMOs of the carbonyl group (banana bonds for C=O are not the LMOs from this treatment). Let n_N, N—H, C_2—N, and C_5—N be LMOs associated with the NH group. Other LMOs are similarly defined. n_H is essentially a π-type LMO. Then n-CMO $\cong .48(n_O - n_{O'}) - .37(C_2{-}N) - .41(C_1{-}C_3)$, π-CMO $\cong .57(n_N) - .43(C_5{-}C_7) - .43(C_5{-}H_{15}) + .29(C{=}O(\pi)) + .28(C_2{-}N_1)$, and π*-CMO $\cong .82(C{=}O(\pi)) - .37(C_3{-}H_9) - .24(C_3{-}H_8)$. The CMOs are fairly delocalized, compared to the LMOs, which themselves have small "tails" on other atoms. Since π-CMO is so delocalized onto the asymmetric C_5, the π → π* transition ([R] = 13.5) should be most affected in rotational strength: more than n → π* ([R] = —7.3). Here [R] is in units of 1.08×10^{40} R (cgs). The MINDO/3 calculated values for [R] are —8.83 (n → π*) and 3.45 (π → π*) for the planar molecule. For the nonplanar molecule, the respective values are calculated to be 3.45 and 4.84. The accuracy is obviously not very good, but the signs are correct. With more accurate wavefunctions, this approach will allow the understanding of CD and ORD spectra in terms of fairly transferable LMOs of molecular fragments. This is an extremely difficult type of property to understand physically, in nontrivial molecules, and this is only a beginning.

Robb, Haines, and Csizmadia [52] have given a theoretical definition of the "size" of LMO electron pairs, and then considered the stereochemical implications in terms of nonbonded interactions. The size is defined by a sphere of radius $\langle \varphi_i^2 | r_{12}^2 | \varphi_i^2 \rangle^{1/2}$, using the center-of-charge (coc) as the origin. The stereochemistry is then defined by the position of the coc and the second moments of an energy-localized LMO. This definition of size is extended to the size of a whole functional group and the directional characteristics of its charge distribution. The molecules treated in this early work where NH_4^+, NH_3, NH_2^-, NH^{-2}, N^{-3}, CO, and FCH_2—OH. This seems a good way to define the "size" of an electron pair. Another definition could involve the (roughly spherical) LMO density surface where appreciable "nonbonded repulsions" begin to develop.

5 Conculsions

LMOs are an important conceptual and quantitative link between chemical intuition and quantum theory for molecules and solids. They are much more visualizable than CMOs in general, particularly when they are equivalent MOs due to symmetry. The CMOs can be usefully visualized as symmetry MO combinations of LMOs when symmetry allows, and approximately so for small deviations from molecular symmetry. Even for open shell states, the closed shell portion of the wavefunction can be usefully described with LMOs. For energy-localized LMOs the wavefunction comes quite close to the simple Hartree-product one in which electrons and LMOs are interchangeable concepts. Even so, the "true" exchange interactions between the LMOs are important to energy surfaces, etc., and are understandable as the orbital-product self-energy. Where the LMOs mutually penetrate, there is a nontrivial probability of interchange of electrons, both of like-spin pairs and unlike-spin pairs, with the latter having no energy effect at the SCF level. The principle components (hybrid AOs) of LMOs are fairly transferable between similar bonds, lone pairs and inner-shells. Their Fock matrix elements can be transferred between similar environments, allowing the approximate calculation of SCF wavefunctions for very large molecules without the heavy computation of electron repulsion integrals. Even more importantly, this transferability applied to the excited pair-natural-orbitals (PNOs) which are so useful in CI calculations of the electron correlation energy. Optimization of basis functions for LMOs and PNOs of small molecules, with similar bonds and lone pairs, can save greatly the optimization effort necessary for larger molecules.

LMOs will allow the understanding of many molecular properties in terms of bond and lone pair components (and functional group components), and the interaction of these components, which will be similar in similarly bonded molecules. These properties include molecular binding energies, NMR shielding and coupling constants, electronic spectra and ionization potentials via diagonalization of the LMO Fock matrix, vibrational spectra via the analogous use of characteristic bond-stretch and bond-bending LMO contributions, and even ORD and CD spectra. Compton scattering is also understandable from Fourier transforms of the LMOs. Even for an activated complex, much of the system can be described via LMOs, and the remainder inherently delocalized MOs may be nearly the same for the given type of reaction (of a type of functional group) regardless of the "side chains" involved. This allows the quantifying of chemical intuition about activated complexes.

The "tails" (minor components) of the orthogonal LMOs are also very important, as they reflect the environmental effects on a given bond or lone pair, although the major HAO components reflect this as well. Several comparatively inexpensive methods have developed for calculation of LMOs approximately, using only a rough approximation for their environment. The major problem here is a certain loss in accuracy when the strict variational principle is given up. However, even the very simple "point charge models", based on the ideas of LMOs, have had much success.

6 References

1. (a) Edmiston C, Ruedenberg K (1963) Rev Mod Phys 35: 457; ibid. (1965) J Chem Phys 43: S97; ibid. (1966) Quantum Theory of atoms, molecules, and the solid state, Academic Press, New York, p 263. (b) Ruedenberg K (1965) In: Sinanoglu O (ed) Modern quantum chemistry. Academic Press, New York, p 85 (Part 1). (c) Fock V (1930) Z Physik 61: 126. (d) Nesbet RK (1962) J Chem Phys 36: 1518 (Sect VII). (e) England W (1973) J Chem Phys 58: 5182; (1971) IJQC 5: 683.
2. (a) Coulson CA (1942) Trans Faraday Soc 38: 433. (b) Lennard-Jones JE, Pople JA (1950) Proc Roy Soc London A202: 166. (c) Lennard-Jones JE (1949) Proc Roy Soc London A198: 14. (d) Hall GG, Lennard-Jones JE (1950) Proc Roy Soc A202: 155; Hall GG (1951) Proc Roy Soc London A205: 541
3. (a) Edmiston C, Krauss M (1965) J Chem Phys 42: 1119; ibid. (1966) J Chem Phys 45: 1833; ibid. (1968) J Chem Phys 49: 192; Edmiston C, Doolittle J, Murphy K, Tang KC, Willson W (1970) J Chem Phys. 52: 3419. (b) Meyer W (1973) J Chem Phys 58: 1017; Ahirichs R, Driessler F (1975) Theor Chim Acta Berlin 36: 275
4. (a) Lowdin PO (1959) Adv Chem Phys, Interscience Pub Inc, New York II: 27; Hurley AC, Lennard-Jones, Pole JA (1953) Proc Roy Soc London A220: 446. (b) Sinanoglu O (1962) J Chem Phys 36: 706, 3198; Nesbet RK (1965) Adv Chem Phys 9: 321; ibid (1969) 14: 1; Edmiston C (1963) J Chem Phys 39: 2394
5. (a) Edmiston C, Bartleson J, Jarvie J (1986) J Am Chem Soc 108: 3593. (b) Tang K, Edmiston C (1970) J Chem Phys 52: 997. (c) Nesbet RK (1963) Rev Mod Phys 35: 552
6. Davidson E, Jones LL (1962) J Chem Phys 37: 2966
7. Weinbaum, (1933) J Chem Phys 1: 593
8. Meyer W (1973) J Chem Phys 58: 1017; (1976) ibid. 64: 2901; ibid. (1984) Advanced theories and computational approaches to the electronic structure of molecules. D Reidel Publ Co, Holland, p 19
9. Dunning TH (1984) Advanced theories and computational approaches to the electronic structure of molecules. D Reidel Publ Co, Holland, p 67
10. (a) Boys SF (1960) Rev Mod Phys. 32: 300. (b) Boys SF (1966) Quantum theory of atoms, molecules, and the solid state. Academic Press, New York, p 253
11. (a) Kleier DA, Halgren TA, Hall Jr JH, Lipscomb WN (1974) J Chem Phys 61: 3905. (b) Von Niessen W (1972) J Chem Phys 56: 4249; ibid. (1972) Theoret Chim Acta 27: 9
12. Magnasco V, and Perico A (1967) J Chem Phys 47: 971
13. Mulliken RS (1960) Rev Mod Phys 32: 232
14. Rajzmann M, Brenier B, Purcell KF (1987) Theor Chim Acta 72: 13
15. Leonard JM Luken WL (1984) Int J Quantum Chem 25: 355
16. Trindle C, Sinanoglu O (1968) J Chem Phys 49: 65
17. (a) Stoll H, Wagenblast G, Preuss H (1980) Theoret Chim Acta Berlin 57: 169. (b) Payne PW (1977) J Am Chem Soc 99: 3787. (c) Edmiston C (1988) J Mol Phys Structure (Theochem) 169: 331

18. Daudel R, Stephens ME, Csizmadia IG, Kozmutza C, Kapuy E (1977) Int J of Quant Chem 11: 665–683
19. (a) Edmiston C, Lindner P (1973) Int J of Quant Chem 7: 309. (b) Dunning TH (1984) Advanced theories and computational approaches to the electronic structure of molecules. D Reidel Publ Co, Dordrecht, Holland, p 67
20. Edmiston C, Krauss M J (1965) Chem Phys 42: 1119; ibid (1966) 45: 1833
21. Gillespe R, Nyholm R (1957) Q Rev Chem Soc 11: 339
22. Jarvie J, Willson W, Doolittle J, Edmiston C (1973) J Chem Phys 59: 3020
23. England W, Ruedenberg K (1973) J Amer Chem Soc 95: 8769
24. England W, Salmon LS, Ruedenberg K (1971) Fortschr Chem Forsch 23: 31
25. (a) Weinstein H, Pauncz R, Cohen M (1971) In: Bates DR (ed) Adv Atomic and Mol Phys, Academic Press, New York 7: 97 (b) Millie PH, Levy B, Berthier G (1975) In: Localization and delocalization in quantum chemistry. D. Reidel Publ Co, Holland I: 59
26. Caldwell D, Eyring H (1978) Adv in Quant Chem 11: 93
27. Kapuy E, Kozmutza C, Daudel R (1980) Theoret Chim Acta Berlin 56: 259
28. Ruedenberg K, (1962) Rev Mod Phys 34: 326; Edmiston C, Ruedenberg K (1964) J Phys Chem 68: 1628; Feinberg MJ, Ruedenberg K (1971) Chem Phys. 54: 1495; Wilson CW, Goddard WA (1972) Theor Chim Acta 26: 195
29. (a) Kapuy E, Kozmutza C, Daudel R, Stephens ME (1979) Theoret Chim Acta Berlin 53: 147. (b) ibid. (1978) Theoret Chim Acta Berlin 50: 31
30. England W, Gordon MS, Ruedenberg K (1975) Theoret Chim Acta Berlin 37: 177
31. Walsh AD (1953) J Chem Soc 22660, 2266, 2288, 2296, 2301
32. Hoffman R, Woodward RB (1986) Acc Chem Res 1: 17
33. Parr RG (1964) J Chem Phys 40: 3726
34. Bartell LS (1968) J Chem Educ 45: 754; Pearson RG (1969) J Chem Soc 91: 1252, (4947); Pearson RG (1976) Symmetry rules for chemical reactions. Wiley, New York
35. Edmiston C, Jarvie J, Bartleson J (1986) J Chem Phys 84: 6907
36. Pauling L (1960) Nature of the chemical bond, 3rd edn. Cornell University Press, Ithaca, New York
37. (a) Sanderson RT (1986) J Chem Educ; Sanderson RT (1983) Polar Covalence. Academic, New York; Sanderson RT (1867) Chemical bonds and bond energy. Academic, New York. (b) Edmiston C (1988) J Chem Ed 65: 219
38. Fliszar S (1983) Charge distributions and chemical effects. Springer-Verlag, New York; England W, Ruedenberg K (1973) J Am Chem Soc 95: 8769
39. (a) Slater JC (1968) Quantum theory of matter, 2nd edn, McGraw-Hill, New York, p 406; (b) Ruedenberg K (1962) Rev Mod Phys 34: 326; (c) Wilson CW, Goddard WA (1972) Theor Chim Acta 26: 195
40. Heitler W, London F (1927) Z Physik 44: 455
41. Pauling L (1928) Chem Rev 5: 173
42. Niessen W von (1975) Theoret Chim Acta Berlin 38: 9
43. Cottrell TM (1954) The strengths of chemical bonds. Butterworths, London
44. Allen TL, Shull H (1961) J Chem Phys 35: 1644; ibid. (1962) J Chem Phys 66: 2281
45. (a) O'Leary B, Duke BJ, Eilers JE (1975) Advan Quant Chem 9: 1; Duke BJ, O'Leary B (1983) Theoret Chim Acta Berlin 62: 223. (b) Deplus A, Leroy G, Peeters D (1974) Theor. Chim. Acta 36: 109. (c) Degand P, Leroy G, Peeters D (1973) Theor Chim Acta 30: 243
46. Minato T, Yamabe S (1985) J Am Chem Soc 107: 4621
47. Pancir J, Turecek F (1984) Chem Phys 87: 223
48. Gleiter R, Schafer W, Sakurai H (1985) 107: 3046
49. Schmidt MW, Cordon MS (1985) J Am Chem Soc 107: 1922
50. Fujimoto H, Yamasaki T, Hataue I, Koga N (1985) J Phys Chem 89: 779
51. Akagi K, Yamabe T, Kato H, Imamura A, Fukui K (1980) J Am Chem Soc 102: 5157
52. Robb MA, Haines WJ, Csizmadia IG (1973) J Am Chem Soc 95: 42
53. Pipek J, Mezey PG (1989) J Chem Phys 90: 4916

Properties of the Fermi Hole and Electronic Localization

W. L. Luken

IBM Corporation, Engineering Graphics Products, Dept. 30KH/MS 966, Neighborhood Road, Kingston, New York 12401, USA

The Fermi hole can be associated with many aspects of molecular structure, including the geometries, charge distributions, and reactivities of molecules. The shape of the Fermi hole may be computed from the first and second order reduced density matrices. The properties of the Fermi hole include the Fermi hole natural orbitals, the Fermi orbitals of an SCF wavefunction, and the Fermi hole mobility function which characterizes the dependence of the Fermi hole on the position of a probe electron. The Fermi hole mobility function has the dimensions of energy, and the values of this function can be used to partition a molecule into loges containing localized electrons. The Fermi hole can be used to transform a set of canonical SCF molecular orbitals into localized orbitals. The Fermi hole may also be used to characterize the electronic structure of unusual molecules such as the [1.1.1] propellane molecule, C_5H_6. Based on the properties of the Fermi hole, the axial carbon atoms of this molecule are found to be joined by a sigma bond composed mainly of p-sigma atomic orbitals on these atoms.

1 Introduction

The total number of electrons in an atom or molecule may be indicated by the number N. The distribution of these electrons in an atom, molecule, or crystal is usually described by the electronic density function $\varrho(\vec{r})$ which specifes the probability density for having an electron located at any position $\vec{r}$. In particular, $\varrho(\vec{r})$ specifies the probability density for having any one of these electrons at $\vec{r}$, averaged over all possible distributions of the remaining N-1 electrons. This distribution may be calculated using quantum mechanics, and it can be measured using X-ray diffraction.

If one of the electrons is located at $\vec{r}_1$, then the distribution of the N-1 other electrons may be described by the conditional density function $\varrho_c(\vec{r}_2; \vec{r}_1)$. Because of the electronic repulsion between the electrons and exchange correlation effects, the conditional density function differs considerably from $\varrho(\vec{r}_2)$, especially when $\vec{r}_2$ is near $\vec{r}_1$.

The effects of the electronic repulsion, or Coulomb correlation, and the effects of exchange correlation both tend to diminish the values of the conditional density function relative to the overall electronic density function when $\vec{r}_2$ approaches $\vec{r}_1$. Consequently, each electron is surrounded by a correlation hole in which the probability of finding a second electron is less than that indicated by the electronic density function $\varrho(\vec{r})$. Of these effects, exchange correlation usually has a much larger influence than Coulomb correlation, and the effect of exchange correlation is called the Fermi hole. The additional smaller effects arising from the repulsion between the electronic charges is called the Coulomb hole.

The importance of the Fermi hole to molecular structure has been known for a long time. In some treatments of molecular structure, however, the effects of the Fermi hole are described in terms of exclusion effects or exchange correlation, without mentioning the Fermi hole by name. This practice tends to obscure the role of the Fermi hole in determining the structure and properties of molecules.

J. W. Linnett, for example, presented a comprehensive method for the qualitative description of molecular structure in the book, "The Electronic Structure of Molecules, A New Approach" [1]. This work builds on the classical work of G. N. Lewis describing the role of the "shared pair of electrons" in chemical binding [2]. The fundamental premises for Linnett's work, as presented in Chapter 1 of this book, are derived largely from an analysis of the "spin correlation" in two model systems. The first of these is based on the radial distributions of the electrons in the 1s2s electronic states of the helium atom. The second model is composed of uncharged spin 1/2 particles confined to a circle. The conclusions drawn from these analyses are then generalized to describe the structure and properties of a wide range of molecules including organic molecules, inorganic molecules, free radicals, and excited states.

Although the Fermi hole is not mentioned by name, the analyses of spin correlation presented by Linnett are equivalent to analyses of the Fermi holes in these nodel systems, and the conclusions obtained by Linnett regarding the electronic structure of molecules are representative of the conclusions which can be reached from studies of the Fermi holes in molecules.

The bulk of the material presented by Linnett is highly qualitative and a posteriori arguments are frequently required to resolve ambiguities. In many cases, it is not possible to choose between alternative explanations of the electronic structure of molecules based on qualitative arguments alone. The principal reason for relying on empiricism, as indicated by Linnett, was the difficulty in obtaining accurate numerical calculations of the electronic structure of molecules at the time when Linnett's book was published.

Since the time of Linnett's work, the ability to calculate electronic wavefunctions has improved dramatically [3]. Very accurate calculations have become routine for small molecules, and the size of the molecules for which reliable calculations can be performed continues to increase. Consequently, it should no longer be necessary to make guesses about the electronic structure of many molecules. Characteristics such as charge distributions, as well as the sizes and shapes of Fermi holes, can be determined from easily calculated wavefunctions.

The following sections of this article provide a review of early work on the Fermi hole, first in the area of solid state physics, and then in the areas of atomic and molecular structure. This is followed by a summary of recent progress in establishing the properties of the Fermi hole in molecules, including the natural representation of the Fermi hole, the Fermi hole mobility function, and the use of the Fermi hole for transforming canonical SCF molecular orbitals to localized molecular orbitals. The use of the Fermi hole as a probe of molecular structure is then demonstrated with a study of the electronic structure of the [1.1.1] propellane C_5H_6.

2 The Fermi Hole in Solid State Physics

The effect of electronic exchange on electronic distributions was first investigated more than 50 years ago [4–7]. Wigner and Seitz showed that electronic exchange not only prevents two electrons with the same spin from occupying the same position, but each electron is surrounded by a region in which the probability of finding a second electron is less than the mean density for the system as a whole. Furthermore, they presented an explicit expression for this distribution in a uniform density free electron gas. This distribution may be written as

$$\varrho_c(\vec{r}_1; \vec{r}_2) = \varrho P(r_{12}/r_s) , \tag{1}$$

where $\varrho = 1/V$ is the average electronic density, $V = 4\pi r_s^3/3$ is the average volume per electron, and $r_{12} = |\vec{r}_2 - \vec{r}_1|$. The function $\varrho_c(\vec{r}_1; \vec{r}_2)$ is the conditional probability for finding an electron at $\vec{r}_1$ given the presence of an electron at $\vec{r}_2$, and $P(r)$ is a correlation factor given by

$$P(r) = 1 - \varrho\, |f(r)|^2 , \tag{2}$$

where

$$f(r) = \frac{r \cos(r) - \sin(r)}{r^3 \sqrt{2\varrho}} . \tag{3}$$

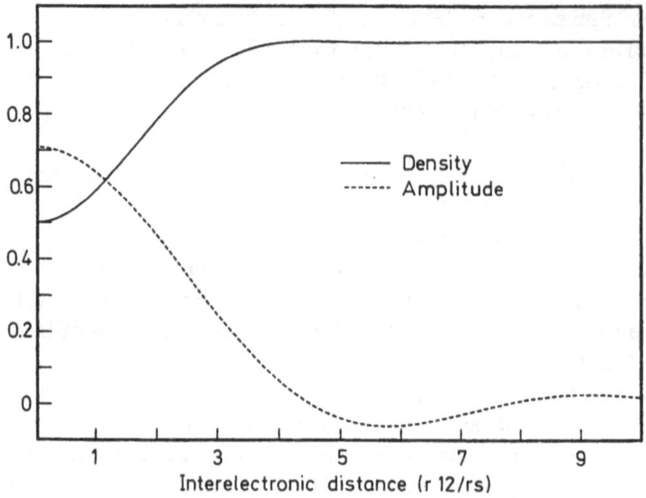

Fig. 1. Exchange correlation in a free electron gas. The solid line indicates the ratio of the conditional electronic density to the average electronic density for a uniform density free electron gas. The conditional density is the difference between the average density and the Fermi hole. The dashed line indicates the amplitude of the corresponding Fermi orbital, and r_s is the radius of a sphere containing one electron at the average density.

The shape of $P(r)$ is shown in Fig. 1. As r approaches zero, $P(r)$ approaches 1/2, and the slope of $P(r)$ approaches zero. The value of $P(r)$ at $r = 0$ indicates the total exclusion of electrons of like spin, which account for half of the total electron density, and the absence of any effect on the electrons of unlike spin.

As r increases from zero to four, the value of $P(r)$ increases monotonically from 0.5 to nearly one. For r greater than four, $P(r)$ oscillates between 1 and $1 - 9/2r^4$ with a period of approximately π.

The correlation factor $P(r)$ is always less than or equal to one. Consequently, the conditional density, ϱ_c is less than or equal to the average density. The difference between the average density and the conditional density represents the effect of electronic exchange on the distribution of electrons. This difference is the Fermi hole,

$$\Delta(\vec{r}_1; \vec{r}_2) = \varrho - \varrho_c(\vec{r}_1; \vec{r}_2), \tag{4}$$

which defines a spherical region around each electron from which other electrons are effectively excluded. The radius of this region may be characterized by the value of r at which the Fermi hole falls to half of its maximum value. This distance is approximetaly $1.8 r_s$. This is very close to the mean distance between nearest neighbors in hexagonal close packed lattice, which is given by

$$d = r_s \left(\frac{4\pi \sqrt{2}}{3} \right)^{1/3} = 1.8094\, r_s \tag{5}$$

Consequently, the entire volume occupied by a uniform density free electron gas is effectively filled with Fermi holes. This suggests that the free electron gas resembles a liquid more than a gas, with each electron at the center of a sphere having a radius of approximately $0.9 r_s$. Under these conditions, no electron can move past its nearest neighbors without passing through the Fermi hole of one of the neighboring electrons.

Because of the Fermi hole, the average electronic repulsion between electrons of like spin is less than it would be in the absence of exchange correlation. This yields a reduction in potential energy equal to the repulsion between each electron and its Fermi hole. For a uniform density free electron gas, the resulting "exchange erngy" is proportional to $\varrho^{4/3}$.

The effects of exchange correlation on the electronic structure of atoms, molecules, and crystals may be treated using local exchange potentials related to the results obtained for the free electron gas [5, 8]. Such treatments, in effect, assume that the electrons constitute a gas with a varying density, and the properties of this gas are equivalent to those of a uniform density free electron gas over small (local) regions. A spherical Fermi hole is assumed to remain centered on each electron, with a radius which depends on the density.

The results derived from the Fermi hole in a uniform density free electron gas can be expected to provide a realistic description of the exchange effects in an atom, molecule, or crystal as long as the electronic density is slowly varying over a scale of distances comparable to the radius of the Fermi hole. The Fermi hole, however, is large, and electronic densities usually vary rapidly over the scale of the Fermi hole. Consequently, in order to obtain satisfactory results, it is usually necessary to consider additional terms beyond the simple $\varrho^{4/3}$ term indicated above.

The resulting "local density approximations" have been widely used in atomic physics and solid state physics [9]. These methods have been also been applied to molecular structure, but with less success than other methods. The following sections will examine the properties of the Fermi hole in molecules, and demonstrate some of the differences between the Fermi holes in molecules and those found in the uniform density free electron gas.

3 Definitions of the Fermi Hole

The properties of the Fermi hole in atoms and molecules have been studied by several laboratories. These studies can be divided into two groups based on differing definitions of the Fermi hole. These definitions may be characterized as the wavefunction based definition and the state based definition.

In the state based approach, correlation effects such as the Fermi hole and the Coulomb hole are considered to be physical properties of individual quantum states [10–17]. Each quantum state may be represented by an exact wavefunction, or by various approximate wavefunctions. Each such wavefunction determines a set of reduced density matrices from which the corresponding Fermi holes and other properties may be determined. Certain properties may vanish for certain classes of approximations, but these are consequences of the definitions of the properties, and not parts of the definitions. Examples of these definitions and their consequences are given following a description of the wavefunction based approach.

In the wavefunction based approach, the Fermi hole and the Coulomb hole are considered to be characteristics of particular types of wavefunctions, rather than quantum states [18–24]. A Fermi hole or Coulomb hole may be associated with a quantum state only if that state can be represented by certain types of wavefunctions. This approach is described below.

3.1 Wavefunction Based Definitions

In the wavefunction based approach, the Fermi hole is defined as the difference between the interelectronic distribution functions for the Hartree and Hartree-Fock wavefunctions [18],

$$\Delta_F(r_{12}) = \varrho_{HF}(r_{12}) - \varrho_H(r_{12}) \tag{6}$$

The Coulomb hole, likewise, is defined as the difference between the interelectronic distribution functions for the Hartree-Fock and exact wavefunctions [19],

$$\Delta_C(r_{12}) = \varrho_{ex}(r_{12}) - \varrho_{HF}(r_{12}) \tag{7}$$

The latter definition resembles the definition of the correlation energy as the difference between the exact energy of an atom or molecule and the corresponding Hartree-Fock energy [25].

Although presented as functions of a single variable, the interelectronic distance r_{12}, these definitions may easily be generalized to functions of $\vec{r}_1$ and $\vec{r}_2$. It would be more appropriate to refer the functions defined above as a reduced Fermi hole and a reduced Coulomb hole. The use of such reduced functions preserves a similarity to the results of Wigner and Seitz who represented the exchange correlation in a free electron gas with a function of r_{12} [6]. This practice implies an expectation that the Fermi hole and Coulomb hole in atoms and molecules will share certain key properties with the corresponding functions for a free electron gas. For example, it may be imagined that Fermi holes and Coulomb holes are always at least roughly spherical and roughly independent of $\vec{r}_1$ for any r_{12}.

The wavefunction based approach has several serious deficiencies. First of all, these definitions assert that the Fermi hole and Coulomb hole, like the correlation energy, are not physical properties of electronic states, but merely artifacts of certain approximations. Secondly, these definitions make it impossible to specify the Fermi hole or Coulomb hole for an arbitrary individual wavefunction. These definitions each require two wavefunctions, and these must be two particular wavefunctions: a Hartree-Fock wavefunction, and either a Hartree wavefunction or a an exact eigenfunction of a many-electron Hamiltonian. The Hartree-Fock wavefunction cannot be any self-consistent field wavefunction, but should have an energy at or near the Hartree-Fock limit. The definition of the "Hartree wavefunction" is ambiguous, meaning either a variationally optimized wavefunction of the Hartree (product) form, or a Hartree product of Hartree-Fock spin-orbitals [18].

Exact or nearly eigenfunctions of the many-electron Hamiltonian are not available for any except a few very simple atoms and molecules. Accurate solutions of the Hartree equations are not available because there is no other use for such wavefunctions. Accurate solutions of the Hartree-Fock equations are available for many states of many atoms [26–28], and for a number of linear molecules [29], but such wavefunctions are generally not obtainable for non-linear molecules. For many electronic states, especially for excited states, the "Hartree-Fock wavefunction" is either undefined, not uniquely defined, or defined in a manner which differs from the closed shell Hartree-Fock wavefunction.

Because of the difficulties involved in obtaining suitable wavefunctions, it has proven impractical to apply the wavefunction based definitions to anything beyond the helium and beryllium atoms [20–24]. This approach has consequently provided no insight into the electronic structure of molecules.

3.2 State Based Definitions

In the state based approach, the Fermi hole and Coulomb hole are each defined in terms of a single wavefunction. This wavefunction is assumed to represent a quantum state of a many electron system. The distribution of electrons in this state can be described by the reduced density matrices determined by this wavefunction [30–35]. Thus, the first order reduced density matrix for any N electron wavefunction Ψ is given by

$$P_1(1;1') = \langle \Psi(1,2,\dots N) \mid \Psi(1',2,\dots N) \rangle_{2\dots N} \tag{8}$$

where the numbers 1, 2, ... N represent the space and spin coordinates associated with electrons 1, 2, ... N, and the subscript on the integral indicates the coordinates over which the integration is to be performed. Likewise, the corresponding second order reduced density matrix is given by

$$P_2(1,2;1',2') = \left(\frac{N(N-1)}{2}\right) \langle \Psi(1,2,3,\dots N) \mid \Psi(1',2',3,\dots N) \rangle_{3\dots N}, \tag{9}$$

The diagonal parts of these matrices are defined by equating coordinate 1 with 1′ and 2 with 2′. If Ψ is determined by a spin-free Hamiltonian, then the diagonal part of P_1 and P_2 may be written as:

$$P_1(1) = P_1(1;1) = \sum_{s=\alpha,\beta} \varrho_s(\vec{r}_1)\, S_s(1), \tag{10}$$

and

$$P_2(1,2) = P_2(1,2;1,2) = \sum_{s=\alpha,\beta} \sum_{t=\alpha,\beta} \varrho_{st}(\vec{r}_1,\vec{r}_2)\, S_s(1)\, S_t(2), \tag{11}$$

where r_1 is the spatial coordinate for electron 1, $S_\alpha(1)$ is a spin function for electron 1, and so forth.

The spatial function $\varrho_s(\vec{r})$ specifies the probability density for finding an electron of spin s at position $\vec{r}$. The electronic density function $\varrho(\vec{r})$ is the probability density for finding an electron at $\vec{r}$ without regard to spin. This is given by

$$\varrho(\vec{r}) = \varrho_\alpha(\vec{r}) + \varrho_\beta(\vec{r}) . \tag{12}$$

The spatial function $\varrho_{st}(\vec{r}_1, \vec{r}_2)$ specifies the probability density for simultaneously finding one electron of spin s at position $\vec{r}_1$ and a second electron with spin t at position $\vec{r}_2$. The probability density for finding on electron at $\vec{r}_1$ and a second electron at $\vec{r}_2$ without regard to spin is given by

$$\varrho_2(\vec{r}_1, \vec{r}_2) = \sum_{s=\alpha,\beta} \sum_{t=\alpha,\beta} \varrho_{st}(\vec{r}_1, \vec{r}_2) . \tag{13}$$

The influence of an electron with spin t at position $\vec{r}_2$ on the probability of finding an electron with spin s at position $\vec{r}_1$ is indicated by the conditional density function

$$\varrho_{cst}(\vec{r}_1; \vec{r}_2) = 2\varrho_{st}(\vec{r}_1, \vec{r}_2)/\varrho_t(\vec{r}_2) . \tag{14}$$

This function specifies the conditional probability density for finding an electron at $\vec{r}_1$ with spin s, given the presence of an electron at $\vec{r}_2$ with spin t. The conditional probability density for finding an electron at $\vec{r}_1$ with any spin, given the presence of an electron at $\vec{r}_2$ with any spin is given by

$$\varrho_c(\vec{r}_1; \vec{r}_2) = 2\varrho_2(\vec{r}_1, \vec{r}_2)/\varrho(\vec{r}_2) . \tag{15}$$

The factor of 2 in Eqs. (14) and (15) results from using the normalization convention of Löwdin [31]. This factor is not needed when using the convention proposed by McWeeny [32, 33, 35] period.

Because of the Coulombic repulsion between pairs of electrons, and exchange effects, the conditional densities ϱ_c and ϱ_{cst} tend to be less than the one-electron denstities $\varrho(\vec{r})$ and $\varrho_s(\vec{r})$ when $\vec{r}_1$ is close to $\vec{r}_2$. The resulting dimunition of the probability of finding an electron in the vicinity of $\vec{r}_2$ is a correlation hole. The magnitude of the correlation hole for an electron of spin s in the presence of an electron of spin t is given by

$$\Delta_{st}(\vec{r}_1; \vec{r}_2) = \varrho_s(\vec{r}_1) - \varrho_{cst}(\vec{r}_1; \vec{r}_2) . \tag{16}$$

Likewise, the magnitude of the correlation hole for an electron of any spin in the presence of a second electron of any spin is given by

$$\Delta(\vec{r}_1; \vec{r}_2) = \varrho(\vec{r}_1) - \varrho_c(\vec{r}_1; \vec{r}_2) . \tag{17}$$

The correlation holes specified by Eqs. (16) and (17) have been defined so as to be mostly positive when r_{12} is small. This differs from the definitions used in certain early work [10–17] by a factor of -1. With the definitions given here, the depth of a correlation hole is described as a positive number, where the definitions used elsewhere describe the same hole in terms of a negative height.

Equation (16) specifies the correlation holes for pairs of electrons having either like spins or unlike spins. The exchange effect, however, acts only on pairs of electrons of like spin. Consequently, the correlation effects represented by $\Delta_{\alpha\beta}$ and $\Delta_{\beta\alpha}$ are entirely the results of the Coulombic interactions, and these functions may be identified as Coulomb holes.

The correlation holes $\Delta_{\alpha\alpha}$ and $\Delta_{\beta\beta}$ represent the interactions between pairs of electrons of like spins and include the influence of both exchange correlation and Coulombic interactions. These functions are defined as the Fermi holes for the quantum state represented by the wavefunction Ψ.

The spin-free correlation hole defined by Eq. (17) represents a composite of the Fermi hole and the Coulomb hole. The effects of electronic exchange, however, are usually much stronger than the Coulomb repulsion. Consequently, the spin-free correlation hole tends to be dominated by the Fermi hole.

In the case of Hartree-Fock wavefunction, or any other wavefunction composed of a single Slater determinant, the Coulomb holes $\Delta_{\alpha\beta}$ and $\Delta_{\beta\alpha}$ vanish for all values of r_1 and r_2. In these cases, the spin-free correlation hole is equivalent to an average of the Fermi holes $\Delta_{\alpha\alpha}$ and $\Delta_{\beta\beta}$, and the Fermi holes represent only the effects of exchange correlation.

In the case of a singlet electronic state, or other electronic state with $M_S = 0$, the distribution of electrons with α spin is identical to that of electrons with β spin. In this case, $\varrho_\alpha = \varrho_\beta$. Likewise, $\Delta_{\alpha\beta} = \Delta_{\beta\alpha}$ and $\Delta_{\alpha\alpha} = \Delta_{\beta\beta}$. Consequently, the spin-free correlation hole for such a state is simply the sum of the Coulomb hole and the Fermi hole,

$$\Delta(\vec{r}_1; \vec{r}_2) = \Delta_{\alpha\alpha}(\vec{r}_1; \vec{r}_2) + \Delta_{\alpha\beta}(\vec{r}_1; \vec{r}_2) . \tag{18}$$

Furthermore, if this state is represented by a wavefunction composed of a single determinant, such as any closed shell self-consistent field (SCF) wavefunction, then the spin-free correlation hole is identical to the corresponding Fermi hole.

The Fermi holes and Coulomb holes defined by Eqs. (16) and (17) are considered to be functions of one electron ($\vec{r}_1$) which are parametrically dependent on the location of a second electron indicated by $\vec{r}_2$. This second electron will be called the "probe electron" in the following sections.

When integrated over all values of $\vec{r}_1$, the Coulomb holes, $\Delta_{\alpha\beta}$ and $\Delta_{\beta\alpha}$, yield a result of zero. This result reflects a cancellation between regions where the Coulomb hole is positive, usually for $\vec{r}_1$ near $\vec{r}_2$, and regions where the Coulomb hole is negative. These correspond to regions where the conditional probability density is diminished or enhanced compared to the ϱ_α or ϱ_β, and represent the extent to which electron 1 is pushed away from the probe electron.

Integration of the Fermi holes $\Delta_{\alpha\alpha}$ and $\Delta_{\beta\beta}$, and the composite correlation hole Δ, over all values of $\vec{r}_1$, however, yield a value of one. Thus, instead of simply pushing the other electrons away, as is found for the Coulomb holes, the Fermi holes appear to represent the complete annihilation of one electron's worth of density. This 'lost' density represents the contribution from the probe electron.

4 The Size and Shape of the Fermi Hole

During the past twenty years, there have been many advances in the ability to calculate accurate electronic wavefunctions for molecules [3]. Until recently, however, there had been little progress in visualization of the Fermi hole in molecules. Judging from the work of Slater [8] and Coulson [18], the Fermi holes in atoms and molecules were expected to resemble those in the free electron gas. That is, the Fermi hole was expected to be roughly spherical and centered on the probe electron. As the probe electron moved, the Fermi hole was expected to follow it like a shadow.

More recently, a series of investigations have shown that the Fermi hole in a molecule is quite different from the Fermi hole in a free electron gas [36–42]. For example, the Fermi hole in a molecule is almost never even roughly spherical, unless the probe electron is located on a nucleus other than that of a hydrogen atom. The Fermi hole is rarely centered on the probe electron, and the Fermi hole follows the probe electron more like a rabbit, lagging behind then jumping ahead.

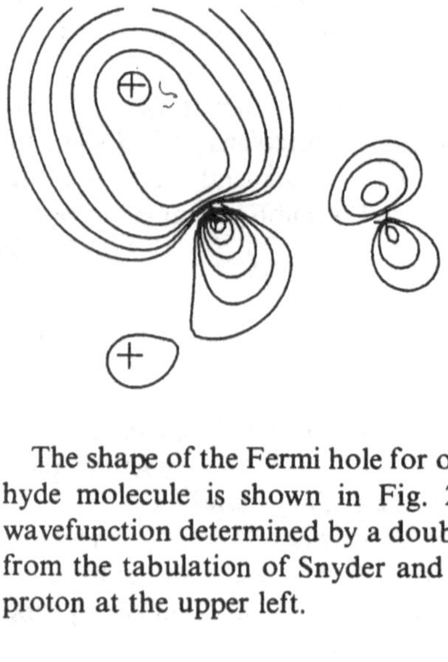

Fig. 2. A Fermi hole for a double zeta SCF wavefunction for the formaldehyde molecule. The positions of the protons (upper, lower left), the carbon nucleus (near center), and the oxygen nucleus (right of center) are each indicated by (+). The probe electron is located on the upper proton. The contour lines indicated electronic densities of 0.005, 0.01, 0.02, 0.04, 0.08, and 0.16 electrons per cubic bohr.

The shape of the Fermi hole for one position of the probe electron in the formaldehyde molecule is shown in Fig. 2. This figure is based on a closed shell SCF wavefunction determined by a double-zeta basis set [43, 44]. The geometry was taken from the tabulation of Snyder and Basch [45]. The probe electron is located on the proton at the upper left.

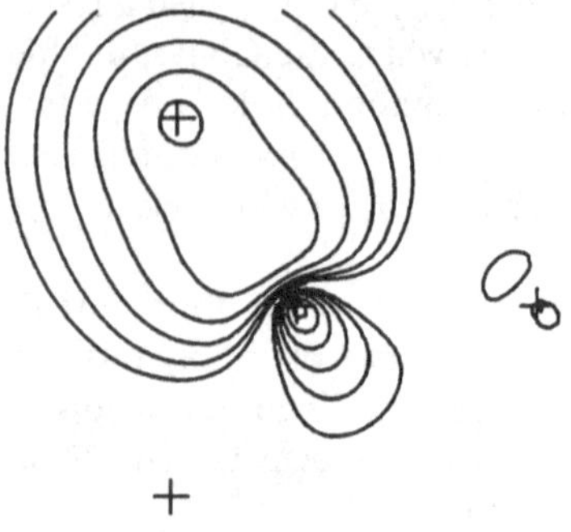

Fig. 3. One of the localized molecular orbitals obtained from a double zeta SCF wavefunction for the formaldehyde molecule. The positions of the nuclei and contour levels are the same as those indicated for Fig. 2. This LMO represents a C—H bond.

The Fermi hole shown in Fig. 2 can be seen to resemble the localized molecular orbital (LMO) associated with a C—H bond shown in Fig. 3. If the probe electron is placed midway between the carbon and oxygen atoms, the Fermi hole resembles a C—O bond LMO [42]. When the probe electron is located on or near the carbon or oxygen nucleus, the Fermi hole is much smaller and nearly spherical, centered on the nucleus and not on the probe electron. In general, if the probe electron is placed at any point dominated by a single LMO, the Fermi hole tends to resemble the density determined by that LMO, including any nodal surfaces which may pass through the LMO. Similar results have been obtained for several molecules, including water [37, 38], ethane [39], methanol [42], and other molecules [41].

The fact that the size of a Fermi hole is similar to the size of an LMO located in the same region of a molecule is understandable because:

1. A Fermi hole and an LMO each represent exactly one electron worth of density, and
2. The Fermi hole and an LMO both have densities near the upper limit of one half of the total density (in a closed shell state).

The similarities between the shapes of Fermi holes and the shapes of LMOs, however, is quite remarkable. The LMOs, for example, are usually determined as members of an orthonormal set, in which case the mutual orthogonality of the LMOs requires that most of the LMOs have nodal surfaces. The Fermi hole is not constrained to be orthogonal to anything, but it adopts the same nodal surfaces seen in the orthonormal LMOs. This supports earlier studies which have concluded that the shapes of LMOs are not strongly influenced by orthogonality requirements [46–48].

The properties of the Fermi hole are more remarkable when the probe electron is not in a region dominated by a single LMO. It is not necessary to place the probe electron in the middle of an LMO. The probe electron can be placed anywhere in a molecule (or even far outside a molecule). If the probe electron moves along a path from the center of one LMO to the center of a second LMO, then the Fermi hole must evolve from from something resembling the first LMO into something resembling the second LMO. If the electronic distribution in a molecule resembled anything like a non-uniform density free electron gas, then this evolution should be gradual, keeping the Fermi hole a least roughly centered on the probe electron.

The actual behavior of the Fermi hole in a molecule, as first shown using animated computer graphics [36], differs markedly from that of a free-electron gas. As the probe electron passes from the center of one LMO to the center of a second LMO, the Fermi hole exhibits three phases:

1. As the probe electron moves away from the center of the first LMO, the Fermi hole does almost nothing. Thus, the Fermi hole effectively lags behind as the probe electron moves away.
2. As the probe electron passes through the intermediate region shared by both LMOs, the Fermi hole suddenly leaps ahead, changing rapidly from a shape resembling the first LMO into a shape resembling the second LMO.
3. As the probe electron approaches the center of the second LMO, the Fermi hole is once again dormant. The shape of the Fermi hole changes very little as the probe electron catches up with the Fermi hole.

This behavior, first observed through visualization of the Fermi hole in a water molecule, can be quantified through the Fermi hole mobility function. This function, which is discussed in a later section of this work, measures the sensitivity of the Fermi hole to the position of the probe electron.

The Fermi orbital for a uniform density free electron gas is spherically symmetric with the radial dependence given by Eq (3). The shape of this function is indicated by the dashed curve in Fig. 1. The presence of radial nodes near $r_{12} = 4.5r_s$ and $7.7r_s$ can be seen in this figure. The Fermi orbital has maximum of $\sqrt{\varrho/2}$ at $r_{12} = 0$, and, as r_{12} approaches zero, the Fermi orbital is proportional to $1 - ar_{12}^2$, where a is a positive constant (1/10). This is consistent with the appropriate cusp condition for coalescence of identical particles [49, 50]. In this case, as r_{12} approaches zero, the wavefunction approaches zero in proportion to r_{12}, the diagonal part of the second order reduced density matrix approaches zero as r_{12}^2, and the Fermi hole approaches $\varrho/2$ minus r_{12}^2 times a positive constant.

The Fermi hole takes on an especially simple form for a singlet state of a two electron wavefunction, such as the ground electronic state of a helium atom or a hydrogen molecule. The two electron density functions $\varrho_{\alpha\alpha}$ and $\varrho_{\beta\beta}$ for these wavefunctions vanish because there is only one electron of each spin. If the position of this electron is specified, the probability of finding another one anywhere else is zero. In this case, the Fermi hole is given by

$$\Delta_{\alpha\alpha}(\vec{r}_1;\vec{r}_2) = \Delta_{\beta\beta}(\vec{r}_1;\vec{r}_2) = \varrho(\vec{r}_1)/2\,, \tag{19}$$

which is independent of $\vec{r}_2$. This condition applies to correlated wavefunctions as well as to the closed shell SCF wavefunctions for these states.

This is a limiting case of the behavior found in larger atoms and molecules. For a totally isolated pair of electrons, the Fermi hole is completely independent of the position of the probe electron. This suggests that if the Fermi hole in a larger system is nearly independent of the position of the probe electron when the probe electron is in a certain region, then that region must be occupied by an electron which is relatively isolated from other electrons.

5 Natural Representations of Correlation Holes

The first order reduced density matrix can be represented with products of members of any complete set of one electron functions,

$$\varrho(\vec{r},\vec{r}') = \sum_{\mu,\nu} d_{\mu\nu}\,\chi_\mu(\vec{r})\,\chi_\nu(\vec{r}')\,. \tag{20}$$

The coefficients $d_{\mu\nu}$ form the elements of a matrix **d** which is often called a "density matrix" leading to confusion with $\varrho(\vec{r},\vec{r}')$. If the basis functions $\chi_i(\vec{r})$ are subjected to a suitable transformation, the first order reduced density matrix may be represented in the "natural" form,

$$\varrho(\vec{r},\vec{r}') = \sum_{\mu} n_\mu\,\varphi_\mu(\vec{r})\,\varphi_\mu(\vec{r}')\,. \tag{21}$$

The functions $\varphi_{\mu'}$ are orthonormal linear combinations of the basis functions χ_{μ}, and they are defined as the "natural" orbitals of $\varrho(\vec{r}, \vec{r}')$. The factors n_{μ} are called the natural orbital occupation numbers, and they may be interpreted as indicating the number of electrons represented by each of the corresponding natural orbitals. This representation is unique, except for unitary transformations among natural orbitals with degenerate occupation numbers, and it provides the most rapidly converging representation of $\varrho(\vec{r}, \vec{r}')$.

The concept of natural orbitals is meaningless for closed shell SCF wavefunctions because all of the occupied molecular orbitals in such a wavefunction have equal occupation numbers of 2. This result is invariant to any unitary transformation of the occupied molecular orbtials. Consequently, the canonical SCF MOs, localized SCF MOs, and any other orthonormal combination of the occupied SCF MOs are all equally "natural." Natural orbitals, however, have been very useful in the calculation of correlated wavefunctions [51–54].

Similar techniques have also been applied to the second order reduced density matrix. The result is a representation of the second order reduced density matrix in terms of "natural geminals," each of which is a function of two electrons. These functions have not been found to be very useful or interesting [35].

It is also possible to define natural representations of the Fermi hole and Coulomb hole. These are given by [38]

$$\Delta_{st}(\vec{r}_1; \vec{r}_2) = \sum_{\mu} n_{\mu}^{(st)}(\vec{r}_2)\, |g_{\mu}^{(st)}(\vec{r}_1; \vec{r}_2)|^2 . \tag{22}$$

Likewise, the natural representation of the composite correlation hole is given by [38]

$$\Delta(\vec{r}_1; \vec{r}_2) = \sum_{\mu} n_{\mu}^{(c)}(\vec{r}_2)\, |g_{\mu}^{(c)}(\vec{r}_1; \vec{r}_2)|^2 . \tag{23}$$

The functions $g_{\mu}^{(st)}$ and $n_{\mu}^{(c)}$ are the correlation hole natural orbitals, and the quantities $n_{\mu}^{(st)}$ and $n_{\mu}^{(c)}$ are the corresponding correlation hole natural orbital occupation numbers. The occupation numbers for the Fermi holes, $\Delta_{\alpha\alpha}$ and $\Delta_{\beta\beta}$, and the

Table 1. Correlation hole natural orbital occupation numbers. These results were determined by a correlated wavefunction for the water molecule, with the probe electron located on one of the protons.

$n^{(\alpha\beta)}(H)$	$n^{(\alpha\alpha)}(H)$	$n^{(c)}(H)$
0.12302	0.98861	1.01512
0.04273	0.01482	0.04883
0.02514	0.01188	0.03996
0.01930	0.00752	0.02654
0.00045	0.00592	0.00224
0.00023	0.00045	0.00044
−0.01904	−0.00383	−0.00885
−0.02534	−0.01085	−0.03553
−0.04335	−0.01461	−0.03996
−0.12309		−0.04879

composite correlation hole sum to unity, and those for the Coulomb holes, $\Delta_{\alpha\beta}$ and $\Delta_{\beta\alpha}$ sum to zero.

Values of the correlation hole natural orbital occupation numbers $n_\mu^{(\alpha\beta)}$, $n_\mu^{(\alpha\alpha)}$, and $n_\mu^{(c)}$ for a probe electron located on one of the protons of a water molecule are shown in Table I. These values were determined by a multiconfigurational wavefunction based on a double-zeta basi set of atomic orbitals and interaction optimized virtual orbitals [38, 51]. This wavefunction includes 85% of the correlation energy obtainable within this double-zeta basis set.

In the case of a closed shell SCF wavefunction, the Coulomb hole vanishes everywhere, and the occupation numbers of the Coulomb hole all vanish. The existence and magnitudes of non-zero occupation numbers for the Coulomb hole may be used to characterize the magnitude and distribution of correlation effects in correlated wavefunctions. The ability to investigate these quantities is limited, in part, by the lack of adequate visualization tools. That is, it is much easier to compute these values and tabulate them than it is to represent them in a meaningful manner which reveals, characterizes, or exploits any potentially useful properties.

The occupation numbers of the Fermi hole and the composite correlation hole for a closed shell SCF wavefunction all vanish except for one which equals unity. In this case, the natural representation of the Fermi hole is the absolute square of the Fermi orbital [37–40]

$$\Delta_{\dot{a}}(\vec{r}_1;\vec{r}_2) = |f(\vec{r}_1;\vec{r}_2)|^2, \tag{24}$$

and the Fermi orbitals is given by

$$f(\vec{r}_1;\vec{r}_2) = \sqrt{2/\varrho(r_2)}\sum_i g_i(\vec{r}_1)\, g_i(\vec{r}_2), \tag{25}$$

where the orbitals $g_i(\vec{r})$ are either the canonical SCF molecular orbitals or any set related to the canonical SCF molecular orbitals by a unitary transformation. The Fermi orbital $f(\vec{r}_1;\vec{r}_2)$ is a function of $\vec{r}_1$ which is parametrically dependent upon the position of a probe electron located at $\vec{r}_2$.

In the case of an isolated pair of electrons, the Fermi hole is equal to half of the total density (Eq. (19)), in which case the Fermi hole natural orbital occupation numbers must equal half of the ordinary natural orbital occupation numbers, independently of the position of the probe electron.

The Fermi hole, the Coulomb hole, and the composite correlation hole have been derived from the second order reduced density matrix. Like the second order reduced density matrix, these functions depend on the coordinates of two electrons. The hole functions, however, must not be confused with geminals, and the correlation hole natural orbitals must not be confused with natural geminals.

The diagonal part of the second order reduced density matrix $\varrho_2(\vec{r}_1, \vec{r}_2)$ depends on two sets of electronic coordinates in an equivalent or symmetric manner. This function is symmetric with respect to exchange of these coordinates because all electrons are identical particles. Natural geminals are antisymmetric with respect to exchange of these coordinates because electrons are fermions. The second order reduced density matrix has the dimensionality of density squared, and the dimensionality of a geminal is the square of an orbital amplitude, which accidently coincides with the dimensionality of electronic density. The occupation numbers of natural geminals are constants which apply to all positions of either electron. The norm and other properties of these functions require integration over the coordinates of both sets of electronic coordinates.

The Fermi hole, Coulomb hole, and composite correlation hole also depend on two sets of electronic coordinates, but the role of the second set of coordinates is in no way equivalent to the role of the first set of coordinates. The dependence of $\Delta(\vec{r}_1;\vec{r}_2)$ or $g_\mu(\vec{r}_1;\vec{r}_2)$ on $\vec{r}_1$ is equivalent to the dependence of the electronic density $\varrho(\vec{r})$ or a molecular orbital $\varphi(\vec{r})$ on the electronic coordinate $\vec{r}$. The dependence of $\Delta(\vec{r}_1;\vec{r}_2)$ or $f(\vec{r}_1;\vec{r}_2)$ on $\vec{r}_2$; as indicated by the semicolon, is a parametric dependence similar to the dependence of an electronic wavefunction on the nuclear coordinates in the Born-Oppenheimer approximation.

Consequently, the Fermi hole, the Coulomb hole, and $\Delta(\vec{r}_1;\vec{r}_2)$ have the dimensionality of electronic density, equivalent to $\varrho(\vec{r})$. The natural orbitals representing these holes, including the Fermi orbitals, have the dimensions of orbital amplitude, equivalent to that of any other orbital, including atomic orbtials, natural orbitals, SCF MOs, LMOs, etc. The natural orbital occupation numbers for these functions depend on the position of the probe electron, $\vec{r}_2$. The norms and other properties of these functions are defined with respect to integration over $\vec{r}_1$ with $\vec{r}_2$ treated as a constant.

6 The Fermi Hole Mobility Function

As noted in Sect. 4, the Fermi hole in a molecule responds to changes in the probe electron more like a rabbit than like a shadow. That is, molecules are found to possess regions where the Fermi hole is insensitive to the position of the probe electron, and these regions are separated by regions where the Fermi hole is very sensitive to the position of the probe electron. The sensitivity of the Fermi hole to the position of the probe electron in a closed shell SCF wavefunction is measured by the Fermi hole mobility function [39],

$$F(\vec{r}) = F_x(\vec{r}) + F_y(\vec{r}) + F_z(\vec{r})\,, \tag{26}$$

where

$$F_v(\vec{r}) = \lim_{\varepsilon\to 0}\left[1 - \frac{\langle f(\vec{t};\vec{r})\,|\,f(\vec{t};\vec{r}+\varepsilon\vec{s})\rangle_{\vec{t}}}{\varepsilon^2}\right], \tag{27}$$

and s vector is a unit vector defined by the coordinate v. The Fermi hole mobility function may be expressed in terms of the SCF molecular orbitals by combining Eq. (25) with Eq. (27). The result is

$$F_v(r) = \left(\frac{2}{\varrho^2}\right)\sum_{i>j}\left[g_i\frac{\partial g_j}{\partial v} - g_j\frac{\partial g_i}{\partial v}\right]^2, \tag{28}$$

for v = x, y, and z.

The function $F(\vec{r})$ has the dimensions of energy. In earlier work [39, 40, 42], the dimensionality of the this function had been described as $(1/\text{distance})^2$, based on the classical interpretation of the quantities in Eq. (28). According to quantum mechanics, however, the derivatives in this equation must be interpreted as momenta. Consequently, $F(\vec{r})$ is proportional to the square of momentum, which is equivalent to energy (hartrees) in atomic units (with electron mass = 1).

The interpretation of $F(\vec{r})$ as an energy is demonstrated by the case of two helium atoms labelled A and B, and separated by a large distance along the x axis. The SCF orbitals may be approximated by those of the isolated atoms, and the sum in Eq. (28) has only a single term. Near nucleus A, $F(\vec{r})$ is extremely small because the amplitude and derivative of the orbital from atom B are much smaller than those from atom A. At the midpoint of the line segment joining the two nuclei, $\varrho_A = \varrho_B$, and the orbitals are proportional to $\exp(-2\varepsilon_2 r)$, where ε_1 is the orbital energy for the helium atom 1s orbital. In this case, the Fermi hole mobility function takes on the value

$$F((\vec{r}_A + \vec{r}_B)/2) = -\varepsilon_1 \,. \tag{29}$$

Consequently, the value of the Fermi hole mobility function at this point equals the energy barrier which must be crossed for an electron to move from one nucleus to the other.

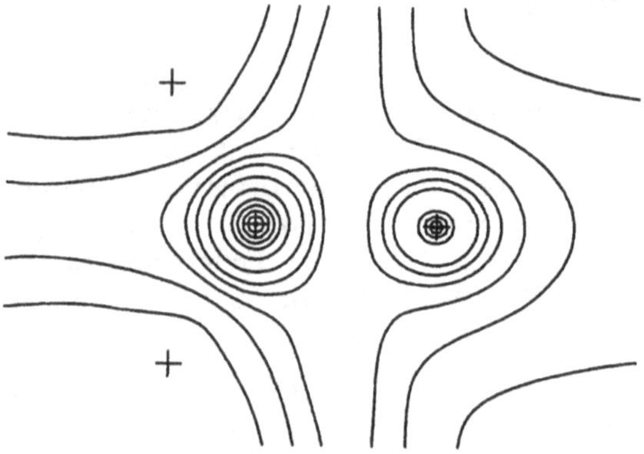

Fig. 4. The Fermi hole mobility function for a double zeta SCF wavefunction for formaldehyde. The positions of the nuclei are indicated as in Fig. 2. The contour levels indicate values of 0.1, 0.25, 0.5, 1.0, 2.0, and 5.0 atomic units (hartrees), with the smallest values occuring nearest each of the nuclei, and the largest values encircling the carbon and oxygen nuclei.

The values of the Fermi hole mobility function in a plane passing through a formaldehyde molecule are represented by contours in Fig. 4. The value of r indicates the position of a probe electron. When the probe electron is in a region where the Fermi hole is insensitive to the position of the probe electron, corresponding to a relatively isolated electron (or ab pair of electrons), then this function has a relatively small value. This condition is most evident when the probe electron is near any of the nuclei.

When the probe electron is in a region separating two relatively isolated electrons (or pairs of electrons) then this function has a relatively large value. This conditions is indicated by the ring of high values surrounding the heavy nuclei, and the smaller ridges separating the protons from each other and other parts of the molecule.

The Fermi hole mobility function for a molecule may be compared to that of an idealized electrical conductor by calculating this function for a uniform density free electron gas. In this case, the value of the Fermi hole mobility function may be obtained by evaluating Eq. (27) based on the Fermi hole indicated by Eq. (3). The result is a constant (independent of $\vec{r}$) determined by the density of the electrons.

An approximation to the Fermi hole mobility function for a free electron gas may be obtained by replacing the Fermi orbital of a free electron gas with a spherical gaussian orbital having maximum density (orbital amplitude squared) equal to half of the total density p. The resulting value of the Fermi hole mobility function is given by [39]

$$F_0(\varrho) = \left(\frac{3\pi}{4}\right)\left(\frac{\varrho}{2}\right)^{2/3} \tag{30}$$

The Fermi hole in a uniform density free electron gas is assumed to follow the motion of the probe electron. That is,

$$\Delta(\vec{r}_1; \vec{r}_2 + \vec{s}) = \Delta(\vec{r}_1 - \vec{s}; \vec{r}_2)\,. \tag{31}$$

If the Fermi hole in a molecule were to follow the probe electron in the same manner, then the Fermi hole mobility function would be given by

$$T(\vec{r}) = T_x(\vec{r}) + T_y(\vec{r}) + T_z(\vec{r})\,, \tag{32}$$

where

$$T_v(\vec{r}) = \lim_{\varepsilon \to 0}\left[1 - \frac{\langle f(\vec{t}\,; \vec{r}) \,|\, f(\vec{t} - \varepsilon\vec{s}\,; \vec{r})\rangle_t}{\varepsilon^2}\right]. \tag{33}$$

For a uniform density free-electron has, $F(\vec{r}) = T(\vec{r})$, as a consequence of Eq. (31). For an atom or molecule, however, $F(\vec{r}) - T(\vec{r})$ will be greater than zero or less than zero according to whether the change in the Fermi hole is greater or less than the effect of shifting the Fermi hole along with the probe electron.

When the limit in Eq. (33) is evaluated, $T(\vec{r})$ is found to be the kinetic energy of the Fermi hole,

$$T(\vec{r}) = \langle \mathrm{f}(\vec{t}; \vec{r})| - (1/2)\ \nabla^2\ |f(\vec{t}; \vec{r})\rangle_t\,, \tag{34}$$

The kinetic energy of the Fermi hole can be expressed in terms of a set of SCF MOs by combining Eq. (25) with Equation (34). The result is

$$T(\vec{r}) = \left(\frac{2}{\varrho}\right) \sum_{ij} g_i(\vec{r})\, g_j(\vec{r})\, T_{ij}\,, \tag{35}$$

where T_{ij} is the kinetic energy integral for orbitals g_i and g_j,

$$T_{ij} = \langle g_i| - (1/2)\ \nabla^2\ |g_j\rangle\,. \tag{36}$$

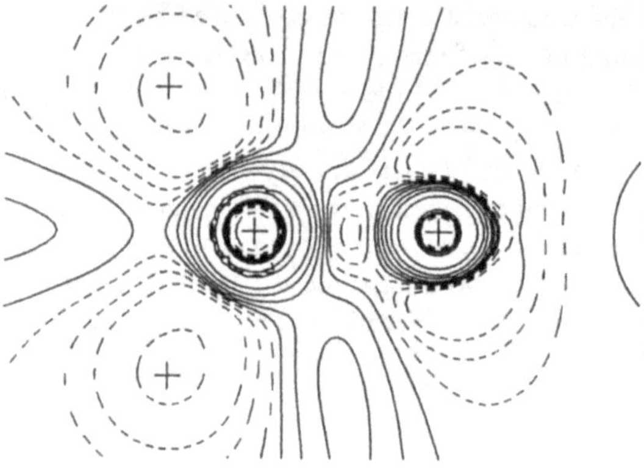

Fig. 5. The difference between the Fermi hole mobility function for formaldehyde (based on a double zeta wavefunction) and F_0(p). The solid lines indicate the same values defined in Fig. 4. The broken lines indicate values of 0.0, —0.1, —0.25, —0.5, —1.0, —2.0, and —1.0 hartrees.

In addition, when integrated over all values of $\vec{r}$, $T(\vec{r})$ yields the total kinetic energy of the electrons in a molecule:

$$\int T(\vec{r})\,\varrho(\vec{r})\,d\vec{r} = 2\sum_i T_{ii} = T\,. \tag{37}$$

Therefore, $T(\vec{r})$ represents a local kinetic energy function, as well as the kinetic energy of the Fermi hole.

If $F(\vec{r}) - T(\vec{r})$ were to be plotted as a function of $\vec{r}$, then an isolated pair of electrons should appear as a region of negative values, and the barriers between isolated pairs of electrons should appear as positive values. Although such figures have not yet been produced, similar results have been obtained using $F_0(\varrho)$ in place of $T(\vec{r})$ [39–42].

The difference between $F(\vec{r})$ and $F_0(\varrho)$ for formaldehyde is shown in Fig. 5. This figure, based on the same wavefunction used to determine Fig. 4, shows large negative values around each of the nuclei. These indicate that, when the probe electron is near a nucleus, the dependence of the Fermi hole on the position of the probe electron is much smaller in the molecule than it is in a free electron gas of the same density. Additional negative contours can be seen near the oxygen atom in regions traditionally

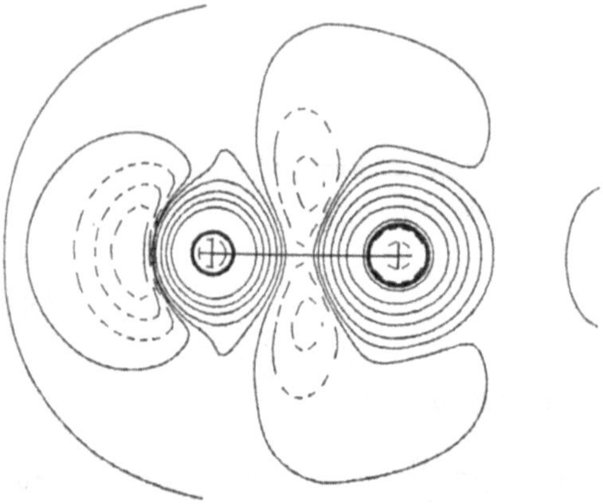

Fig. 6. The difference between the Fermi hole mobility function for formaldehyde (based on an STO-3G SCF wavefunction) and $F_0(\varrho)$. This figure represents a plane containing the carbon and oxygen atoms, perpendicular to the plane of all four nuclei. The solid lines indicate values of 0.1, 0.2, 0.4, 1.0, 2.0, and 4.0 hartrees. The broken lines indicate values of 0.0, —0.1, —0.2, —0.4, —1.0, and —10.0 hartrees.

associated with "lone pair" electrons [1, 2]. Negative values are also found between the carbon atom and the oxygen atom.

The difference between $F(\vec{r})$ and $F_0(\varrho)$ in the plane perpendicular to the plane of the nuclei is shown in Fig. 6. This figure is based on an STO-3G wavefunction instead of a double zeta wavefunction [40]. In this case, it can be seen that the negative values of $F(\vec{r}) - F_0(\varrho)$ between the carbon atom and the oxygen atom are only a section through a small bridge connecing two larger regions above and below the plane of the nuclei. The negative values seen in the right side of Fig. 6 represent a bridge between the two lone pair regions on the oxygen atom.

Fig. 5 and 6 clearly identify eight regions where the shape of the Fermi hole is relatively independent of the position of the probe electron (four nuclei, two lone pair regions, and two regions out of the plane of the nuclei). When the probe electron is located in any of these eight regions, the Fermi hole adopts a relatively stable shape. This shape, however, is not necessarily similar to the shape of the region of stability. When the probe electron is located near a proton, for example, the shape of the Fermi hole resembles a C—H bond LMO, as shown in Fig. 2 and 3. When the probe electron is located in one of the lone pair regions, the Fermi hole resembles a lone pair LMO [40]. When the Fermi hole is located in one of the negative regions above or below a line between the carbon and oxygen atoms, the Fermi hole resembles one of the two LMOs representing the C—O double bond (i.e. tau or "banana" bonds).

The eight regions of negative values of $F(\vec{r}) - F_0(\varrho)$ identified in Figs. 5 and 6 fully account for all eight pairs of electrons which define the electronic structure of the formaldehyde molecule. These regions tend to be separated from each other by regions in which $F(\vec{r})$ is greater than $F_0(\varrho)$. The principal exceptions are the regions associated with the oxygen lone pairs, and the regions associated with the C—O bonds. Similar results have been found for other molecules including ethane [39], methanol [42], and a cyclic enone $C_4H_4O_2$ [40]. In each case, the number of negative local minima in $F(\vec{r}) - F_0(\varrho)$ equals the number of electron pairs.

The regions with $F(\vec{r}) < F_0(\varrho)$ are very similar to loges [13, 55–7]. The relations between loges and the Fermi hole have been discussed by Bader and Stephens [13], who noted that the extent to which electrons may be localized in space is determined solely by the properties of the Fermi hole.

Regions where $F(\vec{r}) < F_0(\varrho)$ form the boundaries between loges. Large values of $F(\vec{r})$ separating two regions with $F(\vec{r}) < F_0(\varrho)$ imply that the regions are strongly isolated. For example, the carbon and oxygen nuclei are surrounded by shells having very large values of $F(\vec{r})$, indicating that the innershell electrons are strongly isolated from the valence electrons.

Small values of $F(\vec{r})$ imply a weak barrier providing little or no isolation. For example, the two lone pair regions on the oxygen atom in formaldehyde are joined by a wide area having $F(\vec{r}) < F_0(\varrho)$. Consequently, it may be more accurate to describe these regions as a single region holding four electrons, rather than two regions each holding a single electron apir.

Although the Fermi hole mobility function has several interesting properties, the formulation presented here is deficient in that it is only suitable for closed shell SCF or other single determinantal wavefunctions. The Fermi hole, however, does not depend on the existence of single determinantal wavefuxctions, and the definition of the Fermi hole adopted here applies to any wavefunction, including the exact correlated eigenfunctions of a many electron hamiltonian. Consequently, it should be possible to define the Fermi hole mobility function in a more general manner not limited to single determinantal wavefunctions.

7 Localized Orbitals Based on the Fermi Hole

It has long been known that it is possible to transform the canonical SCF molecular orbitals of a molecule into a set of localized molecular orbitals by means of a unitary transformation. Each of these orbitals may usually be associated with a chemical bond, a lone pair atomic orbital, or an innershell atomic orbital. The energy, charge density, and all other observable properties determined by the SCF wavefunction are not affected by these transformations. Consequently, these transformations are usually defined in terms of unobservable criteria such as maximization of intraorbital electronic repulsion [58, 59] and maximum separation of orbital centroids [60–62]. Although the resulting orbitals have a striking resemblence to bonds, lone pairs, and so forth, the physical significance of these orbitals may be questioned because they are defined by unobservable criteria.

In spite of possible questions concering their physical significance, several techniques have been developed for determining the transformations which produce these orbitals [58–67]. These methods generally depend on iterative sequences of unitary transformations based on appropriate combinations of integrals over the transformed molecular orbitals. Each iteration may require a substantial computational effort beyond that required to calculate the canonical SCF molecular orbitals, and some cases may require many iterations. In the case of the electronic repulsion criteria [58, 59], the computational effort required to determine this transformation can exceed that required by the original SCF calculation by a large factor.

As is shown below, the properties of the Fermi hole may also be used to transform a set of canonical SCF MOs into localized MOs. Unlike the traditional methods, however, the transformation presented here is determined by a physically observable quantity, the Fermi hole. In addition, this transformation is determined by an explicit expression which defines the localized orbitals directly in terms of the canonical SCF MOs. Consequently, it is not necessary to repeat an iterative calculation until the orbitals converge on a set of localized MOs. The Fermi hole transformation also requires no integrals, except for overlap integrals needed to maintain orthogonality, and these may be eliminated if a non-orthogonal set of localized orbitals is needed [47, 48]. Furthermore, with some additional approximations, the Fermi hole transformation may be simplified to the point where SCF MOs of polyatomic molecules may be converted into at least a partial set of localized MOs using only a hand held calculator.

When the probe electron is located in a region where $F(\vec{r})$ is less than $F_0(\varrho)$, the Fermi orbital is found to resemble a localized orbital determined by conventional methods [39–42]. Consequently, Eq. (25) provides a direct relationship between a set of canonical SCF orbitals $g_i(\vec{r})$ and a localized orbital $f_j(\vec{r}) = f(\vec{r}; \vec{r}_j)$, where $\vec{r}_j$ is a point in a region where $F(\vec{r}_j) < F_0(\varrho(\vec{r}_j))$. The same conclusions are expected from comparisons of $F(\vec{r})$ to the quantity $T(\vec{r})$ defined by Eq. (32), but numerical tests of this prediction are not available.

In order to transform a set of N canonical SCF orbitals into a set of N localized orbitals, it is necessary only to select N points $\vec{r}_j$, j = 1 to N, each of which is located in a region where $F(\vec{r}_i)$ is small. Ideally, each of these points should be near a minimum of $F(\vec{r})$, or $F(\vec{r}) - T(\vec{r})$, or $F(\vec{r}) - F_0(\varrho)$. This condition, however, is not critical because the Fermi hole is relatively insensitive to the position of the probe electron when the probe electron is located in one of these regions.

A set of N Fermi orbitals determined by Eq. (25) is generally not orthonormal. The overlap between a pair of Fermi orbitals, however, is usually very small, and a set of N Fermi orbitals may easily be converted into an orthonormal set of localized orbitals by means of symmetric orthogonalization [66, 68]. The resulting unitary transformation is given by

$$\mathbf{U} = (\mathbf{TT}^{+})^{-1/2}\,\mathbf{T}\,, \tag{38}$$

where

$$T_{ij} = g_j(r_i)/\sqrt{\varrho(r_i)/2}\,. \tag{39}$$

The transformation of a set of canonical SCF orbitals to a set of localized orbitals based on Eqs. (38) and (39) has been demonstrated for each of three molecules [41]. The first example, a cyclic conjugated enone, represents a simple case where conventional methods are not expected to have any special difficulties. The second example, methyl acetylene, is a molecule for which conventional methods have serious convergence problems [69]. The third example, boron trifluoride, is a pathological case for the orbital centroid criterion, with a number of local maxima and saddle points in the criterion of localization.

In each case, the first step in the application of this method is the selection of the set of N points. This set always includes the locations of all of the nuclei in the molecule. For atoms other than hydrogen, the resulting Fermi orbitals are similar to innershell localized orbitals. When the probe electron is located on a hydrogen atom, the Fermi orbitals is similar to an R—H bond orbital.

Additional points for the probe electron may usually determined based on the molecular geometry. The midpoint between two bonded atoms (other than hydrogen) tends to yield a Fermi orbital resembling a single bond. Multiple bonds may be represented with two or three points located roughly one to two bohr a point midway between the multiply bonded atoms, along lines perpendicular to a line joining the nuclei. Likewise, lone pair orbitals may be determined by points located roughly one bohr from the nucleus of an atom which is expected to possess lone pair orbitals.

In order to transform the eleven canonical SCF molecular orbitals of methylacetylene into a set of localized orbitals based on the Fermi hole, it is necessary to select four points in addition to the locations of the nuclei. One of these represents the C—C single bond, and three points are needed for the triple bond. The four C—H bonds are represented by points located on the protons. Transformation of the 22 SCF orbitals of the furenone, $C_4H_4O_2$, requires twelve points in addition to the ten nuclei. Boron trifluoride requires twelve points in addition to the locations of its four nuclei.

In each case, the localized molecular orbitals determined by the Fermi hole were found to be very close to those determined by the orbital centroid criterion. For the furenone molecule, for example, each of the localized orbitals determined by the Fermi hole method was found to have an overlap of 0.994 to 0.999 with one of the localized orbitals determined by the orbital centroid criterion. The remaining (off-diagonal) overlap integrals between these two sets of localized orbitals were found to have a root mean square (RMS) value of 0.011734. Likewise, the RMS off-diagonal overlap between the localized orbitals determined by the Fermi hole and those determined by the orbital centroid criterion for methylacetylene was found to be 0.012874.

Equations (25) and (39), like the other preceding equations, refer to orbitals amplitiudes without regard to the type or quality of any basis set of primitive functions used to construct a wavefunction. This reflects the fact that the Fermi hole is a fundemantal physical property whose existence is not related to any approximations or numerical techniques used to calculate specific wavefunctions. Most wavefunctions, however, are calculated on the basis of a set of one-electron functions $\chi_\mu(\vec{r})$ centered on each of the atoms. In this case, the canonical SCF MOs are defined by specific linear combinations of these functions,

$$\varphi_i(\vec{r}) = \sum_\mu C_{i\mu}\chi_\mu(\vec{r})\,. \tag{40}$$

In certain places, especially when the set of functions χ_μ is of "minimal" quality, a single function χ_μ may dominate a small region of space. That is, $|\chi_\mu(\vec{r})| \gg |\chi_\nu(\vec{r})|$ for all $\nu \neq \mu$ within a volume ω_μ. In this case, the electronic density within this volume is determined almost exclusively by this basis function, so

$$\varrho(\vec{r}) \approx 2\chi_\mu(\vec{r})^2 \left[\sum_i^{occ} C_{i\mu}^2\right]. \tag{41}$$

Likewise, the amplitude of an orbital, when evaluated within this region may be approximated as,

$$g_i(\vec{r}) \approx C_{i\mu}\chi_\mu(\vec{r})\,. \tag{42}$$

Consequently, if the probe electron is located within a region dominated by a single basis function $\chi_\mu(\vec{r})$, the Fermi orbital may be approximated by

$$f(\vec{r}\,;\vec{r}_\mu) \approx \sum_\nu d_{\mu\nu}\chi_\nu(\vec{r})\,, \tag{43}$$

where

$$d_{\mu\nu} = \sum_i^{occ} C_{i\mu}C_{i\nu}/C_\mu\,, \tag{44}$$

and

$$C_\mu = \sqrt{\sum_i^{occ} C_{i\mu}^2}\,. \tag{45}$$

Therefore, if the values of $C_{i\mu}$ are presented in a table in which each column represents an occupied molecular orbital (i), and each row represents an atomic basis function χ_μ, then the following steps are sufficient to determine a single localized orbital:

1. Select a row (u) corresponding to a basis function χ_μ which dominates some region of space.
2. Compute C_μ as the square root of the sum of the squares of the values of $C_{i\mu}$ in this row.
3. Divide each coefficient $C_{i\mu}$ in this row by the value of C_μ. Call these values $d_{i\mu}$.
4. Compute the inner product of the values $d_{i\mu}$ with each row of the table ($C_{i\nu}$). The resulting values, $d_{\mu\nu}$, are the coefficients of each basis function χ_ν in a localized MO which is concentrated in the region dominated by χ_μ.

The results obtained by applying this method to a double zeta SCF wavefunction for the formaldehyde molecule are shown in Table II. These values are based on the first basis function, an s orbital on one of the hydrogen atoms. The resulting localized orbital is a C—H bond orbital. The corresponding localized orbitals as determined by the orbital centroid criterion, the energy criterion, and the Fermi hole (Eqs. (38) and (39)) are also given for comparison. In spite of the enormous variation in the computational effort required to determine each of the columns in Table II, there is very little difference between these columns.

Table 2. Localized molecular orbitals for formaldehyde determined by various methods. Each column indicates a CH bond based on a double zeta SCF wavefunction.

Atom	Basis Function	Centroid Criterion	Energy Criterion	Fermi Orbital	Equation 43
Hydr-1	1s-1	−0.316928	−0.314568	−0.322143	−0.322375
	1s-2	−0.236648	−0.230457	−0.261079	−0.265507
Hydr-2	1s-1	0.035858	0.034761	0.071915	0.076629
	1s-2	0.100335	0.096456	0.137049	0.144429
Carbon	1s-1	0.003703	0.002126	0.004446	0.004403
	1s-2	0.084699	0.055305	0.096618	0.095710
	2s-1	−0.266482	−0.272481	−0.226817	−0.225636
	2s-2	−0.183512	−0.184118	−0.161461	−0.159095
	2px-1	−0.382002	−0.385259	−0.404347	−0.402413
	2px-2	−0.044686	−0.046227	−0.041872	−0.040015
	2py-1	0.0	0.0	0.0	0.0
	2py-2	0.0	0.0	0.0	0.0
	2pz-1	0.244099	0.234529	0.229381	0.221925
	2pz-2	0.026402	0.027216	0.022138	0.022120
Oxygen	1s-1	−0.000505	−0.000221	−0.000580	−0.000559
	1s-2	−0.011706	−0.005869	−0.013087	−0.012609
	2s-1	0.032921	0.025798	0.031761	0.030492
	2s-2	0.027234	0.024482	0.020760	0.021335
	2px-1	0.054412	0.029499	0.138829	0.165050
	2px-2	0.079138	0.067326	0.123740	0.136377
	2py-1	0.0	0.0	0.0	0.0
	2py-2	0.0	0.0	0.0	0.0
	2pz-1	−0.053241	−0.042901	−0.064928	−0.058898
	2pz-2	−0.017322	−0.015548	−0.019532	−0.018039

The results shown in Table II show only one localized MO, and not a full set of localized MOs. The procedure defined by Eqs. (43) to (45) does not necessarily provide a practical method for computing a full set of localized MOs, and in many cases it may not be possible to determine a full set of localized MOs based on these equations because it may be impossible to identify a basis function which dominates each of the regions required to define a full set of localized MOs. Nevertheless, this example demonstrates how a tiny amount of computation based on the Fermi hole can yield results remarkably similar to those of far more complex calculations based on other concepts.

8 Electronic Structure of the [1.1.1] Propellane Molecule

The electronic structure of most ordinary molecules like water, formaldehyde, and ethane are considered to be well understood, at least in their ground electronic states and equilibrium geometries. There are many molecules and molecule-like systems, however, for which the electronic structure is not easily explained by simple or conventional models. These include

1. Metal atom clusters,
2. Boron hydrides,
3. Hypervalent molecules,
4. Highly strained organic molecules,
5. Organic conductors,
6. Electron transfer reactions,
7. Reactive intermediates and transition states in chemical reactions,
8. Electronically excited states, especially those involved in photochemical reactions.

In each of these categories there are a number of outstanding cases, and for each case there is at least one and often several competing explanations of the electronic structure. In many cases, the existing explanations are less than completely satisfactory. The qualitative explanations, for example, are often highly speculative. Quantitative numerical studies, on the other hand, may be flawed by inadequate basis sets or the treatment of electron correlation.

The [1.1.1] propellane molecule, or tricyclo [1.1.0] pentane, C_5H_6, is a good example of a molecule whose electronic structure is not easily explained with conventional methods. This is a highly strained organic molecule composed of three CH_2 groups bridging a pair of carbon atoms lying on a C_3 symmetry axis. The principal questions regarding the electronic structure of this molecule are concerned with the existence and character of any bond or bonds connecting the two axial carbon atoms (C1 and C3).

This molecule was synthesized several years ago, and was found to be remarkably stable [70–72]. There have been a number of theoretical studies of this molecule, but the results of these studies have been inconsistent and inconclusive. For example, an investigation based on extended Hückel wavefunctions concluded that [73]

1. The [1.1.1] propellane with a C1—C3 bond is more stable than the corresponding di-radical structure.

2. The equilibrium C1—C3 separation is only slightly larger than a normal C—C single bond.
3. The HOMO of the [1.1.1] propellane is slightly anti-bonding, in spite of being symmetric with respect to a plane perpendicular to the C1—C3 axis.

Similar results, including a C1—C3 separation of 1.60 Å, were obtained from calculations based on 4-31G SCF wavefunctions [74]. A contour map of the resulting HOMO is shown in Fig. 7. Subsequent calculations based on 6-31G* SCF wavefunctions found an equilibrium C1—C3 separation of 1.543 Å, equal to that of a normal C—C single bond [75, 76].

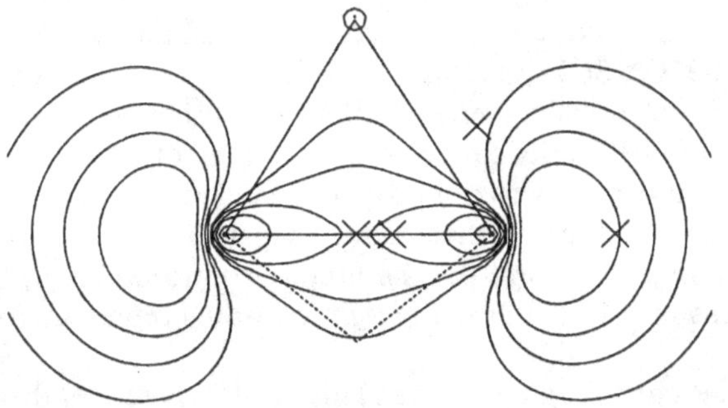

Fig. 7. The charge density for the HOMO from the 4-31G SCF wavefunction for the [1.1.1] propellane. Solid lines are drawn between the nuclei of atoms C1 (left) C2 (top), and C3 (right). Dashed lines are drawn from the nuclei of atoms C1 and C3 to the projected position of atoms C4 and C5 in the plane of atoms C1, C2, and C3. The four Xs indicate the probe electron positions used to determine Fig. 12 to 15. The contour lines indicate density values of 0.004, 0.001, 0.02, 0.04, and 0.1 electrons per cubic bohr.

The symmetry of the HOMO implies a bond between the axial carbon atoms, but the small amplitude at the C1—C3 midpoint implies a very weak bond. The outward hybridization seen in Fig. 7 could even imply anti-bonding characteristics. The shape of the HOMO, however, is not a reliable indication of C1—C3 bonding because this is one of the canonical SCF molecular orbitals (CMOs). CMOs are characteristically delocalized, and individual CMOs rarely posess a one-to-one association with chemical bonds.

Several additional characteristics of the 4-31G SCF wavefunction were also determined in an effort to characterize the interaction between the C1 and C3 carbon atoms [74]. These characteristics include the C1—C3 overlap population, the electron density between these atoms, and the results of transforming the CMOs to localized molecular orbitals. These characteristics all failed to reveral the presence of a chemical bond between the C1 and C3 atoms.

The total overlap population between the C1 and C3 atoms, for example, was found to be sizeable and negative. Population analyses, however, are very unreliable, especially for extended basis sets, because they are based on the assumption that the orbital amplitudes resulting from each orbital basis function ('atomic orbital') can be completely and unambiguously assigned to the point (atom) on which the basis function is centered.

The electron density distribution determined by the 4-31G SCF wavefunction is shown in Fig. 8. The electron density between the C1 and C3 atoms is only about

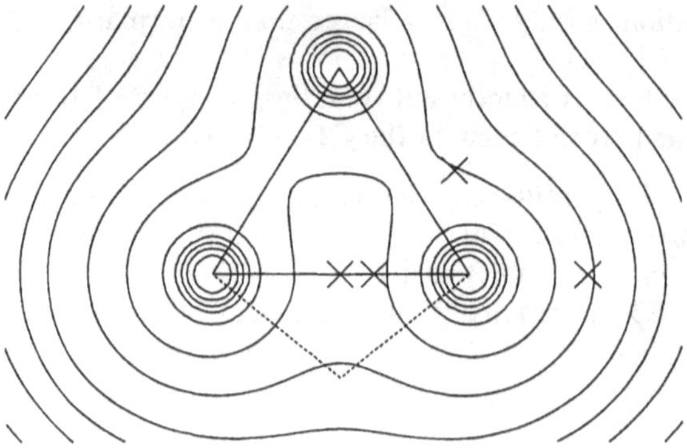

Fig. 8. The total electronic density for the 4-31G SCF wavefunction for the [1.1.1] propellane. The contour lines indicate density values of 0.004, 0.01, 0.02, 0.04, 0.1, 0.2, 0.4, 1.0, 2.0, 4.0, and 10.0 electrons per cubic bohr.

half of that found between atoms C1 and C2, or between C2 and C3. Similar results have been obtained from 6-31G* SCF wavefunctions [75, 76]. Deformation density plots defined by subtracting spherically averaged electron densities for each atom from the electron density of the molecule show a net loss of electrons between the C1 and C3 atoms [77]. These results do not provide a valid representation of the region between these atoms because (a) this region is penalized by contributions from five atoms, rather than two, and (b) the atomic contributions are treated as spherically symmetric, in spite of the unusually tight crowding represented by these atoms.

The 4-31G SCF molecular orbitals have been transformed into LMOs based on the energy criterion [58, 59]. According to Newton and Schulman [74], the resulting LMO associated with the C1 and C3 atoms is hybridized outward, indicating a very weak or anti-bonding interaction between these atoms. The corresponding LMO determined by the orbital centroid criterion [60–62, 66] is shown in Fig. 9. This LMO is concentrated between the C1 and C3 atoms, hybridized inwardly, and suggests a strong C1—C3 bond. These results are all opposed to the results obtained from the energy criterion. In this case, the energy criterion has clearly failed to determine a well localized set of LMOs.

Additional work based on 6-31G* SCF wavefunctions has included calculations of enthalpies of hydrogenation for the [1.1.1] propellane and related molecules

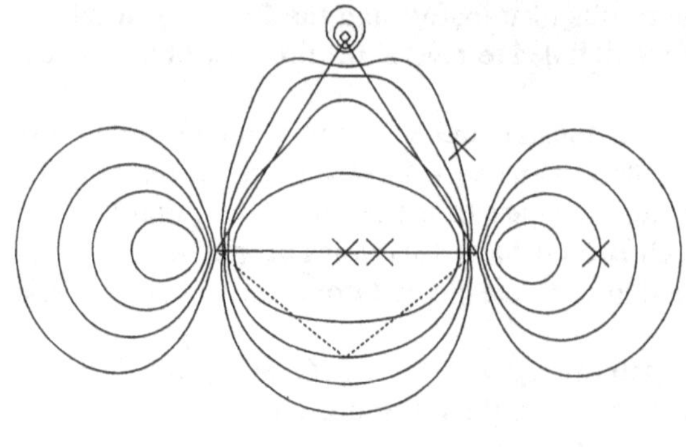

Fig. 9. The C1-C3 bond LMO determined by the orbital centroid criterion for the 4-31G SCF wavefunction for the [1.1.1] propellane. The contour lines indicate density values of 0.004, 0.01, 0.02 and 0.04 electrons per cubic bohr.

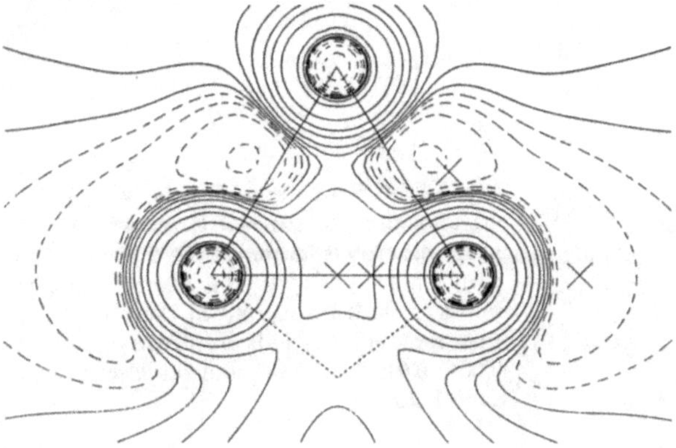

Fig. 10. The difference between $F(\vec{r})$ and $F_0(\varrho)$ for the 4-31G SCF wavefunction of the [1.1.1] propellane. The solid contour lines indicate values of 0.04, 0.1, 0.2, 0.4, 1, 2, and 4 hartrees. The broken contour lines indicate values of 0.0, −0.004, −0.1, −0.2, −0.4, −1, −2, −4, and −10 hartrees. The contour lines for values less than −0.2 are found only within 0.2 Å of the nuclei.

[75, 76]. Based on these results, the [1.1.1] propellane was found to be more stable than the corresponding (C1—C3) diradical by 65 kcal/mole, indicating a strong C1—C3 interaction.

The difference between the Fermi hole mobility function $F(\vec{r})$ and $F_0(\varrho)$ for the [1.1.1] propellane is shown in Fig. 10. This figure shows five regions in which this function is negative. Three small regions are centered on each of the carbon nuclei in this plane. One region crosses the line between atoms C1 and C2, and extends out beyond the left of atom C1. A similar region crosses the line between atoms C2 and C3, and extends out beyond the right side of atom C3.

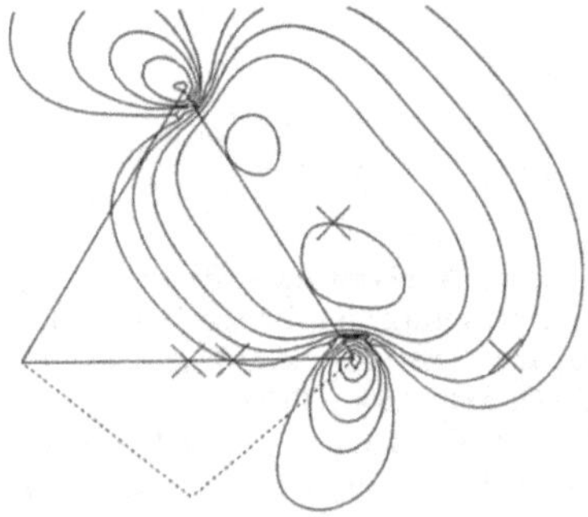

Fig. 11. The C2-C3 bond LMO determined by the orbital centroid criterion for the 4-31G SCF wavefunction for the [1.1.1] propellane. The contour lines indicate density values of 0.004, 0.01, 0.02, 0.04, 0.1, and 0.2 electrons per cubic bohr.

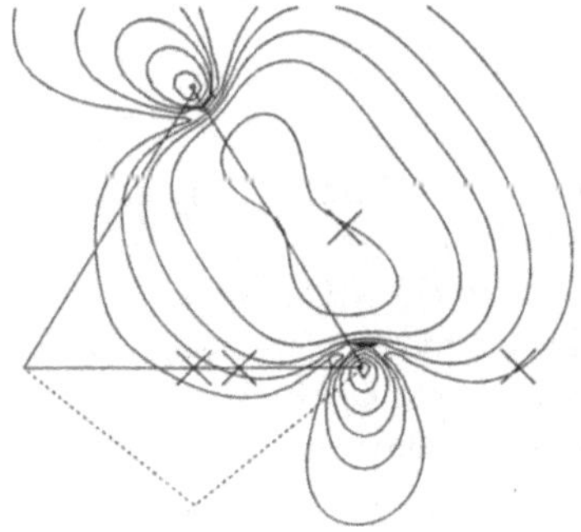

Fig. 12. A Fermi hole for the 4-31G SCF wavefunction for the [1.1.1] propellane. The probe electron is located at the centroid of the LMO shown in Fig. 11. This point is indicated by the X not located on the C1-C3 axis. The contour lines indicate density values of 0.004, 0.001, 0.02, 0.04, 0.1, and 0.2 electrons per cubic bohr.

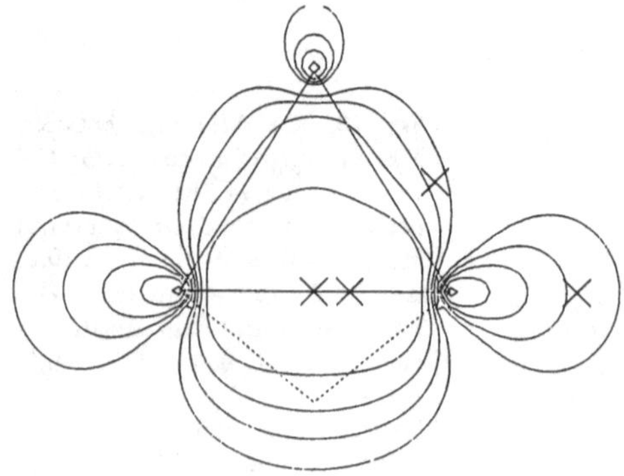

Fig. 13. A Fermi hole for the 4-31G SCF wavefunction for the [1.1.1] propellane. The probe electron is located at the centroid of the LMO shown in Fig. 9. This point is indicated by the X at the midpoint of the C1-C3 axis. The contour lines indicate density values of 0.004, 0.01, 0.02, 0.04, and 0.1 electrons per cubic bohr.

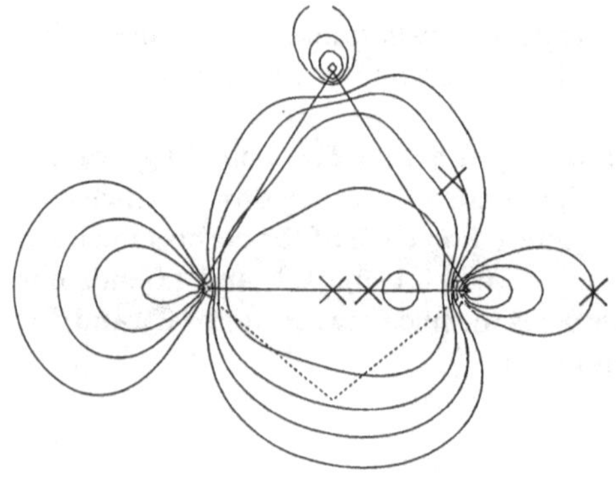

Fig. 14. A Fermi hole for the 4-31G SCF wavefunction for the [1.1.1] propellane. The probe electron is located 0.4 a_0 (0.21 Å) to the right of the C1-C3 midpoint. This point is indicated by the X between the C1-C3 midpoint and atom C3. The contour lines indicate density values of 0.004, 0.01, 0.02, 0.04, and 0.1 electrons per cubic bohr.

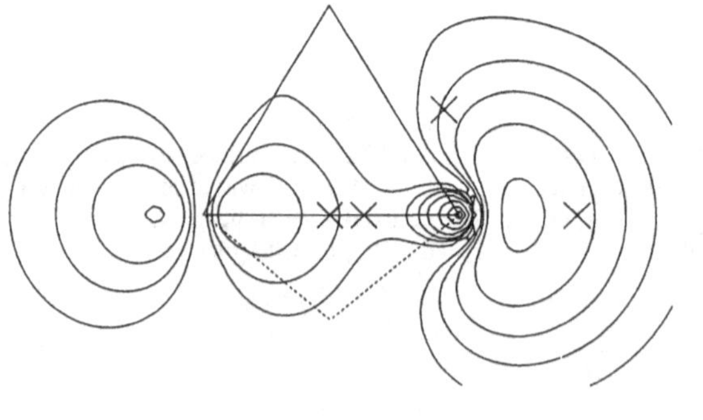

Fig. 15. A Fermi hole for the 4-31G SCF wavefunction for the [1.1.1] propellane. The probe electron is on the C1-C3 axis, 1.39 a_0 (0.73 Å) to the right of atom C3. This point is indicated by the X to the right of atom C3. The contour lines indicate density values of 0.004, 0.01, 0.02, 0.04, 0.1, 0.2, and 0.4 electrons per cubic bohr.

The negative region which crosses the line between atoms C2 and C3 possesses a minimum approximately 0.2 Å away from the midpoint of the C2—C3 line. This mimimum is very close to the centroid of the orbital centroid LMO associated with the C2—C3 bond. The centroid of this LMO is indicated by an X between atoms C2 and C3 in Fig. 7 to 15. The corresponding LMO is shown in Fig. 11. The Fermi hole determined by placing the probe electron at the centroid of the C2—C3 bond LMO is shown in Fig. 12. The Fermi hole in Fig. 12 is nearly identical to the LMO shown in Fig. 11, and both figures indicate orbitals with almost perfect sp^2 hybridization. This is consistent with the conclusions obtained from an investigation based on the "iterative maximum overlap method" [78].

Figure 10 fails to show a negative region between atoms C1 and C3. Such a region would be expected to represent a pair of electrons forming a C1—C3 bond. Instead, there are two small secondary minima in the negative regions crossing the C1—C2 and C2C3 axes. In spite of this, the shape of the Fermi hole is very persistent when the probe electron is located on the C_3 axis, as shown in Figs. 13, 14, and 15. These figures indicate the shapes of the Fermi hole when the probe electron is located at the three positions on the C_3 axis marked by Xs in Figs. 7 to 15.

If the probe electron is located at the midpoint of the line joining atoms C1 and C3, the Fermi hole, as shown in Fig. 13, is very similar to the corresponding LMO determined by the orbital centroid criterion (Fig. 9). The Fermi orbital represented by this Fermi hole shows inward hybridization on both axial carbon atoms, suggesting a strong sigma bond between these atoms. This observation is reinforced by Fig. 14 which shows that the shape of this Fermi hole changes only by a small amount when the probe electron is shifted as much as 0.21 Å towards an axial carbon atom. In this case, as the probe electron moved to the right (towards atom C3), the central portion of the Fermi hole also shifted to the right, but the outer lobes have shifted to the left. Thus the left lobe of the Fermi hole shown in Fig. 14 is actually larger than that shown in Fig. 13, in spite of the fact that the probe electron has moved away from this lobe.

If the probe electron is located within roughly 0.5 Å of a carbon atom, the Fermi hole quickly changes into an innershell orbital. When the probe electron is located along the C_3 axis beyond the innershell region, then the Fermi hole once again exhibits signs of a C1 to C3 sigma bond, as shown in Fig. 15. The shape of the Fermi hole in Fig. 15 has changed significantly in comparison to the shapes shown in Figs. 13 and 14. In spite of this, the Fermi hole of Fig. 15 shows large values on the left side of the C1 atom even though the probe electron is located 0.73 Å to the right of the C3 atom and over 2.3 Å away from the C1 nucleus. The size of left lobe in Fig. 15 is nearly equal to that shown in Fig. 14, and the hybridization at C1 has shifted to essentially pure p.

Figures 13 to 15 indicate a relatively conventional, though distorted, sigma bond joining atoms C1 and C3. This bond possesses relatively high p character at the axial carbon atoms, leading to increased electronic densities along the C_3 axis outside the region between these atoms, lower electronic densities between these atoms, and a diminished binding energy compared to a normal carbon carbon single bond. This description is consistent with the corresponding LMO determined by the orbital centroid criterion, as well as with the results of other calculations [78].

The only exception to this description of the C1—C3 sigma bond in the [1.1.1] propellane is the failure to find a region with negative values of $F(\vec{r}) - F_0(\varrho)$ between these atoms, as shown in Fig. 10. It is likely that this is caused by using $F_0(\varrho)$ instead of the kinetic energy function, $T(\vec{r})$ given by Eq. (35). That is, the function $F(\vec{r}) - T(\vec{r})$ may provide a better representation of the electronic structure of a molecule than is provided by $F(\vec{r}) - F_0(\varrho)$. It is highly desirable to have contour maps of $F(\vec{r}) - T(\vec{r})$ for comparison to Figs. 5, 6, and 10, but such figures are not yet available.

9 Conclusions

The effects of the Pauli exclusion principle and electronic repulsion on the electronic wavefunctions of atoms molecules and solid state structures may be represented by the Fermi hole, the Coulomb hole, and composite correlation holes. Each of these may be described as a function of one electron which is parametrically dependent on the location of a second electron. This second electron is called the probe electron.

The Fermi hole provides a versatile tool for analyzing the electronic structure of atoms and molecules. The usefulness of the Fermi hole was demonstrated over fifty years age in studies of electrical conductors in solid state physics. The results of this work include a formula for the effect of electronic exchange on the electronic repulsion energy of a uniform density free electron gas. This formula has been very influential in the development of local density functional theories of atomic and molecular structure, including the $X\alpha$ SCF methods.

The properties of the Fermi hole have been of great value in the qualitative description of the electronic structure of molecules, as is demonstrated by the work of Linnett. Based on the consequences of the Fermi hole in a few simple models, Linnett was able to characterize the electronic structure of a wide variety of molecules without the need for complex calculations.

Until recently, there had been little progress in establishing or exploiting the properties of the Fermi hole in quantitative studies of molecular structure. Such progress requires practical definitions of the Fermi hole and related quantities such as the Coulomb hole. The definitions based on the work of McWeeny have proven to be the most useful. This approach defines the Fermi hole as a physically observable quantity which is determined by the first and second order reduced density matrices.

Recent work on the properties of the Fermi hole has included the introduction of the Fermi hole natural orbitals, Fermi orbitals, the Fermi hole mobility function, and transformation of canonical SCF molecular orbitals to localized molecular orbitals based on the Fermi hole.

Each of the correlation holes possesses a natural representation determined by a set of correlation hole natural orbitals and their occupation numbers, each of which depends on the position of the probe electron. In the case of a wavefunction composed of a single Slater determinant, the natural representation of the Fermi hole reduces to the square of a single orbital called a Fermi orbital.

Each molecule possesses regions in which the Fermi hole is relatively insensitive to the position of the probe electron. These regions are separated by regions in which the Fermi hole is more sensitive to the position of the probe electron. When the probe electron is located in a region where the Fermi hole is insensitive to the position of the probe electron, the Fermi hole resembles a localized molecular orbital.

The sensitivity of the Fermi hole to the position of the probe electron may be characterized by the Fermi hole mobility function, $F(\vec{r})$. This function has the form of a type of local energy function. In regions where the value of this function is less than the local kinetic energy, $T(\vec{r})$ the Fermi hole lags behind the motions of the probe electron, indicating the presence of a highly localized electron or pair of electrons. Regions where the Fermi hole mobility function exceeds the local kinetic energy correspond to boundaries between localized electrons, or loge boundaries. A conductor is expected to possess extended regions where the Fermi hole mobility function is nearly equal to the local kinetic energy, and an electrical insulator is expected to be composed of regions in which the values of the Fermi hole mobility function are alternately greater than or less than the local kinetic energy.

The local kinetic energy may be approximated by a local density function, $F_0(\varrho)$, based on the properties of a uniform density free-electron gas. This function is determined by approximating the Fermi hole in a free-electron gas with a gaussian function. Plots of $F(\vec{r}) - F_0(\varrho)$ indicate the locations of localized pairs of electrons such as chemical bonds, lone pairs, and innershell orbitals, in many molecules.

The properties of the Fermi hole may be used to transform canonical SCF molecular orbitals into a set of localized SCF molecular orbitals. The transformation matrix for converting the canonical SCF molecular orbitals into localized orbitals is given by a simple explicit function of the SCF orbital amplitudes. Except for orthogonalization, this method requires no integrals and no iterative transformations. The localized orbitals obtained from this method are very similar to the localized orbitals determined by the orbital centroid criterion. The effort required to determine this transformation is much smaller than that required by other methods such as the energy criterion [58, 59] or the orbital centroid criterion [60–62, 66]. With some additional approximations, certain individual localized orbitals may be determined by hand, based on the coefficients of basis functions (e.g. atomic orbitals) in a set of SCF molecular orbitals.

The properties of the Fermi hole may be most useful in characterizing the electronic structure of unusual molecules such as the [1.1.1] propellane. Conventional techniques including localized orbitals based on the energy criterion and deformation density plots have failed to provide evidence of a chemical bond between the two axial carbon atoms in this highly strained molecule. Plots of the Fermi hole show a sigma bond between these atoms based on atomic orbitals having a high p character. A contour map of $F(\vec{r}) - F_0(\varrho)$ did not reveal the central portion of this bond, indicating that it may be necessary to use the local kinetic energy $T(\vec{r})$ instead of the free-electron gas approximation provided by $F_0(\varrho)$.

In spite of the recent progress in establishing properties of the Fermi hole and applying these properties to practical problems, much remains to be learned about these properties. For example, it would be very interesting to produce contour maps of the local kinetic energy $T(\vec{r})$ defined by Eqs. (32) to (35), as well as contour maps of $F(\vec{r}) - T(\vec{r})$ for a variety of molecules. This latter function may provide a better representation of the electronic structure of a molecule than is obtained from $F(\vec{r}) - F_0(\varrho)$.

The definition of the Fermi hole may be applied to any wavefunction, including exact eigenfunctions of the electonic hamiltonian, approximate correlated wavefunctions, and single determinant SCF wavefunctions. The definitions of the Fermi hole mobility function and the local kinetic energy, however, can only be applied to single determinant wavefunctions such as closed shell SCF wavefunctions. It must be possible to generalize these definitions to encompass arbitrary correlated wavefunctions, including the corresponding exact eigenfunctions.

The Fermi hole natural orbital occupation numbers, and the Coulomb hole natural orbital occulation numbers are distributed in space, but these distributions have never been investigated, except for a single point in a water molecule [38]. For example, it would be interesting to investigate the distribution of the Coulomb hole natural orbital occupation numbers as a function of internuclear separation for the H_2 molecule.

The Fermi hole mobility function and the local kinetic energy provide descriptions of the extent to which electrons are localized or confined to limited regions of a molecule. The ability of the Fermi hole to identify the locations of localized electrons including lone pairs and chemical bonds has been demonstrated with several molecules. These have included simple cases like ethane, methanol, and formaldehyde, as well as a more complex case represented by the [1.1.1] propellane. There are, however, a wide variety of molecules and molecule-like systems which possess unusual electronic structures. Several categories of these systems have been identified in Sect. 8, and there are likely to be other categories not covered in this list. The properties of the Fermi hole can provide a great improvement over existing alternatives for characterizing the electronic structure of these systems.

In addition to static characteristics, the properties of the Fermi hole can also be applied to certain dynamic characteristics of the electronic structure of molecules. The static characteristics include specification of where the electrons are located when they are confined to resticted regions. In such cases, the electrons often form the chemical bonds which hold a molecule together. The dynamic characteristics include descriptions of how electrons move from one region to another when such movement is possible.

The properties of the Fermi hole may be used to distinguish insulators from conductors on a microscopic scale [42]. Consequently, it may be possible to use these descriptions to identify regions within molecules which form good, or poor pathways for the motions or conduction of electrons. Such tools could be very useful in a wide range of applications, including the determination of the rates and mechanisms of electron transfer reactions, electrochemical processes, properties of organic conductors, and properties of atomic and molecular scale electrical components.

Acknowledgements: The author wishes to thank Profs. O. Sinanoğlu (Yale), R. G. Parr (N. Carolina), R. McWeeny (Pisa), G. Petersson (Wesleyan), K. B. Wiberg (Yale), L. C. Allen (Princeton), and Dr. E. Clementi (Kingston) for discussions and encouragement in the course of the work presented here.

10 References

1. Linnett JW (1962) The electronic structure of molecules. Methuen, New York
2. Lewis GN (1929) Valence and the structure of atoms and molecules. Chemical Catalog Co.
3. Clementi E (1985) J. Phys. Chem. 89: 4426
4. Bloch F (1929) Z. Phys. 57: 545
5. Dirac PAM (1930) Proc. Cambridge Phil. Soc. 26: 376
6. Wigner EP, Seitz F (1933) Phys. Rev. 43: 804; (1934) 46: 509
7. Slater JC (1934) Rev. Mod. Phys. 6: 209
8. Slater JC (1951) Phys. Rev. 81: 385
9. For a survey of local density approximations, see Dahl JP, Avery J (eds) (1984) Local Density Approximations in Quantum Chemistry and Solid State Physics. (Plenum, New York)

10. Maslen, VW (1956) Proc. Phys. Soc. A69: 734
11. McWeeny R (1960) Rev. Mod. Phys. 32: 335
12. Sperber G (1971) Int. J. Quantum Chem. 5: 177, 189; (1972) 6: 881
13. Bader RFW, Stephens ME (1975) J. Am. Chem. Soc. 97: 7391
14. Besson MA, Suard M (1978) Int. J. Quantum Chem. 10: 151
15. Doggett G (1977) Mol. Phys. 34: 1739
16. Cooper IL, Pounder CNM (1978) Theor. Chim. Acta (Berlin) 47: 51
17. Cooper IL, Pounder CNM (1980) Int. J. Quantum Chem. 17: 759
18. Boyd RJ, Coulson CA (1974) J. Phys. B 7: 1805
19. Coulson CA, Neilson AH (1961) Proc. Phys. Soc. (London), 78: 831
20. Boyd RJ, Katriel J (1974) Int. J. Quantum Chem. 8: 255
21. Banyard KE, Mashat MM (1977) J. Chem. Phys. 67: 1405
22. Banyard KE, Mobbs RJ (1981) J. Chem. Phys. 75: 3433
23. Banyard KE, Youngman PK (1982) J. Phys. B 15: 853
24. Boyd RJ, Yee MC (1982) J. Chem. Phys. 77: 3578
25. Löwdin PO (1959) Adv. Chem. Phys. 2: 207
26. Clementi E (1965) IBM J. Research and Develop Suppl. 9
27. Clementi E, Roetti C (1974) Atomic and Nuclear Data Tables 14: 177
28. Luken WL, Sinanoglu O (1976) Atomic Data and Nuclear Data Tables 18: 525
29. McLean AD, Yoshimine M (1967) IBM J. Research Develop. Suppl. 11
30. Husimi K (1940) Proc. Phys. Math. Soc. Japan, 22: 264
31. Löwdin P-O (1955) Phys. Rev. 97: 1474
32. McWeeny R (1955) Proc. Phys. Soc. London, Ser. A 232: 114
33. McWeeny R (1967) Int. J. Quantum Chem. Symp. 1: 351
34. Kutzelnigg W, Del Re G, Berthier G (1968) Phys. Rev. 172: 49
35. McWeeny R, Sutcliffe BT (1969) Methods of Molecular Quantum Mechanics. Academic, New York
36. Luken WL, Beratan DN (1980) Electron Correlation and the Chemical Bond. 16 mm movie, (Freewater Productions, Durham, NC); also available on VHS (NTSC) video tape. Contact the author for information on availability
37. Luken WL, Beratan DN (1982) Theor. Chim. Acta (Berlin) 61: 265
38. Luken WL (1982) Int. J. Quantum Chem. 22: 889
39. Luken WL Culberson JC (1982) Int. J. Quant. Chem. Symp. 16: 265
40. Luken WL, Culberson JC (1984) in: Dahl JP, Avery J (eds) Local Density Approximations in Quantum Chemistry and Solid State Physics. Plenum, New York
41. Luken WL, Culberson JC (1984) Theor. Chim. Acta (Berlin) 66: 279
42. Luken WL (1984) Croatica Chemica Acta 57: 1289
43. Huzinaga S (1965) J. Chem. Phys. 42: 1293
44. Dunning TH (1970) J. Chem. Phys. 53: 2823
45. Snyder L, Basch H (1972) Molecular Wavefunctions and Properties. Wiley-Interscience, New York
46. England W, Gordon MS (1971) J. Am. Chem. Soc. 93: 4649; (1972) 94: 4818
47. Levy M (1976) J. Chem. Phys. 65: 2473
48. Sundberg KR, Bicerano J, Lipscomb WN (1979) J. Chem. Phys., 71: 1515
49. Pack RT, Byers Brown W (1966) J. Chem. Phys. 45: 556
50. Kato T (1957) Commun. Pure Appl. Math. 10: 151
51. Luken WL, Seiders BAB (1985) Chem. Phys. 92: 235
52. Kutzelnigg W (1963) Theor. Chim. Acta (Berlin) 1: 327
53. McWeeny R, Kutzelnigg W (1968) Int. J. Quantum Chem. 2: 187
54. Meyer W (1973) J. Chem. Phys. 58: 1017
55. Daudel R (1953) Compt. Rend. Acad. Sci. 237: 601
56. Daudel R, Brion H, Odoit S (1955) J. Chem. Phys. 23: 2080
57. Ludeña EV (1975) Int. J. Quantum Chem. 9: 1069
58. Edmiston C, Ruedenberg K (1963) Rev. Mod. Phys. 35: 457
59. Edmiston C, Ruedenberg K (1965) J. Chem. Phys. 43: S 97
60. Boys SF (1966) in: Löwdin P-O (ed) Quantum Theory of Atoms, Molecules and the Solid State Academic, New York, p 253

61. Boys SF (1960) Rev. Mod. Phys. 32: 306
62. Foster JM, Boys SF (1960) Rev. Mod. Phys. 32: 300
63. Magnasco V, Perico A (1967) J. Chem. Phys. 47: 971
64. Verwoerd WS (1979) Chem. Phys. 44: 151
65. Claxton TA (1980) Chem. Phys. 52: 23
66. Leonard JM, Luken WL (1982) Theor. Chim. Acta (Berlin) 62: 107
67. Leonard JM, Luken WL (1984) Int. J. Quantum Chem 25: 355
68. Löwdin P-O (1950) J. Chem. Phys. 18: 365
69. Kleier DA, Halgren TA, Hall JH, Lipscomb W (1975) J. Chem. Phys. 61: 3905
70. Walker FH, Wiberg KB, Michl J (1982) J. Amer. Chem. Soc., 104: 2056
71. Wiberg KB, Walker FH (1982) J. Amer. Chem. Soc., 104: 5239
72. Michl J, Radziszewski GJ, Downing JW, Wiberg KB, Miller RD, Kovacic P, Jawdosiuk M, Bonacic-Koutecky V (1983) Pure Appl. Chem. 55: 315
73. Stohrer WD, Hoffmann R (1972) J. Amer. Chem. Soc., 94: 779
74. Newton MD, Schulman JM (1973) J. Amer. Chem. Soc., 94: 773
75. Wiberg KB (1983) J. Amer. Chem. Soc., 105: 1227
76. Wiberg KB, Wendoloski JJ (1982) J. Amer. Chem. Soc., 104: 5679
77. Jackson JE, Allen LC (1984) J. Amer. Chem. Soc. 106: 591
78. Vuckovic Lj, Vujisic Lj (1984) Croat. Chem. Acta, 57: 801

The Diatomics-in-Molecules Method and the Chemical Bond

P. J. Kuntz
Hahn-Meitner-Institut Berlin GmbH; Glienicker Straße 100, D-1000 Berlin 39

The diatomics-in-molecules (DIM) method leads to a picture of chemical bonding which resembles valence bond (VB) theory to a large extent. While VB and molecular orbital (MO) theories are phrased in terms of electronic structure directly, DIM uses atoms and diatomic fragments as the fundamental units, expressing perhaps more naturally than other methods the concept of bond transferability so essential to the chemist's view of bonding and chemical reaction. The DIM method is formulated here in a general way and is related to the London formula and the LEPS approach, which for years have been the bedrock for theories of the dynamics of chemical reaction. The familiar concepts of bonding — valence, directionality, bond polarity, charge-transfer, and resonance — are discussed from the DIM point of view. The differences from the usual VB theory are pointed out. By way of example, the bonding in small molecules and in ionic clusters is discussed in detail. Applications to organic chemistry and surface chemistry are mentioned. The use of bonding concepts in the design of DIM models is discussed in general and applied to a model for inelastic scattering of OH^- ions from argon.

1 Introduction

The method of diatomics-in-molecules (DIM) [1] is a little over 25 years old and, although applied to many problems in chemistry [2–208], it has only occasionally contributed to disucssions of the chemical bond. This may at first sight seem strange, since DIM is based essentially on valence bond (VB) theory, which is the chemist's traditional method of picturing the electronic structure of molecules and chemical bonding. Upon further reflection, however, there are several factors contributing to the neglect of DIM in this regard. Introduced at a time when largescale ab initio calculations were just becoming feasible, DIM bore the stigma of semi-empiricism. Quite possibly, the similarity in name to "Atoms-in-molecules" (AIM) [209], which had suffered several set-backs [210, 211], may have accentuated this distaste. (DIM in fact is only superficially related to AIM, although the GDIM [83, 94, 102, 103, 151] formulation mixes both methods; AIM is also well-suited to supplying the diatomic fragment information needed as input to DIM [127, 152, 207].) In the early papers, the DIM models were derived from very small basis sets and were often restricted to special geometrical configurations; they lacked the necessary breadth for a discussion of bonding in general.

More recent papers exploit the method's ability to model efficiently the potential energy surfaces (PES) of both ground and excited states over a large range of configuration space, a feature of paramount importance for studies of chemical dynamics and reaction kinetics. Unfortunately, this pragmatic approach tends to ignore the concepts of bonding: DIM is viewed simply as one of the tools the theoretician has at his disposal for representing potential surfaces rather than as a theoretical construct capable of describing chemical bonding or, at the very least, informed by concepts related to bonding.

It would seem, however, that there can be no essential objection to a discussion of bonding in terms of DIM. In fact, the method has a unique vantage point in this respect — it expresses the properties of polyatomic molecules in terms of fragment atoms and diatomic molecules instead of electrons and nuclei. It is therefore closely related to what chemists do and have done for years. Indeed, chemistry would be a hopeless morass of unrelated facts if molecules could not be considered as made up of functional groups having typical, transferable properties. Thus, the design of DIM models must respect the nature of the bonding both in the polyatomic molecule and in the diatomic fragments comprising it — the electronic structure of the fragments cannot be ignored. It is essential to a good DIM model that it *start with* some concept of chemical bonding; only then is it possible to draw conclusions from DIM about the bonding in the polyatomic.

The consistent treatment of diatomics and polyatomics in the above sense forms the main thrust of this article, which emphasises the effects of spin-coupling, spatial symmetry, and charge-transfer — three aspects determining the nature of the bonding in DIM. Some simple molecules will be treated in order to illustrate these points separately and then a few examples will show how the method describes the bonding in more complex molecules or in extended systems such as clusters and surfaces. Finally, the bonding concepts inherent in the DIM formulation will be used to suggest a plausible model for a dynamics calculation, thus illustrating the application of the ideas discussed in this article.

2 Early DIM: The London Formula

2.1 Historical Background

A fore-runner of diatomics-in-molecules, the LEP (London-Eyring-Polanyi) formula is as old as reaction dynamics itself [108]. This formula, and the Sato modification of it (LEPS [4–6, 214]), is widely used in the study of reaction dynamics to represent the potential energy surface of a transfer reaction involving atoms in doublet states:

$$\text{A (doublet)} + \text{BC (singlet)} \rightarrow \text{AB (singlet)} + \text{C (doublet)}\,.$$

The basic formula was first given by F. London [212] for the hydrogen atom transfer reaction (A = B = C ≡ H). The standard text-book derivation of this formula [213, 214] starts with a valence-bond (VB) minimum basis set model for H_3:

$$\psi_1^{VB} = g(|\bar{a}bc| - |a\bar{b}c|)$$

$$\psi_2^{VB} = g(|a\bar{b}c| - |ab\bar{c}|)\,,$$

where $g \equiv \frac{1}{\sqrt{2}}$ and a, b, c are 1s orbitals on the centres A, B, and C respectively. Many approximations are made to the integrals arising out of the matrix elements of the triatomic hamiltonian with these basis functions and, after much algebra, the London formula emerges.

One year after his original paper on DIM [1], Ellison [4] demonstrated that the London formula, as applied in the LEPS manner [5, 6], is equivalent to a DIM treatment with one ^2S state on each H-atom — the same basis used in the VB derivation.

2.2 A generalised London Equation from DIM

The DIM derivation of the London formula is straight-forward and compact, a great improvement on the usual derivation. This economy brings with it a greater insight and, indeed, allows generalisations to other simple systems. For example, a formula for triatomic systems involving atoms in S-states is [215]

$$E = Q - J\,,$$

with

$$Q = Q_{AB} + Q_{BC} + Q_{AC}\,,$$

and

$$J(n) = \sqrt{J_{AB}^2 + J_{BC}^2 + J_{AC}^2 - \frac{2}{n} J_{BC}(J_{AB} + J_{AC}) - \frac{2}{n}\left(n - \frac{2}{n}\right) J_{AB}J_{AC}}\,.$$

Here, E, Q, and J each depend on the three internuclear distances, (R_{AB}, R_{BC}, R_{AC}); the quantities Q_{XY} and J_{XY} depend only on the internuclear distance R_{XY}, as do S_{XY} and T_{XY}:

$$Q_{XY} = \frac{1}{2}(S_{XY} + T_{XY})$$

$$J_{XY} = \frac{1}{2}(S_{XY} - T_{XY})$$

for XY = AB, BC, and AC. Here, S and T are interaction energies for particular states of the XY diatomic fragments. This formula can be used to describe either the doublet potential energy surfaces arising from the atomic states $A(^2S)$, $B(^nS)$, $C(^nS)$ or the surface with multiplicity n arising from $A(^nS)$, $B(^2S)$, $C(^2S)$. In either case, S_{AB} and T_{AB} refer to interactions in the diatomic states with multiplicity $(n - 1)$ and $(n + 1)$ respectively, with analogous meanings for S_{AC} and T_{AC}. S_{BC} and T_{BC} refer to the singlet and triplet states of BC. The original London formula is recovered by setting $n = 2$. The reaction

$$N(^4S_u) + H_2 \rightarrow NH(^3\Sigma^+) + H(^2S_g)$$

could, for example, be described by putting $n = 4$.

The London formula is one of the simplest expressions for a PES obtainable by DIM and it is a good illustration of how the method leads to non-additive threebody interactions from diatomic fragment input — more of that later. From the point of view of chemical bonding, the formula was a great triumph, for it gave a good qualitative picture of the notion that the new bonds are formed at the same times as the old ones break, explaining the fact that the barrier to reaction can be much smaller than the energy required to break one of the diatomic bonds. The DIM derivation of this formula leads at once to the LEP(S) formulation utilising the singlet and triplet diatomic interactions. Normally, the best available potential curves should be used to obtain these interactions.

2.3 The LEPS Formula as a Parallel to the DIM Procedure

In the semi-empirical application of LEPS some of the fragment curves are allowed to vary in order to manipulate the shape of the triatomic PES. (Note that variation of the Sato parameters in the LEPS approach is equivalent to varying the interaction strength of the triplet curves.) Also, the formula is routinely applied to atoms in states other than S-states, allowing a much wider range of application. This is in fact strikingly parallel to the way in which DIM is often applied to more complex cases:

A model is defined through the choice of a suitable VB polyatomic basis set (not too large).

The hamiltonian matrix elements in this basis are approximated by sums of

atomic energies and diatomic fragment interaction energies, suitably weighted.
The eigenvalues of this matrix as a function of geometrical configuration are taken as the polyatomic potential energy surfaces.
Variation of some of the diatomic fragment input allows manipulation of the polyatomic PES.

In the last 20 years, the LEPS method has had countless applications to kinetics and chemical dynamics. DIM has just as great a potential, for it encompasses all that LEPS can do and can be applied at a better level of approximation to a much wider class of molecules, at the same time taking into account the nature of the chemical bond.

3 Formulation of DIM

3.1 General Exposition

There are several ways of presenting the theory. Ellison's original paper [1] is, in my opinion, still the definitive statement of the method and is essential reading for anyone intending to apply the method. Others have improved the presentation somewhat and generalised the method but the original paper remains the best introduction. Kuntz and Roach [23] and Tully [25, 27, 62, 98] present a more compact derivation in matrix notation and, in addition, explicitly include atomic functions other than S-states in arbitrary geometrical configurations. Steiner et al. [28] and Kuntz [82] develop a systematic scheme for writing down the polyatomic matrix elements directly — this is particularly appropriate for programming by computer and is the approach followed here. Eaker and Parr [52, 61] emphasise the spin-coupling aspect of the problem, as does the elegant paper by Pickup [26]. Faist and Muckerman [90, 91] recast the theory in terms of projection operators. Roach and Gimzewski [96], applying the method to a π-electron theory of conjugated hydrocarbons, develop rules for generating the matrix elements for this case; Radlein [135] and Olson and Garrison [147] present similar rules.

DIM distinguishes itself from other approximate quantum mechanical methods in that no calculation of integrals is necessary in the evaluation of the hamiltonian matrix elements of the polyatomic molecule, for these are expressed entirely in terms of the matrix elements of diatomic and atomic hamiltonians in a corresponding basis. This is possible because both the polyatomic and diatomic bases are constructed from a common set of composite functions which are anti-symmetrised products of atomic functions:

$$\mathrm{X}_M = \mathscr{A} \prod_{i=1}^{\mathrm{N}} \varphi_{k_i}^{\alpha_i},$$

for $M = 1, N_X$, the number of polyatomic composite basis functions. $\mathscr{A}$ is the antisymmetriser for all the electrons, i labels the N atomic centres, M the composite basis functions, k_i the atomic functions on centre i, and α_i the chemical species on the centre i (e.g. H, O^+, etc.). Note that α_i and k_i depend on the index of the composite, M, and that the superscript α_i implicitly fixes the number of electrons associated with centre i, n_{α_i}, for composite function M. The hamiltonian matrix in the composite basis, $\boldsymbol{B}_X$, can be expressed in terms of diatomic and atomic fragment matrices in the related fragment composite basis by utilising the (exact) partitioning of the polyatomic hamiltonian operator:

$$\check{H} = \sum_{i=1}^{N-1} \sum_{j=i+1}^{N} \check{H}_{ij} - (N-2) \sum_{i=1}^{N} \check{H}_i .$$

The $M'M$ matrix element of $\boldsymbol{B}_X (= \langle X_{M'} | \check{H} | X_M \rangle)$ is expressed as a sum of fragment terms by applying the partitioning of the polyatomic hamiltonian appropriate to the electron distribution implied by the index M, (which specifies the superscripts α_i and thereby the number of electrons associated with centre i, n_{α_i}):

$$\langle X_{M'} | \check{H} | X_M \rangle = \langle X_{M'} | \sum_{i=1}^{N-1} \sum_{j=i+1}^{N} \check{H}_{ij} - (N-2) \sum_{i=1}^{N} \check{H}_i | X_M \rangle .$$

Ellison [1] shows that the term involving fragment ij is equivalent to $\langle X_{M'} | \mathscr{A}_{ij}^{(ij)} X_M^{ij} \check{H}_{ij} \chi_m^{ij} \rangle$, where

$$\chi_m^{ij} \equiv \mathscr{A}^{(ij)} \varphi_{k_i}^{\alpha_i}(m)\, \varphi_{k_j}^{\alpha_j}(m)$$

and

$$X_M^{ij} \equiv \mathscr{A}_{ij} \prod_{\substack{l=1 \\ l \neq i,j}}^{N} \varphi_{k_l}^{\alpha_l}(M) .$$

$\mathscr{A}^{(ij)}$ is the antisymmetriser for the electrons of fragment ij, $\mathscr{A}_{ij}$ the antisymmetriser for the remaining electrons, and $\mathscr{A}_{ij}^{(ij)}$ the anti-symmetriser involving the permutations between electrons in the fragment ij and the rest of the molecule. The χ_m^{ij} are the composite basis functions for the diatomic species $\alpha_i \alpha_j$. Again, n_{α_i} and k_i depend on the *diatomic* composite index, m, which in turn depends on M. The main approximation in DIM is that the composites $\{\chi_m^{ij}\}$ form a *complete* set, so that $\check{H}_{ij} \chi_m^{ij}$ may be replaced by a linear combination of the functions χ:

$$\check{H}_{ij} \chi_m^{ij} = \sum_{m''} \chi_{m''}^{ij} b_\chi^{ij}(m'', m) ,$$

where the coefficients are the hamiltonian matrix elements in the diatomic composite basis:

$$b_\chi^{ij}(m'', m) = \langle \chi_{m''}^{ij} | \check{H}_{ij} | \chi_m^{ij} \rangle .$$

On substituting this approximation into the original fragment matrix element, and noting that $\mathscr{A}^{(ij)}_{ij} X^{ij}_{M} \chi^{ij}_{m''}$ must correspond to one of the polyatomic basis functions, $X_{M''}$, (since they result from a direct product of atomic functions), we obtain

$$\langle X_{M'} | \check{H}_{ij} | X_{M} \rangle \rightarrow \sum_{m''} \langle X_{M'} | X_{M''(m'')} \rangle \, b^{ij}_{\chi}(m'', m) \, P^{ij}(M', M'') .$$

$P^{ij}(M', M'') = (-)^{\Delta e \Delta q}$, where Δe is the number of electrons associated with the functions $\varphi^{\alpha_l}_{k_l}(M')$ between centres i and j ($i < l < j$) and Δq is the difference in the number of electrons associated with functions $\varphi^{\alpha_i}_{k_i}(M')$ and $\varphi^{\alpha_i}_{k_i}(M'')$; this factor is therefore 1 unless there is charge-transfer between centres i and j associated with the basis functions indexed by M' and M''. The functions φ are chosen to be orthogonal when all centres are infinitely separated from each other:

$$\langle \varphi^{\alpha_i}_{k_i} | \varphi^{\alpha_j}_{k_j} \rangle = \delta_{\alpha_i \alpha_j} \delta_{k_i k_j} .$$

It is customary to assume that this relation also holds for finite distances between the centres; this is called the no-overlap approximation. Since each $X_{M''}$ $(= \mathscr{A}^{(ij)}_{ij} X^{ij}_{M} \chi^{ij}_{m''})$ is constructed from X^{ij}_{M}, there are no matrix elements between polyatomic composites for which $X^{ij}_{M'}$ differs from X^{ij}_{M}. Let

$$\Delta^{ij}_{M'M} = \prod_{\substack{l=1 \\ l \neq i, j}}^{N} \delta_{\alpha_l \alpha'_l} \delta_{k_l k'_l}$$

and

$$\delta^{ij}_{M'M''} = \delta_{\alpha'_i \alpha''_i} \delta_{\alpha'_j \alpha''_j} \delta_{k'_i k''_i} \delta_{k'_j k''_j} .$$

The fragment matrix element then becomes

$$B^{ij}_{X}(M', M) = \Delta^{ij}_{M'M} \sum_{m''} \delta^{ij}_{M'M''} b^{ij}_{\chi}(m'', m) \, P^{ij}(M', M'') ,$$

where the index M'' depends upon m''. The atomic matrices are

$$B^{i}_{X}(M', M) = \Delta^{i}_{M'M} \sum_{m''} \delta_{\alpha'_i \alpha''_i} \delta_{k'_i k''_i} b^{i}_{\varphi}(m'', m) ,$$

where

$$\Delta^{i}_{M'M} = \prod_{\substack{l=1 \\ l \neq i}}^{N} \delta_{\alpha_l \alpha'_l} \delta_{k_l k'_l} .$$

It therefore follows that a knowledge of the atomic fragment matrices, $\boldsymbol{b}^{\alpha_i}_{\varphi}$, and diatomic fragment matrices, $\boldsymbol{b}^{\alpha_{ij}}_{\chi}$, suffices to construct the polyatomic hamiltonian matrix:

$$\boldsymbol{B}_{X} = \sum_{i=1}^{N-1} \sum_{j=i+1}^{N} \boldsymbol{B}^{ij}_{X} - (N-2) \sum_{i=1}^{N} \boldsymbol{B}^{i}_{X} .$$

This is the working equation for DIM. It is analogous to the partitioning of the hamiltonian but is not exact. A DIM model is constructed by specifying the basis functions $\boldsymbol{\varphi}^{\alpha_i}$ for each centre i and then constructing the diatomic and polyatomic bases as direct products of these. The DIM hamiltonian matrix is then constructed.

There are two considerations which complicate the above procedure: 1) the bases $\boldsymbol{\chi}$ and $\boldsymbol{X}$ are not very convenient — one would like to exploit the symmetry (spin and spatial) of the polyatomic and diatomic molecules; 2) the polyatomic basis is defined relative to a fixed coordinate system, so that the bond axis of the diatomic fragments does not lie along the z-axis of the polyatomic but rather takes up a direction specified by the angles ω, the rotations required to bring the z-axis of the polyatomic coordinate system into coincidence with the internuclear distance vector, $\boldsymbol{r}_{ij}$. This means that the fragment matrices $\boldsymbol{b}_{\chi}^{\alpha_{ij}}$ need to be calculated so as to correspond to the geometrical configuration of the polyatomic molecule. Normally, the diatomic hamiltonian matrices, $\boldsymbol{b}_{\chi'}$, are defined in a basis $\boldsymbol{\chi}'$ in which the bond axis of the molecule is the z-axis ($\omega = 0$). The two bases are related by a rotation:

$$\boldsymbol{\chi}' = \boldsymbol{\chi}\boldsymbol{R}(\omega)\,,$$

and the relation between the two types of diatomic fragment matrix is

$$\boldsymbol{b}_{\chi} = \boldsymbol{R}\boldsymbol{b}_{\chi'}\tilde{\boldsymbol{R}}\,.$$

The functions $\{\varphi^{\alpha}\}$ are chosen such that the rotation matrices $\boldsymbol{R}$ have real elements.

It is convenient to transform the polyatomic basis, $\boldsymbol{X}$, so that the individual functions have a particular spin and spatial symmetry (it is assumed that the direct product basis set, $\{\boldsymbol{\varphi}^{\alpha} \otimes \boldsymbol{\varphi}^{\beta} \otimes \ldots\}$, furnishes enough functions to do this. Denoting these symmetry functions by $\boldsymbol{\Psi}$, we have

$$\boldsymbol{\Psi} = \boldsymbol{X}\boldsymbol{T}$$

$$\boldsymbol{B}_{\Psi} = \tilde{\boldsymbol{T}}\boldsymbol{B}_{X}\boldsymbol{T}.$$

Similarily, the diatomic basis can be transformed into symmetry functions:

$$\boldsymbol{\psi}^{\alpha_{ij}} = \boldsymbol{\chi}^{\alpha_{ij}}\boldsymbol{t}^{\alpha_{ij}}$$

$$\boldsymbol{b}_{\chi}^{\alpha_{ij}} = \boldsymbol{t}^{\alpha_{ij}}\boldsymbol{b}_{\psi}^{\alpha_{ij}}\tilde{\boldsymbol{t}}^{\alpha_{ij}}\,.$$

The equations presented here suffice to construct the polyatomic hamiltonian in the symmetry-adapted basis, $\boldsymbol{B}_{\Psi}$, from the diatomic and atomic fragment matrices, $\boldsymbol{b}_{\psi}^{\alpha_{ij}}$ and $\boldsymbol{b}_{\varphi}^{\alpha_i}$ respectively. The latter constitute the numerical input to the DIM method which, in addition to the mappings:

$$\boldsymbol{B}_{X}^{ij} \leftarrow \boldsymbol{b}_{\psi}^{\alpha_{ij}}$$

$$\boldsymbol{B}_{X}^{i} \leftarrow \boldsymbol{b}_{\varphi}^{\alpha_i}\,,$$

specifies $\boldsymbol{B}_{\Psi}$, which is diagonalised at each desired geometrical configuration to obtain the potential energy functions of the polyatomic molecule:

$$\boldsymbol{BC} = \boldsymbol{CE} \, .$$

i.e. the diagonal eigenvalue matrix, $\boldsymbol{E}$, is determined by diagonalising $\boldsymbol{B}$:

$$\boldsymbol{E} = \tilde{\boldsymbol{C}}\boldsymbol{BC}$$

$$\tilde{\boldsymbol{C}} = \boldsymbol{C}^{-1} \, .$$

The formulation here may appear somewhat complicated but is well-suited for programming the DIM method for the computer. This has in fact been done for the general case. There are two separate programs: the master program and the potential energy subroutines. The latter program can be appended to some main program (e.g. a trajectory program) which requires information about the PES (e.g. potential energy, derivatives, non-adiabatic coupling, etc.) for a given geometrical configuration. These subroutines require two data files in order to function: the B-file, containing the diatomic and atomic fragment matrices, $\boldsymbol{b}_{\Psi}^{\alpha_{ij}}$ and $\boldsymbol{b}_{\varphi}^{\alpha_i}$, respectively, and the D-file, containing the information for the construction of $\boldsymbol{B}_{\Psi}$ from the fragment matrices. The B-file consists of the individual matrix elements of the fragments at selected values of the internuclear distance, r; in addition, it contains information about the long-range behaviour of these elements (for $r \to \infty$) and about the energy of the various atomic states. The D-file contains the mappings and the transformation matrix $\boldsymbol{T}$ so that $\boldsymbol{B}_{\Psi}$ can be constructed from the information in the B-file. The B-file must be constructed laboriously from valence bond (or MO) calculations and semi-empirical data; the D-file is constructed in a few seconds from the master program.

The master program is made up of two principal sections: the model designer and the D-file generator. The model designer is an interactive program which produces the list of polyatomic composite functions, $\boldsymbol{X}$, which define the DIM model structure. The basic input to this program is kept as short as possible and consists of the attributes of the polyatomic molecule (no. of centres, chemical element on each centre, total charge, and spin symmetry) and the specification of the spin and spatial symmetries of the atomic functions, $\boldsymbol{\varphi}^{\alpha_i}$, which define the DIM basis. If no restrictions are given, the designer proceeds to construct the list of atomic symmetries on each centre from which the DIM polyatomic basis will be constructed; from this list of "state-groups", the direct product composite basis functions, $\boldsymbol{X}$, for the polyatomic molecule are generated (these are very near to VB structures, but without spin-coupling).

As an example, consider the molecule H_2O in the $^1A'$ state. Allowing the atomic symmetries $O(^1D_g)$, $O(^3P_g)$, $O^-(^2P_u)$, $O^{--}(^1S_g)$, $H(^2S_g)$, and $H^+(^1S_g)$, one obtains the following state-groups:

1. $O(^3P_g)\ H(^2S_g)\ H(^2S_g))$
2. $O(^1D_g)\ H(^2S_g)\ H(^2S_g))$
3. $O^-(^2P_u)\ H(^2S_g)\ H^+(^1S_g))$
4. $O^-(^2P_u)\ H^+(^1S_g)\ H(^2S_g))$
5. $O^{--}(^1S_g)\ H^+(^1S_g)\ H^+(^1S_g))$

These state-groups yield 1, 3, 2, 2, and 1 (i.e. a total of 9) polyatomic functions $\boldsymbol{\Psi}$, respectively.

Since the DIM model is ultimately destined for use in a dynamical calculation, where the subprograms will be called many times, it is very important to restrict the size of the basis set as much as possible. The designer allows for three basic restrictions: 1) in the distribution of electrons among the centres, 2) in the number of spin-coupled functions, and 3) in the number of functions in the state-group list. Implementation of these restrictions requires some previous knowledge of the polyatomic wave function for the states of interest. For example, it is known from VB calculations [216] that only a small subset of spin-couplings need be taken into account for many molecules; this reduces the size of the DIM model enormously. Bonding concepts are a big help in suggesting the nature of these restrictions.

The program to generate the D-file starts from the list of polyatomic composite functions. First, it generates the list of diatomic fragments for which information is needed. Then, for each such fragment, it makes a list of diatomic composites and the functions $\psi^{\alpha ij}$ which correspond to them, putting this information into a skeleton B-file, which must later be fleshed out (with the addition of the matrix elements referred to above) by the user. The D-file itself is then written out; it consists of the specific algebraic formulae for the elements of the polyatomic hamiltonian, $\boldsymbol{B}_{\Psi}$, in terms of the diatomic fragment elements and rotation matrix elements. The generating program removes much of the labour associated with a DIM calculation. The master program performs a great deal of matrix algebra to arrive at its results, but it should be borne in mind that this is done only once; the formula implicit in the D-file is very direct, ensuring that the DIM hamiltonian matrix can later be computed with a minimum of operations.

3.2 An Example: H_3

The above rather complicated exposition can be elucidated by applying it to H_3 using the minimum basis set of one ^2S function on each centre. No charge-transfer and no rotations are involved. The functions corresponding to spin quantum number $M_S = \frac{1}{2}$ are denoted by a, b, and c; those with $M_S = -\frac{1}{2}$ by $\bar{a}$, $\bar{b}$, and $\bar{c}$. From these, four functions for each diatomic fragment and three functions for the triatomic can be constructed (see Tables 1a to 1c). The fragment matrices are constructed from the eigenvectors $\boldsymbol{c}^{ij}$ and eigenvalues $\boldsymbol{e}^{ij}$:

$$\boldsymbol{b} = \boldsymbol{c}\boldsymbol{e}\boldsymbol{c}^{-1} .$$

Table 1a. Atomic Basis for DIM H_3

Function index k	Centre 1	Centre 2	Centre 3
1	a	b	c
2	$\bar{a}$	$\bar{b}$	$\bar{c}$

Table 1b. Diatomic Basis for DIM H_3: Centres 1 and 3

m_{13}	k_1	k_3	χ_m^{13}	c^{13}				b^{13}			
1	1	1	$\|ac\|$	1	0	0	0	T	0	0	0
2	2	1	$\|\bar{a}c\|$	0	$\frac{1}{\sqrt{2}}$	$\frac{1}{\sqrt{2}}$	0	0	Q	$-J$	0
3	1	2	$\|a\bar{c}\|$	0	$-\frac{1}{\sqrt{2}}$	$\frac{1}{\sqrt{2}}$	0	0	$-J$	Q	0
4	2	2	$\|\bar{a}\bar{c}\|$	0	0	0	1	0	0	0	T
				T	S	T	T				

Table 1: $Q = \frac{1}{2}(S + T)$, $J = \frac{1}{2}(S - T)$

Table 1c. Triatomic Basis and Genealogy for $H_3\left(M_S = \frac{1}{2}\right)$

$(ij)\ k \rightarrow$		(12)	3	(13)	2	(23)	1
M	X_M	m_{12}	k_3	m_{13}	k_2	m_{23}	k_1
1	$\|ab\bar{c}\|$	1	2	3	1	3	1
2	$\|a\bar{b}c\|$	3	1	1	2	2	1
3	$\|\bar{a}bc\|$	2	1	2	1	1	2

To construct the matrices B_X^{ij}, it is convenient to refer to the genealogy of the polyatomic composites (Table 1c). For example, in forming B_X^{13}, the indices m and k in the column headed (13) 2 are appropriate. It is clear that the diagonal elements (1,1), (2,2), and (3,3) correspond to the fragment matrix elements $b^{13}(3,3)$, $b^{13}(1,1)$, and $b^{13}(2,2)$ respectively; i.e. to Q, T, and Q, respectively (T = triplet state of H_2; S = singlet state). The only non-zero off-diagonal element is (1,3) because of the requirement that k_2 be the same for both polyatomic composites; this element corresponds to $b^{13}(3,2) = -J$. The other diatomic fragment matrices are constructed in an analogous fashion. The atomic matrices are diagonal:

$$B_X^i(M', M) = \delta_{M'M} E_H ,$$

where E_H is the energy of the ground state of the H-atom and can be set to zero. The DIM hamiltonian matrix in the composite basis is

$$\begin{pmatrix} T_{12} + Q_{13} + Q_{23} & -J_{23} & -J_{13} \\ -J_{23} & T_{13} + Q_{12} + Q_{23} & -J_{12} \\ -J_{13} & -J_{12} & T_{23} + Q_{13} + Q_{12} \end{pmatrix}$$

The eigenvalues of this matrix are the DIM energies for H_3.

A simplification is effected by transforming the basis to yield spin-coupled polyatomic functions $\boldsymbol{B} = \tilde{\boldsymbol{T}}\boldsymbol{B}_X\boldsymbol{T}$, with

$$\boldsymbol{T} = \begin{pmatrix} f & 0 & f/g \\ f & g & -gf \\ f & -g & -gf \end{pmatrix},$$

where $g = \frac{1}{\sqrt{2}}$ and $f = \frac{1}{\sqrt{3}}$. The transformed matrix blocks into a 1×1 part corresponding to the quartet state with eigenvalue $T_{12} + T_{23} + T_{13}$ and the 2×2 block for the doublet:

$$\boldsymbol{B}_{\Psi}(^2A') = \begin{pmatrix} S_{12} + Q_{23} + Q_{13} - \frac{1}{2}(J_{23} + J_{13}) & (g^2/f)(J_{13} - J_{23}) \\ (g^2/f)(J_{13} - J_{23}) & T_{12} + Q_{23} + Q_{13} + \frac{1}{2}(J_{13} + J_{23}) \end{pmatrix}.$$

The eigenvalues of this matrix are given by the London formula.

4 Concepts of Bonding and DIM

4.1 Preliminary Remarks

Since it considers neither functional groups nor orbitals as such (but see Sect. 6.5), DIM is seldom invoked in discussing the chemical bond. To understand the significance of DIM theory with respect to bonding, it must be borne in mind that DIM couches explanations in terms of *diatomic* interactions and *not* in terms of orbitals. Furthermore, because the method does not satisfy the variational principle, increasing the number of VB structures does not necessarily lower the absolute energy. Rather, the basis serves to specify a mathematical structure which, among other things, must properly describe all of the diatomic and atomic fragments implied by it. In other words, the fragments act as *restraints* on the minimisation implicit in diagonalising the hamiltonian matrix — the greater the size of the basis, the more fragments and hence the more restraints.

Each fragment contribution to the hamiltonian matrix, $\boldsymbol{B}_X^{ij}$, if separately diagonalised, must yield precisely the diatomic energies of the fragment *ij* (perhaps shifted by the atomic energies of the remaining centres). Diagonalisation of the full matrix, $\boldsymbol{B}_X$, produces energies which can be viewed as a *compromise* between the various diatomic interactions, complicated of course by the presence of many off-diagonal elements. This compromise is dictated mainly by the requirements of spin-coupling, the rotational properties of the atomic functions, and the effect of bond polarisation (ionicity). All of these act simultaneously in general but it is instructive to consider first some simple systems which exhibit each of these features one at a time.

4.2 Spin-coupling and Valence

Valence, or the limitation on the ability of a given atom to form an arbitrary number of bonds with other atoms, exhibits itself in DIM through the stringent requirements of spin-coupling. In this section, we examine several simple molecules to illustrate how this comes about. Let us begin with a cluster of Ar atoms, a system with trivial spin-coupling, since all atoms are in singlet states. In this case, only one polyatomic composite function is necessary: the product of ^1S functions from each centre. Each fragment matrix, $\boldsymbol{B}_X^{ij}$, consists of a single element, the interaction energy between two Ar atoms at the distance R; i.e. the DIM energy is simply a sum of pair potentials. In one sense, Ar could be viewed as having an infinite valence, since there is nothing in the model preventing the Ar atoms from interacting with as many neighbours as possible: there is no "bond saturation", each interaction being fully independent of the existence of all the others. Of course, there is no "chemical bond" in Ar_2, so there is no "chemical bond" in the DIM clusters either, only van-der-Waals attraction. This situation changes dramatically as soon as any sort of spin-coupling is introduced.

Again, H_3 serves as a good starting point. As seen in the preceding section, the matrix $\boldsymbol{T}$ relating $\boldsymbol{B}_X$ to $\boldsymbol{B}_\Psi$ defines spin-coupled polyatomic basis functions. Those corresponding to doublet spin are

$$\psi_1 = g(|a\bar{b}c| - |\bar{a}bc|)$$

$$\psi_2 = f[|ab\bar{c}|/g - g(|a\bar{b}c| + |\bar{a}bc|)]\,.$$

Notice that this choice of functions corresponds to first coupling atoms 1 and 2 to form either a singlet or triplet and then coupling the result to atom 3 to form an overall doublet:

$$\psi_1 = \mathscr{A}_{ij}^{(ij)}\chi_S^{12}\varphi^3$$

$$\psi_2 = \frac{1}{\sqrt{3}}\mathscr{A}_{ij}^{(ij)}(\sqrt{2}\,\chi_{T_1}^{12}\bar{\varphi}^3 - \chi_{T_0}^{12}\varphi^3)\,,$$

where T_1 and T_0 refer to the triplet functions with spin-quantum number $M_S = 1$ and 0 respectively. In accordance with the linear algebra implied by the DIM assumptions, ψ_1 and ψ_2 are eigenvectors of the fragment hamiltonian $\check{H}_{12}$ in the sense that $\check{H}_{12}\psi_1 = S\psi_1$ and $\check{H}_{12}\psi_2 = T\psi_2$. i.e. the optimal situation from the point of view of the fragment 12 would be to have a (singlet) pair-bond between atoms 1 and 2: H_1—H_2.

The same argument would, of course, hold for fragments 13 and 23, since the basis $\boldsymbol{\Psi}$ could have been chosen so as to couple one of these pairs first. It is, however, impossible to arrange for a pair-bond in all three fragments *simultaneously*, for suppose the triatomic wave function were in fact equal to ψ_1, meaning there would be an (optimal) singlet interaction in the 12 fragment, then fragments 23 and 13 would of necessity involve a *mixture* of singlet and triplet interactions. This can be seen by considering the expectation value of the operator $\check{H}_{23}$ (for example) with ψ_1, which can be read from element 11 of $\boldsymbol{B}_\Psi$ from the previous section:

$$\langle\psi_1|\,\check{H}_{23}\,|\psi_1\rangle = Q_{23} - \frac{1}{2}J_{23} = \frac{1}{4}(S_{23} + 3T_{23})\,.$$

i.e. there is mainly *triplet* interaction between atoms 2 and 3, and since the triplet curve is repulsive, we may say that optimal bonding in fragment 12 implies *antibonding* in fragment 23. Hence, ignoring the 13 bond, ψ_1 could be represented by the symbol

$$H_1 - H_2 \cdot\!\!-\!\!\cdot H_3\,.$$

On the other hand, the function ψ_2, for which there is anti-bonding (T_{12}) in the 12 fragment, exhibits more bonding in the 23 fragment, for

$$\langle\psi_2|\,H_{23}\,|\psi_2\rangle = Q_{23} + \frac{1}{2}J_{23} = \frac{1}{4}(3S_{23} + T_{23})\,,$$

so that ψ_2 can be repesented by

$$H_1 \cdots H_2 -\!\cdot\!- H_3\,.$$

Clearly, sacrificing some of the bonding in the 12 fragment greatly reduces the repulsion in the 13 and 23 fragments, so that a compromise between ψ_1 and ψ_2 produces the lowest energy.

For linear symmetrical H_3, the wave function must have either g- or u-character. If atom 3 is taken to be in the middle, ψ_2 has u-character and hence corresponds to the ground-state with energy $T_{12} + \frac{1}{2}(3S + T)$, where S and T are the singlet and triplet interactions in the 13 and 23 fragments. The transition state may therefore be represented by

$$\cdot H_1 -\!\cdot\!- H_3 -\!\cdot\!- H_2\,\cdot\,.$$

Thus, DIM displays the concept of valence in a strikingly direct manner and shows how the bonding in the transition state is distributed between the nearest neighbours and the anti-bonding between the most distant atoms.

Summarising, the concept of valence emerges in DIM because of two conditions: 1) structures which lead predominantly to one type of spin-coupling in a particular fragment pair imply a definite spin-coupling in the other fragments. 2) In general, not all the spin-coupling are associated with bonding. The second point is very important: were both the singlet and triplet interactions of a bonding nature, there would be no need to strike a compromise; H_3 would be a very strongly bound molecule.

V-valent atoms can be modelled by including at least one basis function with a spin multiplicity ($V + 1$). The Be atom, for example, would be divalent if a triplet function were associated with it. The ground state is a singlet arising from the $1s^2 2s^2$ electron configuration but there are low-lying ^3P_u and ^1P_u states from the $1s^2 2s2p$ configuration; in practice, all three of these states would have to be included in the DIM basis, since they lie so close together. As valency increases, DIM models become more complex and the basis much too large. It is then better to view the multi-valent atom as being in some "valence state" rather than being associated with definite spectroscopic states.

The tetravalent carbon atom could be modelled by a quintet "valence" state. A trivial 1×1 model results from combining $C(^5S_u)$ with four H-atoms in ^2S states. The H-atoms interact with each other via the triplet state of H_2 and the C—H interaction formally corresponds to a $^4\Sigma^-$ state. In contrast to the H_3 case, the C-atom can bond simultaneously with four H-Atoms; the energy is just the sum of the pair interactions: $4D_{C-H} + 6T$. Since all the H-atoms are equivalent, this model would lead to tetrahedral methane, provided that the $^4\Sigma^-$ interaction were powerful enough to offset the repulsive contribution, $6T$. In such an oversimplified model, the interactions would best be treated as effective ones in order to yield a good qualitative representation of the molecules CH_4, CH_3 + H, (CH_3 would be planar in this approximation) and, with some luck, $CH_2(^3\Sigma_g^-)$ + H + H. Note that this model would be incapable of describing singlet methylene, or triplet methylene and a hydrogen *molecule* simultaneously.

Although this model is very primitive, it could be extended to apply to the reaction

$$H + CH_4 \rightarrow H \cdots (CH_3) \cdots H \rightarrow CH_4 + H .$$

Addition of the extra hydrogen atom produces a 5×5 model with the additional fragment interactions $H_2(^1\Sigma_g^+)$ and $CH(^6\Sigma^-)$. To give a consistent picture, the former must be bonding and the latter anti-bonding. Such an augmented model could also cope with the abstraction reaction

$$H + CH_4 \rightarrow H_2 + CH_3 .$$

This product channel could also be avoided, if that were desired, by restricting the H-atoms to triplet interactions only; a simpler, 2×2 model results, which, for dynamical studies, might be preferred from the computational point of view. In any case, even such simple models are capable of reproducing bond saturation and of ensuring the full equivalence of like elements in a molecule.

4.3 Directionality

So far, only functions with S-type symmetry on each centre have been considered. Functions with P, D, F, . . . symmetry complicate the DIM procedure in that their behaviour under rotation must be considered. The fragment matrices refer to a basis fixed with respect to the bond axis of the diatomic molecule (the local z-axis). We have already seen that in general a rotation is required to bring the diatomic

Table 2: $S \equiv AH(^2\Sigma^+)$ interaction, $P \equiv AH(^2\Pi)$ interaction. Q and J as in Table 1

Table 2a. DIM Atomic Basis for $A(^3P)\ H_2\ {}^1A'$

Function index k	Centre 1	Centre 2	Centre 3
1	z_0	b	c
2	$\bar{z}$	$\bar{b}$	$\bar{c}$
3	z		
4	x_0		
5	$\bar{x}$		
6	x		

Table 2b. DIM Diatomic Basis for $A(^3P)H_2\ {}^1A'$: Centres 1 and 2

Molecule	m_{12}	k_1	k_2	χ_m^{12}	b^{12}			
AH	1	2	1	$\|\bar{z}b\|$	$\frac{2}{3}S$	$-\frac{\sqrt{2}}{3}S$	0	0
	2	1	2	$\|z_0\bar{b}\|$	$-\frac{\sqrt{2}}{3}S$	$\frac{1}{3}S$	0	0
	3	5	1	$\|\bar{x}b\|$	0	0	$\frac{2}{3}P$	$-\frac{\sqrt{2}}{3}P$
	4	4	2	$\|x_0\bar{b}\|$	0	0	$-\frac{\sqrt{2}}{3}P$	$\frac{1}{3}P$
HH	1	2	2	$\|\bar{b}\bar{c}\|$	T	0	0	
	2	2	1	$\|\bar{b}c\|$	0	Q	$-J$	
	3	1	2	$\|b\bar{c}\|$	0	$-J$	Q	

Table 2c. Triatomic Basis and Genealogy for AH_2

$(ij)\ k \rightarrow$		(12)	3	(13)	2	(23)	1
M	X_M	m_{12}	k_3	m_{13}	k_2	m_{23}	k_1
1	$\|\bar{z}bc\|$	1	1	1	1	1	2
2	$\|z_0\bar{b}c\|$	2	1	−2	2	2	1
3	$\|z_0b\bar{c}\|$	−2	2	2	1	3	1
4	$\|z\bar{b}\bar{c}\|$	−1	2	−1	2	1	3
5	$\|\bar{x}bc\|$	3	1	3	1	1	5
6	$\|x_0\bar{b}c\|$	4	1	−4	2	2	4
7	$\|x_0b\bar{c}\|$	−4	2	4	1	3	4
8	$\|x\bar{b}\bar{c}\|$	−3	2	−3	2	1	6

coordinate system into coincidence with the polyatomic system. This results in fragment matrices (in the diatomic composite basis) which depend on the orientation of the diatomic with the polyatomic.

As an example, consider the molecule AH_2 for the case that A is in a ^3P_u state and the two hydrogen atoms are in ^2S_g states. Adopting the convention that atom A lies in the origin, B on the positive z-axis, and C in the positive xz-plane leads to three pairs of atomic functions on centre A: (z_0, x_0) for $M_S = 0$, (z, x) for

$M_S = 1$, and $(\bar{z}, \bar{x})$ for $M_S = -1$. In each pair, one function transforms (under rotation) as z and the other as x. The basis functions are shown in Tables 2a to 2c along with the corresponding $\boldsymbol{b}$-matrix elements; T is the interaction in the triplet state of H_2, S the AH($^2\Sigma^+$) interaction, and P the AH($^2\Pi$) interaction. Q and J are defined as in Table 1. The construction of the matrices $\boldsymbol{B}_X^{12}$ and $\boldsymbol{B}_X^{23}$ follows in a straight-forward manner from the genealogy table 2c. For example, the element $B^{12}(3, 4) = (-)^2 b^{AH}(2, 1)$, which has the value $-\frac{\sqrt{2}}{3} S$; the element $B^{23}(6, 7) = b^{HH}(2, 3)$, which is equal to $-J$. In forming $\boldsymbol{B}^{13}$, however, it is necessary to use a matrix $\boldsymbol{b}^{AH}(\beta)$ which has been transformed by a rotation:

$$\boldsymbol{b}(\beta) = \boldsymbol{R}(\beta)\, \boldsymbol{b} \boldsymbol{R}^{-1}(\beta)\,.$$

The angle β is obtained from the coordinates (x, z) of atom 3: $\beta = \cos^{-1}(z/R_{13})$, where $R_{13} = \sqrt{x^2 + z^2}$. For this case,

$$\boldsymbol{R}(\beta) = \begin{pmatrix} \cos\beta & 0 & -\sin\beta & 0 \\ 0 & \cos\beta & 0 & -\sin\beta \\ \sin\beta & 0 & \cos\beta & 0 \\ 0 & \sin\beta & 0 & \cos\beta \end{pmatrix}$$

i.e. a matrix mixing the basis functions z with x.

The rotated matrix is

$$\boldsymbol{b}(\beta) = \begin{pmatrix} \frac{2}{3}(Sc^2 + Ps^2) & -\frac{\sqrt{2}}{3}(Sc^2 + Ps^2) & \frac{2}{3}sc(S-P) & -\frac{\sqrt{2}}{3}sc(S-P) \\ -\frac{\sqrt{2}}{3}(Sc^2 + Ps^2) & \frac{1}{3}(Sc^2 + Ps^2) & -\frac{\sqrt{2}}{3}sc(S-P) & \frac{1}{3}sc(S-P) \\ \frac{2}{3}sc(S-P) & -\frac{\sqrt{2}}{3}sc(S-P) & \frac{2}{3}(Ss^2 + Pc^2) & -\frac{\sqrt{2}}{3}(Ss^2 + Pc^2) \\ -\frac{\sqrt{2}}{3}sc(S-P) & \frac{1}{3}sc(S-P) & -\frac{\sqrt{2}}{3}(Ss^2 + Pc^2) & \frac{1}{3}(Ss^2 + Pc^2) \end{pmatrix},$$

where the rows and columns are indexed by m_{13} (analogous to m_{12} in Table 2b) and $c \equiv \cos\beta$, $s \equiv \sin\beta$. From the genealogy table, $B^{13}(3, 5) = b^{AH}(\beta)\,(2, 3)$ and has the value $-\frac{\sqrt{2}}{3}(S - P)\sin\beta\cos\beta$. The construction of the complete matrix is a purely mechanical exercise.

For the discussion here, we require a hamiltonian matrix in a basis such that atoms 1 and 2 (A and H) are coupled so as to yield a singlet state of AH_2. There are two such functions:

$$\Psi_1 = \frac{1}{\sqrt{3}}\left[X_1 - \frac{1}{\sqrt{2}}(X_2 + X_3) + X_4\right]$$

$$\Psi_2 = \frac{1}{\sqrt{3}}\left[X_5 - \frac{1}{\sqrt{2}}(X_6 + X_7) + X_8\right].$$

The matrix elements in this basis are

$$B_{11} = S_{12} + (S\cos^2\beta + P\sin^2\beta)_{13} + T_{23}$$

$$B_{12} = \sin\beta\cos\beta\,(S - P)_{13}$$

$$B_{22} = P_{12} + (S\sin^2\beta + P\cos^2\beta)_{13} + T_{23}$$

Notice that the operators $\check{H}_{12}$ and $\check{H}_{23}$ are *both* diagonal: each function shows pure anti-bonding in the HH pair (T_{23}); function Ψ_1 corresponds to a pure Σ interaction in the pair 12 and Ψ_2 to pure Π. The 13 interaction depends on the angle β: it is purely Σ for the linear molecule ($\beta = 0$ or π), and purely Π for the right-angled case $\left(\beta = \frac{\pi}{2}\right)$. In this special sense, the bonding is directional but it must be emphasised that the other interactions (here between atoms 2 and 3) may exert even more influence in determining the bond angle in the end.

The eigenvalues of the matrix are

$$E_{\pm}(\beta) = \frac{1}{2}(S_{12} + P_{12}) + \frac{1}{2}(S_{13} + P_{13}) + T_{23} \pm J,$$

with

$$J = \frac{1}{2}\sqrt{(S_{12} - P_{12}) + 2(S_{12} - P_{12})(S_{13} - P_{13})\cos 2\beta}\,.$$

For the linear and right-angled cases, these become

$$E_+(0) = S_{12} + S_{13} + T_{23}$$

$$E_-(0) = P_{12} + P_{13} + T_{23}$$

$$E_+\left(\frac{\pi}{2}\right) = S_{12} + P_{13} + T_{23}$$

$$E_-\left(\frac{\pi}{2}\right) = P_{12} + S_{13} + T_{23}\,.$$

The linear configuration is favoured, for if the sigma interaction is bonding and the pi anti-bonding, then $E_+(0)$ is lower than $E_\pm\left(\frac{\pi}{2}\right)$; if the pi interaction is more bonding than the sigma, then $E_-(0)$ is lower than $E_\pm\left(\frac{\pi}{2}\right)$. If the sigma and pi interactions are identical, the J-term vanishes, so that the results are the same as for a molecule where A is in a ^3S state: $E = S_{12} + S_{13} + T_{23}$. In this case, the molecule would be linear in order to minimise the H—H repulsion. The directionality does have an influence on the shape of the potential energy surface, however, for this example shows that if the two types of interaction are not the same, then it is impossible for the system to benefit from AH binding equally well in all configurations; i.e. even though atom A is divalent (spin-coupling), two strong AH bonds exist only for the linear molecule.

The directional nature of the P-state has greater influence on the bond angle if atoms 2 and 3 are not identical (A + BC). It might then be conceivable that for AB the sigma interaction is bonding and pi anti-bonding but for AC the converse is true. The contribution $(S_{12} + P_{13})$ would then be lower than $(S_{12} + S_{13})$ or $(P_{12} + P_{13})$. Of course, T_{23} would still be anti-bonding but a bent molecule would be feasible.

The phrase "directional bonding" conjures up the idea of orbital hybridisation. This concept is not so easily applied to DIM and has not been investigated in great depth. There are two approaches which might be fruitful. If one is prepared to deal with very large models, enough states on the central atom could be included to allow the construction of hybrid orbitals. For the carbon atom (sp^3 hybridisation), one would ideally need to consider states from several electron configurations: ^5S, ^3D, ^1D from sp^3; ^3P, ^1D from p^4; and ^3P, ^1D from s^2p^2. There is little evidence to suggest that the game would be worth the candle.

A better way may be to devise a "valence state" of the atom in question. The spin coupling and rotational transformation properties could be chosen so that the state would behave in accordance with the qualitative results of simple orbital theories. A multiplicity $(V + 1)$ would ensure a valence of V; directional properties could also be effected by allowing the energy of the state (i.e. the energy which appears in the atomic fragments) to depend upon the orientation of the atom in the rest of the molecule. In any case, the atomic fragment matrices would possess off-diagonal elements in general, since the valence state would not correlate with any particular spectroscopic state of the isolated atom.

4.4 Charge-Transfer, Polar Bonds, Ionicity

Polar and ionic bonds are introduced into DIM via ionic, charge-transfer structures. These are essential to describe molecules where a charge can reside on different centres associated with the same atom: H_3^+, Ar_n^+. Equally important, but often neglected, is the case where there is bond polarity in a heteronuclear diatomic fragment. It is essential that this facet of the bonding be taken into consideration in the polyatomic.

A particularly simple system with which to illustrate charge-transfer is HeH_2^+.

Table 3a. DIM Atomic Basis for $HeH_2^+(^2A')$

Function index k	Centre 1 (He)	Centre 2 (H)	Centre 3 (H)
1	S	b	c
2		$\bar{b}$	$\bar{c}$

Table 3b. DIM Diatomic Basis for $HeH_2^+(^2A')$

Molecule	m_{12}	k_1	k_2	χ_m^{12}	b^{12}			
HeH	1	1	1	$\lvert Sb\rvert$	D			
HeH^+	2	1	—	$\lvert S\rvert$	S			
H_2^+	3	1	—	$\lvert b\rvert$	Q	J		
	4	—	1	$\lvert c\rvert$	J	Q		
	5	1	—	$\lvert\bar{b}\rvert$			Q	J
	6	—	1	$\lvert\bar{c}\rvert$			J	Q

Table 3: $Q \equiv \frac{1}{2}(G + U)$, $J \equiv \frac{1}{2}(G - U)$

Table 3c. Triatomic Basis and Genealogy for HeH_2^+

(ij) $k \rightarrow$		(12)	3	(13)	2	(23)	1
M	X_M	m_{12}	k_3	m_{13}	k_2	m_{23}	k_1
1	$\lvert Sb\rvert$	1		2	1	3	1
2	$\lvert Sc\rvert$	2	1	1		4	1

A minimum basis set model arises from the structures $He(^1S)\,H(^2S)H^+$ and $He(^1S)\,H^+H(^2S)$, shown in Tables 3a to 3c. This time, the fragment operators for HeH, ($\check{H}_{12}$ and $\check{H}_{13}$) are diagonal in the composite basis but $\check{H}_{23}$ is not. The HeH^+ and $H_2^+(^2\Sigma_g^+)$ interactions are bonding, while the HeH and $H_2^+(^2\Sigma_u^+)$ are anti-bonding. Diagonalisation of the DIM hamiltonian yields the eigenvalues:

$$E_\pm = Q_{12} + Q_{13} + Q_{23} \pm \sqrt{(J_{12} - J_{13})^2 + J_{23}^2}\,,$$

where $Q_{12} = \frac{1}{2}(S_{12} + D_{12})$, and $J_{12} = \frac{1}{2}(S_{12} - D_{12})$, with a similar definition for the pair 13. S_{ij} and D_{ij} are the interactions in the ground states of HeH^+ and HeH respectively. The definitions for the 23 pair are obtained by replacing S by G and D by U, the g- and u-interactions between H^+ and H. In C_{2v} configurations ($R_{12} = R_{13}$), $E_+ = S + D + G$ and $E_- = S + D + U$. From this it is clear that the lowest energy in a C_{2v} configuration corresponds to H_2^+ at its (diatomic) equilibrium position, with HeH at the distance which minimises $(S + D)$. The latter is fairly large in HeH_2^+, so that the minimum corresponds to an H_2^+ molecule in its ground state with a He atom bound to it by the long-range ion-induced dipole forces.

In the linear, unsymmetrical configuration, He—H—H, the two VB structures are no longer equally weighted. The structure with the proton in the middle favours

Table 4. Triatomic Basis and Genealogy for H_3^+

$(ij)\ k \rightarrow$		(12)	3	(13)	2	(23)	1
M	X_M	m_{12}	k_3	m_{13}	k_2	m_{23}	k_1
1	$\|\bar{a}b\|$	2	1	5	1	3	2
2	$\|a\bar{b}\|$	3	1	3	2	5	1
3	$\|\bar{b}c\|$	6	1	4	2	2	1
4	$\|b\bar{c}\|$	4	2	6	1	3	1
5	$\|\bar{a}c\|$	5	1	2	1	4	2
6	$\|a\bar{c}\|$	3	2	3	1	6	1

Table 4: The indices m_{ij} for H_2 refer to Table 2, those for H_2^+ to Table 3

the bonding in HeH^+ over the HeH repulsion; of course, a part of the H_2^+ binding energy must be relinquished — the same sort of compromise that arose between singlet and triplet interactions in the discussion of spin-coupling. Their roles are taken over here by the *g*- and *u*-interactions. Numerical results [24] yield a binding energy of about 14 kJ mol^{-1} with respect to He + H_2^+, which is not much different from the C_{2v} case. The important point here is that charge delocalisation in the H_2^+ fragment leads to delocalisation in the triatomic, since the latter can avail itself of the considerable binding in H_2^+ *only* if there is a distribution of charge over the two equivalent atoms.

These effects are even more dramatic for H_3^+, for it is possible for the charge to reside on all three centres — or is it? For this molecule, there are six composite functions, shown in the genealogy table 4. From these, three singlet and three triplet functions can be constructed. The singlets are

$$\Psi_1 = \frac{1}{\sqrt{2}}(X_1 - X_2)$$

$$\Psi_2 = \frac{1}{\sqrt{2}}(X_3 - X_4)$$

$$\Psi_3 = \frac{1}{\sqrt{2}}(X_5 - X_6).$$

The hamiltonian matrix in this basis is

$$\begin{pmatrix} S_{12} + Q_{13} + Q_{23} & J_{13} & J_{23} \\ J_{13} & S_{23} + Q_{12} + Q_{13} & J_{12} \\ J_{23} & J_{12} & S_{13} + Q_{23} + Q_{12} \end{pmatrix}.$$

Notice that only the singlet H_2 interaction appears and that the elements involving H_2^+ are positioned in the same manner in both the polyatomic and diatomic matrices, implying that the polyatomic energy for equilateral triangle configurations lies quite low, since it is possible, by allowing the charge to distribute itself over all three atoms, to exploit the ground state interaction of H_2^+ between all pairs *simultaneously*. The A_1' state-function for D_{3h} configurations is $\Psi = \frac{1}{\sqrt{3}}(\Psi_1 + \Psi_2 + \Psi_3)$; the expectation value of the hamiltonian in this state (see the elements of the matrix above) is

$$E(^1A_1') = \frac{1}{3}(S_{12} + S_{13} + S_{23}) + \frac{2}{3}(G_{12} + G_{13} + G_{23}).$$

All of these terms correspond to binding, so that the molecule in the equilateral triangle configuration is very stable [19–22]. This result is obtained for other molecules whose hamiltonian matrix is similar: Li_3^+ [32].

The concept of resonance structures, stemming from VB theory, is of course automatically taken care of by DIM, since the electronic energy is obtained by diagonalising a matrix. In the case of H_3^+, each of the three structures is associated with the same energy, differing from the other two only in the placement of the positive charge; i.e. these are resonance structures. Application of DIM to aromatic molecules would lead to the usual VB resonance structures in a similar way. Note that resonance in H_2^+ is part of the *input* to DIM in this case. In general, it could also happen that off-diagonal elements produce a large resonance effect between two or more structures of nearly equal energy, even if no resonance were associated with any individual fragment. In any case, the concept of resonance arises naturally in the DIM approach.

Having seen that resonance associated with the strong bonding in the diatomic leads to very strong binding in the triatomic, we might expect this effect to persist for larger ions, but this does not seem to be the case — Polák [51] has shown that the tetramer ion is bound by only 1.25 kJ mol^{-1} with respect to H + H_3^+, and that the minimum occurs at a highly unsymmetrical configuration, corresponding to a neutral H atom bound weakly to the H_3^+. Thus, it would appear that charge delocalisation no longer brings with it the reward of a greatly enhanced binding — valence saturation has set in. Why is this so? The answer lies in the introduction of anti-bonding via the $H_2(^3\Sigma_u^+)$ interaction. In the trimer ion, each VB structure (e.g. HHH^+) involves only *bonding* between the neutral centres (the triplet interaction, T, is missing from the hamiltonian matrix). We have already seen from the analysis of the spin-coupling in H_3 that optimal interaction within one HH pair implies anti-bonding between the atoms from this pair and any external atoms; hence, when the *fourth* H atom is brought close to the trimer ion, it experiences anti-bonding. i.e. *each* of the VB structures for the tetramer ion involves anti-bonding of the fourth H with the other two neutral H's and no amount of charge delocalisation can improve this situation. In fact, as Polák has found out, the charge prefers to reside on only three of the four atoms so that at least the large binding energy associated with the trimer ion is retained. This same qualitative result is found for rare-gas ionic clusters, which are discussed in Sect. 6.3 in detail.

In heteronuclear molecules, charge-transfer is associated with bond polarity. Since DIM is based upon a VB formalism, ionic structures can be used to effect polar bonds: the basis of covalent functions is augmented by ionic structures capable of leading to the desired polarity. The amount of ionic character assigned to the diatomic species (input!) determines the nature of the bonding in the polyatomic. For example, the molecule H_2F in a minimum basis description requires four functions arising from the state-group $F(^2P_u)\,H(^2S_g)\,H(^2S_g)$. Two additional state-groups could be added to introduce bond polarity: $F^-(^1S_g)\,H^+H(^2S_g)$ and $F^-(^1S_g)\,H(^2S_g)\,H^+$, each yielding one extra function. The ionicity (percentage ionic character) of HF enters implicitly via the elements of the HF fragment matrices.

To see the relationship between bond polarity and the fragment matrix, it is instructive to return briefly to the H_2^+ molecule, where "polarity", or more accurately, charge-transfer, is determined completely by symmetry. In Table 3b, the matrix $\boldsymbol{b}_\chi^{12}$ in the composite basis is not diagonal. Diagonalisation leads to a basis, $\boldsymbol{\psi}$, corresponding directly to the $^2\Sigma_g^+$ and $^2\Sigma_u^+$ eigenstates of H_2^+:

$$\boldsymbol{\psi} = \boldsymbol{\varphi c}\,,$$

where in this case,

$$\boldsymbol{c} = \begin{pmatrix} \frac{1}{\sqrt{2}} & \frac{1}{\sqrt{2}} \\ \frac{1}{\sqrt{2}} & -\frac{1}{\sqrt{2}} \end{pmatrix}.$$

Alternatively, the matrix $\boldsymbol{b}$ in the composite basis can be viewed as arising from a *diagonal* energy matrix, $\boldsymbol{e}$, transformed by a matrix of "mixing coefficients", $\boldsymbol{c}$:

$$\boldsymbol{b} = \boldsymbol{cec}^{-1}\,.$$

In this particular case, $\boldsymbol{e}$ has the diagonal elements (G, U). The elements of $\boldsymbol{c}$ then determine the contribution of each composite function to the wave function of each diatomic state.

For homonuclear diatomics, $\boldsymbol{c}$ is fixed by symmetry considerations (the wave function must belong to some irreducible representation of the molecular symmetry group) and is the same for all internuclear distances. In the heteronuclear case, however, the ionicity is free to vary with distance. In the augmented model for HF, for example, only the $^1\Sigma^+$ fragment manifold can have ionic character. Denoting the covalent composite by φ_C and the ionic by φ_I, the fragment wave functions are

$$(\psi_1, \psi_2) = (\varphi_C, \varphi_I)\begin{pmatrix} \cos\theta & -\sin\theta \\ \sin\theta & \cos\theta \end{pmatrix},$$

where θ is a parameter which depends upon the internuclear distance R such that $\lim_{R\to\infty} \theta = 0$. In practice, all the necessary information is contained in the matrix $\boldsymbol{c}$ which is obtained by diagonalising $\boldsymbol{b}$ at each internuclear distance for which ab initio input exists.

Just as the charge distribution in H_2^+ determines the bonding in H_3^+ and H_4^+, so bond polarity in each heteronuclear diatomic fragment determines the nature of the bonding in the polyatomic. The idea is precisely analogous to the discussion of spin-coupling and directional interactions. Suppose that, at the configuration in question, the HF bond were predominantly ionic. It follows that the structure F^-H^+H would be heavily weighted in the polyatomic wave function in order to exploit the stong binding in HF. For this structure, the expectation value of the hamiltonian operator for the HH fragment would then be related to the interaction in H_2^+ and not to H_2. Similarly, the third fragment would involve the interaction in the HF^- molecule. Also, the covalent structure FHH would correspond to binding in H_2 but the FH interaction would have to belong to an *excited* state of HF, since by assumption, the ground state is ionic.

It is worth while at this point, to reiterate the earlier remark that augmenting the polyatomic basis functions does not necessarily lead to a lowering of the energy because the method is always constrained to reach a compromise compatable with the nature of the bonding in the diatomic fragments. This statement is perhaps better understood in the light of the above discussion. Why should one increase the basis at all then? Because we should rightly expect that the basis actually reflect the nature of the bonding in the diatomic molecule — if it is known that a fragment is highly ionic, this should not be ignored. DIM models which do ignore such features ought to be considered inadequate. Such models can, of course easily be set up: in the case of HF, one would simply require that the ground state be associated with the covalent VB structures over the whole range of internuclear distance. In the case of HF one might get away with this; after all, the LEPS equation has been used for years without seriously compromising the results of dynamical calculations based upon it. One should not, however, expect to be able to ignore ionic structures in the NaCl molecule!

One particular area where charge-transfer structures are essential is the chemistry of molecular ions, where ground and excited (charge-transfer) states often lie close together: $Ar^+ + H_2$ and $Ar + H_2^+$ for example. An even crasser example is any molecule where the OH^+ ion is involved: O and H have almost identical ionisation potentials, so that the structures corresponding to $O^+(^4S_u) + H(^2S_g)$ and $O(^3P_g) + H^+$ interact very strongly, even at fairly large internuclear distances [138]. A DIM model employing only the first structure produces a barrier of about 92 kJ mol^{-1} for the reaction [116]

$$O^+(^4S_u) + H_2 \rightarrow OH^+(^3\Sigma^+) + H(^2S)\,.$$

Inclusion of the other structure and incorporation of mixing coefficients from VB calculations causes the barrier to disappear entirely [217]; i.e. a huge qualitative effect hinges on the selection of the basis set — a rather nasty situation, which brings us directly to the next topic: how to use concepts of bonding to design DIM models.

5 Implications of Bonding Concepts for DIM Model Design

From the foregoing discussion, it is clear that DIM contains some of the usual chemical bonding concepts: valence, covalency, ionicity, resonance, charge-transfer, and bond polarity. Some semblance of directional bonding to a central atom is also present. The major difference from the usual theories is the central hypothesis of DIM that everything can be accounted for in terms of atomic and diatomic fragments. Thus, the interpretation of bonding in terms of electronic structure (e.g. orbital hybridisation, valence-shell-electron-pair-repulsion theory [218]) is missing or only indirectly realised in DIM. Of course, DIM can be made exact by increasing the basis. It can be made to approach MO or VB theory arbitrarily closely and must therefore *formally* be capable of yielding the same description of the chemical bond. This viewpoint is artificial, however, for the resulting "concepts" would be so complicated as to be worthless. In practice, DIM, is only applied to problems where a small basis is acceptable. It is meet and proper that concepts of bonding peculiar to DIM be gleaned from such simple systems.

Another approach is to view DIM as a method of projecting the assumed properties of diatomic fragments onto the polyatomics — one gets out what one puts in, or, in other words, the diatomic fragments supply the "axioms", the polyatomic eigenvectors the "theorems". Thus formulated, the central question in choosing a DIM model should be: how can the known ideas about bonding in some particular molecule be applied to the selection of a small but representative and qualitatively adequate basis? Qualitatively adequate means that the results for the polyatomic states of interest are insensitive to the addition of more basis functions.

Since one would be hard pressed to specify a general algorithm leading to such a basis, concepts of bonding are indispensible in approaching this central question. With their help, it is possible to hazard a guess as to what sort of basis constitutes an acceptable model. The guessed model implies a set of fragments, each with a corresponding diatomic basis. Each such fragment basis can be compared with an ab initio description of the same fragment in order to decide whether or not to modify the original model and carry out a further examination. In this way, a satisfactory model can be found iteratively.

Table 5. Diatomic Fragments for BeFH Models

Fragment manifold	No. of Basis Functions		
	BeFH-12 (covalent)	BeFH-18 (cov + ionic)	BeFH-11 (optimal)
$BeF(^2\Sigma^+)$	5	7	5
$BeF(^2\Delta)$	2	2	1
$BeF(^2\Pi)$	5	6	4
$BeF^-(^1\Sigma^+)$	0	2	1
$BeF^-(^1\Pi)$	0	1	0
$HF(^1\Sigma^+)$	1	2	2
$HF(^3\Sigma^+)$	1	1	1
$HF(^1\Pi)$	1	1	1
$HF(^3\Pi)$	1	1	1
$HF^-(^2\Sigma^+)$	0	1	1
$BeH(^2\Sigma^+)$	3	3	2
$BeH(^2\Pi)$	2	2	1
$BeH^+(^1\Sigma^+)$	0	4	3
$BeH^+(^1\Pi)$	0	2	1

Table 5: The optimal model omits $Be(^1P)$ states

The procedure is best illustrated by an example, the BeFH molecule in the $^1A'$ state. The minimum basis arising from the single state-group Be(^1S_g) F(^2P_u) H(^2S_g) is certainly inadequate, for although Be + HF could be described, the divalent character of Be could not emerge from the spin-coupling implied by this group of atomic states. As we have seen, a triplet state of Be is needed. In agreement with conventional wisdom, the $1s^22s2p$ electron configuration must be brought into play; two states, ^3P_u and ^1P_u arise, yielding a tentative DIM model defined by three state-groups. We shall denote this model as BeFH-12 because 12 spin-coupled polyatomic basis functions $\boldsymbol{\Psi}$ arise. The diatomic fragment states implied by this model together with the associated number of diatomic basis functions are listed in Table 5.

VB computations of the fragments using this basis are disappointing, especially for the BeF and BeH manifolds: for the ground state, BeF($X^2\Sigma^+$), there is no binding at all, where there ought to be a dissociation energy of 386 kJ mol^{-1}. The interaction in the BeF manifold obtained from VB calculations with a covalent, ionic, and mixed basis is shown in Fig. 1, which suggests that the inclusion of ionic structures Be^+F^- and H^+F^- would improve matters. This requires 4 additional state groups: $Be^+(^2S_g)\,F^-H$, $Be^+(^2P_u)\,F^-H$, $Be(^1S_g)\,F^-H^+$, and $Be(^1P_u)\,F^-H^+$. These yield 1, 2, 1, and 2 extra basis functions respectively (model BeFH-18). The augmented diatomic bases are considerably more in line with the ab initio calculations and describe the bonding in BeF and BeH much better (see Fig. 1).

So far, the discussion of model design has been very qualitative. Indeed, the first step, the proposal of a tentative model, must remain so, being based partly on experience, partly on intuition. The second step, however, involves *comparison* with good quality diatomic computations; hence, some sort of quantitative *criterion* is required as a measure of the quality of the model. Such an assessment is achieved with the help of projection operators.

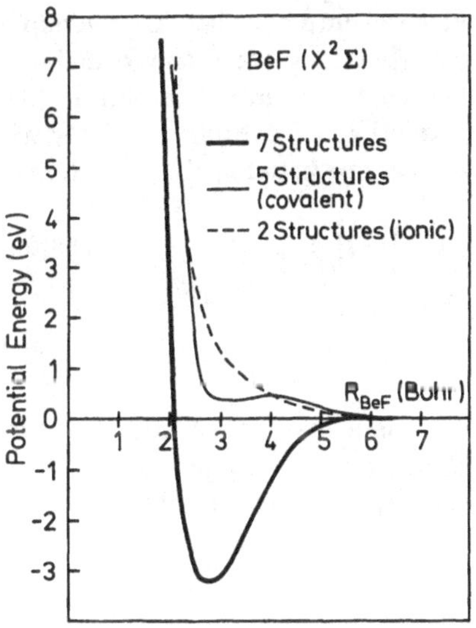

Fig. 1. Fragment interaction energy as estimated by VB calculations using three different bases. (1 eV = 96.487 kJ mol^{-1}; 1 bohr = 52.9 pm). *Roach AC et al.* (1982) *Mol. Phys. 45: 853*

In the spirit of DIM, we assume that, for every conceivable diatomic fragment manifold, we have at our disposal the exact wave functions for all the states belonging to that manifold; e.g. for BeF, we have the wave functions for all states in the $^2\Sigma^+$, $^2\Pi$, $^2\Delta$ ... manifolds. Let the states for one such manifold be denoted by the set $\{|E_i\rangle, i = 1, 2, \ldots N\}$. $|E_i\rangle$ is the state vector corresponding to the eigenvalue E_i. It is convenient to introduce a *row vector* notation for these: $|\boldsymbol{E}\rangle \equiv (|E_1\rangle, |E_2\rangle, \ldots |E_N\rangle)$, so that a *matrix* consisting of the *numbers* $\langle E_i| \check{\mathrm{H}} |E_j\rangle$, is conveniently represented by the symbol $\langle \boldsymbol{E}| \check{\mathrm{H}} |\boldsymbol{E}\rangle$. The projector onto the space spanned by the $\{|E_i\rangle\}$ is then

$$\check{\mathrm{P}}_E = |\boldsymbol{E}\rangle \langle \boldsymbol{E}| \equiv \sum_{i=1}^{N} |E_i\rangle \langle E_i| .$$

We assume that the vectors are orthonormal: $\langle \boldsymbol{E} \mid \boldsymbol{E}\rangle = \boldsymbol{1}_{N\times N}$. The projector associated with the DIM diatomic basis for the manifold in question, $\{|\Psi_k\rangle, k = 1, n\}$, is

$$\check{P}_\Psi = |\boldsymbol{\Psi}\rangle \langle \boldsymbol{\Psi}| ,$$

where

$$|\boldsymbol{\Psi}\rangle = |\boldsymbol{\psi}\rangle \, \boldsymbol{S}^{-\frac{1}{2}}$$

$$\boldsymbol{S} = \langle \boldsymbol{\psi} \mid \boldsymbol{\psi}\rangle ,$$

$\{\psi_i, i = 1, n\}$ being a set of functions having the spin and spatial symmetry of the manifold in question and constructed from the set of diatomic composite functions, χ, implied by the DIM model. The functions $|\Psi_k\rangle$, $|E_i\rangle$, and their associated projectors can be used to assess the adequacy of a particular DIM model and to construct the required diatomic input matrices.

As an aside, it is well to emphasise here that the overlap matrix, $\boldsymbol{S}$, enters these calculations, even though the approach followed here employs the "no-overlap" approximation, which assumes that the *polyatomic* functions are orthonormal. In constructing the *diatomic* matrices, overlap must be taken into account if the projection analysis is to be successful. The introduction of overlap at this level does not de-stabilise the DIM calculation, as does overlap between the polyatomic functions [28].

An (arbitrary) criterion for basis set adequacy [110] exploits the projection *matrix*

$$\boldsymbol{P}_E = \langle \boldsymbol{E}| \check{\mathrm{P}}_\Psi |\boldsymbol{E}\rangle = \boldsymbol{W}\tilde{\boldsymbol{W}},$$

with $\boldsymbol{W}_{N\times n} = \langle \boldsymbol{E} \mid \boldsymbol{\Psi}\rangle$. The diagonal element, $(\boldsymbol{P}_E)_{ii}$, is a measure of how well the state $|E_i\rangle$ is represented by the DIM basis, for if the representation is good, then $|E_i\rangle$ can be expressed as a linear combination of the $|\Psi_k\rangle$ plus a small correction:

$$|E_i\rangle = \sum_{k=1}^{n} d_k \, |\Psi_k\rangle + a_i \, |\Delta_i\rangle ,$$

where $|\Delta_i\rangle$ is a vector orthogonal to the $\{|\Psi_k\rangle\}$ and a_i is a number (hopefully) small compared with 1. Substituting this expansion into the expression for $\boldsymbol{P}_E$ yields the relation $(\boldsymbol{P}_E)_{ii} = 1 - a_i^2$. Hence, the DIM basis may be considered adequate if, for selected, important, diatomic states, the numbers $(\boldsymbol{P}_E)_{ii}$ are greater than some prescribed value, say 0.85. The qualifiers "selected", "important", are necessary, for it is not to be expected that *all* states $|E_i\rangle$, can be adequately described by the DIM basis at *all* internuclear distances.

Once a model has been judged satisfactory, it is necessary to compute the fragment matrices. This is also done with the help of the projector $\check{\mathrm{P}}_E$. In practice, the set $\{|E_i\rangle, i = 1, N\}$ available to us is finite, so that, in general, the space spanned by the $|\Psi_k\rangle$ will not lie completely within that spanned by the $|E_i\rangle$. This is a slight blemish, which can be removed cosmetically by defining effective DIM vectors, $|\boldsymbol{\Psi}_{\mathrm{DIM}}\rangle$, which lie in the space spanned by the $|E_i\rangle$:

$$|\boldsymbol{\Psi}_{\mathrm{DIM}}\rangle \equiv \check{\mathrm{P}}_E\, |\check{\boldsymbol{\Psi}}\rangle\, P_{\Psi}^{-\frac{1}{2}},$$

where $\boldsymbol{P}_{\Psi} \equiv \langle \boldsymbol{\Psi}|\, \check{\mathrm{P}}_E\, |\boldsymbol{\Psi}\rangle = \tilde{\boldsymbol{W}}\boldsymbol{W}$, i.e.

$$|\boldsymbol{\Psi}_{\mathrm{DIM}}\rangle = |\boldsymbol{E}\rangle\, \boldsymbol{Q}\,,$$

where $\boldsymbol{Q} = \boldsymbol{W}\boldsymbol{P}_{\Psi}^{-1/2}$. The vectors $\boldsymbol{\Psi}_{\mathrm{DIM}}$ are orthonormal, since $\tilde{\boldsymbol{Q}}\boldsymbol{Q} = \boldsymbol{1}_{n\times n}$. Also, they lie in the space spanned by $|\boldsymbol{E}\rangle$, since

$$\check{\mathrm{P}}_E\, |\, \boldsymbol{\Psi}_{\mathrm{DIM}}\rangle = |\boldsymbol{E}\rangle\, \langle \boldsymbol{E}\, |\, \boldsymbol{E}\rangle\, \boldsymbol{Q} = |\boldsymbol{E}\rangle\, \boldsymbol{Q} = |\boldsymbol{\Psi}_{\mathrm{DIM}}\rangle\,.$$

The matrix $\boldsymbol{T}_E$ of an operator $\check{\mathrm{T}}$ is defined to be

$$\langle \boldsymbol{\Psi}_{\mathrm{DIM}}|\, \check{\mathrm{P}}_E \check{\mathrm{T}} \check{\mathrm{P}}_E\, |\boldsymbol{\Psi}_{\mathrm{DIM}}\rangle = \tilde{\boldsymbol{Q}}\, \boldsymbol{T}_E \boldsymbol{Q}\,,$$

where $\boldsymbol{T}_E \equiv \langle \boldsymbol{E}|\, \check{\mathrm{T}}\, |\boldsymbol{E}\rangle$.

For the case that the operator is the hamiltonian, $\check{\mathrm{H}}$, the matrix is just the diagonal eigenvalue matrix $\boldsymbol{E}$. The DIM input matrix is then

$$\boldsymbol{b}_{n\times n} = \tilde{\boldsymbol{Q}}_{n\times N} \boldsymbol{E}_{N\times N} \boldsymbol{Q}_{N\times n}\,.$$

This equation shows directly how the (small) DIM input matrix can be recovered from a large ab initio calculation. The latter need not be a VB calculation; it is only necessary to be able to construct the projection matrices that go into defining $\boldsymbol{Q}$. In practice, it is rare that the matrix so constructed can be used directly, for the eigenstates, $|E_i\rangle$, are not exact. In most applications of DIM, the best available diatomic interactions (usually from spectroscopy) should be used; i.e. the matrix $\boldsymbol{b}_{n\times n}$ must be corrected empirically. As detailed discussion of this aspect would be out of place here and the reader is referred to the literature [111, 138, 170].

Applying the selection criterion to the BeFH-18 model confirms its adequacy for the ground state potential surface. As a bonus, the analysis suggests that the state-groups with Be(^1P) are superfluous. The basis can be reduced by 7 functions, resulting in a more compact model, BeFH-11. Such a result implies a great saving in computer time if the model is needed (directly) in dynamical calculations.

6 Selected Examples

6.1 BeFH

We are now in a position to discuss the bonding in several molecules. The BeFH molecule provides a good starting point because it makes use of all the ideas discussed above: spin-coupling, directional effects, and ionic structures. The papers on which this discussion is based [106–108, 110–111] deal mainly with the reaction Be + FH → BeF + H, for which points on a collinear ab initio surface are available [219]. Trajectory calculations on a spline fit to these and on an LEPS surface had been used to criticise the latter, which did not yield results similar to those from the ab initio surface [219]. The implication here was that the LEPS formula and the rules of thumb based upon it (Polanyi's rules [220]) may be invalid for whole classes of reactions. Our work on BeFH was motivated by the thought that, although an LEPS model is inadequate, it might still be possible to develop alternative models by using DIM with a basis set appropriate to BeFH. It must be remembered that LEPS surfaces come from a DIM model for covalent, univalent atoms in *S*-states;. The BeFH system is anything but this!

Having developed "adequate" DIM models in the sense discussed in Sect. 5, we can now discuss the main results. All adequate models agree that the lowest energy corresponds to a $^1\Sigma^+$ state in the linear configuration F—Be—H. This obtains regardless of whether the input matrices are taken directly form extended VB calculations, or whether semi-empirical corrections are introduced. Of the 18 basis functions in the largest model, only 8 contribute significantly to the ground state wave function. Of the remaining functions, 3 involve Be(^1P), which the basis function selection criterion ear-marked as being of negligible importance, and the rest belong to another symmetry. The most important structure (with a weight of 70%), denoted by H—Be$^+$F$^-$, has the H atom interacting with the Be$^+$(^2S) and Be$^+$(^2P) states via the ground and first excited $^1\Sigma^+$ states of BeH$^+$, both of which are bound. This binding, along with the fact that the ionic structures are so heavily weighted in the BeF($^2\Sigma^+$) fragment, accounts for the very large weight of this structure.

Next in importance (10%) is the covalent structure symbolised by H ~ Be ~ F, which denotes the interaction of the Be(^3P_u) state simultaneously with the H and F atoms via the $^2\Pi$ and $^2\Sigma^+$ states respectively; H and F interact via the $^3\Pi$ state. Of similar weight (8%) is the covalent structure symbolised by H—Be—F, where Be(^3P_u) interacts with H and F via $^2\Sigma^+$ interactions and H and F interact via the $^3\Sigma^+$ state. The remaining two contributions, each weighted by about 3%, involve the Be(^1S) state in a non-bonding structure, H · Be · F, and an ionic structure, H$^+$BeF$^-$. The fragment interactions are all $^1\Sigma^+$. Table 6 summarises these results. The reason why the molecule is linear is partly that the first three structures (total weight 88%), exhibit non-bonding or anti-bonding between H and F: bending the molecule incurs extra repulsion. In addition, in a bent molecule, Be$^+$ would be forced by structure 1 to interact with H via an anti-bonding $^1\Pi$ state.

In the reaction Be + HF → BeF + H, the reagents are described by structures 4 and 5, the products mainly by 1, 2, and 3. The barrier to reaction arises because of

Table 6. Important Structures for Linear BeFH and HBeF

Structure	Fragment Interactions BeF		BeH		HF	
1 H — Be$^+$F$^-$	$^2\Sigma^+$	(bonding)	$^1\Sigma^+$	(bonding)	$^2\Sigma^+$	(nonbonding)
2 H ~ Be ~ F	$^2\Sigma^+$	?	$^2\Pi$	(bonding)	$^3\Pi$	(antibonding)
3 H — Be — F	$^2\Sigma^+$	?	$^2\Sigma^+$	(bonding)	$^3\Sigma^+$	(antibonding)
4 H · B · F	$^2\Sigma^+$	(antibonding)	$^2\Sigma^+$	(antibonding)	$^1\Sigma^+$	(bonding)
5 H$^+$ Be · F$^-$	$^1\Sigma^+$	(antibonding)	$^1\Sigma^+$	?	$^1\Sigma^+$	(bonding)

Table 6: The symbol "?" means that the assignment of the structure to a single type of interaction is difficult

the need to change over from the one description to the other. At the saddle point on the collinear surface for Be—F—H, the wave function is a mixture of structures 1, 4, and 5.

It is instructive to consider what happens when the ionic structures are ignored (an "inadequate" model). Leaving out the structures 1 and 5 from model BeFH-11 results in a 7-function model for which the lowest energy, only 19 kJ mol^{-1} below that for BeF + H, (cf around 194 kJ mol^{-1} for model BeFH-11), corresponds to a *bent* molecule with a F—Be—H angle of 125 deg. This time, the covalent structure H ~ Be ~ F has 70% of the weight, followed by H—Be—F (15%) and H · Be · F (12%). The structure H ~ Be ~ F does not exhibit as much directionality as H — Be$^+$F$^-$, so that bending is easier in this model. The main point is that the properties of the molecule change markedly in going from the inadequate 7-function model to the adequate 11-function model, whereas addition of superfluous basis functions leaves the results almost unaltered. This emphasises the importance of finding a model which reflects the known bonding characteristics.

6.2 H_2O

The molecule which has been most extensively investigated is H_2O. It served as the practical example in Ellisons's original paper [1] and, although only two structures, O: H—H and H—O—H, were allowed, the atomisation energy of the molecule ($H_2O \rightarrow O(^3P_g) + H + H$) was very well estimated. A serious drawback of the two-structure model as presented in that paper is that it does not explicity allow for rotation of the fragment matrices: the model is valid only for linear or right-angled configurations. This restriction was later removed by Tully [27], who, with the same covalent structures, obtained very reasonable results for the ground state equilibrium position. There then followed a more ambitious calculation of the entire singlet and triplet potential energy surfaces [93, 113–114], again using a covalent basis yielding workable potential surfaces for studying the dynamics of reaction by means of classical trajectories.

Our own contributions to H_2O, [170, 193], were prompted by a somewhat different question: would the selection criterion and projection operator approach to constructing the matrix elements lead to a satisfactory model for H_2O, not only for the ground state, but for all of the surfaces correlating with the separated-atom limits $O(^3P_g)$ (or $O(^1D_g)$) + H + H? To address this problem, the experience gained from the BeFH work was applied *systematically* to H_2O, always keeping in mind the implications of the bonding in the fragments for the polyatomic molecule.

Obviously of major importance are the fragments OH and H_2. Even a cursory examination of the $OH(X^2\Pi)$ state reveals what turns out to be the dominant feature of our DIM models of H_2O: the highly *ionic* nature of the bonds. OH has a large dipole moment and one that, near the equilibrium position, increases linearly with bond length, implying that the VB structure O^-H^+ is an essential ingredient in describing this molecule. The leading configuration of the MO wave function [221] corresponds to such a structure and indeed, VB calculations [170] show this structure to dominate the wave function at small to intermediate distances.

Incorporation of O^-H^+ structures into the polyatomic basis requires information about OH^-, and VB calculations of this ion show $O^{--}H^+$ to be important [170]. As a consequence, the fully ionic structure $O^{--}H^+H^+$ must also be present in the polyatomic basis for H_2O. Application of the basis function selection criterion to the various fragment manifolds implies that the smallest self-consistent DIM model of the $^1A'$ state of H_2O is a 9-basis function model corresponding to the state-groups displayed in Sect. 3.1. The complete model is specified by the values of 19 independent diatomic matrix elements as a function of internuclear distance; these are divided among 10 fragment manifolds: $OH(^2\Pi)$, $OH(^2\Sigma^+)$, $OH(^2\Delta)$, $H_2(^3\Sigma_u^+)$, $H_2(^1\Sigma_g^+)$, $OH^-(^1\Sigma^+)$, $OH^-(^1\Pi)$, $H_2^+(^2\Sigma_g^+)$, $H_2^+(^2\Sigma_u^+)$, and H_2^{++}. Of these, the H_2, H_2^+, and H_2^{++} states as well as the lowest states of the $OH(^2\Pi)$ and $OH^-(^1\Sigma^+)$ manifolds are fixed to match data in the literature. The remaining elements come from applying the projection technique to VB calculations [170].

This model, which one may regard as lying as close to the non-empirical limit as one may get in DIM, reproduces all of the separated atom limits and the following atom + diatomic limits: $O(^3P_g) + H_2(^3\Sigma_u^+)$, $O(^1D_g) + H_2(X^1\Sigma_g^+)$, $O^-(^2P_u) + H_2^+$, $OH(X^2\Pi)$ + H, and $OH^-(^1\Sigma^+) + H^+$. As such, it satisfies the minimum requirements for interpolating between these limits. The potential surfaces in the regions of triatomic interaction have a number of deficiencies but they do predict a strongly-bound (900 kJ mol^{-1}) water molecule with respect to atomisation (experiment: 975 kJ mol^{-1}); the H—O—H angle is 96.5 deg (experiment: 104.5 deg). (Through adjustment of excited state interactions and mixing coefficients, it is possible to obtain a much better description of the PES over large regions of configuration space.) The contributions of the VB structures to the total wave function in the unadjusted model are shown in Table 7 for the C_{2v} minimum, for the minimum in the symmetric linear $^1\Sigma_g^+$ state, and for the asymptotic limits $OH(X^2\Pi)$ + H and $OH^-(X^1\Sigma^+)$. The latter two entries are, of course, *input* and come from the VB calculations. Note that according to the VB description (after projection onto the DIM space), OH is 48% ionic and OH^- is represented as O^-H (63%) and $O^{--}H^+$ (37%). This leads to a wave function for the H_2O minimum energy configuration in which the O atom is represented by 21% $O(^3P_g)$, 8% $O(^1D_g)$, 54% O^-, and 17% O^{--}; i.e. ionic structures have a weight of 71%. This is hardly surprising, given the large ionic component of the diatomic fragments. Also, the linear symmetric molecule H—O—H ($^1\Sigma_g^+$), which is not very far removed from the C_{2v} minimum, cannot involve the $O(^3P_g)$ basis functions, since the $H_2(^3\Sigma_u^+)$

Table 7. Weight of Structure for H_2O

Structure	$HOH(X^1A_1)$	$HOH(^1\Sigma_g^+)$	$OH^{(-)} + H^{(+)}$
$H{-}O(^3P_g){-}H$	21%	0%	41%
$O(^1D_g)\ H{-}H$	8%	7%	11%
$H{-}O^-(^2P_u)\ H^+$	27%	29%	48%
$H^+\ O^-(^2P_u){-}H$	27%	29%	(63%)
$O^{--}H^+H^+$	17%	35%	(37%)

interaction cannot couple with them to form a g-state. Here, as well, the diatomic fragment correlating with the $O(^1D_g)$ state, the $OH(^2\Sigma^+)$ state, is very ionic in nature, so that the contribution of the $O(^1D_g)$ basis functions is minimal; i.e. the ionic contributions to the ground $^1A'$ state are large over a considerable region of configuration space.

The normal, text-book explanation of bonding in H_2O is in terms of sp^3 hybrid orbitals, the bond angle being less than the tetrahedral angle (109.5 deg) because of repulsion between lone pairs in two of the sp^3 orbitals. The DIM description is not even remotely connected with this picture: the bond angle, for example, results from a very delicate balance between many effects, including the H_2^+ interactions and the repulsion in H_2^{++}. In going from the C_{2v} minimum to the symmetric linear minimum, the weight of the O^{--} structure doubles, O^- remains roughly constant,

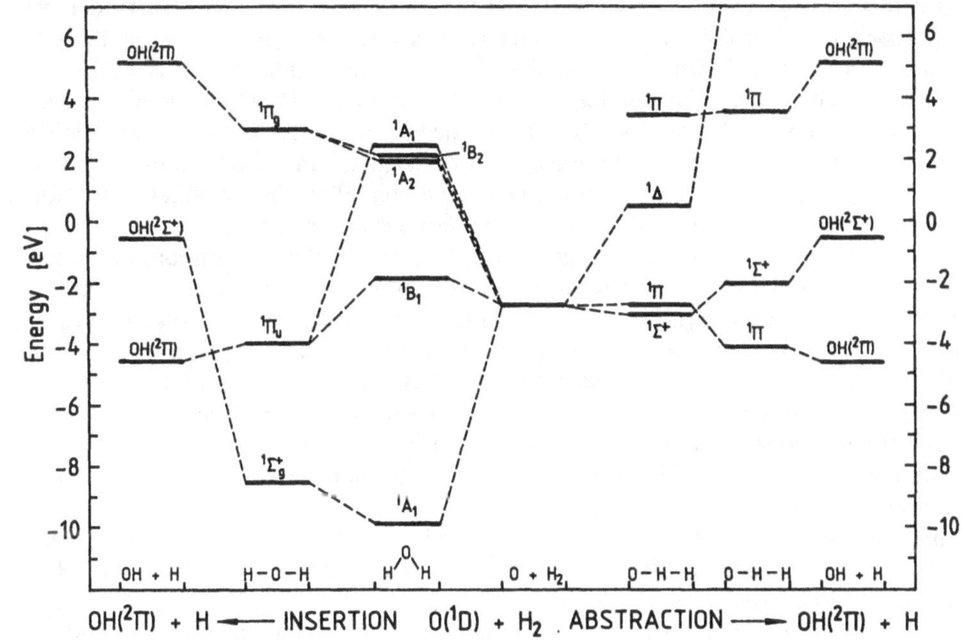

Fig. 2. A DIM state correlation diagram for singlet water. (1 eV = 96.487 kJ mol⁻¹). *Polák R et al. (1987) J. Chem. Phys. 87: 2863*

$O(^3P_g)$ disappears entirely, and $O(^1D_g)$ shows a tendency to decrease slightly: two much long-range repulsion between the two H^+ ions in the C_{2v} configuration is avoided by introducing $O(^3P_g)$ character, for the $H_2(^3\Sigma_u^+)$ repulsion only becomes steep at short range. Of course, the H—H interaction is only part of a rather complicated story.

Whether you are fascinated or disturbed by this picture will depend on your point of view. Those who use DIM usually do not care about the implications for bonding — to them, the method is simply a tool to tackle the more important, dynamics problems. Those who do not use DIM probably shy away from the complexity of the picture, since they are used to thinking in terms of simpler rules. The point I should like to emphasise is that this picture is the *logical consequence* of trying to build up large molecules from *diatomic* fragments. Directional valence and hybrid orbitals were developed in the first instance to explain the bonding in multivalent atoms; these concepts are somewhat foreign to DIM, since they are not really needed in the description of *diatomics*, where there is only *one* direction. These aspects of normal bonding theory do not emerge from DIM, nor should we expect them to.

The DIM description of H_2O can also provide a qualitative explanation for the behaviour of the three surfaces correlating with $O(^1D_g) + H_2$. (See Fig. 2.) In collinear configurations, the $^1\Pi$ surface is most sensitive, not to the ground state $OH(X^2\Pi)$, but to the $OH(^2\Pi)$ state which dissociates to $O(^1D_g) + H_2$. The latter exhibits no bonding, so that the O—H—H $^1\Pi$ surface is repulsive in the entrance channel, as is the $^1\Delta$ surface to an even greater extent. The most important fragment for the $^1\Sigma^+$ surface corresponds to the bonding state $OH(^2\Sigma^+)$; hence, this surface is attractive as the O approaches H_2. In the exit channel, however, these surfaces must approach the $OH(X^2\Pi)$ and $^2\Sigma^+$ states respectively. i.e. the ordering in the exit channel is $^1\Pi < {^1\Sigma^+}$, which implies that the two surfaces intersect. This qualitative feature, predicted by the bonding concepts inherent in the DIM model, is confirmed by ab initio calculations [222]. In fact, the feature was present in earlier ab initio calculations but was not clearly recognised or emphasised. This is another reason for paying attention to bonding concepts in the design of models: qualitative features may emerge, stimulating further research by more exact methods.

Support for this sort of application of DIM to the description of simple molecules comes from an independent investigation into DIM models appropriate for the NH_2 molecule [192]. AIM calculations of the NH fragment were used to suggest an adequate model and to obtain the fragment matrices [152]. As for H_2O, ionic structures are of paramount importance. In addition, the lowest state in the sysmmetric collinear H—N—H configuration has the correct symmetry $(^2\Pi_u)$ only if the ^4P_g state arising from the $2s2p^4$ configuration of N is included in the model. The behaviour of the lowest three states of the triatomic molecule is qualitatively correct over the full range of nuclear configurations. The picture of the bonding in this molecule parallels the discussion on H_2O. AIM calculations on the OH fragments and associated matrices also confirm the VB calculations upon which the H_2O discussion was based [207].

Before concluding the discussion of H_2O, it is well to contemplate, as we did for BeFH, what would happen if the important structures, in this case the ionic ones, were left out? The numerical results consequent upon doing this are easily ascertained by referring to all previous DIM papers in the literature: as long as only the ground-state energy is required, the results are not too bad. In the WMF2 model of Whitlock et al. [93, 114] [$O(^3P_g)$, $O(^1D_g)$, $O(^1S_g)$ + H + H], a ground state surface for the reaction $O(^1D_g) + H_2 \rightarrow OH + H$ yields a deep C_{2v} minimum but a large barrier to inversion (bending of H—O—H). The shape of the surface can be manipulated through alteration of the diatomic fragments not needed to describe the reactants (H_2) or products ($OH(X^2\Pi)$). These are chosen such that the shape of the lowest surface in the entrance channel is controlled by the $OH(^2\Delta)$ fragment. This is unfortunate because it leads to an incorrect assignment of electronic symmetry to linear O—H—H and H—O—H. In the former, for example, there is an intersection between the $^1\Pi$ and $^1\Delta$ surfaces instead of between $^1\Pi$ and $^1\Sigma^+$.

6.3 Singly Ionised Rare Gas Clusters

6.3.1 General Formulation

Application of DIM to this area is interesting but problematical. On the one hand, there is the possibility of charge transfer and local strong binding; on the other hand, most of the interactions are quite weak, so that the final result may prove to be to sensitive to the input. As an example, we consider singly-ionised rare-gas clusters. This case is treated here in some detail, in particular the application to trimer ions.

The simplest DIM model is obtained by admitting only one electronic state for each atomic centre: neutral atoms are assumed to be in the ^1S_g state and the positive ions are assumed to be in the ^2P_u state. The positive charge is allowed to reside on any of the centres so that there are N state-groups possible for a singly ionised N-atom cluster (Rg stands for rare gas):

$$
\begin{array}{ll}
1. & Rg(^1S)\,Rg(^1S)\ \ldots\ Rg(^1S)\,Rg^+(^2P) \\
2. & Rg(^1S)\,Rg(^1S)\ \ldots\ Rg^+(^2P)\,Rg(^1S) \\
 & \\
N-1 & Rg(^1S)\,Rg^+(^2P)\ \ldots\ Rg(^1S)\,Rg(^1S) \\
N & Rg^+(^2P)\,Rg(^1S)\ \ldots\ Rg(^1S)\,Rg(^1S)\,.
\end{array}
$$

Corresponding to each of the state-groups there are 3 polyatomic basis functions, since the ^2P state is 3-fold degenerate. Denoting the three ^2P-functions as x, y, z, and the $Rg(^1S)$ function by s, the 3 basis functions for a particular state group k are

$$
\begin{aligned}
kx &: \ ss \ldots x \ldots ss \\
ky &: \ ss \ldots y \ldots ss \\
kz &: \ ss \ldots z \ldots ss
\end{aligned}
$$

In all, there are $3N$ polyatomic basis functions for a singly-ionised N-atom cluster.

The DIM hamiltonian matrix $\boldsymbol{B}$ is written as a sum of pair contributions, $\boldsymbol{B}^{ij}_X$, and single-centre contributions, $\boldsymbol{B}^i_X$, where i and j label the atomic centres: The atomic contributions, $\boldsymbol{B}^i_X$, are diagonal and the diagonal elements B^i_{kk} are either $E(\mathrm{Rg})$ or $E(\mathrm{Rg}^+)$, so that adopting the convention that $E(\mathrm{Rg}) \equiv 0$, the sum over i simply yields the ionisation potential of Rg, $I(\mathrm{Rg})$; i.e. the atomic terms yield the same contribution, $-(N-2) \cdot I(\mathrm{Rg})$, to *each* diagonal element of the DIM hamiltonian matrix, and so can be ignored.

The symmetric pair-matrices, $\boldsymbol{B}^{ij}_X$, consist mostly of zeros except for elements between those basis functions having the positive charge on atom i or j, (there are exactly 6 such functions: ix, iy, iz, jx, jy, and jz.) and diagonal elements for all the remaining functions. The value of the elements is expressed in terms of the diatomic fragment energies:

$$\mathrm{Rg}_2(^1\Sigma_g^+) \equiv S$$
$$\mathrm{Rg}_2^+(^2\Sigma_g^+) \equiv G$$
$$\mathrm{Rg}_2^+(X^2\Sigma_u^+) \equiv U$$
$$\mathrm{Rg}_2^+(^2\Pi_g) \equiv \bar{G}$$
$$\mathrm{Rg}_2^+(^2\Pi_u) \equiv \bar{U}\,.$$

For this case, the auxiliary quantities must be defined as $Q = \frac{1}{2}(U + G)$ and $J = \frac{1}{2}(U - G)$, with similar definitions for $\bar{Q}$ and $\bar{J}$ in terms of $\bar{U}$ and $\bar{G}$. (This is because the P-state has u-symmetry). If atoms i and j lie on the z-axis, the 6×6 block corresponding to the functions iy, iz, ix, jy, jz, jx is

$$\boldsymbol{B}^{ij}(0,0) = \begin{array}{c} \\ iy \\ iz \\ ix \\ jy \\ jz \\ jx \end{array} \begin{array}{c} \begin{array}{cccccc} iy & iz & ix & jy & jz & jx \end{array} \\ \begin{pmatrix} \bar{Q}_{ij} & 0 & 0 & \bar{J}_{ij} & 0 & 0 \\ 0 & Q_{ij} & 0 & 0 & J_{ij} & 0 \\ 0 & 0 & \bar{Q}_{ij} & 0 & 0 & \bar{J}_{ij} \\ \bar{J}_{ij} & 0 & 0 & \bar{Q}_{ij} & 0 & 0 \\ 0 & J_{ij} & 0 & 0 & Q_{ij} & 0 \\ 0 & 0 & \bar{J}_{ij} & 0 & 0 & \bar{Q}_{ij} \end{pmatrix} \end{array},$$

where Q_{ij} and J_{ij} are obtained from the diatomic fragment potential curves at the internuclear distance r_{ij}, and (0,0) denotes the dependence of the matrix on rotational angles (vide infra). The remaining diagonal elements of $\boldsymbol{B}^{ij}$ have the value S.

For the case that the vector $\boldsymbol{r}_{ij}$ is not parallel to the z-axis, the above matrix must be constructed from the rotated diatomic fragment matrices; this is equivalent to transforming $\boldsymbol{B}^{ij}(0,0)$ by a rotation matrix:

$$\boldsymbol{B}^{ij}(\alpha, \beta) = \boldsymbol{R}(\alpha, \beta) \cdot \boldsymbol{B}^{ij}(0,0) \cdot \tilde{\boldsymbol{R}}(\alpha, \beta)$$

where here we have defined

$$R(\alpha, \beta) = 1_{3N-6} \oplus R^{(i)}(\alpha, \beta) \oplus R^{(j)}(\alpha, \beta)$$

and $R^{(i)}(\alpha, \beta)$ is a 3×3 matrix connecting the functions iy, iz, and iy:

$$R^{(i)}(\alpha, \beta) \equiv \begin{matrix} \\ iy' \\ iz' \\ ix' \end{matrix} \begin{matrix} iy & iz & ix \\ \left(\begin{matrix} \cos\alpha \\ 0 \\ -\sin\alpha \end{matrix}\right. & \begin{matrix} \sin\alpha\sin\beta \\ \cos\beta \\ \cos\alpha\sin\beta \end{matrix} & \left.\begin{matrix} \sin\alpha\cos\beta \\ -\sin\beta \\ \cos\alpha\cos\beta \end{matrix}\right) \end{matrix}$$

The matrix $R^{(j)}(\alpha, \beta)$ is a similar 3×3 block connecting jy, jz, and jx. Carrying out the matrix multiplication yields a symmetric matrix which is diagonal except for the 6×6 block already mentioned. This has the structure

$$B^{ij}(\alpha, \beta) = \begin{pmatrix} M & N \\ N & M \end{pmatrix}$$

where M and N are both 3×3 symmetric matrices which are conveniently written in terms of a matrix $W(a, b)$

$$W_{11} \equiv W_{yy} = a(\cos^2\alpha + \sin^2\alpha\cos^2\beta) + b\sin^2\alpha\sin^2\beta$$

$$W_{21} \equiv W_{zy} = (b - a)\sin\alpha\sin\beta\cos\beta$$

$$W_{31} \equiv W_{xy} = (b - a)\sin\alpha\cos\alpha\sin^2\beta$$

$$W_{22} \equiv W_{zz} = a\sin^2\beta + b\cos^2\beta$$

$$W_{32} \equiv W_{xz} = (b - a)\cos\alpha\sin\beta\cos\beta$$

$$W_{33} \equiv W_{xx} = a(\sin^2\alpha + \cos^2\alpha\cos^2\beta) + b\cos^2\alpha\sin^2\beta$$

In terms of W, the matrices M and N are $M = W(\bar{Q}, Q)$ and $N = W(\bar{J}, J)$. The angles β and α defining the transformation are the angles of the polar coordinates of the relative position vector r_{ij}:

$$r_{ij} = (x_j - x_i)\,\hat{i} + (y_j - y_i)\,\hat{j} + (z_j - z_i)\,\hat{k}\,,$$

so that $\cos\beta = (z_j - z_i)/R$ and $\alpha = \tan^{-1}[(y_j - y_i)/(x_j - x_i)]$.

6.3.2 Trimers With C_s Symmetry

For $N = 3$, the equations simplify somewhat, since the 3 centres can be chosen so as to lie in the xz-plane, which has the effect of seperating the basis functions into two sets, one with A' symmetry (6 functions) and one with A'' symmetry (3 functions). For each state group k, the function ky belongs to the A'' representation; kx and kz belong to the A' representation. In C_s configurations, the DIM matrix B blocks off into a 6×6 A' block and a 3×3 A'' block.

The rotation matrices are also simplified, because only rotations in the xz-plane need be considered; i.e. $\alpha = 0$ so that the matrix $\boldsymbol{R}^{(i)}$ becomes:

$$
\begin{array}{c c}
 & \begin{array}{ccc} iy & iz & ix \end{array} \\
\boldsymbol{R}^{(i)}(0, \beta) = \begin{array}{c} iy \\ iz \\ ix \end{array} & \begin{pmatrix} 1 & 0 & 0 \\ 0 & \cos\beta & -\sin\beta \\ 0 & \sin\beta & \cos\beta \end{pmatrix}
\end{array}
$$

When the 9×9 DIM matrix $\boldsymbol{B}$ is constructed from rotation matrices of this type, it becomes blocked into the 6×6 A' and 3×3 A'' parts described above. Using the convention that atom 1 is at the origin, atom 2 lies on the z-axis, and atom 3 lies in the xz-plane and defining β_i as the angle that r_{i3} makes with the z-axis, the two blocks of the matrix are as follows:

$$\boldsymbol{B}(A') = \boldsymbol{B}_{diag} + \boldsymbol{L}$$

Here, $\boldsymbol{B}_{diag}$ is a diagonal matrix with elements $(S_{12}, S_{12}, S_{13}, S_{13}, S_{23}, S_{23})$ and

$$
\boldsymbol{L} = \begin{array}{c} \\ 3z \\ 3x \\ 2z \\ 2x \\ 1z \\ 1x \end{array}
\begin{array}{c}
\begin{array}{cccccc} 3z & 3x & 2z & 2x & 1z & 1x \end{array} \\
\begin{pmatrix}
q_1 + q_2 & p_1 + p_2 & j_2 & k_2 & j_1 & k_1 \\
p_1 + p_2 & \bar{q}_1 + \bar{q}_2 & k_2 & \bar{j}_2 & k_1 & \bar{j}_1 \\
j_2 & k_2 & Q_{12} + q_2 & p_2 & J_{12} & 0 \\
k_2 & \bar{j}_2 & p_2 & \bar{Q}_{12} + \bar{q}_2 & 0 & \bar{J}_{12} \\
j_1 & k_1 & J_{12} & 0 & Q_{12} + q_1 & p_1 \\
k_1 & \bar{j}_1 & 0 & \bar{J}_{12} & p_1 & \bar{Q}_{12} + \bar{q}_1
\end{pmatrix}
\end{array} .
$$

In the above matrix, the symbols are defined as follows:

$$
\begin{aligned}
q_i &= \bar{Q}_{i3} \sin^2 \beta_i + Q_{i3} \cos^2 \beta_i \\
\bar{q}_i &= Q_{i3} \sin^2 \beta_i + \bar{Q}_{i3} \cos^2 \beta_i \\
p_i &= (Q_{i3} - \bar{Q}_{i3}) \sin \beta_i \cos \beta_i \\
k_i &= (J_{i3} - \bar{J}_{i3}) \sin \beta_i \cos \beta_i \\
j_i &= \bar{J}_{i3} \sin^2 \beta_i + J_{i3} \cos^2 \beta_i \\
\bar{j}_i &= J_{i3} \sin^2 \beta_i + \bar{J}_{i3} \cos^2 \beta_i
\end{aligned}
$$

The matrix for the A'' basis functions is

$$
\boldsymbol{B}(A'') = \begin{array}{c} \\ 3y \\ 2y \\ 1y \end{array}
\begin{array}{c}
\begin{array}{ccc} 3y & 2y & 1y \end{array} \\
\begin{pmatrix}
S_{12} + \bar{Q}_{13} + \bar{Q}_{23} & \bar{J}_{23} & \bar{J}_{13} \\
\bar{J}_{23} & \bar{Q}_{12} + S_{13} + \bar{Q}_{23} & \bar{J}_{12} \\
\bar{J}_{13} & \bar{J}_{12} & \bar{Q}_{12} + \bar{Q}_{13} + S_{23}
\end{pmatrix}
\end{array} .
$$

6.3.3 Trimers With C_{2v} Symmetry

We now consider the configurations for which $r_{13} = r_{23}$; this condition implies that $\beta_2 = \pi - \beta_1$ so that $\cos\beta_2 = -\cos\beta_1$ and $\sin\beta_2 = \sin\beta_1$. In addition, $Q_{13} = Q_{23}$, $J_{13} = J_{23}$, $S_{13} = S_{23}$ with similar results for the barred symbols. Accordingly, the above expressions for the matrices can be simplified through use of the following notation:

$$\beta \equiv \beta_1 = \pi - \beta_2$$

$$S \equiv S_{13} = S_{23}$$

$$q \equiv q_1 = q_2$$

$$j \equiv j_1 = j_2$$

$$k \equiv k_1 = -k_2$$

$$p \equiv p_1 = -p_2$$

with similar definitions of $\bar{q}$ and $\bar{j}$. Introduction of these restrictions into the matrix expressions above results in A'' and A' blocks which are somewhat simplified:

$$\mathbf{L} = \begin{array}{c} \\ 3z \\ 3x \\ 2z \\ 2x \\ 1z \\ 1x \end{array} \begin{array}{c} \begin{array}{cccccc} 3z & 3x & 2z & 2x & 1z & 1x \end{array} \\ \begin{pmatrix} 2q & 0 & j & -k & j & k \\ 0 & 2\bar{q} & -k & \bar{j} & k & \bar{j} \\ j & -k & Q_{12}+q & -p & J_{12} & 0 \\ -k & \bar{j} & -p & \bar{Q}_{12}+\bar{q} & 0 & \bar{J}_{12} \\ j & k & J_{12} & 0 & Q_{12}+q & p \\ k & \bar{j} & 0 & \bar{J}_{12} & p & \bar{Q}_{12}+\bar{q} \end{pmatrix} \end{array}.$$

$$\mathbf{B}(A'') = \begin{array}{c} \\ 3y \\ 2y \\ 1y \end{array} \begin{array}{c} \begin{array}{ccc} 3y & 2y & 1y \end{array} \\ \begin{pmatrix} S_{12}+2\bar{Q} & \bar{J} & \bar{J} \\ \bar{J} & \bar{Q}_{12}+S+\bar{Q} & \bar{J}_{12} \\ \bar{J} & \bar{J}_{12} & \bar{Q}_{12}+\bar{Q}+S \end{pmatrix} \end{array}$$

A greater simplification results if the basis functions are transformed so as to belong to the irreducible representations of the C_{2v} group. The functions $3x$, $3y$, and $3z$ already belong to such representations (A_1, B_2, and B_1 respectively); to obtain symmetry functions from the remaining basis functions it is necessary to take linear combinations of the functions on the centres 1 and 2: $1z + 2z$, (B_2); $1z - 2z$, (A_1); $1x + 2x$, (A_1); $1x - 2x$, (B_2); $1y + 2y$, (B_1); and $1y - 2y$, (A_2). The

DIM matrices must be subjected to a similarity transformation: $\tilde{T}BT = B(C_{2v})$. The transformation matrix for the A' block is (letting $g \equiv 1/\sqrt{2}$)

$$
T(A') = \begin{array}{c} \\ 3z \\ 3x \\ 2z \\ 2x \\ 1z \\ 1x \end{array}
\begin{array}{c} \begin{array}{cccccc} 3_A & x_A & z_A & 3_B & x_B & z_B \end{array} \\
\begin{pmatrix} 0 & 0 & 0 & 1 & 0 & 0 \\ 1 & 0 & 0 & 0 & 0 & 0 \\ 0 & 0 & -g & 0 & 0 & g \\ 0 & g & 0 & 0 & -g & 0 \\ 0 & 0 & g & 0 & 0 & g \\ 0 & g & 0 & 0 & g & 0 \end{pmatrix} \end{array} .
$$

The transformation matrix for the A'' block is

$$
T(A'') = \begin{array}{c} \\ 3y \\ 2y \\ 1y \end{array}
\begin{array}{c} \begin{array}{ccc} 3_B & y_B & y_A \end{array} \\
\begin{pmatrix} 1 & 0 & 0 \\ 0 & g & -g \\ 0 & g & g \end{pmatrix} \end{array} .
$$

The similarity transformations lead to symmetry-adapted DIM B-matrices. The $^2B_1$ matrix resulting from transformation of the A'' functions is

$$
B(^2B_1) = \begin{array}{c} \\ 3_B \\ y_B \end{array}
\begin{array}{c} \begin{array}{cc} 3_B & y_B \end{array} \\
\begin{pmatrix} S_{12} + 2\bar{Q} & \sqrt{2}\bar{J} \\ \sqrt{2}\bar{J} & \bar{U}_{12} + S + \bar{Q} \end{pmatrix} \end{array} .
$$

The $^2A_2$ matrix resulting from the A$''$ functions is

$$
B(^2A_2) = \bar{G}_{12} + S + \bar{Q} .
$$

The (symmetric) $^2A_1$ matrix resulting from the A' functions is

$$
B(^2A_1) = \begin{array}{c} \\ 3_A \\ x_A \\ z_A \end{array}
\begin{array}{c} \begin{array}{ccc} 3_A & x_A & z_A \end{array} \\
\begin{pmatrix} S_{12} + 2\bar{q} & & \\ \sqrt{2}(J \sin^2\beta + \bar{J}\cos^2\beta) & \bar{U}_{12} + S + \bar{q} & \\ \sqrt{2}(J - \bar{J}) \sin\beta\cos\beta & \sqrt{2}(Q - \bar{Q}) \sin\beta\cos\beta & G_{12} + S + q \end{pmatrix} \end{array}
$$

The (symmetric) $^2B_2$ matrix resulting from the A$'$ functions is

$$
B(^2B_2) = \begin{array}{c} \\ 3_B \\ x_B \\ z_B \end{array}
\begin{array}{c} \begin{array}{ccc} 3_B & x_B & z_B \end{array} \\
\begin{pmatrix} S_{12} + 2\bar{q} & & \\ \sqrt{2}(J - \bar{J}) \sin\beta\cos\beta & \bar{G}_{12} + S + \bar{q} & \\ \sqrt{2}(J \cos^2\beta + \bar{J}\sin^2\beta) & \sqrt{2}(Q - \bar{Q}) \sin\beta\cos\beta & U_{12} + S + q \end{pmatrix} \end{array} .
$$

6.3.4 Trimers In Linear Configurations

Setting $\beta = 0$ results in a further simplification since the matrices separate into Σ and Π blocks. The B_1 matrix yields a 2×2 matrix for the Π_u states; the A_1 matrix yields a 2×2 Π_u-block corresponding to the B_1 matrix and a Σ_g^+ state with energy $(G_{12} + S + Q)$. The B_2 matrix yields a Π_g state with energy $(\bar{G}_{12} + S + \bar{Q})$ and a 2×2 Σ_u^+ block with the same structure as the ${}^2B_1$ 2×2 matrix but with unbarred symbols replacing the barred ones. The A_2 matrix yields a component of the Π_g state.

6.3.5 Trimers With D_{3h} Symmetry

In D_{3h} configurations, the three internuclear distances are equal $(r_{12} = r_{13} = r_{23})$ so that $\beta = 60°$. Moreover, the fragment functions Q_{ij}, J_{ij}, S_{ij}, $\bar{Q}_{ij}$, $\bar{J}_{ij}$, etc. may be written without their subscripts, since the internuclear distances are equal to each other. The 6×6 block corresponding to the A' functions is conveniently written as

$$\boldsymbol{B}(A') = \boldsymbol{B}_{diag} + \boldsymbol{L}$$

where $\boldsymbol{B}_{diag}$ is a 6×6 diagonal matrix with all diagonal elements equal to $S + Q + \bar{Q}$, and

$$\boldsymbol{L} = \begin{array}{c} \\ 3z \\ 3x \\ 2z \\ 2x \\ 1z \\ 1x \end{array} \begin{array}{c} \begin{array}{cccccc} 3z & 3x & 2z & 2x & 1z & 1x \end{array} \\ \begin{pmatrix} \frac{1}{2}(\bar{Q}-Q) & 0 & \frac{1}{4}(J+3\bar{J}) & \frac{\sqrt{3}}{4}(\bar{J}-J) & \frac{1}{4}(J+3\bar{J}) & \frac{\sqrt{3}}{4}(J-\bar{J}) \\ 0 & \frac{1}{2}(Q-\bar{Q}) & \frac{\sqrt{3}}{4}(\bar{J}-J) & \frac{1}{4}(3J+\bar{J}) & \frac{\sqrt{3}}{4}(J-\bar{J}) & \frac{1}{4}(3J+\bar{J}) \\ \frac{1}{4}(J+3\bar{J}) & \frac{\sqrt{3}}{4}(\bar{J}-J) & \frac{1}{4}(Q-\bar{Q}) & \frac{\sqrt{3}}{4}(\bar{Q}-Q) & J & 0 \\ \frac{\sqrt{3}}{4}(\bar{J}-J) & \frac{1}{4}(3J+\bar{J}) & \frac{\sqrt{3}}{4}(\bar{Q}-Q) & \frac{1}{4}(\bar{Q}-Q) & 0 & \bar{J} \\ \frac{1}{4}(J+3\bar{J}) & \frac{\sqrt{3}}{4}(J-\bar{J}) & J & 0 & \frac{1}{4}(Q-\bar{Q}) & \frac{\sqrt{3}}{4}(Q-\bar{Q}) \\ \frac{\sqrt{3}}{4}(J-\bar{J}) & \frac{1}{4}(3J+\bar{J}) & 0 & \bar{J} & \frac{\sqrt{3}}{4}(Q-\bar{Q}) & \frac{1}{4}(\bar{Q}-Q) \end{pmatrix} \end{array}.$$

The 3×3 block corresponding to the A'' functions is

$$\boldsymbol{B}(A'') = \begin{array}{c} \\ 3y \\ 2y \\ 1y \end{array} \begin{array}{c} \begin{array}{ccc} 3y & 2y & 1y \end{array} \\ \begin{pmatrix} S+2\bar{Q} & \bar{J} & \bar{J} \\ \bar{J} & S+2\bar{Q} & \bar{J} \\ \bar{J} & \bar{J} & S+2\bar{Q} \end{pmatrix} \end{array}.$$

The construction of a symmetry-adapted basis is effected by transformation with the following matrices:

$$T(A') = \begin{array}{c} \\ 3z \\ 3x \\ 2z \\ 2x \\ 1z \\ 1x \end{array} \begin{array}{c} \begin{array}{cccccc} z_A & x_A & z_{E1} & x_{E1} & z_{E2} & x_{E2} \end{array} \\ \begin{pmatrix} \frac{1}{\sqrt{3}} & 0 & \frac{2}{\sqrt{6}} & 0 & 0 & 0 \\ 0 & \frac{1}{\sqrt{3}} & 0 & 0 & \frac{2}{\sqrt{6}} & 0 \\ -\frac{1}{2\sqrt{3}} & \frac{1}{2} & \frac{1}{2\sqrt{6}} & -\frac{1}{2\sqrt{2}} & -\frac{1}{2\sqrt{2}} & \frac{\sqrt{3}}{2\sqrt{2}} \\ -\frac{1}{2} & -\frac{1}{2\sqrt{3}} & \frac{1}{2\sqrt{2}} & -\frac{\sqrt{3}}{2\sqrt{2}} & \frac{1}{2\sqrt{6}} & -\frac{1}{2\sqrt{2}} \\ -\frac{1}{2\sqrt{3}} & -\frac{1}{2} & \frac{1}{2\sqrt{6}} & \frac{1}{2\sqrt{2}} & \frac{1}{2\sqrt{2}} & \frac{\sqrt{3}}{2\sqrt{2}} \\ \frac{1}{2} & -\frac{1}{2\sqrt{3}} & -\frac{1}{2\sqrt{2}} & -\frac{\sqrt{3}}{2\sqrt{2}} & \frac{1}{2\sqrt{6}} & -\frac{1}{2\sqrt{2}} \end{pmatrix} \end{array}$$

$$T(A'') = \begin{array}{c} \\ 3y \\ 2y \\ 1y \end{array} \begin{array}{c} \begin{array}{ccc} y_A & y_{E1} & y_{E2} \end{array} \\ \begin{pmatrix} f/g & 2f & 0 \\ f/g & -f & g \\ f/g & -f & -g \end{pmatrix} \end{array}, \qquad f = \frac{1}{\sqrt{6}}, \qquad g = \frac{1}{\sqrt{2}}.$$

The matrix $T(A')$ was obtained by first rotating the x and z coordinates on centre 1 by $-120°$ and on centre 2 by $+120°$ according to the following transformation:

$$(x', z') = (x, z)\begin{pmatrix} \cos\theta & -\sin\theta \\ \sin\theta & \cos\theta \end{pmatrix}$$

The new basis functions $3x$, $2x'$, $1x'$ and $3z$, $2z'$, $1z'$ were then transformed by the matrix $T(A'') \otimes T(A'')$.

The DIM B-matrices in the D_{3h} symmetry-adapted basis functions are considerably simplified. The $T(A'')$ transformation yields a diagonal matrix having the following elements:

$$B(A_2'') = S + 2\bar{Q} + 2\bar{J},$$

$$B(E'') = S + 2\bar{Q} - \bar{J}, \qquad \text{(doubly degenerate)}.$$

The $T(A')$ transformations yield two 1×1 blocks of A_2' and A_1' symmetry respectively, and two identical 2×2 blocks of E' symmetry. The A' block is

$$\mathbf{B}(A_1', A_2') = \begin{array}{c} \\ z_A \\ x_A \end{array} \begin{array}{c} \begin{array}{cc} z_A(A_1') & x_A(A_2') \end{array} \\ \begin{pmatrix} S + \frac{1}{2}(U + 3\bar{G}) & 0 \\ 0 & S + \frac{1}{2}(3G + \bar{U}) \end{pmatrix} \end{array}.$$

The doubly-degenerate E' block is

$$\mathbf{B}(E') = \begin{array}{c} \\ z_{E1}, x_{E1} \\ x_{E2}, z_{E2} \end{array} \begin{array}{c} \begin{array}{cc} z_{E1}, x_{E1} & z_{E2}, x_{E2} \end{array} \\ \begin{pmatrix} S + \frac{1}{2}(Q + 3\bar{Q}) + \frac{1}{4}(3\bar{J} - J) & \pm\frac{3}{4}(J + \bar{J}) \\ \pm\frac{3}{4}(J + \bar{J}) & S + \frac{1}{2}(3Q + \bar{Q}) + \frac{1}{4}(3J - \bar{J}) \end{pmatrix} \end{array}.$$

For 1×1 and 2×2 matrices, the eigenvalues and eigenvectors can be written out explicitly. The eigenvalues of a 2×2 matrix with diagonal elements (a, b) and off-diagonal element c are

$$E_i = \frac{1}{2}(a + b) \pm \sqrt{\left(\frac{a - b}{2}\right)^2 + c^2}$$

The eigenvectors may be written in terms of a parameter Θ defining the unitary matrix which diagonalises the 2×2 matrix

$$\begin{pmatrix} a & c \\ c & b \end{pmatrix}\begin{pmatrix} \cos\Theta & -\sin\Theta \\ \sin\Theta & \cos\Theta \end{pmatrix} = \begin{pmatrix} \cos\Theta & -\sin\Theta \\ \sin\Theta & \cos\Theta \end{pmatrix}\begin{pmatrix} E_1 & 0 \\ 0 & E_2 \end{pmatrix}$$

$$\Theta = \frac{1}{2}\tan^{-1}\left(\frac{2c}{a - b}\right)$$

In terms of Θ the eigenvalues are

$$E_1 = a\cos^2\Theta + b\sin^2\Theta + c\sin 2\Theta$$
$$E_2 = a\sin^2\Theta + b\cos^2\Theta - c\sin 2\Theta$$

The matrices for C_{2v} and linear configurations have the structure typified by the 2×2 $^2B_1$ block:

$$a = S_{12} + 2Q$$
$$b = S + U_{12} + Q$$
$$c = \sqrt{2}\,J$$

The eigenvalues are

$$E_i = \frac{1}{2}(S + S_{12} + U_{12} + 3Q) \pm \frac{1}{2}\sqrt{(S_{12} - U_{12} + Q - S)^2 + 8J^2}$$

The eigenvector parameter is

$$\Theta = \frac{1}{2}\tan^{-1}\left[\frac{2\sqrt{2}J}{S_{12} - U_{12} + Q - S}\right]$$

For D_{3h} configurations, the eigenvalues of A_1' and A_2' symmetry are

$$E(A_1') = S + \frac{1}{2}(U + 3\bar{G})$$

$$E(A_2') = S + \frac{1}{2}(3G + \bar{U})$$

The matrix elements for the E' blocks have the following structure:

$$a = S + \frac{1}{2}(Q + 3\bar{Q}) + \frac{1}{4}(3\bar{J} - J)$$

$$b = S + \frac{1}{2}(Q + 3\bar{Q}) + \frac{1}{4}(3J - \bar{J})$$

$$c = \frac{3}{4}(J + \bar{J})$$

The eigenvalues are

$$E_i = S + (Q + \bar{Q}) + \frac{1}{4}(J + \bar{J}) \pm \sqrt{(\bar{U} - U)^2 + \frac{9}{16}(J + \bar{J})^2}$$

As an example, the above procedure is applied to Ar clusters [206]. The DIM models for all clusters of a particular noble gas are fully defined only when the atomic and diatomic fragment matrices are specified. These were defined by taking the points of the curves U, G, $\bar{U}$, and $\bar{G}$ for Ar_2^+ from the ab initio computations of Böhmer and Peyerimhoff [223]. The $Ar_2(^1\Sigma_g^+)$ interactions were taken from Watts [224]. The trimer ions are found to be most stable in the $^2\Sigma_u^+$ state of the linear symmetric molecule. The ground ($^2\Sigma_u^+$) state is bound by 133 kJ mol^{-1} at R = 261 pm (cf. Böhmer and Peyerimhoff who get 130 kJ mol^{-1} at R = 262 pm); the $^2\Pi_u$ state is very weakly bound at long range and the other states are essentially repulsive in character.

In the linear symmetric $^2\Sigma_u^+$ state, the eigenvector is characterised by the parameter $\Theta \approx 45°$; i.e. the linear molecule has a charge distribution corresponding to the VB-structures Ar—Ar$^+$—Ar (50%), Ar$^+$—Ar—Ar (25%), and Ar—Ar—Ar$^+$

(25%). This charge distribution is very insensitive to the internuclear distance in the molecule. The same charge distribution at the equilibrium distance is obtained from ab initio calculations [223]. This rather remarkable behaviour can be explained by a closer examination of the equation defining the parameter Θ. The numerator of the argument of $\tan^{-1}$ is the energy splitting between the (repulsive) ${}^2\Sigma_g^+$ and (bound) ${}^2\Sigma_u^+$ states of Ar_2^+ at a distance where the interaction is considerable; the denominator consists of two interactions (U_{12} and S_{12}) which are negligibly small, (since the distance between the outer atoms 1 and 2 is large), and $(Q - S)$, the difference between the average energy of the Ar_2^+ ${}^2\Sigma_u^+$ and ${}^2\Sigma_g^+$ states, Q, and the energy of the Ar_2 ${}^1\Sigma_g^+$ state at a short range, S. Numerical calculations show that the term $(Q - S)$ is small in comparison with J over a large range of internuclear distance; this in turn leads to an argument of $\tan^{-1}$ which corresponds to an angle of magnitude approximately $\pi/2$, with the result that Θ must be near $\pi/4$ (45°). We therefore conclude that the behaviour of the wave function is partly accidental (in that Q and S are of similar value) and partly due to the large role of resonance in the rare gas dimer ions ($|J| \gg |Q|$). Note that this resonance is not nearly so large in the Π-states of Ar_2^+, which is partly the reason why the trimer ion is not bent. For D_{3h} configurations, the ground state at intermediate distance is doubly degenerate and correlates with the ${}^2A_1$ and ${}^2B_2$ states in the C_{2v} irreducible representations. In the neighbourhood of the D_{3h} configuration, the ${}^2A_1$ state has a shallow (local)

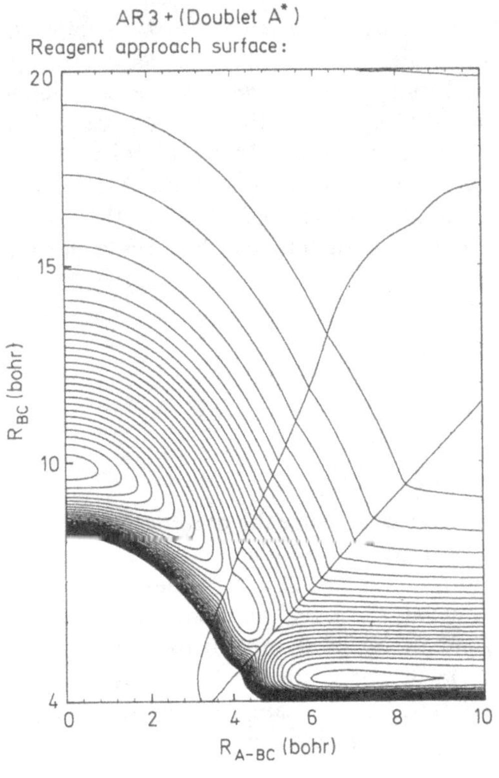

Fig. 3. DIM potential energy surface for Ar_3^+ in C_{2v} configurations. The abscissa is the distance from atom A to the mid-point of BC; the ordinate is the internuclear distance in the BC molecule.
(1 bohr = 52.9 pm). *Kuntz PJ, Valldorf J (1988) Z. Phys. D. 8: 195*

minimum which lies considerably higher then the global minimum of the linear $^2\Sigma_u^+$ state (which correlates with the $^2B_2$ state) (see Fig. 3).

For $N > 3$, ionic clusters consist of a small ionic molecule (trimer, or tetramer) surrounded by shells of nearly neutral atoms. The trimer ion is particularly stable and forms the core of clusters up to size 13, where the first shell of neutrals is complete. With the filling of the second shell, there is a tendency for the central ion to change from a trimer to a tetramer and for a small fraction of the positive charge to spread out over the rest of the atoms in the cluster. The clusters are very floppy, there being many geometrical configurations with an energy near the lowest energy found.

Improvements to the DIM model can be effected in two ways: increasing the basis, or incorporating (small) 3-body effects into the polyatomic matrix elements. The latter procedure has been carried out [225], where account was taken of the induced-dipole induced-dipole forces between neutrals under the influence of a neighbouring ion. Strictly speaking, this is no longer a DIM procedure; however, it may point the way to a systematic way of representing the PES: instead of expanding the PES itself in a many-body expansion [226, 227], one should first obtain a model hamiltonian for the system and then introduce the many-body effects into the matrix elements.

It is interesting to compare the above case with He_3^+ and with H_3^+. The former differs from the above treatment in that the ions are in S-states and not P-states; this has the effect that the quantity J must be defined as $J = \frac{1}{2}(G - U)$ but apart from this, the results are similar. The charge distribution in the symmetric linear molecule ($^2\Sigma_g^+$) is much the same as before; however, the difference in energy between the C_{2v} and linear configurations is much smaller, because the directional restrictions arising from the P-states are missing in the He system. One might wonder why H_3^+ and He_3^+ behave so differently, since the DIM hamiltonian matrix has exactly the same structure. However, the roles of the g- and u-states are switched in these molecules: in H_2^+, the g-state is the ground state and the u-state is repulsive. For He the reverse is true. In D_{3h} configurations, the eigenvalues for both molecules are

$$E(A') = S + 2G$$

$$E(E') = S + \frac{1}{2}(3U + G) .$$

For H_3^+, both of the terms S and G are attractive — hence the deep minimum. For He_3^+, S is only weakly attractive and G is repulsive, so that the E' state is the ground state, even though it is made up of a mixture of G and U interactions. For He, more binding can be realised in a distorted, C_{2v}, configuration. The difference in the two systems is partly a symmetry effect and partly due to the fact that the S-state interaction is so much weaker for He—He than for H—H. This is again an instance of how the properties of the diatomic, in this case the symmetry of the ground state, influence the qualitative nature of the bonding in the polyatomic.

6.4 Surface Chemistry

This is a relatively new area for DIM and it is very complex. Two main appoaches have been reported. In the first, the surface is modelled by a small cluster of atoms, so that the whole treatment amounts to a computation for a large molecule. An example of this type of approach is the study of the interaction between H atoms and a Li_9 cluster [117] as well as the study of the dissociative adsorption of H_2 on a Li_8 cluster [115, 140].

Charge-transfer in the neighbourhood of a surface is perhaps somewhat more amenable to the method, since this process is similar to the harpooning mechanism of M. Polanyi [228] which, years ago, was so successfully applied to gas phase reactions in bulk and under beam conditions. Again, the main interest here is in dynamical calculations which can be carried out in a small diabatic basis. The diabatic states are chosen to reflect the system before and after the charge transfer. The system W_5Na has been studied in this way, again with the tungsten surface represented by a small cluster [157, 178, 182, 187].

Another approach incorporates the entire physical surface into the DIM model in an approximate way. This has been applied to the formation of Na^+ and Na_2^+ ions from Na* (^2P_u) atoms in the neighbourhood of a tungsten surface *partially covered* with sodium [164, 179, 186]. Provided the covering is less than a monolayer, the adsorbed sodium can be considered to be a layer of Na^+ ions on the surface. Since each ion induces an associated image charge in the metal, the covered surface can be approximated as an infinitely extended dipole layer. This layer determines the position of ionic potential energy surfaces relative to neutral ones and as such influences the position of "avoided crossings" between the adiabatic potentials. A DIM model can be constructed by considering the entire tungsten surface to be an "atom" and then adding $(N - 1)$ Na^+ ions and one sodium atom. In addition to the Na_2^+ diatomic fragment interactions (in ground and excited states), it is important to incorporate the interaction between a Na^+ ion and the entire dipole layer into the Na^+—W "diatomic fragment".

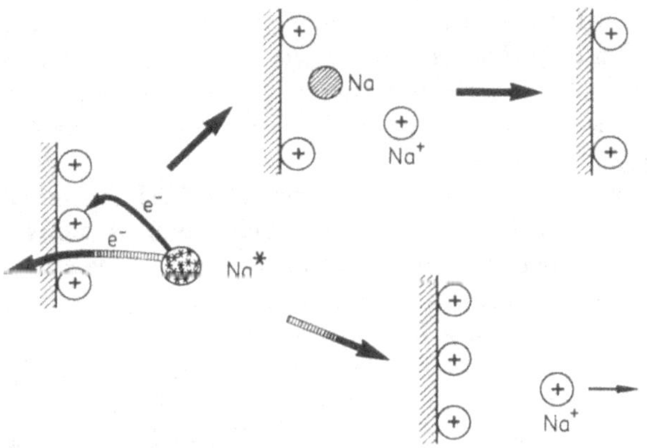

Fig. 4. Schematic illustration of ionisation of excited Na in the neighbourhood of a tungsten surface. *Kuntz PJ (1985) Chem. Phys. 94: 371*

Such a model yields a physically realistic description of the processes occurring in the neighbourhood of the surface. In particular, examination of the DIM polyatomic wave functions leads to a simple harpooning picture for the production of ions: if the electron from $Na^*(^2P_u)$ jumps into the metal surface, Na^+ ions are formed. If, however, the electron jumps onto one of the adsorbed ions, the latter is repelled from the surface and joins the incident sodium atom (now an ion) to form Na_2^+. This process is illustrated schematically in Fig. 4.

Such a model can also explain the reverse reaction, the production of excited sodium atoms from Na^+ ions incident on a partially covered surface [186, 229]. Admittedly, the simplest DIM models omit features such as the presence of a continuum of states associated with the ionic curves. The models can, however, be refined to include such effects.

It is not yet certain whether or not DIM is well-suited to describing the surface structure itself and those things which depend upon it. It is more likely to be of use in describing processes on or near the surface (catalysis) or, even better, on adsorbed layers on the surface. A surface covered with a monolayer of Cs does not lead to the production of excited Li atoms from incident Li^+ ions but if, subsequently, oxygen is added to the system, excited Li atoms are observed [230]. Such effects could readily be investigated with DIM models involving Cs, Cs^+, O^-, Li^+, $Li(^2S)$, and $Li^*(^2P)$ basis functions. In fact, DIM may be one of the few methods available for such purposes.

6.5 Molecules-in-Molecules

In many chemical reactions, whole groups of atoms remain relatively intact, so that there is no need to dissect these into their diatomic components as the normal DIM procedure does. A considerable simplication of the basis set results if such groups are treated as single entities. This can be done in the normal framework by restricting the number of spin-couplings. For example, an ethyl group obtained by coupling two quintet C atoms to five doublet H atoms would require many basis functions. Some of these correspond to VB structures where two of the H atoms are coupled to form a singlet. For most cases, such structures are not important and one could do very well by allowing only that structure where the H-atoms interact with each other via a triplet (repulsive) potential energy curve. This approach effectively reduces the group in question to a single entity while still retaining the possiblility of describing the internal motion.

A disadvantage of the above approach is that the properties of the individual groups are still determined by *diatomic* interactions within the group. A simpler and sometimes more effective way of parring a model down to manageable size is to treat the sub-sets of atoms as the basic entities ("atoms") of the DIM method *from the very start.* Roach and Gimzewski show how this is done in a paper on VB π-electron theory applied to conjugated hydrocarbons [96]. The basic units, the "π-atoms", are taken to be the CH_2, CH, or C species obtained on breaking all C—C bonds in the molecule. For butadiene, for example, $CH_2=CH—CH=CH_2$ breaks up into the "atoms" CH_2 + CH + CH + CH_2. Each choice of a pair of these defines one of six "π-diatomics" needed as input. The fragment information is obtained empirically from the lowest singlet (N) and triplet (T) states of planar ethylene as a function of the C—C bond distance. The CH bond energy, D_{CH}, and the dissociation energy of the N-state, D_N, are chosen to reproduce the experimental enthalpy of atomisation of ethylene, $(D_N + 4D_{CH})$, and to lie close to published values of mean bond enthalpies. In all, only 7 empirical parameters are needed to specify the fragment input.

Small, workable models for the molecules benzene, naphthalene, azulene, styrene, butadiene, and the allyl-, benzyl-, and triphenylmethyl-radicals are easily constructed using this approach. Enthalpies of atomisation, excitation energies and bond lengths agree remarkably well with experiment in most of these cases, the largest discrepancy occurring for butadiene, which requires additional ionic structures in order to improve the excitation energy.

This work is very encouraging, showing the way in which the basic idea of DIM can be applied to large molecules without incurring models of intractable size. While it is true that the π-systems are special cases and that there is some arbitrariness in assigning an energy to the basic units as their internal configurations are altered, nevertheless, these systems provide an excellent example of the use of bonding concepts within a DIM framework to achieve a unified, self-consistent description of a whole class of molecules from a small amount of empirical input. The work is a gem in the DIM literature.

6.6 Models for Dynamics Based on DIM Bonding Concepts: Ar + OH^-

The $ArOH^-$ molecule will now be used to illustrate the use of bonding concepts in the design of DIM models at the qualitative level. To my knowledge, no DIM calculations exist for this molecule but a model potential would be of interest in studying theoretically the inelastic scattering of OH^- ions on Ar, for which experiments have already been reported [231]. The discussion here is therefore qualitative and speculative but serves to show how the bonding concepts outlined in this article are applied to the design of a DIM model for a concrete example. The discussion is restricted to the ground state only, and no rearrangement channels are considered; i.e. chemical reaction is a latent possibility in the DIM model but this fact is not explicitly emphasised in the design.

First, consider what sort of model could be envisaged *without* appealing to DIM [231]. A rough-and-ready potential surface for studying the dynamics of non-reactive or dissociative processes results from describing the OH^- molecule by a morse potential (i.e. OH^- is strongly bound) and the Ar—OH^- interaction by the sum of two pair potentials, one for Ar—O and one for Ar—H. If the negative charge is assumed to reside on the O atom, then the latter two potentials consist of interactions between Ar and O^- and between Ar and H. The latter has a very weak van-der-Waals attraction at long range but is strongly repulsive at short range [23]. The former is presumably also mainly repulsive but would have a stronger medium-to-long range attractive component due to the ion-induced dipole forces between Ar and O^-.

Dynamical calculations based on such a potential would probably show the results to be dominated by two main effects: 1) Upon approaching Ar, the OH^-

molecule would tend to orient itself to favour the configuration Ar—O—H over Ar—H—O, since the Ar—O interaction in the entrance channel is attractive over a greater range of internuclear distance than is the Ar—H interaction. 2) Since the foregoing effect ensures that the H-atom is pushed out of the way, the scattering event would be dominated by the interaction between the two heavy atoms, Ar and O^-. It is unlikely that there would be significant vibrational excitation of OH^-.

A DIM treatment leads to a fundamentally different picture. Referring back to the section of H_2O, we know from the analysis of the OH^- wave function that the structure $O^{--}H^+$ is very important near the OH^- equilibrium distance; hence, the structure $ArO^{--}H^+$ must be included along with ArO^-H in the DIM model. This, in turn, requires diatomic input for the fragments ArO^{--} and ArH^+ as well as additional information about the OH^- fragment. Now the Ar—O^{--} interaction is probably similar to that between Ar—Ne plus the ion-induced dipole component (attractive); i.e. it is similar to Ar—O^-. The Ar—H^+ interaction, on the other hand, corresponds to the strongly bound (386 kJ mol^{-1}) ground state of this molecule-ion. The DIM potential energy surface would therefore tend to favour the configuration Ar—H—O over Ar—O—H. Furthermore, the approach of Ar would tend not only to attract the proton to a position between Ar and O but also to weaken the OH^- bond to compensate for the strengthening of the Ar—H^+ bond: there would quite possibly be incipient chemical reaction. But this is exactly the sort of situation which favours vibrational excitation. Clearly the dynamics on such a potential surface would be markedly different from that for the additive pair-potentials. The extent of the difference depends, of course, on the relative importance of the various DIM basis functions and, at the moment, there are no calculations at our disposal to estimate this.

As a final remark, it is clear that the same considerations apply to the uncharged molecule ArOH, for the projection analysis of the VB wave function for OH in its ground state showed the structure O^-H^+ to be very important. Accordingly, the structure ArO^-H^+ must be included in the DIM model of this system. At high collision energy, there is quite possibly a direct pathway from the reagents Ar + OH to ionic products $ArH^+ + O^-$. This is all somewhat speculative but calculations on these two systems could be carried out fairly easily and would probably confirm the broad picture outlined here. If so, the DIM models would provide a very good start to the study of the inelastic and dissociative processes. The DIM procedure together with a few rudimentary bonding concepts provides a way of thinking about these systems and, at the same time, suggests new experiments.

7 Concluding Remarks

The DIM description of polyatomic molecules does not attempt to answer questions such as "What is a chemical Bond?", (an inappropriate question in any case [232]), nor does it clarify why such a bond exists at all (another inappropriate question?), for DIM *starts* with an assumed complete knowledge of all atoms and diatomic molecules. Since bonding in diatomics already exists, any rationalisations of polyatomic bonding are ultimately expressed in terms of diatomic behaviour: the H_2O molecule is stable because the spin-coupling appropriate to the ground state allows for an exploitation of the bonding in the $X^2\Pi$ state of *both* OH fragments *simultaneously*. The molecule has a large dipole moment because the $OH(X^2\Pi)$ wave function at the appropriate internuclear distance contains the structure O^-H^+ with a large weight. In itself, DIM is not a theory of binding directly in terms of electronic structure. Nonetheless, it takes account of such theories *indirectly* through the assumed electronic structure of the diatomic fragments.

As a result of this indirectness, the description of bonding is a somewhat complicated interplay between spin-coupling, charge-transfer, and the rotational properties of the atomic functions. The first two features are more easily visualised than the third, which, despite one's instinctive feeling, is often not the main factor in controlling the disposition of bonds about a central atom: bond directionality does not arise in a simple way. DIM is closely related to a valence bond (VB) description of a molecule but does not quite stand on its own feet, since a VB description of the fragments is needed as input to the method.

Extensions of DIM to larger molecules is hampered by the large basis sets required to handle the spin-coupling and the rotational properties of the direct-product basis. It is here that information from other theories can help out. Omitting many of the spin-couplings at the design stage of a model is certainly possible but the decision about which structures to leave out can only be made on the basis of some theory external to DIM. A full VB treatment of the ground state of benzene would require very many spin-couplings but only 5 of these are really needed to get a good semi-quantitative description [216]; modern VB theory could therefore be of enormous help to DIM in selecting the most important couplings in molecules where the choice is not so obvious.

Work on ways to eliminate the many angular momentum states is not as far advanced. At the moment, two avenues appear attractive. First, the need for many atomic basis functions (e.g. in describing hybrid orbitals) could be obviated by defining certain atoms to be in "valence" states having simple rotational properties assigned to them. This idea could be supplemented by a second step: taking small groups of atoms to be the primary entities in the theory [96]. The latter approach would allow application to organic molecules, for example.

I feel it is best to view DIM not as a theory of bonding in itself, but rather as a method of modelling polyatomic molecules in accordance with the concepts of bonding implicit in other theories (VB and MO). The examples discussed here show that failure to take account of these concepts leads to inconsistent models, which are no more than artificial methods of interpolating the potential energy surfaces connecting the various diatomic fragments. Compliance with these concepts, on the other hand, sometimes uncovers subtle properties of the real molecules (as seen, for example, in the behaviour of the potential surfaces of $O(^1D) + H_2$ in the entrance channel), which can stimulate further work by more exact methods. In this spirit, DIM can promote a fruitful application of the concepts of bonding to the description, not just of stable molecules, but also to the more exciting areas of chemical reactions and the nature of the transition state.

8 References

1. Ellison FO (1963) J. Am. Chem. Soc. 85: 3540
2. Ellison FO, Huff NT, Patel JC (1963) J. Am. Chem. Soc. 85: 3544
3. Ellison FO, Patel JC (1964) J. Am. Chem. Soc. 86: 2115
4. Ellison FO (1964) J. Chem. Phys. 41: 2198
5. Cashion JK, Herschbach DR (1964) J. Chem. Phys. 40: 2358
6. Cashion JK, Herschbach DR (1964) J. Chem. Phys. 41: 2199
7. Ellison FO, Huff NT (1965) J. Chem. Phys. 42: 364
8. Pfeiffer GV, Ellison FO (1965) J. Chem. Phys. 43: 3405
9. Pfeiffer GV, Huff NT, Greenawalt EM, Ellison FO (1967) J. Chem. Phys. 46: 821
10. Wu AA, Ellison FO (1967) J. Chem. Phys. 47: 1458
11. Wu AA, Ellison FO (1968) J. Chem. Phys. 48: 727
12. Companion AL (1968) J. Chem. Phys. 48: 1186
13. Wu AA, Ellison FO (1968) 48: 1491
14. Wu AA, Ellison FO (1968) 48: 5032
15. Abrams RB, Patel JC, Ellison FO (1968) 49: 450
16. Companion AL, Steible DJ Jr, Starshak AJ (1968) J. Chem. Phys. 49: 3637
17. Companion AL (1969) J. Chem. Phys. 50: 1165
18. Tyndall JR, Companion AL (1970) J. Chem. Phys. 52: 2036
19. Preston RK, Tully JC (1971) J. Chem. Phys. 54: 4297
20. Tully JC, Preston RK (1971) J. Chem. Phys. 55: 562
21. Huang JJ (1971) J. Chem. Phys. 55: 5136
22. Krenos J, Preston RK, Wolfgang R, Tully JC (1971) Chem. Phys. Lett. 10: 17
23. Kuntz PJ, Roach AC (1972) J. Chem. Soc. Faraday Trans. II. 68: 259
24. Kuntz PJ (1972) Chem. Phys. Lett. 16: 581
25. Tully JC (1973) Ber. Bunsenges. Phys. Chem. 77: 557
26. Pickup BT (1973) Proc. R. Soc. London. A333: 69
27. Tully JC (1973) J. Chem. Phys. 58: 1396
28. Steiner E, Certain PR, Kuntz PJ (1973) J. Chem. Phys. 59: 47
29. Bauschlicher CW Jr, O'Neil SV, Preston RK, Schaefer HF III, Bender CF (1973) J. Chem. Phys. 59: 1286
30. Preston RK, Cross RJ Jr (1973) J. Chem. Phys. 59: 3616
31. Tully JC (1973) J. Chem. Phys. 59: 5122
32. Ellison FO, Delledonne MJ (1973) J. Chem. Phys. 59: 6179
33. Holloway TT, O'Brien TJ (1974) Bull. Am. Phys. Soc. 19: 260
34. Chapman S, Preston RK (1974) J. Chem. Phys. 60: 650
35. Krenos JR, Preston RK, Wolfgang R, Tully JC (1974) J. Chem. Phys. 60: 1634
36. Tully JC (1974) J. Chem. Phys. 60: 3042
37. Lin YW, George TF, Morokuma K (1974) J. Chem. Phys. 60: 4311
38. Lin YW, George TF, Morokuma K (1975) Chem. Phys. Lett. 30: 49
39. Kuntz PJ, Whitton WN (1975) Chem. Phys. Lett. 34: 340
40. Zimmermann IH, George TF (1975) Chem. Phys. 7: 323
41. Gelb A, Jordan KD, Silbey R (1975) Chem. Phys. 9: 175
42. Wilson CW Jr (1975) J. Chem. Phys. 62: 4842
43. Zimmermann IH, George TF (1975) J. Chem. Phys. 63: 2109
44. Kuntz PJ (1976) in: Miller WH (ed) Dynamics of molecular collisions, Part B, Plenum, New York, p 53
45. Miller DL, Wyatt RE (1976) Chem. Phys. Lett. 38: 410
46. Baer M, Top AH (1976) Chem. Phys. Lett. 39: 134
47. DeVries PL, George TF (1976) Chem. Phys. Lett. 43: 391
48. Viehland LA, Mason EA, Steven TH, Monchick L (1976) Chem. Phys. Lett. 44: 360
49. Krenos JR, Lehmann KK, Tully JC, Hierl PM, Smith GP (1976) Chem. Phys. 16: 109
50. Kuntz PJ, Whitton WN (1976) Chem. Phys. 16: 301
51. Polák R (1976) Chem. Phys. 16: 353
52. Eaker CW, Parr CA (1976) J. Chem. Phys. 64: 1322

53. Baer M, Top AH (1976) J. Chem. Phys. 64: 3078
54. Tully JC (1976) J. Chem. Phys. 64: 3182
55. Whitton WN, Kuntz PJ (1976) J. Chem. Phys. 64: 3624
56. Collins FS, Cross RJ Jr (1976) J. Chem. Phys. 65: 644
57. Tully JC, Truesdale CM (1976) J. Chem. Phys. 65: 1002
58. Chapman FM Jr, Hayes EF (1976) J. Chem. Phys. 65: 1032
59. Preston RK, Cohen JS (1976) J. Chem. Phys. 65: 1589
60. Hayes EF, Siu AKQ, Chapman FM Jr, Matcha RL (1976) J. Chem. Phys. 65: 1901
61. Eaker CW, Parr CA (1976) J. Chem. Phys. 65: 5155
62. Tully JC (1977) in: Segal GA (ed) Semiempirical methods of electronic structure calculation, vol 7A, Plenum, New York, p 173
63. Baer M, Beswick JA (1977) Chem. Phys. Lett. 51: 360
64. Top AH, Baer M (1977) Chem. Phys. 25: 1
65. Havemann U, Pacak V, Hermann Z, Schneider F, Zuhrt C, Zülicke L (1977) Chem. Phys. 28: 147
66. Micha DA (1977) J. Chem. Phys. 66: 1255
67. Sathyamurthy N, Duff JW, Stroud C, Raff LM (1977) J. Chem. Phys. 67: 3563
68. Schubert JG (1978) Diatomics-in-molecules studies of CH_2^+. Report WIS-TCI-591 of the Theoretical Chemistry Institute of the University of Wisconsin, Madison WI
69. Stepanov NF, Nemukhin AV (1978) Opt. Spectroscopy 45: 1193
70. Dahl JP (1978) Int. J. Quantum Chem. 14: 191
71. Osherov VI, Polujanov LV (1978) Theor. Chim. Acta 49: 123
72. Polák R, Vojtík J, Schneider F (1978) Chem. Phys. Lett. 53: 117
73. Companion AL (1978) Chem. Phys. Lett. 56: 500
74. Poppe D (1978) Chem. Phys. 35: 151
75. Stine JR, Muckerman JT (1978) J. Chem. Phys. 68: 185
76. Kendrick J, Kuntz PJ, Hillier IH (1978) J. Chem. Phys. 68: 2373
77. Truhlar DG, Horowitz CJ (1978) J. Chem. Phys. 68: 2466
78. Eaker CW (1978) J. Chem. Phys. 69: 1453
79. Matcha RL, Pettitt BM, Meier PF, Pendergast P (1978) J. Chem. Phys. 69: 2264
80. Huestis DL, Schlotter NE (1978) J. Chem. Phys. 69: 3100
81. Nemukhin AV, Stepanov NF (1979) Int. J. Quant. Chem. 15: 49
82. Kuntz PJ (1979) in: Bernstein RB (ed) Atom-molecule collison theory: A guide for the experimentalist. Plenum, New York, p 79
83. Wu AA (1979) Mol. Phys. 38: 843
84. Baer M, Beswick JA (1979) Phys. Rev. A 19: 1559
85. Nemukhin AV, Stepanov NF (1979) Chem. Phys. Lett. 63: 396
86. Schneider F, Zülicke L (1979) Chem. Phys. Lett. 67: 491
87. Kuntz PJ, Kendrick J, Whitton W (1979) Chem. Phys. 38: 147
88. Kuntz PJ, Doran M, Hillier IH (1979) Chem. Phys. 43: 197
89. Kendrick J, Kuntz PJ (1979) J. Chem. Phys. 70: 736
90. Faist MB, Muckerman JT (1979) J. Chem. Phys. 71: 225
91. Faist MB, Muckerman JT (1979) J. Chem. Phys. 71: 233
92. Meier PF, Hayes EF, Ellison FO (1979) J. Chem. Phys. 71: 1948
93. Whitlock PA, Muckerman JT, Kroger PM (1980) in: Truhlar DG (ed) Potential energy surfaces and dynamics calculations, Plenum, New York, p 551
94. Wu AA (1980) Mol. Phys. 39: 1287
95. Zaitsevskii AV, Nemukhin AV, Stepanov NF (1980) Mol. Phys. 41: 377
96. Roach AC, Gimzewski E (1980) J. Chem. Phys 73: 1294
97. Isaacson AD, Muckerman JT (1980) J. Chem. Phys. 73: 1729
98. Tully JC (1980) Adv. Chem. Phys. 42: 63
99. Nemukhin AV, Stepanov NF (1980) J. Molec. Struct. 67: 81
100. Micha DA (1977) J. Chem. Phys. 66: 1255
101. Polishchuk AY, Shpil'rain EE, Yakubov IT (1980) High Temp. 17: 987
102. Wu AA (1981) Mol. Phys. 42: 379
103. Wu AA (1981) Mol. Phys. 43: 1459
104. Last I (1981) Chem. Phys. 55: 237

105. Eaker CW, Allard LR (1981) J. Chem. Phys. 74: 1821
106. Kuntz PJ, Roach AC (1981) J. Chem. Phys. 74: 3420
107. Roach AC, Kuntz PJ (1981) J. Chem. Phys. 74: 3435
108. Kuntz PJ (1982) Ber. Bunsenges. Phys. Chem. 86: 367
109. Kuntz PJ, Chang CC (1982) J. Phys. Chem. 86: 1212
110. Schreiber JL, Kuntz PJ (1982) J. Chem. Phys. 76: 1872
111. Kuntz PJ, Schreiber JL (1982) J. Chem. Phys. 76: 4120
112. Truhlar DJ, Duff JW, Blais NC, Tully JC, Garrett BC (1982) J. Chem. Phys. 77: 764
113. Kroger PM, Muckerman JT (1981) Radiochim. Acta. 28: 215
114. Whitlock PA, Muckerman JT, Fisher ER (1982) J. Chem. Phys. 76: 4468
115. Vojtík J, Fiser J (1982) Chem. Phys. Lett. 86: 312
116. Ellison FO, McCandlish ML (1982) J. Phys. B 15: 229
117. Vojtík J, Fiser J (1982) Surf. Sci. 121: 111
118. Vojtík J, Paidarová I (1982) Chem. Phys. Lett. 91: 280
119. Lee KT, Bowman JM, Wagner AF, Schatz GC (1982) J. Chem. Phys. 76: 3563
120. Raynor S, Herschbach DR (1982) J. Phys. Chem. 86: 1214
121. Kuntz PJ, Sizun M (1982) J. Phys. B 15: 511
122. Clary DC (1982) Chem. Phys. 71: 117
123. Richtsmeier SC, Eades RA, Dixon DA, Gole JL (1982) in: Gole JL, Stwalley WC (eds) Metal bonding and interactions in high temperature systems. ACS, Washington, p 177
124. Richtsmeier SC, Hendewerk ML, Dixon DA, Gole JL (1982) J. Phys. Chem. 86: 3932
125. Richtsmeier SC, Dixon DA, Gole JL (1982) J. Phys. Chem. 86: 3937
126. Richtsmeier SC, Dixon DA, Gole JL (1982) J. Phys. Chem. 86: 3942
127. Paidarová I, Vojtík J, Polák R (1982) Chem. Phys. 72: 119
128. Ljudkovski SV, Nemukhin AV, Stepanov NF (1983) J. Mol. Struct. 13: 403
129. Vojtík J, Krtková A, Polák R (1983) Theor. Chim. Acta 63: 235
130. Koshi M, Ito H, Matsui H (1983) Chem. Phys. Lett. 103: 180
131. Clary DC (1983) Chem. Phys. 81: 379
132. Polishchik AY (1983) Theor. & Exp. Chem. 19: 119
133. Clary DC, Garrett BC, Truhlar DG (1983) J. Chem. Phys. 78: 777
134. Blais NC, Truhlar DG, Garrett, BC (1983) J. Chem. Phys. 78: 2956
135. Radlein DSA (1983) J. Chem. Phys. 78: 3084
136. Ellison FO (1983) J. Chem. Phys. 78: 5024
137. Schneider F, Zülicke L, Polák R, Vojtík J (1983) Chem. Phys. 76: 259
138. Kuntz PJ, Chang CC (1983) Chem. Phys. 75: 79
139. Vojtík J, Paidarová I (1983) Chem. Phys. Lett. 99: 93
140. Vojtík J, Savrda J, Fiser J (1983) Chem. Phys. Lett. 97: 397
141. Duggan JJ, Grice R (1983) J. Chem. Phys. 78: 3842
142. Duggan JJ, Grice R (1984) J. Chem. Soc. Faraday Trans. II. 80: 729
143. Duggan JJ, Grice R (1984) J. Chem. Soc. Faraday Trans. II. 80: 739
144. Duggan JJ, Grice R (1984) J. Chem. Soc. Faraday Trans. II. 80: 809
145. Duggan JJ, Grice R (1984) J. Chem. Soc. Faraday Trans. II. 80: 795
146. Olson JA, Micha DA (1984) J. Chem. Phys. 80: 2602
147. Olson JA, Garrison BJ (1984) J. Chem. Phys. 81: 1355
148. Paidarová I, Vojtík J (1984) Chem. Phys. 84: 225
149. Schneider F, Zülicke L, Polák R, Vojtík J (1984) Chem. Phys. 84: 217
150. Schneider F, Zülicke L, Polák R, Vojtík J (1984) Chem. Phys. Lett. 105: 608
151. Wu AA (1984) Mol. Phys. 51: 1
152. Polák R, Vojtík J (1984) Chem. Phys. 87: 273
153. Clary DC, Connor JNL (1984) J. Phys. Chem. 88: 2758
154. Mayne HR, Polanyi JC, Sathyamurthy N, Raynor S (1984) J. Phys. Chem. 88: 4064
155. Vojtík J, Paidarová I (1984) Chem. Phys. Lett. 103: 305
156. Hirst DM (1984) J. Phys. B: At. Mol. Phys. 17: 505
157. Olson JA, Garrison BJ (1985) J. Chem. Phys. 83: 1392
158. Richtsmeier SC, Jagger T, Gole JL, Dixon DA (1985) Chem. Phys. Lett. 117: 274
159. Garret BC, Truhlar DG (1985) J. Chem. Phys. 82: 4543
160. Tucker SC, Truhlar DG, Garrett BC, Isaacson AD (1985) J. Chem. Phys. 82: 4102

161. Kuntz PJ, Polák R (1985) Chem. Phys. 99: 405
162. Chang CC, Kuntz PJ (1985) Chem. Phys. 99: 73
163. Vojtík J (1985) Int. J. Quant. Chem. 28: 593
164. Kuntz PJ (1985) Chem. Phys. 94: 371
165. Chapman S (1985) J. Chem. Phys. 82: 4033
166. Vojtík J, Krtková, Polák R (1985) Theor. Chim. Acta 66: 405
167. Kimura M (1985) Phys. Rev. A 32: 802
168. Kimura M, Chapman S, Lane NF (1985) in: Coggiola MJ, Huestis DL, Saxon RP (eds) Electronic and atomic collisions. 14th ICPEAC Abstracts of Contributed Papers. North-Holland, Amsterdam, p 582
169. Kimura M (1985) in: Coggiola MJ, Huestis DL, Saxon RP (eds) Electronic and atomic collisions. 14th ICPEAC Abstracts of Contributed Papers. North-Holland, Amsterdam, p 583
170. Polák R, Paidarová I, Kuntz PJ (1985) J. Chem. Phys. 82: 2352
171. Maslová ME, Polishchuk AY (1985) Theor. & Exp. Chem. 21: 84
172. Vojtík J (1985) Int. J. Quantum Chem. 28: 943
173. Schneider F, Polák R, Vojtík J (1985) Chem. Phys. 99: 265
174. Eaker CW, Muzyka L (1985) Chem. Phys. Lett. 119: 169
175. Eaker CW (1985) J. Phys. Chem. 89: 2612
176. Hesslich J, Kuntz PJ (1986) Z. Phys. D. 2: 251
177. Roach AC, Kuntz PJ (1986) J. Chem. Phys. 84: 822
178. Olson JA, Garrison BJ (1986) J. Vac. Sci. Technol. A. 4: 1222
179. Kuntz PJ (1986) Int. J. Quantum Chem. 29: 1105
180. Vojtík J, Paidarová I (1986) Chem. Phys. 108: 115
181. Gislason EA, Sizun M (1986) Chem. Phys. 107: 311
182. Olson JA, Garrison BJ (1986) Nucl. Instrum. Methods Phys. Res. B. 14: 414
183. Kuntz PJ (1986) J. Phys. B. 19: 1731
184. Schneider F, Zülicke L, Polák R, Vojtík J (1986) Sov. J. Chem. Phys. 3: 1931
185. Polák R (1986) Chem. Phys. 103: 277
186. Kuntz PJ, Lacmann K (1987) Surf. Sci. 192: 859
187. Olson JA, Garrison BJ (1987) J. Phys. Chem. 91: 1430
188. Schlier C, Nowotny U, Teloy E (1987) Chem. Phys. 111: 401
189. Firth NC, Grice R (1987) J. Chem. Soc. Faraday Trans. II 83: 1011
190. Firth NC, Grice R (1987) J. Chem. Soc. Faraday Trans. II 83: 1023
191. Firth NC, Grice R (1987) J. Chem. Soc. Faraday Trans. II 83: 1029
192. Polák R, Vojtík J (1987) Chem. Phys. 114: 43
193. Polák R, Paidarová I, Kuntz PJ (1987) J. Chem. Phys. 87: 2863
194. Eaker CW (1987) J. Chem. Phys. 87: 4532
195. Badenhoop JK, Schatz GC, Eaker CW (1987) J. Chem. Phys. 87: 5317
196. Nehmukhin AV, Lyudkovskii SV, Stepanov NF (1987) Vestn. Mosk. Univ. Ser. 2: Khim. 28: 332
197. Baer M, Nakamura H (1987) J. Chem. Phys. 87: 4651
198. Last I, George TF (1987) J. Chem. Phys. 87: 1183
199. Vojtík J, Paidarová I, Schneider F (1987) Chem. Phys. 114: 369
200. Noorbatcha I, Thareja S, Sathyamurthy N (1987) J. Phys. Chem. 91: 2171
201. Schlier CG, Vix U (1987) J. Chem. Phys. 113: 211
202. Vojtík J, Spirko V, Jensen P (1986) Collect. Czech. Chem. Commun. 51: 2057
203. Gianturco FA, Schneider F (1987) Chem. Phys. 111: 113
204. Gianturco FA, Schneider F (1986) Chem. Phys. Lett. 129: 481
205. Gadea FX, Kuntz PJ (1988) Mol. Phys. 63: 27
206. Kuntz PJ, Valldorf J (1988) Z. Phys. D 8: 195
207. Polák R, Kuntz PJ (1988) Mol. Phys. 63: 865
208. Kuntz PJ, Niefer B, Sloan J (1988) J. Chem. Phys. 88: 3629
209. Moffit W (1951) Proc. Roy. Soc. (London) A210: 245
210. Coulson CA (1960) Revs. Mod. Phys. 32: 170
211. Parr RG (1964) The quantum theory of molecular electronic structure. Benjamin, New York, p 108
212. London F (1929) Z. Electrochem. 35: 552

213. Glasstone S, Laidler KJ, Eyring H (1941) The theory of rate processes. McGraw-Hill, New York, p 67
214. Johnston HS (1966) Gas phase reaction rate theory. Ronald Press, New York
215. Kuntz PJ (1985) in: Baer M (ed) Theory of chemical reaction dynamics, vol I p 71
216. Gerratt J (1987) Chem. in Britain 23: 327
217. Mahling S, Whitton WN (private communication)
218. Gilespie RJ (1963) J. Chem. Ed. 40: 295
219. Schor H, Chapman S, Green S, Zare RN (1978) J. Chem. Phys. 69: 3790
220. Polanyi JC, Schreiber JL (1974) in Jost W (ed) Physical chemistry, An advanced treatise, VIA. Kinetics of gas reactions, chap 6, Academic, New York
221. Werner HJ, Rosmus P, Reinsch EA (1983) J. Chem. Phys. 79: 905
222. Meier U, Peyerimhoff S (1987) private communication
223. Böhmer U, Peyerimhoff S (1986) Z. Phys. D 3: 195
224. Watts RO, McGee IJ (1976) Liquid state chemical physics. Wiley, New York, chap 7
225. Amarouche M, Durand G, Malrieu JP (1988) J. Chem. Phys. 88: 1010
226. Carter S, Mills IM, Murrell JN (1979) Mol. Phys. 37: 1885
227. Varandas AJC, Murrell JN (1977) Faraday Disc. Chem. Soc. 62: 92
228. Polanyi M (1932) Atomic reactions. Williams and Norgate, London.
229. Hermann J, Gehring J, Kempter V (1986) Surf. Sci. 171: 377
230. Kempter V (1987) Private communication
231. Herbst E, Mulholland KA, Champion RL, Doverspike LD (1977) J. Chem. Phys. 67: 5074
232. Popper KR (1972) Objective knowledge. Oxford University Press, p 195

Calculation of Electron Correlations by Using Local Operators

P. Fulde
Max-Planck-Institut für Festkörperforschung, 7000 Stuttgart 80, FRG

The use of local operators in electron-correlation calculations is reviewed. It is put in perspective with other methods based on canonical or localized orbitals, respectively. Results for atoms, molecules and solids are presented. For molecules containing first row atoms only, simple analytic expressions are derived for the inter- as well as intraatomic correlation energy contributions to the ground-state energy. It is also shown how the method of local operators can be applied to calculating completely delocalized quasiparticle excitations in e.g. semiconductors.

1 Introduction

The understanding and computation of electronic correlations is one of the frontiers of theoretical chemistry and solid state physics [1–3]. Because of their mutual Coulomb repulsion the dynamics of electrons in atoms, molecules or solids is a complex many-body problem. Dealing with it requires therefore drastic simplifying approximations. The problem becomes relatively simple when it is assumed that the electrons move independently of each other. In that case the total wave function of the electronic system reduces to an antisymmetrized product of single-electron wave functions which often is written in form of a Slater determinant. The one-particle functions are obtained by solving a single-particle Schrödinger equation. The result of this simplification is that the electron repulsions are taken into account only on the average. They enter the potential of the one-electron Schrödinger equation by averaging over the positions of the other electrons. When the former is determined self-consistently, the independent electron approximation goes over into the self-consistent field (SCF) approximation. It is well known that within the independent electron approximation the SCF eigenfunctions and eigenvalues are the best solutions of the full many-electron Hamiltonian.

In reality, however, an electron is not moving in accordance with where the other electrons are on the *average*, but rather where they are *actually*. In order to reduce the Coulomb repulsions as much as possible a correlated motion of all electrons must take place. It prevents them from coming too close to each other and therefore decreases their mutual repulsions. In effect each electron carries around itself a correlation hole, i.e. a spatial region within which the probability is depleted, as compared with the independent electron case, to find other electrons. It is precisely this correlation hole which is neglected when the SCF approximation is made. Therefore electrons can approach each other too closely and the Coulomb repulsion is overrated. The concept of the correlation hole is crucial for understanding, why local operators are particularly suitable for describing electron correlations. Therefore we shall introduce this concept in some more detail.

A missing correlation hole shows up clearly in the pair-distribution function $g(\boldsymbol{r}, \boldsymbol{r}')$. The latter is defined by the probability of finding an electron at position $(\boldsymbol{r} + \boldsymbol{r}')$ provided that there is one at point $\boldsymbol{r}'$, *relative* to the probability of finding a particle at position $(\boldsymbol{r} + \boldsymbol{r}')$ without that constrain. The pair-distribution function is closely related to the instantaneous density-density correlation function

$$S(\boldsymbol{r}, \boldsymbol{r}') = \frac{1}{\varrho(\boldsymbol{r} + \boldsymbol{r}')} \langle\varphi|\, \hat{\varrho}(\boldsymbol{r} + \boldsymbol{r}')\, \hat{\varrho}(\boldsymbol{r}')\, |\varphi\rangle \tag{1.1}$$

where $|\varphi\rangle$ is the state of the system. We have introduced the density operator

$$\hat{\varrho}(\boldsymbol{r}) = \sum_{i=1}^{N} \delta(\boldsymbol{r} - \boldsymbol{r}_i) \tag{1.2a}$$

where $\boldsymbol{r}_i$ denote the positions of the electrons. Furthermore

$$\varrho(\boldsymbol{r}) = \langle\varphi|\, \hat{\varrho}(\boldsymbol{r})\, |\varphi\rangle\,. \tag{1.2b}$$

The pair-distribution function can be expressed in terms of $S(\boldsymbol{r}, \boldsymbol{r}')$ as

$$g(\boldsymbol{r}, \boldsymbol{r}') = \frac{1}{\varrho(\boldsymbol{r}')} [S(\boldsymbol{r}, \boldsymbol{r}') - \delta(\boldsymbol{r} - \boldsymbol{r}')] . \tag{1.3}$$

Thereby the δ-function subtracts the correlation of an electron with itself. The function $g(\boldsymbol{r}, \boldsymbol{r}')$ is easily calculated for a homogeneous electron gas when the SCF approximation is made. In that case $g(\boldsymbol{r}, \boldsymbol{r}')$ depends on the relative coordinate only and one finds [4]

$$g_{SCF}(r) = 1 - \frac{9}{2}\left[\frac{\sin k_F r - k_F r \cos k_F r}{(k_F r)^3}\right]^2 \tag{1.4}$$

where k_F is the Fermi momentum. This function is graphically displayed in Fig. 1a. The oscillations in $g_{SCF}(r)$ result from the presence of a Fermi surface in momentum space. They are too small in order to be seen in the figure. At $r = 0$ the pair distribution function drops to 0.5. This is a consequence of the Pauli principle. It prevents electrons with equal spins from occupying the same position in space, while electrons with anti-parallel spin may come arbitrarily close. Therefore $g_{SCF}(r)$ shows the exchange hole which every electron carries around itself. The energy associated with it is the exchange energy. The moving electron with the exchange hole around it is the simplest case of a quasiparticle. When $g(r)$ is calculated for the *exact* ground state $|\psi_0\rangle$ instead for $|\varphi_{SCF}\rangle$, it has a form which is schematically shown in Fig. 1b. One notices the depletion of $g(r)$ for small values of r. It is due to electron

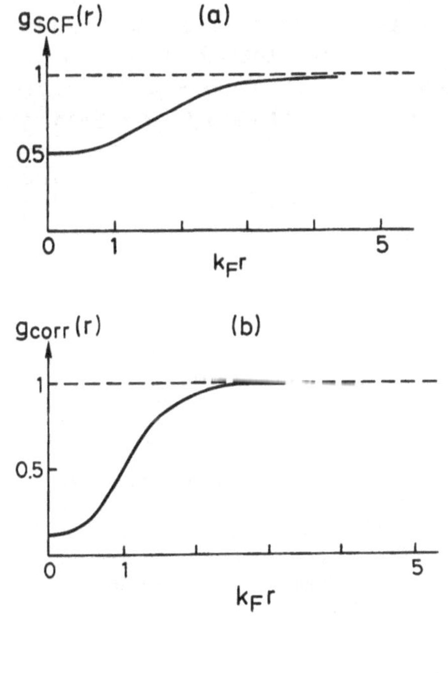

Fig. 1a and b. Pair-distribution function g(r) for the homogeneous electron gas: **a)** in SCF approximation. **b)** including electron correlations

correlations and represents the correlation hole. The energy associated with it is the correlation energy. The latter is defined as the difference between the exact and the SCF energy. From the definition of $g(\mathbf{r}, \mathbf{r}')$ it follows in a straightforward manner that the following sum rule holds

$$\int d^3r' \, \varrho(\mathbf{r}') \, (g(\mathbf{r}, \mathbf{r}') - 1) = -1 \tag{1.5}$$

The depletion in density around an electron due to the exchange-correlation hole corresponds therefore exactly to one particle. The above considerations show that the correlation hole is *local* in character. Therefore it is natural to describe it by using local functions. The more local the functions are which are used and the larger their number is, the better one can model the correlation hole and compute the correlation energy.

Let us consider the problem of electron correlations from yet another point of view. When electrons are treated within the SCF approximation the charge fluctuations at an atomic site of a molecule or solid come out too large. This feature is closely

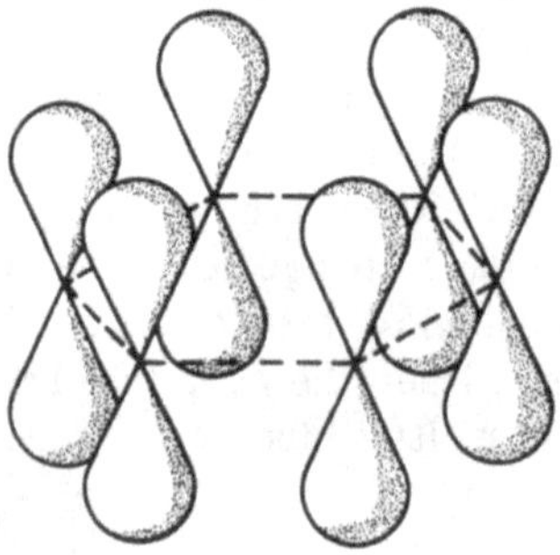

Fig. 2. Schematic plot of the p_z orbitals in C_6H_6

related to the shortcomings of the pair distribution function and is best seen by using a minimal set of basis functions. For the purpose of demonstration we consider the π electrons of a benzene molecule C_6H_6. We describe them by a set of basis functions for which we choose the (real) p_z functions of the valence electrons at the different carbon sites (see Fig. 2). The molecular orbitals (MO's) $\chi_\mu(\mathbf{r})$ are given by

$$\chi_\mu(\mathbf{r}) = \sum_{n=1}^{6} d_{\mu n} p_z(n) \, . \tag{1.6}$$

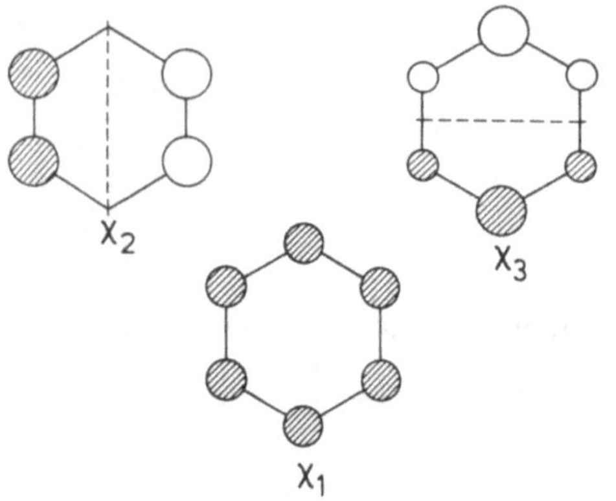

Fig. 3. MO's of lowest energy for π electrons in C_6H_6. The sizes of the circles indicate the relative weight of the contributions of the $p_z(n)$ orbitals. Shaded and empty circles refer to positive and negative coefficients, respectively

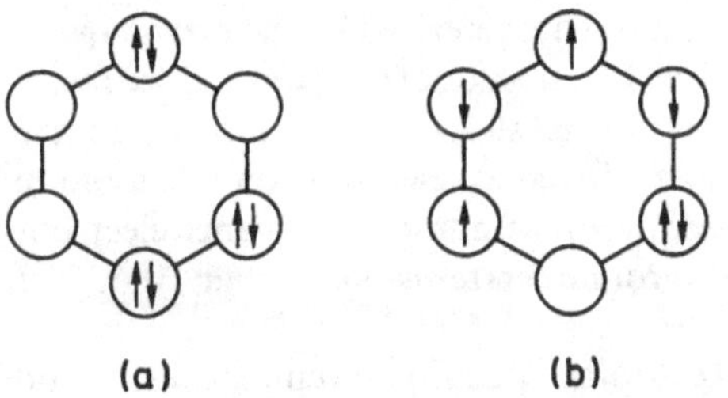

Fig. 4a and b. Two configurations of the π electrons in C_6H_6 : **a**) is an "unfavourable" configuration from the point of Coulomb repulsions, while **b**) is a "favourable" configuration

The index n lables the different C sites. The three MO's which are occupied in the ground state of the molecule are schematically shown in Fig. 3. The sizes of the circles around the C sites indicate the relative sizes of the coefficients $d_{\mu n}$. Shaded and empty circles refer to $d_{\mu n} > 0$ and $d_{\mu n} < 0$, respectively. In the ground state $|\varphi_{SCF}\rangle$ each MO is doubly occupied. Imagine $|\varphi_{SCF}\rangle$ expressed in terms of products of functions $p_z(n)$. This can be done by using the form (1.6) when constructing the Slater determinant of the ground state. Then a large number of different configurations results. Two of them are shown in Fig. 4. The electronic Coulomb repulsion is very different in the two cases, namely much larger in case (a) than in (b). Therefore we call (a) an "unfavourable" and (b) a "favourable" configuration. Configuration (a) contains large charge fluctuations. Only empty or doubly occupied sites are present despite of the fact, that on the average there is one π electron per site. We can characterize the shortcomings of the SCF approximation by noticing that in the SCF ground state $|\varphi_{SCF}\rangle$ the charge fluctuations are too large. The unfavourable configurations have too large weights. They are forced on $|\varphi_{SCF}\rangle$ by the requirement that it can be written in form of a single Slater determinant. Correlations suppress partially all those configurations contained in $|\varphi_{SCF}\rangle$, which correspond to large charge fluctuations.

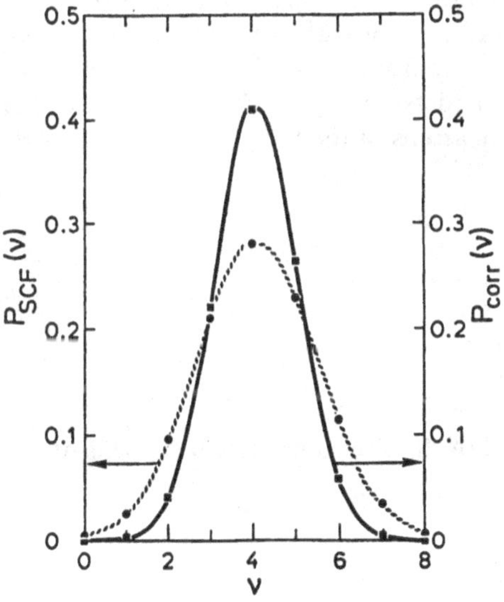

Fig. 5. Probability $P(\nu)$ of finding ν valence electrons on a C atom in C_2H_2 for the SCF and for the correlated ground state (from Ref. [5])

Charge fluctuations e.g. on an atomic site, can be characterized by the mean-square deviation of the charges from their average value. Electronic correlations reduce therefore those mean-square deviations. An example is shown in Fig. 5 where the probability is plotted of finding a given number of valence electrons on a C atom in C_2H_2. One notices that the mean-square deviations in the number of valence electrons are considerably less when the correlated ground state is used than they are within the SCF ground state.

The problem of dealing with electron correlations, which quantum chemistry and solid-state theory is facing, can be reduced to the following: Finding an efficient and accurate way of describing the correlation hole (in particular for small values of $\boldsymbol{r}$) or, alternatively, finding an efficient and accurate way of reducing electronic charge- and similarly spin-fluctuations.

In most cases molecular calculations are done within a basis set of L nonorthogonal functions $f_i(\boldsymbol{r})$. For the $f_i(\boldsymbol{r})$ usually Gaussian-type orbitals are used which are centered on different atoms. The creation and annihilation operators for electrons with spin σ in orbital $f_i(\boldsymbol{r})$ are $\hat{a}_{i\sigma}^+$ and $\hat{a}_{i\sigma}$ respectively. They fulfill the anticommutation relations $[\hat{a}_{i\sigma}^+, \hat{a}_{j\sigma}]_+ = S_{ij}\delta_{\sigma\sigma}$, where S is the overlap matrix, i.e. $\langle f_i \mid f_j \rangle = S_{ij}$.

Let us consider a system of N electrons for which one wants to compute the ground-state wave function $|\psi_0\rangle$. The dimension of the Hilbert space, within which $|\psi_0\rangle$ has to be found, is of order $\binom{2L}{N}$. It equals the number of different N-electron configurations which can be constructed. For practical calculations this number is usually much too large. Therefore the Hilbert space within which one wants to compute approximately $|\psi_0\rangle$ must be drastically reduced. In the following we shall discuss a number of different methods for selecting an appropriate subspace within which the ground-state problem can be solved. The aim is thereby to find a subspace with dimensions as small as possible, when results of a given quality are required. In order to achieve this, consequent use must be made of the fact that the correlation hole around an electron is a fairly local object. This suggests working with local orbitals or localized orbitals. The differences between the two are pointed out in the next section. In the first case it is more convenient to work with local *operators* which generate the local orbitals rather than with the orbitals themselves. After having compared the different theoretical approaches we shall discuss a number of results and show how they improve our understanding of chemical bonds. The main emphasis will be on molecules involving atoms of the first row. But also solids will be discussed.

2 Comparison of Different Methods

We start out by expressing the Hamiltonian H of the electronic system in terms of the basis set. It can be written in the form

$$H = \sum_{ij\sigma} \varepsilon_{ij} a_{i\sigma}^+ a_{j\sigma} + \frac{1}{2} \sum_{ijkl\sigma\sigma'} V_{ijkl} a_{i\sigma}^+ a_{k\sigma'}^+ a_{l\sigma'} a_{j\sigma} . \tag{2.1}$$

The first term includes the potential due to the nuclei, which the electrons experience and their kinetic energy. The second term describes the electron-electron interactions. The operators $a^+_{i\sigma}$, $a_{i\sigma}$ are adjoint to the operators $\hat{a}^+_{i\sigma}$, $\hat{a}_{i\sigma}$. Their commutation relations are therefore $[a^+_{i\sigma}, a_{j\sigma'}]_+ = S^{-1}_{ij}\delta_{\sigma\sigma'}$. Furthermore, the relation holds $[a^+_{i\sigma}, \hat{a}_{j\sigma'}]_+ = \delta_{ij}\delta_{\sigma\sigma'}$. In the independent electron approximation H is replaced by

$$H_{SCF} = \sum_{ij\sigma} f_{ij} a^+_{i\sigma} a_{j\sigma} \tag{2.2}$$

where the Fock matrix is defined in the well-known way as

$$f_{ij} = \varepsilon_{ij} + \sum_{kl\sigma'} \left(V_{ijkl} - \frac{1}{2} V_{ilkj} \right) \langle a^+_{k\sigma'} a_{l\sigma'} \rangle \tag{2.3}$$

and $\langle ... \rangle$ implies taking an average with respect to the SCF ground state $|\varphi_{SCF}\rangle$, i.e. $\langle ... \rangle = \langle \varphi_{SCF}| ... |\varphi_{SCF}\rangle$. The latter is an antisymmetrized product of the one-electron states $|\varphi_{\mu\sigma}\rangle = c^+_{\mu\sigma}|0\rangle$ (here $|0\rangle$ denotes the vacuum). The $|\varphi_{\mu\sigma}\rangle$ are eigenstates of H_{SCF}, i.e.

$$H_{SCF} |\varphi_{\mu\sigma}\rangle = \varepsilon_\mu |\varphi_{\mu\sigma}\rangle . \tag{2.4}$$

Therefore one can write

$$|\varphi_{SCF}\rangle = \prod_{\mu,\sigma}^{N} c^+_{\mu\sigma} |0\rangle . \tag{2.5}$$

We shall distinguish between occupied and unoccupied one-electron states by using Greek and Latin letters, respectively. Furthermore we shall consider in the following closed shell system only.

Formally, the exact ground-state wave function $|\psi_0\rangle$ can be written within the full or complete Hilbert space as

$$\begin{aligned} |\psi_0\rangle &= |\varphi_{SCF}\rangle + \sum_{i\mu} \alpha^i_\mu |\varphi^i_\mu\rangle + \sum_{\substack{i<j \\ \mu<\nu}} \alpha^{ij}_{\mu\nu} |\varphi^{ij}_{\mu\nu}\rangle + \dots \\ &= \left(1 + \sum_{i\mu} \alpha^i_\mu c^+_i c_\mu + \sum_{\substack{i<j \\ \mu<\nu}} \alpha^{ij}_{\mu\nu} c^+_i c^+_j c_\nu c_\mu + \dots \right) |\varphi_{SCF}\rangle . \end{aligned} \tag{2.6}$$

In order to avoid excessive labeling, we include here and sometimes in the following, the spin index σ in the other indices i, μ etc. Thus $|\psi_0\rangle$ consists of a sum of terms in which an increasing number of electrons have been moved from occupied to unoccupied states of $|\varphi_{SCF}\rangle$. One notices that all possible configurations are included this way (full CI expansion). The expansion (2.6) is of little use and, as pointed out above, the Hilbert space must be drastically reduced. One obvious choice is to use the one which is spanned by $|\varphi_{SCF}\rangle$, $|\varphi^i_\mu\rangle = c^+_i c_\mu |\varphi_{SCF}\rangle$ and those $|\varphi^{ij}_{\mu\nu}\rangle = c^+_i c^+_j c_\nu c_\mu |\varphi_{SCF}\rangle$ which couple directly to $|\varphi_{SCF}\rangle$, i.e. for which $\langle \varphi^{ij}_{\mu\nu}| H |\varphi_{SCF}\rangle \neq 0$. This is called CI-SD, i.e. one is considering a configuration space containing single and double substitutions. But even that space is too large in order to diagonalize H in it and find $|\psi_0\rangle$, because the number of two-particle excitations $|\varphi_{\mu\nu}^{ij}\rangle$ is of order $N^2(L-N)^2/4$. Therefore the Hilbert space must be further reduced. There are several ways of doing this. They are discussed in the following subsections. But before doing this we want to mention briefly the so-called size consistency problem. It is well known that a wave function $|\psi_0\rangle$ of a system when calculated within the space of two-particle excitations only, contains a decreasing portion of the electron correlations as the size of the system increases. This is intuitively obvious. Assume a system of A noninteracting atoms. In order to describe the correlations of electrons on a single atom one needs to admixe two-particle excitations to the SCF ground state of that atom. Since the total wave function is the product of the wave functions for the different atoms, it must contain A × 2-particle excitations. Keeping 2-particle excitations only is therefore not sufficient. The problem can be solved by writing for the correlated ground-state wave function instead of the form (2.6) the following one

$$|\psi_0\rangle = \exp\Big[\sum_{i\mu} \alpha_\mu^i c_i^+ c_\mu + \sum_{ij\mu\nu} \alpha_{\mu\nu}^{ij} c_i^+ c_j^+ c_\mu c_\nu + \dots\Big] |\varphi_{SCF}\rangle\,. \tag{2.7}$$

This is the ansatz used in the coupled-cluster method [6–7]. When going back to the problem of A noninteracting atoms, one notices that the form (2.7) factorizes, as required, i.e. $\exp\left(\sum_i A_i\right) |\varphi_{SCF}\rangle = \prod_i \exp A_i\, |\varphi_{SCF}\rangle$, where A_i refers to electrons on atom i. An alternative form for $|\psi_0\rangle$ is presented in the next section.

2.1 Canonical Orbital Methods

The simplest way of reducing the dimensions of the Hilbert space $\mathscr{H}$ is by working with selected single and double excitations which are obtained by applying some threshold (T) criteria. Within such a space $\mathscr{H}(T)$ the eigenvalues $E_n(T)$ and eigenfunctions $\psi_n(T)$ can be calculated. The index $n = 0$ corresponds to the ground state. The effect of the left-out configurations can be computed as follows [8, 9]. Let us take one of them which we label by $|\varphi_\tau\rangle$ and assume that its influence is sufficiently weak, so that the part of the improved eigenfunctions which lies in $\mathscr{H}(T)$ remains unchanged (except for a renormalization factor). In that case the effect of $|\varphi_\tau\rangle$ on e.g. the ground state is found from solving a simple 2 × 2 matrix problem with coupling matrix element $H_{0,\tau}(T) = \langle\psi_0(T)|\, H\, |\varphi_\tau\rangle$. It leads to an improved ground-state wave function and energy. The effect of all excluded configurations on the ground state energy is then estimated to be

$$E_0 = E_0(T) + \lambda \sum_\tau \frac{|H_{0,\tau}(T)|^2}{E_0(T) - E_\tau} \tag{2.8}$$

where $E_\tau = \langle\varphi_\tau|\, H\, |\varphi_\tau\rangle$. The correction factor λ is introduced because setting $\lambda = 1$ would overestimate the correction of the energy. The value of λ is obtained by doing the calculations for two different thresholds T_1 and T_2.

Another possible selection of states is by introducing the concept of pair-natural orbitals [10–12] (PNO's). It uses the fact that the number of pair holes $c_\mu c_\nu |\varphi_{SCF}\rangle$ is $N(N-1)/2$ and therefore relatively small. The aim is therefore to limit for a given pair of holes the number of important excitations. For that purpose one introduces the pair states

$$|\psi_{\mu\nu}\rangle = |\varphi_{SCF}\rangle + \sum_{i<j} \alpha^{ij}_{\mu\nu} |\varphi^{ij}_{\mu\nu}\rangle \tag{2.9}$$

(compare with Eq. (2.6)). By requiring that $E_{\mu\nu} = \langle\psi_{\mu\nu}| H |\psi_{\mu\nu}\rangle \langle\psi_{\mu\nu} | \psi_{\mu\nu}\rangle^{-1}$ is stationary, one obtains the set of equations

$$\langle\varphi^{ij}_{\mu\nu}| H |\varphi_{SCF}\rangle + \sum_{k<l} \langle\varphi^{ij}_{\mu\nu}| H |\varphi^{kl}_{\mu\nu}\rangle \alpha^{kl}_{\mu\nu} = (E_{SCF} + \varepsilon_{\mu\nu}) \alpha^{ij}_{\mu\nu} . \tag{2.10}$$

Here we have set $E_{\mu\nu} = E_{SCF} + \varepsilon_{\mu\nu}$. When we introduce one index for a pair only, i.e. $|\varphi^{P}_{\omega}\rangle$ for $|\varphi^{kl}_{\mu\nu}\rangle$, then the PNO's are defined as those superpositions of pair substitutions which diagonalize $\langle\varphi^{P}_{\omega}| H |\varphi^{P'}_{\omega}\rangle$. It turns out that in many cases a relatively small number of PNO's per pair of holes is required, say M, in order to obtain satisfactory results for the correlation energy. The dimension of the Hilbert space is thereby reduced to order $N(N-1) M/2$. Note that the PNO's belonging to different pairs are generally not orthogonal to each other. Also note that within that space also couplings between different pairs are taken into account when the correlation energy calculations are performed.

Nowhere in the above considerations have we used that the correlation hole of an electron is fairly localized. Instead, the reduced Hilbert space is spanned by states which are constructed from CMO's, which extend over the molecule. Therefore it takes many of those two-particle excited states in order to construct with high accuracy the correlation hole of the electrons. Expressed differently, the method is only slowly convergent, i.e. the Hilbert space must be extended considerably in order to achieve high accuracy.

2.2 Localized Orbital Methods

A more appropriate way of selecting configurations is by working with *localized* molecular orbitals (LMO's). We denote the corresponding creation operators by $\tilde{c}^+_i$, $\tilde{c}^+_\mu$ etc. In terms of them the SCF ground state takes the form

$$|\varphi_{SCF}\rangle = \prod_{\mu=1}^{N} \tilde{c}^+_\mu |0\rangle . \tag{2.11}$$

The $\tilde{c}^+_\mu$ are obtained from the c^+_μ by a unitary transformation U, which operates in the space spanned by the *occupied* CMO's, i.e.

$$\tilde{c}^+_\mu = \sum_{\nu=1}^{N} U_{\mu\nu} c^+_\nu . \tag{2.12}$$

In order to find the matrix U, two different localization procedures are widely used. They are due to Edmiston and Ruedenberg [13], and to Foster and Boys [14], respectively. The aim is thereby to limit each LMO to the smallest possible volume within the molecule and at the same time to have the LMO's spaced as far apart as possible. The requirement is either to maximize the energy of self-repulsion of the LMO's [13] or to maximize the sum over the squared distance between the centers of charge of the different LMO's [14]. The larger the self-repulsion of a LMO is, the smaller is the volume to which the orbital is confined. On the other hand, when the sum over the squared distances of the charge centers is maximum, the LMO's are kept as far apart as possible. Generally the Foster-Boys method is easier to handle numerically.

The same procedure which has been described for the occupied orbitals could in principle be applied to the unoccupied orbitals as well. A sufficiently reduced Hilbert space would be spanned by states of the form $\tilde{c}_i^+ \tilde{c}_j^+ \tilde{c}_\mu \tilde{c}_\nu \, |\varphi_{SCF}\rangle$ with orbitals μ, ν, i, j in the immediate spatial neighborhood of each other.

A step in the right direction is the Local Correlation method [15–17]. It combines ideas of the so-called "direct CI" method [18, 1b] with the physical picture of a local correlation hole. Some previous suggestion in this direction are due to Hurley [19]. Starting point are again the hole states $\tilde{c}_\mu \tilde{c}_\nu \, |\varphi_{SCF}\rangle$, but the requirement is given up, that the virtual states, into which the electrons are excited are orthogonal with respect to each other. For example, one may use for them basis set functions $f_i(\boldsymbol{r})$, or more precisely those parts of them which are orthogonal to the occupied LMO's. This has also the advantage that one avoids a transformation of the V_{ijkl} into another basis. The $f_i(\boldsymbol{r})$ are, of course, well localized. Because of the local character of the correlation holes only those functions $f_i(\boldsymbol{r})$ have to be taken into account which are situated in the direct neighborhood of the hole states. For various applications of the method we refer to Refs. [15–17].

One can go one step further and pose the following question. For a given accuracy of the correlation energy, what is the minimum number of functions $\chi_\nu^i(\boldsymbol{r})$ in virtual space which is required to achieve this? How are those functions to be constructed? Expressed differently, let $(\tilde{b}_{\nu\sigma}^i)^+$ denote the creation operator for electrons with spin σ in orbital $\chi_\nu^i(\boldsymbol{r})$. How large have M_i^{tot}, M_j^{tot} to be, so that the Hilbert space spanned by $(\tilde{b}_{\nu\sigma}^i)^+ \, (\tilde{b}_{\mu\sigma'}^j)^+ \, \tilde{c}_{\mu\sigma'} \tilde{c}_{\mu\sigma'} \tilde{c} \, |\varphi_{SCF}\rangle$ (i = 1, ... , M_i^{tot}; j = 1, ... , M_j^{tot} and the spin indices are shown explicitely) yields this accuracy? This problem was solved in Ref. [20]. An energy criterion was applied for optimizing, the functions $\chi_\nu^i(\boldsymbol{r})$. By using CH_4 and SiH_4 as an example, it was shown that their number is considerably less than that of PNO's, which are required to obtain the same accuracy. The functions $\chi_\nu^i(\boldsymbol{r})$ are local and allow for a simple physical interpretation. For more details and a plot of them we refer to Ref. [20].

Reducing the dimensions of Hilbert space by working with LMO's is more desirable than working with CMO's, because they are more apt for describing the local correlation hole around the electrons. However, the LMO's are not ideal either because of their limited degree of localizations. Firstly, there are the orthogonalization tails which those functions have. They result from the requirement that the occupied LMO's are orthogonal to each other. Furthermore, one expects the LMO's to be better localized in a molecule like CH_4 than in C_6H_6, where also the π electrons must be considered. In CH_4 one has well defined bonds and the LMO's essentially agree with them. On the other hand, the π electrons in C_6H_6 are delocalized. Three localized MO's can be constructed for them only by breaking the symmetry of the molecule. In that case they are concentrated on alternate positions on the ring. Even then their fall-off is slower than in molecules with well-localized bonds. This problem is well-known in solids. For systems with partially filled conduction bands it is not possible to construct well-localized Wannier functions, which are the analogue of the LMO's. The extreme is the case of the homogeneous electrons gas. In that case the Wannier functions fall off asymptoticly like r^{-2} for large values of r and not exponentially. This is so, because they have to be constructed from plane-wave states with momenta $|k| < k_F$, where k_F is the Fermi momentum. Note that in C_6H_6 the π states are half filled only. Because of a finite spacing of the energy levels, the localization problem is not as severe as in a metal.

Finally we want to discuss briefly the method of Perturbative-Configuration-Interaction using Localized-Orbitals (PCILO) [21]. It has the merit of being very simple with a right physical picture behind it. The prize is a limited accuracy, but this is not so important here. The PCILO method has been formulated for a minimal basis set $f_i(\boldsymbol{r})$ for which a zero-differential overlap (ZDO) approximation is made. It reduces the number of interaction matrix elements V_{ijkl} decisively. For each chemical bond i a bonding and antibonding orbital is constructed. Let us denote the corresponding creation operators by $B_{i\sigma}^+$ and $A_{i\sigma}^+$, respectively. The ground state is assumed to be of the form

$$|\varphi_0\rangle = \prod_{i\sigma} B_{i\sigma}^+ |0\rangle \tag{2.13}$$

and does not have SCF quality. This approximation is often called Bond Orbital Approximation (BOA) [22]. The Hilbert space within which the correlated ground state $|\psi_0\rangle$ is calculated is spanned by the states $A_{i\sigma}^+ B_{i\sigma} |\varphi_0\rangle$ and $A_{i\sigma}^+ A_{j\sigma'}^+ B_{j\sigma'} B_{i\sigma} |\varphi_0\rangle$, respectively. Furthermore, even within that space H is not diagonalized but computed perturbationally instead. The PCILO method clearly takes advantage of the local character of a correlation hole by working with local bond excitations. Because of its simplicity, one can derive analytic expressions for various correlation-energy contributions and obtain good order of magnitude estimates. We shall refer to this method in later sections.

2.3 Use of Local Operators

The logical consequence of the above is to free oneself completely from any orthogonality requirement when particle excitations are taken into account [23, 24]. This refers to the initial state as well as to the final states. It enables one to work with optimally local functions and has the advantage that only a minimum of configurations are required for generating the correlation hole around an electron. The Hilbert space can be reduced to a minimum for any required accuracy of the results. It turns out to be very convenient to work with local *operators*, which correspond to the local functions, instead of working with the functions themselves

[23, 24]. Because correlations are described in terms of these local operators the method has been called Local Ansatz.

In order to see how the generation of correlation holes is achieved, let us assume that the molecular volume is covered by meshes of local functions $g_l(\boldsymbol{r})$. This is symbolically shown in Fig. 6. We can expand them in terms of the basis set as

$$g_l(\boldsymbol{r}) = \sum_m \gamma_{lm} f_m(\boldsymbol{r}) . \tag{2.14}$$

They need not be orthogonal. The operators for creating or annihilating an electron with spin σ in the local orbital $g_l(\boldsymbol{r})$ are denoted by $b_{l\sigma}^+$ and $b_{l\sigma}$, respectively. Eq. (2.14) is then written in equivalent form as

$$b_{l\sigma}^+ = \sum_m \gamma_{lm} \hat{a}_{m\sigma}^+ . \tag{2.15}$$

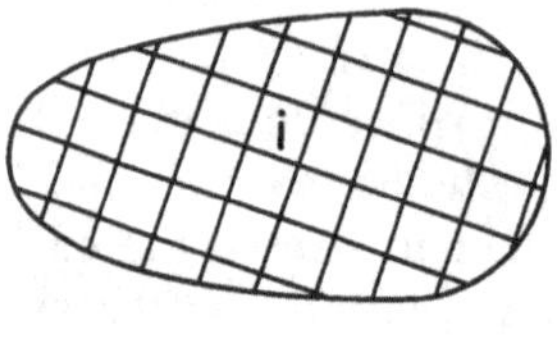

Fig. 6. Schematic picture of the subdivision of the molecular volume into (nonorthogonal) segments. To each of them corresponds a local function $g_i(r)$

One can define number operators $n_l = \sum_\sigma n_{l\sigma}$ with $n_{l\sigma} = b_{l\sigma}^+ b_{l\sigma}$ and spin operators $\boldsymbol{s}_l = \frac{1}{2} \sum_{\alpha\beta} b_{l\alpha}^+ \boldsymbol{\sigma}_{\alpha\beta} b_{l\beta}$ for electrons in local orbitals $g_l(\boldsymbol{r})$. The components of $\boldsymbol{\sigma}$ are the Pauli matrices. The selection criterion for the states, which span the reduced Hilbert space is the following: Only those two-particle excitations are considered, which are contained in states of the form $O_{kl} |\varphi_{SCF}\rangle$ with

$$O_{kl} = \begin{cases} n_{l\sigma} n_{l-\sigma} \delta_{kl} \\ n_k n_l \\ \boldsymbol{s}_k \boldsymbol{s}_l \end{cases} \tag{2.16}$$

When expressed in terms of the b operators one has, for example

$$n_{l\sigma} n_{l-\sigma} |\varphi_{SCF}\rangle = b_{l-\sigma}^+ b_{l\sigma}^+ b_{l\sigma} b_{l-\sigma} |\varphi_{SCF}\rangle . \tag{2.17}$$

The operators $b_{l\sigma}^+$ (and similarly $b_{l\sigma}$) can be expanded in terms of the operators $c_{\mu\sigma}^+$ and $c_{i\sigma}^+$ as follows

$$b_{l\sigma}^+ = \sum_{\mu=1}^{N/2} u_{l\mu} c_{\mu\sigma}^+ + \sum_{i=(N/2)+1}^{L} v_{li} c_{\sigma i}^+ . \tag{2.18}$$

This corresponds to expanding the $g_l(\boldsymbol{r})$ in terms of the CMO's. When (2.18) is set into Eq. (2.17) one notices that $n_{l\sigma} n_{l-\sigma} |\varphi_{SCF}\rangle$ contains a part which is proportional to $|\varphi_{SCF}\rangle$, and in addition one-electron excitations $|\varphi_\mu^i\rangle$ as well as two-electron excitations. Only the latter are of interest here. The two-particle excitation part is of the form

$$n_{l\sigma}n_{l-\sigma}|\varphi_{SCF}\rangle \xrightarrow[\text{2-part. excit.}]{} \sum_{ij\mu\nu} v_{li}v_{lj}u_{l\mu}u_{l\nu}|\varphi^{i\bar{j}}_{\mu\bar{\nu}}\rangle \tag{2.19}$$

The bar over an index implies that the spin is opposite to that of the corresponding other index, e.g. $(i, \bar{j}) \rightarrow (i\sigma, j-\sigma)$. This shows that the states $n_{l\sigma}n_{l-\sigma}|\varphi_{SCF}\rangle$ are weighted superpositions of the $|\varphi^{ij}_{\mu\nu}\rangle$. The great advantage is that the state $n_{l\sigma}n_{l-\sigma}|\varphi_{SCF}\rangle$ is distinct from $|\varphi_{SCF}\rangle$ only in a spatial region given by the extent of the local function $g_l(\boldsymbol{r})$. When we choose for the latter an atomic-like orbital it is much better localized than e.g. a LMO. Therefore by using local operators $n_{i\sigma}$ and $\boldsymbol{s}_{i\sigma}$ one can generate the correlation hole around each electron in an optimal way. An important point is that one needs only of order of 7N functions $g_i(\boldsymbol{r})$ in order to span, by means of (2.16), a sufficiently large Hilbert space [24]. This subspace allows for recovering of the order of 95% of the total correlation energy. The latter is the energy which one obtains in the spaced spanned by *all* $|\varphi^{ij}_{\mu\nu}\rangle$.

In order to avoid misunderstandings it should also be pointed out that states of the form $O_{kl}|\varphi_{SCF}\rangle$ can be always made to span the space of *all* two-particle excitations $|\varphi^{ij}_{\tau}\rangle$. In that case, however, the functions $g_l(\boldsymbol{r})$ are generally no longer localized and their number has to be of order NL.

3 The Local Ansatz

In this section we want to discuss in more detail the method of the Local Ansatz (LA) [23, 24] and to give a physical interpretation of it. Thereby the amphasis will be on the ground state, but also the problem of excited states is discussed. The results of the method, when applied to atoms, molecules and solids are discussed in the next section.

As pointed out before, the main idea of the Local Ansatz is to work with optimally local functions $g_i(\boldsymbol{r})$ when generating the correlation hole around the electrons. Their proper choice is a vital part of the method.

We start out again from the SCF ground state $|\varphi_{SCF}\rangle$ and make the following ansatz for the correlated ground state $|\psi_0\rangle$

$$|\psi_0\rangle = e^S|\varphi_{SCF}\rangle\,. \tag{3.1}$$

The operator S generates charge (or density) correlations as well as spin correlations. It consists of the form

$$S = -\sum_{mn}\eta_{mn}O_{mn} \tag{3.2}$$

and we want the operators O_{mn} to take the forms (2.16). But as pointed out before, when these operators act on $|\varphi_{SCF}\rangle$ they do not only generate two-particle excitations but also one-particle excitations, and a term which is proportional to $|\varphi_{SCF}\rangle$ itself. In order to eliminate the latter we introduce

$$O_{mn} = O'_{mn} - \langle O'_{mn} \rangle \tag{3.3}$$

and identify O'_{mn} with

$$O'_{mn} = \begin{cases} n_{m\uparrow} n_{m\downarrow} \delta_{mn} + n_m n_n (1 - \delta_{mn}) \\ s_m s_n \end{cases} \tag{3.4}$$

The index m refers to the orbital $g_m(\boldsymbol{r})$. The expression (3.3) for O_{mn} leaves us still with the one-particle excitations. They should not be tied to the two-particle excitations and are eliminated by forbidding contractions within O_{mn} operators when expectation values involving them are evaluated. This restriction implies immediately $\langle O'_{mn} \rangle = 0$ and makes the distinction between O_{mn} and $O_{m'n}$ become obsolete.

In order to understand better the effect of the operator exp S in Eq. (3.1) let us consider for the pupose of illustration the π electrons in C_6H_6 (see Fig. 2). We identify the functions $g_i(\boldsymbol{r})$ with the $p_z(i)$ orbitals of the different sites i. Consider the operator $\exp(-\eta n_{i=0\uparrow} n_{0\downarrow})$. When applied to $|\varphi_{SCF}\rangle$ it picks out that parts of it, in which the orbital $p_z(i = 0)$ is occupied by two electrons. When $\eta > 0$ the state $\exp(-\eta n_{0\uparrow} n_{0\downarrow}) |\varphi_{SCF}\rangle$ has a reduced probability amplitude for finding two electrons in orbital $p_z(0)$ as compared with $|\varphi_{SCF}\rangle$. This is precisely what correlations are expected to do, i.e. they partially suppress "unfavourable" configurations like the one with two electrons in orbital $p_z(0)$.

Similarly, when the operator $\exp[-\eta n_m n_n]$ is applied to $|\varphi_{SCF}\rangle$ it picks out that part of it, in which electrons are found in orbital $p_z(m)$ as well as in orbital $p_z(n)$. Depending on the sign of η, the amplitude of that part is either deminished or enhanced. Similar arguments hold for $\exp[-\eta s_m s_n] |\varphi_{SCF}\rangle$. For $\eta > 0$ that part of $|\varphi_{SCF}\rangle$, in which two electrons with antiparallel spins are in orbital $p_z(m)$ and $p_z(n)$, respectively, is enhanced while the part with two electrons of equal spin in orbitals $p_z(m)$ and $p_z(n)$ is decreased in amplitude. This makes the way transparent in which exp S generates a correlation hole around the electrons.

In the above example we have used for the functions $g_i(\boldsymbol{r})$ atomic orbitals (here $p_z(i)$). This way one can reduce e.g. charge fluctuations between *different* atoms (see Fig. 4). But in order to describe correlations between electrons on the

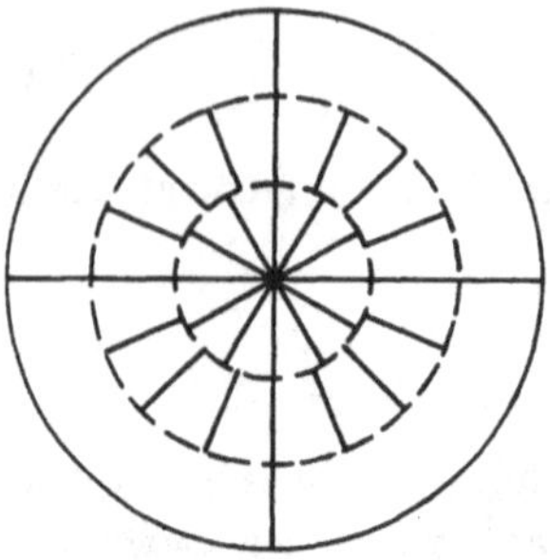

Fig. 7. Segmentation of the atomic volume. The segments represent (nonorthogonal) GTO hybrid functions with different exponential decrease

same atom one needs better localized functions. The meshes shown in Fig. 6 must be finer. An example is shown in Fig. 7 for the Ne atom. The different segments refer to different local functions $g_i(\boldsymbol{r})$. They are nonorthogonal. In practice those functions consist of three sets of hybrid functions with different exponential decrease. In order to construct 20 hybrids like for the middle shell of Fig. 7 one needs 1s, 3p, 5d, 7f and 4g basis functions. Finding the optimal segmentation of an atomic volume can be a tedious optimization process. However, once it has been found it can be used whenever that atom is part of a molecule.

The parameters η_{mn} in Eq. (3.2) are determined from the requirement that the energy

$$E = \frac{\langle\psi_0| H |\psi_0\rangle}{\langle\psi_0 | \psi_0\rangle} = \frac{\langle e^S H e^S\rangle}{\langle e^S e^S\rangle} \tag{3.5}$$

is minimized. By applying a linked cluster theorem [25] one can rewrite this expression in the form

$$E = \langle e^S H e^S\rangle_c \,. \tag{3.6}$$

The index c implies that only "connected" contractions between operators are taken when the expectation value is evaluated. Operators are contacted in such a way that the expectation value does not factorize into disconnected parts.

Unfortunately it is not possible to evaluate (3.6) without further approximations. In the following we shall use the approximation exp S = 1 + S. In that case E reduces to

$$E = E_{SCF} - 2 \sum_{mn} \eta_{mn} \langle O_{mn} H\rangle_c + \sum_{mn} \sum_{m'n'} \eta_{mn} \langle O_{mn} H O_{m'n'}\rangle_c \, \eta_{m'n'} \,. \tag{3.7}$$

With $E_{SCF} = \langle H\rangle$ the different expectation values are easy to evaluate when $|\varphi_{SCF}\rangle$ and the operators O_{mn} are expressed in terms of the basis set. For details we refer to Ref. [24].

Minimizing Eq. (3.7) results in the following set of equations for η_{mn}

$$\sum_{m'n'} \langle O_{mn} H O_{m'n'}\rangle_c \, \eta_{m'n'} = \langle O_{mn} H\rangle_c \,. \tag{3.8}$$

The correlation energy $E_{corr} = E - E_{SCF}$ of the system can therefore be written as

$$E_{corr} = -\sum_{mn} \eta_{mn} \langle O_{mn} H\rangle_c = \sum_{mn} \varepsilon_{corr}(m, n) \tag{3.9}$$

It is composed of contributions $\varepsilon_{corr}(m, n)$ which correspond to different local operators O_{mn}. When the η_{mn} are known, the ground state $|\psi_0\rangle$ in the form of Eq. (3.1) is known.

The above concept for calculating the correlated ground-state wave function can be generalized to excited states. Correlations in the excited states of molecules are often very different from the ones in the ground state. In order to demonstrate this let us consider diamond or Si as a macroscopic molecule. When an electron is moved from the occupied valence band to the unoccupied conduction band it leaves a hole state behind. The excited electron and the hole are usually well separated in space because of the large dimensions of the system. We neglect for the moment exciton like effects, which bind the electron to the hole. To good approximation one can therefore consider separately the state $c^+_{i\sigma}\,|\varphi_{SCF}\rangle$ with an electron added to the ground state, and the hole state $c_{\mu\sigma}\,|\varphi_{SCF}\rangle$. Considering the first one only, one can make again the ansatz

$$|\psi^{ex}_{i\sigma}\rangle = e^S\, c^+_{i\sigma}\,|\varphi_{SCF}\rangle\,. \tag{3.10}$$

However, there is now an essential difference to the ground state problem. Since there is an additional electron present, the system is locally no longer charge neutral. Charge neutrality is of course reestablished when the hole is included, but the latter is spatially well separated from the additional electron. Since the system is locally not charge neutral, the additional electron will polarize its surroundings, i.e. the bonds in its neighborhood (see Fig. 8). When it moves through the solid the polarization cloud is moving with it. The electron together with its polarization cloud forms a quasiparticle. The same arguments apply to the hole state.

What one wants to calculate is the energy of the quasiparticles. Due to the above, the operator exp S in Eq. (3.10) must generate a polarization cloud around the additional electron and it must also describe the correlations of the remaining electrons. The latter will resemble closely those in the ground state. In generalization of Eq. (3.2) the operator S in Eq. (3.10) is therefore of the form [26]

$$S = -\sum_{mn}\left[\eta^{(0)}_{mn} O_{mn} P_{mn}(i) + \eta^{(e)}_{mn} O_{mn} Q_{mn}(i)\right]. \tag{3.11}$$

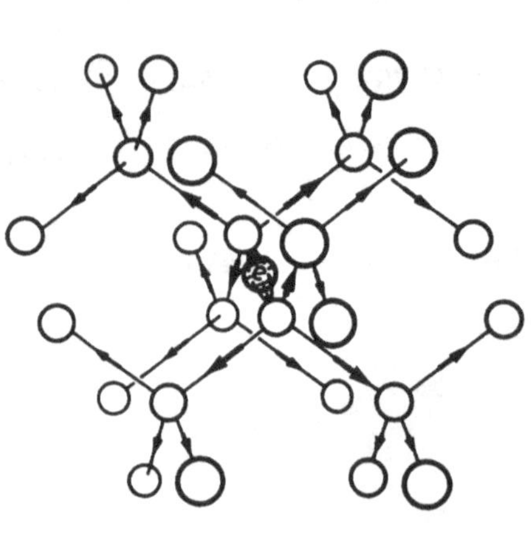

Fig. 8. Bond polarizations in the vicinity of an electron in the conduction band of e.g. diamond

The operators O_{mn} are the same as before. The $P_{mn}(i)$ and $Q_{mn}(i)$ are projection operators. They fulfil the equation

$$P_{mn}(i) + Q_{mn}(i) = 1 \tag{3.12}$$

and are defined by the following relations

$$[O_{mn}P_{mn}(i), c_{i\sigma}^{+}]_{-} = 0\,. \tag{3.13}$$

The physical content of these relations is easily seen. The operator $O_{mn}Q_{mn}(i)$ generates the polarization cloud of the electron $c_{i\sigma}^{+}$ while $O_{mn}P_{mn}(i)$ generates the correlations of the remaining electrons. Because of the relations (3.13) it is

$$O_{mn}P_{mn}(i)\, c_{i\sigma}^{+}\, |\varphi_{SCF}\rangle = c_{i\sigma}^{+}O_{mn}\, |\varphi_{SCF}\rangle\,. \tag{3.14}$$

The operator O_{mn} acts therefore on $|\varphi_{SCF}\rangle$ like in the ground state. It generates what we call ground-state correlations. The operator $O_{mn}Q_{mn}(i)$ on the other hand, does not commute with $c_{i\sigma}^{+}$. Its only influence on $|\varphi_{SCF}\rangle$ is that it generates single-particle excitations (relaxation or ΔSCF processes). Consequently, $O_{mn}Q_{mn}(i)$ "sticks" to the additional electron and generates a correlation hole around it, of which the polarization cloud is part of it.

The energy of the (N + 1) electron state (3.10) is conveniently written as

$$\begin{aligned} E_i^{N+1} &= \langle\psi_{i\sigma}^{ex}|\, H\, |\psi_{i\sigma}^{ex}\rangle\, \langle\psi_{i\sigma}^{ex}|\psi_{i\sigma}^{ex}\rangle^{-1} \\ &= \varepsilon_i + E_N^0 \end{aligned} \tag{3.15}$$

where E_N^0 is the ground state energy of the N electron system. By using a linked cluster expansion one can show [26] that the energy ε_i of the additional electron can be written as

$$\varepsilon_i = \frac{\langle c_{i\sigma}\, e^{S+}\, H\, e^{S}\, c_{i\sigma}^{+}\rangle_c}{\langle c_{i\sigma}\, e^{S+}\, e^{S}\, c_{i\sigma}^{+}\rangle_c} + [\langle e^{S+}\, H\, e^{S}\rangle_c - \langle e^{S_0}\, H\, e^{S_0}\rangle_c]\,. \tag{3.16}$$

The operator S_0 in the last term is given by Eq. (3.1) with η_{mn} parameters determined from a ground-state calculation. The parameters $\eta_{mn}^{(0)}$, $\eta_{mn}^{(e)}$ in Eq. (3.16) are obtained by minimizing E_i^{N+1}. As the size of a molecule becomes very large (as in a polymer), the term in brackets in Eq. (3.16) approaches zero. In that case the parameters $\eta_{mn}^{(0)}$ approach the values obtained from a ground-state calculation. The reason for that is simple. The additional electron with its polarization cloud influences less and less the correlations of the remaining N electrons as the size of the system increases.

The above considerations demonstrate that the concept of local operators can be applied not only to ground-state but also to excited-state calculations. A severe limitation is merely the expansion $\exp S = 1 + S\cdot$ when the expectation values (3.6) or (3.16) are evaluated. As in any CI or coupled electron pair calculation, such an expansion limits the calculations to not too strongly correlated electronic systems. As long as molecules are involved containing atoms of the second or third row only, this approximation is usually satisfied. Noticable exceptions are molecules like O_3 in which the π electrons are strongly correlated. However, when atoms with d (or f) electrons are involved, the above expansion is often clearly insufficient. Going beyond it by taking, for example, also terms of order $\langle S^2HS\rangle_c$ or $\langle S^2HS^2\rangle_c$ into account increases considerably the complexity of the calculation. For truly strongly correlated systems this is not sufficient either, and the method of starting from $|\varphi_{SCF}\rangle$ and building in correlations by an operator exp S must be questioned. A quantitative treatment of molecules or solids with strongly correlated electrons remains a challenging problem.

4 Results for Atoms and Molecules

We want to apply the above formalism and the physical picture which goes with it to atoms, molecules and in the next section also to solids. The main emphasis will be on molecules, and thereby in particular on finding simple rules which govern the different correlation contributions. Results of ab initio calculations will be presented [27]. But we want also to show that most of them can be obtained by much simpler means, namely by what we will call a semiempirical approach to electron correlations. Thereby one computes the *inter*atomic correlations by the Local Ansatz method, using a minimal basis set only. The *intra*atomic correlations are added by making use of an "atoms in molecules" approach. The quality of the results turns out to be surprisingly good in cases where a comparison with ab initio results is available. One advantage of the semiempirical approach is, e.g. that the correlation energy of almost any hydrocarbon molecule can be calculated.

4.1 Atoms

In order to show to which accuracy the correlation energy is obtained, when local operators of the form (3.4) are used, we consider the Ne atom. We use a basis set consisting of 5s(9s), 3p(5p), 3d, 1f GTO's. For a more accurate specification we refer to Ref. [24]. The atomic volume is subdivided as shown in Fig. 7. Three sets of functions $g_i(\boldsymbol{r})$ are used with different exponential decrease from the center of the atom. The set with the fastest decrease (inner set) consists of 12 non orthogonal hybrid functions. The set with the next fastest decrease (middle set) consists of 20 hybrids, while the set with the slowest decrease (outer set) consists of 4 hybrids. Finding the optimal form of the functions $g_i(\boldsymbol{r})$ is a straightforward but tedious process. This has been considered as a disadvantage of the method, mainly because this step is not completely free of arbitrariness. We do not share this point of view, because the final results for the correlation energy are very insensitive to details of the optimizing procedure.

A comparison of the calculated correlation energy $E_{corr} = E - E_{SCF}$ for the valence shell (i.e. valence-core and core-core electron correlations) with that of a corresponding CEPA (coupled electron pair approximation) calculation gives very close agreement. The values are $E_{corr} = 7.07$ eV and 7.29 eV, respectively. There are 21 different η parameters in the calculation. It is of interest to find out from which of them the largest contributions to E_{corr} result. For example, one can show that the most important correlations are due to the operators $n_i n_i$ and $n_i n_j$ when they refer to functions $g_i(\boldsymbol{r})$, $g_j(\boldsymbol{r})$ belonging to the middle set. On the other hand, one loses only less than 1% of the correlation energy when one neglects density correlations within the outer set and in addition spin correlations, except when they refer to the middle set. The net conclusion is that density correlations are much more important than spin correlations and that the former are largest when the electrons are in a spatial region described by the middle set.

Of interest is also that one should be able to establish a connection to Linnett's double quartet theory. With respect to Ne this theory predicts that the 8 valence electrons should preferably sit with alternating spins at the corners of a cube. Indeed, the calculations within the Local Ansatz yield spin-spin correlations which fluctuate strongly with distance. This would seem to support Linnett's point of view [28]. However, more detailed investigations are still lacking.

4.2 Molecules

We start out by discussing the ab initio calculations, before we digress to the extremely simple semiempirical correlation calculations. With respect to the ab initio calculations, one wants to know what the simplifications are, as compared with standard quantum chemistry calculations, when local operators are used and furthermore what the quality of their results is. For that purpose we shall consider the molecule CH_4 in more detail. One of the aims is to demonstrate the additivity of various correlation contributions. This is one of the main advantages of using local operators for correlation calculations.

We use a basis set of 6s(9s), 4p(5p), 1d GTO's for the C atom and sets of 3s(4s), 1p GTO's for the different H atoms, respectively. Let us begin with the interatomic correlations, which reduce the charge fluctuations between the different atoms. For the corresponding functions $g_i(\boldsymbol{r})$ one chooses the atomic parts of the four localized bond orbitals. Expressed differently, one first expands the LMO's in terms of the basis set, i.e.

$$\tilde{c}^{+}_{\mu\sigma} = \sum_{i,A} \alpha_{\mu i}(A)\, \hat{a}^{+}_{i\sigma}(A) \tag{4.1}$$

where the index A specifies at which atom A the basis set function $f_i(r)$ is centered. Then one identifies the operators $b^{+}_{i\sigma}$ (see Eq. (2.15)) with the largest contribution, which an atom makes to $\tilde{c}^{+}_{\mu\sigma}$, i.e.

$$b^{+}_{A\sigma}(\mu) = C_A(\mu) \sum_{i} \alpha_{\mu i}(A)\, \hat{a}^{+}_{i\sigma}(A)\,. \tag{4.2}$$

Here $C_A(\mu)$ is a normalization factor of order unity.

Therefore for CH_4 the functions $g_i(\boldsymbol{r})$ take the form of four sp^3 hybrid functions at the C atom and an s like function at each of the H atoms. Once they have

Table 1. Correlation energy for the ground state of CH_4 by using the Local Ansatz and for a CI calculation with the same basis set [29]. A better comparison would be with a CEPA calculation. One notices the additivity of the H atom contributions

Type of correlation energy (c.e.)	Local Ansatz [eV]	CI [eV]
Total	4.90	4.90
interatomic c.e.	1.76	
+ intraatomic c.e. C atom	3.91	
+ intraatomic c.e. *one* H atom	4.16	
+ intraatomic c.e. *second* H atom	4.40	
+ intraatomic c.e. *third* H atom	4.65	

been found, the operators O_{mn} are constructed according to Eqs. (3.3–4). With the help of the operators $n_i n_i$ occupancies of H atoms by two electrons are partially suppressed. Spin correlations of the form $\boldsymbol{s}_i \boldsymbol{s}_j$ between electrons on different H atoms imply spin correlations among electrons on the C atom. They ensure that for large atomic distances the $2p^2$ electrons at the C atom form a triplet state according to Hund's rule. The calculated interatomic correlation energy is listed in Table 1. Next, intraatomic correlations are added. For the C atom three sets of functions $g_i(\boldsymbol{r})$ are determined, similarly as it was done for the Ne atom. Due to the smaller basis set the numbers of hybrid functions in the inner, middle and outer set are 4, 12 and 4, respectively (see for comparison Fig. 7). The optimization of the functions $g_i(\boldsymbol{r})$ has been described in Ref. [24]. As has been pointed out before, this procedure is not completely free of arbitrariness, but the final results are very insensitive to the optimizing procedure. Also, once the $g_i(\boldsymbol{r})$ for the C atom have been found, one may use them whenever that atom is part of a molecule. Therefore the work of finding them has to be done only once.

Despite of this, the situation with respect to finding the functions $g_i(\boldsymbol{r})$ for intraatomic correlations remains open for improvements. One possible way of avoiding the problem is simply by including all local two-particle excitations $\hat{a}^+_{i\sigma}(A)$ $\hat{a}^+_{j\sigma'}(A)$ $\hat{a}_{k\sigma'}(A)$ $\hat{a}_{l\sigma}(A)$ in the calculations which couple to the SCF fround state. The index A refers to basis set functions $f_i(\boldsymbol{r})$, which are centered at site A. Such a program is presently being developed.

Similar arguments hold true for intraatomic correlations on different H atoms. Again, three sets of functions $g_i(\boldsymbol{r})$ are constructed. Here they consist of four sp^3 hybrids each. One may use the same functions as for the computation of intraatomic correlations in the H_2 molecule. Finally one can also construct operators $n_i n_j$ (or $\boldsymbol{s}_i \boldsymbol{s}_j$) for which n_i refers to an *inter*atomic local orbital $g_i(\boldsymbol{r})$, while n_j refers to an *intra*-atomic local orbital $g_j(\boldsymbol{r})$ at the same site.

In Table 1 it is shown how the correlation energy increases when one includes successively the intraatomic correlation operators O_{mn} for the C atom and for the different H atoms. Whenever one H atom is added the correlation energy increases by a fixed amount. This demonstrates the additivity of the correlation contributions of the different atoms. One notices also that the final result agrees very closely with that of a CI calculation done with the same basis set. A better comparison would be with a CEPA calculation, which should yield a slightly larger value for the correlation energy than does the CI calculation. But no such calculation is available. Finally we want to comment on the total computational effort which is required for a correlation energy calculation of a molecule by means of the Local Ansatz. The most costly computational step is a transformation of the interaction matrix V_{ijkl} into the basis of the $g_i(\mathbf{r})$. It involves of the order of $M^4L/8$ multiplications. M is the size of the basis set within which SCF calculations are done, while L ($\simeq 2M$ to $4M$) is the basis set for the correlation calculations. This is much less than the L^5 step which one encounters in CEPA or CI calculations.

4.3 Simplified Correlation Calculations for Molecules

Although the Local Ansatz is a very economical method for doing ab initio correlation-energy calculations it is still limited to molecules of medium size. It is therefore of interest to exploit the possibility of performing simplified correlation calculations, which can be applied also to large molecules. This can be accomplished by performing semiempirical SCF calculations onto which one implements correlation calculations [5, 30]. Although they do not have ab initio accuracy, they provide considerable insight into the nature of electron correlations.

The underlying idea of the semiempirical approach to electron correlations is to calculate the interatomic correlation energy within the Local Ansatz by using a minimal basis set. The intraatomic correlations are included by using an "atoms in molecule" approach. This is possible due to the additivity of different correlation energy contributions as pointed out above [24]. The same conclusions were reached by Lievin et al. [31] and we shall use some of their ideas. In the following we shall consider only molecules containing atoms of the first row.

4.3.1 Interatomic Correlations

We begin by describing the computations of interatomic correlations. They are done in close analogy to the one described in section III except that we introduce additional simplifications. One is to assume that the overlap matrix S_{ij} is diagonal. In that case the operators $a_{i\sigma}^+$ and $\hat{a}_{i\sigma}^+$ are the same. Furthermore, for the computation of the matrix elements ε_{ij} and V_{ijkl} of the Hamiltonian (2.1) the INDO (intermediate neglect of differential overlap) scheme is used. It implies considerable simplifications for the matrix elements V_{ijkl}. Only the Coulomb integrals $V_{iikk} = J_{ik}$ are kept, as well as the one-center exchange integrals $V_{ikik} = V_{ikki} = K_{ik}$, for which the functions $f_i(\mathbf{r})$ and $f_k(\mathbf{r})$ are centered on the same atom. In order to ensure rotational invariance, certain relations between the one-center Coulomb and exchange integrals must hold. For more details, how the values for the ε_{ij}, J_{ik} and K_{ik} are chosen we refer to the original literature [30]. It is perhaps worth stressing that in order to reproduce correctly the results of high-quality ab initio calculations for the interatomic correlation energy, one must use *un*screened on-site Coulomb integrals. With the Hamiltonian being specified, the CMO's are calculated and the ground-state wave function $|\varphi_0\rangle$ is constructed. Again, only closed shell systems are considered. From the CMO's one derives the LMO's by means of the Forster-Boys localization procedure. Very often they are essentially localized on two atoms and correspond to the classical chemical bonds. In that case one determines from them the corresponding hybrids for each atom and after a symmetric Löwdin $S^{-1/2}$ transformation identifies them with the functions $g_i(\boldsymbol{r})$. When the derived LMO's are not well localized, one first has to project them onto the different atoms and then find the $g_i(\boldsymbol{r})$ as described in Ref. [35]. For example, the $g_i(\boldsymbol{r})$ which are used in order to describe the correlations of π electrons in C_6H_6 are simply the $p_z(i)$ orbitals. By applying the Local Ansatz the interatomic correlation energies can be calculated, once the $g_i(\boldsymbol{r})$ are known. Because of Eq. (3.9) one can decompose the total interatomic correlation energy into different contributions which are associated with different bonds. At this place it should be pointed out that for the special case of completely localized orbitals the present calculation scheme resembles closely that of the PCILO method [21]. One difference is that the PCILO method uses a perturbational expansion while here we apply a variational calculation. The operators $O_{ii} = n_i n_i$ and $O_{ij} = n_i n_j$, when acting on $|\varphi_0\rangle$, correspond in the PCILO method to a double excitations within a bond and single excitations in two different bonds, respectively.

The method has been applied to calculate the interatomic correlation energy for a large number of molecules involving C, N and H atoms. Some of the molecules were of considerable size. A number of interesting relations were found. We want to discuss them in the following. Let us begin with σ bonds. The (intrabond) correlation energy associated with a σ bond was found to depend only on the bond length d and on the type of bond (e.g. C—H, C—N, N—H etc.). Examples are shown in Figs. 9 and 10.

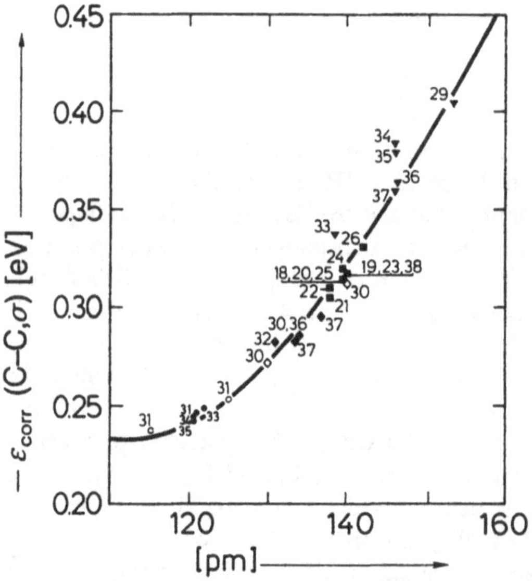

Fig. 9. Interatomic correlation energy within C—C σ bonds as function of bond length for different molecules. The latter are listed in the text according to their number. The solid line corresponds to Eq. (4.3) (from Ref. [30]).

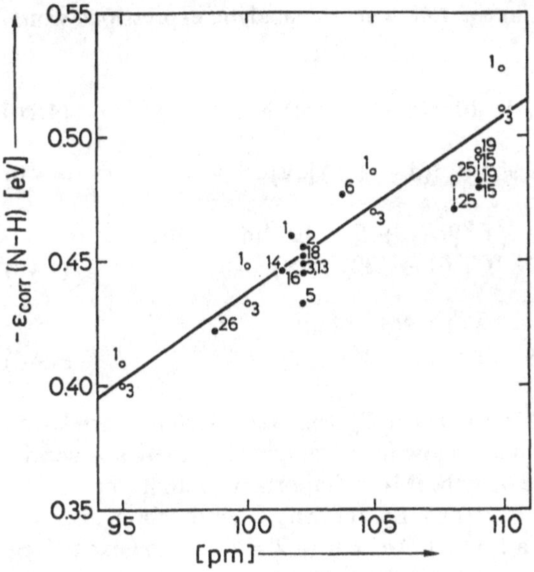

Fig. 10. Interatomic correlation energy for N—H bonds in different molecules as function of bond length. For molecules 1 and 3 also nonequilibrium bond lengths were considered (open circles); for molecules 15, 19 and 25 $-\varepsilon_{corr}(N{-}H)\,(1-\alpha_p^2)^{-5/2}$ is shown by open circles. The solid line corresponds to Eq. (4.4a) (from Ref. [30])

The numbers in the drawings refer to the following closed shell molecules. 1: NH_3; 2: N_2H_4; 3: N_2H_2; 4: N_2; 5: N_3H; 6: NH; 7: N_2O_4; 8: N_2O; 9: N_2F_4; 10: HCN; 11: CN^-; 12: $(CH_3)_3N$; 13: CH_3NH_2; 14: CH_3NH_2; 15: $CH_3NH_3{}^+$; 16: CH_2NH; 17: $(CH_2N)_2$; 18: $(C_6H_5)NH_2$; 19: $(C_6H_5)NH_3^+$; 20: $(C_6H_5)CN$; 21–23: six-membered dinitrogen heterocycles, i.e. $C_4N_2H_4$ isomers: pyridazine (21, $C_4N_2H_4$ I) pyrimidine (22, $C_4N_2H_4$ II) and pyrazine (23, $C_4N_2H_4$ III); 24: C_5H_5N; 25: $C_5H_5NH^+$; 26: pyrrol; 27: purine; 28: CH_4; 29: C_2H_6; 30: C_2H_4; 31: C_2H_2; 32: allene; 33: diacetylene; 34: propyne; 35: C_4H_6; 36: butadiene; 37: transhexatriene; 38: benzene; 39: azulene; 40: naphthalene.

One notices that in the case of a C—C σ bond the data for the different molecules fall onto a universal curve. It is given by the equation

$$\begin{aligned}\varepsilon_{corr}(C{-}C;\sigma) &= -0.27 - 3.0\cdot 10^{-3}(d-130)\\ &\quad - 1.0\cdot 10^{-4}(d-130)^2\ [eV]\end{aligned} \tag{4.30}$$

when d is measured in pm. For the N—H bonds the data scatter somewhat more, but still lie within 10% on a universal curve. The scattering can be partially accounted for by correcting for the bond polarity α_p. The latter is defined by $\alpha_p = |1 - \bar{n}_H|$ where $\bar{n}_H$ is the occupancy of the H site. It can be shown that a finite bond polarity diminishes the correlation energy of a (fictitous) nonpolar bond by a factor of $(1-\alpha_p)^{5/2}$. Therefore we show also in Fig. 10 data of the form $[\varepsilon_{corr}(N{-}H)]\cdot(1-\alpha_p^2)^{-5/2}$. It is seen that this reduces the scatter of the data. Similar results are found for other σ bonds. We list in the following the analytic expressions which fit the numerical results

$$\varepsilon_{corr}(N{-}H) = -0.44 - 7.1 \cdot 10^{-3}(d - 100)\ [eV] \tag{4.4a}$$

$$\varepsilon_{corr}(C{-}H) = -0.39 - 5.9 \cdot 10^{-3}(d - 100)\ [eV] \tag{4.4b}$$

$$\varepsilon_{corr}(C{-}N;\sigma) = -(1 - \alpha_p^2)^{5/2}[0.31 + 5.7 \cdot 10^{-3}(d - 130) + 1.30 \cdot 10^{-4}(d - 130)^2]\ [eV] \tag{4.4c}$$

$$\varepsilon_{corr}(N{-}N;\sigma) = -0.37 - 7.8 \cdot 10^{-3}(d - 130) - 1.35 \cdot 10^{-4}(d - 130)^2\ [eV]\,. \tag{4.4d}$$

One exception to the above is the N—Nσ bond in N_2O_4. Here the calculated correlation energy is by 0.5 eV smaller than it would follow from applying Eq. (4.4d). A special feature of that bond is its extreme length. More important, within the INDO scheme one finds that it contains 2.3 electrons. By applying a bond-orbital approximation (BOA) one would argue that a fraction $\Delta n = n - 2 = 0.3$ of the excitations from bonding into antibonding states cannot take place as compared with the case of two electrons in the bond. This suggests that $-\varepsilon_{corr}(N{-}N;\sigma)\cdot(1-\Delta n)^{-1}$ should be identified with the right hand side of Eq. (4.4d). This would raise the value for the bond by 0.37 eV. For a discussion of the remaining difference we refer to Ref. [30].

Until now we have not yet considered interbond correlations of σ electrons. They are described by operators of the form $O_{ij} = n_i n_j$ with i and j referring to hybrids, which belong to neighboring bonds. The corresponding correlation energy contributions are relatively small, as can be seen from Table 2. They depend more sensitively on the actual form of the hybrids $g_i(\mathbf{r})$ than the intrabond correlation energy. One notices that the interbond correlation energy is larger for sp^3 than for sp^2 hybrids.

Table 2. Interbond correlation energy contributions $\varepsilon_{corr}^{(i,j)}$ in eV for a pair of bonds which join at a central C or N atom. sp^3, sp^2, and sp refers to the hybridization on the C (or N) atom (from Ref. [30])

	sp^3	sp^2	sp
H—C—H	0.010	0.004	—
H—C—C	0.021	0.009	0.000
H—C—N	0.017	0.006	0.000
H—N—H	0.025	0.015	—
H—N—C	0.044	0.023	—
H—N—N	0.041	0.021	—
C—C—C	—	0.018	0.003
C—C—N	—	0.020	—
N—C—N	—	0.011	—
C—N—C	0.077	0.041	—
C—N—N	—	0.045	—
N—N—N	—	—	0.010

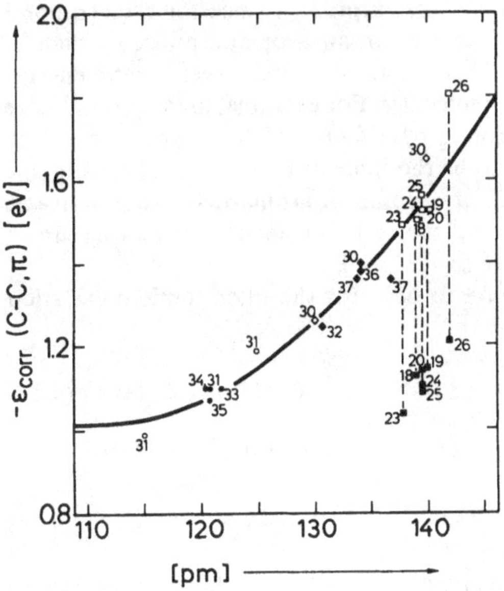

Fig. 11. Interatomic correlation energy contribution of a pair of π electrons in a C—C π bond as function of bond length. Unfilled symbols refer to nonequilibrium bond lengths. Renormalized correlation energies for aromatic systems (18–20, 23–26) are shown by unfilled squares. For molecules 23 and 26 additional corrections due to an enhanced π-electron density were included. The solid line corresponds to Eq. (4.5c) (from Ref. [30])

Next we consider π bonds. In the presence of a π bond there is also a contribution to ε_{corr} which results from correlations between π and σ electrons. When there are two π bonds present there is also an additional contribution π/π' between the two. The various contributions can be computed by the Local Ansatz technique. By adding them one can associate a correlation energy with each π bond (or pair of π electrons) and check whether the data falls onto a universal curve. The results for C—Cπ and C—Nπ bonds are shown in Figs. 11 and 12. One notices that in both cases there is a group of aromatic molecules, which have a smaller correlation energy than

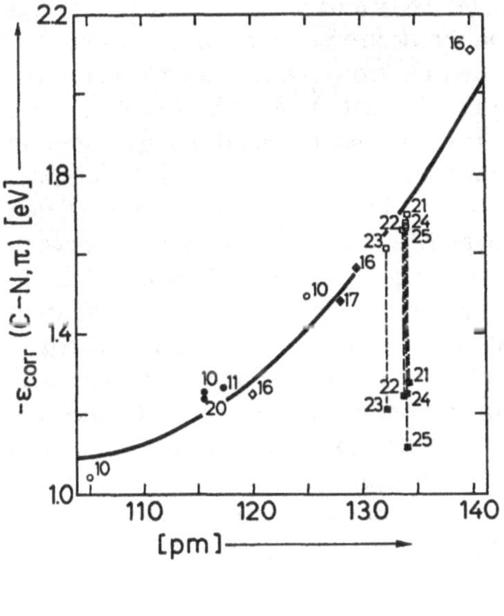

Fig. 12. The same as in Fig. 11 but for C—N π bonds. Renormalized energies for aromatic molecules (21–25) are shown by unfilled squares. For molecule 25 an additional correction due to bond polarity is included. The solid line corresponds to Eq. (4.5b) (from Ref. [30])

would follow from the bond length. This is not surprising, since for a given bond length the correlation energy is certainly smaller in an aromatic molecule than in a molecule with well-localized π bonds. This is related to the larger kinetic energy, which the π electrons have in an aromatic molecules. For example, in the ground state of C_6H_6 the kinetic energy of a π electron is by a factor of 4/3 larger than e.g. in C_2H_4, when the bond length is assumed to be the same in both cases. Therefore one expects that multiplication of ε_{corr} by 4/3 for π bonds in aromatic benzene derivates should bring the value closer to the universal curve. This is indeed the case as can be seen 11 and 12. For more details we refer to Ref. [30].

The following simple relations we found to hold for the interatomic correlation energy associated with π bonds

$$\tilde{\varepsilon}_{corr}(N{-}N;\pi) = -1.99{-}4.64\cdot 10^{-2}(d-130) - 6.63\cdot 10^{-4}(d-130)\ [eV] \quad (4.5a)$$

$$\tilde{\varepsilon}_{corr}(C{-}N;\pi) = -1.57{-}3.47\cdot 10^{-2}(d-130) - 6.28\cdot 10^{-4}(d-130)\ [eV] \quad (4.5b)$$

$$\tilde{\varepsilon}_{corr}(C{-}C;\pi) = -1.57{-}3.47\cdot 10^{-2}(d-130) - 6.28\cdot 10^{-4}(d-130)\ [eV]\,. \quad (4.5c)$$

$\tilde{\varepsilon}_{corr}$ and ε_{corr} are the same except for aromatic molecules for which $\tilde{\varepsilon}_{corr} = (4/3)\,\varepsilon_{corr}$.

The above considerations enable us to define the quantity Σ_i as a measure of the strength of interatomic electron correlations. The following is chosen for Σ_i.

$$\Sigma_i = \frac{\langle\varphi_{SCF}|\,\Delta n_i^2\,|\varphi_{SCF}\rangle - \langle\psi_0|\,\Delta n_i^2\,|\psi_0\rangle}{\langle\varphi_{SCF}|\,\Delta n_i^2\,|\varphi_{SCF}\rangle}\,. \quad (4.6)$$

Here $\Delta n_i = n_i - \langle n_i\rangle$ describes the deviation of an orbital occupancy from its average value. The mean-square deviations are computed for the correlated and uncorrelated ground-state wave function, respectively. The index i refers to the different hybrid functions $g_i(r)$ which are used for the description of interatomic correlations. Σ_i describes the reductions in the mean-square deviations of the electronic charge distribution. It tells us how strongly in a bond the electronic charge fluctuates around its mean value as compared with uncorrelated electrons. In a Heitler-London description of the H_2 molecule the charges do not fluctuate at all. There is always one electron at each hydrogen site. In that case $\Sigma = 1$. On the other hand, when the ground state $|\psi_0\rangle$ of H_2 is described by an antisymmetrized product of occupied molecules orbitals, the fluctuations are largest and $\Sigma = 0$. This is the limit of uncorrelated electrons. Knowing the mean-square deviations of the charges is equivalent to knowing the interatomic correlations. The correlation-strength parameter Σ is therefore an important quantity. It has to be known in addition to the charge distribution $\langle n_i\rangle$ in order to characterize a bond. Shown in Fig. 13 are the computed Σ values for a number of bonds [32]. One notices a number of interesting features. Generally Σ is larger for π bonds than for σ bonds. This is expected, since the overlap between orbitals forming the bond is smaller in the first case. One notices also that there are groups of bonds with comparable Σ values. For example, different σ bonds which are part of double bonds form such a group (with $\Sigma \simeq 0.14$) and so do π bonds which belong to triple bonds (with $\Sigma \simeq 0.27$). It is also seen that

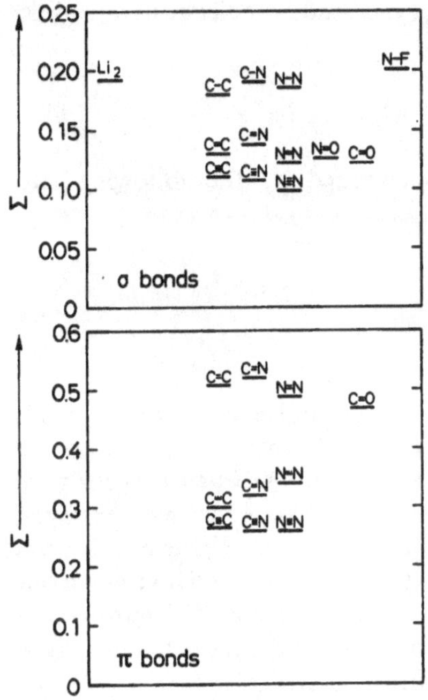

Fig. 13. Correlation strength parameter Σ for a number of bonds formed by first-row atoms. Single, double, triple and aromatic bonds are indicated by single, double, triple and dotted over solid lines, respectively (from Ref. [32])

π bonds in aromatic molecules have a smaller π value ($\Sigma \simeq 0.33$) than π bonds in non-aromatic molecules ($\Sigma \simeq 0.5$). Furthermore, the correlation strengths in triple bonds are generally smaller than the ones in double bonds. This is due to the smaller bond length which causes an increase in the overlap between the orbitals forming the bond.

4.3.2 Intraatromic Correlations

After having discussed interatomic correlations we turn towards an efficient calculation of intraatomic correlations. There is, of course, some arbitrariness when one is dividing the electronic correlations into an interatomic and intraatomic part. We regard all correlations which can be described within a minimal basis set as interatomic. They contain also e.g. a s^2p^2 excitation at a carbon site. This excitation is also present in a single atom and therefore could have been labeled intraatomic as well.

In the following use is made of experience gained with ab initio calculations on small molecules. Thereby it was found that to good approximation intraatomic correlations can be treated separately for each atom of a molecule. Interference effects are very small and may be neglected.

Starting point is the ground state $|\psi_0\rangle$ expressed in terms of a minimal basis set. It contains the interatomic correlations in form of the sansatz (3.1). Next, a population analysis of $|\psi_0\rangle$ is done with respect to each atom of the molecule. This is accomplished by assuming, that the probability $P_A(\nu)$ of finding ν electrons at a given atomic site A is given by a Gaussian distribution

$$P_A(\nu) = (2\pi\beta^A)^{-1/2} \exp[-(\nu - \nu_0^A)^2/2\beta^A] . \tag{4.7}$$

Here ν_0^A and β_A are parameters which are determined by the following two relations

$$\begin{aligned} \sum_\nu \nu P_A(\nu) &= \bar{n}_A \\ \sum_\nu \nu^2 P_A(\nu) &= \overline{n_A^2} . \end{aligned} \tag{4.8}$$

The first equation defines the average valence electron number $\bar{n}_A$ at site A, while the second equation defines the second moment of this electron number. Both, i.e. $\bar{n}_A$ and $\overline{n_A^2}$ can be determined from $|\psi_0\rangle$. For a fixed number ν of valence electrons at site A, there exist different configurations i with weights $w_i(\nu, A)$. For example, for $\nu = 5$ the configurations s^2p^3, sp^4, p^5 have different relative weights at a C or N site, respectively. We shall assume that these relative weights do not change when one goes over from $|\varphi_0\rangle$ to $|\psi_0\rangle$. The $P_A(\nu)$ themselves are, of course, different for the uncorrelated and correlated ground state (see Fig. 5). Leaving the $w_i(\nu, A)$ unchanged is an approximation which is discussed in more detail in Ref. [33].

Each of the different configurations i consists of different terms. We shall assume that they enter with weights according to their respective degeneracies, e.g. $s^2p^3 = (1/20)(4 \cdot {}^4S + 10 \cdot {}^2D + 6 \cdot {}^2P)$. This assumption is not strictly fulfilled even for $|\varphi_0\rangle$, when the chemical environment of site A is of low symmetry. Furthermore it underestimates the effects of Hund's rule correlations contained in $|\psi_0\rangle$, in particular in π electron systems. The Hund's rule ground state should have a larger weight than given by its degeneracy. For the sake of simplicity we shall neglect those complications in the following. The error introduced by this approximation is expected to be less than 0.2 eV per C or N atom, respectively.

The intraatomic correlation energy of a molecule is then given according to Lievin et al. [31] as

$$E_{corr}^{intra} = \sum_A \sum_{\nu i} P_A(\nu)\, w_i(\nu, A)\, \varepsilon_{\nu i}^{corr}(A) . \tag{4.9}$$

The $\varepsilon_{\nu i}^{corr}(A)$ are the correlation energies of an atom A with ν valence electrons in different configurations. They include the effect of the core electrons. They have been tabulated for atoms of the first row [33, 34]. The energies are accurately known only for neutral and positively charged atoms. The ones for negative ions are only estimated and therefore may lead to some errors when used in Eq. (4.9). Furthermore, excitations of the form $s^2p^\nu \to s^0p^{\nu+2}$ ($\nu = 0, \dots 4$) are already contained in the minimal basis set and must be substracted out here.

The intraatomic correlation energy has been computed according to Eq. (4.9) for a large number of molecules consisting of first-row atoms. Thereby it has been found that the results can be well described by the simple algebraic expression

$$E_{corr}^{intra}(A) = -(0.119 + 1.020\, r_p^A)(\bar{n}_A)^{1.23} . \quad [\mathrm{eV}] \tag{4.10}$$

Table 3. Intraatomic correlation energy for a H and a C (or N) atom in small molecules. Calculations were done by a population analysis (PA) and by applying the analytic expressions (4.10, 4.11) instead (AM). In the latter case the electron densities $\bar{n}_H$ and $\bar{n}_{C/N}$ as well as the fraction r_p of p electrons at the C(N) atom are required (from Ref. [30])

	H atom			C/N atom			
	$\bar{n}_H$	AM	PA	$\bar{n}_{C/N}$	r_p	AM	PA
CH_4	1.034	0.233	0.230	3.865	0.752	4.67	4.71
C_2H_6	1.030	0.231	0.231	3.911	0.762	4.79	4.82
C_2H_4	1.016	0.224	0.224	3.967	0.749	4.81	4.83
C_2H_2	0.961	0.197	0.196	4.039	0.732	4.82	4.84
NH_3	1.006	0.219	0.221	4.982	0.770	6.52	6.50
N_2H_4	1.002	0.217	0.220	4.977	0.769	6.52	6.52
N_2H_2	0.967	0.200	0.203	5.034	0.728	6.29	6.28
N_2	—	—	—	5.000	0.660	5.73	5.78

Here $r_p^A = n_p^A/\bar{n}_A$ is the fraction of p electrons n_p^A among the $\bar{n}_A = n_s^A + n_p^A$ valence electrons at a first-row atom A in a molecule. This relation shows that a p electron is stronger correlated by more than 1 eV than a s electron. For a H atom, the computed intraatomic correlation energy follows the algebraic from

$$E_{corr}^{intra}(H) = -0.216(\bar{n}_H)^{2.30}. \quad [eV] \tag{4.11}$$

In Table 3 a comparison is made between the correlation energy of a number of small molecules, when the analytic expressions (4.10) and (4.11) are used and when a population analysis with subsequent application of Eq. (4.9) is made. One notices close agreement in both cases.

Together with Eqs. (4.3–5) the relations (4.10) and (4.11) constitute a set of simple expressions from which the total correlation energy of a molecule involving C, N and H atoms can be computed.

4.3.3 Total Valence-Electron Correlation Energies

Listed in Table 4 are the results of inter- plus intraatomic correlation energies for a number of different molecules, when calculation are done within the semiempirical scheme. Also shown are the experimental correlation energies as determined by Cremer [35]. It is seen that the overall agreement is good. For CH_4 the calculated values are by 5% too small, while for C_2H_2 they are by 4% too large. It can be shown that the errors result thereby from the intratomic part. For more details we refer to Ref. [30].

The above computational scheme can be used to obtain better insight into the role, which various correlation contributions play. For example, it is well known that electron correlations make sizeable contributions to molecular binding energies. For small molecules like N_2 or HCN this contribution is almost 4.5 eV. The analytic expressions (4.4; 4.5) and (4.10; 4.11) enable one to tell how these numbers come about. The interatomic correlation energies can be easily calculated and one can

Table 4. Semiempirical correlation energy calculation for a number of molecules (in eV). Calculations are done by means of the Local Ansatz and a population analysis (LA + PM) and by using the analytic expressions instead (LA + AM). Also shown are the experimental correlation energies (Expt) as given in Ref. [37]

No.	Molecule	INDO		
		LA + PA	AM	Expt
28	CH_4	7.57	7.46	7.97
29	C_2H_6	14.28	14.21	14.97
30	C_2H_4	14.06	13.96	13.69
31	C_2H_2	13.38	13.28	12.82
1	NH_3	8.62	8.61	9.09
2	N_2H_4	16.48	16.46	16.95
3	N_2H_2	16.02	16.06	15.81
4	N_2	14.59	14.37	14.69
16	CH_3NH_2	15.38	15.30	15.84
14	CH_2NH	15.07	15.02	14.78
10	HCN	14.08	13.90	13.77
21	$C_4N_2H_4$ (I)	40.52	40.53	—
22	$C_4N_2H_4$ (II)	40.49	40.49	—
23	$C_4N_2H_4$ (III)	40.31	40.35	—
27	purine	61.56	61.30	—
38	C_6H_6	38.90	38.82	—
39	azulene	63.92	63.68	—
40	naphthalene	63.88	63.66	—

study the different contributions to it. The changes in the intraatomic correlation energies as compared with the free atom case are mainly due to conversions of s into p electrons at C sites and to contributions from H sites. Similarly one can study problems like the correlation contribution to the inversion-barrier height in NH_3. With the help of a slide rule one can make a simple and rather precise estimate of it. One merely has to know the changes in the bond length and in the p electron count at the N site, when the planar instead of the equilibrium geometry is used. Everything else follows from the analytic expressions given above. For more details see Ref. [30].

5 Applications to Solids

In this section a description is given of the applications of local operators to correlation calculations for solids. This is of interest for several reasons. First of all, it is likely that solids will turn out to be a fertile field for quantum chemistry calculations in the future. There is no reason, why one should not treat molecules and solids by the same methods. It is also of interest to study which of the molecular features remain in a solid. This holds true in particular for the correlation energy. In the following we want to describe ground state calculations for semiconductors, before we consider excited states. In all these cases we find it mandantory to work with local operators.

5.1 Ground State Calculations

Ab initio ground state correlation calculations exist at present only for very few systems, i. e. for diamond [36] and more recently also for silicon and graphite. Thereby the main problem has been the SCF part of the calculations, due to the slow convergence of the non-local exchange. In contrast, the correlation energy part of the calculations was without major difficulties. In addition to the initio work, calculations are also available on a semiempirical level [37, 38] in analogy to those for molecules described in Sect. 4.3. Although they lack ab initio accuracy, they reproduce in a simple way the main results. Finally, there have been also model calculations performed for transition metals with the aim of elucidating the role of d electron correlations in the formation of magnetic order. It should be pointed out that local operators have been applied rather early in model treatments of correlations in transition metals by Mott [39], by Gutzwiller [40], and by Hubbard [41] (for reviews see Ref. [42, 43]).

5.1.1 Ab initio Calculations for Diamond

The SCF ground-state wave function is calculated by using a basis set of 4s(7s) and 2p(3p) GTO's. It is expressed in terms of the one-particle density matrix, which fixes the Fock matrix (see Eq. (2.3)). The interatomic correlation energy was computed by choosing for the $g_i(\boldsymbol{r})$ the atomic parts of the bond orbitals after orthogonalizing them with respect to each other and to the 1s core states. They consist of four tetrahedral hybrid functions which can be extracted from the density matrix. The different correlation energy contributions are listed in Table 5. In Fig. 14 it is shown in which hybrids electrons are, which are correlated (through operators of the form $n_i n_j$ and $\boldsymbol{s}_i \boldsymbol{s}_j$). The correlation energy falls off rather fast with increasing bond separations. This is a consequence of the van der Waals type of interactions, which take place between different bonds and which determine the interbond correlation energy. Further details can be found in Ref. [36]. The intraatomic correlations are calculated by using a spatial partitioning of the atomic volume

Table 5. Different interatomic correlation-energy contributions (in eV per unit cell) (from Ref. [36])

Form of correlation	Notation	Correlation energy contribution [eV]
a) $n_i n_i$ (correlations within a bond)	$E^{(1)}_{corr}$	−1.52
b) $n_i n_j$, $\boldsymbol{s}_i \boldsymbol{s}_j$ (correlations between neighboring bonds)	$2\,E^{(2)}_{cor}$	0.84
c) $n_i n_j$, $\boldsymbol{s}_i \boldsymbol{s}_j$ (correlations between next-nearest neighbor bonds)	$24\,E^{(3)}_{corr} + 12\,E^{(4)}_{corr}$	−0.11
d) $n_i n_j$, $\boldsymbol{s}_i \boldsymbol{s}_j$ (correlations between more distant bonds, estimated)		−0.05 ± 0.03
Total interatom. correl. energy		−2.52 ± 0.03

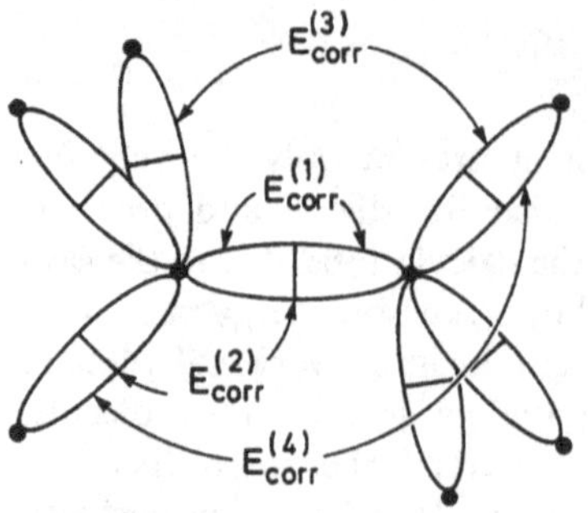

Fig. 14. Correlations between electrons in different hybrids of a diamond lattice

as used previously for the C atom e.g. in CH_4. It corresponds to a 4-fold, 12-fold and 4-fold subdivision of the three shells shown in Fig. 7. This requires inclusion of a set of d-functions into the basis set. The following intraatomic correlation energy for the valence electrons are found per unit cell,

$$E_{corr}^{intra} = -3.68 \qquad [eV/uc]. \tag{5.1}$$

From calculations for a single C atom it is known, how large the defect in the correlation energy is due to to the limited size of the basis set. Furthermore, by studying small molecules, one knows how this defect transfers when the C atom is part of a molecule. It turns out that a basis set of the above size yields 73% of the correlation energy limit [36]. Since the Local Ansatz misses typically 5% of the correlation energy within a given basis set, one may extrapolate the above findings and estimate the following limiting value for the correlation energy of the valence electrons [36]

$$E_{corr}^{limit} = -8.82 \pm 0.16 \qquad [eV/uc]. \tag{5.2}$$

An analysis reveals that correlations contribute approximately 25% to the binding energy of diamond.

5.1.2 Semiempirical Correlation Calculations

The correlation calculations for covalent semiconductors become extremely simple, and in fact can be done analytically, when the bond-orbital approximation (BOA) is made [37]. In that case all calculations are done within the space spanned by the bonding and antibonding combinations of orthogonal atomic hybrids, which form bonds. The corresponding creation operators are $B_{I\sigma}^+$ and $A_{I\sigma}^+$, respectively (I is the bond index). The SCF ground state is (compare with Eq. (2.13))

$$|\varphi_{SCF}\rangle = \prod_{I\sigma} B_{I\sigma}^+ |0\rangle \tag{5.3}$$

and there are two electrons in each bond. The operator S (see Eq. (3.2)) takes the form

$$S = -\eta_0 \sum_I S_I - \eta_1 \sum_{\langle IJ \rangle} S_{IJ} - \dots \tag{5.4}$$

where I and J are *bond* indices. The operator S_I ($= S_{II}$) describes correlations within bond I, i.e. $S_I = A^+_{I\sigma}A^+_{I-\sigma}B_{I-\sigma}B_\sigma$: S_{IJ} describes correlations between nearest-neighbor bonds as indicated by $\langle IJ \rangle$, and is of the form $S_{IJ} = \sum_{\sigma\sigma'} A^+_{I\sigma}A^+_{J'\sigma'}B_{J'\sigma'}B_{I\sigma}$.

It can be shown that η_0 and η_1 are of the simple form

$$\eta_0 = (V_0^D - 12\eta_1 V_1^D)(2t_0)^{-1}$$
$$\eta_1 = V_1^D(1 - \eta_0)(2t_0\alpha)^{-1} \tag{5.5}$$

where $V_0^D = V_{A_IB_IA_IB_I}$ and $V_1^D = V_{A_IB_IA_JB_J} - \frac{1}{2} V_{A_IB_JA_JB_I}$ are interaction matrix elements in *bond* representation. They are dipole-dipole interaction matrix elements, averaged over spins, for bond I and between bonds I and J, respectively. Furthermore, t_0 is the bare hopping matrix element between the two hybrids which form a bond. The constant α can be considered as being due to screening. It is given by

$$\alpha = 1 + V_0^D(2t_0)^{-1} + 2V_1^D t_0^{-1}. \tag{5.6}$$

The interatomic correlation energy per unit cell, which results from the first two terms in the expansion of S is found to be

$$E_{corr}^{inter} = -2\eta_0 V_0^D - 24\eta_1 V_1^D, \tag{5.7}$$

with η_0 and η_1 given by Eqs. (5.5). In order to obtain explicit results, one must know V_0^D, V_1^D and t_0 for a system. These quantities are known for diamond, and we assume that we can scale them with the respective bond length d according to

$$t_0^x = t_0^c \left(\frac{d_c}{d_x}\right)^{1.4}$$
$$(V_i^D)^x = (V_i^D)^c \left(\frac{d_c}{d_x}\right)^{1.4}, \qquad (i = 0,1) \tag{5.8}$$

The index C stands for diamond while x = Si, Ge and α-Sn. The resulting values for E_{corr}^{inter} are listed in Table 6.

Table 6. Calculated interatomic (or bond) correlation energies per unit cell (in eV) (from Ref. [37])

	C	Si	Ge	α-Sn
$-2\eta_0^{st}V_0^D$	1.8	1.2	1.2	1.1
$-24\eta_1^{st}V_1^D$	0.8	0.4	0.4	0.4
$-E_{korr}^{inter}$	2.6	1.6	1.6	1.5

The interatomic correlations must be supplemented by the intraatomic ones. This is done by making use of the "atoms in molecules" approach, which was discussed before in Sect. 4.3. The population analysis for a C site is made by using Eq. (4.7). It is $\nu_0^C = 4$ and one finds

$$\beta_{SCF}^{c} = \langle \varphi_{SCF} | (\Delta n)^2 | \varphi_{SCF} \rangle = 2$$

$$\begin{aligned} \beta_{corr}^{c} &= \langle \psi_0 | (\Delta n)^2 | \psi_0 \rangle_c \\ &= 2(1 - \eta_0 - 6\eta_1) \end{aligned} \tag{5.9}$$

with $\Delta n = n - 4$. For diamond one finds $\beta_{corr}^C = 1.25$, which shows that interatomic correlations reduce the charge fluctuations considerably. The intraatomic correlation energy per unit cell is then given by

$$E_{corr}^{intra} = 2 \sum_{\nu} P_c^{corr}(\nu)\, \varepsilon_{\nu}^{corr} \tag{5.10}$$

where ε_{ν}^{corr} is the correlation energy for ν valence electrons on an atom. It is assumed that the relative weights of the different configurations are the same as on the C atom in CH_4. By subtracting the correlation energy of two isolated C atoms one obtains the intraatomic correlation energy contribution to the binding energy. Such an analysis has been done only for diamond, since atomic calculations of the required form are available only for the C atom [33].

5.2 Excited States

In Sect. 3, it was shown how the method of the Local Ansatz can be extended to excited states. A challenging problem is thereby the delocalization of the excited states in a solid. Therefore one has to describe the correlation hole of the moving quasiparticles. The Local Ansatz is an alternative to the more customary standard Green's function techniques [44] which deal with electronic eigenstates of the momentum. With those states it is difficult to compute on an ab initio basis the short range part of the correlation hole of the quasiparticle. However by making appropriate model assumptions the Green's function technique can be very succesful [45–47].

Ab initio calculations of excited states within the Local Ansatz have been attempted for diamond [48]. However, they have been done only with respect to interatomic correlations. The intraatomic correlation part is very elaborate due to relaxation effects, which occur when an electron is added (or subtracted) to the system. The interatomic correlations give by far the largest contribution to the reduction of the energy gap from its SCF value. This is so, because it is the build-up of a polarization cloud around the electron in the conduction band (or hole in the valence band) which leads to that reduction, and this is an interatomic correlation effect. We shall not describe the ab initio calculations in detail but refer instead to the original literature [48]. Instead we want to dwell some more on the semiempirical correlation calculation scheme [38], which was applied before in Sect. 5.1. It shows in a simple and transparent way the influence of electron correlations on the quasiparticle energy spectrum of semiconductors. For that purpose we add first an electron to the conduction band of the system, so that the state is of the form

$$|\varphi_{\mathbf{k}c\sigma}\rangle = A^{+}_{\mathbf{k}c\sigma} \mid \varphi_{SCF}\rangle \tag{5.11}$$

where $|\varphi_{SCF}\rangle$ is given by Eq. (5.3). Here $A^{+}_{\mathbf{k}c\sigma}$ (c is a band index) takes the role of the operator c_i^+ in Eq. (3.10). One may expand $A^{+}_{\mathbf{k}c\sigma}$ in terms of the operators $A^{+}_{I\sigma}$ in bond representation

$$A^{+}_{\mathbf{k}c\sigma} = N_B^{-1/2} \sum_{I} \alpha_{\mathbf{k}c}(I)\, A^{+}_{I\sigma} \tag{5.12}$$

where N_B is the number of bonds. The operator S is written as

$$S = S^{(e)} + S^{(0)} = -\sum_{IJ} \eta^{(e)}_{IJ} S^{(e)}_{IJ} - \sum_{IJ} \eta^{(0)}_{IJ} S^{(0)}_{IJ} \tag{5.13}$$

(compare with Eq. (3.11)).

The operators $S^{(e)}_{IJ}$ corresponds to $O_{IJ}Q_{IJ}(\boldsymbol{k}, c)$ and is of the form

$$S^{(e)}_{IJ} = \frac{1}{4} \sum_{\sigma\sigma'} A^{+}_{J\sigma'} B_{J\sigma'} A^{+}_{I\sigma} B_{I\sigma} \tag{5.14}$$

while the operator $S^{(0)}_{IJ}$ corresponds to $O_{IJ}P_{IJ}(\boldsymbol{k}, c)$ and is

$$S^{(0)}_{IJ} = \frac{1}{4} \sum_{\sigma\sigma'} A^{+}_{J\sigma'} B_{J\sigma'} A^{+}_{I\sigma} B_{I\sigma} \tag{5.15}$$

The operator $A^{+}_{I\sigma}$ exp $(S^{(0)}) \mid \varphi_{SCF}\rangle$ introduces agåin electron correlations within and between bonds, like in the ground state, with one modification, though, Due to the presence of the additional electron, some of the ground state correlations are blocked. This results in a *loss* of ground-state correlation energy. When the operator $S^{(e)}$ is applied on $A^{+}_{I\sigma} |\varphi_{SCF}\rangle$ one obtains

$$S^{(e)} A^{+}_{I\sigma} |\varphi_{SCF}\rangle = -\frac{1}{4} \sum_{J\sigma'} \eta^{(e)}_{IJ} A^{+}_{J\sigma'} B_{J\sigma'} A^{+}_{I\sigma} |\varphi_{SCF}\rangle\,. \tag{5.16}$$

It is seen that $S^{(e)}$ generates dipoles (by means of $A^{+}_{J\sigma'} B_{J\sigma'}$) in the bonds which surround bond I. This way a long-ranged polarization cloud can be created around the extra electron $A^{+}_{I\sigma}$ in that bond. It reduces considerably the energy which is required for putting the electron into the conduction band.

The energy of the extra follows from Eq. (3.16). It is approximated by

$$\varepsilon_{c\sigma}(\boldsymbol{k}) = \frac{\langle A_{\mathbf{k}c\sigma}\, e^{S^+}\, H\, e^{S}\, A^{+}_{\mathbf{k}c\sigma}\rangle_c}{\langle A_{\mathbf{k}c\sigma}\, e^{S^+}\, e^{S}\, A^{+}_{\mathbf{k}c\sigma}\rangle_c}\,. \tag{5.17}$$

The parameters η_{IJ} are obtained by minimizing this expression. As before, the approximation $\exp S \simeq 1 + S$ is made when the expectation value is calculated. The correlation energy $\varepsilon_{c\delta}^{corr}(\boldsymbol{k})$ consists of two parts, i.e.

$$\varepsilon^{corr}(\boldsymbol{k}) = \varepsilon_{c\sigma}^{\pi}(\boldsymbol{k}) + \varepsilon_{c\sigma}^{\eta}(\boldsymbol{k}) \tag{5.18}$$

One part represents the energy *gain* due to the polarization of the bonds by the extra electron. It is

$$\varepsilon_{c\sigma}^{\eta}(\boldsymbol{k}) = -\sum_{IJ} \eta_{IJ}^{(e)} \langle A_{kc\sigma} H_{res} S_{IJ}^{(e)} A_{kc\sigma}^{+} \rangle_c \tag{5.19}$$

where $H_{res} = H - H_{SCF}$.

The other part represents the *loss* of ground-state correlation energy due to the extra electron. One finds

$$\varepsilon_{c\sigma}^{\eta}(\boldsymbol{k}) = -\sum_{IJ} \eta_{IJ}^{(0)} \langle A_{kc\sigma} H_{res} S_{IJ}^{(0)} A_{kc\sigma}^{+} \rangle_c\,. \tag{5.20}$$

The $\eta_{IJ}^{(0)}$ are the same as found from the ground-state calculation (see Eqs. (5.5). Therefore the upper script (0) will be omitted in the following. Not surprisingly, one finds that

$$\varepsilon_{c\sigma}^{\eta}(\boldsymbol{k}) = -\frac{1}{4}(-2V_0^D\eta^0 - 24V_1^D\eta_1) \tag{5.21}$$

which is just one fourth of the ground-state correlation energy per unit cell (see Eq. (5.7)). This means that the correlation energy for one bond is lost, due to blocking, when the additional electron is present. For $\varepsilon_{c\sigma}^{\pi}(\boldsymbol{k})$ one finds

$$\varepsilon_{c\sigma}(\boldsymbol{k}) = -\frac{\pi_c^2\left(1 - \frac{1}{2}\eta_0 - \eta_1\right)}{\bar{E}_A + 2T_0 - 2V_{sc} - \varepsilon_{c\sigma}(\boldsymbol{k})}\,. \tag{5.22}$$

The following quantities appear in this expression,

$$\pi_c^2 = 2\sum_{J(\neq I)} \left(V_{A_I A_I A_J B_I} - \frac{1}{2} V_{A_I B_J A_J A_I}\right)^2. \tag{5.23}$$

As before, the V's are interaction matrix elements in bond representation. One can show that π_c^2 is a sum of squared polarization matrix elements, which for a diamond type of lattice turns out to be of the form

$$\pi_c = \lambda^2 \frac{1.373}{d^2} 10^7\ [\mathrm{eV}]^2 \qquad (\text{d in pm})\,. \tag{5.24}$$

As before, d is the bond length while $\lambda = p/d$ is the ratio between the effective dipole length of a bond p and the bond length. It can be calculated from the interaction matrix $V_{A_IA_JB_JB_I}$ (see Ref. [38]). The quantity E_A is the mean energy of the conduction electron band, i.e.

$$\bar{E}_A = \langle A_{J\sigma} H_{SCF} A_{J\sigma}^+ \rangle . \tag{5.25}$$

Furthermore, $2T_0$ is the bonding-antibonding splitting in SCF approximation and $V_{sc} = -(V_{A_IB_IA_IB_I} - \frac{1}{2} V_{A_IA_IB_IB_I})$. The same calculations, which have been outlined for an added electron, can be repeated for a hole which is added to the valence band. In that case $\bar{E}_A$ is replaced by $\bar{E}_B$, which is the mean energy of the valence band. A list of values for the different quantities is given in Table 7 for diamond, silicon and germanium. Thereby it has been assumed that $p = a_0/4$, where a_0 is the lattice constant. This implies $\lambda = 3^{-1/2}$. This choice is not free of arbitrariness. It has been used before, though [49].

At his stage we want to comment on possible basis set effects. The above correlation calculations were done by using a minimal basis set, i.e. by working with sp^3 hybrid functions. When one allows for larger basis sets the correlation energy is expected to change. For example, the polarizability of the bonds will increase while the loss of ground-state correlations will decrease. Investigations of the dielectric function [50] and of optical properties [49, 51] suggest that basis-set effects are relatively small, i.e. in diamond in the 1–2 eV regime. Correlations, which cannot be described within a minimal basis set were called before intraatomic. For the ground state, they can be estimated by the "atoms in molecules" method. A similar estimate for excited states is more complex but can be done also. Instead of doing this, we take here a more pragmatic point of view. We assume that their effects can be incorporated in the quantity λ, for which an appropriate choice is required. In order to apply the above theory to diamond, Si and Ge one uses results of recent SCF calculations [52] for the three systems and determines $\varepsilon_{n\sigma}^{SCF}(\mathbf{k})$. The correlation energies are calculated according to Eqs. (5.18) and (5.21, 5.22) and by applying corresponding expressions for an added hole. By using the ab initio SCF results and inserting them into the denominators of Eqs. (5.22) one is going beyond the BOA. The result of the calculations for Si is shown in Fig. 15, where the SCF energy bands and the ones including electron correlations are shown. In addition

Table 7. Numerical values for various parameters entering the theory. $2T = \bar{E}_A - \bar{E}_B$ and E_{corr}^{inter} is defined by Eq. (5.7) (see also Eq. (5.21))

	C	Si	Ge
η_0	0.20	0.24	0.25
η_1	0.029	0.030	0.032
E_{corr}^{inter}	− 2.6	−1.6	−1.6
$\bar{E}_A$	21.8	12.8	11.7
$\bar{E}_B$	−13.3	−8.6	−8.3
V_{sc}	8.7	6.2	6.0
λ^2	1/3	1/3	1/3

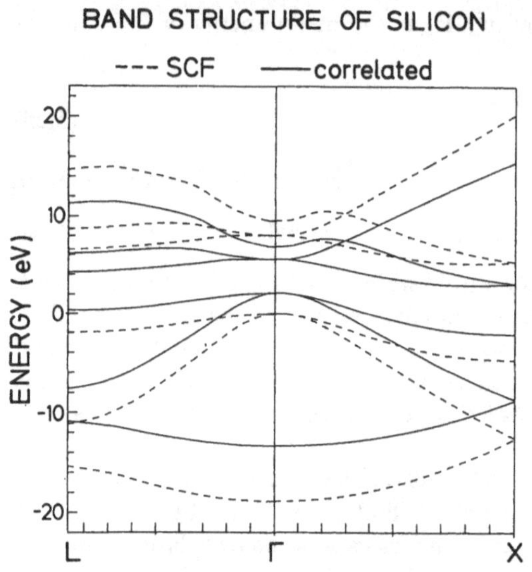

Fig. 15. Band structure for Si within the SCF approximation (dashed lines) and after correlations have been taken into account (solid lines) (from Ref. [38])

to narrowing substantially the energy gap between the valence and conduction band, correlations lead also to a reduction of the band width. Table 8 contains a comparison of the experimental direct band gaps and the valence band widths with those of a SCF calculation, and after correlation effects have been included. One notices a good overall agreement, but it should be stressed that the particular choice of λ is important for that.

Table 8. Comparison of the experimental direct gaps and valence-band width with those of a SCF calculation and including correlations (in eV) (from Ref. [38]) [a] Ref. [53], [b] Ref. [54], [c] Ref. [55], [d] Ref. [56], [e] Ref. [57]

	C	Si	Ge
W_{val}^{SCF}	29.9[a]	18.9[a]	18.2[a]
$E_{gap}^{SCF}(\Gamma)$	14.2[a]	8.0[a]	5.1[a]
W_{val}^{theor}	24.6	15.2	14.4
$E_{gap}^{theor}(\Gamma)$	7.2	3.0	0.44
W_{val}^{expt}	24.2 ± 1[b]	12.5 ± 0.6[c] 14.7 ± 0.5	13.0 ± 0.5[d]
$E_{vap}^{expt}(\Gamma)$	7.4	3.1[c], 3.3[c]	0.9[c]

Acknowledgement. I am deeply indebted to Dr. H. Stoll and Dr. G. Stollhoff for numerous discussions and helpful suggestions.

6 References

1. a. Kutzelnigg W (1977) in: Schaefer HF III (ed) Methods of electronic structure theory. Plenum, New York, p 129
   b. Meyer W (1977) in: Schaefer HF III (ed) Methods of electronic structure theory. Plenum, New York, p 413
   c. Buenker RJ, Peyerimhoff SD, Bruna PJ (1981) in: Csizmadin IG and Daudel R (eds) Computational theoretical organic chemistry. Reidel, Dordrecht, p 183
   d. Zirz C and Ahlrichs R (1979) in: Guest MF and Wilson S (eds) Proceedings of the Daresbury Study Weekend, Science Research Council, Warrington
   e. Ahlrichs R (1979) Comp. Phys. Comm. 17: 31
   f. Leroy G, Sana M, Burke LA, Nguyen (1980) in: Daudel R et al (eds) Quantum theory of chemical reactions, Vol. 1. Reidel, Dordrecht, p 91
   g. Malrieu JP (1977) in: Segal GA (ed) Modern theoretical chemistry, vol 7. Plenum, New York, p 69
2. Szabo A, Ostlund NS (1982) Modern quantum chemistry: Introduction to advaned electronic structure theory. Macmillan, New York
3. Mahan GD (1981) Many-particle physics, Plenum, New York
4. see e.g. Pines D and Nozières P (1966) The theory of quantum liquids. Benjamin, New York
5. Pfirsch F, Böhm MC, Fulde P (1985) Z. Phys. B60: 171
6. Cizek J (1969) Adv. Chem. Phys. 14: 35
7. Kümmel H, Lührmann KH, Zabolitzky JG (1978) Phys. Lett. C36: 1 and references therein.
8. Buenker RJ and Peyerimhoff SD (1983) in: Löwdin P-O and Pullman (eds) New horizons of quantum chemistry. Reidel, Dordrecht, p 183
9. Bruna PJ and Peyerimhoff SD (1987) in: Lawley KP (ed) Ab initio methods in quantum chemistry — I (Advances in chemical physics, vol 67). John Wiley, Chichester, p 1
10. Edmiston C and Krauss M (1965) J. Chem. Phys. 42: 1119
11. Ahlrichs R and Kutzelnigg W (1968) J. Chem. Phys. 49: 1819
12. Meyer W (1977) in: Schaefer HF III (ed) Methods of electronic structure theory, Plenum, New York
13. Edmiston C and Ruedenberg K (1963) Rev. Mod. Phys. 35: 457
14. Foster JM and Boys SF (1960) Rev. Mod. Phys. 32: 300
15. Pulay P (1983) Chem. Phys. Lett. 100: 151
16. Pulay P and Saebo S (1986) Theor. Chim. Acta 69: 357
17. Saebo S and Pulay P (1987) J. Chem. Phys. 86: 914
18. Roos BO and Siegbahn PEM (1977) in: Schaefer HF III (ed) Methods of electronic structure theory. Plenum, New York, p 277
19. Hurley AC (1976) Electron correlations in small molecules. Academic, London
20. Dieterich K and Fulde P (1987) J. Chem. Phys. 87: 2976
21. Diner S, Malrieu JP, Claverie P (1969) Theor. Chim. Acta 13: 1
22. Harrison W (1980) Electronic structure and the properties of solids. Freeman, San Francisco
23. Stollhoff G and Fulde P (1977) Z. Phys. B26: 251 and (1978) ibid 29: 231
24. Stollhoff G and Fulde P (1980) J. Chem. Phys. 73: 4548
25. Horsch P and Fulde P (1979) Z. Phys. B36: 23
26. Horsch S, Horsch P, Fulde P (1983) Phys. Rev. B28: 5977
27. For a review see: Stollhoff G (1988) Adv. Phys., to be published
28. Linnett JW (1964) The Electronic structure of molecules. Wiley, New York
29. Meyer W (1973) J. Chem. Phys. 58: 1017
30. Oleś AM, Pfirsch F, Fulde P, Böhm MC (1986) J. Chem Phys. 85: 5183
31. Lievin J, Breulet J, Clercq P, Metz JY (1982) Theor. Chim. Acta 61: 512
32. Oleś AM, Pfirsch F, Fulde P and Böhm MC (1987) Z. Phys. B66: 359
33. Verhaegen G and Moser CM (1970) J. Phys. B3: 478
34. Declaux PG, Moser CM, Verhaegen G (1971) J. Phys. B4: 296
35. Cremer D (1982) J. Comput. Chem. 3: 165
36. Stollhoff G and Bohnen KP (1988) Phys. Rev. B37, 4678 and references therein
37. Borrmann W and Fulde P (1985) Phys. Rev. B31: 7800

38. Borrmann W and Fulde P (1987) Phys. Rev. B35: 9569
39. Mott NF (1949) Proc. Phys. Soc. 62: 416, 342, 358
40. Gutzwiller M (1963) Phys. Rev. Lett. 10: 159
41. Hubbard J (1963) Proc. Roy. Soc. A 276: 238
42. Friedel J (1969) in: Ziman JM (ed) The physics of metals, 1. Electrons. Cambridge University Press, London
43. Fulde P, Kakehashi Y, Stollhoff G (1987) in: Capellmann H (ed) Metallic magnetism. Springer, Berlin Heidelberg New York (Topics in Current Physics, vol 42)
44. see e.g. Abrikosov AA, Gorkov LP, Dzyaloshinski IE (1963) Methods of quantum theory in statistical physics. Prentice Hall, Englewood Cliffs NJ
45. Hedin L and Lundqvist S (1969) in: Seitz F, Turnbull D, Ehrenreich H (eds) Solid state physics 23: 1
46. Hybertsen MS and Louie SG (1986) Phys. Rev. B34: 5390
47. von der Linden W and Horsch P (1988) Phys. Rev. B37, 8351 (1988)
48. Horsch S, Horsch P, Fulde P (1984) Phys. Rev. B29: 1870
49. Strinati G, Mattausch HJ, Hanke W (1982) Phys. Rev. B25: 2867
50. Breuer NE (1975) Phys. Rev. B12: 1487
51. Turner RD and Inkson JC (1976) J. Phys. C9: 3583
52. von der Linden W, Bohnen KP, Fulde P (1986) Phys. Rev. B34: 1063
53. von der Linden W and Horsch P (private communication)
54. McFeely FR, Kowalczyk SR, Ley L, Cavell RG, Pollak RA, Shirley DA (1974) Phys. Rev. B9: 5268
55. Landoldt-Börnstein (1982) Numerical data and functional relationships. Springer, Berlin Heidelberg New York (New Series, Group III, vol. 17a)
56. Ley L, Kowalczyk S, Pollak R, Shirley DA (1972) Phys. Rev. Lett. 29: 1088
57. Straub D, Ley L, Himpsel FJ (1985) Phys. Rev. Lett. 54: 142

The Concept of the Chemical Bond in Solids

M. Grodzicki*

I. Institute of Theoretical Physics, University of Hamburg, Jungiusstrasse 9, D-2000 Hamburg 36, FRG

* Present address: University of New Orleans, Dept. of Chemistry, New Orleans, Louisiana 70148

1 Introduction

The description of the electronic structure of solids was based almost exclusively on the (delocalized) band picture until the end of the 1960s. The band theory of solids was founded in 1928 with the famous paper by Felix Bloch: "Über die Quantenmechanik der Elektronen in Kristallgittern" (On Quantum Mechanics of Electrons in Crystal Lattices) [1]. The theorems proved therein, and the related physical concepts became the basis of the entire subsequent work on this topic, as e.g. the band picture itself and the Bloch functions. The further development of constructing solutions for the Bloch problem, i.e. the solution of the Schrödinger equation for the local periodic potential, took its course along several different lines. Their common and most remarkable property when compared to parallel developments in quantum chemistry is the fact that, for two reasons, the Hartree-Fock method never played any significant role in treating the band structure problem. The first and more practical one is the prohibitive amount of computational effort easily recognized when looking at the expense of even small molecule calculations. The second and much more important conceptual one is that the density of states at the Fermi level vanishes for a free electron gas in the Hartree-Fock approximation [2] and the Fermi velocity becomes infinite [3]. Bearing in mind that electrons in normal metals behave almost like free-electron, together with the knowledge that almost all electronic properties of solids are related to the electrons in the direct vicinity of the Fermi surface, the Hartree-Fock approximation leads to obviously disastrous consequences in solids. A third important aspect compared with modern quantum chemistry is the minor role of the LCAO approximation in the periodic potential problem since such basis functions satisfy "atomic" boundary conditions being insignificant in an ideal crystal. Though already introduced in Bloch's paper and despite its great explanatory power, the LCAO approach has not been used in first-principles calculations apart from very few exceptions, but has served merely as the basis for constructing parametrized model Hamiltonians. For these three reasons, it is not surprising that the development of band structure methods, in its earlier stage, followed lines completely different from the techniques of solving the molecular problem.

The situation only changed around 1970 initially caused by the increasing interest in nonperiodic systems (such as impurities, alloys and amorphous materials, surfaces and interfaces, adsorption), but also by treating more complicated substances with big unit cells; for Bloch's theorem relates the phases of the wave function only in different cells, and when these become so large that intracell interactions play the decisive role compared with the intercell ones, molecular orbital methods are more suitable. Therefore, a revived interest in molecular orbital or tight-binding methods and chemical bonding concepts could be observed [4], which was accompanied by the formulation of the molecular problem in terms and concepts that have been used so far only in band theory as, e.g., the pseudopotential [5] and the local density functional formalism [6].

There is, however, another good reason for the development of methods analyzing the electronic structure of solids in terms of chemical bonding. The band picture — as another typical characteristic — establishes unifying, but quite abstract principles and concepts for the description of the electronic structure of periodic systems. Hence, the band picture is not specific in the sense that it covers the whole range of solids from simple metals to noble gases and even molecular crystals without taking into account their quite different physical nature. Accordingly, the connection of the bandpicture to phenomenological concepts as bonds, atoms, atomic and ionic radii etc. is not obvious. Moreover, band theory contains the information that the total energy of the solid is lower than that of the separated atoms, but it does not make intelligible the various interactions in a natural way leading to the phenomenon of cohesion, i.e. why atoms tend to form a solid.

An analysis of this type of questions requires the construction of "bridging concepts" between the abstract theoretical level of band theory and the phenomenological level of empirical terms and classifications. In the case of electronic structure theories such bridging concepts are often supplied by an appropriate interpretation and partitioning of the total charge density. For instance, investigating the electron density distribution in crystals by experimental [7] or theoretical methods exhibits significant differences reproducing — in a natural way — the empirical classification system of covalent, metallic, ionic and van der Waals crystals (cf. Fig. 1). These differences apparently correspond to different types of chemical bonding, and suggest the separate discussion of these classes when more specific conclusions about the interactions in the respective crystal types are desired. Thus, even though the bond

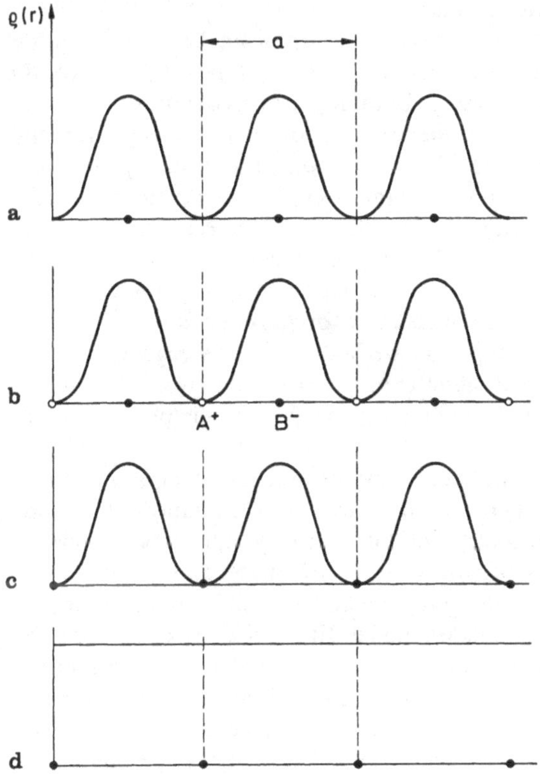

Fig. 1a–d. Idealized valence electron density distributions $\varrho(r)$ for the main types of solids, a is the lattice constant. (**a**) Van der Waals crystals: the electron density remains localized at the atoms; (**b**) ionic crystals: the electrons are transferred to the anion B^-; (**c**) covalent crystals: the density is concentrated in the bonds between the atoms; (**d**) metallic crystals: the density is completely delocalized over and constant within the whole metal

picture is less universal, it facilitates the understanding of many physical properties of solids by constructing special approximate schemes for the respective interactions based on physical intuition, and it offers simpler models — though of limited applicability — on a less abstract level more closely related to empirical needs.

The usual way of understanding the origin of cohesion in solids in terms of the chemical bond consists in constructing the attractive and repulsive interactions leading to the minimum in total energy. In the bond picture this is accomplished by discussing the origin of the different types of chemical bonding in solids. However, any investigation of this kind is faced with a serious problem: the interactions cannot be inferred unequivocally from empirical facts, but additional and independent theoretical assumptions are compulsory from which those data have to be deduced comparable with experiments. This of course implies ambiguities in the physical interpretation of the respective mathematical expression since a successful prediction of empirical data does not neceesarily prove the underlying physics to be correct. Consequently, there is always a jeopardy of arbitrariness in such a discussion, and there may exist more than one possibility to ponder over the same physical effect, so that an alternative description may be correct as well. Fortunately, we are meanwhile in the position to remove some of these ambiguities and to decide by detailed calculations whether a mechanism suggested for the origin of bonding is physically correct or not. Nevertheless, many textbooks on solid state physics as well as on quantum chemistry still offer interpretations of the various types of the chemical bond, that are at least misleading if not definitely wrong in the sense that they are contradictory to the results of these accurate calculations.

Accordingly, the main purpose of this article is to concentrate on the correct physical interpretation of the interactions in the various types of solids leading to the phenomenon of chemical bonding or cohesion, but not the proposal of new methods for calculating the electronic structure of solids in terms of bonds for three reasons. Firstly, the practical application of chemical bond concepts to solid state problems is presented exhaustively in a recent textbook by Harrison [8] whereas the interpretational aspects are not dealt with. Secondly, a description of the electronic structure based on the concept of the chemical bond will serve predominantly as a tool for qualitative models, and under this aspect, the understanding of the physical nature of the mechanisms underlying the formation of the chemical bond is inevitable. Otherwise, the development of qualitative or semiempirical methods will easily degenerate into a pure fitting procedure without physical significance, serving merely as instruments for predicting some numbers but without giving any insight into the physical nature of the problem.

Finally, it must be emphasized that the possibility of understanding a number of physical properties of solids in the bond picture does not at all imply that a solid is merely something like a big molecule. On the contrary, the most important developments in the quantum theory of solids are based in recognition of the fact that, by idealizing the solid as an infinite many-particle system, a great number of properties emerge that are otherwise hidden under finite size effects. From this point of view, the distinction between a molecule and a solid is of an essentially qualitative nature with important experimentally confirmed consequences, whereas it is merely quantitative under the aspect of description in terms of chemical bonds. These qualitatively new and different phenomena are collective or cooperative in the sense that they involve the strong interaction of a large number of particles, thus, precluding the reduction to properties of a composed finite particle system. This is the reason that modern solid state physics is mainly based on the concept of the "elementary excitation in periodic structures" but not on the concept of the chemical bond.

2 Statistical and Dynamical Correlations

The paradigmatic type of the chemical bond is, without any doubt, the covalent bond, i.e. understanding the chemical bond at all, means understanding the covalent bond first. The identification of this type of bonding with the electron pair bond has strongly influenced the interpretation of the first quantum mechanical treatment of the covalent bond; accordingly, it was assumed that this phenomenon has to be considered as a two-particle effect or, more specifically, as an "exchange effect". Although it has been generally accepted by now that the chemical bond is a predominantly one-electron effect, some misinterpretations of the chemical bond are eventually based on certain misunderstandings of genuine quantum mechanical concepts. These misunderstandings arise from the subsequent "tacit assumptions" forming the basis of the theoretical methodology of physics [9, 10], and having their origin in the principles of Newtonian mechanics:

— for each observable event (= effect), a "mechanism" can be constructed from which that effect can be deduced. Then the mechanism is said to explain the effect;
— if two (or more) events are correlated, i.e., they regularly appear together or successively, a mechanism representable as a physical interaction exists that can be assmed to be the cause of the correlation.

These assumptions constitute a very successful heuristical basis for theoretical reasoning in science, but they are by no means universally valid, i.e. there exist correlations that cannot be interpreted in this way. Let us consider the following example [11] as a simple case. It is well known that the probability is 1/6 to throw a number between 1 and 6 with a (homogeneous) die. Taking two dice, however, and looking at the probability distribution for the results of the sum of both, not all combinations turn out to be equally probable, because there exist more combinatorial possibilities of obtaining 5 points than, e.g. 2. The resulting distribution can be easily calculated and is displayed in Table 1. Hence, at the first sight, it looks as if there exists an "invisible hand" being responsible for the "effect" that 7 occurs more often than all other combinations.

Table 1. Probability distribution W of the sum of two dices

sum	2	3	4	5	6	7	8	9	10	11	12
$W \cdot 36$	1	2	3	4	5	6	5	4	3	2	1

According to the above-mentioned tacit assumptions, we automatically ask for a (physical) interaction that might be responsible for this correlation. The answer to this simple example is, of course, "no", and we do not have any problems to comprehend it because we know the reason for this unequal distribution. Such correlations which cannot be interpreted or "explained" by physical or dynamical interactions will be called statistical correlations. They can be characterized by the following properties:

(S1) The correlation is independent of the distance of the two events. It does not make any difference, for instance, whether or not the two dice are in the same shaker, or one die is thrown in Hamburg and the other one in New York. The result of the correlation as described above will be always the same.
(S2) The correlation is independent of time, i.e., it does not make a difference whether both dice are thrown together at the same time or successively.
(S3) The two events that are statistically correlated are dynamically independent in the following sense. Throwing a certain number, e.g. a 5 with one die, does not increase the probability of getting a 2 with the other one only because the sum of 7 has the highest probability. On the contrary, the probability of throwing any number with the second die remains 1/6.

Analogous statistical correlations occur as a characteristical feature also in quantum mechanics, their nature, however, being not so evident. This can be demonstrated with two non-interacting electrons in a one-dimensional potential box with infinite walls [12]. The only requirement that the electrons have to fulfill is to obey the Pauli principle according to their indistinguishability. Solving the Schrödinger equation for this simplified system, yields the following result for the probability distribution as a function of the interelectronic distance d. In the singlet ground state it is constant, i.e. the motion of the electrons is uncorrelated and they are completely independent. On the other hand, this is not the case in the first excited singlet state and in the triplet ground state which are accidentally degenerate. In the singlet state the maximum of the probability occurs at $d = 0$, i.e. the electrons tend to condense, whereas in the triplet state the maximum appears at 0.37a where a is the length of the potential box. Thus, in the first case it looks as if there existed an "attractive interaction" between the electrons, while in the second one a "repulsive interaction" attempts to keep them apart from one another. According to the initial assumption of noninteracting electrons, however, there does not exist any physical force or interaction, but the two electrons are correlated exclusively due to their indistinguishability, i.e. the Pauli principle.

Another statistical correlation of basic importance for the quantum theory of the chemical bond is caused by the superposition principle. As generally known, the probabilities for two (dynamically) independent events are uncorrelated in classical statistics, i.e. the total probability is the sum or product of the individual probabilities. This is, however, not the case in quantum mechanics as is usually demonstrated with the help of the "double-slit-Gedankenexperiment". Even though the electrons are assumed to be dynamically independent, there must exist certain correlations leading to the usual interference pattern. Such correlations are caused by the phase coherence of the respective wave functions implying that quantum mechanical particles as electrons and photons can "interfere with themselves". For, when this coherence is destroyed, the interference pattern vanishes and the classical probability distribution of uncorrelated events is obtained. These correlations due to phase coherence can be simulated by adding the probability amplitudes instead of the probabilities themselves, as it would be the case for independent systems in classical statistics. Proceeding in accordance with this rule, i.e. the superposition principle, the total probability of phase correlated systems contains an additional term, viz. the superposition or interference term, that plays the crucial role in the understanding of the formation of the chemical bond.

By generalizing these examples one might, for statistically correlated properties, point out that the knowledge of such a property for one system does not allow any prognosis about the respective property of the other system, and that they are in this sense dynamically independent. On the other hand, dynamical correlations, i.e. correlations that can be understood and interpreted as the result of physical interactions have the subsequent properties:

(D1) The correlation decreases with an increase of the distance of the two events or systems that are correlated.

(D2) Dynamically correlated events are depending on time, in the sense, that a change of a physical property in one system will influence the other system only with a certain retardation.

(D3) Changes of state in one system will generally influence also the state of the other one. Thus, the knowledge about the relevant properties of one system often facilitates predictions about the correlated properties of the other system.

These considerations elucidate the fact that correlations, due to the superposition and to the Pauli principle, are statistical correlations and must not be interpreted in a dynamical way. This result shall now be applied to the physical interpretation of the Hartree-Fock equation forming principally the basis of all one-electron methods in quantum chemistry. When starting with the Hartree approximation where the Pauli-principle is not taken into account, one obtains with the wavefunction ("Hartree-product")

$$\Psi(\vec{r}_1, \ldots, \vec{r}_N) = \psi_1(\vec{r}_1) \ldots \psi_N(\vec{r}_N) \tag{1}$$

by independent variation of the orbitals $\psi_i(\vec{r})$ the Hartree-equation:

$$-\left[\frac{\Delta}{2} + \frac{Z}{r}\right]\psi_i(\vec{r}) + \left[\sum_{\substack{k=1\\(k\neq i)}}^{N} \int \frac{|\psi_k(\vec{r}')|^2\, d^3r'}{|\vec{r}-\vec{r}'|}\right]\psi_i(\vec{r}) = \varepsilon_i \psi_i(\vec{r}) \tag{2}$$

which can be given the form

$$h\psi_i(\vec{r}) - v_{XHi}(\vec{r})\, \psi_i(\vec{r}) = \varepsilon_i \psi_i(\vec{r}) \tag{3}$$

$$h = -\frac{\Delta}{2} - \frac{Z}{r} + \int \frac{\varrho(\vec{r}')\, d^3r'}{|\vec{r}-\vec{r}'|}; \qquad \varrho(\vec{r}) = \sum_{k=1}^{N} |\psi_k(\vec{r})|^2 \tag{4}$$

$$v_{XHi}(\vec{r}) = \int \frac{|\psi_i(\vec{r}')|^2\, d^3r'}{|\vec{r}-\vec{r}'|}. \tag{5}$$

Eq. (3) is easily interpreted in such a way that the orbital-independent one-electron Hamiltonian h is corrected for the self-interaction v_{XHi} of the orbital under consideration, sometimes called the "Hartree-exchange" term.

Including, in the next step, the Pauli-principle means that the Hartree-product, Eq. (1), has to be replaced with a Slater-determinant being the totally anti-symmetrized combination of spin-orbitals. The respective one-electron Schrödinger-equation is the Hartree-Fock-equation which may be written in a form analogous to Eq. (3)

$$(h - v_{XHFi})\,\psi_i(\vec{r}) = \varepsilon_i \psi_i(\vec{r}) \tag{6}$$

$$\begin{aligned} v_{XHFi}(\vec{r}) &= \sum_{k,\,II} \int \frac{\psi_k^*(\vec{r}\,')\,\psi_i^*(\vec{r})\,\psi_i(\vec{r}\,')\,\psi_k(\vec{r})\,d^3r\,d^3r'}{|\vec{r}-\vec{r}\,'|\cdot|\psi_i(\vec{r})|^2} \\ &=: \int \frac{|\Psi_{II}(\vec{r},\vec{r}\,')|^2\,d^3r\,d^3r'}{|\vec{r}-\vec{r}\,'|}\,. \end{aligned} \tag{7}$$

Now, due to the Pauli principle, an additional contribution to the Coulomb energy appears. It presents conceptual difficulties since it does not have a classical counterpart so that its physical interpretation is not obvious. The usual interpretation is based on the classical analogue of resonance phenomena in a system or coupled oscillators exchanging their energy. Similarly, the additional term in Eq. (7) has been interpreted as resulting from the exchange of two electrons at $\vec{r}$, $\vec{r}\,'$ between two orbitals $\psi_i(\vec{r})$ and $\psi_k(\vec{r}\,')$. This is, however, a poor analogy since it is not a physical quantity that oscillates here, but the probability of a given configuration. This "electron exchange" is thus not a real physical process and may not be interpreted, e.g. as an effect of the quantum mechanical tunneling of an electron between two states. Moreover, the frequently used picture of an exchange force acting between two electrons of parallel spins is misledading. As emphasized in the discussion of the two electrons in a box, there is no physical force keeping electrons of the same spin apart, and this "exchange force" is a purely imaginary force in the sense that it can do no work, for instance. It is therefore analogous to the imaginary forces in classical mechanics that are introduced to restrict the motion of a particle to a surface or to describe the fact that two bodies cannot penetrate each other ("Zwangskräfte"), and this last example supplies a useful analogy to the exchange force between two electrons of the same spin [13]. To illustrate this, we remind of the similar structure of Eqs. (3) and (6), and that the Hartree-Fock-exchange $v_{XHFi}(\vec{r})$ contains the self-interaction $v_{XHi}(\vec{r})$ for $k = i$. Thus, when the two electrons approach each other very closely, i.e. $|\vec{r} - \vec{r}\,'| \to 0$, the probability $|\Psi_{II}(\vec{r}, \vec{r}\,')|^2$ vanishes creating an exchange- or Fermi-hole around each electron. This hole can be understood as the quantum mechanical analogue to the classical principle that two bodies cannot be in the same place. The exclusion principle is thus static and not kinetic in its character.

Another frequently given interpretation of the Hartree-Fock-exchange term is based on the division of the exchange integrals in diagonal and nondiagonal terms. This view, however, is also unsatisfactory since such a division depends on the choice of the basis orbitals, and is not invariant under unitary transformations. The reason for this misinterpretation lies in the identification of electrons with orbitals and does not take into account that due to superposition and antisymmetrization all electrons make use of all orbitals [13].

Further problems with the physical interpretation of the exchange term arise when two electrons are separated after having undergone an interaction. Assuming that the total wavefunction has to be antisymmetric for all times and distances after the interaction, the motion of the two electrons remains correlated even when orbitals do not overlap anymore. This „everlasting correlation" poses some problems of understanding at first sight, but nothing is odd with this correlation when taking into consideration that the exchange interaction is not due to a dynamical but to a statistical correlation, i.e. when one does not search for a dynamical interaction to explain this correlation. There exist also several other aspects in the physical interpretation of the chemical bond related to statistical correlations, as we shall see in the next sections.

3 The Covalent Bond

The year of birth of quantum chemistry can be dated back to 1927 when Heitler and London published their famous paper on the first quantum mechanical treatment of the H_2 molecule [14], suggesting the purely quantum mechanical exchange interaction between the two electrons as the completely new mechanism for the origin of the chemical bond. This interpretation was directly oriented on the subsequent formal consideration leading to a bonding state, i.e. to a state with lower energy. For large distances, where the physical interaction between the two atoms is negligible, the molecular wavefunction can be written as the product of the respective atomic orbitals (cf. Fig. 2):

$$\Psi_I(\vec{r}_1, \vec{r}_2) \propto \varphi_1(\vec{r}_1 - \vec{R}_A) \cdot \varphi_2(\vec{r}_2 - \vec{R}_B) \tag{8}$$

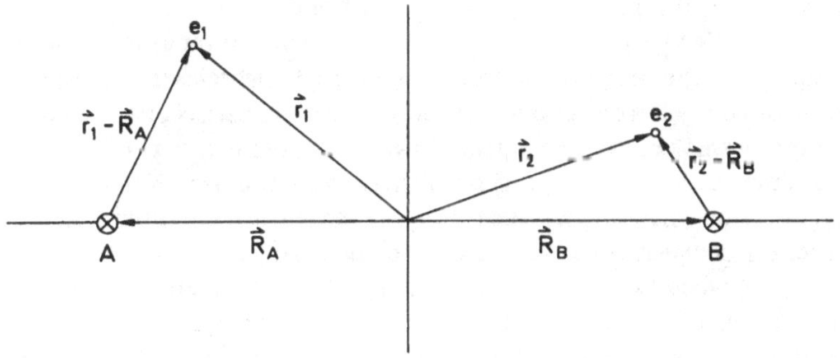

Fig. 2. Geometrical parameters in the H_2 molecule

For smaller distances, however, this is not a reasonable description because, due to the indistinguishability of the two electrons, e_1 can be at proton B and e_2 at proton A with the same probability. Hence Ψ_I is degenerate with the wavefunction

$$\Psi_{II}(\vec{r}_1, \vec{r}_2) \propto \varphi_1(\vec{r}_2 - \vec{R}_A) \cdot \varphi_2(\vec{r}_1 - \vec{R}_B) \quad (9)$$

Solving this problem by perturbation theory for degenerate states, i.e. diagonalizing the Hamiltonian

$$\begin{pmatrix} \langle\Psi_I|\mathcal{H}|\Psi_I\rangle & \langle\Psi_I|\mathcal{H}|\Psi_{II}\rangle \\ \langle\Psi_{II}|\mathcal{H}|\Psi_I\rangle & \langle\Psi_{II}|\mathcal{H}|\Psi_{II}\rangle \end{pmatrix}$$

yields the following result for the molecular wavefunction

$$\Psi^{\pm}(\vec{r}_1, \vec{r}_2) \propto \Psi_I(\vec{r}_1, \vec{r}_2) \pm \Psi_{II}(\vec{r}_1, \vec{r}_2) \quad (10)$$

where, in fact, the solution $\Psi^{+}(\vec{r}_1, \vec{r}_2)$ corresponds to a bonding state and, according to the Pauli-principle, both electrons have antiparallel spins. The direct pictorial interpretation of the derivation and result of this calculation has convinced Heitler and London that the phenomenological electron-pair-bond concept of the chemists is intimately connected with the quantum mechanical exchange interaction, which leads to the stabilization of the spin-saturated electron-pair constituting the bond.

This point of view was extended and summarized one year later in two separate papers of London [15] and Heitler [16], the former being the author who stressed the following properties of the electron-pair bond:
- the very existence of chemical bonding depends on the exchange energy and the concept of electron spin;
- the saturation property of the electron-pair bond turns out to be a consequence of spin-pairing;
- the valency of an atom equals the number of unpaired electrons (or holes) in the valence shell;
- hypervalency, e.g. in pentavelent phosphorus, can be understood by the inclusion of excited atomic states.

The last point became the basis for Pauling's hybridization concept that will be treated later. Eventually, London also discusses the ionic bond, giving some hints about a later explanation of the polarity of heteropolar bonds.

On the other hand, Heitler [16] tried to specify this concept by quantification in relating the bond energy to exchange integrals: "Zwei freie Leuchtelektronen die zu zwei verschiedenen Atomen gehören, können vermöge ihrer Austauschenergien eine Anziehung der Atome bewirken. Je eine solche Austauschenergie J_Q zwischen zwei Atomen kann als eine Valenzbindung, symbolisiert durch einen Valenzstrich, gedeutet werden." (Two free valence electrons belonging to two different atoms enable an attractive interaction between these atoms due to their exchange energies. Such an exchange energy J_Q between two atoms can be interpreted as a valence bond symbolized by a "valence dash".) Especially these two papers became, for a certain period, the basis of the quantum mechanical interpretation of the chemical bond, and the concepts developed in this work exerted a strong influence on following quantum chemical activities, mainly because it was seen as the quantitative basis of the Lewis' electron-pair model of the chemical bond [17]. The long survival of this interpretation is remarkable for several reasons:

- as early as 1931, F. Hund [18] emphasized the insignificance of exchange integrals for the chemical bond and showed that singlet-pairing is energetically less favourable than triplet-pairing if the latter case is possible without violating the Pauli-principle (Hund's rule) being, e.g., the reason for the paramagnetism of O_2;
- also the H_2^+-ion is stable with a bonding energy of 2.65 eV even though only a single electron contributes to the bond;
- an alternative suggestion for the mechanism leading to the covalent bond was given in 1933 by H. Hellmann [19] based on the uncertainty relation, and coming very close to the modern interpretation: "... für Valenzelektronen, die einen unabgesättigten Spin mitbringen und deshalb bei Absättigung der Spins denselben Raum besetzen dürfen, bedeutet die Übergangsmöglichkeit zum anderen Atom eine Ausbreitung im Koordinatenraum, damit eine Schrumpfung im Impulsraum und *ein Absinken der kinetischen Energie*." (... for valence electrons which carry an unsaturated spin and, therefore, may occupy the same space due to their opposite spins, the possibility of a transition to another atom means a spreading in coordinate space, i.e. a shrinking in momentum space and the decrease of kinetic energy.)

However, such an interpretation of the covalent bond as an effect of the decrease in kinetic energy has not been accepted because it apparently contradicts the virial theorem. It has been argued though that the binding energy $E_B = \Delta T + \Delta V$ of a bound state is negative, and since, according to the virial theorem $\Delta V = -2\,\Delta T$, the change ΔT in kinetic energy cannot be less than zero because in this case ΔV would increase twice as much, yielding a positive energy change, i.e. no bonding. The mechanism for the formation of the covalent bond would, on the contrary, be the decrease of the potential energy according to the virial theorem. Consequently, the second physical interpretation of the chemical bond was based on the argument that in the bonding region additional charge is accumulated being then attracted by both nuclei leading to a decrease in potential energy. This argument

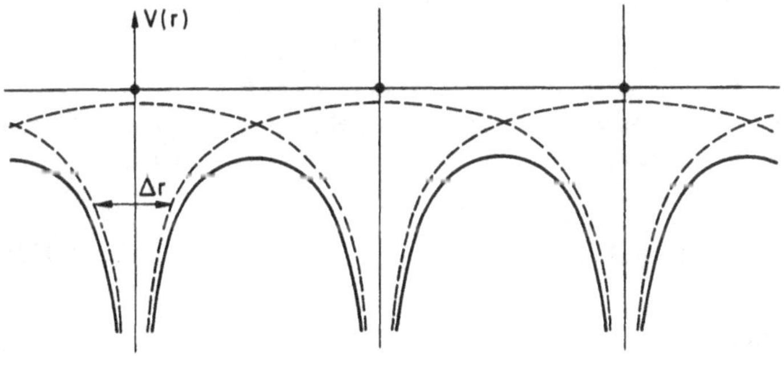

Fig. 3. Shape of the molecular potential (*solid line*) arising from the superposition of atomic potentials (*dashed line*). Δr denotes the uncertainty in position of an atomic electron

ignores, however, two important facts making it essentially wrong. Firstly, imaging the shape of the potential in molecules or solids (cf. Fig. 3), it can immediately be seen that, with respect to the potential energy, the bonding region is energetically the most unfavourable region for the electrons. Secondly, it has to be considered that bond formation does not produce any additional amount of electron density, but merely leads to a reorganization of charge density, i.e. an increase in the bonding region must be accompanied by a corresponding decrease somewhere else. Yet it is well known from electron density difference plots [20] that the density decreases in the region close to the nuclei. Thus, bond formation is related to a transfer of electron density from regions close to the nuclei, being more favourable for the electrons with respect to potential energy, to the less favourable bonding regions. Such a process must consequently be connected to an increase in potential energy, but not to a decrease.

In spite of these apparent contradictions, the interpretation of the covalent bond as an effect of the decrease in potential energy became the commonly accepted concept for understanding bond formation, as it is presented even nowadays in most of the textbooks on quantum chemistry and solid state physics. This is once more surprising because the correct interpretation, reconciling Hellmann's interpretation with the virial theorem, dates back to 1962 when K. Ruedenberg in one of the benchmark papers of quantum chemistry about "The Physical Nature of the Chemical Bond" [13] presented a detailed analysis of the various contributions to bonding. Nevertheless, even 25 years later, the conclusions drawn in this paper cannot be denoted as common knowledge within the physical and chemical scientific communities. We shall thus briefly repeat the main results of his investigation starting from the H_2^+-ion demonstrating first of all that unlike the original Heitler-London interpretation the chemical bond is predominantly a single-particle effect. For a more detailed discussion we refer, apart from the original article [13], to Refs. 21–23, to the contribution of Kutzelnigg in this volume, or to the textbooks of Kutzelnigg [24] or Klessinger [25].

Let be $\varphi_\nu(\vec{r})$ a wave function located at the atomic site ν with the corresponding probability density $\chi_\nu(\vec{r}) = |\varphi_\nu(\vec{r})|^2$. The classical molecular probability density is just the sum of the atomic ones, i.e.

$$\chi^{\rm mol}_{\rm class}(\vec{r}) = \frac{1}{2}[\chi_\nu(\vec{r} - \vec{R}_\nu) + \chi_\mu(\vec{r} - \vec{R}_\mu)] \tag{11}$$

On the other hand, a quantum mechanical treatment requires the superposition of amplitudes instead of densities to account for the phase coherence. Hence, one has to start with a molecular wave function assumed here in LCAO representation

$$\psi_\pm(\vec{r}) = N_\pm^{-1/2}[\varphi_\nu(\vec{r} - \vec{R}_\nu) \pm \varphi_\mu(\vec{r} - \vec{R}_\mu)] \tag{12}$$

with the normalization constant

$$N_\pm = 2(1 \pm S_{\mu\nu}) \tag{13}$$

where $S_{\mu\nu}$ denotes the overlap integral. The quantum mechanical probability density for the bonding state $\psi_+(\vec{r})$ can then be written as

$$\chi_{qm}^{mol}(\vec{r}) = |\psi_+(\vec{r})|^2 = \chi_{klass}^{mol}(\vec{r}) + \chi_{int}(\vec{r}) \tag{14}$$

emphasizing that the quantum mechanical and the classical density differ from each other by the purely quantum mechanical interference or difference density

$$\chi_{int}(\vec{r}) = \frac{\varphi_\nu^*(\vec{r})\,\varphi_\mu(\vec{r}) - S_{\mu\nu}\chi_{class}^{mol}(\vec{r})}{1 + S_{\mu\nu}} \tag{15}$$

It can easily be proven that the interference density integrated over whole space vanishes in accordance with the fact that bond formation corresponds merely to a reorganization of the charge density. Moreover, the interference density for the bonding state turns out to be positive in the bonding region yielding the commonly known result of charge accumulation in the bond leading to the conclusion that "... the interference density is responsible for the origin of the specific and substantial stabilization energy which cannot be understood in terms of a quasi-classical picture and which gives rise to the phenomena of 'covalent chemical binding'" [26].

So far the analysis does not say anything about the mechanism connected with this process of stabilization, and it is the merit Ruedenberg and coworkers [13, 22, 23] to have shown quantitatively that it is, in fact, the decrease of kinetic energy in the bonding region accompanying the accumulation of charge due to quantum mechanical interference. To demonstrate this, it is necessary to split the expectation values of kinetic and potential energy changes into their components parallel and perpendicular to the bond direction and to discuss them separately. In this way, it is also possible to resolve the seeming contradiction to the virial theorem, referring to an early result of Finkelstein and Horowitz [27] who showed that the hydrogen atomic orbitals appear to be modified in the molecule since a lower value for the total energy is obtained by using slightly contracted orbitals in comparison with atomic ones. This result is interpreted by Ruedenberg in such a way that the atoms are prepared for bonding ("promotion") and that the promoted, i.e. the contracted orbitals interfere. This contraction leads to a significant increase of kinetic and decrease of potential energy whereas its net contribution to the binding energy is inessential. In conclusion, the virial theorem is not violated because the kinetic energy increases perpendicular to the bond direction due to promotion while the interference has nearly no effect there. On the other hand, the interference effect with the decrease of kinetic energy dominates the increase due to promotion in bond direction yielding a net contribution to stabilizing the molecule. Moreover, it becomes evident why the virial theorem cannot be directly applied in order to understand the phenomenon of covalent bonding. In its quantum mechanical modification it relates expectation values over whole space, but such expectation values are insignificant for the discussion of bond formation, e.g. the interference density integrated over the space vanishes. What is really important for bond formation are the changes in the bonding region, and here the interpretation of Hellmann [19] and Ruedenberg [13] completely applies.

We wish to conclude our analysis with an additional remark on the concept of promotion since it has recently been criticized as not being the origin of the binding in H_2^+ [28]. Although Ruedenberg never claimed that the origin of binding in H_2^+ is due to promotion but to interference and the criticism is therefore inadequate, it comprises, nevertheless, some interesting aspects because it contributes to a better understanding of the meaning of the promotion process. The critique is based on a result obtained by Pettifor [29] expanding the exact wave function of H_2^+ in spherical harmonics around the proton position. The first — spherically symmetric — term has the same shape as the atomic orbital, i.e. it appears uncontracted. The steeper decay in the outer region beyond the nucleus results solely from the admixture of the aspherical terms with higher l-values. This means that the atomic orbital does not undergo a contraction but is actually lopsided due to the presence of the other proton resulting in a slower decay into the bonding region and in a steeper one into the outer parts of the molecule. Thus, the deformation of the wave function appears to be essentially angular-dependent instead of being a simple radial contraction. In consideration of this result, Ruedenberg's concept of promotion can be understood easily as the simulation of an environmental effect — angular distortion of the atomic orbital — by modification of a corresponding atomic quantity, i.e. orbital contraction. From the methodological point of view it is, however, common practice in physics to incorporate external influences on a particle by modifying some of its properties, leading e.g. to the quasiparticle concept.

For instance, the Bloch-electron can be interpreted as a quasiparticle containing the interaction of the electron with the static lattice, or a polaron can be seen as a Bloch-electron modified by its interactions with the lattice vibrations. In analogy, the contracted orbital can be viewed as an atomic orbital modified by the influence of the electrostatic field of the neighbouring proton. One must only be aware of the fact, that the process of promotion does not, in some way or other, correspond to an element of physical reality, but is just the simulation of an external influence and, in a strict sense, is thus fictitious.

In conclusion, according to Ruedenberg's analysis the formation of the covalent bond can be understood as a result of different steps whereby the partition itself is, of course, fictitious:

(1) The quasiclassical interaction: it results from the superposition of atomic probability densities. Around the equilibrium distance it is merely connected with small energy changes but is strongly repulsive for shorter distances.
(2) The quantum mechanical interference: it leads to an accumulation of charge for bound states in the bonding region, connected with a significant decrease of kinetic energy. This is the decisive mechanism during the formation of the covalent bond.
(3) The promotion of the atomic orbitals: it causes an intraatomic change of the kinetic and potential energy without a significant influence on the binding energy. It is more important for the correct shape of the wave function at small distances, and emphasizes its origin due to external influences.
(4) In molecules with several electrons, the electron-electron interaction has to be taken into account. It has a destabilizing effect and is denoted as the penetration interaction by Ruedenberg.

(5) In polar molecules, an additional Coulomb interaction occurs due to partial charge transfer of valence electrons between the constituent atoms. This contribution is always stabilizing and will be of the same order of magnitude as the interference contribution if the atoms have large electronegativity differences.

To give the gist of the whole story, we can distinguish three stages in the development of the physical interpretation of the chemical bond, namely the exchange picture of Heitler and London, the concept of potential energy decrease, and eventually the mechanism of kinetic energy decrease in the bonding region. Remarkably, these changes in the physical interpretation did not affect the formalism in the sense that they led to improved computational schemes so that they were not of vital interest for the practice of electronic structure calculations. Most likely, this is the reason that these developments have been widely ignored. Moreover, this analysis illustrates that a working mathematical scheme by no means allows unambiguous conclusions about the physical interpretation of the respective facts of reality. Therefore, the successful prediction of experimental data must not tempt us to believe that also the physical model, as the conceptual basis of the formal description, is correct. The subsequent sections will support this claim by even more striking examples.

4 The Metallic Bond

Many physical properties of normal metals can, at least qualitatively, be described and interpreted in the frame of a model of free or only slightly disturbed electrons (electron-gas model). The electronic properties of metals are therefore usually regarded as typical solid state phenomena and the "metallic bond" does not seem to be accessible to concepts based on the bond picture; hence, it is generally believed as being out of the scope of genuine chemical bonding theory. On the other hand, there exist certain types of chemical bonds, such as the π-bonds in organic chemistry and bonds in electron-deficient molecules (e.g. boron compounds) bearing much resemblance with the metallic bond. Under this aspect, the metallic bond can be viewed as the extreme case of an electron-deficient or unsaturated covalent bond; thus, meaningful questions connected with the bond concept remain to be solved, especially what kind of mechanism leads to the metallic bond or, more precisely, what kind of attractive and repulsive interactions contribute to the cohesion energy in metals.

In view of the fact that the valence electrons in normal metals forming the conduction band are quite well described by the free or only weakly disturbed electron gas model, the expectation value of the potential energy is correspondingly small for the typical metals. Therefore, we should expect that, even more pronounced than in the covalent bond, changes in the kinetic energy should constitute the dominating mechanism in forming the metallic bond. This picture is most easily visualized by considering the electrons as completely delocalized over the whole metal leading to a strong decrease of kinetic energy. Although this is basically correct, one should be cautious in applying oversimplified pictures when dealing with the physical

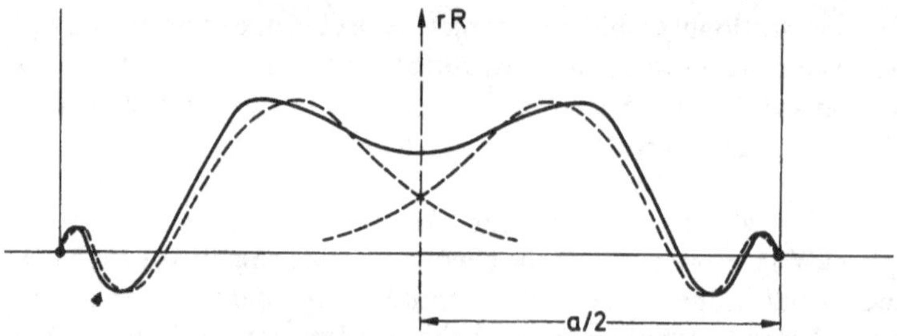

Fig. 4. Contraction of an atomic Na(*3s*)-wavefunction (*dashed line*) when subjected to the Wigner-Seitz boundary conditions. a is the lattice constant

interpretation of chemical bonding, as we have seen in the previous discussion of the covalent bond. Therefore, we will provide a more careful analysis of the mechanisms leading to the metallic bond.

The first hint for an interpretation based on the changes of kinetic energy was given by Wigner and Seitz [30] in 1933, with their treatment of the electronic structure of sodium. According to the periodicity of an ideal crystal, it can be thought of as being partitioned in unit cells, and the Bloch wavefunction, i.e. the crystal orbital in one cell has to match with the corresponding one in the neighbouring cell with vanishing first derivative. This matching condition leads to a reduction of the curvature of the Bloch orbital in the bonding region compared with the respective atomic orbital resulting eventually in a decrease of the kinetic energy of electrons from the *bonding region* as sketched in Fig. 4. Accordingly, the deformation of the wave function due to the zero slope at the cell boundary will also influence the inner part of the orbital by — figuratively speaking — pressing it more into the core region. We have, therefore, the same phenomenon of contraction of the wave function as resulting from the analysis of the covalent bond. Moreover, it is evident that Bloch orbitals built up of atomic orbitals do not satisfy the virial theorem. The wrong conclusion that they should satisfy the virial theorem, was drawn by Friedel [31] only because of neglecting overlap integrals between neighbouring atoms. Namely, it can be shown that the change in kinetic energy in tight-binding approximation with next-neighbour overlap is for the bonding state [32]

$$\Delta T = -(S_{\mu\nu} U_\nu - E_B) \tag{16}$$

where U_ν is the intraatomic matrix element

$$U_\nu = \int |\varphi_\nu(\vec{r})|^2 \, V_\nu^{at}(r) \, r^2 \, dr \tag{17}$$

and E_B is the binding energy. For zero overlap, Friedel's result is obtained that the change in kinetic energy is exactly the opposite of the binding energy and positive; but the first term in (16) is not negligible because the intraatomic matrix element (17) has a typical order of magnitude of several Ry. Once again, the virial theorem is only satisfied when the orbital contraction is taken into account as in the discussion of the covalent bond. To sum up, the existence of both the covalent and the metallic bond can be deduced from essentially the same mechanism, namely the decrease of kinetic energy in the bond region accompanied by a contraction of the orbitals involved in binding.

In the next step, we wish to search for local or transferable quantities, i.e. we wish to bridge the gap between the perfectly delocalized free electron gas model and a local picture. For this purpose, it is helpful to investigate, why the free electron gas model works so well. For the following reason, this is not at all self-evident. The potential in a metal, e.g. along a direction of maximum atomic density, has a form as sketched in Fig. 3: it is rather deep, close to the nuclei and flat in the interstitial region. According to such a potential, it is difficult to comprehend why the valence electrons should behave as free electrons. For, each time when approaching the nucleus they should stay there for some time due to the deep potential well, losing their free electron character. The solution to this problem is provided by one of the fundamental concepts in the electronic structure theory of solids, namely the pseudopotential. The basic, though incomplete, understanding of its action is immediately achieved by the pictorial argument that the core electrons prevent the valence electrons from coming too close to the nucleus. It is then argued that this is due to the Pauli principle: assuming that all core levels are occupied, the valence electrons cannot go into a core level and, hence, they are kept away from the core region. Although this argument is principally correct, it is not the full truth. Namely, first of all one has to take into account that the valence orbitals must be orthogonal to all core orbitals which is, in the first instance, a purely mathematical restriction to prevent the "variational collapse". Such an orthogonality requirement leads automatically to a low probability amplitude of the valence orbitals in regions where the core amplitude is large, i.e. electrons occupying the respective orbitals tend to avoid each other. Therefore, the mutual orthogonality acts like a "repulsive interaction" which is, of course, not a dynamical interaction (cf. Sect. 2) and, moreover, is by no means related to the Pauli-principle. For, also the orbitals in one-electron theories have to be orthogonal to each other, independently of a possible antisymmetrization when constructing a many-particle wave function from the orbitals.

However, at the same time the orthogonality condition causes that the valence orbitals have nonvanishing amplitudes in the core and, consequently, they have — in spite of the Pauli principle — a probability density in the core region that turns out to be nonnegligible. This can easily be proven by computing r^n-expectation values of the valence orbitals where the domain of integration is restricted to the core region. The decisive reason, why nevertheless the deep potential of the core does not affect the valence electrons can be related again to the kinetic energy. Namely, the gain in potential energy is almost perfectly compensated by a corresponding increase of kinetic energy also reflected in the oscillations of the valence orbital near the nucleus. Quantitatively, this may be expressed by the statement that the expectation value of the Hamiltonian with respect to a valence orbital and integrated over the core region vanishes in good approximation:

$$\langle H \rangle_{\text{core}} = \int_{\text{core}} \psi^*_{\text{val}}(\vec{r})\, H \psi_{\text{val}}(\vec{r})\, d^3r \approx 0 \tag{18}$$

This is the famous cancellation theorem [33] giving the actual reason that the deep potential in the core region is effectively ignored by the valence electrons. To summarize, the valence electrons do not see the core, not because their probability density vanishes there, but because there is a negligible contribution to the total energy. Consequently, as far as just energies are discussed or compared, one can simulate the combined effect of the Pauli principle, of the orthogonality condition and of the cancellation theorem by constructing a "pseudopotential" excluding the valence electrons from the core. In turn, this formal construction should not be physically misinterpreted in such a way that the core region is actually forbidden for the valence electrons.

The fundamental significance of pseudopotentials with regard to technical as well as to conceptual aspects cannot be overestimated and is generally recognized in solid state physics. It is, however, surprising that the explanatory power of the pseudopotential concept, especially for qualitative reasonings, has been widely ignored in chemistry, e.g. it has not been mentioned, much less discussed in almost all the standard textbooks on quantum chemistry. Two examples for the value of the pseudopotential concept will be given subsequently. The first concerns the use of Slater-functions [34] as node-free basis orbitals in many semiempirical valence-electron-only methods. Solving the pseudopotential Schrödinger equation, it can be shown that a suitable choice of the pseudopotential yields as solutions "pseudoorbitals" being node-free in the core region. The Slater-type orbitals can thus be interpreted as the solutions of an appropriate pseudopotential equation. The second example is a consequence of the cancellation theorem, namely, that size and structure of the core do not play a role for many properties of the valence electrons as far as they are outside the core. This is the fundamental basis for understanding the periodicity of atomic properties and the chemical similarity of atoms within a group of the periodic system. Only within the frame of the pseudopotential concept is it therefore possible to introduce a term like "halogen atom" in a meaningful way, and to bridge the gap between a rigorous quantum mechanical description and the phenomenological concepts of empirical chemistry.

The last example also shows clearly that the pseudopotentials are transferable to a high degree and this transferability can even be extended to the whole atom. Of course, we do not expect the orbitals themselves of the conduction electrons to be localizable, but it is nevertheless possible to construct such local quantities. Namely, the gas of the conduction electrons tends to distribute itself in a way to screen the positive charges of the ion-cores forming the metal lattice. The metal may then be viewed as an assembly of electrically neutral objects "ion-core + screening-charge" and this picture has lead to the concept of only weakly interacting "pseudoatoms" [35] or "renormalized atoms" [36]. An empirical hint for the reasonability of this picture may be seen in the fact that the volume and the cohesion energy of a normal metal do not change very much when it is melted. Such a behaviour indicates that the total energy of the metal depends mainly on the atomic volume and far less on the details of the crystal structure. This mathematical concept, however, should not be misunderstood in that way as if the total charge density of the conduction electrons is actually well localized around a single atom in the metal. On the contrary, recent experimental charge density studies of metals [37] exhibit clearly a distribution of the conduction electrons as expected from the free electron gas model. But why does a metal then appear sometimes as a collection of impenetrable touching spheres? The answer is as follows. First, the kinetic energy of the conduction electrons acts as the resistance against collapsing of the metal atoms long before their cores start to touch each other. Secondly, the interactions within the ensemble of the electrons and the ion cores are essentially independent of direction. For both reasons, the metal can be viewed as consisting of pseudoatoms in many cases. Hence, the pseudoatom concept is another example of the problems one is faced with when trying to derive physical interpretations from formally correct descriptions of a physical system. Nevertheless, the combination of the two limiting pictures of the free electrons and the pseudoatoms can be taken for justifying the physical basis of a widely used potential model that considers every atom as surrounded by a sphere embedded in the constant potential of the free-electron-gas model. This construction of the potential within a metal is called "muffin-tin potential" [38], including originally the assumption that the spheres surrounding the atoms should not overlap for mathematical reasons, viz. to obtain single-valued wave functions. Later on, it became evident that introducing overlapping spheres is more advantageous, i.e. enlarging them to spacefilling Wigner-Seitz spheres but keeping the mathematical formalism of the nonoverlapping spheres. By the help of this potential model an appropriate set of basis functions can be constructed, namely the "muffin-tin orbitals" [39–41] which can be utilized to derive transferable matrix elements for electronic structure calculations. Such schemes facilitating numerical and enabling analytical calculations of electronic states, altogether, have been developed by several groups. Although this formalism is of some interest for molecular calculations, too, we will not go into details but refer to the literature [42–45].

Instead of that, we wish to discuss another concept closely related to the concept of pseudoatoms, which is of particular advantage for large systems with many close lying eigenvalues, i.e. with many nearly degenerate eigenstates $\psi_k(\vec{r})$. In such a system, the individual eigenstates are in general unstable under small perturbations which may be physically irrelevant in the sense that they do not produce any observable change in the whole system. Apparently, this makes a description of physical properties in terms of wave functions highly inappropriate and much of the relevant information is contained in another quantity, namely the "local density of states" [46]:

$$n(E, \vec{r}) = n(E)\, |\psi_E(\vec{r})|^2 \tag{19}$$

Here, n(E) is the total density of (energy) states of the system under consideration, and the local density can be interpreted as the total density modulated by the probability density of the state with energy E. For a small system with discrete energy levels E_k, Eq. (19) can be transformed into a form more familiar to quantum chemists by the following operations. The function n(E) decomposes then into a sum of δ-functions:

$$n(E) = \sum_k \delta(E - E_k) \tag{20}$$

yielding for the local density of states

$$n(E, \vec{r}) = \sum_k \delta(E - E_k)\, |\psi_k(\vec{r})|^2 \tag{21}$$

Introducing the LCAO approximation

$$\psi_k(\vec{r}) = \sum_{\mu i} c^{\mu}_{ik} \varphi^{(\mu)}_i(\vec{r} - \vec{R}_\mu) \tag{22}$$

with the atomic spin-orbital $\varphi^{(\mu)}_i(\vec{r} - \vec{R}_\mu)$ at atomic site μ into Eq. (21), we obtain

$$\begin{aligned} n(E, \vec{r}) &= \sum_k \sum_{\substack{\mu\nu \\ ij}} c^{\mu *}_{ik} c^{\nu}_{jk} \delta(E - E_k) \varphi^{(\mu)*}_i(\vec{r} - \vec{R}_\mu) \varphi^{(\nu)}_j(\vec{r} - \vec{R}_\nu) \\ &= \sum_{\substack{\mu\nu \\ ij}} P^{\mu\nu}_{ij}(E) \varphi^{(\mu)*}_i(\vec{r} - \vec{R}_\mu) \varphi^{(\nu)}_j(\vec{r} - \vec{R}_\nu) \end{aligned} \tag{23}$$

with the bond-order matrix

$$P^{\mu\nu}_{ij}(E) = \sum_k \delta(E - E_k) c^{\mu *}_{ik} c^{\nu}_{jk} \tag{24}$$

The total density of states can then be written as

$$n(E) = \int n(E, \vec{r}) \, d^3r = \sum_{\substack{\mu\nu \\ ij}} P^{\mu\nu}_{ij}(E) S^{\mu\nu}_{ij} \tag{25}$$

with the overlap matrix $S^{\mu\nu}_{ij}$, and may be decomposed into (atomic site) local densities of states

$$n^{\mu}_i(E) = P^{\mu\mu}_{ii}(E) + \sum_{\substack{\mu \neq \nu \\ j}} P^{\mu\nu}_{ij}(E) S^{\mu\nu}_{ij} \tag{26}$$

when a Mulliken population analysis [47] is applied. Integrating additionally over E, the usual gross orbital occupation numbers are obtained. In this way, a local quantity which is closely related with the concept of the pseudoatom is obtained. Moreover, we shall demonstrate in Sect. 7 that the occupation numbers may serve as one tool to bridge the gap between a delocalized and a localized description of the electronic structure.

5 Van der Waals Bonding

Although the Van der Waals bond is not a typical chemical bond in that it is made up of a considerable reorganization of electrons in the valence shell, we wish to analyse briefly the mechanisms leading to attractive and repulsive interactions between neutral closed-shell atoms because they represent another type of a typical quantum mechanical many-body effect. This fact, however, is not intelligible from the discussion of the Van der Waals bond in many textbooks of solid state physics but is concealed under the subsequent phenomenological description.

The starting point of this pseudoexplanation is the known $1/R^6$-dependence of the attractive interaction between nonoverlapping atoms, showing the same distance dependence as a dipole induced by an external dipole-field. Now imagine a rare gas atom as the paradigmatic species exhibiting the Van der Waals interaction. Such an atom is a closed shell system and, due to the spherical charge density, does not have any dipole or quadrupole moment. Hence, it cannot produce any dipole-field, and for this reason it has been argued, in the picture of the Bohr atom, that there exists an instantaneous dipole moment of order $p_1 \approx a_0 e$ because the electron revolves around the nucleus, being once on the one side of the nucleus and once on the other side. This motion produces an electric field $E \sim p_1/R^3$ which will polarize a second rare gas atom with polarizability α, thus inducing a dipole moment $p_2 = \alpha E \sim \alpha p_1/R^3$. Since the interaction between two dipoles decreases as $1/R^3$, one obtains an estimate for the interaction energy

$$|U| \approx p_1 p_2/R^3 \approx \alpha(a_0 e)^2/R^6 \tag{27}$$

showing the correct distance dependence. Then it is claimed that the attractive force between two neutral atoms arises from their instantaneos dipole moments.

Even though the mathematical result is correct, it becomes immediately apparent that the physical interpretation will most likely be false. For, it is based on the quasiclassical picture of the Bohr atom including the assumption that electrons in an atom move on definite orbits. Whereas this model may occasionally be applied for deriving formal results, it results in well-known serious errors if it is used for supplying physical interpretations and should, for several reasons, not be applied for such a purpose. Firstly, since the "explanation" of the Van der Waals interaction as given above is explicitly based on definite electronic orbits, it cannot represent the correct physical mechanism. Secondly, such an "explanation" would be on the same level as a model "explaining" the stability of the hydrogen atom as the consequence of the balance between the electrostatic attraction and the centrifugal force acting on the electron. Finally, the application of such a picture should be avoided on account of pedagogical reasons because it provides a mechanism for the dispersion interaction that is essentially the same as for the inductive interactions, i.e. those interactions arising from the polarization of a neutral atom by a molecule with a permanent dipole moment. Whilst these inductive interactions can be reduced to the behaviour of a single atom in an external field, this is impossible in case of the dispersion interactions.

Therefore, we will describe the nonfictitious physical mechanism leading to the condensation of neutral spherically symmetric atoms into the solid state. For this purpose, let us first imagine what will happen when we continuously increase the distance between the atoms in a metal. In this case, the bands will become increasingly narrower because the overlap will be more and more reduced, but within the one-electron picture the metal will, even for large interatomic distances, remain a conductor. Such a result is hard to be believed, and actually it will not happen but it is an artefact due to the neglect of the nonlocal part of the electron-electron interaction, mainly the Coulomb correlation. This many electron effect causes a phase transition from the metallic to the insulating state (Mott transition) at a certain interatomic distance dependent on the electron density [48]. Thereby, the Coulomb correlations become important especially in narrow bands since narrow bands mean weak dynamical interactions, and then the statistical correlations between the electrons become dominant.

The situation turns out to be analogous in the rare-gas crystals; in order to emphasize this aspect, let us make two additional idealizing assumptions. Firstly, the overlap between wave functions on different atoms shall be negligible so that no quantum mechanical interference is involved. Secondly, because of the zero overlap between the atoms, the wavefunction for the whole system shall be constructed as the simple nonsymmetrized product of the wavefunctions of the isolated system. As one consequence, the zero order Hamiltonian does not have the full symmetry of the exact Hamiltonian, but it underlines that possible correlations between the electrons are neither due to the superposition principle nor to the Pauli principle. Furthermore, on the ground of these assumptions, the interaction between the two atoms can be calculated by ordinary Schrödinger perturbation theory. The first treatment by such a second-order perturbation theory was provided by London in 1930 [49]. The physical interpretation of the result is that the atoms are subjected to a virtual transition to an excited state, *simultaneously* and not separately. Due to this coherence in the probability density of the electrons in both atoms, there exists a time-averaged dipole-dipole interaction, even though the time-averaged dipole moment of the isolated atom is zero, i.e. an electron located at a certain atom reacts upon the instantaneous rather than the average position of the electrons located at a neighbouring atom. Therefore, the "true" source of the attractive R^{-6}-interaction is the correlation between the electrons in the two atoms also present when the respective charge densities do not overlap. Hence, the electrons are correlated in a similar way as the noninteracting two electrons in the box discussed in Sect. 2. The nature of the dispersion interaction as a correlation effect is also displayed by analyzing the additivity of the Van der Waals forces [50]: "The Coulomb interaction correlates the dipole moments p_1 and p_2 with their mutual distance in such a way as to give the directional factor

$$-3 \cdot (\vec{p}_1 \cdot \vec{R}) \cdot (\vec{p}_2 \cdot \vec{R}) + (\vec{p}_1 \cdot \vec{p}_2) \cdot |\vec{R}|^2 \tag{28}$$

as function of direction its minimal value, $-2|\vec{p}_1| \cdot |\vec{p}_2| \cdot |\vec{R}|^2$, *simultaneously* for all pairs. Thus the nonstatic dipole moments $\vec{p}_i$ do not depend on the *single* molecules, but only on the *pairs* of molecules which interact with each other". In this sense, the Van der Waals interaction is a typical quantum mechanical many-body effect, i.e. London's perturbational treatment exhibits them as being a property arising only from an interaction but not from the superimposed properties of single atoms.

At the same time, however, London also showed that it is possible tò mathematically express the energy gain caused by two-electron-correlations in terms of virtual one electron excitations. This mathematical simplification can be correlated with an idealization in the respective physical system of the two interacting atoms. For, one may replace each atom by a dipole oscillator the characteristic frequencies of which correspond to the allowed electronic excitation energies. This result is consistent with the results from perturbation theory if the average quantum energy of the dipole oscillators is identified with their zero point energy. From this analysis, the origin of the simple picture derived from Bohr's theory can easily be understood [51].

If the dipole model connects the electronic correlation energy with the energy states of dipole oscillators caused by photon absorption and emission, it is also feasible to interpret the effect of this photon field as inducing spontaneous dipole fluctuations. The field produced by these fluctuations induces a reaction field on a neighbouring polarizable atom also viewed as a dipole which in turn lowers the energy of the spontaneous fluctuations. This quasi-classical dipole model considerably reduces the complexity of the whole matter and opens the possibility not just for eliminating the quantum mechanical many-body formalism but also for expressing the interaction energy in such a way that it becomes comparable with experimental data. Consequently, on the one hand, such a procedure can bridge the gap between theory and experiment, and it is for this reason useful and methodologically legitimate. On the other hand, it is clear that such a formal simplification may — under no circumstances — be interpreted as a part of physical reality, i.e. the model picture of two interacting fluctuating dipoles may not be regarded as physically real.

The dipole oscillator model contains another interesting feature. While the electrons obey the Fermi statistics, the oscillators behave as Bose particles. Applying Bose statistics to the set of oscillators, a temperature dependence — not inherent in the original perturbation formulation — is introduced. Thus, the dipole model originally constructed to simplify the quantum mechanical description yields additional information which can be proven by experiment to be even qualitatively correct. Does this feature give the model a touch of reality? Apparently not. It illustrates solely that the same effect can be modelled or simulated by different model assumptions. Hence, the content of reality of a certain model is not determined by the deduced experimental data but by independent and mature theoretical reflections.

6 Ionic Bonding

The ionic bond as the bond between ions of opposite charge might be generally considered as the most comprehensible type of bonding, seeming to bear no ambiguities in the interpretation of the underlying mechanisms and their physical reality. But even in this case the matter is by far not as simple as it appears at first sight. To elucidate this claim, let us briefly discuss the bonding between two other particles of opposite charge, namely the positively charged proton and the negatively charged electron leading to a stable hydrogen atom. The pseudo-classical picture of the Bohr atom "explains" the stability as the balance between the electrostatic attraction and the centrifugal force acting on the electron. Evidently, such a picture cannot resemble the physical reality since it is based on the assumption that an electron in an atom revolves on definite orbits in contradiction to the principles of quantum mechanics. Accordingly, the quantum mechanical mechanism for the stability of the hydrogen atom is substantially different. While the Coulomb interaction as the attractive mechanism remains unchanged, the repulsive mechanism is now deduced from the uncertainty principle, i.e. its origin is genuinely quantum mechanical. The qualitative argument is as follows. The Coulomb potential of the hydrogen atom has a shape as displayed in Fig. 3. For r approaching zero the potential tends to $-\infty$ being the reason that an electron viewed as a classical particle would collapse into the proton. However, according to the uncertainty relation $\Delta r \cdot \Delta p \gtrsim \hbar/2$, a decreasing Δr requires an increasing Δp leading to an increased kinetic energy for the following reason. The uncertainty of the momentum operator p is by definition $(\Delta p)^2 = \langle p^2 \rangle - \langle p \rangle^2$ yielding $(\Delta p)^2 \leqq \langle p^2 \rangle$ because $\langle p \rangle^2 \geqq 0$. Taking into account that the expectation value of the kinetic energy, $\langle T \rangle$, is proportional to $\langle p^2 \rangle$, an increasing Δp results in an increasing $\langle T \rangle$. The Bohr radius is then the distance at which the Coulomb attraction is balanced by the kinetic energy comprehensible, thus, as a repulsive interaction. In conclusion, it is again the kinetic energy playing an important role — now as a repulsive interaction — in understanding the stable ground state of the hydrogen atom. This insight will also serve as a guide for interpreting the repulsive mechanism for the ionic bond.

In the subsequent analysis of the ionic bond we will restrict ourselves to the alkali halides MX as the paradigmatic representatives of ionic crystals. Starting with the attractive interaction, the physical interpretation seems to be completely clear, viz. as the electrostatic interaction between the closed shell ions M^+ and X^-. The corresponding binding energy is called the Madelung energy and can easily be calculated analytically, apart from some well-known purely technical problems, yielding

$$E(Mad) = -\alpha \cdot (Ze)^2/R \qquad (29)$$

where α is the Madelung constant, Z is the ionic charge +1 in this case, and R is the distance between M and X.

However, quantum mechanical calculations clearly exhibit [52] that there is not a complete charge transfer even in the most polar alkali halide CsF, and the effective charges in LiI are even less than $|0.5e|$. Moreover, a charge density analysis gives overlap charges amounting approximately 0.4e being in the same order of magnitude or even larger than in the purely covalent diatomic halogens X_2. We may thus expect considerable covalent contributions to the Madelung energy though they cannot be separated from the electrostatic point charge interaction in an unambiguous way. For instance, assume the effective charges of MX as $Q = \pm 0.7e$ in Eq. (29). Since, however, these effective charges depend on the type of population analysis they are constructed from, they are not well-defined quantities. Accordingly, the separation of the Madelung energy in ionic and covalent parts bears the same defect although an estimate of the lower bound for covalent contributions to the binding energy is certainly given by the amount arising from the overlap charge between the halogen and the alkali atom. Following this rule, it has to be concluded that covalent contributions cannot be considered as small corrections of the attractive ionic interaction in the alkali halides [52]. Additionally, one has to introduce a finite size correction for the ions [53] when the point-charge approximation is removed, being also inherent in this simple model.

Let us turn next to the physical interpretation of the repulsive interaction. In analogy to the rare-gas atoms as the standard model of a closed shell species, this interaction is being interpreted as the repulsion between the closed shells of the ions M^+ and X^- due to the Pauli principle when they start to overlap as it is the case in the rare-gas crystals. For the simplest systems such as two He or Ne atoms, this interaction can be computed with a high degree of accuracy giving qualitatively an exponentially decreasing repulsive interaction. This result leads to the mathematical expression for the repulsive potential,

$$U_{rep}(r) = B \exp(-r/\varrho) \tag{30}$$

going back to Born and Mayer [54]. The parameter B is a measure for the strength of the interaction and ϱ characterizes its range. Both parameters have to be determined from experimental data, e.g. from the bulk modulus and the equilibrium distance, with the result that, for the alkali halides, ϱ is approximately independent of the special compound and is in this sense transferable or universal. This deliberates it from being a pure fit parameter and ascribes to it a touch of physical reality. Furthermore, the energy of cohesion following from Eq. (30)

$$|E_{coh}| = \frac{\alpha e^2}{R}(1 - \varrho/R) \tag{31}$$

can be compared with experimental values and agrees remarkably well with them. This result and further successful applications of this model [55] seem to confirm the standard physical picture of the alkali halides as consisting essentially of closed shell ions. But what is really real on the physical basis of this model, particularly in view of the fact that we know, as already mentioned that the alkali halides do not strictly consist of closed shell ions. I wish to claim that it is less the empirical support for this model but mainly the analogy with the closed shell rare-gas crystals which has lead to this belief. To support this point of view, a brief excursion to physical methodology is necessary.

Without any doubt, arguing by analogy exhibits a common and legitimate methodological practice in physics. A certain quantity is calculated exactly or with high accuracy for simple model systems. In the next step, the functional form, being in this case the exponential decrease of the repulsive interaction, Eq. (30), is transferred to more complicated systems and will be modified, if necessary, by introducing some parameters to correct for the "nonideality". Such a procedure is not at all questionable but bears a problematical aspect. For, very often the transfer of the mathematical expression is connected with a simultaneous transfer of the physical interpretation which is generally inadmissible and can lead to more or less serious misinterpretations. Exactly this occurs here since the assumption of a closed shell system is appropriate for the simple model system of the rare-gas atoms, but it is no longer valid for the alkali halides as already mentioned above. Thus, when the repulsive interaction cannot be identified with the closed shell repulsion we have to ask whether another mechanism might exist from which a short-range repulsive interaction of the Born-Mayer type, Eq. (30), could be deduced.

According to the previous analysis of the covalent and metallic bond as well as of the stability of the hydrogen atom, one may suspect that once again the kinetic energy may supply such a mechanism. This is in fact the case as can be proven by computing the expectation value of the kinetic energy in bond direction in the diatomic molecule MX as a function of the bond distance [52]. Due to the additional charge transfer resulting in the strongly attractive Coulomb interaction, the equilibrium distance is small enough to lie now distinctly in the repulsive part of the curve for the bond component of the kinetic energy. Thus, unlike in the covalent bond, the kinetic energy acts here indeed like a repulsion. This quantitative result can also be illustrated by a qualitative analysis based on the uncertainty relation together with the simple model assumption that only the valence electron of the alkali atom is free to move between the two atoms. In the first step, we estimate the binding energy of the alkali halide molecule assuming that the attraction is due to the Coulomb-interaction whereas the repulsion is caused by the kinetic energy $p^2/2m$ of the alkali valence electron. The uncertainty in the position of the electron equals approximately the bond distance R so that we obtain for the binding energy by using the uncertainty relation

$$E_B = \hbar^2/(2mR^2) - e^2/R \tag{32}$$

The cohesion energy per ion-pair of the crystal is accordingly

$$E_{coh} = -\alpha e^2/R + \beta\hbar^2/(2mR^2) \tag{33}$$

restricting the repulsive interaction to next neighbours, their influence being described by the factor β. Equation (33) can be directly compared with the Born-Mayer value, Eq. (31), giving the result

$$E_{coh} = -(\alpha e^2/R)\cdot(1 - \sigma/R); \qquad \sigma = \beta\hbar^2/(2me^2\alpha) \approx 0.27\beta/\alpha\,[\text{Å}] \tag{34}$$

The parameter σ describes the range of the interaction. The argument leading to Eq. (33) is valid only at the equilibrium distance but the functional form of this repulsive interaction can also be determined within the frame of this model by discussing the transition-matrix-element of the kinetic energy of the valence electron between the alkali- and the halogen-atom:

$$T_{MX} = (\hbar^2/2m) \int \varphi^*_{Xp}(\vec{r} - \vec{R}_X)\,\Delta\,\varphi_{Ms}(\vec{r} - \vec{R}_M)\,d^3r \tag{35}$$

Since the atomic wave-functions decay as $\exp(-\zeta r)$, it is easily proven that the integral T_{MX} is proportional to an exponential of the distance R_{MX} as well.

This analysis may be extended by constructing other types of interactions that may be thought of as occuring during the formation of the ionic bond. The result will be a conglomeration of attractive and repulsive interactions as:

a) Coulomb interaction between point charges: attractive and contributing to the conventional Madelung energy;
b) partial covalent bonding contributing also to the Madelung energy and in general not separable from a) around the equilibrium distance;
c) finite size correction to the Madelung energy;
d) polarization contributions, proportional to $1/R^4$ arising e.g. from the polarization of the halogen valence shell by the alkali cation;
e) increase of kinetic energy for smaller distances;

f) "Pauli repulsion" due to orthogonalization effects;
g) "Pauli repulsion" due to exchange- and correlation-effects;
h) attractive and repulsive Van der Waals interactions;
i) many-body interactions;
k) repulsion between neighbours of higher order;
l) zero-point vibrations.

Without claiming this list to be complete, the problem connected with analyzing interactions by comparison with empirical information becomes evident. It is impossible to say anything about the relative importance of the various contributions to the energy of cohesion and even quantum mechanical calculations cannot solve this problem unequivocally. Moreover, a careful analysis of the various contributions will show that they partially rely on different models leading either to contradictions or to abundancies when just a different name is given to the same interaction.

To sum up our analysis, the practical success of the Born-Mayer scheme in predicting experimental data does not prove anything about the physical reality of the underlying assumptions entering the model and the interactions, i.e. the close agreement between experimentally and theoretically derived data does not confirm the conclusion that the alkali halides are actually consisting of closed-shell ions. In particular, the actual repulsive interaction ought not to be identified with the repulsion between closed-shell ions. This does not mean, of course, that there exist any objections against the formal treatment of alkalihalides on the level of the Born-Mayer model. The criticism only concerns the rash conclusions regarding the physical nature of bonding in the ionic crystals as well as the actual interactions; conclusions that are based solely on the efficiency of the model in predicting some experimental data. To provide an analogous example, nowadays nobody would consider the epicycles of Ptolemy as physically real just because it was possible to predict planetary positions with sufficient accuracy over several centuries. By generalizing these examples, the agreement of theoretical deductions with experimental values does not tell us anything definite about the physical reality of the underlying model assumptions. If it is at all possible to gain any knowledge about them, the ionic bond has to be explored and analyzed on the theoretical level in a similar way [52] as it was accomplished for the covalent bond by Ruedenberg if serious misinterpretations and wrong conclusions are to be avoided.

7 Hybridization and Localization

A very attractive feature of the chemical bond picture in solids consists of the possibility of expressing physical properties in terms of local bonds instead of bands being delocalized over the crystal. The construction of localized electron-pair bonds is closely related to the concept of hybridization going back to Pauling [56]. Unfortunately, the term "hybridization" is used in physics and chemistry with differing meanings which has led to several misinterpretations. For this reason, we will start by analyzing the various meanings as far as they are of significance here. To this end, we have to distinguish primarily between two basically different meanings which we will call dynamical and formal hybridization.

A typical example for a dynamical hybridization is the $s - d$ hybridization in transition metals [38] where a comparatively narrow band of *d*-states lies within a broad *s*-band. This type of hybridization can be characterized as follows:

1. It corresponds to a real physical (or dynamical) interaction producing physical effects directly accessible to experiments, such as electrical resistivity in this example.
2. The interaction takes place (in the one-electron approximation) only between states of the same symmetry, i.e. between states transforming according to the same irreducible representation of the respective symmetry group.
3. The mixing coefficients between the hybridizing states are determined by the strength of the respective dynamical interaction.

The dynamical hybridization is obviously not restricted to a special choice of the basis orbitals, i.e. to a localized description, and is thus a meaningful concept, also inherent within MO- or band-theory. A special case of dynamical hybridization is the shifting of (accidental) degeneracies between states of the same symmetry. Another important effect of the dynamical hybridization is the bonding in the alkaline earth metals: whereas the diatomic molecules are only very weakly bonded, the respective metals have large cohesion energies.

While the dynamical hybridization is apparently the result of a quantum mechanical treatment of a physical system, documented by the deduced mixing coefficients, the term "formal hybridization" should be understood as a procedure for fixing the mixing coefficients previously according to some more or less plausible reasoning, i.e. they appear as an *assumption* but not as a *result* of a molecular calculation. The meaning of this definition is most easily explained with the help of the paradigmatic case of this type of hybridization, viz. the construction of sp^3 orbitals for an atom in tetrahedral environment as, e.g. carbon in methane. As can be shown by group theoretical arguments, one can form the following four hybrid orbitals from the $2s$ and $2p$ orbitals of carbon:

$$\begin{aligned} h_1 &= as + b(p_x + p_y + p_z) \\ h_2 &= as + b(-p_x - p_y + p_z) \\ h_3 &= as + b(-p_x + p_y - p_z) \\ h_4 &= as + b(p_x - p_y - p_z) \end{aligned} \tag{7.1}$$

The basic idea of the "formal" hybridization concept consists in fixing the mixing coefficients to $a = b$ ($= 1/2$ if normalization of h_i is taken into account) instead of assuming them as variational parameters for an actual quantum chemical calculation. Apart from this practical aspect of avoiding laborious computations, such a hybridization picture offers conceptual advantages of great importance which are responsible for the fact that this concept did not become obsolete with the advent of supercomputers enabling numerical calculations with high degree of accuracy. Some of these advantages of the hybridization concept are:

— the directional dependence of the covalent bond can be simulated in an easy and vivid manner;
— it provides a sound theoretical basis for the phenomenological-empirical concepts as the electron-pair bond, the lone-pair, and the "atom-in-a-molecule" conception, at all;
— it matches very closely the experience that many aspects of the chemical bond are amenable to a local description, at least in zero-order approximation, expressed by additivity rules for numerous molecular properties.

Nevertheless, the formal hybridization has — as any empirically derived concept — certain limitations that must not be overlooked. In particular, it is not possible to draw from the geometrical structure any definite conclusion about the respective electronic structure, as could be suggested by the example of tetrahedral coordination. To clarify this point, another simple example may serve as a warning against such premature inferences. The bonding angle in water is known to be 104.5°, being rather close to the tetrahedral angle of about 109.5°. Moreover, two lone-pairs are located at the oxygen atom. This has led some people to develop the model that water can be understood as an "incomplete tetrahedron" with two corners being occupied by H atoms whereas the other two corners correspond to the oxygen lone-pairs. This is, of course, one possible arrangement that can be inferred from the hybridization concept but it is not the only one. Another guess utilizes the MO-result that one oxygen lone-pair is perpendicular to the molecular plane, e.g. the electrons in the oxygen $2p_z$-orbital if the molecule is in the xy-plane, and is not involved in any hybridization with the oxygen $2s$, $2p_x$, $2p_y$ orbitals. Knowing about this, one can try next to simulate the structure by a (planar) sp^2 hybridization corresponding to the geometrical structure of an equilateral triangle. However, this model would lead to a bonding angle of 120° so that one has to invent an argument for the shrinking of the angle by about 15°. This is occasionally supplied by the ad-hoc assumption that the lone-pair "needs more space" than the electron-pairs in the two O—H bonds so that there exists an additional repulsion between the lone-pair and the bonds, eventually leading to this reduced bond angle. This argument sounds rather artificial and it certainly means an overstressing of the concept of formal hybridization. We do not want to continue with these attempts to derive the interactions within the molecule from the geometrical arrangement; the purpose was merely to emphasize the fact that the geometry is a result of the interactions within the molecule, the origin of which has to be figured out independently, and not the other way round.

Similar arguments apply to the case of tetrahedral sp^3 hybridization because any other choice of a and b in Eq. (7) is compatible with tetrahedral symmetry as well. Therefore, it may seem at first sight as if the choice $a = b$ is a pure ad-hoc assumption, and as if this formal hybridization were, consequently, merely a mathematical procedure without any relation to the actual physical situation. This is, however, not necessarily true. For the reliability of this assumption gets support in some cases by comparing the initially assumed s^1p^3 occupation of the atomic orbitals with the corresponding results of quantum chemical calculations. Thereby, one possible connection is being established via a population analysis of the computed charge density [47] yielding atomic orbital occupation numbers which can be directly compared with the mixing scheme of the hybrid atomic orbitals. Let us discuss the series of the tetrahedrally bonded hybrides and halides of C, Si, Ge and Sn, in order to illustrate this connection. Performing an SCC—Xα molecular-orbital calculation [57] together with a population analysis [58], slightly modified in comparison with the Mulliken scheme, the charge density distribution can be characterized by effective charges Q, atomic orbital occupations n_s, n_p and by overlap charges n_{AB} as displayed in Table 2. The following trends are remarkable: whereas in CH_4 the atomic orbital occupation numbers of the carbon atom are very close to the sp^3 assumption of the hybridization scheme, increasing deviations can be observed through the series down to SnH_4 or to CBr_4.

Table 2. Charge density at the central atom of tetrahedral group 4 compounds

	Q	n_s	n_p	n_{AB}	d(Å)
CH_4	−0.19	1.06	3.13	0.82	1.0935
SiH_4	0.45	1.15	2.40	0.77	1.4811
GeH_4	0.61	1.15	2.24	0.74	1.5251
SnH_4	0.76	1.22	2.02	0.72	1.7108
CF_4	0.70	1.00	2.30	0.68	1.323
CCl_4	0.45	1.14	2.40	0.60	1.7667
CBr_4	0.27	1.36	2.37	0.52	2.12
$SnCl_4$	1.34	0.91	1.75	0.50	2.2808

The deviations are caused by essentially three mechanisms: the first one is the apparently decreasing tendency of the *s* orbitals to participate in bonding from C to Sn; the second one is the increasing polarity of the bonds, and the third one the increasing bond distance given in the last column of Table 2.

Particularly the last point is immediately intelligible. For with increasing bond distance d, e.g. in CH_4, the interaction decreases leading eventually in the limit $d \rightarrow \infty$ to the s^2p^2 occupation of the free carbon atom in the ground state. Thus, on one side the numerical results demonstrate convincingly that tetrahedral symmetry does not fix the actual "mixing coefficients" of the atomic orbitals in such way that this symmetry necessarily requires occupation numbers leading exactly to sp^3 hybrids. On the other side, it is also seen that the choice of an sp^3 occupation comes considerably closer to the final result than the s^2p^2 occupation of the atomic ground state. However, the actual values s^xp^y are determined by the whole body of intramolecular interactions and can only be obtained as the result of a numerical calculation. In this sense, the phenomenon of the mixing of atomic orbitals is an example of dynamical hybridization because it is not anymore and ad-hoc assumption but a result deduced from some (dynamical) interaction processes.

Finally, by comparing how closely the ad-hoc assumptions of formal hybridization match the results of the dynamical hybridization process, a number of empirical rules can be established that express the conditions for the applicability of the formal hybridization scheme. These rules are the following:

a) hybridization between atomic orbitals of different angular momentum is favoured if the radial shape of the involved wavefunctions is similar [59];
b) hybridization is favoured with decreasing difference in the respective atomic orbital energies;
c) the tendency to hybridization decreases with increasing bond distance;
d) the hybridization concept becomes less applicable with increasing bond polarity.

The tendencies expressed in these rules can be confirmed by corresponding model calculations. Thereby it turns out that the first condition of similar radial shape is by far the most important one. Accordingly, the elements of the first period are those for which the hybridization concept is most appropriate, in the sense that it is coming close to physical reality. For, the radial shape of the 2*s*- and 2*p*-orbitals is very similar due to the small core consisting only of the 1*s*-electrons. With increasing core size this similarity diminishes leading to the observed decreasing tendency of the atomic orbitals to hybridize [59]. Furthermore, the conditions a)–d) emphasize that particularly the hydrocarbons are to be expected to fall within the range of applicability of the hybridization concept: we meet similar radial shape, short bond distances and low polarity in these compounds. The results obtained from considerations based on the hybridization concept match very well with comparable quantities as obtained from accurate quantum-chemical calculations, and they consequently exhibit close resemblance to physical reality.

The situation changes, however, when turning next to the other domain of the formal hybridization concept, viz. the transition metal complexes. Starting with a simple example, we firstly compare $SiCl_4$ and $TiCl_4$. Both molecules have tetrahedral symmetry and can serve again as a test for the validity of the classical example of sp^3- and sd^3-hybridization, respectively. By analyzing the electronic configuration at the central atom, we can ascertain that neither Si nor Ti exhibit a typical sp^3- or sd^3-hybridization configuration but a population analysis resulting from a molecular orbital calculation yields the configurations $3s^{1.0}3p^{1.8}$ for Si and $4s^{0.3}3d^{2.6}$ for Ti, respectively [60]. This result can easily be understood by recalling the requirements for the applicability of formal hybridization. First of all, the radial shape of the atomic orbitals involved in the hybridization process is considerably dissimilar, especially in Ti. Next, the high polarity of the bonds in both molecules leads to a significant loss of electrons at the central atom making any conclusions about occupation numbers from ad-hoc assumptions questionable altogether. Finally, it can be shown [60] that the main bond characteristics in $TiCl_4$ are even obtained without the inclusion of the 4*s*-function at Ti so that one cannot draw also any conclusions from geometrical similarities about the local electronic configuration at one of the atoms.

These trends are even more distinct when comparing analogous *hypervalent* main-group and transition metal compounds. To emphasize this point, we have investigated the electronic structure of several group 5 and group 15 pentahalides [60]. All of them possess the same geometrical structure of a trigonal bipyramid leading in the hybridization picture to sp^3d or sd^3p configurations at the central atom [61]. To examine whether this assumption has a physical basis we are presenting the results of a charge density analysis as derived from an SCC-Xα calculation in Table 3. These results can be briefly summarized as follows:

Table 3. Charge density analysis of pentahalides EX_5

	effective charges			overlap population			
	E	X(eq)	X(ax)	E — eq	E — ax	Δ	n_d
PF_5	1.33	−0.24	−0.31	0.69	0.58	0.11	0.21
AsF_5	1.54	−0.29	−0.34	0.68	0.62	0.06	0.11
VF_5	1.67	−0.33	−0.34	0.63	0.61	0.02	2.29
PCl_5	1.09	−0.16	−0.30	0.60	0.46	0.14	0.18
$SbCl_5$	1.39	−0.24	−0.33	0.67	0.60	0.07	0.03
$NbCl_5$	1.47	−0.27	−0.33	0.81	0.77	0.04	2.65

1. While the effective charges of the halogen atoms in axial position are almost equal for all of the six compounds investigated, significant differences exist with respect to the equatorial halogens. This tendency is even more distinctly reflected in the overlap population providing a qualitative measure for the covalent contribution to the bond. For, the differences Δ between respective equatorial and axial populations (sixth column of Table 3) turn out to be smaller by almost an order of magnitude in the transition metal compounds. This means that in these molecules the five bonds are approximately of the same strength while in the main-group compounds the axial bonds are significantly weaker and, for this reason, they are differing in character.

2. As regards the differences in the electronic configuration of the central atoms, the most noteworthy result is that the occupation numbers of the higher atomic orbitals at the central atoms are surprisingly small having a typical order of magnitude of 0.15 for the *d*-orbital of the main-group elements, and of 0.3 and 0.4 for the *s*- and *p*-functions at the transition metals, respectively. Whereas, in the former cases, the electronic configuration at the central atom is that of a typical *s*-*p*-bond with a very low direct *d*-participation, the *s*- and *p*-orbitals of the transition metals contribute scarcely to the bond because — due to the interaction with the ligand orbitals — they are shifted into the region of the empty antibonding states. This effect is also well-known from band structure calculations when comparing, e.g. metallic vanadium [62] with vanadium monoxide [63] or V_2O_3 [64]. While the 4*s*-band in vanadium lies about 6 eV below the 3*d*-band at the center of the Brillouin-zone, it is repelled in the oxides, because of metal-oxygen interaction, beyond the Fermi level and remains unoccupied.

3. The inclusion of *d*-orbitals for the main-group elements leads to a slight increase of charge in the bonding region, i.e. a strengthening of the bond and a reduced bond polarity. Therefore, the inclusion of *d*-functions is essential for quantitative investigations of moments of the electronic density and properties such as dissociation energies or force constants depending critically on bond strengths. Notwithstanding, the qualitative overall features of the electronic structure are quite well reproduced without the *d*-functions. This is confirmed by the small *d*-orbital occupation numbers (last column of Table 3) and by the fact that the photoelectron spectra can be fairly well reproduced without *d*-functions on the central atom. Analogous conclusions are valid for the transition metals, i.e. the qualitative

structure of the electronic charge density is described already by the *d*-orbitals whereas the *s*- and *p*-functions are non-essential.

4. Therefore, the bonding properties of the main-group systems are best described by a combination of three covalent pair bonds with the equatorial halogens and by an axial delocalized four-electron-three-center-bond. In this case, the formal hybridization picture based on five equivalent electron pair bonds with significant *d*-orbital participation is apparently inappropriate. The situation is a little bit different for the transition metal compounds because the five bonds are approximately of the same character and allow a description by five equivalent electron-pair bonds. Nevertheless, the actual electronic configuration at the central atom is far away from the ad-hoc assumption of an sd^3p configuration of the hybridization picture. At the same time, however, it becomes intelligible why the model works as far as properties of the *d*-electrons are concerned. For the actual *d*-occupation numbers, as derived from molecular orbital calculations, are in the same order of magnitude as those of the hybridization model, whereas the loss of electrons migrating to the electronegative ligands goes at the expense of the *s*- and *p*-electrons.

5. Finally, both series of pentahalides, the transition metal as well as the main-group species, exhibit approximately the same differences between axial and equatorial bond distances. Hence, these differences do not indicate the character of the bonds, regardless of whether two- or three-center bonds are concerned, or whether a *d*-orbital participation is significant in case of the main-group elements.

By summarizing these results, it can be said that reasonings about the electronic structure and bonding properties based on superficial empirical similarities as, e.g. of the geometrical structure are not justified especially if transition metals are compared with their respective main-group counterparts. Furthermore, when comparing the ad-hoc assumptions of the formal hybridization concept about the local electronic structure of an atom in a molecule with corresponding results from molecular orbital calculations, the general experience shows firstly that a realistic description is supplied for less polar compounds of elements of the first period, particularly for the hydrocarbons. Secondly, the physical significance of the hybridization picture is less clear for the transition metal compounds because its efficiency results from the fortuitous cancellation of two erroneous assumptions, namely the covalent nature of the transition metal complexes and the large occupation numbers of *s*- and *p*-orbitals. Hence, hybridization is still a working scheme for qualitative discussions, as far as only properties of the *d*-electrons are involved. Finally, the concept is essentially inappropriate for polar compounds of heavier main-group elements where such a cancellation is not present any longer.

8 Conclusions

This essay has been an attempt to illustrate that the bond concept may be useful for analysing several aspects of the electronic structure of solids. We have seen that even the electronic structure of metals may be interpreted in concepts closely resembling a bondlike description. It has been shown that also in the bond picture unifying principles exist that supply the basis for understanding and interpreting the quantum mechanics of the various types of chemical bonds in solids, namely the uncertainty relation, the change in the kinetic energy during bond formation and the presence of statistical correlations; the latter emphasize that not every change in energy is related to a force in the sense of Newtonian dynamics.

Furthermore, several simplifying approaches to special types of chemical bonding in solids have been discussed, such as the Born model for the ionic bond, the dipole oscillator model for the Van der Waals bond, the concept of pseudoatoms in metals and the various hybridization schemes. It has been demonstrated that such simple model considerations and semiempirical methods are, in the first place, valuable for deriving data directly comparable with measurements. One may ask, however, if those methods are really necessary and if it were not preferable to concentrate on the improvement of rigorous quantum mechanical calculations. The appropriate answer supporting the point of view presented in this essay was given by Wigner and Seitz more than thirty years ago [65]:

> "If one had a great calculating machine, one might apply it to the problem of solving the Schrödinger equation for each metal and obtain thereby the interesting physical quantities, such as the cohesive energy, the lattice constant, and similar parameters. It is not clear, however, that a great deal would be gained by this. Presumably the results would agree with the experimentally determined quantities and nothing vastly new would be learned from the calculation. It would be preferable instead to have a vivid picture of the behaviour of the wave functions, a simple description of the essence of the factors which determine cohesion and an understanding of the origins of variation in properties from metal to metal."

Consequently, semiempirical approaches are even necessary for interpreting rigorous quantum mechanical calculations in a chemically significant way as well as for bridging the gap between the general theoretical principles and the inductively constructed phenomenological concepts. For instance, even one of the most fundamental assumptions of solid state physics and quantum chemistry, viz. the existence of a fixed and well-defined crystalline or molecular structure, cannot be derived from quantum mechanical principles since it is contradictory to the uncertainty relation. Accordingly, all concepts based on the assumption of a definite structure, i.e. almost the whole body of solid state physics and quantum chemistry must be regarded as not being rigorous in a strictly theoretical sense.

However, there also exists some danger connected with using semiempirical methods concerning the physical interpretation; by the critical analysis of some semiempirical approaches we have elucidated that the agreement between theoretically derived and measured data does not allow any definite conclusion to be made about how close the underlying models and concepts match the physical reality of the physical systems or interactions under consideration. More generally speaking, the main purpose of this article has been the demonstration that conclusions about physical reality on the ground of semiempirical methods and inductively developed empirical concepts are misleading if not impossible for three reasons. First, the development of such concepts and models by induction requires simple examples: it is very unlikely that the chemical bond concept would have been created by investigating, e.g. intermetallic compounds. Generalizations from inductive reasonings are therefore always subject to limitations and can hamper deeper insights when transferred to other fields. Then they become artificial in the sense that the reality is "pressed" into a concept of too limited validity. Secondly, models and concepts for physical systems and interactions are very often constructed by analogy to known or simpler systems; accordingly, these conceptions are in general not deducible from the formalism from which the experimentally available date are derived, but they are, on the contrary, independent and additional ingredients. Thirdly, the success of a simplified model or method may be based on the fortuitous cancellation of unknown errors masking the relation to physical reality. This should always be kept in mind when semiempirical methods or strongly idealized concepts are to be physically interpreted: if at all, only a thorough theoretical analysis, not the practical success, yields insight into the physical reality.

This point of view is admittedly based on a specific conception of what has to be considered as physically real. If physical reality is being identified with observations and experimental data only, an analysis based on theoretical conceptions is unnecessary if not prejudicial because it is mostly concerned with unobservable terms. If, however, physical reality is also ascribed to models, as e.g. the atomic model, or to mechanisms, such as gravitational or electromagnetic forces, even though they are not directly observable but merely indirectly via their effects, a theoretical analysis is compulsory. To express it more directly: if the aim of physical theories were just restricted to the prognosis of experimental data, there would be no need for theories but the development and application of algorithms and empirical correlations would be already sufficient or even more useful. The reason why we prefer in most cases to work with theories is not due to their superior predictive power but to their interpretational value usually not inherent in algorithms and empirical correlations. In other words, we are convinced we can learn from theories something about the very reality hidden behind the observable phenomena. To illustrate that this is principally possible, although within certain limitations and with a mass of errors, detours and dead ends, was the main objective of our analysis of the chemical bond concept in solids.

9 References

1. Bloch F (1928) Z. Phys. 52: 555
2. Kittel C (1967) Quantum Theory of Solids. Wiley, New York, p 92
3. Pines D (1955) Sol. St. Phys. 1: 373 and 410
4. Ehrenreich H, Seitz F, Turnbull D (eds) Sol. St. Phys. (1980) vol 35
5. Ehrenreich H, Seitz F, Turnbull D (eds) Sol. St. Phys. (1970) vol 24
6. Callaway J, March NH (1984) Sol. St. Phys. 38: 136
7. Coppens P, Stevens ED (1977) Adv. Quant. Chem. 10: 1
8. Harrison WA (1980) Electronic Structure and the Properties of Solids. WH Freeman, San Francisco
9. Grodzicki M (1986) In: Bamme A, Berger W, Kotzmann E (eds), Anything goes — Science everywhere? Profil, München, p 135
10. Grodzicki M: Die Wirklichkeit der Physik — Konstruktion oder Entdeckung? (to be published)
11. Pagels HR (1982) The cosmic code, Simon and Schuster, New York

12. Braunbek W (1954) Z. Naturf. 9a: 959
13. Ruedenberg K (1962) Rev. Mod. Phys. 34: 326
14. Heitler W, London F (1927) Z. Phys. 44: 455
15. London F (1928) Z. Phys. 46: 455
16. Heitler W (1928) Z. Phys. 51: 805
17. Lewis GN (1916) J. Am. Chem. Soc. 38: 762
18. Hund F (1931) Z. Phys. 73: 1 and 565
19. Hellmann H (1933) Z. Phys. 85: 180
20. Bader RFW, Henneker WH, Cade PE (1967) J. Chem. Phys. 46: 3341
21. Kutzelnigg W (1973) Angew. Chem. 85: 551
22. Feinberg MJ, Ruedenberg K, Mehler EL (1970) Adv. Quant. Chem. 5: 28
23. Feinberg MJ, Ruedenberg K (1971) J. Chem. Phys. 54: 1495
24. Kutzelnigg W (1978) Einführung in die Theoretische Chemie, vol 2. Verlag Chemie, Weinheim
25. Klessinger M (1982) Elektronenstruktur organischer Moleküle, Verlag Chemie, Weinheim
26. Ref. 13, p 342
27. Finkelstein BN, Horowitz GE (1928) Z. Phys. 48: 118
28. Heine V (1980) Sol. St. Phys. 35: 1
29. Pettifor DG (1973) J. Chem. Phys. 59: 4320
30. Wigner E, Seitz F (1933) Phys. Rev. 43: 804
31. Friedel J (1969) In: Ziman JM (ed), The physics of metals, Cambridge University Press, chap 8
32. Pettifor DG (1971) J. Phys. F1: L62
33. Hellmann H (1935) J. Chem. Phys. 3: 61
Phillips JC, Kleinman L (1959) Phys. Rev. 116: 287
34. Slater JC (1930) Phys. Rev. 36: 57
35. Ziman JM (1964) Adv. Phys. 13: 89
36. Gelatt CD, Ehrenreich H, Watson RE (1977) Phys. Rev. B15: 1613
37. Schneider JR, Kretschmer HR (1985) Naturwiss. 72: 191
38. Ziman JM (1972) Sol. St. Phys. 26: 1
39. Andersen OK (1975) Phys. Rev. B12: 3060
40. Andersen OK, Jepsen O (1977) Physica 91B: 317
41. Skriver HL (1984) The LMTO method. Springer, Berlin Heidelberg NewYork
42. Pettifor DG (1977) J. Phys. F7: 613
43. Andersen OK, Klose W, Nohl H (1978) Phys. Rev. B17: 1209
44. Harrison WA, Froyen S (1980) Phys. Rev. B21: 3214
45. Straub GK, Harrison WA (1985) Phys. Rev. B31: 7668
46. Friedel J (1954) Adv. Phys. 3: 446
47. Mulliken RS (1955) J. Chem. Phys. 23: 1833, 1841, 2338 and 2343
48. Mott N (1974) Metal-insulator transitions, Taylor and Francis, London
49. London F (1930) Z. Phys. 63: 245
50. Lieb EH, Thirring WE (1986) Phys. Rev. A34: 40
51. Langbein D (1973) Festkörperprobleme XIII: 85
52. Grodzicki M (to be published)
53. Wilson JW, Heinbockel JH, Outlaw RA (1986) J. Chem. Phys. 84: 543
54. Born M, Mayer JE (1932) Z. Phys. 75: 1
55. Tosi MP (1964) Sol. St. Phys. 16: 1
56. Pauling L (1928) Proc. Natl. Acad. Sci. 14: 359
57. Grodzicki M (1980) J. Phys. B13: 2683
58. Grodzicki M, Männig V, Trautwein A, Friedt JM (1987) J. Phys. B20: 5595
59. Kutzelnigg W (1984) Angew. Chem. 96: 262
60. Grodzicki M, Elbel S (1987) In: Maksic ZB (ed), Molecular structure and bonding, Ellis Horwood, Chichester, p 239
61. Kimball GE (1940) J. Chem. Phys. 8: 188
62. Mattheiss LF (1964) Phys. Rev. 134: A 970
63. Kraan DJ (1973) Phys. Stat. Sol. 60b: 587
64. Grodzicki M, Jepsen O: Sol. St. Comm. (submitted for publication)
65. Wigner EP, Seitz F (1955) Sol. St. Phys. 1: 97

Chemical Implication of Local Features of the Electron Density Distribution

E. Kraka and D. Cremer

Theoretical Chemistry, University of Göteborg, Kemigården 3, S-41296 Göteborg, Sweden

Investigation of the electron density distribution $\varrho(\mathbf{r})$, its associated Laplace field $\nabla^2\varrho(\mathbf{r})$, and the energy density distribution $H(\mathbf{r})$ leads to a unique definition and description of atoms in molecules, chemical bonds, molecular structure, and structural changes. From the local features of $\varrho(\mathbf{r})$, $\nabla^2\varrho(\mathbf{r})$, and $H(\mathbf{r})$ valuable insight into the bonding mechanism and the reactive behavior of a molecule is gained, in particular into distortions of the valence shell of an atom upon bond formation; the order, π-character, polarity, and bent character of a bond; the degree of electron delocalization and bond equalization in a conjugated π-system; the extent of through-bond and through-space interactions; the sites that are prone to electrophilic or nucleophilic attack; and the bonds that can easily be broken in a molecule. Important chemical concepts or models such as π- and σ-aromaticity, homoaromaticity, ring strain, the frontier orbital model or the HSAB concept can be reevaluated on the basis of the density analysis. This is demonstrated for a number of examples ranging from π-delocalization in polyenes and cyclopolyenes, the peculiar stability of three-membered rings, the basicity of N containing aromatic molecules to the bonding in π-complexes. Furthermore, the analysis of $\varrho(\mathbf{r})$, $\nabla^2\varrho(\mathbf{r})$, and $H(\mathbf{r})$ openes fascinating avenues into new areas such as the chemistry of helium organic compounds.

Index of Symbol

A, B	denote atoms
A	degree of bond equalization based on bond orders n
C	Configuration space
D	degree of π-delocalization based on π-character η
a, b	constants in bond order — electron density relationship
d	deviation of bond path from internuclear connection line
d**r**	volume element in three-dimensional space
dω	volume element in n-dimensional space
dσ	surface element
E	molecular electronic energy
F(**r**)	force density
G(**r**)	kinetic energy density
H(**r**)	electronic energy density
H_b	electronic energy density at the bond critical point $\mathbf{r}_b$
K	number of bonds
K_σ, K_π	number of σ and π bonds
L(**r**)	one-particle Lagrange density
l_A^{AB}	value of $-\nabla^2\varrho(\mathbf{r})$ in the valence shell of atom A at position $\mathbf{r}_A^{AB}$ in direction of bond AB
l_A^n	value of $-\nabla^2\varrho(\mathbf{r})$ in the lone pair region of atom A
l_A^h	value of $-\nabla^2\varrho(\mathbf{r})$ at a concentration hole in the valence shell of atom A
N	number of π electrons
n	bond order

n_a	number of maxima $\mathbf{r}_a$ of $\varrho(\mathbf{r})$
n_b	number of bond critical points $\mathbf{r}_b$ of $\varrho(\mathbf{r})$
n_r	number of ring critical points $\mathbf{r}_r$ of $\varrho(\mathbf{r})$
n_c	number of cage critical points $\mathbf{r}_c$ of $\varrho(\mathbf{r})$
$\mathbf{n}(\mathbf{r})$	unit vector normal to surface $S(\mathbf{r})$
M^i_A	ith order moment of atom A
$\hat{O}(\mathbf{r})$	one-electron operator
$O(\Omega_A)$	atomic property, represented by the operator $\hat{O}(\mathbf{r})$
p	distance of a point on the valence shell from the nucleus
Q	degree of π-delocalization
R_b	bond path length
R_e	equilibrium distance between two atoms
$\mathbf{R}$	nuclear configuration (geometry)
$\mathbf{R}_c$	catastrophe point
r	rank of the Hessian matrix (matrix of second derivatives) of $\varrho(\mathbf{r})$
$\mathbf{r}$	position vector in molecular space
$\mathbf{r}_A$	position vector of a critical point of $\nabla^2\varrho(\mathbf{r})$ in the valence shell of atom A
$\mathbf{r}_a$	position vector of a maximum of $\varrho(\mathbf{r})$
$\mathbf{r}_b$	position vector of a bond critical point of $\varrho(\mathbf{r})$
$\mathbf{r}_c$	position vector of a cage critical point of $\varrho(\mathbf{r})$
$\mathbf{r}_r$	position vector of a ring critical point of $\varrho(\mathbf{r})$
$S(\mathbf{r})$	zero-flux surface
S(A, B)	zero-flux surface between atoms A and B
s	signature of the Hessian matrix of $\varrho(\mathbf{r})$
T	kinetic energy
V	potential energy
$V(\mathbf{r})$	potential energy density
$\mathbf{W}$	control space in catastrophe diagram
$\mathbf{v}_1, \mathbf{v}_2, \mathbf{v}_3$	eigenvectors of the Hessian matrix of $\varrho(\mathbf{r})$, direction of the principal curvatures
$\mathbf{u}_i$	normalized eigenvector of the Hessian matrix
u, v, w	control parameters of catastrophe diagram
x, y, z	cartesian coordinates
α	geometrical angle
β	interpath angle
γ	angle between the lines connecting two valence shell critical points with the enclosed nucleus
Γ^1	first order density matrix
$\varepsilon_b, \varepsilon_r$	ellipticity at bond, ring critical point
ε'	effective ellipticity
η	π-character of a bond
$\lambda_1, \lambda_2, \lambda_3$	eigenvalues of the Hessian matrix of $\varrho(\mathbf{r})$, (principal curvatures)
$\lambda_i(A - B)$	curvature of single bond AB
ξ	ratio (ϱ_r/ϱ_b) 100
$\varrho(\mathbf{r})$	spinfree one-particle electron density distribution
ϱ_b	electron density at the bond critical point

ϱ_c	electron density at the cage critical point
ϱ_r	electron density at the ring critical point
$\nabla\varrho(\mathbf{r})$	gradient vector (field) of $\varrho(\mathbf{r})$
$\nabla^2\varrho(\mathbf{r})$	Laplace field of $\varrho(\mathbf{r})$
σ	standard deviation
σ	stress tensor
τ	dihedral angle
Ψ	molecular wavefunction
χ	electronegativity
Ω, Ω_A	subspace, atomic subspace of atom A

1 Introduction

The molecular wave function of a molecule embodies all the information about structure, stability and reactivity of the molecule. However, the wave function Ψ contains this information in such a complex form that is difficult to translate into heuristic concepts and ad hoc models familiar to chemists. One way to decode chemically useful information from Ψ is based on a representation of the wave function by antisymmetrized products of molecular orbitals (MO) [1]. The MOs themselves are formed by a linear combination of atomic orbitals (LCAO) [2], which are the smallest building blocks of the wave function. Hence, it is possible to discuss molecular properties in terms of atomic properties by determining the contribution of individual AOs to the MOs that form the wave function.

The MO approach has been successfully applied to reveal the important role of the valence shell electrons for chemical bonding [3], to distinguish between σ- and π-electrons [4], to describe individual bonds of molecules by hybrid orbitals [5], to draw a connection between Lewis' electron pair model and localized MOs [6], to predict the shape of simple molecules by analyzing MO diagrams [7], to assess the properties of atoms in molecules and chemical bonds in terms of orbital and overlap populations [8], to break up complicated molecules into simpler entities and to describe the union of these entities by perturbational MO (PMO) theory [9], to rationalize the reactivity of molecules by analyzing orbital symmetry [10], to stress the role of the frontier orbitals (FO) in chemical reactions [11], etc. This list of accomplishments achieved by applying MO theory could be easily extended and, therefore, it is not astonishing that most chemists prefer to rationalize the physical and chemical properties of atoms and molecules in terms of the properties of the MOs.

The use of MOs, however, implies some disadvantages that have to do with the fact that MOs (as well as AOs) are nonobservable. Also, MOs can be transformed in many ways (using appropriate unitary transformations) without changing energy, geometry or other molecular properties. A chemical observation that is easy to rationalize with one set of MOs may be difficult to understand when described with a transformed set of MOs. For example, canonical MOs are always delocalized and, therefore, they are not helpful when comparing bond localization in a cycloalkane and in a conjugated cyclopolyene. However, the corresponding localized MOs immediately reveal bond localization in the former and bond delocalization in the latter compound.

Also, many orbital models attribute importance to particular MOs while ignoring the effects of all other MOs. Although this simplified approach is helpful in many cases, its justification is doubtful and application is only possible with advances knowledge of the actual chemical situation. Failures of the approach are corrected by considering the properties of additional MOs. For example, within the FO-model [11], molecular reactivity is considered to be dominated by the properties of HOMO and LUMO. In those cases, where predictions on reactive behavior elucidated from HOMO and LUMO are not in line with experimental observations, the second (third, etc.) HOMO (LUMO) are included into the analysis [12].

Models and concepts based on MO-theory are often easy to understand but difficult to put on a quantitative basis. For example, homoaromaticity is a result of 1,3 overlap between orbitals thus closing a cyclic conjugated system with (4q + 2) π-electrons [13]. If the topology and the geometry of a cyclic system is known potential homoaromaticity can easily be predicted by counting electrons and analyzing MOs in a qualitative way. However, the question how much orbital overlap is necessary to turn on homoaromatic interactions can not be answered in an unambiguous way.

Considering these disadvantages of MO descriptions of molecules it becomes highly desirable to base a better description on a molecular quantity that is

1. a function of the total wave function,
2. invariant under unitary transformations of the MOs,
3. an observable molecular property,
4. directly related to the molecular energy (stability), geometry (structure), and other molecular properties,
5. dissectable into atomic and bond contributions,
6. indicative of the distribution of negative and positive charge in the molecule,
7. a continuous function of a position vector **r** to allow a distinction between different (reactive) sites in the molecule.

The only quantity that fulfills all these requirements is the total one electron density distribution $\varrho(\mathbf{r})$. It is a direct function of Ψ, invariant with regard to unitary transformations of the MOs, and an observable molecular property that can be determined by a combination of X-ray and neutron diffraction experiments [14, 15]. Also it is easy to calculate $\varrho(\mathbf{r})$ once the MOs and the wave function Ψ of a molecule have been determined [16].

The fundamental role of the one electron density $\varrho(\mathbf{r})$ was pointed out first by Hohenberg and Kohn [17]. They showed that the energy of a nondegenerate (electronic) ground state of a molecule is an unique functional of $\varrho(\mathbf{r})$. Therefore, physical and chemical properties of a molecule can be related to $\varrho(\mathbf{r})$.

Bader and co-workers have shown that properties of the electron density distribution $\varrho(\mathbf{r})$ can be used to derive a partitioning of the molecular space into subspaces that are associated with the atoms in a molecule [18–20]. This method has been called "virial partitioning" because the virial theorem remains valid in the subspaces thus obtained. Also, chemical bonds can be described by the properties of $\varrho(\mathbf{r})$ [21–23] and reactive sites in a molecule can be identified by analyzing the properties of $\varrho(\mathbf{r})$ [24–28]. In conclusion, investigation of $\varrho(\mathbf{r})$ provides an improved description of a molecule as compared to the amalysis of particular MOs. This does not mean that the analysis of the MOs should be discarded in favour of the analysis of $\varrho(\mathbf{r})$. On the contrary, information from both MOs and $\varrho(\mathbf{r})$ should be combined in such a way that the most complete and useful description of the molecule emerges. MOs are still to be used for qualitative arguments while the analysis of $\varrho(\mathbf{r})$ proves the basis for

1. defining atoms in molecules and chemical bonds,
2. defining molecular structure and structural changes,
3. quantitatively assessing the properties of atoms in molecules and chemical bonds,
4. specifying chemical models and concepts,
5. relating changes in molecular structure (chemical reactivity) to the local properties of the molecular charge distribution.

We will demonstrate this by briefly reviewing the essential features of the analysis of the electron density distribution $\varrho(\mathbf{r})$, its gradient vector field $\nabla\varrho(\mathbf{r})$, and its Laplace field $\nabla^2\varrho(\mathbf{r})$ in Sects. 2 and 3. Then, based on the local properties of $\varrho(\mathbf{r})$ we will describe π-conjugation (Sect. 4), homoconjugation (Sect. 5), and σ-conjugation (Sect. 7). Selected examples will be given to demonstrate both the feasibility and the usefulness of the analysis of $\varrho(\mathbf{r})$ [29–33]. Special emphasis will be laid on a reconsideration of well-known chemical concepts such as aromaticity and antiaromaticity [29] (Sect. 4), homoaromaticity [34–37] (Sect. 5),. and σ-aromaticity [38–41] (Sect. 7). A quantitative assessment of the latter requires a deeper understanding of the notion of ring strain, of the relationship of three-membered rings and π-complexes, and of the chemical bonding in cyclopropane [38–44] (Sect. 6). Section 8 will be devoted to the description of chemical reactivity as assessed from the properties of $\varrho(\mathbf{r})$ [45–47]. Sections 2 through 9 establish the scenario of carrying out the analysis of $\varrho(\mathbf{r})$ in such a way that allows reevaluation and improvement of chemical concepts, that leads to new insights into molecular properties, and that lays the basis for discovery of novel compounds. The latter point has been demonstrated in a number of investigations which, e.g. dealt with unknown carbon, beryllium compounds [48], the peculiar structure of dications [49], and the question of the existence of helium compounds [50–51]. Since these studies will be reviewed in another article [52], we will only sketch some highlights of this work in the final chapter of this article which contains concluding remarks and an outlook.

2 Analysis of the Electron Density Distribution $\varrho(\mathbf{r})$

Figure 1 shows a perspective drawing of the calculated electron density distribution $\varrho(\mathbf{r})$ of the methyl cation, CH_3^+, with regard to the plane containing the four nuclei. At the positions of the nuclei, $\varrho(\mathbf{r})$ attains maximal values [53]. In off-nucleus

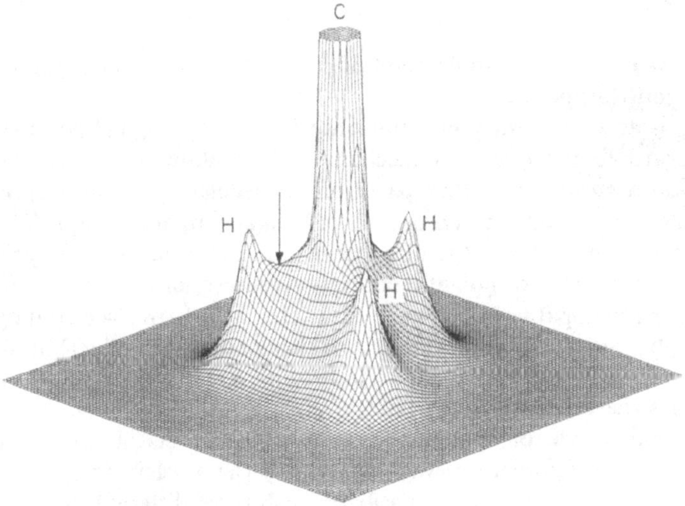

Fig. 1. Perspective drawing of the calculated electron density distribution $\varrho(\mathbf{r})$ of the methyl cation, CH_3^+, with regard to the plane containing the four nuclei. The position of one of the three saddle points between the C and the H atoms is marked by an *arrow*. For a better representation values above a threshold are cut off. (HF/6-31G(*p*, *d*) calculations)

direction, $\varrho(\mathbf{r})$ decreases exponentially and quickly approaches zero for large distances from the nuclei. The same pattern is found for all molecules. It seems that $\varrho(\mathbf{r})$ of a molecule is dominated by the electron density distribution of the various atoms constituting the molecule. In so far, one cannot expect much useful information about molecules when analyzing $\varrho(\mathbf{r})$.

However, a closer look at Fig. 1 reveals that in the bonding region $\varrho(\mathbf{r})$ adopts a distinct behavior that cannot be explained by just superimposing the electron densities of the corresponding atoms. For example, the distribution $\varrho(\mathbf{r})$ possesses relatively large values in the CH bond region. The C and H nuclei are connected by a ridge of maximum electron density like the summits of a mountain range. Any lateral displacement from the path leading along the ridge means a decrease in $\varrho(\mathbf{r})$. Therefore, we speak of a path of maximum electron density (MED path) linking bonded atoms. As can be seen in Fig. 1, there are just three MED paths between the C atom and the three H atoms, which is in line with the fact that CH_3^+ contains three CH bonds.

Also, shown in Fig. 1 is the position $\mathbf{r}_b$ (marked by an arrow) of the minimum of $\varrho(\mathbf{r})$ along the MED path. It corresponds to a maximum of $\varrho(\mathbf{r})$ in all directions perpendicular to the MED path, i.e. it is a saddle point of $\varrho(\mathbf{r})$ in three dimensions. The position $\mathbf{r}_b$ is uniquely defined by the first and second derivatives of $\varrho(\mathbf{r})$, collected in the gradient vector $\nabla\varrho(\mathbf{r})$ and the Laplacian $\nabla^2\varrho(\mathbf{r})$. The value of $\nabla\varrho(\mathbf{r})$ equals zero at $\mathbf{r}_b$. Two of the three eigenvalues of the Hessian matrix of $\varrho(\mathbf{r})$ (matrix of the second derivatives), λ_1 and λ_2 are negative, while the third eigenvalue, λ_3, is positive at $\mathbf{r}_b$. The eigenvalue λ_3 defines the curvature along the MED path, λ_1 and λ_2 the curvatures along the principle axes spanning the plane perpendicular to the MED path. The saddle point $\mathbf{r}_b$ of $\varrho(\mathbf{r})$ can be specified by rank and signature of the Hessian matrix [54]. For $\lambda_i \neq 0$ (i = 1, 2, 3), the rank r is 3 and the signature s is given by

$$s = \sum_i \lambda_i/|\lambda_i| \tag{1}$$

Since $\lambda_1, \lambda_2 < 0$ and $\lambda_3 > 0$ for $\mathbf{r}_b$, the saddle point of $\varrho(\mathbf{r})$ between two bonded atoms is a (3, —1) stationary (critical) point.

The critical point $\mathbf{r}_b$ is an important point, not only from a topological point of view but also with regard to the quantum mechanics of an atom in a molecule [18–20]. It is located on a surface S(r) that partitions molecular space in atomic subspaces. This can be seen when analyzing not only $\varrho(\mathbf{r})$ but also $\nabla\varrho(\mathbf{r})$. The gradient vector $\nabla\varrho(\mathbf{r})$ always points into the direction of maximum increase of $\varrho(\mathbf{r})$. If one evaluates the gradient vector of $\varrho(\mathbf{r})$ at a continuous successions of points **r** in molecular space, then a gradient path of $\nabla\varrho(\mathbf{r})$ will result. This is the path traced out by the gradient vector of $\varrho(\mathbf{r})$. Since it is perpendicular to the contourlines of $\varrho(\mathbf{r})$ at all points in molecular space, it is an orthogonal trajectory of $\varrho(\mathbf{r})$. The collection of all gradient paths defines the gradient vector field of $\varrho(\mathbf{r})$.

Figure 2 shows the gradient vector field of the electron density distribution $\varrho(\mathbf{r})$ of CH_3^+ with regard to the reference plane containing the nuclei. Inspection of the gradient vector field reveals that one can distinguish three different types of trajectories. First, there are those trajectories which terminate at one of the nuclei (type I trajectories). Secondly, there are trajectories that spring at one of the saddle points $\mathbf{r}_b$ between C and the H nuclei and terminate at one of the nuclei next to $\mathbf{r}_b$

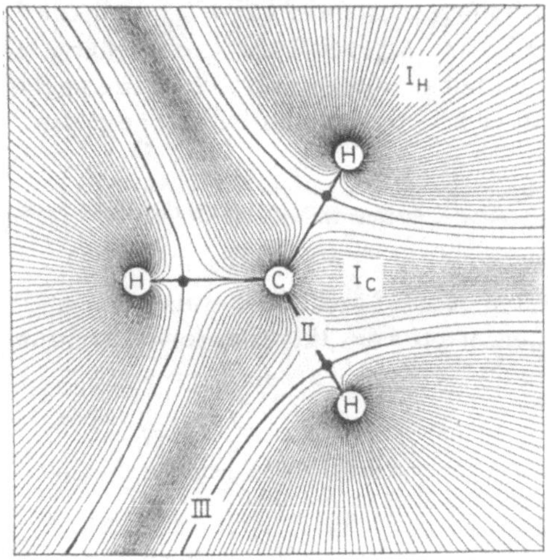

Fig. 2. Gradient vector field $\nabla\varrho(\mathbf{r})$ of the electron density distribution of the methyl cation, CH_3^+, shown for the plane containing the four nuclei. Saddle points $\mathbf{r}_b$ are denoted by *dots*. There are three different types of trajectories. Type $I_H(I_C)$ trajectories start at infinity and end at a H(C) nucleus. Type II trajectories (*heavy lines*) define the MED paths linking the carbon atom to the hydrogen atoms. Type III trajectories form the three zero-flux surfaces $S(\mathbf{r})$ between the C and H atoms. In the two dimensional display only their traces can be seen. Type III trajectories spring at infinity and terminate at the saddle points. (HF/6-31G(p, d) calculations)

(type II trajectories). For each saddle point $\mathbf{r}_b$ just two type II trajectories exist which define the MED path linking bonded atoms (heavy lines). Finally, there are trajectories that originate at infinity and terminate at the saddle point $\mathbf{r}_b$ (type IH trajectories). In three dimensions, these trajectories define the limiting surface $S(\mathbf{r})$ separating bonded atoms. In the two-dimensional display of Fig. 2, only traces of the three surfaces $S(\mathbf{r})$ that separate C and H atoms can be seen. There are no trajectories passing through $S(\mathbf{r})$ (Fig. 2), i.e. the flux of $\nabla\varrho(\mathbf{r})$ vanishes for all surface points:

$$\nabla\varrho(\mathbf{r})\,\mathbf{n}(\mathbf{r}) = 0 \qquad \mathbf{r} \in S(\mathbf{r})\,, \tag{2}$$

where $\mathbf{n}(\mathbf{r})$ is the unit vector normal to the surface $S(\mathbf{r})$. The surface $S(\mathbf{r})$ is called zero-flux surface [18–20].

2.1 Definition of an Atom in a Molecule

As discussed for CH_3^+ (Fig. 2), the zero-flux surfaces $S(\mathbf{r})$ partition molecular space into subspaces, each of which contains exactly one and only one atomic nucleus [55]. Since this observation has been made for many molecules [21–23], it is reasonable to consider the subspaces defined by the zero-flux surfaces as *atomic subspaces*. All type I trajectories terminating at a given nucleus form the basin Ω. As was first proposed by Bader and co-workers [18–20], the union of a nucleus and its associated basin defines the *atom in a molecule*.

The subspaces given by Eq. (2) are unique since quantum mechanical laws such as the virial theorem are valid in a subspace Ω [19]. If one associates Ω with an atom one can speak of a "virial atom". The dissection of the molecules space into atomic subspaces is called the "virial partitioning" method [19]. With the definition of an atom in a molecule, it is possible to determine the properties of atoms in a molecule. For example, the space Ω_A that is filled by the gradient paths terminating at one atomic nucleus is the volume of the corresponding atom A.

If $\varrho(\mathbf{r})$ is integrated over Ω_A, the atomic monopole moment M_A^0 will be obtained. The difference between the atomic number Z_A and M_A^0 yields the atomic charge. If $\mathbf{r}^n\varrho(\mathbf{r})$ is integrated over Ω_A for n = 1, 2, ... atomic dipole, quadrupole, . . . moments will result. Any atomic property $O(\Omega_A)$ that can be represented by an one-electron operator $\hat{O}$ is given by:

$$O(\Omega_A) = \int_{\Omega} \hat{O}(\mathbf{r})\, \varrho(\mathbf{r})\, d\mathbf{r}\,. \tag{3}$$

The definition of atoms in molecules leads to a detailed and quantitative description of their properties (see, Table 1). In addition, it provides the basis for a definition of the chemical bond.

Table 1. Description of atoms in molecules, chemical bonds, and molecular structure in terms of the properties of $\varrho(\mathbf{r})$

Terms used in Density Analysis	Comment	Chemical Term
1. General definitions		
Nucleus + Basin	Basin Ω_A is formed by all gradient paths that terminate at the nucleus of atom A.	Atom A
Zero-flux Surface	Surface S (AB) is formed by all gradient paths that terminate at the saddle point $\mathbf{r}_b$ between atoms A and B.	Interatomic surface between atoms A and B
Bond path	*Necessary condition*: Existence of a maximum electron density (MED) path linking bonded atoms A and B. *Sufficient condition*: Negative energy density $H(\mathbf{r}_b)$ at the minimum of the MED path.	Covalent bond AB
Molecular graph	Network of MED (bond) paths for a given geometry **R**.	Molecule
Molecular structure	Class of all equivalent molecular graphs.	Molecular structure
2. Properties of an atom in a molecule		
$\int_{\Omega} d\mathbf{r} = V$	Volume V of basin Ω.	Atomic volume
$\int_{\Omega} \mathbf{r}^n\, \varrho(\mathbf{r})\, d\mathbf{r} = M^n$	Moments M^n of basin Ω n = 0:	Atomic monopole moment
	n = 1:	Atomic dipole moment
	n = 2:	Atomic quadrupole moment
$O(\Omega) = \int_{\Omega} \hat{O}(\mathbf{r})\, \varrho \mathbf{r}\, d\mathbf{r}$	Any atomic property $O(\Omega)$ which can be represented by a one-electron operator $\hat{O}(\mathbf{r})$ can be obtained by integrating $\hat{O}(\mathbf{r})\ \varrho(\mathbf{r})$ over the atomic volume Ω.	Atomic property O
3. Bond properties of covalent bonds		
Bond path length R_b	$R_b \geqq R_e$; R_b is larger than the geometrical distance R_e for bent bonds.	Bond length
Interpath angle β	β is (normally) larger (or smaller) than the geometrical angle α in strained rings.	Bond angle
Bond order n	Evaluated from ϱ_b according to Eq. 8.	Bond order

Bond ellipticity ε	Defined by the curvature λ_1 and λ_2 of ϱ_b perpendicular to the bond according to Eq. 9.	π-Character
Shift parameter Δ	Deviation of $\mathbf{r}_b$ from the geometrical center of the bond path.	Bond polarity
Distance d	Deviation of bond path from the interatomic connection line (together with R_b and β).	Bent bond character

2.2 From Atoms to Chemical Bonds

Analyzing $\varrho(\mathbf{r})$ reveals that there are as many (3, —1) saddle points, zero-flux surfaces, and MED paths as there are bonds in a molecule. For example, in CH_3^+ (Figs. 1 and 2) only the C and H atoms are connected by MED paths but not the H atoms among themselves. If a bond is broken in a molecule the corresponding MED path will vanish in the electron density distribution. It will reappear if the bond is formed again [21]. These observations have led to the conclusion that there is a one to one correspondence between MED paths, saddle points $\mathbf{r}_b$, and zero-flux surfaces S on the one hand and chemical bonds on the other hand. As a consequence, the existence of a saddle point $\mathbf{r}_b$ accompanied by the existence of a MED path linking the two nuclei of the adjoining atomic subspaces can be considered to be a necessary condition for the existence of a chemical bond [22, 26–28].

However, not every MED path represents a covalent chemical bond. MED paths are also found in case of van der Waals complexes, hydrogen bonds or electrostatic (ionic) interactions [22], i.e. in all cases of closed shell interactions. In order to uniquely define a covalent bond and to distinguish it from ionic and other bonds, a second, sufficient condition has to complement the first one. According to the variation principle the molecular charge distribution is such as to minimize the energy and bonding is intimately connected with an energy gain relative to the separated atoms. Therefore it seems reasonable to base a sufficient condition for chemical bonding on energetic grounds.

It has been shown that the grain in energy, ΔE, resulting from bond formation is due to a complex interplay of kinetic and potential energy contributions [56–59]. When two bonded atoms share their electrons in a chemical bond, the space which the electrons can occupy has been increased with regard to that of the isolated atoms. The electrons *delocalize* and, as a consequence, the uncertainty of position of each electron increases. According to the Heisenberg uncertainty principle, this leads to a decrease of the momentum uncertainty. Since the kinetic energy is proportional to the square of the momentum uncertainty [60], delocalization of electrons causes a drop in the kinetic energy and, hence, a stabilization of the molecule [61].

In the case of the H_2 molecule, the form of the molecular wavefunction Ψ is characterized by the bonding σ_g MO that results from an in-phase overlap of the $1s$ AOs. Along the bond axis (z-axis), the σ_g MO takes the form qualitatively shown in Fig. 3a. Bonding overlap implies a flattening of the wavefunction in the bond

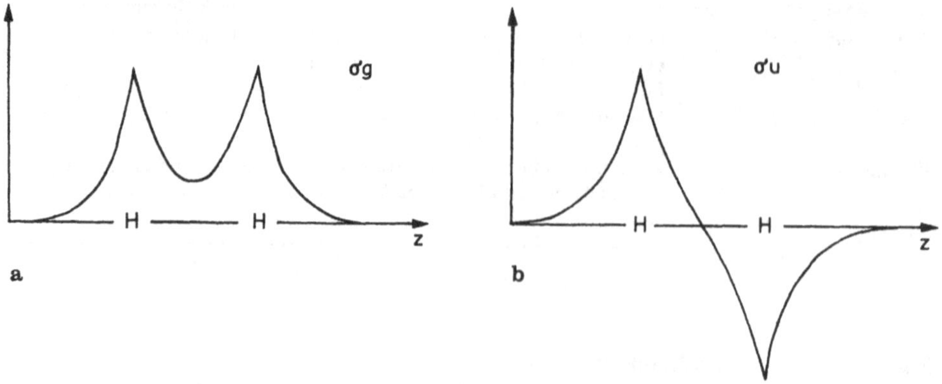

Fig. 3a, b. Schematic representation of **a** the bonding σ_g MO and **b** the antibonding σ_u MO of the H_2 molecule shown along the bond axis (z-axis)

region reflected by a decrease in $\partial\Psi/\partial z$ and an increase in $\partial\Psi/\partial x$ and $\partial\Psi/\partial y$. Since the kinetic energy parallel to the H—H bond may be expressed as a function of $\partial\Psi/\partial z$, bonding overlap and delocalization of electrons leads to a reduction of the kinetic energy in the direction of the bond [57–59]. At the same time, components of the kinetic energy perpendicular to the bond increase yielding an overall, though minimal, increase of the kinetic energy.

Similar observations can be made for the three components of $\varrho(\mathbf{r})$ namely, $\varrho(x)$, $\varrho(y)$, and $\varrho(z)$. Delocalization of charge into the bonding region (occupation of the ϱ_g MO) causes a decrease in the curvature λ_3 of $\varrho(\mathbf{r})$ along the bond path while the two curvatures perpendicular to the bond path, λ_1 and λ_2, increase. Since $\lambda_3 > 0$ and λ_1, $\lambda_2 < 0$, their sum becomes more negative, which is indicative of a concentration of electronic charge in the bonding region (see Sect. 3). These observations suggest that covalent bonding is accompanied by a decrease of the kinetic energy (parallel to the bond) with respect to the free atoms and by a concentration of negative charge in the bonding region [62].

If heavy atoms are bonded, the situation will be far more complicated. Occupied antibonding orbitals cause an increase rather than a decrease of $\nabla\Psi$ (see, e.g. Fig. 3b) and, hence, an increase in the kinetic energy parallel to the bond. At the same time, the sum of the curvatures λ_i may become positive, indicative of a depletion of charge in the bonding region. Studies on A_2 and AH molecules [20, 63], suggest that there are regions with a drop in the kinetic energy, but they can be located anywhere, even in nonbonding regions.

In order to derive a more general model of the chemical bond valid for systems with and without hydrogen atoms, Ruedenberg [64] has proposed a conceptual partitioning of the binding energy, which is the difference of the total molecular energy and the sum of the energies of the atoms in their ground states. According to this scheme, the binding energy comprises five contributions that result from

1. promotion energy,
2. quasi-classical interaction energy,
3. sharing penetration,
4. sharing interference energy, and
5. charge transfer energy.

Although this partitioning of the binding energy provides valuable insight into the mechanism of bonding of small molecules, it is difficult to apply in general. This has to do with the choice of the atomic ground state, and some arbitrariness in the construction of the hybrid orbitals, which are necessary for the evaluation of the promotion energy. For larger molecules, an inflation of numbers will impede a clear description of chemical bonds [65]. Also, it remains to be seen how this description performs in the case of nonclassical bonds. — Other schemes have been proposed to describe chemical bonds by partitioning the energy [66, 67]. However, none of these descriptions is simultaneously simple, generally applicable, and free of shortcomings.

2.3 Distinction Between Covalent Bonds and Closed Shell Interactions

In view of the complications inherent to the various bonding analyses that are based on the total energy, it is better to turn to local properties of the energy and analyze them in a way similar to that employed for the analysis of the electron density distribution $\varrho(\mathbf{r})$. For this purpose, the energy density $H(\mathbf{r})$ has been defined [22, 26, 28]:

$$H(\mathbf{r}) = G(\mathbf{r}) + V(\mathbf{r}) \tag{4}$$

$G(\mathbf{r})$ is a local kinetic energy density [68] given by:

$$G(\mathbf{r}) = 1/2\, \nabla\nabla' \Gamma^1(\mathbf{r};\mathbf{r}')|_{\mathbf{r}=\mathbf{r}'}, \tag{5}$$

with $\Gamma^1(\mathbf{r};\mathbf{r}')|_{\mathbf{r}=\mathbf{r}'}$ being the first order density matrix [69]. $V(\mathbf{r})$ is the average field experienced by one electron in a many particle system [70]:

$$V(\mathbf{r}) = N \int [\Psi^*(-\mathbf{r}\nabla\nabla)\, \Psi]\, d\tau' \tag{6}$$

where $d\tau'$ denotes integration over spin coordinates of all N electrons and integration over all N electrons but one. Integrated over the total molecular space, $H(\mathbf{r})$ yields the molecular electronic energy E

$$E = \int H(\mathbf{r})\, d\mathbf{r} = \int G(\mathbf{r})\, d\mathbf{r} + \int V(\mathbf{r})\, d(\mathbf{r}) = T + V \tag{7}$$

$G(\mathbf{r})$ is always positive and $V(\mathbf{r})$ is always negative. The sign of $H(\mathbf{r})$ determines whether accumulation of charge at a point $\mathbf{r}$ is stabilizing $[H(\mathbf{r}) < 0]$ or destabilizing $[H(\mathbf{r}) > 0]$. Investigation of a variety of different chemical bonds [22] has revealed that covalent bonding is characterized by a predominance of the local potential energy density $V(\mathbf{r})$ in the interatomic region reflected by $H(\mathbf{r}_b) \equiv H_b < 0$. For closed shell interaction, van der Waals complexes, or ionic bonds, $H_b > 0$ has been found [22]. Therefore, it is reasonable to base a *sufficient* condition for covalent bonding on the properties of $H(\mathbf{r})$ in the bond region. In view of these considerations, one can speak of a covalent bond between two atoms A and B if [22, 26, 28]

1. a MED path links A and B, i.e. the atomic basins Ω_A and Ω_B are adjoined, only separated by an interatomic surface S(A, B) (*necessary condition*), and

2. the local energy density H(**r**) at the saddle point $\mathbf{r}_b$ is smaller than zero (*sufficient condition*).

If both conditions are fulfilled we will call the MED path a "*bond path*" and the saddle point $\mathbf{r}_b$ a bond critical point [71]. If, however, H_b is $\geqq 0$, there will be a closed shell interaction between A and B leading to either ionic bonding, van der Waals interactions, or other forms of noncovalent bonding.

It is clear that a reduction of the kinetic energy (density) in the bonding region is a prerequisite of a negative H(**r**), i.e. the definition given above is in line with the idea that chemical bonding (mostly) implies a decrease in the kinetec energy in the direction of the bond. But since H(**r**) comprises the effects of both local kinetic and local potential energy, it is also indicative of bonding in those cases where the kinetic energy decrease in the bonding region is disguised by other effects as found for N_2, O_2, and F_2 [72] or cyclopropane [73]. We will corroborate on this point further when discussing the Laplacian of $\varrho(\mathbf{r})$ (see Sect. 3).

2.4 Description of the Covalent Bond

The definition of the covalent bond provides a basis for its characterization in terms of the properties of $\varrho(\mathbf{r})$ along the bond path or at the bond critical point. For example, the value of $\varrho(\mathbf{r})$ at $\mathbf{r}_b$, ϱ_b, can be used to determine a bond order and its position can serve as a measure for the bond polarity. Further, the anisotropy of ϱ_b can be related to the π-character of a bond. In the case of strained molecules, the curvature of the bond path reflects the bent bond character of a bond [22, 23].

For a variety of hydrocarbon compounds with single, double and triple bonds, aromatic, strained and unstrained bonds a linear relationship between ϱ_b and the CC bond distance R has been found [21, 22]. This relationship has been utilized to establish a definition for the bond order n [21] (see Fig. 4a):

$$n = \exp\left[a(\varrho_b - b)\right] . \tag{8}$$

The constants a and b have been determined by assigning Lewis bond orders 1, 2, and 3 to ethane, ethylene, and acetylene. Bond orders for molecules containing nitrogen, oxygen or other hetero atoms can be defined in a similar way [22]. It has to be stressed that a bond order defined in terms of ϱ_b does not suffer from the ambiguities of an analysis of chemical bonding in terms of molecular orbitans. It is simple to evaluate and simple to use [74].

In case of a hetero polar bond, the basin of the more electropositive atom is smaller than that of its bonding partner because it has lost part of its negative charge. As a consequence, the bond critical point $\mathbf{r}_b$ is always closer to the more electropositive atom. For the AH_n series with A = Li, Be, B, C, N, O a linear relationship between the deviation Δ of the bond critical point $\mathbf{r}_b$ from the center of the bond path and the electronegativity difference between A and H has been found [22]. With increasing electronegativity of A the bond critical point moves toward the H atom which is indicative of a continuous compression of the H basin. However, for H_2O a natural compressibility limit of H is reached that cannot be surpassed. Bearing this in mind, the position of the bond critical point can be used as measure of the bond polarity [75].

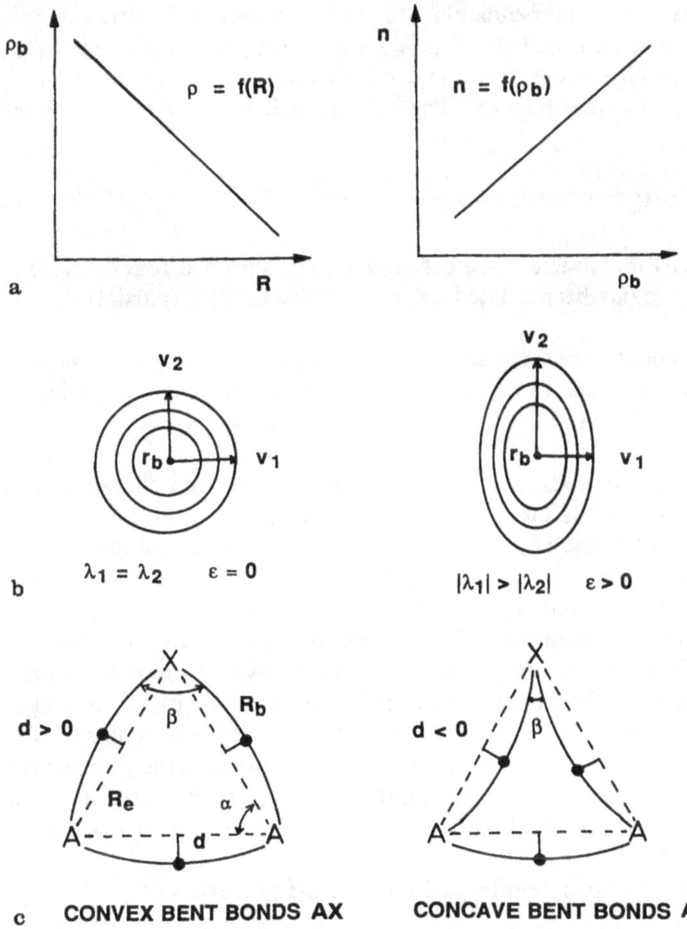

Fig. 4a–c. Description of a covalent bond in terms of the properties of the electron density distribution $\varrho(\mathbf{r})$. **a** Schematic representation of the relationship between the electron density ϱ_b at the bond critical point and the bond length R (left side) and of the relationship between the bond order n and ϱ_b (right side). **b** Qualitative contourline diagrams of $\varrho(\mathbf{r})$ in the plane that contains the bond critical point $\mathbf{r}_b$ and is perpendicular to a CC single bond (left side) and a CC double bond (right side). Ellipticities ε and eigenvectors $\mathbf{v}_1$ and $\mathbf{v}_2$ that indicate the direction of steep (λ_1) and soft curvature (λ_2) are given. $\mathbf{v}_2$ is the major axis. **c** Definition of parameters R_b, d, and β describing the curved MED path (*solid line*) of a bent bond in strained rings. Outwardly (inwardly) curved bond paths denote convex (concave) bent bonds. R_b is larger than the geometrical distance R_e. The angle β is larger (convex bent bonds) or smaller (concave bent bonds) than the geometrical angle α

The density distribution at the bond critical point in the plane perpendicular to the bond may be isotropical or anisotropical [21, 22]. The ratio of the curvatures λ_1 and λ_2 is a measure of the anisotropy of ϱ_b [21]. In case of the CC single bond in ethane, the negative charge distribution at $\mathbf{r}_b$ is isotropical and, hence, $\lambda_1 = \lambda_2$. However, for ethylene (or other hydrocarbons with CC double bonds) the density distribution at $\mathbf{r}_b$ is anisotropical ($\lambda_1 < \lambda_2$). The electron density extends farther in the space above and below the molecular plane, i.e. in the direction of the π MOs, than in the molecular plane perpendicular to the σ bond. The curvature λ_2 associated with the π-direction is called *soft curvature* while λ_1 is called steep curvature. The corresponding eigenvectors of the Hessian matrix, the directions of λ_1 and λ_2, are given by $\mathbf{v}_1$ and $\mathbf{v}_2$ (see Fig. 4b). The ratio λ_1 to λ_2 has been used to define the bond ellipticity ε [21]:

$$\varepsilon = \lambda_1/\lambda_2 - 1 . \tag{9}$$

For bonds with (partial) π-character ε-values larger than zero are found [21, 22]. Therefore, the ellipticity ε has been related to the π-character of a (partial) double bond.

For triple bonds, ε is equal or close to zero since there are now two soft curvatures ($\lambda_1 \approx \lambda_2$). It is useful in this case to define effective ellipticities ε_i' by comparing λ_i (i = 1, 2) with λ_i(A — B), where λ_i(A — B) are the curvatures of the single bond A — B [22].

Another interesting feature of the chemical bond, is its bent bond character [22]. The analysis of $\varrho(\mathbf{r})$ provides a basis for a quantitative description of this phenomena. In case of strained molecules, the bond path may not coincide with the interatomic connection line. It can be inwardly or outwardly curved, as schematically shown in Fig. 4c. The bond critical point $\mathbf{r}_b$ is shifted from the interatomic connection line by a distance d. The shift parameter d has a positive sign in case of outwardly curved bonds (convex bent bonds) and a negative sign in case of inwardly curved bonds (concave bent bonds). For a bent bond, the bond path length R_b is longer than the internuclear distance R_e. Similarly, the interpath angle β is larger (for outwardly curved bonds) or smaller (for inwardly curved bonds) than the geometrical angle α as shown in Fig. 4c. Thus, the bent bond character, i.e. the curvature of a bond path is quantitatively described by the ratio R_b/R_e, the interpath angle β, and the shift parameter d.

Table 1 summarizes the various definitions based on the properties of $\varrho(\mathbf{r})$.

2.5 Definition of Molecular Structure and Structural Changes

Once an atom in a molecule and the chemical bond have been defined, it is straightforward to define a molecule and molecular structure. Before doing this we have to characterize other types of stationary (critical) points of $\varrho(\mathbf{r})$ that can be found in the molecular charge distribution. Apart from the the saddle points $\mathbf{r}_b$ of $\varrho(\mathbf{r})$ in the bond region (bond critical points), there are three other types of stationary points, namely the local maxima $\mathbf{r}_a$ of $\varrho(\mathbf{r})$ that occur at the position of the nuclei [53], ring critical points $\mathbf{r}_r$, and cage critical points $\mathbf{r}_c$. A ring critical point is found in the center of a ring while cage critical points are inside a cage. The number of positive and negative eigenvalues λ_i given by the signature s of the Hessian matrix (see Eq. (1)) for each of these critical points is given in Table 2. Hence, $\mathbf{r}_a$ is a (3, —3) critical point (maximum), $\mathbf{r}_b$ is a (3, —1) critical point (saddle point), $\mathbf{r}_r$ a (3, +1) critical point (saddle point), and $\mathbf{r}_c$ a (3, +3) critical point (minimum). The number and types of gradient vector paths terminating (springing) at each critical point are also given in Table 2. Further, each point can be classified as a l-dimensional attractor (3-l-di-

Table 2. Description of critical (stationary) points of $\varrho(\mathbf{r})$

Notation	$\mathbf{r}_a$	$\mathbf{r}_b$	$\mathbf{r}_r$	$\mathbf{r}_c$
Critical point	Nucleus	Bond[f]	Ring	Cage
Eigenvalues of the Hessian Matrix	$\lambda_1, \lambda_2, \lambda_3 < 0$	$\lambda_1, \lambda_2 < 0, \lambda_3 > 0$	$\lambda_1 < 0, \lambda_2, \lambda_3 > 0$	$\lambda_1, \lambda_2, \lambda_3 > 0$
Rank, Signature[a]	(3, −3)[e]	(3, −1)	(3, +1)	(3, +3)
Mathematical Description	Maximum	Saddle point	Saddle point	Minimum
Topological Description[b]	3-dimensional attractor	2-dimensional attractor	1-dimensional attractor	Repeller
Trajectories[c, d]				
terminating	∞ type I	∞ type III	2 type III	—
originating	—	2 type II	∞ (n_b I + n_b II)	∞ (n_a I) + ∞ (n_b II)
Associated Element[d]	basin Ω	Zero-flux surface S	Edge of n_b surfaces	Corner of n_b surfaces
Chemical Relevance	Nucleus[e]	Bond path[f]	Ring structure	Cage structure

[a] The rank of the Hessian matrix is given by the number of eigenvalues λ_i unequal zero. The signature is defined by Eq. (1).

[b] An attractor "attracts" gradients paths (trajectories). They terminate at the attractor. A repeller "repells" gradient paths. They originate at the repeller. An I-dimensional attractor is at the same time a (3-I)-dimensional repeller.

[c] Type I trajectories fill the basin Ω, type II trajectories form the bond path, and type III trajectories form the zero-flux surface.

[d] n_a denotes the number of adjoining atomic basins, n_b the number of bonds in a ring or in a cage. $n_a \geqq 1$, $n_b(\text{ring}) \geqq 3$, $n_b(\text{cage}) \geqq 6$ where the equal sign would apply to a diatomic molecule, a three-membered ring, and a tetrahedral cage structure.

[e] The nucleus is a pseudo maximum. See Ref. [53].

[f] Since not each point $\mathbf{r}_b$ is a bond critical point, the correct notation would be MED-critical point. For reasons of simplicity, we use a term that best fits in the list of structural elements associated with the critical points of $\varrho(\mathbf{r})$.

mensional repeller) in view of the trajectories being "attracted" (being "repelled") by it. The chemical relevance of each point is also given in Table 2.

The number n and type of critical points that can exist in a molecule is determined by the Poincare-Hopf relationship [54]:

$$n_a - n_b + n_r - n_c = 1\,, \tag{10}$$

Utilizing the structural assignments given in Table 2 for each critical point, n_a denotes the number of nuclei, n_b the number of bonds and closed shell interactions, n_r the number of rings, and n_c the number of cages in a molecule.

Utilizing the definitions given above, a molecule is defined as a collection of 3-dimensional attractors, i.e. atoms, linked by a network of bond paths, i.e. chemical bonds. For a given nuclear configuration **R** (geometry) of the molecule, it is characterized by the molecular graph, i.e. by the critical points of $\varrho(\mathbf{r}, \mathbf{R})$ and the associated MED paths connecting the atomic nuclei. Molecular graphs that possess the same number of MED paths linking the same nuclei form an equivalent class [76]. They represent a single molecular structure, which in turn is associated with a structural region of the configuration space. For a given molecular state, there exists a finite number of classes of equivalent graphs (structures) and associated structural regions. Accordingly, the molecular configuration space is partitioned into a finite number of structural regions. Within a structural region, molecular structure does not change (is structurally stable [77]) with respect to differential changes of geometry). However, at the boundaries between two structural regions, a differential change in geometry leads to a discontinuous change in the molecular graph. Hence, at these boundaries the unstable structures of a molecule are located. A point **R** of nuclear configuration space that is structurally stable is called a regular point. In contrast, a point $\mathbf{R}_c$ on the boundary between two stable structures is called catastrophe point. Point $\mathbf{R}_c$ will be a catastrophe point of bifurcation type if $\varrho(\mathbf{r}, \mathbf{R}_c)$ possesses at least one singularity, i.e. one critical point with a rank less than three; otherwise it is a catastrophe point of conflict type. An example for both cases is given below.

The collection of all structural regions together with their boundaries is called a structure diagram. For a given system, the structure diagram embodies all information concerning possible structures and their possible changes. Therefore, it should contain information to elucidate the mechanism of possible chemical reactions.

A change in structure necessitates passage of the system through a catastrophe point. This is a discontinuous process which can be best described with catastrophe theory [78]. For example, with catastrophe theory, one can predict that for the isomerization reaction

$$\mathrm{ABC} \rightarrow \mathrm{BCA}$$

there should exist two different reaction mechanisms depending on what kind of catastrophe points are passed during the course of the reaction. The first possible

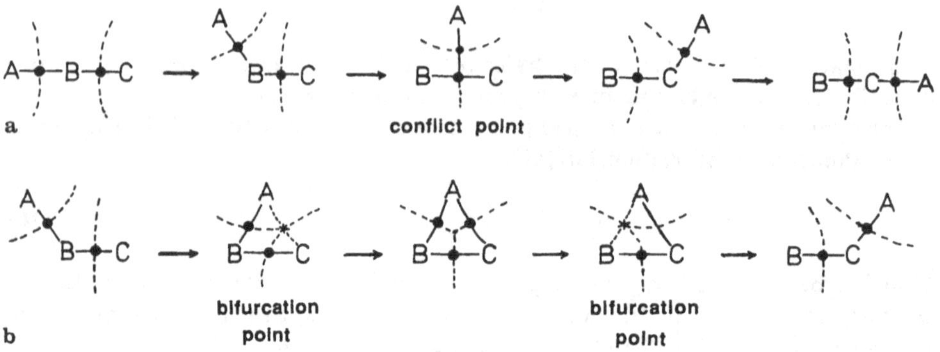

Fig. 5a, b. Qualitative molecular graphs representing the sequence of structural changes during the course of the isomerization reaction ABC → BCA. Bond critical points are denoted by *dots*, MEP paths by *solid lines* and the traces of the zero-flux surfaces by *dashed lines*. **a** Bond switching mechanism. The system passes through one catastrophe point of the conflict type. **b** Mechanism involving ring formation. Singularities in $\varrho(\mathbf{r})$ are denoted by *stars*. The system passes through two catastrophe points of the bifurcation type. A *dashed line* between A and C (B) denotes the onset of bond formation (rupture)

mechanism is a simple switching of bond paths from one atom (attractor) to the other. As shown in Fig. 5a, atom A remains bonded to atom B until a catastrophe point of the conflict type is reached. This conflict point is caused by a direct link of atom A to the bond critical point $\mathbf{r}_b$(B, C). An infinitesimal movement of A in direction of C causes an abrupt switch of the A, (BC) path to nucleus C and the formation of BCA as indicated in Fig. 5a.

The second possible mechanism involves ring formation during the course of the reaction. The series of structural changes in this reaction is illustrated in Fig. 5b. Here, atom A moves toward C until a bifurcation catastrophe point is reached. This bifurcation point is caused by the birth of a new bond and ring critical point yielding a singularity in $\varrho(\mathbf{r})$ in form of a (2, 0) critical point. An infinitesimal displacement of A toward C causes this bifurcation point to transform into the bond critical point $\mathbf{r}_b$(AC) and the ring critical point $\mathbf{r}_r$(ABC). A ring structure is formed. When A moves further toward C a second catastrophe point is passed, at which $\mathbf{r}_b$(AB) and $\mathbf{r}_r$(ABC) disappear. Finally, the product BCA is formed. The two bifurcation catastrophe points may correspond to the transition states leading from the reaction and product side to the ring structure, which in turn may be a true intermediate.

It is appealing to associate the catastrophe points in the structure diagram with the transition states of the corresponding reaction. A limited number of investigations suggest that there may be a one-to-one relationship between catastrophe points and transition states encountered in a reaction. There even could exist a relationship between the coordinates of the catastrophe point and the nuclear configuration of the transition state [79]. In this case it would be possible to transform energy propensity rules such as the Hammond postulate into the electron density picture. However, no proof for the existence of a relationship between the number of catastrophe points and the number of transition states or between $\mathbf{R}_c$ and the transition state geometry has been given so far.

3 The Laplace Field of the Electron Density Distribution $\varrho(\mathbf{r})$

The electron density distribution $\varrho(\mathbf{r})$ exhibits a simple pattern. Local maxima only occur at the position of the nuclei and local minima only in the interior of cage structures. Saddle points can be found inside rings and at the MED paths linking the nuclei of neighboring atoms. These properties of $\varrho(\mathbf{r})$ lead to definition of the atom in a molecule, the chemical bond, molecular structure, and structural changes.

However, important features and useful models that describe the electronic structure of atoms and molecules are not reflected in any obvious way by $\varrho(\mathbf{r})$. Noteworthy are the shell structure of the electrons in an atom and the usual dissection of the electrons of a molecule into inner shell, bond, and lone electron pairs. It seems difficult to associate molecular stability and reactivity with these features of the electronic structure within an analysis of $\varrho(\mathbf{r})$. Shell structure and electron pair model, however, can also be extracted from the properties of $\varrho(\mathbf{r})$ if the Laplacian of $\varrho(\mathbf{r})$, $\nabla^2\varrho(\mathbf{r})$, is analyzed.

3.1 Mathematical Meaning of the Laplace Field of a Scalar Function

The Laplacian of any scalar field f(**r**), ∇^2f(**r**), is given by the second derivatives of f(**r**) with regard to **r**. It is obtained as the sum of the eigenvalues or the trace of the rank 3 Hessian matrix of f(**r**) with regard to the three components x, y, z of the vector **r**. The Laplacian is an important property of a scalar field [80]. It is negative where the scalar field concentrates. In particular, ∇^2f(**r**) adopts a minimum where f(**r**) possesses a maximum. As shown for the one-dimensional scalar function f(x) in Fig. 6, the Laplacian of f at a point x provides a measure of the difference between the value of f at x and the average value of f at all neighboring points x + Δx. For $\nabla^2 f(x) < 0$, f(x) is greater than the average value and the opposite holds for $\nabla^2 f(x) > 0$. *Hence the Laplacian* $\nabla^2 f(\mathbf{r})$ *is a measure of local concentration and depletion*

$$\lim_{\Delta x \to 0} \left\{ f(x) - \frac{1}{2}\left[f(x-\Delta x) + f(x+\Delta x) \right] \right\} =$$

$$-\frac{1}{2} \lim_{\Delta x \to 0} \left\{ \left[f(x+\Delta x) - f(x) \right] - \left[f(x) - f(x-\Delta x) \right] \right\} =$$

$$-\frac{1}{2}\left(\frac{d^2 f}{dx^2}\right) dx^2 = -\frac{1}{2} \nabla^2 f(x)\, dx^2$$

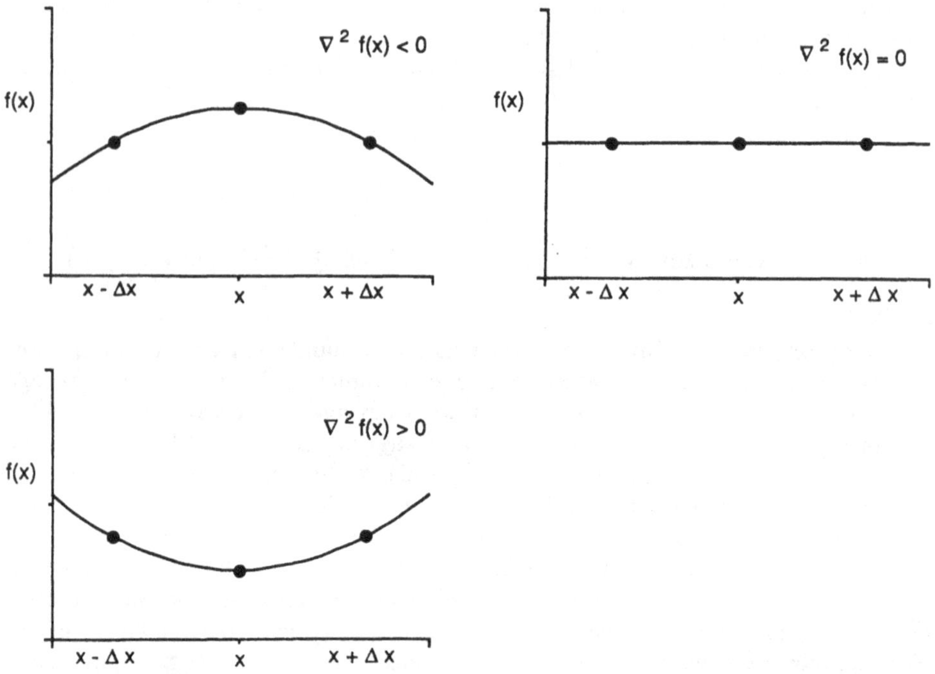

Fig. 6. Definition of the Laplacian ∇^2f(x) of a one-dimensional scalar function f(x) and a schematic representation of the behavior of f(x) in case of $\nabla^2 f(x) = 0$, $\nabla^2 f(x) < 0$, and $\nabla^2 f(x) > 0$

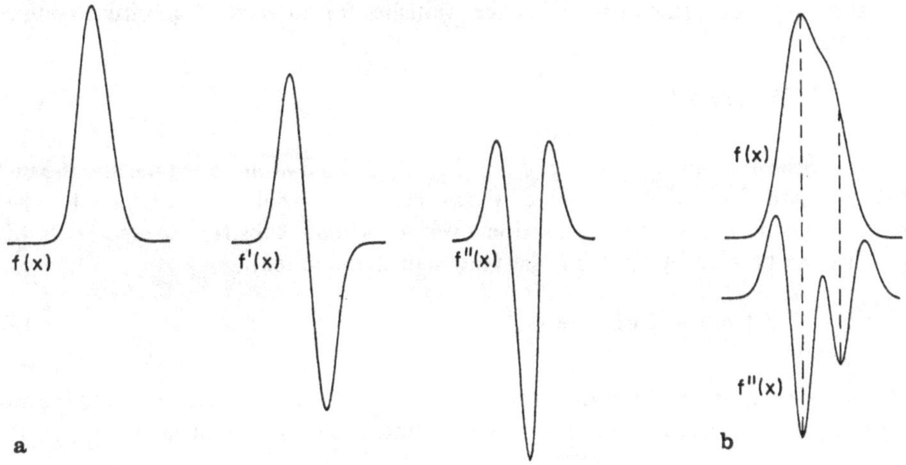

Fig. 7. a Characterization of a spectral band by the first and second derivatives of the enveloping function f(x). **b** Determination of the positions of two overlapping bands in a spectrum

*of the scalar field f(**r**). Since the behavior of f(**r**) at **r** is determined by comparison with an internal standard, namely the average value of all neighboring points this definition of concentration and depletion is completely independent of an arbitrary reference function [81].*

The properties of the Laplacian are used to specify the local behavior of f(**r**). For example, in derivative spectroscopy [82] the maximum of the function f(x) enveloping a spectral band is determined by the position of the minimum of the second derivative $d^2f/dx^2 = f''(x) = \nabla^2 f(x)$, i.e. the minimum in f''(x) determines a maximal concentration of the enveloping function f(x) (Fig. 7a). This looks trivial but turns out to be extremely useful if two bands overlap in a spectrum (Fig. 7b). From f(x) it is not possible to determine the positions of the maxima. On the other hand, the minimum of f''(x) clearly reveals the positions of maximum concentration of f(x).

An investigation of the properties of the Laplace field is particularly useful when a threedimensional function such as $\varrho(\mathbf{r})$ is analyzed. For example, the strong exponential decay of $\varrho(\mathbf{r})$ in the off-nucleus direction disguises any local properties of $\varrho(\mathbf{r})$. These are unearthed by the Laplacian of $\varrho(\mathbf{r})$.

3.2 Physical Meaning of the Laplace Field of $\varrho(\mathbf{r})$

The analysis of the Laplacian of the electron density $\varrho(\mathbf{r})$, $\nabla^2\varrho(\mathbf{r})$, is more than a convenient way of describing the distribution $\varrho(\mathbf{r})$ with an appropriate mathematical tool. The Laplacian $\nabla^2\varrho(\mathbf{r})$ is of fundamental physical significance. It determines the one-particle Lagrangian density L(**r**) of a quantum mechanical system [19]:

$$L(\mathbf{r}) = -\hbar^2/4m\nabla^2\varrho(\mathbf{r})\,. \tag{11}$$

The Lagrangian, integrated over all space, vanishes for an isolated quantum system, e.g. a molecule [19]:

$$\int_{\infty} L(\mathbf{r})\, d\mathbf{r} = 0. \tag{12}$$

For an open quantum system, e.g. an arbitrarily defined atom in a molecule the integral attains a finite nonzero value. However, for a virial atom defined by the zero flux consition (Eq. (2)), integration over the atomic subspace Ω also leads to zero. This can be seen by applying the Gaussian theorem [83]

$$\int_{\Omega} \nabla^2 f\, d\omega = \oint_{S} \nabla f\, d\sigma = 0, \tag{13}$$

($d\omega$ denotes integration of a n-dimensional function f over a volume element Ω and $d\sigma = \mathbf{n}\, ds$ denotes integration over a surface element S) to the Laplacian of $\varrho(\mathbf{r})$:

$$\int_{\Omega} \nabla^2 \varrho(\mathbf{r})\, d\mathbf{r} = \oint_{S} \nabla\varrho(\mathbf{r})\, \mathbf{n}(\mathbf{r})\, d\mathbf{r} = 0. \tag{14}$$

Because of Eqs. (11) and (14) the integral (15) vanishes:

$$\int_{\Omega} L(\mathbf{r})\, d\mathbf{r} = 0. \tag{15}$$

It is because of this common property of the virial atom and the total molecule that a unique quantum mechanical description exists both for the total molecule and for the virial atoms within it [19].

Another important feature of the Laplacian $\nabla^2\varrho(\mathbf{r})$ is that it provides a link between the distribution of $\varrho(\mathbf{r})$ and the forces acting on the system. The force density $F(\mathbf{r})$ defined as:

$$F(\mathbf{r}) = -\nabla\sigma(\mathbf{r}), \tag{16}$$

in which $\sigma(\mathbf{r})$ is the quantum stress tensor first introduced by Pauli [84]:

$$\sigma(\mathbf{r}) = \hbar^2/4m[(\nabla\nabla + \nabla'\nabla') - (\nabla\nabla' + \nabla'\nabla)]\, \Gamma^1(\mathbf{r}, \mathbf{r}')|_{\mathbf{r}=\mathbf{r}'}. \tag{17}$$

Noting the form of $G(\mathbf{r})$ (Eq. (5)) and that of $L(\mathbf{r})$ (Eq. (11)), the following relationship is established:

$$\mathrm{tr}\, \sigma(\mathbf{r}) = -2G(\mathbf{r}) - L(\mathbf{r}). \tag{18}$$

Since the following relationship holds between the trace of the stress tensor and the potential energy density $V(\mathbf{r})$ [85]:

$$V(\mathbf{r}) = \mathrm{tr}\, \sigma(\mathbf{r}), \tag{19}$$

the *local virial theorem* results [19]

$$2G(\mathbf{r}) + V(\mathbf{r}) = -L(\mathbf{r}) = \hbar^2/4m\, \nabla^2\varrho(\mathbf{r}). \tag{20}$$

Because of Eq. (15) integration over total space or an atomic subspace Ω leads to the virial theorem

$$2\int_{\Omega} G(\mathbf{r})\,d\mathbf{r} + \int_{\Omega} V(\mathbf{r})\,d\mathbf{r} = 2T_{\Omega} + V_{\Omega} = 0\,. \tag{21}$$

The local virial theorem locally connects the Laplacian with the kinetic energy density and the potential energy density. Since $G(\mathbf{r})$ is always positive and $V(\mathbf{r})$ is always negative, the sign of the Laplacian of $\varrho(\mathbf{r})$ determines the relative magnitudes of the local values of the potential and kinetic energy densities. In regions with $\nabla^2\varrho(\mathbf{r}) < 0$, the potential energy density is dominant. This leads to a concentration of negative charge. In contrast, in regions with $\nabla^2\varrho(\mathbf{r}) > 0$ the kinetic energy density dominates and causes a depletion of negative charge.

The Laplacian of $\varrho(\mathbf{r})$ is directly related to the local contributions to the electronic energy of a system. The local virial theorem can be used to elucidate the sufficient condition for covalent bonding described in Sect. 2.3. From Eq. (20) it follows that

$$H(\mathbf{r}) = \hbar^2/4m\,\nabla^2\varrho(\mathbf{r}) - 2G(\mathbf{r}) \tag{22}$$

and

$$2H(\mathbf{r}) = \hbar^2/4m\,\nabla^2\varrho(\mathbf{r}) + V(\mathbf{r})\,. \tag{23}$$

Eqs. (22) and (23) show that $H(\mathbf{r})$ will always be negative in the bonding region if $\nabla^2\varrho(\mathbf{r}) < 0$, i.e. if the electrons concentrate in the bonding region. On the other hand, for a homonuclear diatomic molecule there will always be electron concentration in the internuclear region if bonding MOs are occupied. Occupation of antibonding MOs leads to a depletion of negative charge. It depends on the number of occupied (anti)bonding MOs and their shape whether the Laplace concentration, $-\nabla^2\varrho(\mathbf{r})$, is larger or smaller than zero. If there is depletion, *the actual electron density in the internuclear region may still lead to bonding* when the magnitude of the potential energy density $V(\mathbf{r})$ is so large (due to relatively high atomic number and concomitant strong electron, nucleus attraction) that electrons in the bonding region stabilize the system, i.e. decrease the total energy.

3.3 Chemical Meaning of the Laplace Field of $\varrho(\mathbf{r})$

Since negative charge is concentrated in regions where $\nabla^2\varrho(\mathbf{r}) < 0$, it is convenient to refer to $-\nabla^2\varrho(\mathbf{r})$ as the *Laplace concentration of* $\varrho(\mathbf{r})$. The Laplacian concentration exhibits a maximum parallel to a maximum in $\varrho(\mathbf{r})$.

The vanishing of the Laplacian integrated over the total molecular space or over an atomic subspace Ω defined by the zero-flux surface (Eq. (2)) has an important consequence. All fluctuations in $\nabla^2\varrho(\mathbf{r})$ are such that local depletion or concentration of electronic charge cancel each other out for the molecule and for an atom in a molecule as well. If electrons concentrate at one position of the molecule, e.g. at the site of an electron lone pair, negative charge will be depleted at another site of

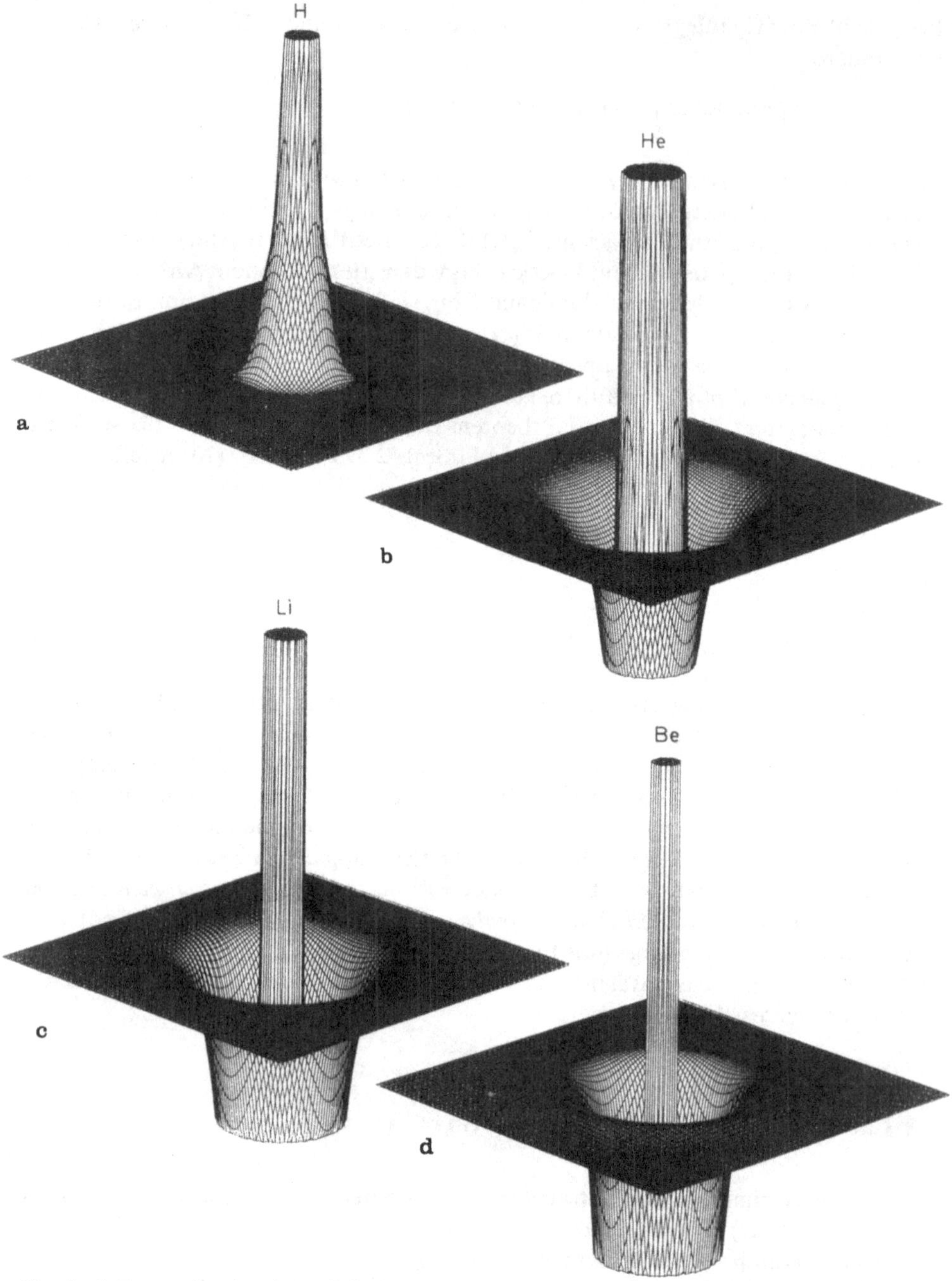

Fig. 8a–j. Perspective drawings of the Laplace concentration $-\nabla^2\varrho(\mathbf{r})$ of H, He, and the first row atoms Li through Ne. The inner concentration peaks are assigned to the 1*s* electrons and the outer concentration spheres to the valence shell electrons. For a better presentation values above and below a threshold are cut off. Different scaling factors are used depending on the number of electrons. **a** H(^{2}S). **b** He(^{1}S). **c** Li(^{2}S). **d** Be(^{1}S). **e** B(^{2}P; $1s^22s^22p_x$) (reference plane is the xy plane). **f** C(^{3}P; $1s^22s^22p_x2p_y$) (in the xz plane). **g** N(^{4}S). **h** O(^{3}P; $1s^22s^22p_x^22p_y2p_z$) (in the xy plane). **i** F(^{2}P; $1s^22s^22p_x^22p_y^22p_z$) (in the xz plane). **j** Ne(^{1}S). (UHF/6-31G(*d*) calculations)

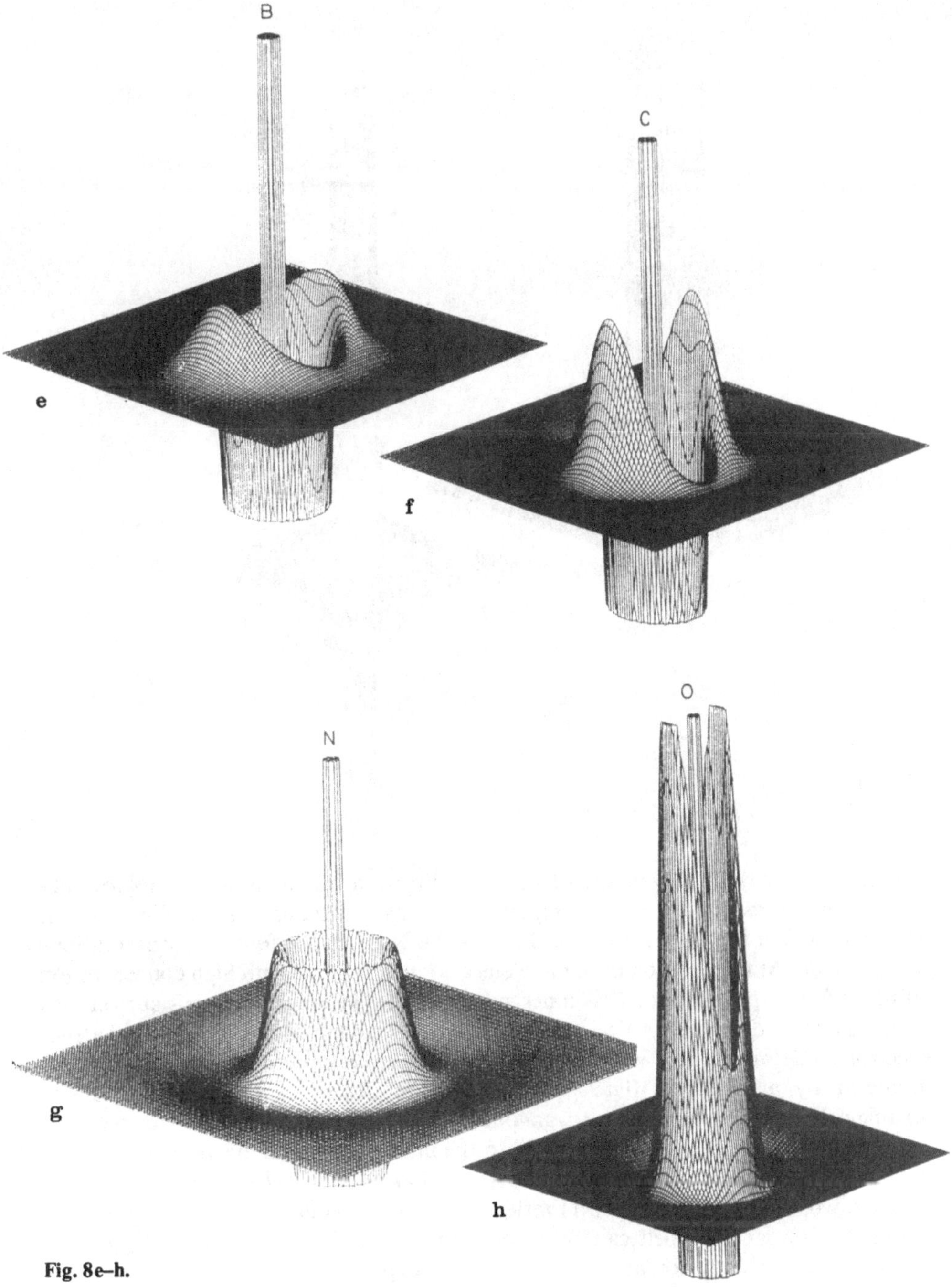

Fig. 8e–h.

the molecule, e.g. at the site of a possible nucleophilic attack. While the charge distribution $\varrho(\mathbf{r})$ of an isolated atom with spherical electron density distribution is characterized by one single local maximum at the position of the nucleus and an

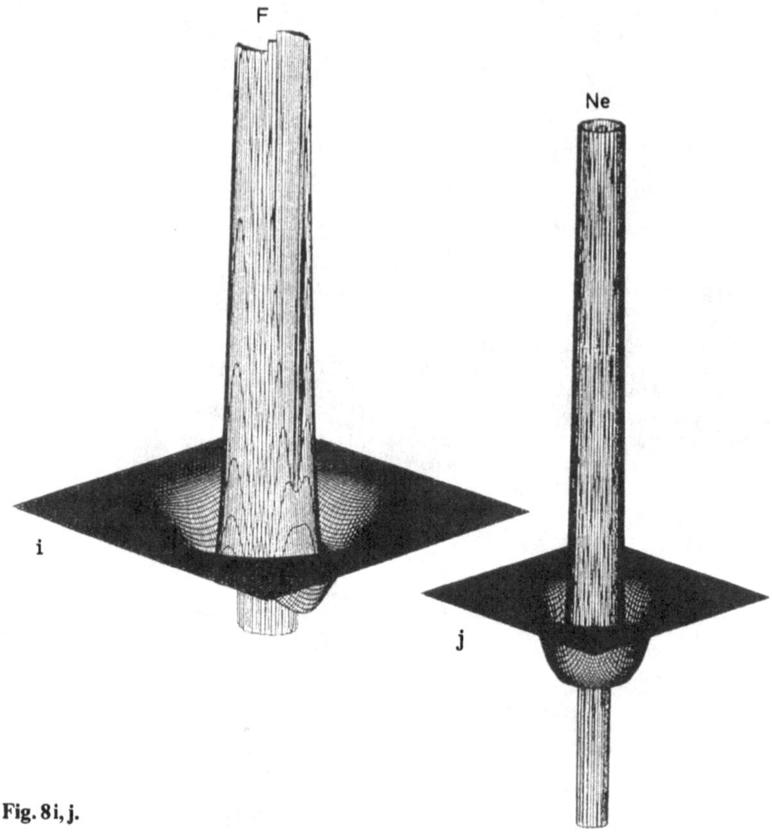

Fig. 8i, j.

exponential decay of $\varrho(\mathbf{r})$ in the off-nucleus direction, the associated Laplace concentration adopts a much more complex pattern as is demonstrated in Fig. 8 for H, He, and the first row atoms Li—Ne. Picking the N atom, the following observations can be made. At the position of the nucleus a compact sphere with high concentration of negative charge is found. This inner sphere of charge concentration is surrounded by a sphere of charge depletion. In the region where one would expect the valence electrons a second sphere with relatively large electron concentration can be found followed by a relatively flat sphere of charge depletion. Obviously, one can distinguish between an inner and an outer sphere of electron concentration. It is appealing to assign the 1*s* electron of N to the inner sphere and the valence electrons to the outer sphere and to speak of *inner shell and valence shell concentrations.* In other words, the Laplacian of $\varrho(\mathbf{r})$ reflects the shell structure of the electrons in an atom. Each quantum shell can be associated with a pair of shells of $\nabla^2\varrho(\mathbf{r})$, one with charge concentration and one with charge depletion.

The fluctuations in the Laplacian of $\varrho(\mathbf{r})$ must vanish if the Laplacian is integrated over the total space of the atom, i.e. spheres of electron concentration and electron depletion are a necessary consequence of the quantum mechanics of the atom. The Laplacian reveals that regions with an excess of potential energy are followed by regions with an excess of kinetic energy. Electrons stay longer in the former than in the latter.

The shell structure is clearly reflected in the case of atoms B to Ne (Fig. 8c–8j). For Li and Be, the valence shell concentration is very diffuse and can only be detected by a quantitative analysis of the Laplacian. Also, the outer sphere of charge depletion is very flat for atoms to the right of the periodic table. It develops within a row and is pronounced for atoms such as He, F, or Ne (see Fig. 8b, i, j).

The inner shell and the valence shell concentration contract with increasing atomic number. If the maximum concentration of $\varrho(\mathbf{r})$ in the valence shell is characterized by the distance p from the position of the nucleus, then p values decrease from 1.3 Å (Li) to 0.3 Å (Ne) in the first row. At the same time $-\nabla^2\varrho(\mathbf{r})$ increases from 0.048 (Li) to 368.368 e Å^{-5} (Ne).

For an isolated atom with an anisotropical charge distribution, as in the case of the atoms B, C, O, and F, the Laplace concentration in the valence sphere is no longer spherical. Local maxima (lumps) and local minima (holes) emerge in the valence shell concentration. For example, in case of the C atom in its ^{3}P ($1s^2\,2s^2\,2p_x\,2p_y$) state (Fig. 8f), the lumps are found in the direction of the occupied *p* orbitals and the holes in the direction of the empty *p* orbital. It seems that the lumps and holes in the valence sphere of the Laplacian reflect the occupation of the valence orbitals. At least, this is suggested by the Laplacian concentration of B(^{2}P), C(^{3}P), O(^{3}P), and F(^{2}P).

3.4 The Laplace Field of a Molecule

When an isolated atom with a spherical charge distribution forms a bond, its valence sphere concentration is distorted in a characteristic way. A pattern of lumps and holes is formed in the valence shell. In particular, local concentration maxima ((3, −3) critical points of $-\nabla^2\varrho(\mathbf{r})$) develop in direction of a bond with a neighboring atom while saddle points ((3, −1) critical points of $-\nabla^2\varrho(\mathbf{r})$) emerge between the bonds. The pattern of maxima and saddle points of the Laplace concentration of an atom is characteristic for the number and type of bonds with neighboring atoms [39]. It can be described by an *atomic graph*, i.e. the network of lines connecting the nucleus with the local maxima in the valence sphere concentration of the Laplacian. A critical point $\mathbf{r}_A$ in the valence sphere can be characterized quantitatively by the value l_A^{AB} of $-\nabla^2\varrho(\mathbf{r})$ at $\mathbf{r}_A^{AB}$, by the distance p from the nucleus, and by the angle γ between the lines connecting two valence sphere critical points with the nucleus.

While the valence sphere concentration of an sp^3 hybridized C(^{5}P) atom is spherical, that of the carbon atom in methane exhibits the pattern schematically given by the atomic graph of Fig. 9a. A perspective drawing of the Laplacian concentration of CH_4 (with respect to the plane containing the C atom and two H atoms) is shown in Fig. 9b. There are four concentration lumps, one along each of the CH bond paths, and six concentration holes, each between two CH bonds. Accordingly, the valence shell concentration of C is best characterized by a tetrahedral atomic graph (Fig. 9a).

The HOMO of methane is dominantly concentrated in the CH bonding region while the LUMO preferentially occupies the regions between the CH bonds (Fig. 10).

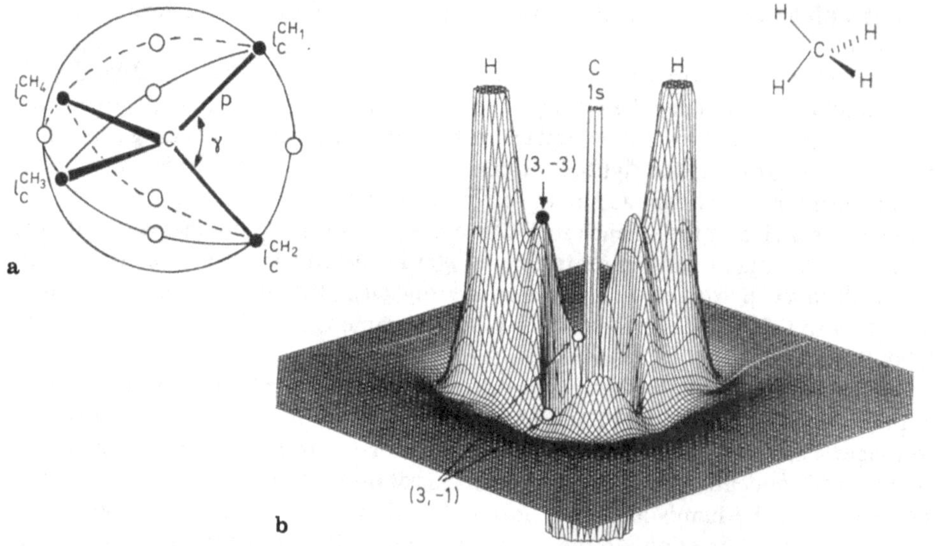

Fig. 9. **a** Schematic representation of the atomic graph of the C-atom in methane. Positions of concentration maxima l_CCH (3, —3) are denoted by *dots*, positions of saddle points (3, —1) are denoted by *open circles*. Distances p and angles γ determine the form of the atomic graph. **b** Perspective drawing of the Laplace concentration $-\nabla^2\varrho(\mathbf{r})$ of methane, depicted in the plane containing the C and two H atoms. Lumps (3, —3) and holes (3, —1) in the valence shell concentration of C are indicated. For a better presentation values above and below a threshold are cut off. (HF/6-31G(*d, p*) calculation)

Obviously, regions with concentration and depletion of negative charge coincide with the regions in which HOMO and LUMO possess their largest amplitude. Thus, in the same way as one can predict the sites of nucleophilic and electrophilic attack in a molecule by an analysis of the frontier orbitals (FOs), an investigation of the Laplacian $\nabla^2\varrho(\mathbf{r})$ yields information about the active sites of the molecule. Sites with

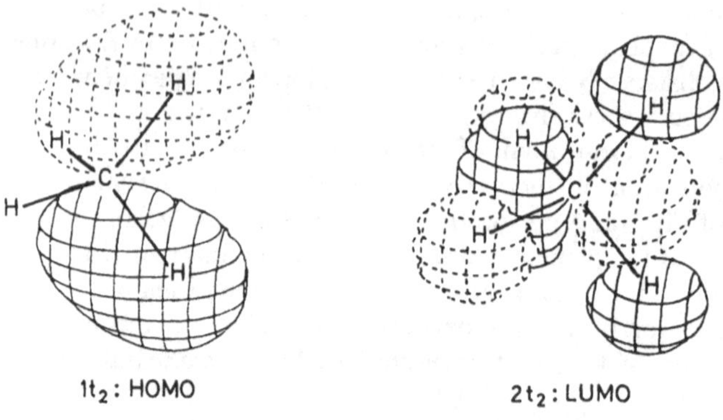

Fig. 10. Perspective drawing of the HOMO and LUMO of methane. Only one component of the $1t_2$ and the $2t_2$ sets are shown

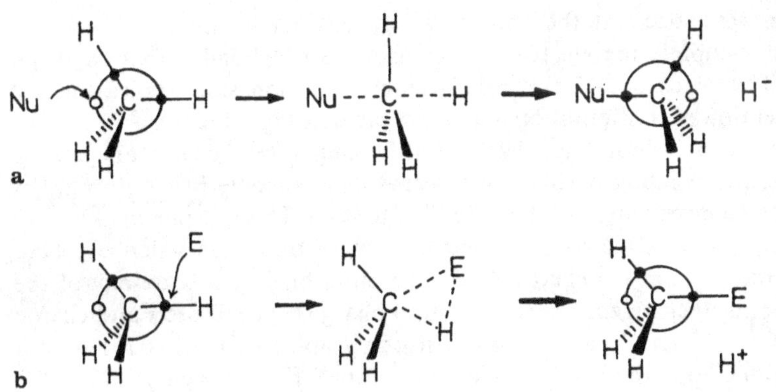

Fig. 11. a Schematic representation of a nucleophilic attack on methane. **b** Schematic representation of an electrophilic attack on methane

concentration holes in the valence sphere are prone to a nucleophilic attack while sites with concentration lumps in the valence sphere are prone to an electrophilic attack. A nucleophile will preferentially attack at a concentration hole between two CH bonds, i.e. the attack will take place at the C nucleus rather than at a CH bond. Formation of a bond weakens a CH bond at the opposite site of the valence sphere and the corresponding H will dissociate. Hence, a nucleophile leads to an inversion of the configuration at C. This is typical of a S_N2 mechanism and is schematically shown in Fig. 11a.

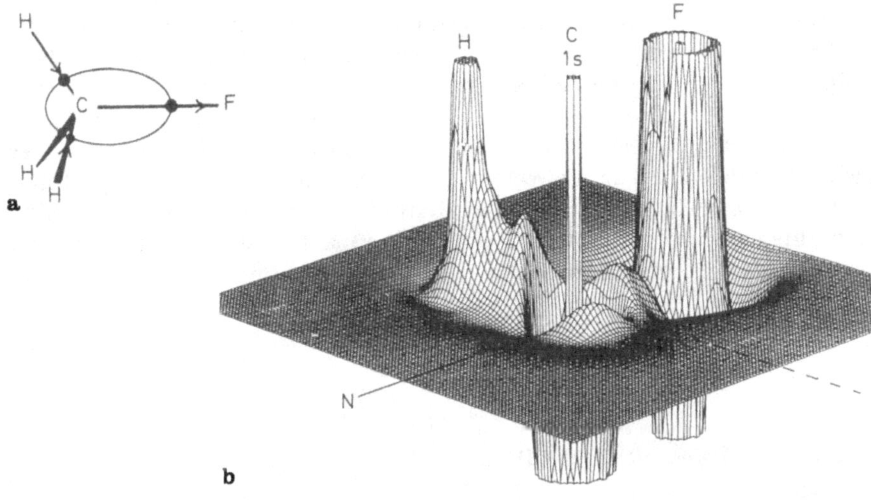

Fig. 12. a Distortion of the valence sphere of the C atom in methylfluoride. The valence sphere is indicated by a large *ellipse*. The direction of the distortion is given by an *arrow*. Propagation of distortion effects is shown for the H atoms. **b** Perspective drawing of the Laplace concentration $-\nabla^2\varrho(\mathbf{r})$ of methylfluoride, depicted in the plane containing the C, the F, and one H atom. Lumps and holes in the valence shell concentration of C are indicated. For a better presentation values above and below a threshold are cut off. (HF/6-31G(d, p) calculation)

An electrophile will attack at the concentration maxima in the CH bond, i.e. contrary to the nucleophile the electrophile attacks the CH bond rather than the carbon nucleus. This leads to a 2-electron 3-center transition state typical of S_E2 reactions with retention of configuration and is indicated in Fig. 11 b.

In the case of heteropolar bonding, charge transfer caused by the electronegativity difference between the bonding partners is reflected by a characteristic deformation of the valence sphere concentration of the atoms in question. This is depicted in Fig. 12 for methylfluoride, CH_3F. Due to the electronegativity of F, the valence sphere concentration of the C atom is pulled into the direction of F. The topology of the atomic graph remains unchanged, but the extent of charge concentration and charge depletion in the CF and CH bond regions is strongly influenced. As a result of a charge transfer from C to F, the bond critical point of the CF bond is only 0.43 Å [86] from the carbon nucleus, i.e. the local charge concentration at carbon has been drawn into the subspace of the fluorine atom. This is in line with the low l_C^{CF} value of 10.50 e Å^{-5} and a large distance value p_C^{CF} of 0.555 Å. In comparison, the corresponding values for the CH bond in methane are $l_C^{CH} = 28.62$ e Å^{-5} and $p_C^{CH} = 0.519$ Å. Since all fluctuations in $\nabla^2\varrho(\mathbf{r})$ have to cancel each other out within the boundary of a bonded atom (compare with Eq. (15)), the decrease of concentration at C in direction of F is accompanied by an increase of charge concentration on the opposite side of the valence sphere (see Fig. 12). The three concentration maxima in the direction of the CH bonds are considerably larger and closer to the C nucleus than the corresponding ones in methane ($l_C^{CH} = 34.32$ e Å^{-5}, $p_C^{CH} = 0.506$ Å). In a chain of bonded atoms, distortions in the valence sphere caused by electronegativity differences between the bonding partners can propagate to some extent along the chain. This is reflected by the atomic graphs of the involved atoms.

Again, the pattern of the Laplace concentration in CH_3F correlates with the form of the corresponding frontier orbitals. Predictions made within the FO model can also be made by inspection of the Laplace concentration. Thus, the analysis of the Laplacian of $\varrho(\mathbf{r})$ provides a link between the MO approach and the electron density approach for the description of molecules and their chemical behavior. Since *the Laplace concentration covers the effects of all occupied MOs*, it does not suffer from the ambiguities of FO theory with regard to the choice of the FOs (HOMO, second HOMO, third HOMO, etc.). The analysis of $-\nabla^2\varrho(\mathbf{r})$ leads to more reliable predictions. In those cases where the FOs are not dominant the shape of the Laplacian should differ from the shape of HOMO and LUMO and, in this way, should provide a basis to foresee the flaws of the FO model [87].

4 Description of π-Conjugation

Conjugation of double bonds leads to π-electron delocalization. Although the distinction between σ- and π-electrons is no longer relevant when discussing $\varrho(\mathbf{r})$, features of the electron density in the bonding region can be related to features of the σ/π-model. A complete investigation of a conjugated bond AB would imply

1. an analysis of $\varrho(\mathbf{r})$ at the bond critical point $\mathbf{r}_b(AB)$;
2. an analysis of $\varrho(\mathbf{r})$ along the bond path AB starting at the valence shell of atom A and ending at the valence shell of atom B;
3. an analysis of $\varrho(\mathbf{r})$ in the internuclear surface S(A, B);
4. an analysis of $\varrho(\mathbf{r})$ in the subspace of A and the subspace of B.

For reasons of simplicity the investigation of $\varrho(\mathbf{r})$ is constrained in most cases to the first step, namely an analysis of ϱ_b, the electron density at the bond critical point $\mathbf{r}_b$. As noted in Sect. 2.4, the bond ellipticity ε can be related to the π-character of the bond. For example, by using the π double bond of ethylene as a suitable reference bond, the ratio η given in per cent gives the π-character of any other formal CC double or single bond once its bond ellipticity has been found:

$$\eta = 100\,\varepsilon(\mathrm{CC})/\varepsilon(\mathrm{ethylene})\,. \tag{24}$$

By definition, η of ethane (bond order n = 1) and ethylene (n = 2) are 0 and 100%. In the case of π-conjugation in polyenes or cyclopolyenes, the π-character η is decreased for the formal double bonds and increased for the formal single bonds. Accordingly, bond orders n < 2 and n > 1 result. Hence, bond order n and π-character η extracted from the properties of ϱ_b allow a quantitative analysis of conjugative effects.

4.1 Conjugation in Polyenes

In *trans*-1,3-butadiene, the π-character of the formal double bonds is decreased to $\eta = 92.5\%$, the bond order to n = 1.89 (compare with Table 3) [88]. π-electrons propagate along the C chain thus inducing partial π-character to the formal CC single bond. This is reflected by $\eta = 15.1\%$ and n = 1.21 for C2—C3. If the ratio

Table 3. Description of CC bonds in conjugated systems and appropriate reference molecules[a]

Molecules	Bond	Bond length R [Å]	Bond order n[b]	Bond Ellipticity ε	π-character η[b] [%]
Ethane	C—C	1.527	1.0	0	0
Ethylene	C=C	1.317	2.0	0.45	100
Benzene	C···C	1.386	1.5	0.23	50
trans-1,3-Butadiene (τ = 180°)					
	C=C	1.323	1.89	0.433	92
	C—C	1.467	1.21	0.070	15
cis-1,3-Butadiene (τ = 0°)					
	C=C	1.323	1.93	0.432	93
	C—C	1.479	1.14	0.066	14
orthogonal 1,3-Butadiene (τ = 90°)					
	C=C	1.319	1.96	0.463	100
	C—C	1.489	1.08	0	0

[a] HF/6-31G(*d*) calculations [30]. For a definition of parameters see text.
[b] All n- and η-values have been normalized to yield the proper number of bonds and electron pairs.

$Q = \eta(C—C)/\eta(C=C)$ is considered to be a quantitative measure for π-delocalization, then one can say that the π electrons of *trans*-1,3-butadiene are delocalized by 16%.

In general, the overlap between π-orbitals is a prerequisite for π-conjugation. Maximal overlap and effective π-delocalization is given by a parallel alignment of π-orbitals. Information that reflects the degree of overlap can also be extracted from ϱ_b. For this purpose, the direction of its major axis, i.e. the direction $\mathbf{v}_2$ of the soft curvature λ_2 is used (see Sect. 2.4). In *trans*-1,3-butadiene all major axes are perpendicular to the molecular plane, i.e. the major axis of the formal single bond is parallel to those of the formal double bonds (see Fig. 13a). We say that the major axes "overlap" completely as indicated by

$$\text{overlap } (\textit{trans}\text{-1,3-butadiene}) = \mathbf{u}_2(\text{C1=C2})\, \mathbf{u}_2(\text{C2—C3}) = 1.0$$

$$\text{with} \quad \mathbf{u}_2 = \mathbf{v}_2/|\mathbf{v}_2| \, .$$

For molecules such as allene or butatriene the major axes of the bond ellipticities of neighboring double bonds are perpendicular to each other, yielding

$$\text{overlap (allene)} = \mathbf{u}_2(\text{C1=C2})\, \mathbf{u}_2(\text{C2=C3}) = 0$$

as is qualitatively shown in Fig. 13c, d. This is in line with the orbital description of π-conjugation.

For *cis*-1,3-butadiene the degree of π-delocalization is slightly lower ($Q = 15\%$) as is indicated by n, ε, and η values given in Table 3. This agrees with the MO description of the cis form. If π-delocalization is gradually disturbed by a rotation at the formal single bond of 1,3-butadiene either from $\tau = 0°$ (*cis* form) or from $\tau = 180°$ (*trans* form) to $\tau = 90°$ (*orthogonal* form), then the ellipticity of the single bond will decrease continuously to zero as is shown in Fig. 14. π-conjugation is no longer possible in the orthogonal form.

Inspection of the bond order n (Table 3), however, reveals that n(C—C) for $\tau = 90°$ is 1.08 rather than 1.0. Obviously, there are other effects than π-conjugation that stabilize the central bond. First of all, C2 and C3 are sp^2-hybridized C atoms and, therefore the CC σ-bond is shorter and stronger than the σ-bond in ethane. Both ϱ_b

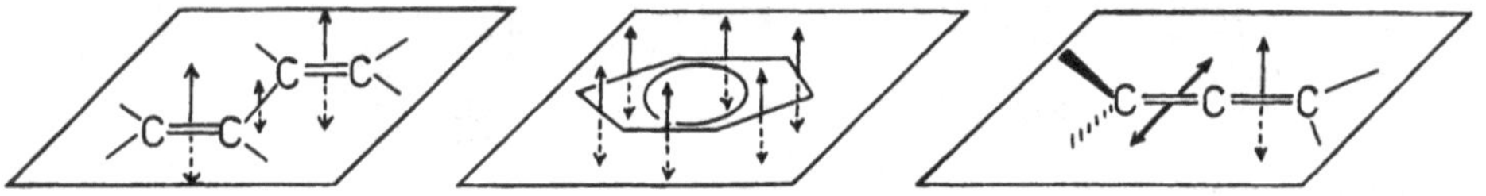

a b c

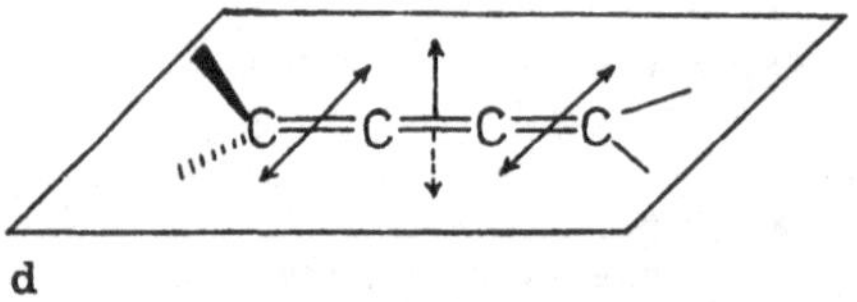

d

Fig. 13a–d. Ellipticities of **a** *trans*-1,3-butadiene, **b** benzene, **c** allene and **d** butatriene. The direction of the major axis is indicated by an *arrow*. The length of the arrow is proportional to the value of ε. (HF/6-31G(*d*) calculations)

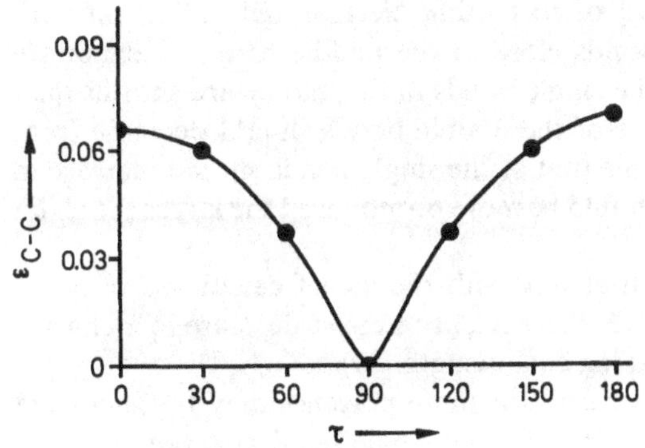

Fig. 14. Dependence of the bond ellipticity ε(C2–C3) in 1,3-butadiene on the dihedral angle τ(C1C2–C3C4). (HF/6-31G(*d*) calculations)

and n reflect this. Secondly, the curvatures λ_1 and λ_2 (—13.1 and —13.0 e Å^{-5}) perpendicular to the bond path are both softer than λ_1 of C2—C3 in the *trans* form (—14.1 e Å^{-5}), i.e. the effective ellipticites ε' (with regard to ε(C2—C3 *trans*) are $<1(\varepsilon' = 0.96)$ as in the case of acetylene (0.86). This is indicative of hyperconjugative effects between C3—H or C3—C4 and the adjacent π-bond C1C2. By utilizing ε, ε', and n values other conjugative effects can be separated from π-effects thus leading to a unified description of conjugation.

Studies on conjugated polyenes have been focused on the question, whether or not the CC bond lengths converge to a common value with increasing chain length, i.e. whether or not electron delocalization increases to 100% for infinitely long polyenes [89]. Ab initio calculations on the geometries of polyenes have led to geo-

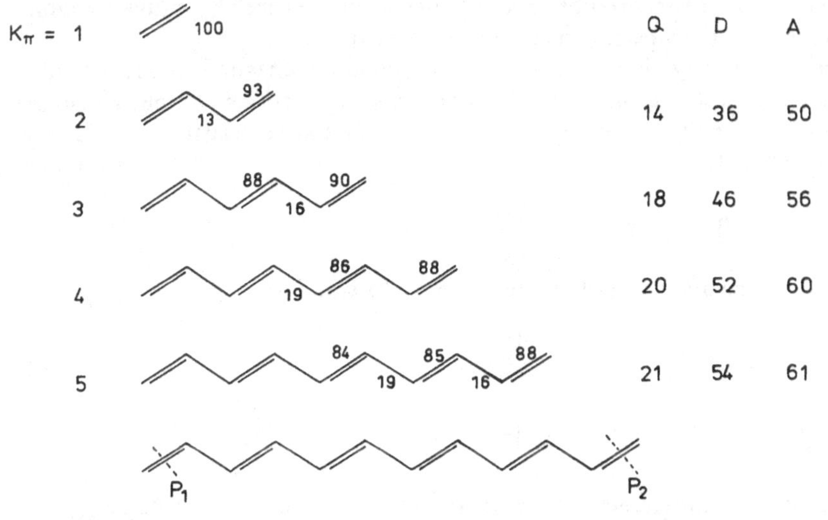

Fig. 15. π-character of polyenes with one to five double bonds. Values are given in per cent (see Eq. (24)) and are normalized to the total number of π-electrons. Delocalization of π-electrons is considered between the points P_1 and P_2, i.e. the terminal CH_2 groups are excluded. The degree of π-delocalization is measured by Q and D (Eq.s (25) and (26)), the degree of bond equalization by A (Eq. (29)). (HF/4-31G calculations from Ref. 30.)

metrical parameters that are indicative of increasing electron delocalization with increasing chain length. The double bonds close to the middle of the C chain are longer than those at the end, whereas the single bonds in the middle are shorter than those at the end. Hence, the π-character of the double bonds should decrease from the end to the middle of the C chain while that of the single bonds should increase in the same direction. Bond equalization should be more pronounced the larger the chain length is.

These findings can be verified quantitatively with the aid of calculated ratios η. According to the η values shown in Fig. 15, there is a progressive decrease in π-character toward the middle bond with increasing chain length [89].

The degree of delocalization Q (in per cent) in these polyenes may be calculated as the ratio of the averaged π-character in a formal single bond and the averaged π-character in a formal double bond:

$$
\begin{aligned}
Q &= 100\, K_\pi/K_\sigma \Big[\sum_i \eta(C{-}C)\Big]\Big/\Big[\sum_j \eta(C{=}C)\Big] \\
&= 100\Big[\sum_i \eta(C{-}C)\Big]\Big/\Big[\sum_j \eta(C{=}C)\Big],
\end{aligned}
\tag{25}
$$

where K_σ is the number of formal single bonds and K_π the number of formal double bonds. Since the number K_π of formal double bonds in a polyene is $K_\sigma + 1$, only properly averaged values η can be compared. On the other hand, the terminal groups in a polyene represent a wall that interrupts π-delocalization. Thus, it is reasonable to exclude the terminal groups and, thereby, the wall effect and to measure delocalization only between P_1 and P_2 as schematically shown in Fig. 15. Hence, the number of π-bonds is $K'_\pi = K_\sigma - 1 + 1/2 + 1/2 = K_\sigma$, i.e. the ratio of formal single and formal double bonds is one as in the case of conjugated monocyclic polyenes. This has been considered in Eq. (25).

Q increases only very slowly with increasing bond equalization. Also, it is often difficult to distinguish e.g. in charged (cyclo) polyenes between formal single and double bonds. Therefore, π-electron delocalization should be measured by the parameter D (in per cent):

$$D = 100(1 - \sigma_\eta) \tag{26}$$

that is based on the average π-character η_{av} and the standard deviation σ_η:

$$\eta_{av} = 1/K \sum_i^K \eta(AB) \tag{27}$$

$$\sigma_\eta = 1/K \sum_i^K [\eta(AB)_i - \eta_{av}]^2/\eta_{av}^2, \tag{28}$$

where K denotes the total number of heavy atom bonds in the conjugated system.

If for some reason ε or η cannot be measured, the bond orders n(AB) can be used to define a bond equalization parameter A:

$$A = 100(1 - 9\sigma_n) \tag{29}$$

with

$$n_{av} = 1/K \sum_{i}^{K} n(AB)_i \tag{30}$$

and

$$\sigma_n = 1/K \sum_{i}^{K} [n(AB)_i - n_{av}]^2/n_{av}^2 . \tag{31}$$

For a system with 100% π-delocalization leading to complete bond equalization as in benzene, $\eta = 50\%$, $\sigma_\eta = 0$, $n = 1.5$, and $\sigma_n = 0$. π-delocalization and bond equalization are zero for $\eta_1 = 0$, $\eta_2 = 100$, $n_1 = 1$ and $n_2 = 2$ as can be verified by applying Eqs. (26)–(31).

Both Q, D, and A values given in Fig. 15 for polyenes up to decapentaene confirm that π-electron delocalization and bond equalization increase with increasing chain length. In Fig. 16, the delocalization parameter D is plotted against the chain length K. Clearly, D approaches a limit value for increasing chain length. However, the available data suggest that convergence to a limit value of 100% is slow or may not be achieved at all.

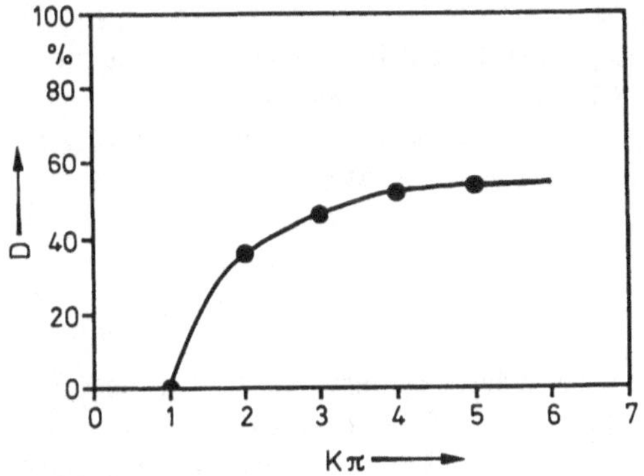

Fig. 16. Dependence of the degree of delocalization D on the number of double bonds K_π in a polyene. (Compare with Fig. 15)

4.2 Conjugation in Cyclic Systems — Aromaticity and Antiaromaticity

Cyclic conjugation involving $(4q + 2)$ π-electrons can lead to aromatic character. Aromaticity is reflected by the energetic, structural, and magnetic properties of the ground state of the compound in question. It can be determined by [90]

1. counting π-electrons and comparing their number with the Hückel $4q + 2$ value,
2. calculating the Dewar resonance energy [91],
3. determining bond alternation or bond equalization parameters, and
4. relating magnetic properties to those of appropriate reference compounds.

All these criteria have been used in the literature. It has been found that criteria 1 and 2 are very useful for chemists while 3 and 4 are more difficult to apply [92].

Since the analysis of $\varrho(\mathbf{r})$ leads to a description of the chemical bond in terms of bond order and bond ellipticity, it is easy to determine the degree of bond equalization A and the degree of π-delocalization D in a potentially aromatic molecule and to use this as a measure of aromaticity.

As noted above, the bond order n covers all bond strengthening effects, not just those due to π-delocalization. For example, for the orthogonal form of 1,3-butadiene an A parameter of 32% is calculated although π-delocalization is no longer possible. Here, A reflects an increase in n(C2C3) due to the shorter $C(sp^2)$—$C(sp^2)$ bond and additional hyperconjugative effects. We note that any definition of aromaticity based on global properties of the bond such as bond length, bond energy, bond order (even though it has been derived from a local property of $\varrho(\mathbf{r})$) can only be used in a very qualitative way. Any quantitative use requires the separation of other effects not due to the delocalization of π-electrons.

The ellipticity ε of a bond and the ε-dependent parameter η may be better suited for a quantitative definition of the aromatic character of a molecule. Utilizing the delocalization parameter D (Eq. (26)) we expect aromatic character when

1. there is π-conjugation in a cyclic system as reflected by the overlap of the bond ellipticities,
2. $(4q + 2)$ π-electrons are delocalized in the conjugated cyclic system and
3. the degree of π-delocalization measured by D is significantly larger than in the appropriate acyclic reference compound(s) used to determine the Dewar resonance energy (*definition of aromaticity*).

It will be reasonable to consider π-overlap as sufficient if the overlap of the major axes of adjacent bonds is larger than 0.7, i.e. the angle enclosed by the major axes is larger than 45°. The number N of π-electrons is obtained by comparing the sum of (normalized) bond orders with the total number K of bonds:

$$N = 2\left(\sum_{i}^{K} n_i - K\right). \tag{32}$$

For $N \pm c = 4q + 2$ with $c \leqq 0.5$, criterion 2 is considered to be fulfilled. For benzene, 6π electrons are delocalized in the ring. All major axes of the CC bonds are perpendicular to the molecular plane, i.e. they overlap completely (see Fig. 13b). Each CC bond possesses the same value for n, ε, and η (Table 3) which is indicative of complete delocalization with D = 100%. Hence benzene is aromatic.

It is interesting to compare π-delocalization for benzene and 1,3-cyclobutadiene. Although the bond ellipticities of the latter also overlap completely (see Fig. 17a) the π-character of the CC bonds is 4 and 96%, and π-delocalization is just Q = 5% (D = 17.7%), i.e. it is lower than in any polyene. The interaction of $4q$ π-electrons leads to a suppression of π-delocalization. Analogous to the definition of aromaticity, we can now give a definition for antiaromaticity. It is found when

1. $4q$ π-electrons are conjugated in a cyclic system and
2. the degree of π-delocalization measured by Q or D is suppressed as compared to that found for the appropriate acyclic reference compound(s) used to determine the Dewar resonance energy (*definition of antiaromaticity*).

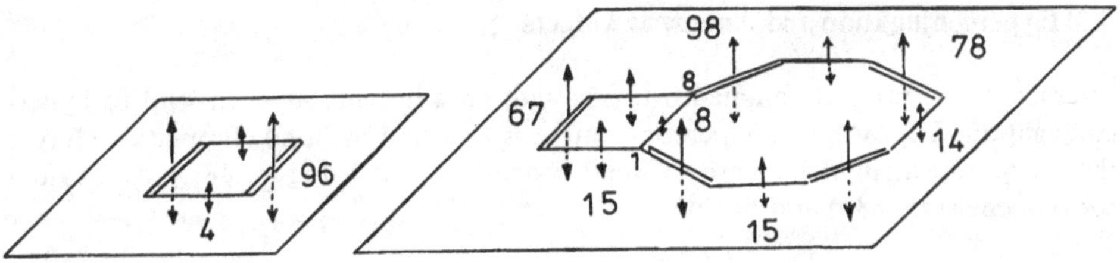

Fig. 17 a, b. Major axes of the bond ellipticities ε (*arows*) and π-character η for **a** 1,3-cyclobutadiene and **b** bicyclo[6.2.0]decapentaene. The length of the arrows is proportional to the value of ε. Values of η in per cent and normalized to the total number of π-electrons. (HF/STO-3G calculations from Ref. [29])

With the ϱ-based definitions of aromaticity and antiaromaticity the electronic nature of cyclopolyenes can be investigated. To demonstrate this we discuss one example, namely bicyclo[6.2.0]decapentaene [29]. This compound has attracted interest by both experimentalists and theoreticians, in particular after its first successful synthesis by Oda and co-workers [93] and, independently, by Vogel and co-workers [94]. The component rings of bicyclo[6.2.0]decapentaene possess $4q$ π-electrons and, therefore, could be destabilized by antiaromatic character. On the other hand, there are $(4q + 2)$ π-electrons in the perimeter of bicyclo[6.2.0]decapentaene, possibly leading to aromatic stabilization provided bicyclo[6.2.0]decapentaene is planar or close to being so. The question is whether peripheral or monocyclic delocalization dominates the electronic nature of bicyclo[6.2.0]decapentaene, whether the π-system is stabilized by aromatic or destabilized by antiaromatic interactions that outweigh any aromatic effects.

Cremer and Schmidt [29] have carried out ab initio calculations on bicyclo[6.2.0]-decapentaene which show that the molecule is essentially planar and that π-overlap is fully developed (Fig. 17b). Bicyclo[6.2.0]decapentaene contains a dimethylene-cyclobutene unit rather than a cyclobutadiene unit in order to avoid 4π electron interactions in the four-membered ring. Investigation of $\varrho(\mathbf{r})$ leads to the η-values shown in Fig. 17b. According to these data the central 1,8-bond of bicyclo[6.2.0]-decapentaene possesses only low π-character ($\eta = 8$), less than the central bond in 1,3-butadiene. Delocalization of π-electrons occurs essentially along the periphery of the bicyclic system. The π-character of the formal CC single bonds (15%) is somewhat higher, that of the formal CC double bonds (84%) is somewhat lower than that of *trans*-1,3-butadiene calculated at the same level of theory (13 and 93.5%) [29]. Hence, delocalization is slightly stronger in the periphery of bicyclo[6.2.0]-decapentaene than in *trans*-1,3-butadiene.

This is confirmed by the calculated delocalization parameter D. For the four-membered ring D is as low as for cyclobutadiene itself (18.5 vs 17.7%) while for the eight-membered ring D is 42.7%. However, D is 52.8% in the periphery of bicyclo[6.2.0]decapentaene, clearly larger than D(*trans*-1,3-butadiene) = 35%.

The energetic consequences of peripheral π-delocalization have been estimated to be 4.4 ± 1.5 kcal mol^{-1}, i.e. bicyclo[6.2.0]decapentaene is slightly aromatic despite the fact thar it contains antiaromatic ring units [29].

4.3 Hyperconjugation and Anomeric Effects

Interactions of properly aligned σ-bonds with an adjacent π-system lead to hyperconjugation. The extent of hyperconjugation is reflected by bond ellipticities ε larger than zero and a slight increase of bond orders n [21–23]. Figure 18 gives ε values for propene (Fig. 18a) and the ethyl cation (Fig. 18b).

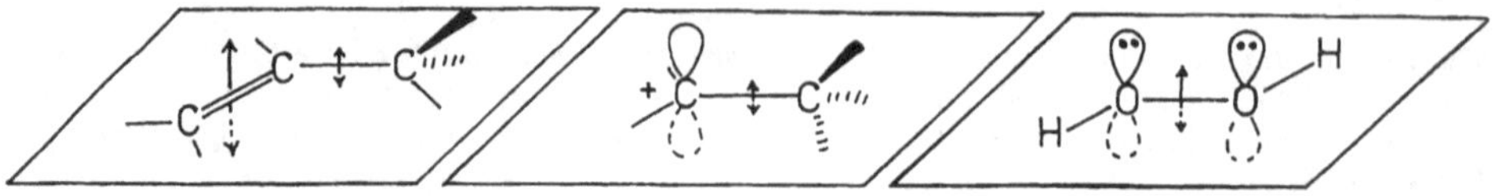

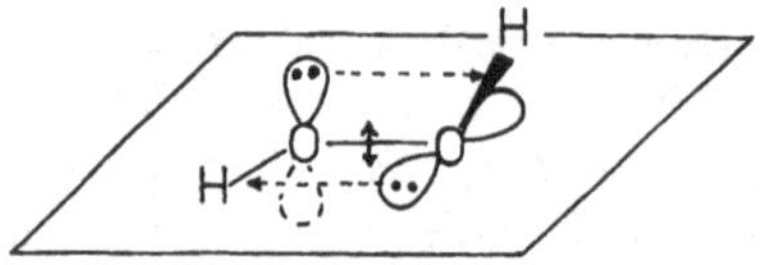

Fig. 18a–d. Bond ellipticities ε of **a** propene, **b** ethyl cation, **c** trans hydrogen peroxide, and **d** skewed hydrogen peroxide (dihedral angle 120°). Interactions between the lone pair orbitals causing the anomeric effect are indicated for skewed H_2O_2. (HF/STO-3G [21] and HF/DZ + P calculations [31])

In the former molecule, the CC single bond possesses an ellipticity of 0.03. The major axis is parallel to that of the adjacent double bond, i.e. both ellipticities overlap completely. The bond order of the double bond is slightly increased (n = 2.1). Its ellipticity is 14% larger than that of ethylene. Obviously, the methyl group of propene donates π-charge into the double bond. Similarly hyperconjugative interactions of the methyl group of the ethyl cation lead to a transfer of negative charge into the empty *p*-orbital. This is indicated by ε(C—C$^+$) = 0.03 and a major axis parallel to the empty *p*-orbital at the positive C.

Anomeric interactions involve delocalization of negative charge from an electron lone pair to the σ* orbital of a vicinal bond that is coplanar with the electron lone pair orbital. As a consequence, the σ-bond connecting the interacting centers adopts partial π-character.

In Fig. 18c and d λ and ε values of *trans* and skewed H_2O_2 (τ = 120°) are shown. In the planar form, the OO bond H_2O_2 possesses π-character due to the occupation of π and π* MOs. This electronic arrangement is unfavorable. Rotation about the OO bond leads to a coplanar arrangement of the lone pair orbitals at the vicinal O. Negative charge delocalizes into the σ*(OH) orbitals as indicated in Fig. 18d. Since this involves both OH bonds and lone pairs, the soft curvature (λ_2) and the steep curvature (λ_1) adjust leading to an electron arrangement similar to that of acetylene. Anomeric interactions are indicated by a nearly cylindrical density distribution at the OO σ-bond. The value of ϱ_b and, hence, the bond order n(OO) increase indicating that the OO bond is strengthened and H_2O_2 becomes more stable in line with experimental and theoretical observations [31]. The description of H_2O_2 demonstrates that anomeric effects similar as hyperconjugative effects can be quantitatively assessed from the properties of $\varrho(\mathbf{r})$ at the bond critical points of a molecule.

This has been used by LaBarge and co-workers to investigate anomeric effects in final ozonides (1,2,3-trioxolanes) [32, 33]. These authors determined the positions of the lone pair concentrations at the various O atoms for a given conformation of the molecule. Then, calculated σ^* MOs were associated with the bonds in α-position to the O atoms and ordered according to their energy. Finally, the number of anomeric interactions, i.e. coplanar alignments of the lone pair concentrations at O and vicinal CX bonds were counted in dependence of the conformation of the final ozonid and weighted by the relative energy of the σ^*(CX) MO. In this way, the most stable conformations of substituted final ozonides were predicted and verified by ab initio calculations [32, 33].

5 Description of Homoconjugation

5.1 π-Character of Three-membered Rings

As discussed in the previous section, σ-bonds can possess a bond ellipticity larger than zero due to hyperconjugative or anomeric effects. While these bond ellipticities generally are very small indicating less than 5% π-character, considerably larger bond ellipticities are calculated for formal single bonds in three-membered rings [21, 38, 39].
Table 4 gives relevant values which show that the π-character of the CC bonds in cyclopropane should be of the same magnitude as that of ethylene even though the bond order is close to n = 1.

There is, however, an important difference between the bond ellipticity in a polyene and in a three-membered ring. As indicated in Fig. 19, the major axes, i.e. the directions of the soft curvatures of the ring bonds, are not perpendicular but parallel to the plane of the carbon nuclei. This means that in the ring plane electron density

Table 4. Bond order n, bond ellipticity ε_b, and degree of surface delocalization[a]

Molecule	CC			CX			Ring		
	ϱ_b [e Å^{-3}]	n	ε_b	ϱ_b [e Å^{-3}]	n	ε_b	ϱ_r [e Å^{-3}]	ξ^b [%]	ε_r^c
Cyclopropane	1.681	1.00	0.489				1.379	82.0	0
Aziridine	1.763	1.06	0.391	1.823	0.97	0.502	1.485	83.1	0.445
Oxirane	1.819	1.08	.0.314	1.771	0.96	0.879	1.533	85.8	0.094
Cyclobutane		1.00	0.001				0.554	33	0
Benzene		1.50	0.231				0.141	7	0

[a] HF/6-31G(*d*) calculations [38]. All bond orders are normalized.
[b] Ratio $(\varrho_r/\varrho_b(\mathrm{av}))$ 100 where ϱ_b(av) is the average of all ϱ_b values.
[c] Ellipticity at the ring critical point.

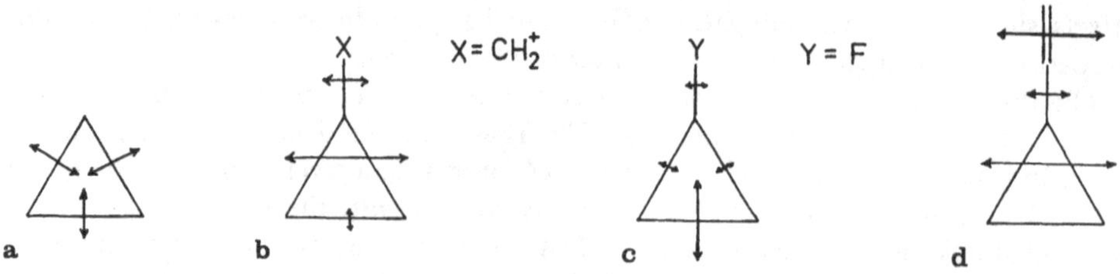

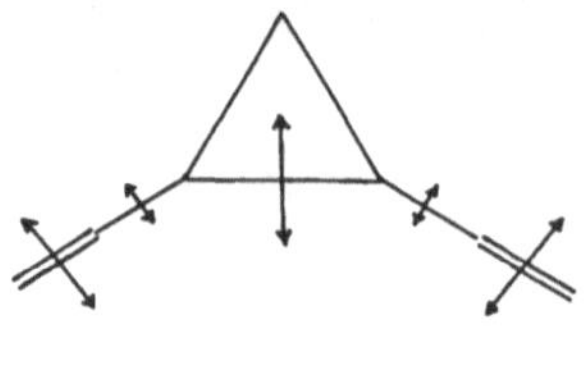

Fig. 19a–e. Surface delocalization in substituted cyclopropanes. Major axes of bond ellipticities are indicated by *arrows*; the direction of surface delocalization in the three-membered ring is given by a larger arrow. **a** Cyclopropane. **b** π-Acceptor substituted cyclopropane. **c** π-Donor substituted cyclopropane. **d** Vinylcyclopropane. **e** *cis*-1,2-Divinylcyclopropane

extends both toward the ring center and toward the outside of the ring. This is unique for three-membered rings. For cyclobutane and larger rings the bond ellipticities are vanishingly small [38] (Table 4). The electron density distribution $\varrho(\mathbf{r})$ in the three-membered ring bonds clearly reveals features of the π-density of an alkene double bond. This is in line with experimental observations that suggest π-character for the three-membered ring bonds [95].

Due to the fact that the electron density extends into the ring, $\varrho(\mathbf{r})$ is considerably larger at the center of the three-membered ring than is usually observed at the center of other rings. For example, the density at the ring critical point of cyclopropane is still 82% of that at the bond critical point $\mathbf{r}_b$(CC) (ξ in Table 4). In cyclobutane (benzene), there is just 33% (7%) of the density at the $\mathbf{r}_b$(CC) points [38]. Hence, cyclopropane as well as other three-membered rings are unique in the sense that electron density is smeared out all over the ring plane in a similar manner as it is along the carbon framework of 1,3-butadiene or benzene. In order to distinguish between the various modes of electron delocalization, we speak of "ribbon delocalization", when electrons delocalize along the "ribbon" of a π-conjugated system (see Sect. 4), and of "surface delocalization" when electrons delocalize in the ring surface [38, 40, 42].

Surface delocalization is characterized by ξ, the percentage of negative charge at the ring critical point relative to that at the bond critical point [96], and the ring critical ellipticity ε_r. The latter parameter is zero for cyclopropane (Table 4) because both λ_{1r} and λ_{2r} possess the same value, i.e. $\varrho(\mathbf{r})$ is isotropical in the plane of the ring at $\mathbf{r}_r$.

For substituted cyclopropanes, interaction between ring and substitutents influence the π-character of the CC bonds [39]. Depending on the nature of the substituent the bond ellipticity of the vicinal CC bonds exceeds that of the distal CC bond or vice versa. For the same reason, $\varrho(\mathbf{r})$ at the ring critical point becomes anisotropical, i.e. $\varepsilon_r > 0$ and $\lambda_1 \neq \lambda_2$. *The major axis of the ring ellipticity ε_r always points to the CC bond(s) with the largest π-character.* Hence, a direction can be assigned to surface delocalization in substituted three-membered rings. This is the direction of the major axis of ε_r [38, 39].

Cremer and Kraka [38, 39] have shown that there is a direct relationship between orbital interactions in substituted cyclopropanes and the various modes of surface delocalization. For example, a π-acceptor substituent that leads to a depopulation of the antibonding Walsh MO of cyclopropane, and, hence, a shortening of the distal bond, causes the direction of surface delocalization to be parallel to the distal bond indicating that the π-character of the vicinal bonds is larger than that of the distal CC bond (Fig. 19b). Similarly, a π-donor substituent causes surface delocalization to be perpendicular to the distal CC bond. In this case, the π-character of the distal bond is larger than that of the vicinal bonds (Fig. 19c).

Due to the π-character of the CC bonds, the cyclopropane unit can both participate and close a gap between two π-conjugated systems. This is indicated in Fig. 19d for vinylcyclopropane. The major axis of the CC double bond is parallel to the direction of surface delocalization. The formal CC σ-bond linking double bond and ring adopts partial π-character. Its major axis overlaps with those of vinyl and cyclopropane, establishing a conjugated system. Obviously, the cyclopropane ring plays the role of a second ethylene unit. Vinylcyclopropane and 1,3-butadiene are isoconjugate.

If the substituent effects of two vinyl groups are superimposed as in *cis*-1,2-divinylcyclopropane, the π-character of the bond C1C2 will be larger than that of bonds C1C3 and C2C3. Surface delocalization takes the direction indicated in Fig. 19e. Again, the cyclopropane ring adopts the role of an ethylene unit that participates in conjugation. 1,2-Divinylcyclopropane is isoconjugate with 1,3,5-hexatriene.

These examples demonstrate that the union of a three-membered ring and one or more double bonds leads to π-conjugation. If two molecular entities that could form a conjugated system are separated by an isolating group X, this can be by-passed by the formation of a three-membered ring, in which X is the apex group and the basal bond connects the conjugated systems. Conjugation of this type has been called homoconjugation [97]. If homoconjugation leads to closure of a ring with $(4q + 2)$ π-electrons, a homoaromatic system will be established. Both homoconjugation and homoaromaticity are consequences of the peculiar properties of three-membered rings in general and cyclopropane in particular. In the same way, as the electronic features of three-membered rings are described with the aid of the properties of $\varrho(\mathbf{r})$, homoconjugation and homoaromaticity can be described [34]. This is shown in the Subsections that follow.

5.2 Homoconjugation

A vinyl group and a carbenium ion that are separated by just one CH_2 group can homoconjugate. The simplest system to possibly show this type of conjugation is the homoallyl cation, $C_4H_7^+$ [97]. The most notable property of the homoallyl cation is its facile interconversion to the cyclopropylcarbinyl cation [98]. This can be anticipated from the electron density distribution of the homoallyl cation as reflected by bond orders and ellipticities [34] (Fig. 20a). The bond C2C3 is a slightly weakened $(n = 0.9)$ σ-bond. Nevertheless, it possesses a substantian ellipticity $(\varepsilon = 0.3)$ with its major axis lying in the plane C1C2C3 and being directed toward the region between

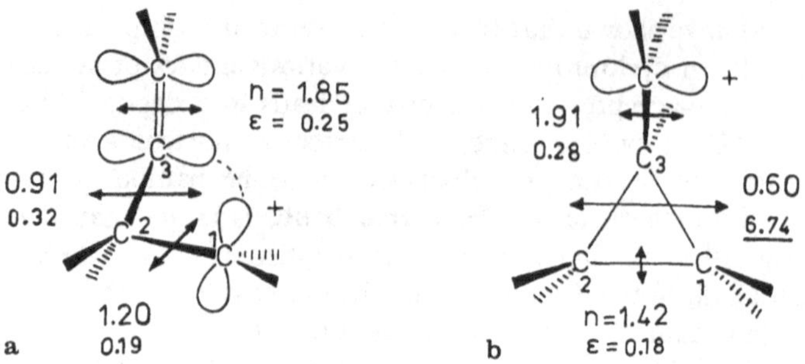

Fig. 20a, b. Bond orders n and bond ellipticities ε of $C_4H_7^+$. **a** Homoallyl cation. **b** Cyclopropylcarbinyl cation. (HF/6-31G(*d*) calculations [34])

C1 and C3. This is indicative of the onset of a C1C3 bond which is realized in the cyclopropylcarbinyl cation.

The cyclopropylcarbinyl cation has two very labile ring bonds (Fig. 20b). The bond orders of C1C3 and C2C3 are just 0.6 and the ellipticities are extremely large ($\varepsilon = 6.7$). The ring critical point lies almost on a line connecting the bond critical points $\mathbf{r}_b$(C1C3) and $\mathbf{r}_b$(C2C3). Its density is marginally smaller than that at the two bond critical points, and in line with this, the negative curvature of ϱ_r is close to zero. There is a trough in the distribution of negative charge in the plane of the ring linking $\mathbf{r}_r$ and the two bond critical points $\mathbf{r}_b$. The trough gives the preferred direction of surface delocalization as parallel to the bond C1C2. Little energy is needed to break the ring structure and to merge the three critical points in the bond critical point $\mathbf{r}_b$(C2C3) of the homoallyl cation.

In the cyclopropylcarbinyl cation, the major axis of the bond ellipticity of bond C3C4 is parallel to the direction of surface delocalization. Three-membered ring and C3C4 bond form a conjugate system in which the cyclopropyl ring takes the part of an ethylene unit. The cyclopropylcarbinyl cation is isoconjugate with the allyl cation. The onset of these interactions are already revealed by the density properties of the homoallyl cation as shown above.

Homoconjugation is generally considered to arise from π-overlap between nonbonded groups or atoms in a conjugated system. In terms of structure theory it implies the formation of a three-membered ring with enhanced bond ellipticities and a direction of surface delocalization that overlaps largely with the major axes of an adjacent π-system. In general, it is difficult to predict homoconjugative interactions just on the basis of orbital considerations (see Sect. 5.3). Therefore, homoconjugation is mostly postulated in an ad hoc manner to rationalize the properties or the reactivity of a given compound. By way of contrast, we will speak of homoconjugation only if conjugation, interrupt by an atom or group X, has been established by the formation of a labile bond with large π-character. A restriction of the term "homoconjugation" in this way helps to clarify the notion of homoaromaticity discussed in the next Subsection.

5.3 Homoaromaticity

The concept of homoaromaticity introduced into chemistry by Saul Winstein [97] is a useful tool to rationalize the chemical behavior of cyclic $(4q + 2)$ π-electron systems in which conjugation is interrupted by a sp^3-hybridized carbon atom or by any other saturated group Y [13]. Under certain circumstances these systems can circumvent the insulating group by 1,3-orbital interactions that close the conjugated system. Then, cyclic conjugation leads to homoaromatic character of the molecule.

1

The question is to which extent 1,3 orbital interactions have to be established to guarantee cyclic conjugation of $4q + 2$ electrons and, hence, homoaromaticity.

There have been many experimental and theoretical investigations to answer this question [13]. Nevertheless, an unequivocal answer could not be found due to the fact that homoaromaticity is only a vaguely defined quantity. This becomes clear when scrutinizing Winstein's definition of homoaromaticity for hydrocarbons [97].

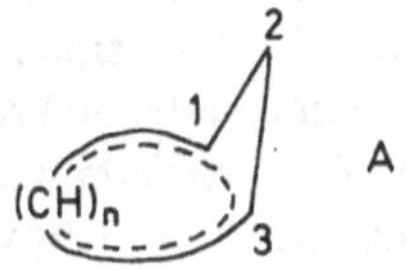

2

The potentially aromatic system A with $(4q + 2)$ π-electrons will be homoaromatic if

1. the system is closed by a 1,3 bond,
2. the bond order of the 1,3 bond is between 0 and 1,
3. orbital overlap between the participating p-AOs at centers 1 and 3 is neither σ nor π, but intermediate between σ and π, and

3

4. the $(4q + 2)$ π-electrons are fully delocalized in the resulting closed cycle.

On the basis of MO therory it is impossible to verify for any homoaromatic compound requirements 1 through 4. Although there have been many attempts to describe bonding in terms of MO theory, this has not led to an unified model of bonding that unequivocally distinguishes bonded and nonbonded interactions, that provides an unique definition of bond order, that defines a minimum overlap that is necessary for bonding or that quantifies the extent of π-delocalization in a cyclic system. The key question whether an overlap of 0.1 or even 0.03 is sufficient to establish a homoaromatic bond cannot be answered by MO-theory [99].

Also, experimental descriptions of potentially homoaromatic systems do not provide an unambiguous assessment of homoaromatic character. Does for example a 1,3-CC distance of 1.8 or 1.7 Å already imply bonding? How is homoaromatic stabilization measured in a system that contains both a strained cyclopropane unit and a conjugated π-system? How can magnetic properties be related to homoaromatic interactions? Despite all attempts to answer these questions, there has been only limited progress toward a quantitative description of homoaromaticity.

The analysis of $\varrho(\mathbf{r})$ provides a basis for a rigorous definition of homoaromaticity [34]. Utilizing the definitions of covalent bonding, bond order, π-character, and π-delocalization presented in Sects. 3 and 4, Winstein's definition of homoaromaticity can be translated into density language.

A potentially aromatic system with $(4q + 2)$ π-electrons will be homoaromatic if

1. the system is closed by a 1,3-bond path with a bond critical point $\mathbf{r}_b(C1, C3)$ and $H_b < 0$;
2. the bond order n (see Eq. 8) of the 1,3 bond is $0 < n < 1$,
3. the π-character of the 1,3-bond as measured by the bond ellipticity ε is larger than that of cyclopropane: $\varepsilon(1,3) > \varepsilon$(cyclopropane), and
4. the major axis of $\varepsilon(1,3)$ overlaps effectively with those of the neighboring bonds.

This is a quantitative definition of homoaromaticity that is generally applicable, not only for hydrocarbons. However, this definition is much more stringent than Winstein's definition. It excludes all those systems with 1,3-interactions that do not lead to a bond path. Hence, it describes homoaromaticity in a restricted sense.

In the following the electron density description of homoaromaticity will be applied to the archetype of all homoaromatic compounds, namely the homotropylium cation [100]. In 1967 Winstein concluded that the available data on the homotropylium cation were in favor of a nonclassical homotropylium structure [97]. Childs and co-workers reached the same conclusion on the basis of a X-ray diffraction study of 2-hydroxyhomotropylium hexachloroantimonate [101]. Ab initio calculations carried out by Haddon [102] led to an equilibrium geometry that is in good agreement with the geometry found in the X-ray diffraction study [101].

Using the ab initio geometry of Haddon the bond orders and ellipticities shown in Fig. 21 have been obtained by Cremer and co-workers [34]. The most important result of this investigation is the observation of a 1,3 bond path with a bond order considerably lower than 1 ($n = 0.71$) and a bond ellipticity that is considerably

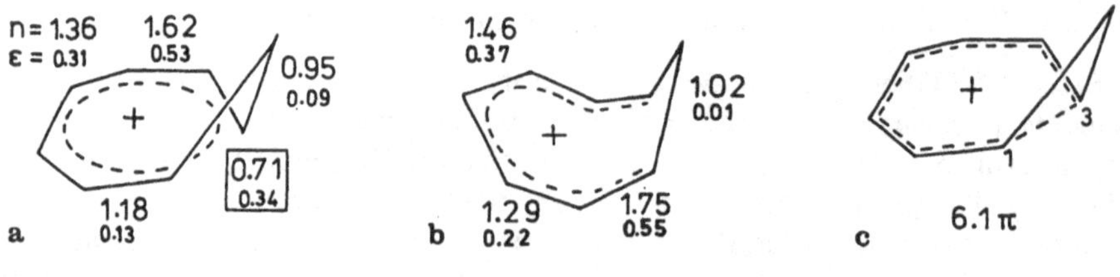

Fig. 21a–c. Bond orders n and bond ellipticities ε of $C_8H_9^+$. **a** Homotropylium cation. **b** Nonhomoaromatic cyclooctatrienyl cation. **c** Number of π-electrons delocalized in the homotropylium cation. (HF/STO-3G calculations [34])

larger than that of cyclopropane obtained at the same level of theoretical accuracy. The corresponding π-character is 46%. Hence, *homotropylium cation possesses a homoaromatic bond.* The bond path is bent below the plane of the seven-membered ring which is in line with a 1,3 overlap that is intermediate between σ and π. For full realization of an aromatic system, all seven bond orders n and ellipticities ε should be equal and all major axes should be parallel and perpendicular to the ringe plane as in the case of the tropylium cation. In the homotropylium cation, the bond orders alternate between 0.7 (C1C3), 1.2 (C3C4), 1.6 (C4C5), and 1.4 (C5C6). The π-character of the ring bonds is 45, 17, 70, and 41 %. The dot products of the unit vectors defining neighboring major axes (see Sect. 4) are 0.89 (C2C3, C3C4). 0.98 (C3C4, C4C5), 0.99 (C4C5, C5C6), and 0.97 (C5C6, C6,C7). Hence, overlap of the major axes is close to 100% [34].

The bond orders in the ring sum to 9.1, a value equivalent to six single bonds and six π-electrons delocalized in the ring. The degree of delocalization D (Eq. (26), Sect. 4) is 78%, much higher than that of any of the polyenes discussed in Sect. 4. From Fig. 21 it becomes obvious that full delocalization (D = 100%) is not achieved due to the low bond order and π-character of the bonds C3C4 and C1C8 that link the homoaromatic bond to the pentadienyl fragment.

According to the definition given above, the homotropylium cation is a true homoaromatic system [34]. In addition, the analysis of $\varrho(\mathbf{r})$ reveals that the homoaromatic bond C1C3 is rather labile. This is a consequence of the proximity of the ring critical point $\mathbf{r}_r$(C1C2C3) and the bond critical point $\mathbf{r}_b$(C1C3). Both points are just 0.44 Å apart and the density at these points differ by only 0.015 e Å^{-3}. The other two three-membered ring bonds, C1C2 and C2C3 possess relatively small ellipticities. Their properties are closer to those of normal σ-bonds rather than to those of cyclopropane bonds. This is also a consequence of the migration of the ring critical point $\mathbf{r}_r$ toward the homoaromatic bond C1C3. Obviously, surface delocalization in the cyclopropane unit is largely reduced in favor of π-character of the bridging bond C1C3.

Any lengthening of the C1C3 distance (1.58 Å) leads to further migration of $\mathbf{r}_r$ toward $\mathbf{r}_b$(C1C3). Coalescence of these points (yielding a catastrophe point) causes a rupture of the homoaromatic bond. A second isomere of $C_8H_9^+$ can exist with a C1C3 distance of 2.324 Å. It is 6.4 kcal mol^{-1} less stable than the homotropylium cation [102], does no longer possess a cyclopropane unit and is best described as a heptatrienyl cation which is part of an eight-membered ring [34]. In conclusion, the analysis of $\varrho(\mathbf{r})$ along the lines established in Sect. 2, facilitates the characterization of potentially homoaromatic compounds. Utilizing the definitions given above a clear distinction between true homoaromatic and nonhomoaromatic systems can be made.

5.4 Do Neutral Homoaromatic Compounds Exist?

In the fifties, even before Winstein developed the concept of homoaromaticity, Turner and co-workers [103] pointed out that according to measured heats of hydrogenation 1,3,5-cyclohepatriene should be stabilized by about 9 kcal mol^{-1}. Later this value was refined to 4 kcal mol^{-1} by Roth and co-workers who analyzed heats of hydrogenation with the aid of force field calculations [104]. The stabilization of 1,3,5-cyclohepatriene was considered to arise from 1,6-homoconjugative interactions and resultant homoaromaticity of a cyclic triene with 6π electrons.

There have been many experimental and theoretical investigations on the question of homoaromaticity in cycloheptatriene, partially supporting partially rejecting this possibility [105]. In the late seventies Christoph, Paquette, Gleiter, Houk, and others raised the principal question whether homoaromatic character is possible in neutral compounds [106, 107]. Christoph and co-workers [106] investigated a select group of trienes, including C_{16}-hexaquinacene, triquinacene, and *cis*^{3}-1,4,7-cyclo-nonatriene, the constituent double bonds of which adopt an arrangement potentially suitable for effective pp-overlap. From their investigation the authors concluded that *one can rule out the likelihood that neutral homoaromatic character will ever be discovered* [106]. Houk and co-workers [107] pointed out that homoconjugative interactions between π-systems should be destabilizing due to closed-shell repulsion and, therefore, the existence of neutral homoaromatic molecules is rather limited [108]. To scrutinize these arguments and to resolve their contradiction with the apparent stabilization of 1,3,5-cycloheptatriene, Cremer and co-workers investigated the latter molecule with the aid of ab initio methods [35, 36].

In its equilibrium geometry 1,3,5-cycloheptatriene adopts a boat conformation with dihedral angles as large as 60°. The distance between C1 and C6 is 2.5 Å. For this form, bond orders and ellipticities typical of a cyclopolyene with partial π delocalization are found (see Fig. 22). There is no indication of homoconjugative 1,6-interactions. 1,3,5-Cycloheptatriene is not a homoaromatic system.

It is, however, striking that despite the nonplanarity of 1,3,5-cycloheptatriene, π-delocalization is a large as in planar 1,3-butadiene. Analysis of the conformation of cycloheptatriene with the aid of Cremer-Pople puckering coordinates [109] revealed that cycloheptatriene in its equilibrium geometry does not represent a pure boat form but a mixed conformation comprising 77% boat and 23% chair form, as schematically shown in Fig. 23. The admixture of the chair form leads to a perturbed boat conformation, which is flattened in the triene part of the seven-membered ring as reflected by folding angles of 27° (stern of the boat) and 49° (bow of the boat) [35]. In this way, π-delocalization is improved and 1,3,5-cycloheptatriene stabilized. Also, hyperconjugative effects reflected by the small ellipticities of bonds C1C7 and C6C7 (ε = 0.04) add to the stability of the system. *There is no need to invoke homoaromaticity to explain the stability of 1,3,5-cyclohepatriene.* However, this finding not necessarily implies tnat homoaromaticity is not possible for a neutral molecule.

Upon decreasing the distance C1C6, the energy of cycloheptatriene increases and electron density is redistributed in the ring. In particular, there is a build up of electron

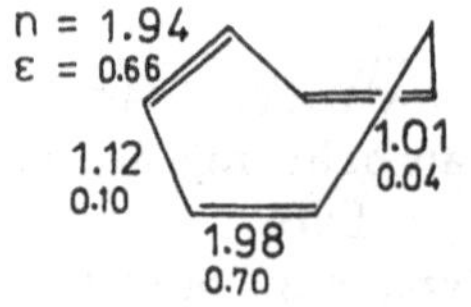

Fig. 22. Bond orders n and bond ellipticities ε of cyclohepatriene. (HF/6-31G(*d*) calculations)

Fig. 23. Representation of the perturbed boat conformation of 1,3,5-cycloheptatriene (side view) as a linear combination of boat and chair form of the seven-membered ring (X = CH_2)

density between C1 and C6 indicating the onset of a bond formation. 1,3,5-Cycloheptatriene rearranges to its valence tautomer norcaradiene. The energy change along the reaction path as determined by ab initio calculations [36], is shown in Fig. 24. According to these calculations the activation energy is 10.7 kcal mol^{-1} and the reaction is endothermic by 5.4 kcal mol^{-1} (Fig. 24).

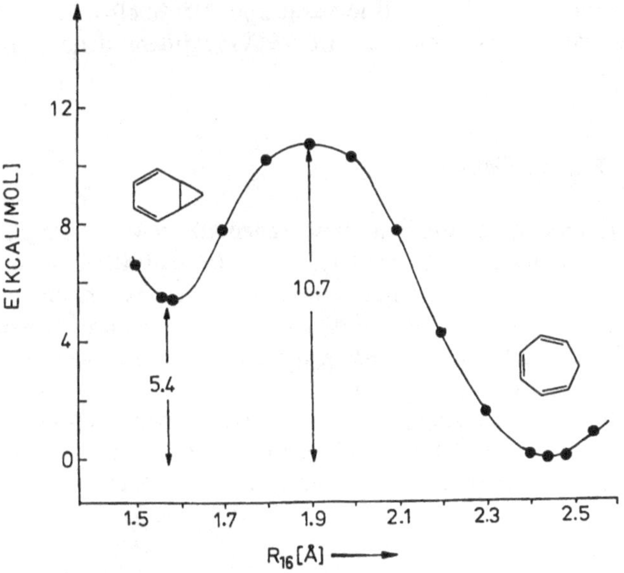

Fig. 24. Potential energy surface calculated for the valence tautomerization of 1,3,5-cycloheptatriene to norcaradiene. As a reaction coordinate the 1,6-distance (R_{16}) has been taken. *Dots* denote calculated energies for fixed R_{16} values. (MP2/6-31G(*d*) calculations)

In view of the fact that norcaradiene contains a strained C_3 ring (the strain energy of cyclopropane is 27.5 kcal mol^{-1} [110], see Sect. 6), the energy gap of just 5 kcal mol^{-1} between the two valence tautomers appears extremely small. An explanation for the surprising stability of norcaradiene emerges from its bond orders and ellipticities as obtained from the analysis of $\varrho(\mathbf{r})$ (see Fig. 25). The bond C1C6 is labile (n = 0.85) with a rather large ellipticity indicating substantial π-character. If one assumes the electrons forming this bond to interact with the other π-electrons of the six-membered ring, an ensemble of six π-electrons will be formed that can delocalize in the ring. According to the definition given above *norcaradiene is a*

Fig. 25. Bond orders n and ellipticities ε of norcaradiene. The number of π-electrons delocalized in the six-membered ring are given

neutral homoaromatic system. Homoaromatic stabilization leads to a partial compensation of the strain induced by the cyclopropane unit. Hence, the energy difference between norcaradiene and cycloheptatriene is less than expected [111].

Neutral homoaromatic compounds exist! Possible candidates are norcaradiene or compounds containing norcaradiene such as bisnorcaradiene (tricyclo[4.4.1.$0^{1,6}$]-undeca-2,4,7,9-tetraene), the valence tautomer of 1,6-methano[10]annulene [36, 112].

5.5 Homoantiaromaticity — Fact or Fiction?

If one extends the concept of homoconjugation in a purely schematic way, one might expect that homoconjugative interactions also lead to cyclic delocalization of $4q$ π-electrons. In this case, one could speak of homoantiaromaticity. However, this is a definition according to pattern probably without any chemical relevance. Since delocalization of ($4q$) π-electrons should lead to destabilization, there is no reason why homoconjugative interactions should take place at all.

Cremer and co-workers [34] have investigated the bicyclo[3.1.0]hexenyl cation (Fig. 26a), which has been described as being homoantiaromatic [113]. An interaction of the allylic 2π system with the cyclopropane unit similarly as encountered in nor-

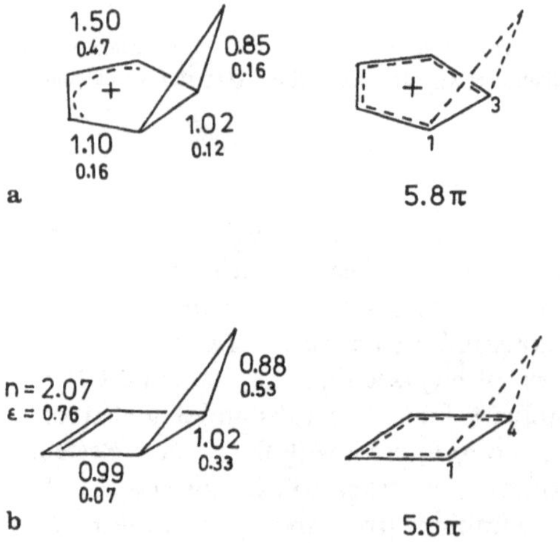

Fig. 26. Bond orders n and bond ellipticities ε of potentially homoantiaromatic molecules. **a** Bicyclo[3.1.0]hexenyl cation. (HF/STO-3G calculations [34]) **b** Bicyclo[2.1.0]pen-2-tene. (HF/6-31G(*d*) calculations) Number of electrons delocalized in the two molecules are given on the right side

caradiene (Section 5.3) would entail destabilizing 4π electron interactions. However, the bicyclo[3.1.0]hexenyl cation avoids homoantiaromatic 4π electron delocalization as is clearly revealed by calculated n and ε values (Fig. 26a). The properties of the bond C1C3, including its length (1.501 Å) [34] are those of a CC bond in an isolated cyclopropane. On the other hand, the two external bonds of the three-membered ring, C1C2 and C2C3, are extended (1.535 Å) and are the most labile bonds in the cation with n being just 0.85. They possess ellipticities whose major axes have substantial overlaps with the corresponding axes of the neighboring bonds in the five-membered ring. Their properties are similar to, but less extreme than, those of the homoaromatic bond in the homotropylium cation.

The analysis of $\varrho(\mathbf{r})$ of bicyclo[3.1.0]hexenyl cation suggests that the ring of the six outer bonds forms a conjugated system. Their bond orders sum to 6.8, equivalent to four single bonds and a π-system of approximately six electrons. The labile character of bonds C1C2 and C2C3 accounts for the perambulatory properties of the cyclopropane ring. For methyl-substituted derivatives of the bicyclo[3.1.0]-hexenyl cation it has been found that C2 and, hence, the cyclopropane ring migrates around the five-membered ring by a series of sigmatropic 1,4-rearrangements [114]. Unlike the homotropylium cation, in which the external bonds of the three-membered ring are effectively moved from the conjugated system, delocalization of negative charge persists over the entire surface of the three-membered ring and this surface is conjugated with the π-system of the five-membered ring. The direction of surface delocalization is parallel rather than perpendicular to the bond C1C3 indicating that the latter is excluded from conjugation.

Similar observations can also be made for other potentially homoantiaromatic compounds. For example, a molecule for which one could expect 4π electron delocalization similar to the 6π delocalization in norcaradiene (Sect. 5.3) is bicyclo[2.1.0]pent-2-ene [113] (Fig. 26b). Again, the external bonds rather than the bridging bond of the three-membered ring are labile as reflected by n = 0.88 and $\varepsilon = 0.53$. Approximately six electrons are delocalized on the perimeter of the six-membered ring.

In conclusion, potentially homoconjugative systems with $(4q)$ π-electrons clearly avoid homoantiaromaticity by adopting an electronic situation with $(4q + 2)$ delocalized electrons. This is possible because surface delocalization in the three-membered ring can be both perpendicular and parallel to the 1,3-bond. It seems that the preferred direction of surface delocalization is always as such that a $(4q + 2)$ system of delocalized π-electrons is formed and a homoantiaromatic system is avoided. The question remains whether homoantiaromaticity will also be avoided if steric factors enforce interactions of $4q$ electrons. These interactions will clearly be destabilizing. In this situation, the molecule will adopt a geometry that minimizes destabilizing interactions, i.e. it will avoid a homoconjugative bond. In view of these considerations *it is unlikely that homoantiaromaticity exists*. This does not exclude that through-space (or through-bond) interactions are still destabilizing to some extent leading to molecular properties different from those expected [115].

5.6 Distinction Between Homoconjugation and Through-Space Interactions

The definition of homoaromaticity given in Sect. 5.3 is based on changes of molecular structure as reflected by the network of bond paths. Compounds that possess a

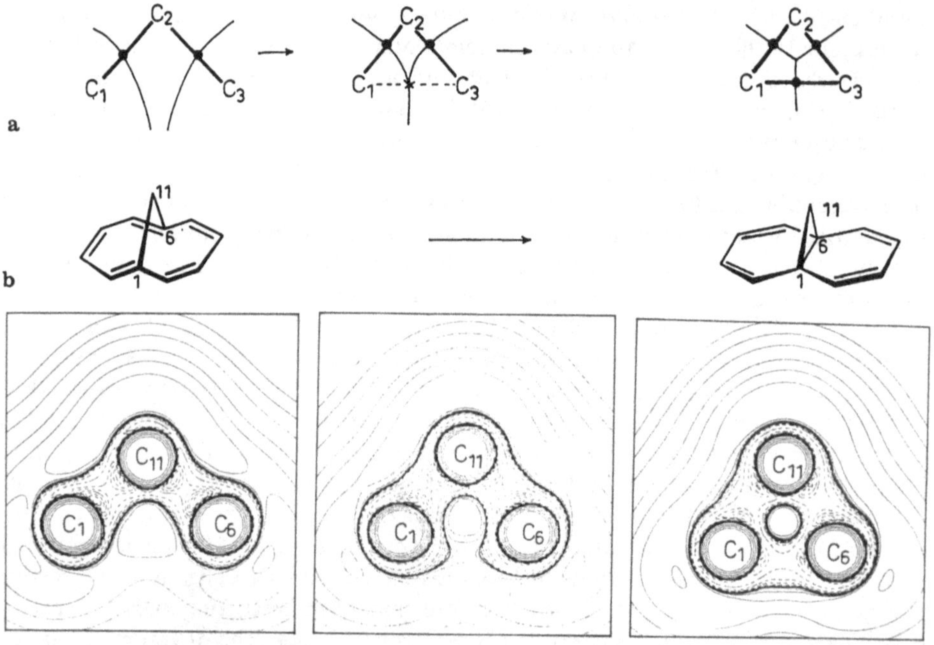

Fig. 27. a Schematic view of the changes of the atomic subspaces upon the approach of C1 and C3 in a potentially homoaromatic system. *Heavy solid lines* denote bond paths, *dots* bond critical points, *a star* a catastrophe point. The traces of the zero-flux surfaces in the plane are given by *light solid lines*. **b** Change of the Laplace concentration (shown in form of contour-line diagrams) for the rearrangement of 1,6-methanol[10]annulene to bisnorcaradiene (tricyclo[4.4.1.0^{1,6}]undeca-2,4,7,9-tetraene). The reference plane contains nuclei C1, C6, and C11. *Dashed contour-lines* denote concentration of charge ($\nabla^2\varrho(\mathbf{r}) < 0$) and *solid contour-lines* denote depletion of charge ($\nabla^2\varrho(\mathbf{r}) > 0$). (HF/6-31G($p$, d) calculations)

homoaromatic bond in a cyclic ($4q + 2$) π-system are homoaromatic in the closer sense. However, there are other compounds that exhibit 1,3-interactions that do not lead to a 1,3-bond path but affect the properties of the molecule. An example for such a case is the cyclobutenyl cation. The spectroscopic properties of this cation are often quoted to emphasize its "homoaromatic" character [13]. Yet, none of the ab initio calculations carried out so far leads to an electron density distribution with a 1,3 homoaromatic bond path [34] despite the fact that other properties of the cation such as geometry and conformation are well reproduced by ab initio theory [116].

The question is which properties of $\varrho(\mathbf{r})$ indicate through-space interactions and the stepwise formation of a chemical bond. Fig. 27a presents a schematic view of how the atomic subspaces Ω change on an approach of C1 and C3. Clearly, the subspace C2 extends less and less in the region between C1 and C3 until the surfaces of C1 and C3 coincide and a bond path between C1 and C3 is formed. At the same time the Laplace concentration between C1 and C3 gradually increases and converges to the one found for a three-membered ring. This is shown in Fig. 27b for the case of the valence tautomerism of 1,6-methano[10]annulene to bisnorcaradiene [36].

An analysis of the Laplace concentration, $-\nabla^2\varrho(\mathbf{r})$, yields information about the

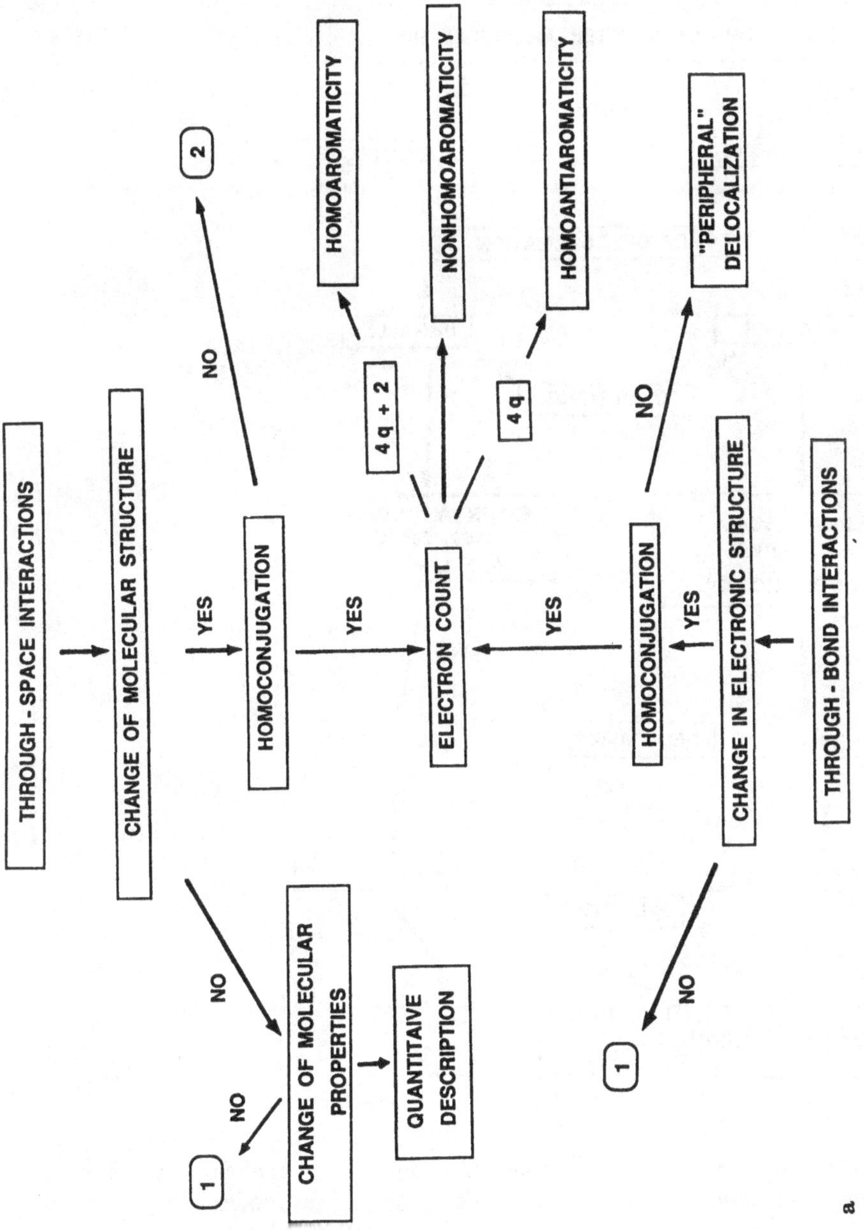

Fig. 28. **a** Flow chart diagram for the distinction of through-space, through-bond, and homoconjugative interactions utilizing the properties of $\varrho(\mathbf{r})$. The numbers have the following meaning. 1: Through-space or through-bond interactions are negligible. 2: Other nonclassical structures are formed. **b** Flow chart diagram showing the two possible modes of conjugation between a three-membered ring and an adjoining π-system

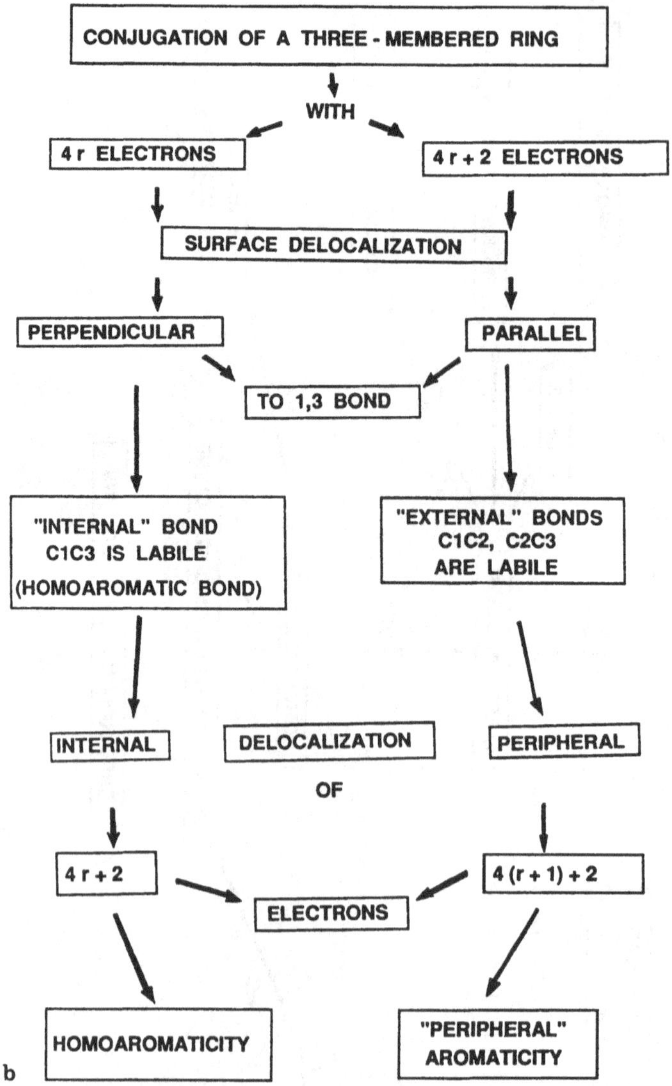

b

extent of through-space interactions and the concomitant changes in the molecular properties. Hence, a clear distinction between the various modes of intermolecular interactions is possible. This is shown in form of a flow chart diagram in Fig. 28a. Through-space interactions can lead to a change in molecular structure, in particular a 1,3 bond may be established due to homoconjugative interactions. An electron count will reveal whether a homoaromatic system is formed. If the molecular structure is not changed, i.e. no new bond path is established, the molecular properties will change to some degree. Strong changes in typical parameters such as NMR shifts have led to the denotation of "homoaromaticity" in the broader sense [13]. Examples are the cyclobutenyl cation or the isoelectronic 1,2-dihydro-borete [37].

Alternatively, homoaromaticity may be caused by through-bond interactions. The latter do not cause a change in molecular structure (if they would change molecular structure, the interaction would have to be considered as through-space interaction), but a change in the electronic structure which in turn may lead to a homoconjugative bond and to a conjugated cyclic system. "Internal" delocalization of $(4q + 2)$ π-electrons involving a 1,3 bond yields homoaromaticity as in norcaradiene while "external" delocalization of $(4q + 2)$ π-electrons involving bonds 1,2 and 2,3 results in a labilization of these bonds and "peripheral aromaticity" (Fig. 28b).

Homoaromaticity requires conjugation of a three-membered ring with an adjoined π-system. A four-membered ring (bridging of two saturated groups) does not lead to homoaromatic character. This is due to the fact that surface delocalization of electrons is found just for three-membered rings and not for larger ones [38, 42].

Depending on the number of π-electrons in the adjoining π-system, the preferred direction of surface delocalization is perpendicular or parallel to the 1,3-bond to yield in any case an $(4q + 2)$ ensemble of delocalized electrons. Only in the former case is the π-character of the 1,3 bond increased and a homoaromatic system formed. Noteworthy is that both ionic and neutral homoaromatic compounds exist. Homoantiaromaticity, however, is not possible.

Obviously, homoconjugation is the result of the exceptional electronic structure of a three-membered ring. One could expect that homoconjugation first leads to a destabilization of the molecule since formation of a three-membered ring results in increased ring strain. Since this is not the case, ring strain must be outweighed or at least be compensated largely by the stabilizing effects of surface delocalization. We will consider this question in Sect. 6.

6 Description of Strain

In general, one distinguishes between four major types of strain [117], namely

1. bond stretching strain,
2. bond angle or Baeyer strain,
3. torsional or Pitzer strain, and
4. nonbonded or Dunitz-Schomaker strain.

Strain is caused by deviations from the strain-free molecular geometry, e.g. by deviations of the bond lengths from ideal values, by deformation of the tetrahedral, trigonal or linear bond angles, by bond eclipsing as opposed to bond staggering, and by nonbonded distances that are smaller than ideal van der Waals distances. Since an extensive discussion of the various types of strain can be found in Ref. 40, we refrain from repeating this discussion. Instead we focus on the question how strain affects the properties of $\varrho(\mathbf{r})$. If this is known, a quantitative description of strained molecules with the properties of $\varrho(\mathbf{r})$ is straightforward.

Actually, the description of an elongated bond in terms of the properties of $\varrho(\mathbf{r})$ is trivial. As discussed in Sect. 2.4 an increase in the bond distance R leads to a linear decrease of ϱ_b, the electron density at the bond critical point $\mathbf{r}_b$. More interesting are the changes in $\varrho(\mathbf{r})$ caused by a compression of bond angles. Angle bending as encountered, e.g. in small cycloalkanes leads to bent bonds. The bend of the ring bonds should reflect the amount of Baeyer strain induced by angle compression. Therefore, a realistic and quantitative description of bent bonds is a prerequisite for an assessment of ring strain in small rings.

6.1 Baeyer Strain — Description of Bent Bonds

The bond paths are images of the covalent bonds of a molecule [22, 23]. Any deviation of the bond path from the internuclear connection line is indicative of bond bending. The bent bond character is quantitatively assessed by determining the bond path length R_b, the interpath angle β, and the shift parameter d (for a definition of these parameters see Sect. 2, Fig. 4c).

In Table 5, bent bond parameters of some strained rings are compared. For cyclopropane, the CC bond paths are 0.009 Å longer than the R_e distances [38, 41]. The shift parameter is positive (0.06 Å, Table 5), reflecting that the bent bonds in cyclopropane are convex, i.e. outwardly curved. The interpath angle β is 19° larger than the geometrical angle α of 60°. The corresponding values for cyclobutane (Table 5) indicate that going from a three- to a four-membered ring leads to a significant decrease in bond bending. For example, β(CCC) in cyclobutane is just 7° larger than α(CCC) (Table 5). Bond bending is negligible for five-membered or larger rings.

If one of the C atoms in cyclopropane is replaced by a more electronegative atom X, the bend of the distal CC bond will increase while that of the vicinal CX bonds decreases, i.e. the CX bonds approach the internuclear connection lines. These trends are reflected by the bent bond parameters given in Table 5 for oxirane (X = O). The corresponding molecular graphs are depicted in Fig. 29.

For a further increase of the electronegativity of X, the CX bond paths curve inwardly into the ring, i.e. the bent bonds become concave and $d < 0$, $\beta < \alpha$ are found

Table 5. Bent bond parameters of strained rings in comparison with geometrical parameters[a]

Molecule	Bond	R_e [Å]	ΔR^b [10^3 Å]	d [Å]	ε	Angle	α	β
Cyclopropane	CC	1.497	9	0.06	0.49	CCC	60	78.8
Oxirane	CC	1.453	23	0.094	0.31	CCO	58.8	72.8
	CO	1.401	3	0.004	0.88	COC	62.4	75.8
Protonated Oxirane	CC	1.446	18	0.087	0.05	CCO	61.1	56.7
	CO	1.498	184	−0.171	2.16	COC	57.7	19.9
Fluoro-ethyl cation	CC	1.445	15	0.083	0.03	CCF	61.9	4.3
	CF	1.533	537	−0.455	$>10^2$	CFC	56.2	0.02
Bicyclobutane	CC[c]	1.484	18	0.089	0.36	CCC[c]	58.8	73.6
	CC	1.513	9	0.056	0.49	CCC	60.6	77.4
Cyclobutane	CC	1.544	3	0.038	<0.01	CCC	88.6	95.6

[a] All values from HF/6-31G(*d*) calculation [38, 43].
[b] $\Delta R = 1000\,(R_b - R_e)$.
[c] Central bond and angle opposite to central bond, respectively.

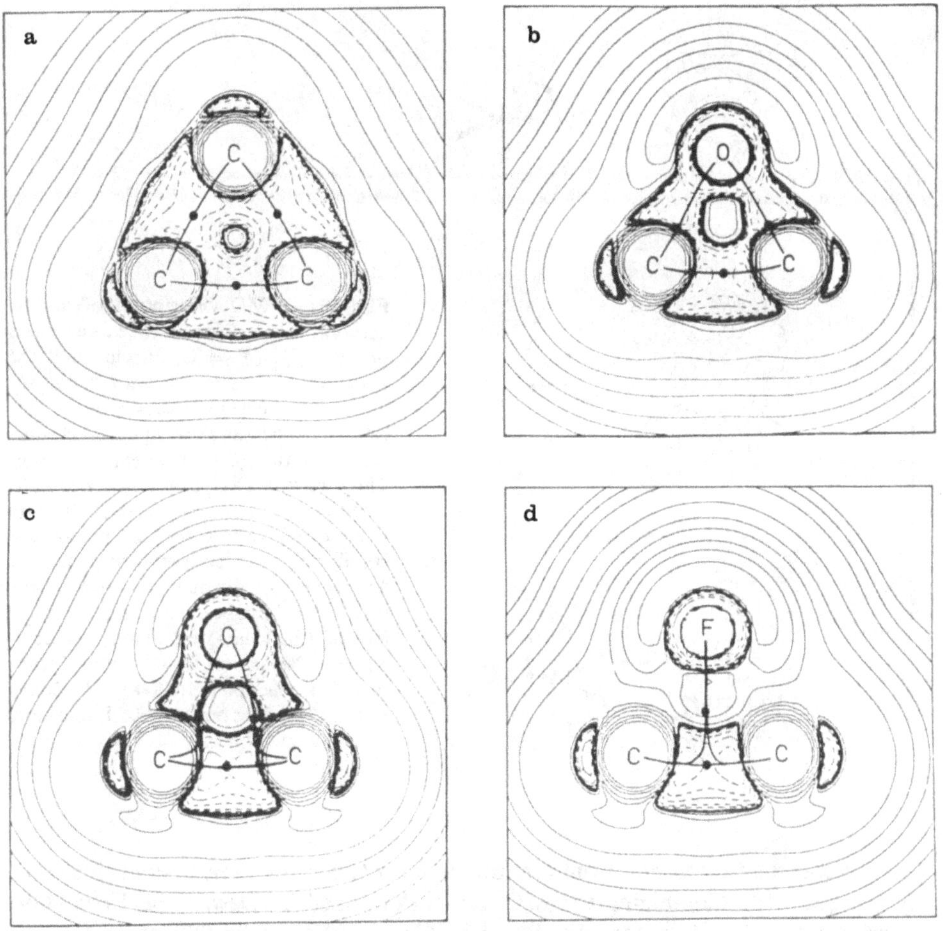

Fig. 29 a–d. Molecular graphs and Laplace concentrations, $-\nabla^2 \varrho(\mathbf{r})$, of **a** cyclopropane, **b** oxirane, **c** protonated oxirane, and **d** halogen-bridged fluoroethyl cation. The reference plane contains the nuclei of the heavy atoms. Bond paths are indicated by *heavy solid lines* and bond critical points by *dots*. *Dashed lines* are in regions where electronic charge is concentrated ($\nabla^2 \varrho(\mathbf{r}) < 0$) and *solid lines* in regions where charge is depleted ($\nabla^2 \varrho(\mathbf{r}) > 0$). Inner shell concentrations are not shown. (HF/6-31G(*d*) calculations [38])

[38]. At the same time the bending of the CC bond is reduced. This is the case for $X = OH^+$ and $X = F^+$ (Table 5 and Fig. 29). For $X = F^+$, the vicinal bond paths are so extremely inwardly curved that they coincide largely. The CFC interpath angle β is just 0.02° and the shift parameter d is −0.455 Å. The corresponding molecular graph (Fig. 29) resembles no longer a three-membered ring but a π-complex [38].

6.2 From Three-membered Rings to π-Complexes — Limits of the Concept of Strain

The molecular graphs given in Fig. 29 reveal that there is a continuous transition from three-membered rings to π-complexes depending on the electronegativity of the constituent atoms. This has been observed not only for C-containing but also for Si-containing three-membered rings [44].

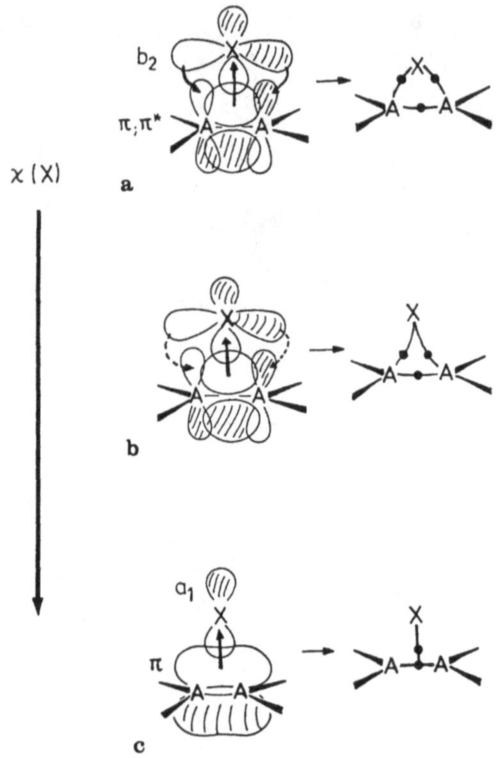

Fig. 30a–c. MO description of donor-acceptor interactions between the basal group A_2H_4 (A = C, Si, etc.) and the apex group X. The relevant orbitals (a_1 and b_2 symmetry) are shown on the left side and the corresponding molecular graphs on the right side of each diagram. The direction of charge transfer is indicated by *arrows* (*dashed arrows* indicate reduced charge transfer). **a** Three-membered ring with convex bent bonds. **b** Three-membered ring with concave bent bonds. **c** π-Complex. The electronegativity χ of X increases from a) to c). Reprinted with Permission from J. Mol. Structure (Theochem), 169 (1988) 531. Copyright (1988) Elsevier Science Publishers, Amsterdam

Dewar was the first to point out a relationship between three-membered rings and π-complexes [118]. Combining the MO model developed by Dewar and Ford [119] and an electron density model developed by Cremer and Kraka [38], the following picture emerges (Fig. 30). The curvature of the bond paths in three-membered rings can be rationalized by donor, acceptor interactions between the group (atom) X and the basal group A=A (with A = CH_2, SiH_2, etc.). The basal group donates electrons from a π MO into a suitable low lying a_1 MO (C_{2v} symmetry assumed) thus establishing a build up of electron density in the center of the ring. Back donation from a high lying b_2 MO into the π* MO of A=A leads to a peripheral accumulation of electron density, mainly responsible for the curvature of the bond paths (Fig. 30).

1. If donation and back-donation are comparable, a three-membered ring with convex bent bonds will result (Fig. 30a).
2. If back-donation is reduced due to the electronegativity χ of X (reduced donor capacity of X), a three-membered ring with concave bent bonds results (Fig. 30b). As a further consequence, the vicinal AX bonds are weakened.
3. In the extreme case, back-donation becomes negligible and the interaction between X and A_2 is characterized by the amount of negative charge transferred from A_2 to X. The T-structure of a π-complex is formed with X being "bound" to the bond critical point $\mathbf{r}_b$(A=A) (Fig. 30c).

The analysis of $\varrho(\mathbf{r})$ of a potential three-membered ring immediately reveals whether the structure is closer to a ring or a π-complex [38]. This is of importance in discussions of strain since it helps to avoid severe misinterpretations: It makes no sense to discuss the "strain" of a π-complex or a ring with large π-complex character (see, e.g., protonated oxirane [40]). For example, concave bending is indicative of interpath angles considerably lower than ideal bond angles. Strain energies derived from these angles, e.g. with the aid of Hooke's law are extremely large, larger than bond dissociation energies. However, these strain energies are derived under the assumption that fully developed bonds are bent, an assumption that is not fulfilled for concave bent bonds, that are partial bonds with a bond order $n < 1$. *The concept of strain does not apply to partial bonds or any nonclassical bond. Therefore, it is useless to discuss the strain of three-membered rings with concave bonds and partial π-complex character [40].*

6.3 Pitzer Strain and Dunitz-Schomaker Strain

Bond eclipsing leads to changes in geometry, energy, and other molecular properties. It is also reflected by changes in $\varrho(\mathbf{r})$, e.g., ϱ_b is increased in the eclipsed bonds contrary to expectations based on an electrostatic model of interacting bond electron pairs [39]. The increase in ϱ_b, however, is a secondary effect caused by a) a significant lengthening of the bond that connects the eclipsed bonds and b) a concomitant small shortening of the latter bonds. Because of this, it is difficult to unearth the actual effect of bond eclipsing on $\varrho(\mathbf{r})$ and to distinguish between the cause and the effect.

A way to show the impact of bond eclipsing on $\varrho(\mathbf{r})$ is to compare the charge distributions in hydrogenfluoride, HF, and methylfluoride, CH_3F [39]. In HF, the nonbonded electron pairs at F are arranged in form of a torus of negative charge. If H is replaced by CH_3, accumulation of electron density in the CH bonds will cause a change of the potential (V), kinetic (G), and total energy (H) densities in both the CF bonding region and the valence sphere of F. Off-nucleus positions that are staggered to the CH bonds possess lower $V(\mathbf{r})$ and, hence, $H(\mathbf{r})$ values than those that are eclipsed (Fig. 31a). Correspondingly, higher ϱ-values are found at points with lower $H(\mathbf{r})$ values. This is illustrated by the difference density plot of the lone pair region at the F atom of CH_3F (Fig. 31b). The preference for staggering of nonbonded charge distributions is obvious and holds also for the Laplace concentration (Fig. 31c). From this and other observations it has been concluded that *bonded and nonbonded charge concentrations try to avoid each other (principle of avoidance of charge concentrations [39])*. If the molecular geometry forces vicinal charge concentrations to be eclipsed, Pitzer strain will result.

Avoidance of charge concentrations is also apparent in the distribution $-\nabla^2\varrho(\mathbf{r})$ of cyclobutane [41]. Dunitz and Schomaker [120], pointed out as early as 1952 that repulsive 1,3-interactions between the C atoms in cyclobutane are responsible for a considerable part of its strain. Destabilizing 1,3 interactions lead to a depletion of electron density inside the cyclobutane ring. This can be visualized by plotting a contourline diagram of the Laplace concentration in the plane of the carbon

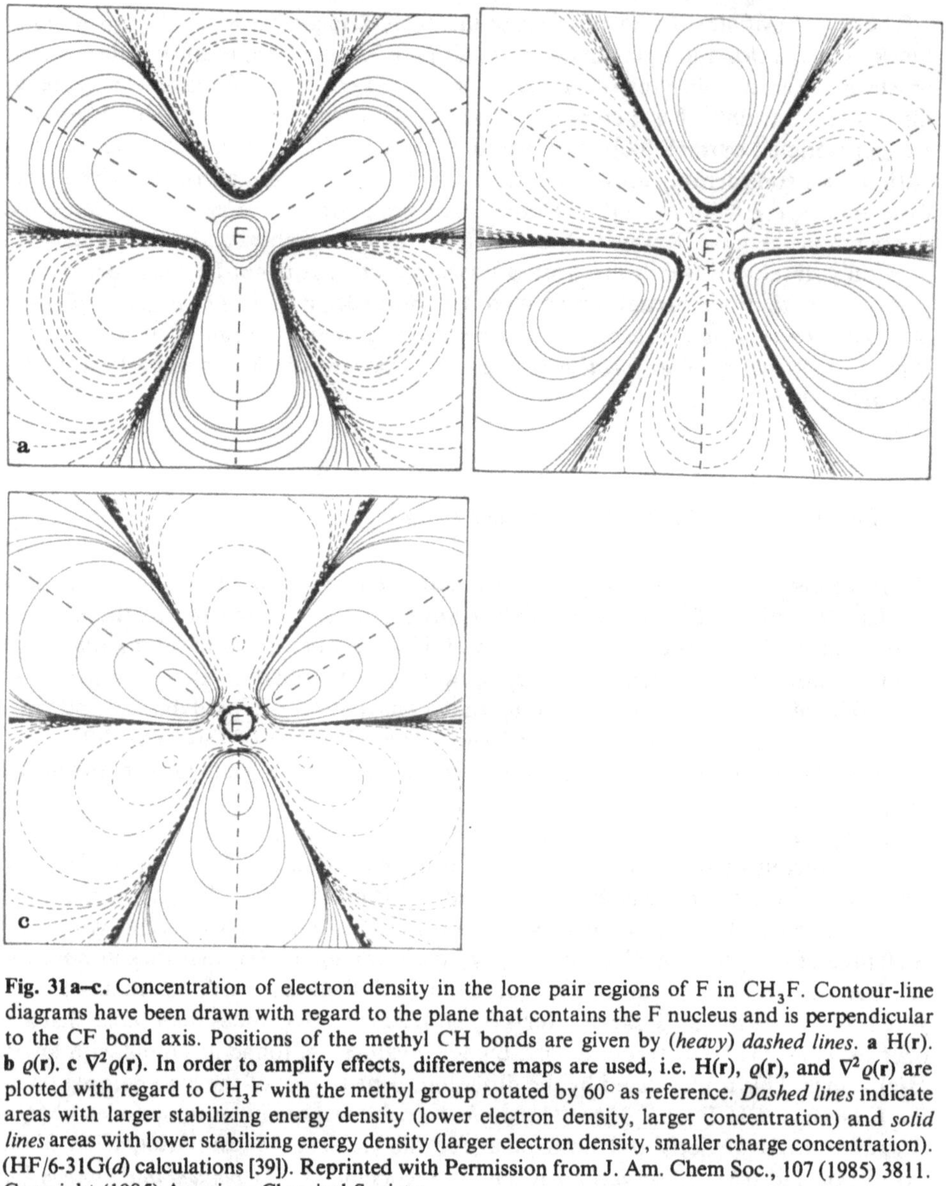

Fig. 31 a–c. Concentration of electron density in the lone pair regions of F in CH_3F. Contour-line diagrams have been drawn with regard to the plane that contains the F nucleus and is perpendicular to the CF bond axis. Positions of the methyl CH bonds are given by (*heavy*) *dashed lines*. **a** H(**r**). **b** $\varrho(\mathbf{r})$. **c** $\nabla^2\varrho(\mathbf{r})$. In order to amplify effects, difference maps are used, i.e. H(**r**), $\varrho(\mathbf{r})$, and $\nabla^2\varrho(\mathbf{r})$ are plotted with regard to CH_3F with the methyl group rotated by 60° as reference. *Dashed lines* indicate areas with larger stabilizing energy density (lower electron density, larger concentration) and *solid lines* areas with lower stabilizing energy density (larger electron density, smaller charge concentration). (HF/6-31G(*d*) calculations [39]). Reprinted with Permission from J. Am. Chem Soc., 107 (1985) 3811. Copyright (1985) American Chemical Society.

nuclei of planar cyclobutane (Fig. 32). Contrary to cyclopropane (compare with the contourline diagram in Fig. 29a) negative charge is depleted in the center of the ring. Repulsive 1,3 interactions between the C atoms seem to impede surface delocalization of electrons in cyclobutane.

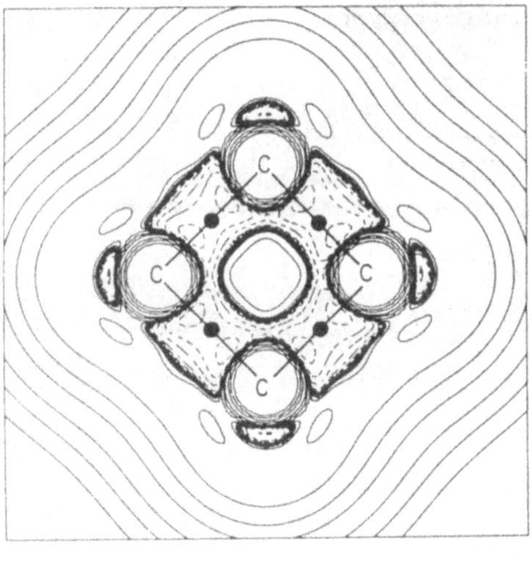

Fig. 32. Molecular graph and contour-line diagram of the Laplace concentration of cyclobutane. Bond paths are indicated by *heavy lines* and bond critical points by *dots*. *Dashed lines* are in regions where electronic charge is concentrated ($\nabla^2\varrho(\mathbf{r}) < 0$) and *solid lines* in regions where charge is depleted ($\nabla^2\varrho(\mathbf{r}) > 0$). Inner shell regions with $\nabla^2\varrho(\mathbf{r}) < 0$ are not shown. (HF/6-31G(*d*) calculations)

6.4 Energetic Consequences of Strain

A dissection of strain into the various contributions provides a basis for the calculation of the strain energy. For this purpose, one has to define a set of suitable reference compounds and to use their properties to define a strain-free molecular state. Cremer and Gauss [41] have carried out such an analysis for cyclopropane and cyclobutane. They proceeded in the following way.

1. Stretching strain energies were estimated from Hooke's law and known k(CC) and k(CH) force constants.
2. Baeyer strain energies were determined with the aid of Hooke's law *utilizing calculated interpath angles β rather than physically meaningless geometrical angles α.* In addition, a CCC bending force constant was derived that does not contain effects arising from 1,3 CC repulsion.
3. Pitzer strain energies were evaluated directly from ab initio calculations using ethane as an appropriate reference compound.
4. The Dunitz-Schomaker strain energy for cyclobutane was taken from appropriately scaled CNDO/2 energies obtained for model calculations with and without C,C-nonbonded repulsion [21].
5. Stabilization effects due to strengthening of the external bonds were directly evaluated. For this purpose $\varrho(\mathbf{r})$ was integrated over the interatomic (zero-flux) surface S(CH) and the resulting integral was related to the bond energy.

Strain energies obtained in this way are listed in Table 6. For cyclobutane, they add up to the conventional strain energy of 27 kcal mol^{-1} determined from thermochemical data [110]. In the case of cyclopropane, however, there is a discrepancy of 16.4 kcal mol^{-1} between the theoretical value (44.4 kcal mol^{-1}, Table 6) and the conventional strain energy (28 kcal mol^{-1} [110]). This energy difference has to be attributed to the stabilizing effects of surface delocalization of σ-electrons [38, 40–42]. Hence, *the energetic consequences of surface delocalization are responsible for the*

Table 6. Ab initio strain energies and stabilization energies of cyclopropane and cyclobutane[a]

Strain	Strain Energy [kcal mol^{-1}]	
	Cyclopropane	Cyclobutane
	Destabilization Energies	
Stretching	0.5	1.0
Baeyer[b]	46.3	13.0
Pitzer	4.0	3.9
Dunitz-Schomaker	0	12.0
Total	50.8	29.9
	Stabilization Energies	
CH Strengthening	6.4	2.8
σ-Delocalization	x = 16.4	0
Conventional strain energy	28.0 = 44.4 + x	27.1

[a] From Ref. [41].
[b] Baeyer strain energy of cyclopropane calculated with Hooke's law (41.3 kcal mol^{-1} [41]) plus energy increase from anharmonicity effects calculated form a bending function with and without a cubic term for an interpath angle $\beta = 79°$ (5 kcal mol^{-1}). Note that the strain energy of propane has been set erronously to 5.1 kcal mol^{-1} in Ref. 41 (Table IX). This energy, however, is compensated by the increase in the CC bond energy relative to that of ethane (Table V, Ref. 41).

puzzling similarity of the strain energies of cyclopropane and cyclobutane. In addition, they explain the existence of homoaromatic compounds as well as the relatively large stability of bicyclic molecules such as norcaradiene or bisnorcaradiene.

7 σ-Conjugation — Modes of σ-Electron Delocalization

7.1 σ-Conjugation

Over the last three decades various authors, in particular Dewar, have repeatedly pointed out the possibility and chemical consequences of σ-conjugative interactions [122, 123]. The concept of σ-conjugation is based on the fact that resonance integrals between different AOs of a given atom will not vanish even if the AOs are orthogonal. This applies also to the resonance integrals between geminal sp^n hybrid orbitals of a C atom, which are considerably larger than that between adjacent 2pπ AOs in the case of conjugated double bonds. Hence interactions between σ-bonds should be at least as strong as interactions between π-bonds.

Utilizing the concept of σ-conjugation Dewar was able to rationalize a number of apparent anomalies in organic chemistry that are difficult to explain in other ways [122]. These examples included, among others, the astonishing stability of three-membered rings, the pyramidal conformation of radicals and biradicals, bond staggering in saturated molecules, the gauche and anomeric effects, and the geometry of triplet carbene.

There always is σ-conjugation in a molecule with three or more atoms. Conjugative effects between adjacent σ-bonds are approximately constant in most cases and, therefore, they can be absorbed in the empirical values of appropriate reference compounds used within the concept of localized bonds. σ-Conjugation should influence the distribution $\varrho(\mathbf{r})$ in a molecule. For example, it has been found for many unstrained molecules in their equilibrium geometry that the bond paths do not coincide with the internuclear connection lines. They are slightly curved toward each other ($\beta < \alpha$, $|\beta - \alpha| < 1°$) despite the fact that coulombic repulsion between the electron bond pairs should cause outwardly curved bent bonds. Bending of normal σ-bonds may be indicative of σ-conjugation. Also the Laplace concentration in the valence shell of an atom that is bonded to at least two neighboring atoms should reflect the degree of σ-conjugation. Probably, valence shell holes between concentration lumps in the direction of the bond paths and the geometrical arrangement of concentration lumps and holes depend on the extent of σ-conjugation.

From a practical point of view, however, a description of σ-conjugative effects in terms of the properties of $\varrho(\mathbf{r})$ is an enterprise without any basis. Since all molecules with three or more atoms always should benefit from σ-conjugation, it is impossible to assess the degree of σ-conjugation by comparison with a suitable reference molecule that does not possess σ-conjugation. A description of σ-conjugation will only be possible in those cases where σ-conjugation causes changes in the molecular properties that are no longer expected on the basis of normal σ-conjugative effects. The latter are absorbed in descriptions based on the localized bond model. If the localized bond model fails to describe the properties of a molecule, it will be reasonable to expect exceptional σ-conjugative effects. Such a case is encountered for three-membered rings [42]. But before discussing these effects it will be useful to describe orbitals and modes of σ-electron delocalization in cyclic systems.

7.2 Ribbon, Surface, and Volume Delocalization of σ-Electrons

While for acyclic molecules it is not difficult to distinguish between σ and π-MOs, this may become problematic for cyclic compounds. Therefore, definitions have been introduced [41], which help to classify orbitals in cyclic systems and which allow a distinction between various modes of electron delocalization. In the following these rules are summarized for cycloalkanes, but a generalization to other N-membered rings is straightforward.

1. If a cycloalkane is considered to be made up of CH_2 entities each possessing two singly occupied orbitals, two sets of orbitals can be distinguished: a) the r-set that consists of radially (toward the ring center) oriented sp^2 hybrid orbitals and b) the *t*-set that consists of tangentially (with regard to the ring perimeter) oriented *p*-orbitals.

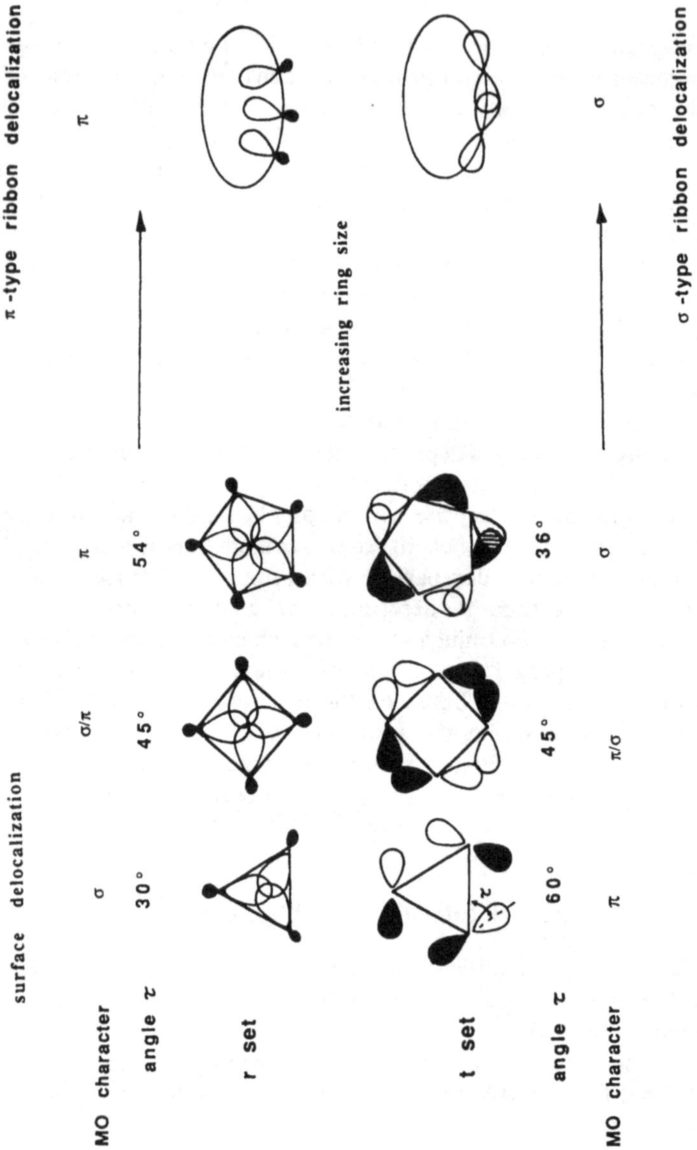

Fig. 33. Characterization of ring orbitals (σ or π) with the aid of the angle τ

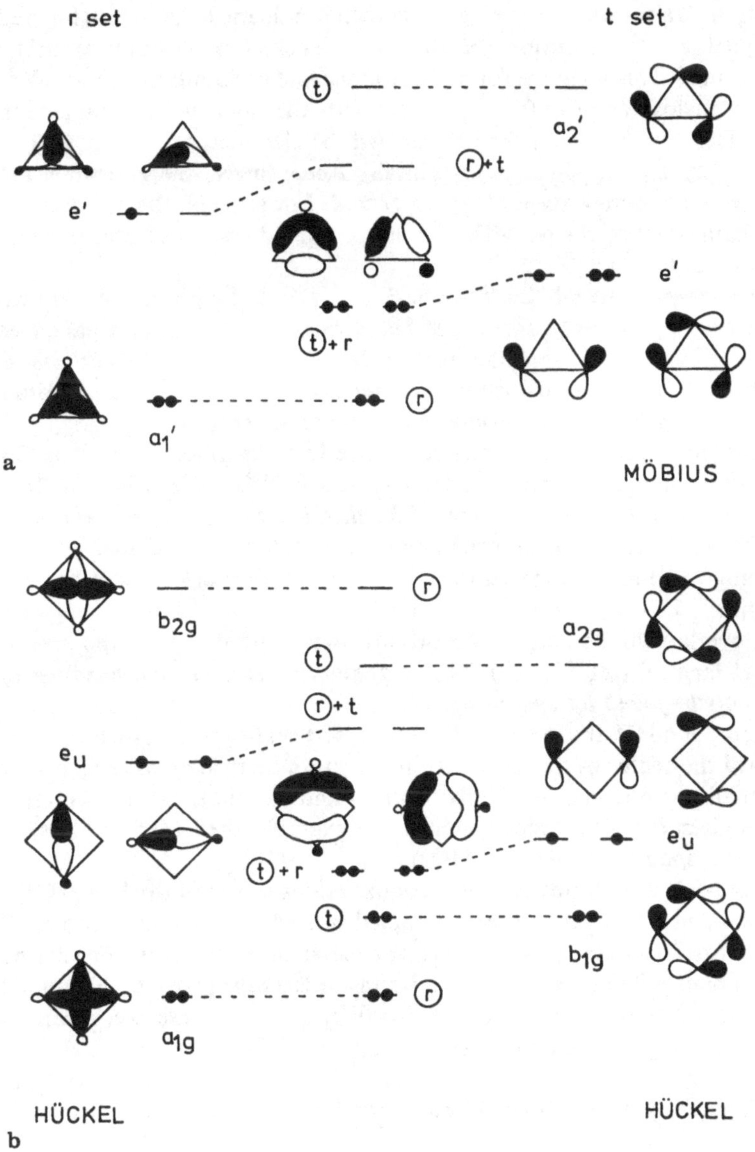

Fig. 34a, b. *r* and *t*-MOs of **a** cyclopropane and **b** cyclobutane. The predominant nature of the final MOs is indicated by *circles*. Reprinted with Permission from J. Am. Chem Soc., 108 (1986) 7467. Copyright (1986) American Chemical Society

The two orbital sets are shown in Fig. 33 for the three-, the four-, and the five-membered ring. The *r*-orbitals always lead to a Hückel system while the *t*-orbitals form a Möbius system for odd N but a Hückel system for even N. The actual framework MOs of the ring are formed by linear combinations of *r*- and *t*-orbitals, respectively. The ring MOs with dominant *r*- or *t*-character (for even N) are occupied by $4q + 2$ electrons while ring MOs with dominant *t*-character (for odd N) are occupied by $4q$ electrons. In this way Hückel- or Möbius-aromatic subshells are formed. This is shown for cyclopropane and cyclobutane in Fig. 34.

2. If *r*-(*t*-)orbitals enclose angles $90° \geqq \tau > 45°$ with the internuclear connection lines of a ring (see Fig. 33), then the orbitals will be classified as π-orbitals. For $45° > \tau \geqq 0°$, they are classified as σ-orbitals. Accordingly, the *r*-orbitals are σ-MOs for $N = 3$ while they are π-MOs for $N > 4$. For large N, the rπ-MOs are topologically equivalent to the pπ-MOs of a π-conjugated system such as a cyclopolyene (see Fig. 33).
   The *t* orbitals correspond to π-MOs for $N = 3$. This is in line with the well-known π-character of the ring bonds of three-membered rings such as cyclopropane (see Section 5.1). For $N > 4$ the *t*-orbitals form σ-MOs (Fig. 33). These definitions collapse in the case of the four-membered ring because both *r*- and *t*-orbitals enclose angles of 45° with the interatomic connection lines (Fig. 33).
3. Electrons occupying *r*-type orbitals can delocalize in the surface of the ring for small N or along the chain (ribbon) of ring atoms if N is large (Fig. 33). It is reasonable to distinguish between *surface delocalization and ribbon delocalization* of electrons [38, 42]. Electrons occupying *t*-type orbitals for $N = 3$ delocalize on a circle enveloping the three-membered ring. For large N, they delocalize along the ribbon of atoms.
   If N is small, overlap will be better for *r*-orbitals than *t*-orbitals while the reverse will be true for large N (see Fig. 33). This suggests that *surface delocalization of electrons can only be found for small rings.*
4. For a cage compound of the type $(CH)_N$ ($N = 4$: tetrahedrane), *r*-orbitals are oriented toward the center of the cage. At each C atom there are two tangentially oriented, mutually orthogonal *p*-orbitals. Both *r*- and *t*-orbitals can be classified using the rules given for monocyclic systems. For example, the *r*-MOs (*t*-MOs) of tetrahedrane correspond to σ-MOs (π-MOs) [42].
   The actual cage MOs are formed as linear combinations of *r*- and *t*-orbitals. Cage MOs with dominant *r*-(*t*-)character are occupied in such a way that "aromatic" subshells are formed similar as in the case of cycloalkanes. Electrons occupying MOs with dominant *r*-character can delocalize inside the cage provided N is small and there is strong overlap between the *r*-orbitals. In this case we speak of *volume delocalization of electrons* [41, 42].

One has to distinguish between three different modes of electron delocalization, namely

i) ribbon delocalization of either σ- or π-electrons,
ii) surface delocalization of σ-electrons or
iii) volume delocalization of σ-electrons.

The three modes of σ-electron delocalization are shown in Fig. 35. The extent of surface delocalization depends on the overlap of the r-orbitals, i.e. it depends on the topology and the geometry of the ring. Inspection of the overlap in cycloalkanes reveals that significant effects resulting from surface delocalization can only be expected for three-membered rings [40, 41]. *Due to their topology, three-mem-*

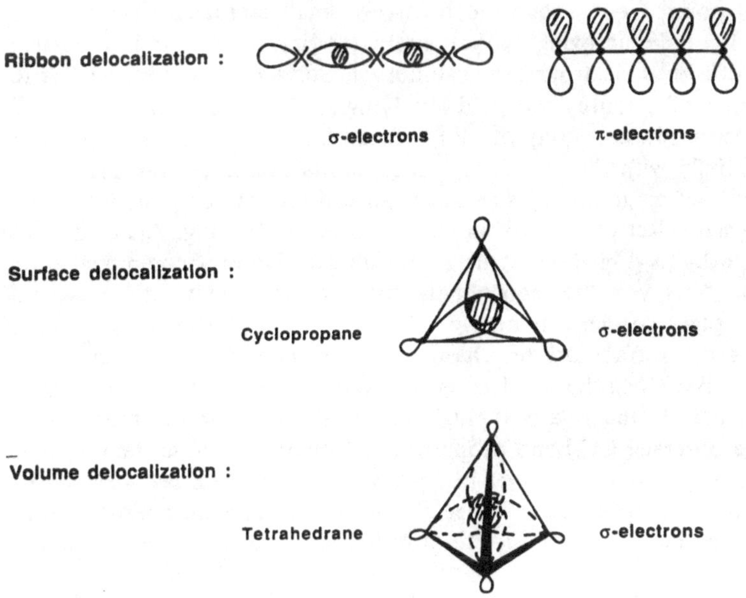

Fig. 35. Possible modes of electron delocalization

bered rings are exceptional since they possess σ-electrons delocalized over the ring surface and π-electrons delocalized in peripheral bent bonds.

Similar considerations apply to volume delocalization. If a cage consisting of three-membered rings is formed (e.g. tetrahedrane), *volume delocalization* of σ-electrons will be possible [41, 124].

7.3 Surface Delocalization and Bonding in Cyclopropane

There have been a number of clues indicating that the properties of cyclopropane are exceptional. Dewar [122] was the first to list these clues and to discuss them in view of σ-electron delocalization, which may cause or, at least, may influence the properties of cyclopropane. Later Cremer and co-workers [38–42] reinvestigated cyclopropane utilizing ab initio methods and the topological analysis of $\varrho(\mathbf{r})$. According to these investigations, cyclopropane differs from other cycloalkanes by

1. its relatively low strain energy (SE) that is almost identical to that of cyclobutane (28 and 27 kcal mol^{-1}, respectively),
2. its relatively high electron density in the center of the ring,
3. its relatively short CC distances,
4. its relatively high CC bond strength,
5. the upfield shifts of its proton and ^{13}C NMR signals,
6. its electronic interactions with substituents,
7. its ability to enhance conjugation in homoaromatic systems.

As was already pointed out above the relatively small strain energy of cyclopropane is a result of destabilizing Baeyer strain (46 kcal mol^{-1}) and stabilizing surface delocalization of σ-electrons (16 kcal mol^{-1}). Surface delocalization has to do with the existence of a doubly occupied low-lying *surface orbital* (a_{1g} in Fig. 34) that stems from the in-phase overlap of all *r*-orbitals inside the ring. Since overlap decreases exponentially with the size of the ring, the nature of the surface orbital changes dramatically when going to larger rings. Already for cyclobutane it does no longer bring sufficient electron density into the center of the ring. Also, the two 1,3 antibonding e_u orbitals (Fig. 34) that are responsible for Dunitz-Schomaker strain reduce the electron density in the center of the ring, thereby yielding the observed depletion of the Laplace concentration (Fig. 32).

For larger rings the surface orbital changes to a ribbon orbital which allows σ-electrons to delocalize along the ring framework. Occupation of the surface orbital in cyclopropane leads to the relatively high electron density inside the ring, is responsible for the observed CC bond properties, and may also cause the magnetic properties of cyclopropane. As discussed in Sect. 5, the interaction between substituents and cyclopropane as well as the special role of the three-membered ring in homoaromatic compounds can be understood in terms of surface delocalization of σ-electrons.

Surface delocalization of σ-electrons in cyclopropane is nicely reflected by the Laplace concentration of the electrons in the plane of the cyclopropane ring. In Fig. 36 a perspective drawing of the theoretically determined Laplace concentration is shown [38, 41]. The corresponding contourline diagram is given in Fig. 29a. Electrons are concentrated (dashed contourlines in Fig. 29a) along the CC bond

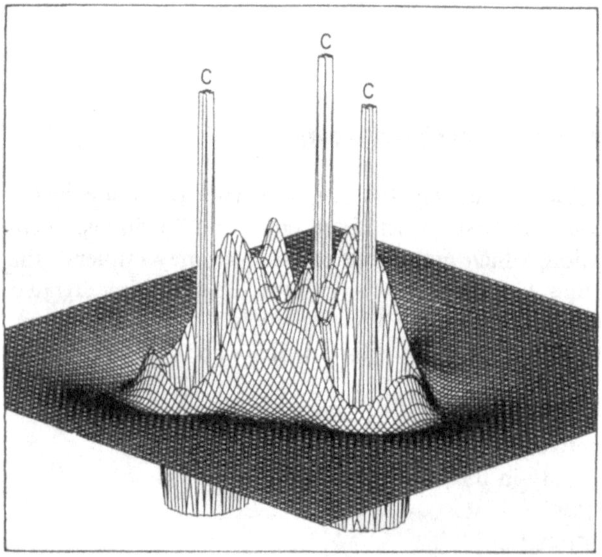

Fig. 36. Perspective drawing of the Laplace concentration $-\nabla^2\varrho(\mathbf{r})$ of cyclopropane shown in the ring plane. Inner shell concentrations are indicated by the atomic symbol C. Laplace values above and below a threshold are cut off. (HF/6-31G(*d*) calculations)

paths (heavy solid lines in Fig. 29a) as well as inside the ring. For other cycloalkanes such as cyclobutane there is depletion rather than concentration of negative charge inside the ring (Fig. 32).

Various other authors have made similar observations when investigating the electron density distribution of cyclopropane. Coulson and Mofitt [125] were the first to note that there is a plateau of relatively high negative charge inside the C_3 ring. In a more recent investigation, Schwarz and co-workers [73] found that the total electron density is increased by 0.16 e $Å^{-3}$ in the center of cyclopropane as compared to the electron density of the promolecule formed by three spherical isolated carbon atoms. Delocalization of σ-electrons occupying the surface orbital leads to a strong reduction of their kinetic energy which in turn triggers enhanced AO contraction at the C atoms. This restores the virial relation, lowers the total energy, and leads to CC bond shortening [73]. Hence, σ-electron delocalization directly influences the length of the ring bonds.

Ahlrichs and Ehrhardt [126] have calculated shared electron numbers for alkenes. While bonding is reflected in these compounds by two-center contributions and negligible contributions from three- and four-center terms, a CCC shared electron number of 0.3 is calculated for cyclopropane, that is indicative of three-center bonding.

Putting all facts together, the following conclusions can be drawn. Bonding in cyclopropane is exceptional since the C atoms are linked by (see Fig. 37)

1. *a central 2-electron 3-center bond ("super-σ bond") and*
2. *two peripheral 4-electron 3-center bonds ("π-bonds").*

Fig. 37. Nonclassical 2-electron 3-center bond and 4-electron 3-center bonds in cyclopropane

The classical structure of cyclopropane is misleading. It does not reflect the high degree of σ-electron delocalization which influences stability, geometry, magnetic properties, etc. of cycloproane.

7.4 Pros and Cons of σ-Aromaticity

Dewar has drawn an analogy between the =HC—CH= groups of a conjugated (cyclo)polyene and the —CH_2— groups of a cycloalkane [122]. He has argued that cyclopropane and benzene are isoconjugate; the latter possesses a system of six delocalized π-electrons, the former a system of six delocalized σ-electrons. Therefore, cyclopropane should be considered to be σ-aromatic and its properties should be seen to result or at least to be influenced by σ-aromaticity.

There is no question that the peculiar properties of cyclopropane in particular and three-membered rings in general are coupled with the unusual bond features encountered in these molecules. The question is only whether such a situation should be described with a term (aromaticity) that is heavily loaded with what has been found for cyclopolyenes with $4q + 2\pi$-electrons. Aromaticity is expected for those systems that can only be represented in form of two or more resonance structures [90–92]. For example, the electronic structure of benzene has to be described as a resonance hybrid of two classical cyclohexatriene structures. It is impossible to anticipate π-delocalization and π-aromaticity of benzene by just looking at one cyclohexatriene structure. It seems that there is a considerable difference between benzene and cyclopropane in so far as the latter molecule is normally represented by just one classical structure. However, does this structure provide a realistic image of bonding and electron delocalization in cyclopropane?

We have already discussed in Sect. 6 that there is a close electronic relationship between three-membered rings and π-complexes. According to this relationship, cyclopropane can be written as a resonance hybrid of three equivalent methylene, ethylene π-complexes as shown in Fig. 38 [118]. Of course, a methylene, ethylene π-complex does not exist but the same is also true for cyclohexatriene. One might argue that resonance structures used to describe π-aromatic compounds are all "classical" in the sense that they possess just single and double bonds. However, this argument can easily be refuted. Cremer and Kraka have demonstrated that the bond

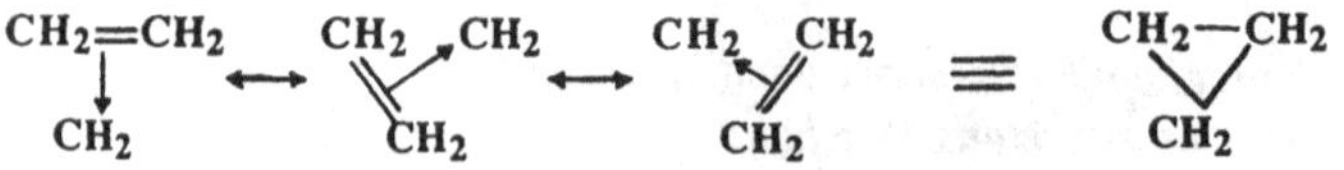

Fig. 38. "Resonance structures" of cyclopropane and shorthand notation

paths, i.e. the paths of maximum electron density between bonded atoms, nicely describe the bonds of three-membered rings and π-complexes [38]. For π-complexes a bond path is found where chemists draw the arrow between basal group, e.g. ethylene, and the apical group, e.g. a halogen cation. Hence, the π-complex notation is as realistic as any other representation of molecular structure and, accordingly, it can be placed at the same level with the "classical" structures.

Actually, the π-complex description of cyclopropane is better suited to reflect its properties than the notation normally used in chemistry. It indicates

1. that there are different types of CC bonds in cyclopropane,
2. the π-character of the CC bonds,
3. the possibility of resonance stabilization,
4. a possible shortening of the CC bonds,
5. increased s-character in the CH bonds and a corresponding CH bond strengthening, etc.

In conclusion, one can consider cylopropane in the same way as a resonance hybrid as this is done for benzene. In this respect, the classical structure of cyclopropane is the shorthand notation for the three equivalent resonance structures shown in Fig. 38.

The discussion clearly shows that a rejection of the term σ-aromaticity on the grounds that it is inconsistent with common understanding of the term aromaticity is not justified. Even if the usual criteria for π-aromaticity [90–92], namely (see Sect. 4.2)

1. $4q + 2$ π-electrons (Hückel rule),
2. the Dewar resonance energy,
3. partial or complete bond equalization, and
4. diatropicity of the NMR signals,

are applied, the term σ-aromaticity cannot be discarded.

1. There are six delocalized σ-electrons that form the ring bonds of cyclopropane. Two of them establish a Hückel-aromatic, the remaining four a Möbius-aromatic system (Fig. 34). In whatever way the electron count is carried out, an aromatic ensemble of electrons is obtained.
2. Calculation of the Dewar resonance energy implies a comparison with a suitable acyclic reference compound that contains the same number of CC σ-bonds. In addition, it requires the separation of energetic effects resulting from ring strain on the one hand and σ-aromaticity on the other hand. This has been done for cyclopropane and a delocalization energy of 16 kcal mol^{-1} has been obtained [41].
3. The CC bonds in cyclopropane are all equivalent. In addition, they are considerably shorter (0.03–0.05 Å) than those of other (cyclo)alkanes.
4. Due to the high s character of the CH bonds, one would expect the ^{1}H-NMR signal for cyclopropane to appear downfield from the signals of the CH_2 protons of alkanes. However, it appears upfield by 1 ppm ($\delta = 0.22$ ppm) [127]. Also, the ^{13}C-NMR signal is shifted by 20 ppm upfield from other aliphatic ^{13}C-NMR signals [128]. It was found that the isotropic shift of —3.8 ppm from Me_4Si is the consequence of that component of the chemical shift tensor for CH_2 that is perpendicular to the ring plane. The value of this component (—36 ppm from Me_4Si) is indicative of strong circulation of electrons in the ring plane [128]. Both the ^{1}H- and the ^{13}C-NMR shifts of cyclopropane suggest σ-electron delocalization in the ring.

Looking at points 1 through 4 in one context, it seems to be appropriate to consider cyclopropane as being σ-aromatic. Accepting σ-aromaticity all properties of cyclopropane are easily rationalized while the rejection of this concept entails the unpleasant task of looking for additional explanations. Certainly, this is reason to lean on an acceptance of the notion of σ-aromaticity. Also, the concept of σ-aromaticity helps to rationalize the properties of substituted cyclopropanes [39], those of the Si, Ge, etc. analogues of cyclopropane [41, 44], and of homoaromatic compounds [34, 38].

8 Chemical Reactivity

To make predictions on the chemical reactivity of a molecule knowledge of its labile bonds and the sites prone to electrophilic or nucleophilic attack is needed. This information can be extracted from the density distribution $\varrho(\mathbf{r})$ of the molecule.

8.1 Chemical Reactivity as Reflected by Properties of $\varrho(\mathbf{r})$

A labile bond is mostly indicated by a low bond order. Also, the shape of the bond path reveals whether a bond can easily be broken or not. A suitable example to demonstrate this is the differing behavior of oxiranes in basic, neutral or acidic media. The addition of nucleophiles to oxiranes to yield 1,2-disubstituted products, which is widely used in synthesis [129], is known to be considerably accelerated in acidic solutions. For an unsymmetrically substituted oxirane the nucleophilic attack

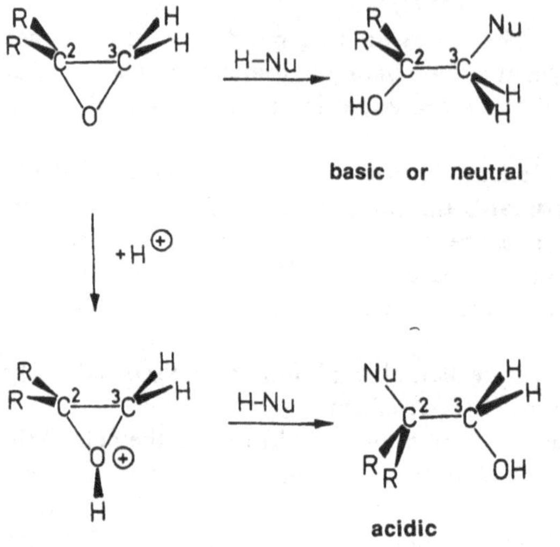

Fig. 39. Nucleophilic ring opening of an epoxide in basic, neutral or acidic solvents

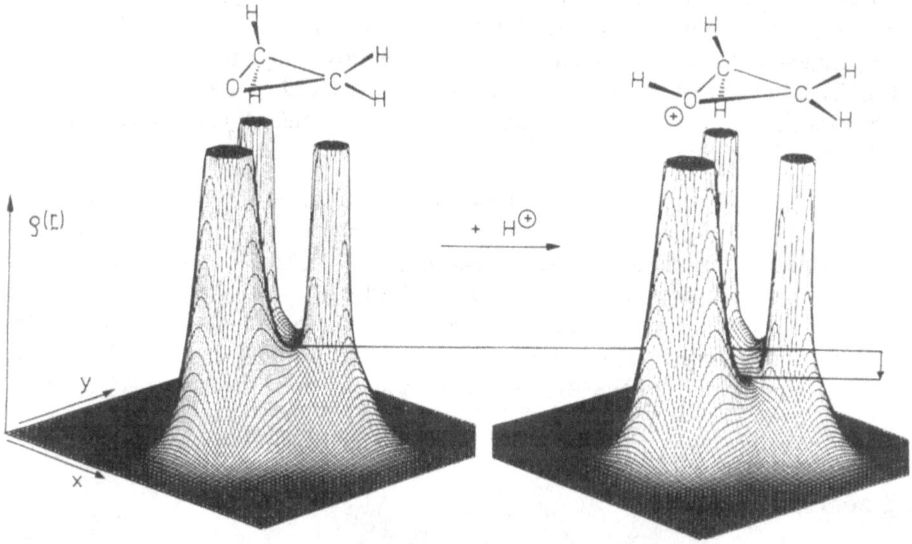

Fig. 40. Perspective drawing of the electron density distribution $\varrho(\mathbf{r})$ of oxirane and protonated oxirane shown with respect to the ring plane. The value of $\varrho(\mathbf{r})$ is cut off above a predetermined value to improve the representation. The weakening of the CO bonds upon protonation is indicated. (HF/6-31G(d) calculations)

occurs predominantly at the sterically less hindered C atom in neutral or basic media while in acid solution surprisingly the higher substituted C atom of oxirane is attacked (compare with Fig. 39). Since ring opening is always accompanied by a Walden inversion, a S_N1 mechanism can be ruled out in acidic media [129].

In Fig. 40, the effect of protonation (or complexation with a Lewis acid) at the O atom of oxirane is shown with the aid of perspective drawings of $\varrho(\mathbf{r})$ obtained for the plane of the three heavy atoms. There is a dramatic decrease of the electron density in the CO bond regions upon protonation. The corresponding bond orders decrease from 1.0 to 0.77 and the bond paths bend inside the ring, indicating that protonated oxirane attains partial π-complex-character (compare with Fig. 29c) [38]. All these observations suggest an acceleration of CO bond rupture in acidic media in line with experimental observations.

The preference of an attack on the sterically more hindered C atom of protonated oxirane seems to depend on three effects [38].

1. By increasing the π-complex character of the three-membered ring the alkyl groups bend backwards toward the CC axis to adopt the position they take in normal π-complexes. In this way, sterical shielding of C2 is reduced.
2. According to calculated bond orders for substituted protonated oxiranes [38] the bond C2O is more facile to bond rupture than the bond C3O.
3. The concentration holes in the valence sphere of C2 (see Fig. 41) are more prone to nucleophilic attack than those at C3.

Hence, the regioselectivity of the reaction can be elucidated from the electron density distribution $\varrho(\mathbf{r})$ of oxirane and protonated oxirane without the need to speculate on the charge distribution in the transition state of the reaction.

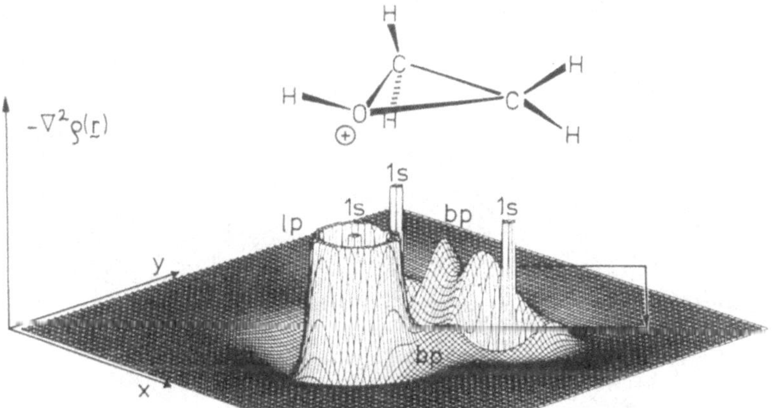

Fig. 41. Perspective drawing of the Laplace concentration $-\nabla^2\varrho(\mathbf{r})$ of protonated oxirane in the plane of the ring. Inner shell (1*s*), bond (bp), and lone pair (lp) concentrations and the difference of the CC and CO bond concentrations are indicated. Laplace values above and below a threshold are cut off. (HF/6-31G(*d*) calculations)

8.2 The Structure Diagram

Reactions as the one discussed in Sect. 8.1 can be seen in a broader context if the structure diagram of the compound in question is analyzed (see Sect. 2.5). The structure diagram of a given system embeds all information concerning possible structures and their possible changes. To discuss structural changes of a molecule such as oxirane a control space **W** has to be defined, which is of the same dimension as the configuration space **C** of the corresponding system (**W** ⊂ **C**). For example, if just changes of the heavy atom bonds in oxirane are considered, **W** will be a three-dimensional control space spanned by the control parameters u, v, w, and applies to any system of the general form A_2X.

In Fig. 42a the structure diagram of A_2X is shown in the plane spanned by the parameters u, v located at a constant value of w. Control parameters u and v are chosen to describe the two independent coordinates of X. For each pair of (u, v) the third degree of freedom of A_2X is considered to be optimized. In this way, v

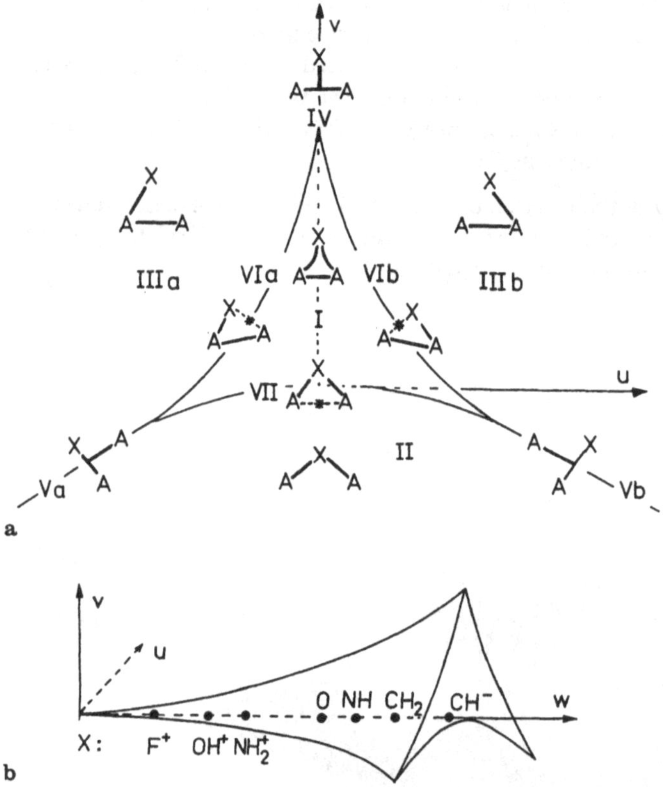

Fig. 42. Structure diagram of system A_2X shown in (u, v, w) space. **a** Control parameter w = constant. For each of the structural regqons I–VII a molecular graph is given. *Asterisks* denote catastrophe points. **b** Three-dimensional display of the bifurcation set of the elliptic umbilic in (u. v, w) space for **w** ≧ 0. The meaning of **w** as chemical control parameter is illustrated by giving the approximate positions of three-membered rings C_2H_4X on the positive **w** axis. Reprinted with Permission from J. Am. Chem Soc., 107 (1985) 3800. Copyright (1985) American Chemical Society

Table 7. Description of structural regions of the A_2X structure diagram

Structural region	I	II	IIIa, IIIb	IV	Va, Vb	VIa, VIb	VII
Number of critical points of $\varrho(\mathbf{r})$[a]							
(3, −3) [atoms]	3	3	3	3	3	3	3
(3, −1) [bonds]	3	2	2	2	2	2	2
(3, +1) [rings]	1	0	0	0	0	0	0
(2,0)[b] [dissolved bonds]	0	0	0	0	0	1	1
Form of molecular graph	3MR[c]	C_{2v} chain	C_1 chain	C_{2v} T-form	C_s T-form	change from C_1 chain	3MR[c] to C_{2v} chain
Structure of A_2X	X A–A	A–X–A	A–A–X	(A_2)–X	(AX)–A		
Catastrophe	none	none	none	conflict	conflict	bifurcation	bifurcation

[a] Associated chemical unit is given in brackets.
[b] Singularity of $\varrho(\mathbf{r})$.
[c] Three-membered ring.

represents the reaction path of a C_{2v} approach of X toward A_2 (decreasing v, a_1 symmetrical displacement of X) while u describes b_2 symmetrical displacements of X from this reaction path.

The topologically stable molecular graphs of A_2X belong to one of the structural regions I, II or III. The boundaries of these stable regions are formed by a set of all catastrophe points corresponding mathematically to the unfolding [130] of the elliptic umbilic catastrophe. Regions IV, and V a, b denote the positions of all conflict catastrophes, regions VI a, b and VII the position of all bifurcation catastrophes (compare with the isomerization mechanism described in Sect. 2.5). In Table 7a complete description of the structural region of the A_2X system, including molecular graphs (I–III) as well as catastrophe graphs (V–VII) is summarized. Transitions from I to II or III correspond to ring opening processes, and approach of X toward A_2 along v results in the formation of a three-membered ring via a π-complex.

Similar structure diagrams are obtained for oxirane and protonated oxirane. For the lather, the area of the hypocycloid-shaped region I is smaller which is reflected by a decrease in the control parameter w. Hence, by the value of w the degree of π-complex character and an assessment of the stability of the three-membered rings can be made. Cremer and Kraka [38] have used w as a chemical control parameter reflecting the electronic nature of both A and X. The larger the electronegativity difference between A and X (for A, X not equal H), the smaller becomes the parameter w. On this basis, various combinations of A (carbon) and X have been placed at positive w values of the control space as is illustrated in Fig. 42b. The most labile three-membered rings with high π-complex character adopt small w-values (e.g. $X = F^+$), a stable three-membered ring such as cyclopropane ($X = CH_2$) a large w value. All catastrophe points leading to acyclic structures form the elliptic umbilic [78].

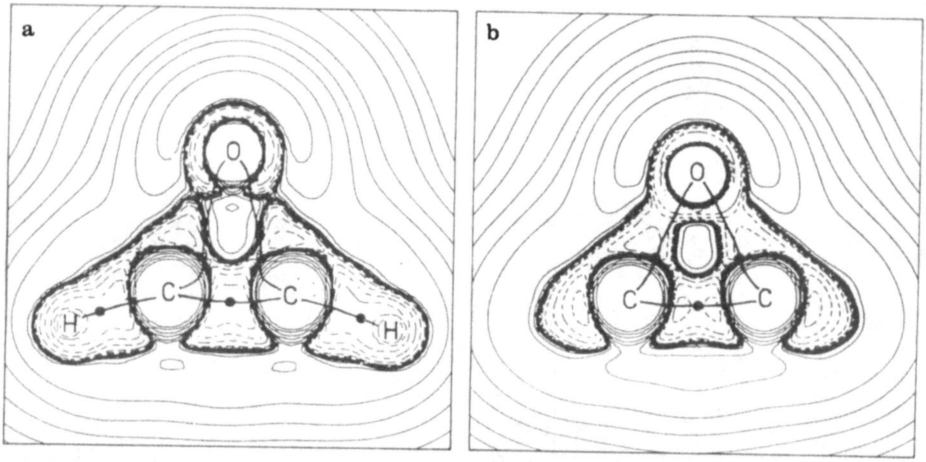

Fig. 43a, b. Molecular graphs and contour-line diagrams of the Laplace concentration $-\nabla^2\varrho(\mathbf{r})$ of **a** oxirene and **b** 1,3-dioxabicyclobutane. For oxirene, the reference plane is identical with the molecular plane. For 1,3-dioxabicyclobutane, the reference plane contains O1, C2, and C4. *Heavy solid lines* denote bond paths, *dashed (solid) contour-lines* are in regions with charge concentration (depletion). (HF/6-31G(d) calculations [43])

With a structure diagram such as that shown in Fig. 42b structural changes of a whole class of molecules can be discussed and a relationship between equilibrium properties and reaction behavior can be established. For example, analysis of $\varrho(\mathbf{r})$ of 1,3-dioxabicyclobutane and oxirene have revealed [43] that these systems possess lower w values and, hence, larger π-complex character than oxirane. This is clearly reflected by the molecular graphs of the two molecules shown in Fig. 43 (compare with Fig. 29b). The structure of oxirene is close to a π-complex between acetylene and oxygen. This can be understood in terms of the donor-acceptor model of three-membered rings discussed in Sect. 6.2. Acetylene is a weaker π-acceptor than ethylene because its π^* MOs are higher in energy than the π^* MO of ethylene. Accordingly, there is a dominance of π-donation from acetylene to a a_1-symmetrical p-orbital at O). As a consequence the CO bonds are labile and the stability of oxirene is low [43].

In the case of 1,3-dioxabicyclobutane, there are donor, acceptor interactions between acetylene and *two* O atoms. Acetylene donates electrons to both O atoms and, therefore, looses more charge than in oxirene. To compensate this loss, π-backdonation is stronger and the CO bonds are more developed (less concave). Hence, 1,3-dioxabicyclobutane takes an intermediate position between oxirane and oxirene. Its π-complex character is weaker and its stability higher than that of oxirene. Similar considerations apply to other bicyclic or unsaturated three-membered rings [43].

While molecular graph, structure diagram, and the value of w in that order represent an increasing wealth of information on the reactivity of the corresponding molecule, it is not clear at the moment whether this information can be used as a basis for predicting reaction dynamics. Future investigations have to show whether the catastrophe points of the structure diagram bear a direct chemical relevance, e.g. by giving the locations of the transition states that have to be passed when going from one stable nuclear configuration to another.

8.3 Chemical Reactivity as Reflected by the Laplace Field of $\varrho(\mathbf{r})$

The determination of a structure diagram is a very costly way to assess the reactivity of a molecule because it requires many calculations before the dissection of **W** into the various structural regions emerges. A simpler way to identify the reactive sites in a molecule is an approach that has amply been applied within MO theory. In Sect. 3 it has already been pointed out that the Laplace concentration in the valence shell of an atom can be related to the form of the frontier orbitals or, more general, to those MOs that are important to predict the course of the reaction. To demonstrate this we will discuss below some simple examples [27].

Figure 44 presents a perspective drawing of $\varrho(\mathbf{r})$ of HCN. Although it is easy to see that N possesses more electrons than C, detailed predictions with regard to the sites most prone to electrophilic or nucleophilic attack are only possible by a quantitative analysis of $\varrho(\mathbf{r})$ including integrations over the three atomic basins Ω. Since this is very time consuming, the analysis of $\varrho(\mathbf{r})$ should be replaced by an inspection of Laplace field.

Figure 45 depicts the Laplace concentration $-\nabla^2\varrho(\mathbf{r})$ of HCN. The concentration

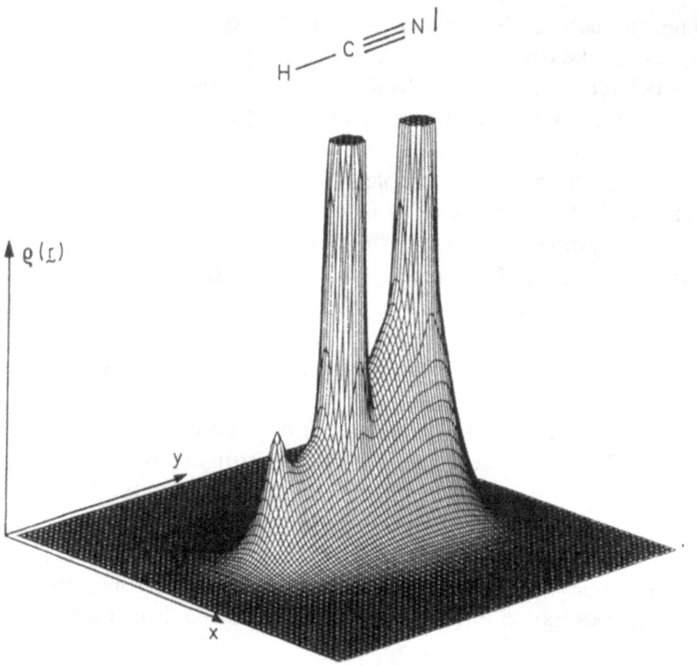

Fig. 44. Perspective drawing of the electron density distribution $\varrho(\mathbf{r})$ of HCN. The reference plane contains the three nuclei. The value of $\varrho(\mathbf{r})$ is cut off above a predetermined value to improve the representation. (HF/6-31G(*d*) calculations)

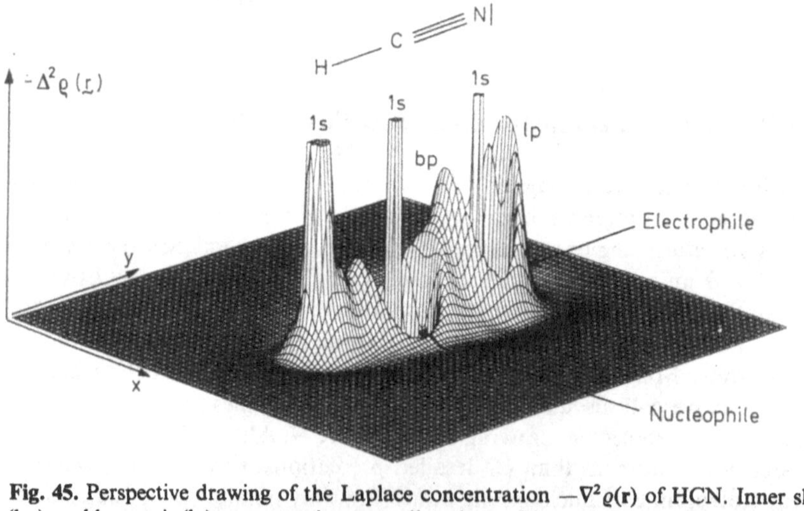

Fig. 45. Perspective drawing of the Laplace concentration $-\nabla^2\varrho(\mathbf{r})$ of HCN. Inner shell (1*s*), bond (bp), and lone pair (lp) concentrations as well as the positions of electrophilic and nucleophilic attack are indicated. The reference plane contains the three nuclei. Laplace values above and below a threshold are cut off. (HF/6-31G(*d*) calculations). Reprinted with Permission from Modelling of Structure and Properties of Molecules, Z. B. Maksic, Ed., published by Ellis Horwood Limited, Chichester, England 1987

lumps in the Laplace distribution can be associated with the inner shell electron pairs (1s), bond electron pairs (bp), and an electron lone pair (lp) at the N atom as indicated in Fig. 45. Clearly, an electrophile will attack at the back of the N atom where the lp concentration is found.

There are distinct holes in the valence shell of both the C and the N atom (Fig. 45). However, the hole at C is significantly deeper than that at the N atom, i.e. the C nucleus is less shielded than the N nucleus. A nucleophile will preferentially attack at the C atom, in line with experimental observations. Hence, inspection of the Laplace concentration of the electrons immediately reveals the reactive sites of a molecule.

This information can also be exploited in a quantitative way. The four aza compounds (Fig. 46) quinoline, isoquinoline, 2,7-methanoaza[10]annulene, and 3,8-methanoaza[10]annulene differ considerably with regard to their basicity although they are all isoconjugate 10π systems [45]. 2,7-Methanoaza[10]annulene is a significantly weaker base ($pk_a = 3.20$) than either pyridine or quinoline ($pk_a = 5.23$ and 4.92, respectively). The perimeter of a bridged [10]annulene is slightly puckered and, as a consequence, a lone-pair orbital of an atom in α-position to the bridge is no longer perpendicular to the π-orbitals. Accordingly, the lone pair orbital and π-orbitals overlap partially and the electron lone pair can delocalize to some extent into the annulene ring thus reducing the electron density, and hence, the basicity of the N-atom in 2,7-methanoaza[10]annulene [45].

This is nicely reflected by the perspective drawings of the Laplace concentrations of quinoline and 2,7-methanoaza[10]annulene taken in the plane defined by the N nucleus and the nuclei of the neighboring atoms (Fig. 47a, b). The lone pair concentration in the bridged annulene is clearly lower that that of quinoline. On the other hand, the lone pair concentration of the 3,8-methanoaza[10]annulene is comparable to that of isoquinoline (Fig. 47d, c).

If the value of $-\nabla^2\varrho(\mathbf{r})$ at the position of the lone pair concentration lumps is plotted against measured pk_a values, a linear relationship will result (Fig. 48). The larger the electron concentration at a lone pair site, the more basic the molecule. Utilizing this relationship the pk_a value of 3,8-methanoaza[10]annulene has been calculated to be 4.8 in good agreement with a value of 4.7 ± 0.5 that was measured *after* the theoretical calculations had beem carried out [45].

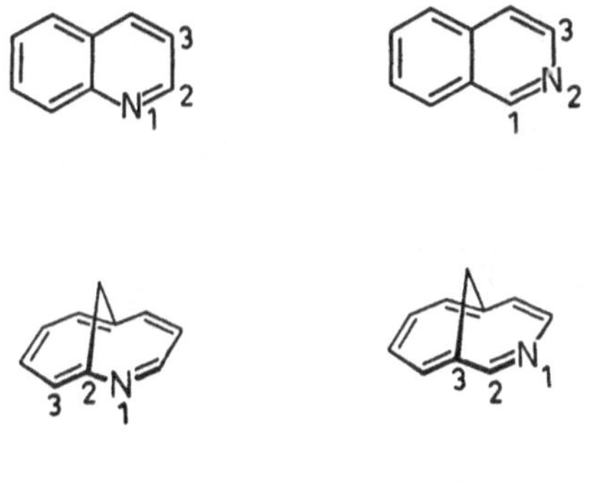

Fig. 46. Quinoline, isoquinoline, 2,7-methanoaza-[10]annulene, and 3,7-methanoaza[10]annulene

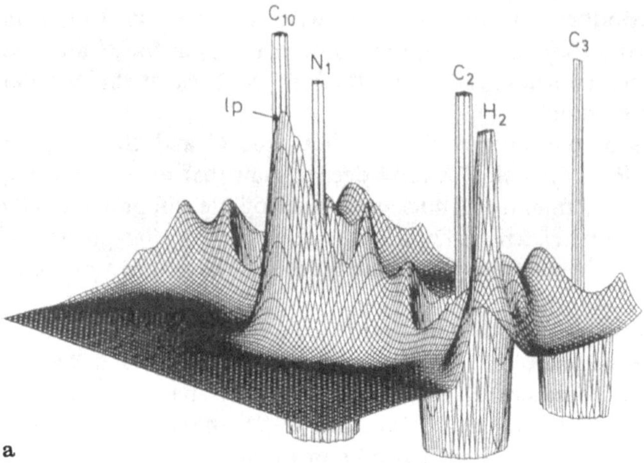

a

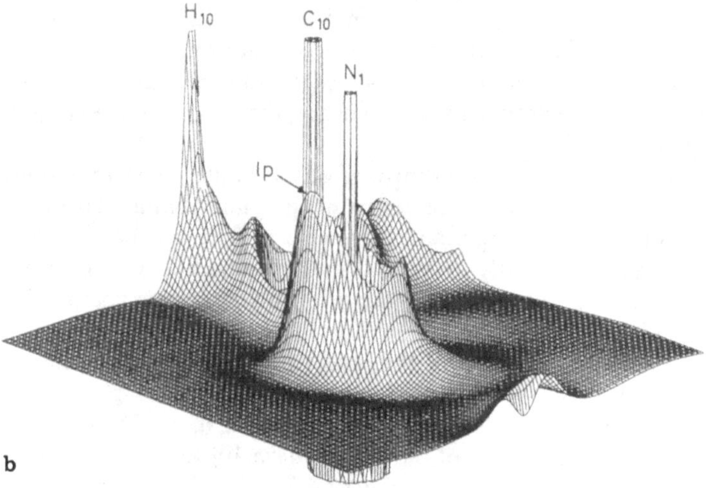

b

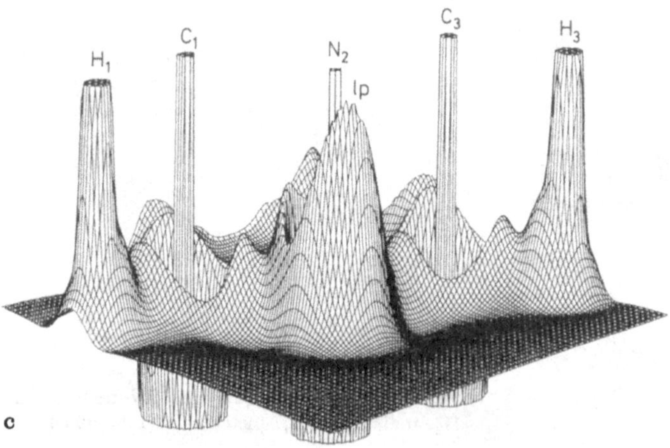

c

Fig. 47 a–c.

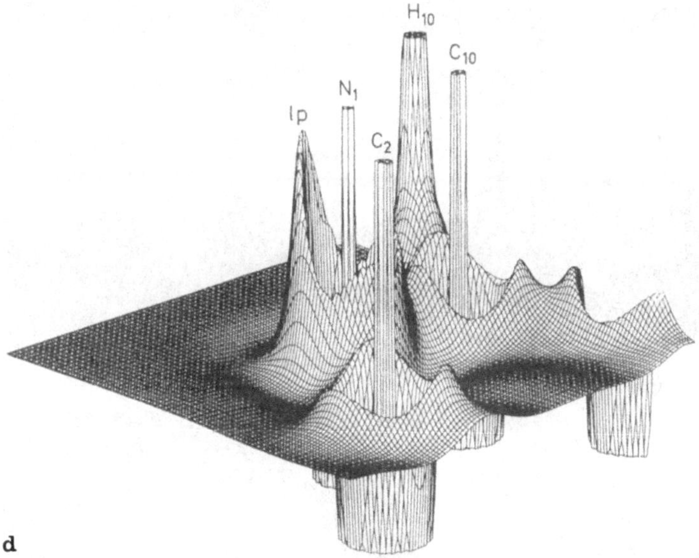

d

Fig. 47 a–d. Perspective drawing of the Laplace concentration $-\nabla^2\varrho(\mathbf{r})$ of **a** quinoline, **b** 2,7-methanoaza[10]annulene, **c** isoquinoline, and **d** 3,7-methanoaza[10]annulene. In each case, the plane containing the N nucleus and those of the neighboring atoms is chosen as a reference plane. Laplace values above and below a threshold are cut off. (HF/4-31G calculations [45])

Pearson has developed the concept of hard and soft acids and bases (HASB concept) [131], which is very useful to classify and to rationalize the reactivity pattern of many molecules. An ambident nucleophile such as the cyanide ion, CN^-, possesses a soft side that reacts with soft electrophile (e.g., alkylhalogenids) and a hard side that reacts with hard electrophiles (e.g., carbenium ions). Hard and soft side of CN^- can be nicely visualized from perspective and contourline diagrams of the Laplace concentration shown in Fig. 49. The concentration lumps at the lone pair side of the N atom is considerably higher than that at the lone pair side of the C atom (Fig. 49a). However, the latter is diffuser and reaches much

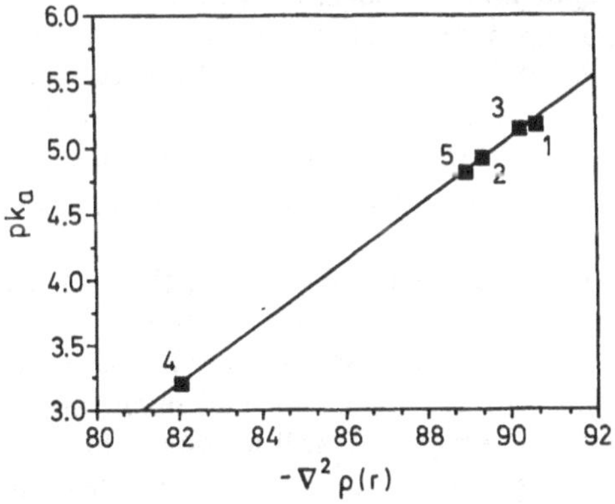

Fig. 48. Relationship between the pK_a value and the Laplace concentration $-\nabla^2\varrho(\mathbf{r})$ in e Å^{-5} at the site of the lone pair electrons of nitrogen. 1: pyridine; 2: quinoline; 3: isoquinoline; 4: 2,7-methanoaza[10]annulene; 5: 3,8-methanoaza[10]annulene. (HF/4-31G calculations [45])

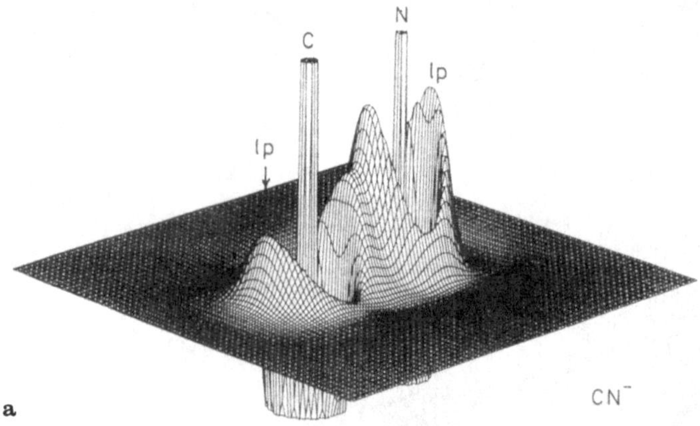

a

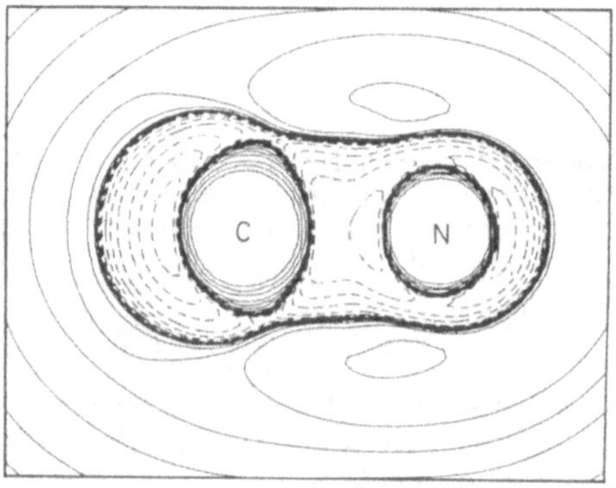

b

Fig. 49 a, b. Perspective drawing **a** and contour-line diagram **b** of the Laplace concentration $-\nabla^2\varrho(\mathbf{r})$ of the cyanide ion, CN^- (HF/6-31G(*d*) calculations)

more into space and, therefore, is earlier "seen" by an approaching electrophile than the more tightly bound lone pair concentration at nitrogen. This example demonstrates that soft and hard sides of ambident nucleophiles can easily be identified by the Laplace concentration.

Holes in the valence sphere concentration correspond to (3, +1) critical points of $-\nabla^2\varrho(\mathbf{r})$. The Laplace concentration at a valence shell critical point is always a maximum in the radial direction, i.e. the eigenvalue λ associated with this direction is always negative. If in addition $-\nabla^2\varrho(\mathbf{r})$ is a maximum in the tangential plane (perpendicular to the radial direction), all three $\lambda < 0$ and a (3, −3) critical point will be encountered, corresponding to a concentration maximum in the bond or in the lone pair region. A minimum of $-\nabla^2\varrho(\mathbf{r})$ in the tangential plane is indicative of a valence shell concentration hole and a (3, +1) critical point of the Laplacian concentration.

Valence shell holes and (3, +1) critical points are found, e.g., where the π^* MOs of alkenes, imino, keto, and other compounds with unsaturated functional groups adopt their largest amplitude. This is shown for ethylene in Fig. 50. The size and depth of the hole provides not only an idea where a nucleophile N might attack but also where an electrophile E might attack. Certainly, E will approach first the "π-concentration" in the bond region yielding a π-complex (Fig. 50). Then, to

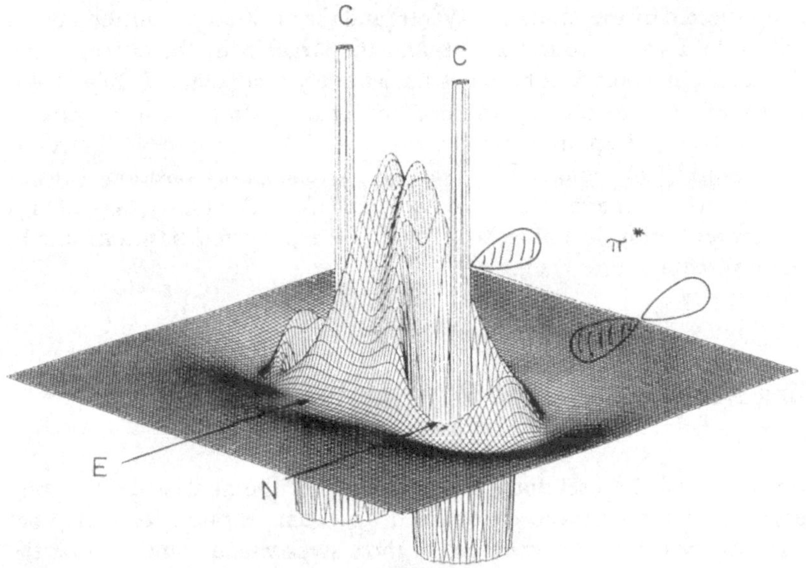

Fig. 50. Perspective drawing of the Laplace concentration $-\nabla^2\varrho(\mathbf{r})$ of ethylene shown with regard to the plane perpendicular to the molecular plane and containing the C nuclei ("π-direction"). Concentration holes in the valence shell can be associated with the π* MO. The positions of electrophilic (E) and nucleophilic attack (N) are indicated. Laplace values above and below a threshold are cut off. (HF/6-31G(*d*) calculations)

establish a bond to one of the atoms of the double bond, it will turn to the atom with the largest π-electron density and the smallest π* concentration hole. In this way, the size of the valence shell concentration hole determines both the site of the nucleophilic and the electrophilic attack.

As an example, the Laplace concentrations at the (3, +1) critical points of the π* holes of benzene, fluorobenzene, and nitrobenzene are listed in Table 8. They possess

Table 8. Description of the concentration holes in the valence shell below and above the molecular plane[a]

Molecule	Critical point	Position[b]	$-\nabla^2\varrho(\mathbf{r})$ [e Å^{-5}]	p [Å]
C_6H_6	C1 (3, +1)		4.32	0.528
C_6H_5F	C1 (3, −1)		8.54	0.511
	C2 (3, +1)	Ortho	5.02	0.524
	C3 (3, +1)	Meta	3.97	0.529
	C4 (3, +1)	Para	4.64	0.526
$C_6H_5NO_2$	C1 (3, −1)		9.35	0.510
	C2 (3, +1)	Ortho	3.20	0.531
	C3 (3, +1)	Meta	4.54	0.527
	C4 (3, +1)	Para	3.23	0.532

[a] HF/4-31G calculations. Ref. [133];
[b] with regard to substituent.

different sizes as reflected by the values of $-\nabla^2 \varrho(\mathbf{r})$ and the position parameter p of the hole. The smaller the Laplace concentration and the larger p is, the deeper is the hole. From the values in Table 8, it becomes immediately clear that a F substituent directs an electrophile (nucleophile) into *para* and *ortho* (*meta*) position, because the holes are less (very) deep in these positions. A NO_2 substituent, however, directs an electrophile (nucleophile) into *meta* (*ortho* and *para*) position. This is completely in line with experimental observations [132]. Obviously, knowledge of the Laplace concentration is sufficient to predict the preferred position of substitution in aromatic compounds [133].

9 Concluding Remark and Outlook

In the previous Sections, the usefulness and feasibility of the analysis of the total electron density distribution have been outlined and demonstrated for a number of examples. The analysis is best carried out in three steps which comprise first the determination of the paths of maximum electron density (MED paths) in a molecule, second the calculation of the properties of electron density $\varrho(\mathbf{r})$ and energy density $H(\mathbf{r})$ at the path critical points $\mathbf{r}_b$ and, finally, the investigation of the Laplace concentration. All Steps of this analysis are easily implemented and require only marginally more computer time than, for example, a Mulliken population analysis, i.e. they require only seconds of computer time [134, 135].

The three-step analysis of the total ectron density distribution leads to

1. an identification of chemical bonds and molecular structure;
2. the distinction between covalent bonds and closed shell interactions (ionic bonds, hydrogen bonds, van der Waals interactions);
3. an assessment of the π-character of a bond;
4. a measure for the degree of π-delocalization and bond equalization in unsaturated systems;
5. knowledge of the distortion of the valence sphere of an atom upon bond formation;
6. the location of concentration lumps arising from inner shell, bond, and lone electron pairs;
7. detection of those sites in the molecule that are prone to nucleophilic or electrophilic attack.

Furthermore, the investigation of the properties of the electron density distribution facilitates the description of through-space and through-bond interactions, in particular π-conjugation, homoconjugation, and σ-conjugation, thereby laying the basis for a clear definition of aromaticity, antiaromaticity, homoaromaticity, σ-aromaticity, etc.

With somewhat more computational effort bond path lengths and interpath angles are calculated. Their knowledge is important in the case of strained molecules, for which the analysis of $\varrho(\mathbf{r})$ leads to a description of bent bonds and their chemical or physical consequences. For example, calculation of the interpath angles helps to quantify the energetic consequences of bond angle or Baeyer strain.

Even more information is obtained by investigating $\varrho(\mathbf{r})$ in the zero-flux surface S(A, B) that separates the atomic subspaces of atoms A and B. Integration of $\varrho(\mathbf{r})$ or H($\mathbf{r}$) over S(A, B) leads to quantities that are measures for the number of electrons forming bond AB or for the bond energy E(AB) [41]. Integration of $\varrho(\mathbf{r})$ and H($\mathbf{r}$) over the basins of A and B gives similar quantities for the atoms. Also, atomic multipole moments, which provide a detailed insight into the electronic structure of a molecule, can be obtained by applying the appropriate one-electron operator and integrating over the atomic subspaces.

However, all calculations which involve integration over a surface or a subspace are costly because they have to be done numerically. The distributions $\varrho(\mathbf{r})$ and H($\mathbf{r}$) like energy, geometry, and other molecular properties depend on the method and the basis set used and, therefore, it is reasonable to carry out integrations only if the molecular wave function has been obtained with a DZ or DZ + P basis set.

The three-step analysis of $\varrho(\mathbf{r})$ has been successfully applied in many cases as was discussed above. Furthermore, it has opened some fascinating avenues that have led to the discovery of novel compounds [48–52]. To illustrate this we briefly discuss here an example taken from our investigations on helium chemistry [50–52].

Helium possesses the highest ionization potential of all elements and, therefore, shows a very moderate tendency of establishing an electron pair bond by donating its electrons to an electron acceptor. A suitable acceptor, however, can be found by bearing in mind 1) that the acceptor ability of an atom or molecule requires a hole in its valence shell concentration and 2) that helium does not possess π-electrons, i.e. it can donate only σ-electrons into a σ-hole of the acceptor. In other words, the bonding

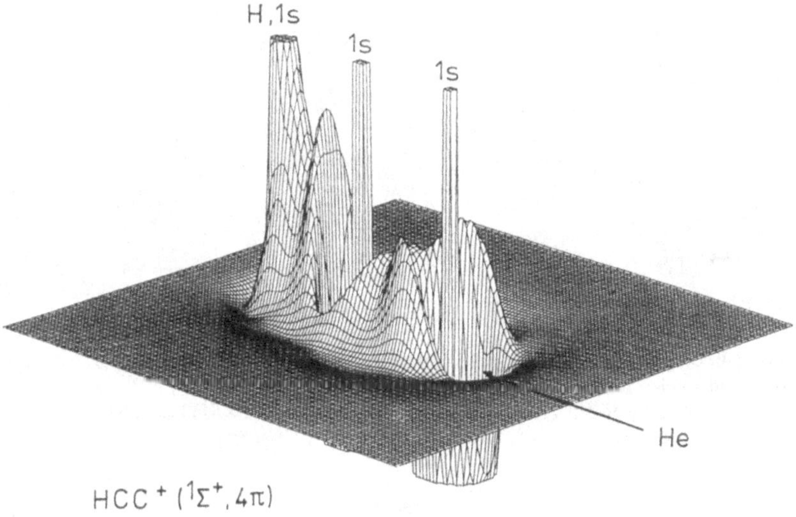

Fig. 51. Perspective drawing of the Laplace concentration $-\nabla^2\varrho(\mathbf{r})$ of the ethinyl cation, HCC^+, in its $^1\Sigma^+(4\pi)$ state. The reference plane contains the three nuclei. The site of the σ-hole to which He binds is indicated by an *arrow*. Laplace values above and below a threshold are cut off. (HF/6-31G(*d, p*) calculations)

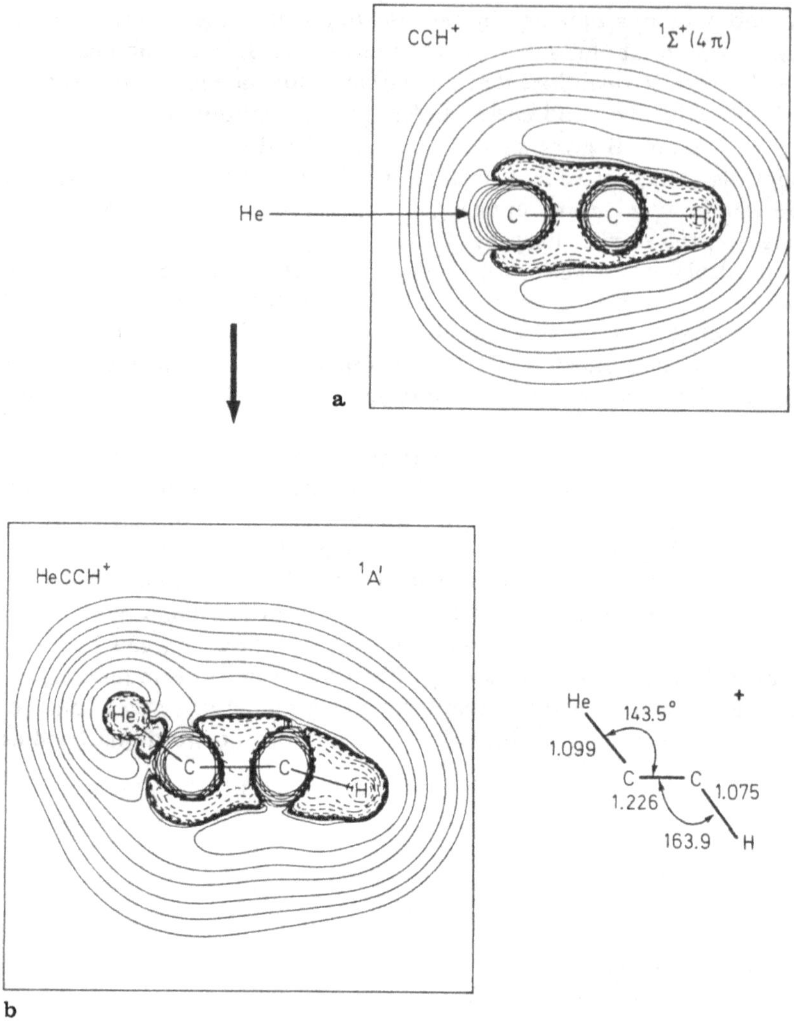

HECCH$^+$ ⟶ HE + CCH$^+$ ($^1\Sigma^+$) 59.3 KCAL/MOL

⟶ HE + CCH$^+$ ($^1\Delta$) 23.7 KCAL/MOL

Fig. 52 a, b. Contour-line diagrams of the Laplace concentration $-\nabla^2\varrho(\mathbf{r})$ of **a** CCH^+($^1\Sigma^+$, 4π) and **b** $HeCCH^+$ ($^1A'$). The nuclei are in the reference plane. Inner-shell concentrations for C are not shown. For $HeCCH^+$ the calculated MP2/6-31G(*d*, *p*) geometry and the MP4(SDTQ)/6-311G(2*df*, 2*dp*) dissociation energies are also given [51]

partner for helium must possess low-lying unoccupied σ-MOs that accept the He electrons.

By checking the Laplace concentration of various carbon cations, it has been found [51] that, among other, the ethinyl cation, CCH^+, should bind helium.

Figure 51 gives the calculated Laplace concentration of CCH^+ in an excited $^1\Sigma^+(4\pi)$ state. The concentration peaks of the 1*s* electrons of H and the two C atoms are indicated. There are holes in the valence sphere of the central C atom that are typical of a π^* LUMO. However, there is an even deeper hole at the terminal C atom in the nonbonding region that is indicative of an unoccupied σ MO. This hole imparts pronounced electron acceptor ability to CCH^+. The electrons of an approaching He atom can be drawn in this hole as indicated in Fig. 51.

Figure 52a gives the contourline diagram of the Laplace concentration of CCH^+ ($^1\Sigma^+$, 4π), which also shows the large σ-hole at the terminal C atom. Upon collision of He and CCH^+, the monoheliumacetylene monocation, $HeCCH^+$, is formed (Fig. 52b). $HeCCH^+$ is nonlinear and stable with regard to decomposition to He and CCH^+ by at least 24 kcal/mol. The contourline diagram of the Laplace concentration reveals that the spherical concentration of He is distorted toward the C atom. It forms a droplet-like appendix in the CHe bonding region that is typical of a *semipolar covalent bond*. This is confirmed by the analysis of H(**r**), which is negative at the bond critical point. Hence, *$HeCCH^+$ is a stable covalently bonded compound* that should be synthesizable. By substituting the H atom by alkyl or aryl groups, a whole class of stable helium organic compounds should be obtained, the monoheliumalkine cations.

This example demonstrates how the analysis of $\varrho(\mathbf{r})$, H(**r**), and $\nabla^2\varrho(\mathbf{r})$ can be used to discover novel compounds or a whole new are in chemistry. Future developments will definitely lead to further examples in this respect. Investigation of atom, atom interactions in nonclassical compounds, in transition metal compounds, in molecules with fluctuating valences, in atom clusters, etc. will probably unravel new interesting features of the bonding mechanism. Also, an analysis of the changes in $\varrho(\mathbf{r})$ and H(**r**) along a reaction path and a comparison with the corresponding changes in the molecular energy present a challenging and probably highly rewarding enterprise.

Acknowledgements: We thank Prof. W. H. E. Schwarz and Prof. T. P. Radhakrishnan for valuable comments on the manuscript. This work was supported by the Deutsche Forschungsgemeinschaft and the Fonds der Chemischen Industrie. Calculations for this article have been carried out on the CRAY XMP 48 of the National Supercomputer Center (NSC), University of Linköping, Sweden. We thank the NCS for a generous allotment of computer time.

10 References

1. (a) Pople JA, Beveridge DL (1970) Approximate molecular orbital theory. McGraw-Hill, New York
(b) Dewar MJS (1969) The molecular orbital theory of organic chemistry. McGraw-Hill, New York
(c) Parr RG (1964) Quantum theory of molecular electronic structure. WA Benjamin, New York
2. Roothan CCJ (1951) Rev. Mod. Phys. 23:69
3. (a) Hoffman R (1963) J. Chem. Phys. 39: 1397
(b) For a review see (1977) Semiempirical methods of electronic structure calculations. In: Segal GA (ed) Modern theoretical chemistry, New York (vols 7 and 8)

4. (a) Salem L (1966) The molecular orbital theory of conjugated systems. WA Benjamin, New York
(b) Heilbronner E, Bock H (1968) Das HMO-Modell und seine Anwendung — Grundlagen und Handhabung, Verlag Chemie, Weinheim
5. (a) Pauling L (1960) The nature of the chemical bond, 3rd edn, Cornell University Press, Ithaca, New York
For a recent review on hybrid orbitals, see (b) Bingel WA, Lüttke W (1981) Angew. Chem. 93: 944; Angew. Chem. Int. Ed. Engl. 20: 899
6. (a) Boys SF (1960) Rev. Mod. Phys. 32: 296
(b) Edminston C, Ruedenberg K (1963) Rev. Mod. Phys. 35: 457
7. See, e.g. Gimarc BM (1979) Molecular structure and bonding — the qualitative molecular orbital approach. Academic, New York
8. Mulliken RS (1955) J. Chem. Phys. 23: 1833; 23: 1841; (1962) 36: 3428
9. Dewar MJS, Dougherty R (1975) The PMO theory of organic chemistry. Plenum, New York
10. (a) Woodward RB, Hoffmann R (1970) Die Erhaltung der Orbitalsymmetrie, Verlag Chemie, Weinheim
(b) Woodward RB, Hoffman R (1969) Angew. Chem. 81: 797
(c) Anh NT (1972) Die Woodward-Hoffmann-Regeln und ihre Anwendungen, Verlag Chemie, Weinheim
11. (a) Fukui K (1982) Angew. Chem. 94: 852
(b) Fleming I (1979) Grenzorbitale und Reaktionen organischer Verbindungen, Verlag Chemie, Weinheim
12. Fukui K, Yonezawa T, Nagata C, Shingu H (1954) J. Chem. Phys. 22: 1433
13. For reviews see: (a) Warner PM (1976) In: Nozoe T, Breslow R, Hafner K, Ito S, Murata I (eds) Topics in nonbenzeniod aromatic chemistry. Hirokawa, Tokyo, vol 2
(b) Paquette LA (1978) Angew. Chem. Int. Ed. Engl. 17: 106
14. Becker P (1980) Electron density and magnetization densities in molecules: Nato advanced study institutes, series B: Phys. Plenum, New York, vol. 48
15. Coppens P, Hall MB (1981) Electron density distributions and the chemical bond, Plenum, New York
16. Breitenstein M, Dannöhl H, Meyer H, Schweig A, Seeger R, Seeger U, Zittlan W (1983) Int. Rev. Phys. Chem. 3: 335
17. Hohenberg P, Kohn W (1964) Phys. Rev. 136B: 864
18. Bader RFW, Nguyen-Dang TT, Tal Y (1981) Rep. Prog. Phys. 44: 893
19. Bader RFW, Nguyen-Dang TT (1981) Adv. Quantum Chem, 14: 63
20. Bader RFW (1981) in: Deb BM (ed) The force concept in chemistry. Van Nostrand Reinhold, NewYork, p 39
21. Bader RFW, Slee TS, Cremer D, Kraka E (1983) J. Am. Chem. Soc. 105: 5061
22. Cremer D, Kraka E (1984) Croat. Chem. Acta 57: 1259
23. Cremer D (1988) In: Maksic ZB (ed) Modelling of structure and properties of molecules. Ellis Horwood, Chichester, England, p 125
24. Bader RFW, Essen H (1984) J. Chem. Phys. 80: 1943
25. (a) Bader RFW, MacDougall PJ, Lau CDH (1984) J. Am. Chem. Soc. 106: 1594
(b) Bader RFW, MacDougall PJ (1985) J. Am. Chem. Soc. 107: 6788
26. Cremer D, Kraka E (1984) Angew. Chem. Int. Ed. Engl. 23: 62
27. (a) Kraka E, Cremer D (1985) Z. Kristallogr. 170: 106
(b) Cremer D, Kraka E (1985) Z. Kristallogr. 170: 32
28. Kraka E (1984) Thesis, University of Cologne, Cologne, West Germany
29. Cremer D, Schmidt T (1985) J. Org. Chem. 50: 2684
30. Cremer D, Bock CW, to be published
31. Cremer D (1983) In: Patai S (ed) The chemistry of functional groups, peroxides, John Wiley, New York, p 1
32. LaBarge MS, Hillig II KW, Kuczkowski RL, Cremer D (1986) J. Phys. Chem. 90: 3092
33. LaBarge MS, Keul H, Kuczkowski RL, Wallasch M, Cremer D (1988) J. Am. Chem. Soc. 110: 2081
34. Cremer D, Kraka E, Slee TS, Bader RWF, Lau CDH (1983) J. Am. Chem. Soc. 105: 5069
35. Cremer D, Dick B, Christen D (1984) J. Mol. Struct. 110: 227
36. Cremer D, Dick B (1982) Angew. Chem. Int. Edt. Engl. 21: 865

37. Cremer D, Gauss J, Schleyer PvR, Budzelaar PHM (1984) Angew. Chem. Int. Edt. Engl. 23: 370
38. Cremer D, Kraka E (1985) J. Am. Chem. Soc. 107: 3800
39. Cremer D, Kraka E (1985) J. Am. Chem. Soc. 107: 3811
40. Cremer D, Kraka E (1988) In: Liebman JF, Greenberg A (eds) Molecular structure and energetics, VCH Publishers, Deerfield Beach, vol 7, p 65
41. Cremer D, Gauss J (1986) J. Am. Chem. Soc. 108: 7467
42. Cremer D (1988) Tetrahedron 44: 7427
43. Budzelaar PHM, Cremer D, Wallasch M, Würthwein M E-U, Schleyer PvR (1987) J. Am. Chem. Soc. 109: 6290
44. Cremer D, Gauss J, Kraka E (1988) J. Mol. Struct. (THEOCHEM) 169: 531
45. Tückmantel W, Andree G, Seidel A, Schmickler H, Lex J, Kraka E, Haug M, Cremer D, Vogel E (1985) Angew. Chem. 97: 592; Angew. Chem. Int. Edt. Engl. 24: 592
46. Cremer D, Bock CW (1986) J. Am. Chem. Soc. 108: 3375
47. Cremer D (1985) Jahrbuch der Akademie der Wissenschaften, Göttingen, p 1
48. Koch W, Frenking G, Gauss J, Cremer D, Sawaryn A, Schleyer PvR (1986) J. Am. Chem. Soc. 108: 5732
49. Koch W, Frenking G, Gauss J, Cremer D (1986) J. Am. Chem. Soc. 108: 5808
50. Koch W, Franking G, Gauss J, Cremer D, Collins JR (1987) J. Am. Chem. Soc. 109: 5917
51. Frenking G, Koch W, Gauss J, Cremer D (1988) J. Am. Chem. Soc. 110: 8007
52. Frenking G, Cremer D (1989) Structure and bonding in press
53. The maximum in $\varrho(\mathbf{r})$ at the nuclear positions is not a true critical point since $\nabla\varrho(\mathbf{r})$ is discontinuous there as a consequence of the cusp condition on Ψ. However, there always exists a function homeomorphic to ϱ that coincides with ϱ almost everywhere and for which the nuclei are (3, —3) critical points.
54. Collard K, Hall GG (1977) Int. J. Quant. Chem. 12: 623
55. This is an observation that has been made in all cases investigated so far. However, at the moment there is no proof that each subspace of a molecule possesses just one and only one atomic nucleus.
56. Hellmann H (1933) Z. Phys. 35: 180
57. Bader RWF, Preston HJT (1969) Int. J. Quant. Chem. 3: 327
58. Feinberg MJ, Ruedenberg K (1971) J. Chem. Phys. 54: 1495
59. Kutzelnigg W (1973) Angew. Chem. 13: 551
60. Nordholm S (1988) J. Chem. Educat. (in press)
61. The kinetic energy T is always >0 and the potential energy V always <0. Since E = T + V, a drop in T leads to a decrease in E and, hence, a stabilization of the molecule
62. We note that the simplified statement "bonding leads to a decrease in the kinetic energy T" often found in text books is not correct. Only the componente of T parallel to the bond decreases; the two components perpendicular to the bond increase
63. Preston HJT (1969) Thesis, McMaster University, Hamilton, Canada
64. (a) Ruedenberg K (1975) in Čhelvet O, Daudel R, Diner S, Malrieu JP (eds) Localization and delocalization in quantum chemistry. D Reidel, Dordrecht vol 1, p 223
(b) Ruedenberg K (1962) Rev. Mod. Phys. 34: 326
65. (a) Edmiston C, Ruedenberg K (1963) J. Phys. Chem. 68: 1628
(b) Layton Jr. EM, Ruedenberg K (1963) J. Phys. Chem. 68: 1654
(c) Rue RR, Ruedenberg K (1963) J. Phys. Chem. 68: 1676
66. (a) Wilson CW, Goddard III WA (1972) Theor. Chim. Acta 26: 195
(b) Goddard III WA, Wilson CW, (1972) Theor. Chim. Acta 26: 211
67. Driessler F, Kutzelnigg W (1976) Theoret. Chim. Acta 43: 1
68. Cohen L (1979) J. Chem. Phys. 70: 788
69. Davidson ER (1976) Reduced density matrices in quantum chemistry. Academic, New York
70. Bader RFW (1980) J. Chem. Phys. 73: 2871
71. We note that Bader (see Refs. 18 through 20) uses these terms differently by calling all MED paths bond paths, not differentiating between covalent bonds and closed shell interactions. This, however, is contrary to general chemical understanding.
72. See, e.g. Ref. 20, p 108
73. Pan DK, Gao JN, Lin HL, Huang MB, Schwarz WHE (1986) Int. J. Quantum Chem. 29: 1154
74. Of course, the analysis of $\varrho(\mathbf{r})$ at a single point $\mathbf{r}_b$ rather than in the whole bonding region is a

simplification. Also, the bond order defined in such a way depends on the theoretical method used. For a detailed discussion see Ref. 22 and 23
75. If bonds between atoms of different rows of the periodic table are considered, the different number of inner shells will have to be taken into account
76. (a) Bader RFW, Tal Y, Anderson SG, Nguyen-Dang TT (1980) Israel J. Chem. 19: 8
(b) Bader RFW, Nguyen-Dang TT, Tal Y (1979) J. Chem. Phys. 70: 4316
77. Note that structural stability of the molecular graph is a mathematical term that is not necessarily related to thermodynamical stability of the molecule
78. (a) Thom R (1975) Structural stability and morphogenesis. Benjamin/Cummings, Reading, MA
(b) Poston T, Stewart I (1981) Catastrophe theory and its applications. Pitma, Boston
79. Investigation of isomerization reactions suggest that $\mathbf{R}_c$ and the location of the transition state are very close. Cremer D (unpublished results)
80. Morse PM, Feshbach H (1953) Methods of theoretical physics. McGraw-Hill, New York, vol 1, p 6
81. Generally, concentration of charge is discussed with the aid of difference density maps, which strongly depend on the proper choice of a reference state. For a discussion see, e.g., Ref. 26
82. See, e.g., Drago RS (1977) Physical methods in chemistry. Saunders, WB Company, London, p 323
83. See, e.g., Arfken G (1985) Mathematical methods for physicists. Academic, New York, p 57
84. See, e.g., Pauli W (1980) General principles of quantum mechanics. Springer Verlag, Berlin Heidelberg New York
85. Bader RFW (1985) Acc. Chem. Res. 18: 9; see also Ref. 18 and references cited therein
86. HF/6-31G(*d*, *p*) calculations, Ref. 28
87. Of course, $\nabla^2\varrho(\mathbf{r})$ depends on the method and the basis set of the ab initio calculation. If $\varrho(\mathbf{r})$ is not adequately described, an analysis of $\nabla^2\varrho(\mathbf{r})$ will be difficult. But the same cbnsiderations apply to an analysis of the energy or other molecular properties obtained by ab initio calculations
88. See Ref. 30. Note that values of Table 3 have been calculated at the HF/6-31G(*d*) level of theory while those of Fig. 15 have been obtained at HF/4-31G. Small differences in η reflect basis set dependency
89. See, e.g. Bock CW, George P, Trachtman M (1984) J. Mol. Struct. (THEOCHEM) 109: 1
90. See, e.g., (a) Garrat PJ (1971) Aromaticity. McGraw-Hill, London
(b) Stevenson GR (1986) in Liebman JF, Grrenberg A (eds) Molecular structure and energetics. VCH Publishers, Deerfield Beach, vol 3, p 57
91. (a) Dewar MJS, Schmeising HN (1959) Tetrahedron 5: 166; (1960) Tetrahedron 11: 96
(b) de LLano C, Dewar MJS (1969) J. Am. Chem. Soc. 91: 789
(c) see also Ref. 1 b
92. (a) Bergmann ED, Pullman B (1971) Aromaticity, pseudo-aromaticity, anti-aromaticity. Proc. Int. Symp, Jerusalem
(b) Lloyd D (1984) Non-benzoid conjugated carbocyclic compounds. Elsevier, Amsterdam
93. Oda M, Oikawa H (1980) Tetrahedron Lett. 21: 107
94. Kawka D, Mues P, Vogel E (1983) Angew. Chem. Int. Ed. Engl. 22: 1003
95. (a) Ferguson LN (1973) Highlights of alicyclic chemistry. Franklin Palisades, NJ (part 1, chap 3)
(b) Charton M (1970) In: Zabicky J (ed) The chemistry of alkenes. Wiley-Interscience, New York, vol 2
(c) Wendisch D (1971) In: Methoden der Organischen Chemie. Houben-Weyl-Müller, Thieme Verlag, Stuttgart vol IV, p 3
96. In Refs. 38 and 39 the symbol η is used. Since in this work η denotes π-character, η has been replaced by ξ
97. (a) Winstein S (1959) J. Am. Chem. Soc. 81: 6524
(b) Winstein S (1969) Q. Rev. Chem. Soc. 23: 141
98. See, e.g.,
(a) Hehre WJ, Hilberty PC (1974) J. Am. Chem. Soc. 96: 302
(b) Hehre WJ, Hilberty PC (1972) J. Am. Chem. Soc. 94: 5917
(c) Levi DA, Blurock ES, Hehre WJ (1979) J. Am. Chem. Soc. 101: 5537
(d) Hanack M, Schneider H-J (1967) Topics Curr. Chem. 554

99. Haddon RC (1987) J. Am. Chem. Soc. 109: 1676. Haddon has defined an orbital overlap criterion for homoconjugation. Although this is very useful to investigate through-space interactions, it cannot provide an answer to the question whether a homoaromatic bond is formed or not. See also Ref, 13b
100. For a review on the homotropylium, see Childs RF (1984) Acc. Chem. Res. 17: 347
101. Childs RF, Varadarajan A, Lock CJL, Faggiani R, Fyfe CA, Wasylishen RE (1982) J. Am. Chem. Soc. 104: 2452
102. (a) Haddon RC (1979) J. Org. Chem. 44: 3608
(b) Haddon RC, Raghavachari KJ (1983) J. Am. Chem. Soc. 105: 118
103. Turner RB, Meader WR, Doering WvE, Knox LH, Mayer JR, Wiley DW (1957) J. Am. Chem. Soc. 79: 4127
104. Roth WR, Böhm M, Lennartz H-W, Vogel E (1983) Angew. Chem. 95: 1011; Int. Ed. Engl. 22:1007
105. See, e.g.
(a) Canrow K (1961) J. Am. Chem. Soc. 83: 2958
(b) Dewar MJS (1974) Pure Appl. Chem. 767
106. (a) Christoph GG, Muthard JL, Paquette LA, Böhm MC, Gleiter R (1978) J. Am. Chem. Soc. 100: 7782
(b) McMurry JE, Haley GJ, Matz JR, Clardy JC, Dyne GV, Gleiter R, Schäfer W (1984) J. Am. Chem. Soc. 106: 5018
(c) For a differing view see Liebman JF, Paquette LA, Petersen JR, Rogers DW (1986) J. Am. Chem. Soc., 108: 8267
107. Houk KN, Gandour RW, Strozier RW, Rondan NG, Paquette LA (1979) J. Am. Chem. Soc. 101: 6797
108. The authors do not rule out neutral homoaromaticity but constrain this to the case in which two interacting π moieties are substituted by strong donors (the first) and strong acceptors (the second), thus benefitting from stabilizing HOMO LUMO interactions
109. (a) Cremer D, Pople JA (1975) J. Am. Chem. Soc. 97: 1358
(b) Essen H, Cremer D (1984) Acta Cryst. B40: 418
(c) Cremer D (1984) Acta Cryst. B40: 498
110. (a) Cox JD, Pilcher G (1970) Thermochemistry of organic and organometallic compounds. Academic, New York
(b) Benson SW, Cruickshank FR, Golden DM, Hangen GR, O'Neal HE, Rodgers AS, Shaw R, Wash R (1969) Chem. Rev. 69: 279
(c) Benson SW (1976) Thermochemical Kinetics, Wiley, London
111. Homoaromaticity of norcaradiene has also been suggested on the basis of MO considerations:
(a) Jorgensen WL (1975) J. Am. Chem. Soc. 97: 3082
(b) Jorgensen WL (1976) J. Am. Chem. Soc. 98: 6784
112. Vogel E, Scholl T, Lex J, Hohlneicher G (1982) Angew. Chem. 94: 878; (1982) Angew. Chem. Suppl. 1882
113. For example, homoantiaromaticity has been suggested by Jorgensen for bicyclo[3.1.0]hexenyl cation and bicyclo[2.1.0]pent-2-ene. See Ref. 111
114. (a) Childs RF, Sakai M, Parrington BD, Winstein S (1974) J. Am. Chem. Soc. 96: 6403
(b) Childs RF, Winstein S (1974) J. Am. Chem. Soc. 96: 6409
(c) Childs RF, Zeya M (1974) J. Am. Chem. Soc. 96: 6418
115. This seems to be the case for 3*H*-cyclo-nona[def]biphenylene that has been investigated by Wilcox CF, Blain DA, Clardy J, Duyne vG, Gleiter R, Eckert-Maksić M (1986) J. Am. Chem. Soc. 108: 7693
116. See, e.g. Haddon RC, Raghâvachari K (1983) J. Am. Chem. Soc. 105: 118 and references cited therein
117. For reviews on strain, see (a) Ref. 40
(b) Greenberg A, Liebman JF (1978) Strained molecules. Academic, New York
(c) Liebman JF, Greenberg A (1976) Chem. Rev. 76: 311
(d) Wiberg K (1986) Angew. Chem. Int. Ed. Engl. 25: 312
118. (a) Dewar MJS (1945) Nature 156: 748
(b) Dewar MJS (1946) J. Chem. Soc. 406: 777
(c) Dewar MJS (1951) Bull. Soc. Chim. Fr. C71

(d) Dewar MJS, Marchand AP (1965) Annu. Rev. Phys. Chem. 16: 321
(e) Walsh AD (1947) Nature 159: 165; Nature 159: 712
119. Dewar MJS, Ford GP (1979) J. Am. Chem. Soc. 101: 183
120. Dunitz JD, Schomaker V (1952) J. Chem. Phys. 20: 1703
121. Bauld NL, Cesak J, Holloway RL (1977) J. Am. Chem. Soc. 99: 8140
122. (a) Dewar MJS (1979) Bull. Soc. Chim. Belg. 88: 957
(b) Dewar MJS, McKee ML (1980) Pure Appl. Chem. 52: 1431
(c) Dewar MJS (1984) J. Am. Chem. Soc. 106: 669
123. (a) Dewar MJS, Petit R (1954) J. Chem. Soc. 1617
(b) Sandorfy C (1955) Canad. J. Chem. 33: 1337
(c) Pople JA, Santry DP (1964) Mol. Phys. 7: 269; 7: 301
124. Cremer D, Reichel F (1989) J. Am. Chem. Soc. (to be published)
125. Coulson CA, Moffit WE (1949) Philos. Mag. 40: 1
126. Ahlrichs R, Ehrhardt C (1985) Chem. Z. 19: 120 and personal communication
127. Emsley JW, Feeney J, Sutcliffe LH (1966) High resolution nuclear magnetic resonance spectroscopy. Pergamon, Oxford, p 690
128. Zilm KW, Beeler AJ, Grant DM, Michl J, Clou T, Allred EL (1981) J. Am. Chem. Soc. 103: 2119
129. (a) Buchanan JG, Sable HZ (1972) in: Thagarajan BS (ed) Selective organic transformations. Wiley, New York, vol 2, p 1
(b) Smith JG (1984) Synthesis 629
130. A family of functions describing a given type of catastrophe is called unfolding [78]
131. (a) Pearson RG (1963) J. Am. Chem. Soc. 85: 3533
(b) Pearson RG, Songstad J (1967) J. Am. Chem. Soc. 89: 1827
(c) Ho T-L (1975) Chem. Rev. 75: 1
132. See, e.g. Ref. 11b, chap 3
133. Cremer D, Bock CW (to be published)
134. For example, the analysis of $\varrho(\mathbf{r})$, $H(\mathbf{r})$, and $\nabla^2\varrho(\mathbf{r})$ for a three-heavy atom system with 70 basis functions carried out with the ab initio package COLOGNE 87 [135] requires 6 s while the Mulliken population analysis requires 5 s of computer time
135. Gauss J, Kraka E, Reichel F, Cremer D, (1988) Universität Köln, COLOGNE 88

Electron Deformation Densities and Chemical Bonding in Transition Metal Complexes

A. A. Low and M. B. Hall
Department of Chemistry Texas A & M University College Station, TX 77843-3255, USA

The relationship between electron density and chemical bonding has been studied both experimentally and theoretically by the use of electron deformation density maps. For transition metal complexes, deformation density maps have proved useful in the study of the nature of metal-ligand bonding, metal-metal bonding and the electronic configuration about the metal. However, puzzling features have been found in the deformation density maps of some transition metal complexes such as the presence of unexpected accumulation peaks or the absence of expected accumulation peaks. A review of reported deformation density maps of transition metal complexes is presented here, and includes a brief discussion of many of these puzzling features.

Abbreviations

STO — Slater type orbitals
GTO — Gaussian type orbitals
SCF — self-consistent field
CI — configuration interaction
LMO — localized molecular orbitals
Bz — Benzene
TPP — tetraphenylporphyrin
OMe — methoxy (OCH_3)
TMP — tetramethylporphyrin
Pc — phthalocyanine
THF — tetrahydrofuran
meim — 2-methyl-imidazole
OEP — octaethylporphyrin
Cp — cyclopentadiene
HOMO — highest occupied molecular orbital
t-Bu — *tert*-butyl group
tmp — 1,3,5-trimethoxyphenyl
mhp — 2-hydroxy-6-methylpyridine

1 Introduction

The study of charge density and its relationship to chemical bonding has been the subject of increasing research over the years [1]. In order to study how electron density changes with bond formation, a difference density sometimes called the electron deformation density is calculated. The electron deformation density is defined as the difference between the total molecular electron density and a reference density called the promolecule, which is usually the sum of the spherically averaged ground state constituent atoms.

$$\Delta\varrho(r) = \varrho_{molecule}(r) - \varrho_{promolecule}(r) \quad (1)$$

$$\varrho_{promolecule}(r) = \sum_{atoms} \varrho_{spherical}(r - r_i) \quad (2)$$

r_i = position of spherical atom i.

Deformation density maps may be derived experimentally through difference Fourier synthesis of x-ray structure determinations. Because the positional parameters calculated in least square fittings in x-ray structures may be biased by aspherical charge distributions about the atoms, positional parameters are usually obtained from single crystal neutron diffraction studies (called the X—N method) or by high order refinement based on high angle x-ray data only (called the X—X method).

In order to reduce "smearing" of electron density by thermal vibrations, the collection of data is performed at low temperatures. Even at low temperatures, thermal smearing of electron density causes some loss of clarity in deformation density maps. Other sources of discrepancy include acentric or disordered structures or systematic errors made during collection or processing of the data.

Theoretical deformation densities may be computed through ab initio SCF molecular orbital calculations. Generally, the basis sets used consist of either Slater-type (STO) or Gaussian-type (GTO) orbitals. A minimal basis set with one STO (or the equivalent combination of GTO's) has been clearly found to be inadequate to describe the charge density in even diatomics such as O_2 and HF [2]. Adding one more STO (or GTO's) per atomic orbital gives a basis set of double (or higher) zeta quality and greatly improves the description of charge density. The addition of polarization functions for unoccupied *d* or *p* orbitals also have been shown to dramatically effect deformation densities [3]. In addition, configuration interaction may be added to account for electron correlation. It is generally agreed that in order to calculate a qualitatively correct deformation density, at least a double zeta quality basis set is required.

Generally, deformation density maps are plotted as two dimensional contour maps in particular planes of the compound in question. For experimentally determined maps, the units of electron density are almost always electrons per cubic angstrom (e Å^{-3}). Theoretically determined maps may also be plotted in electrons per cubic angstrom but they are also plotted in atomic units of density (au) or electrons per atomic unit of length cubed (e (au)$^{-3}$). The atomic unit of length is the Bohr radius, a_0, which is equal to 0.529177 Å. Therefore, an atomic unit of density is equal to 6.748315 e Å^{-3}. See Table I for other conversion factors.

Often, experimental and theoretical deformation deformation density maps are used in conjunction with each other to help understand the bonding in a molecule. Various factors must be accounted for before a comparison can be made. The effects of thermal vibration in experimental maps, called thermal smearing, is significant even at low temperature. Multipole refinement can be used in experimental studies to produce a "static" experimental map. Conversely, methods to simulate thermal motion in theoretical maps have been proposed [4]. In addition, it must be noted that theoretical calculations are usually performed on single isolated molecules where experimental studies are performed on a molecule situated in a crystal lattice. Neighboring molecules in the lattice may have a significant effect on the molecule's structure and the shape of the electron cloud surrounding the molecule.

There are some general features of deformation densities which can be intuitively deduced. If one considers the classical model of covalent bonding based on the H_2 molecule, one would intuitively expect to see accumulation of charge between two covalently bonded atoms. However, for some covalent bonds, little or no accumulation of charge can be seen between bonded atoms. The lack of observable accumulation in covalent bonds has been attributed to several reasons: 1) the rearrangement of the

Table 1. Conversion factors useful in deformation density studies

Length:	1 au = 1 bohr = 1 a_0 = 0.529777 Å = 5.2977×10^{-9} m
Electron Density:	1 au = 1 ea_0^{-3} = 6.748315 eÅ^{-3} = 1.08121×10^{12} C m^{-3}

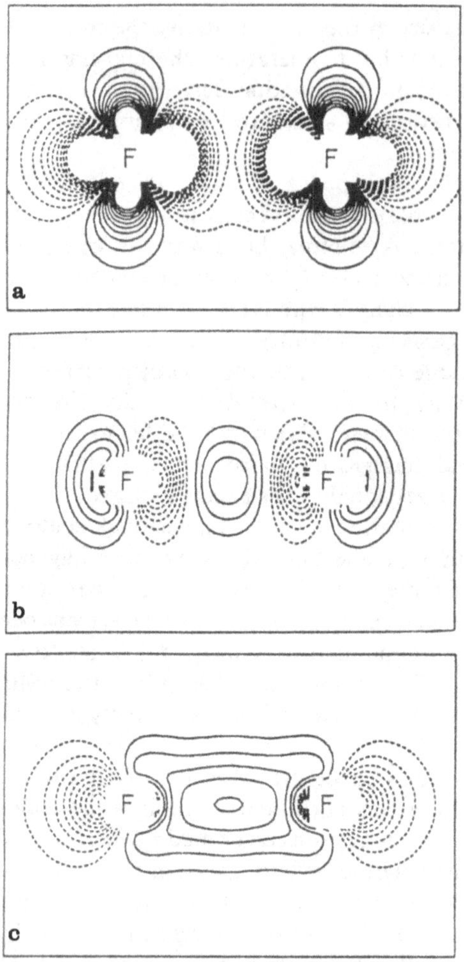

Fig. 1a–c. Theoretical deformation density from GMO—CI molecular density for F_2: **(a)** molecule minus spherical atoms; **(b)** molecule minus oriented atoms; **(c)** molecule minus valence-state hybrid atoms. Positive contours are *solid*, negative contours are *dashed*. The zero contour is omitted. Contour interval of 0.1 e $Å^{-3}$ (Reprinted from [5])

atom's valence shell in preparation for bonding; 2) the formation of bonds to other atoms may result in loss of density along the bond direction and 3) the bond may be a relatively long bond between atoms with compact valence orbitals.

The first effect can be demonstrated clearly in the deformation density map of the F_2 molecule. When the sum of the density of two spherically averaged F atoms is subtracted from the total density calculated from an ab initio, two-configuration, SCF wave function of the F_2 molecule [5], the entire region along the bond axis corresponds to a region of density deficit (see Fig. 1). There are accumulations of density in the π region perpendicular to the axis.

Instead of subtracting spherical fluorine atoms, a more chemically meaningful deformation density could be obtained by subtracting a properly oriented F atom with 1 electron in the p_σ orbital and 2 electrons in the p_π orbitals. If two properly oriented F atoms are subtracted from the total density, accumulation can be found in the bonding region of F_2 and outside the bond region along the bond axis (see Fig. 1b). These features represent bonding density from constructive interference and lone-pair density from polarization and hybridization, respectively.

An analysis of the localized molecular orbitals (LMO's) of F_2 reveals some 2s character in the F_2 bond. This hybridization causes the mainly 2s lone pair orbital to move away from the bonding region. The valence state hybrid orbitals may be obtained by truncation of the LMO's of F_2 [6]. When the density of these valence state hybrid atoms is subtracted from the total density of the F_2 molecule, the density difference due to constructive interference alone can be seen (Fig. 1c).

A similar effect is seen in the experimental and theoretical deformation density maps of H_2O_2 which show very little accumulation of density in the O—O bond. [7] (Fig. 1a). The deformation density in the region of the O—O bond is, in fact, negative. Initially this observation seems rather puzzling until one considers the high occupancy (1.33 electrons) of *p* orbitals in a spherically averaged oxygen atom. Since the *p* orbitals involved in the O—O bond each contribute 1 electron to the bond, it is not surprising that when a sperically averaged *p* orbital of an oxygen atom is subtracted from the molecule, the result is a negative deformation density along the bond axis. When the choice of the promolecule is changed from a collection of spherically averaged atoms to a collection of "prepared" OH fragments, some accumulation of charge in the O—O bond may be observed. Thus, the choice of promolecule must be considered carefully if one wishes to accurately represent the charge density associated with chemical bonding in deformation density maps.

Deformation densities of many transition metal complexes have been studied. Some of the objectives of these studies include the nature of metal-ligand bonding and metal-metal bonding and the ground state electronic configuration about the metal atom. The main thrust of this review will be concerned with studies of transition metal complexes. Many puzzling features have been observed and will be discussed. Included among these features are lack of accumulations due to the second and third reasons mentioned above.

2 Mononuclear Complexes

2.1 $[Co(NH_3)_6]$ $[Co(CN)_6]$

The first experimental deformation density map determined for a transition metal complex was that of the coordination complex, $[Co(NH_3)_6]$ $[Co(CN)_6]$ (shown in Figure 2) [8]. The crystal structure contains discrete $[Co(NH_3)_6]^{+3}$ and $[Co(CN)_6]^{-3}$ ions. The coordination geometry about each cobalt is very close to octahedral. There are weak N—II ··· N≡C hydrogen bonds between cations and anions.

The number of electrons about each cobalt was estimated by direct integration of the electron density in the area within 1.22 Å of the cobalt (covalent radius of Co^{+3}) to be 26.3 ($\pm$.03) electrons for $[Co(NH_3)_6]^{+3}$ and 26.8 ($\pm$.03) for $[Co(CN_6]^{-3}$. This experimentally obtained value of 26.3 electrons in the vicinity of the cobalt atom for $[Co(NH_3)_6]^{+3}$ is very close to the value of 26.4 electron calculated by ab initio Hartree-Fock calculations for $[Co(NH_3)_6]^{+3}$ [9]. This value shows that the cobalt atom has a largely neutralized charge with the positive charge spread out among the hydrogen atoms. This is also borne out by the electron deficiencies found around

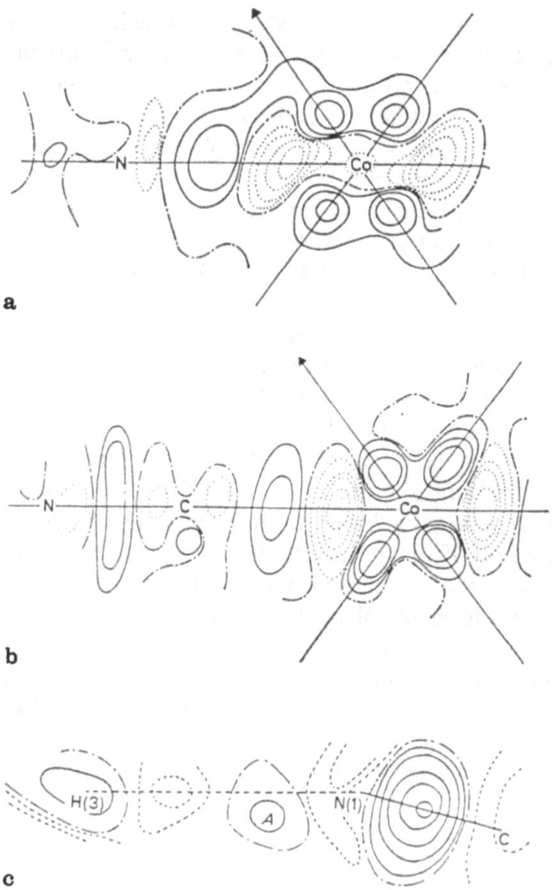

Fig. 2a–c. Deformation density plots from the final difference Fourier syntheses of $[Co(NH_3)_6]$ $[Co(CN)_6]$: **(a)** Section through Co—N bonds and a threefold axis of $[Co(NH_3)_6]^{+3}$, contours of 0.1 e Å^{-3}; **(b)** Section through a Co—C—N bond and a threefold axis of $[Co(CN)_6]^{-3}$, contours of 0.1 e Å^{-3}; **(c)** Section through the close cation-anion H···N=C contact, contours of 0.05 e Å^{-3} (Reprinted with permission from [8])

the hydrogen atoms in the final difference Fourier maps. Some caution, however, must be exercised in comparison of these two values because of the different methods used to obtain the values for both methods. Experimentally, the number of electrons was obtained through direct integration of the volume 1.22 Å from the cobalt. The theoretical number was obtained through Mulliken population analyses. The fact that the values obtained from both methods are similar may just be fortuitous.

The residual electron density along the Co—N bond of $[Co(NH_3)_6]^{+3}$ (see Fig. 2a) shows a maximum between the cobalt and nitrogen atoms with minima on each side. The authors suggest that the electron density has moved to the bond center from the nitrogen lone pair. This same pattern of minimum-maximum-minimum is found for the Co—C bond in $[Co(CN)_6]^{-3}$ (see Fig. 2b). The electron densities in the Co—C and C=N bonds are more elongated perpendicular to the bond than in the Co—N bond in the cation suggesting d_π—p_π interactions.

In the region of the close N—H ... N=C contact, a small peak of accumulation is found off the C—N line, 0.75 Å away from the cyanide nitrogen, in the direction of the hydrogen atom of the ammine. This small peak, labelled as "A" in Fig. 2c, is attributed to the lone pair of the nitrogen atom pointing towards the hydrogen.

In the immediate vicinity of the cobalt atom, well-defined maxima are found on the three-fold axes of the complexes, pointing between the ligands. This pattern of maxima correspond to what is expected for 6 electrons in the t_{2g} orbitals (d_{xy}, d_{xz} and d_{yz}) in an octahedral field. This pattern of electron density accumulation is found frequently about octahedrally coordinated transition metals.

A simple crystal field argument can be used to explain the asphericity of charge in octahedrally coordinated complexes. If one starts with a spherically symmetric ground state transition metal and allows six ligands to approach if from the z, —z, y, —y, x and —x directions, the ligands will interact strongly with the e_g orbitals (d_{z^2} and $d_{x^2-y^2}$) causing them to rise in energy. As a result, electron density will flow from these orbitals into the t_{2g}(d_{yz}, and d_{xz}) orbitals which point between the axes. Thus, density will accumulate along the threefold axes of the octahedron as seen in the standard deformation density maps of most octahedrally coordinated transition metal complexes.

This argument to rationalize the asphericity of charge density in octahedral metal complexes is based solely on nearest neighbor contacts. A recent study has shown that second nearest neighbor forces may also have a significant effect on the pattern of charge density about the metal atom [10]. In this study, the experimental electron deformation density of the complex bis(hydrazine)bis(hydrazinecarboxylato)cobalt(II) was studied. This complex contains two bidentate hydrazinecarboxylates which bind the cobalt with a ligand-metal-ligand angle of less than the ideal 90°. Experimental studies have shown that the larger *d*-electron accumulations generally occur in the larger L—M—L angles [11]. However, as shown in Fig. 3, the larger *d*-electron accumulation occurs in the smaller angle. This larger peak lies on the vector between

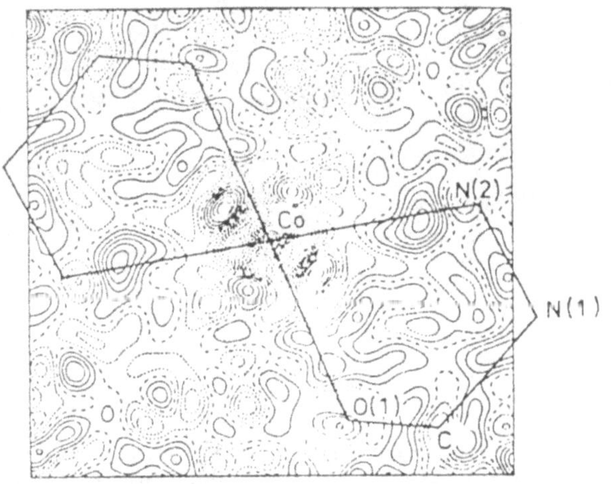

Fig. 3. Deformation density of bis(hydrazine)bis(hydrazinecarboxylato)cobalt(II) in the plane of the cobalt atom and the bidentate ligands. Contour interval of 0.1 e Å^{-3}. Negative contours are *broken* (Reprinted with permission from [10])

the Co atom and the C atom which was calculated from the experimental data to be the most electropositive atom in the structure. This suggests that effects of second nearest neighbors on the pattern of density accumulation about the central metal atom can be significant.

2.2 Benzene Chromium Tricarbonyl

The compound benzene chromium tricarbonyl, [BzCr(CO)$_3$], has been extensively studied along with dibenzene chromium in order to shed light on the nature of the η^6-benzene to metal bonding and its effect on the benzene ring. Initially, there was

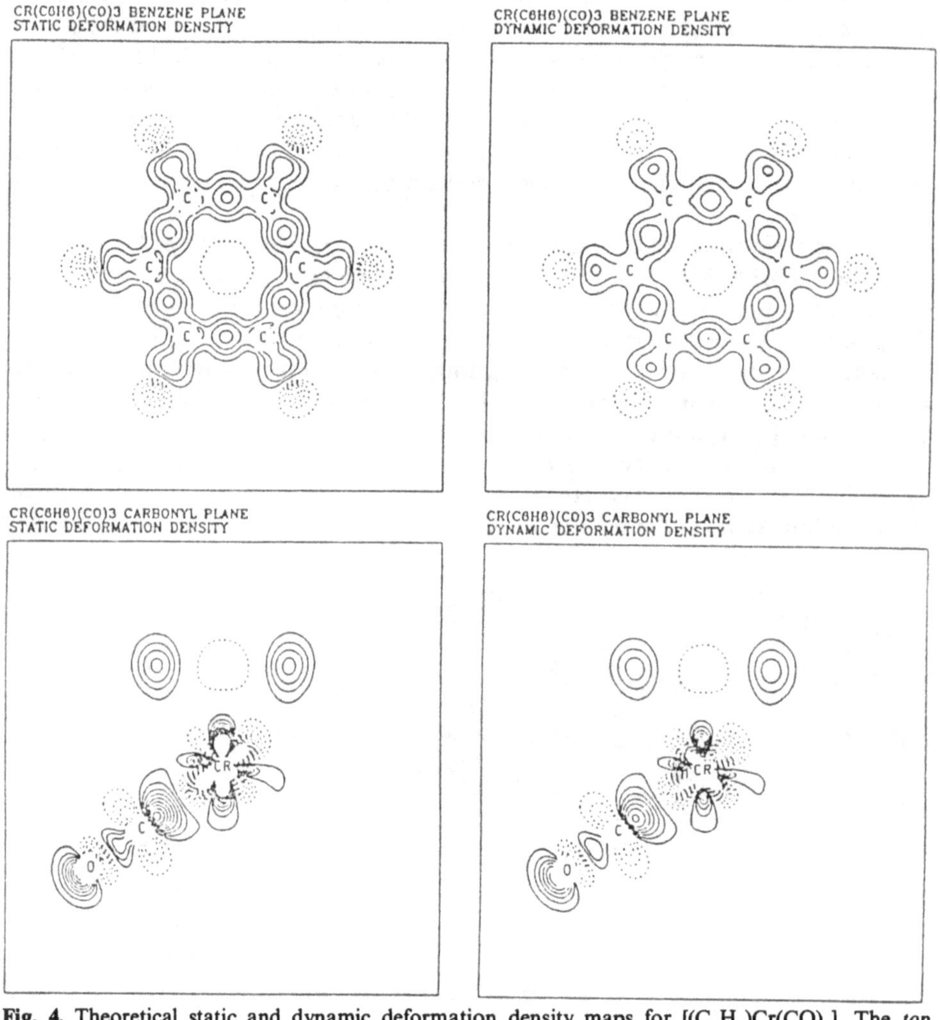

Fig. 4. Theoretical static and dynamic deformation density maps for [(C_6H_6)Cr(CO)$_3$]. The *top* two show the electron density distribution in the plane of the benzene ring while the *bottom* maps show the density in a plane containing a chromium atom and one carbonyl group. Contour intervals of 0.10 e Å^{-3}. Negative contours are *broken*. (Reprinted with permission from [15] Copyright 1985, American Chemical Society)

some controversy as to the symmetry of the benzene ring in dibenzene chromium. It was finally determined to have D_{6h} symmetry [12]. In the compound $[BzCr(CO)_3]$, the symmetry of an isolated molecule is C_{3v}. However, the room temperature x-ray structure determination showed no deviation of the benzene ring from D_{6h} symmetry [13].

In the low temperature x-ray structure determination, done in conjunction with the neutron diffraction study, two types of C—C bonds were found in the benzene ring [14]. With reference to the carbonyls on chromium, the C—C bonds which were *cisoid* to the carbonyls had an average bond length of 1.432 Å while those *transoid* to the carbonyls had an average bond length of 1.406 Å. In addition, the hydrogen atoms of the benzene ring were slightly out of the plane of the benzene ring, bent

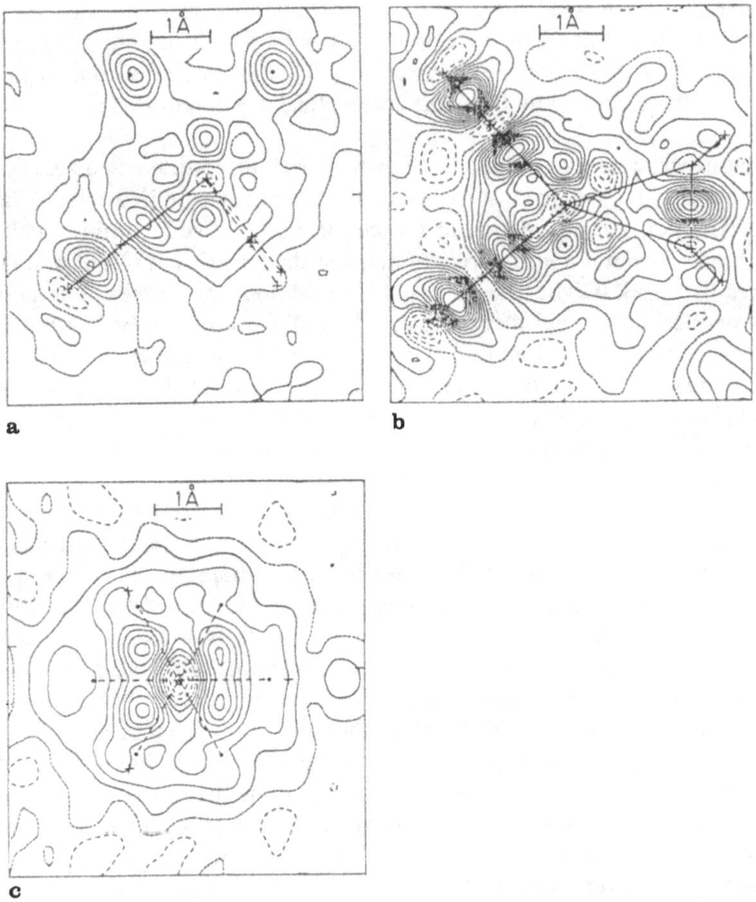

Fig. 5a–d. Experimental electron density of $[(C_6H_6)Cr(CO)_3]$: **(a)** in a plane passing through the chromium atom, the center of the benzene ring and one of the carbonyls, contour intervals of 0.1 e $Å^{-3}$ **(b)** in a plane passing through the chromium atom and a lone C—C bond of the benzene ring, two carbonyls almost lie in the plane, contour intervals of 0.05 e $Å^{-3}$ **(c)** in a plane parallel to the benzene ring containing the Cr atom, contour intervals of 0.05 e $Å^{-3}$. Negative contours are *dashed*, zero contour is *dotted*. (Reprinted with permission from [16] Copyright 1987, American Chemical Society)

0.03 Å in the direction of the chromium atom. Interestingly, in the crystal structure of $[BzCr(CO)_3]$, the molecules do not take full advantage of the C_{3v} symmetry available to them upon crystallization, but instead lie on crystallographic mirror planes.

The theoretical deformation density map of $[BzCr(CO)_3]$ was computed assuming the C_{3v} symmetry of the complex [15]. These deformation density maps are shown in Fig. 4. In addition to the static theoretical maps, a technique to incorporate thermal smearing was used to produce dynamic deformation maps. In Fig. 4, the top maps show the deformation density in the plane of the benzene ring. The bottom two maps show the density in the plane containing the chromium atom, a carbonyl, and passing through the midpoints of two C—C bonds of the benzene ring.

The deformation density near the chromium atom shows a pattern of 8 lobes with 4 lobes of accumulation separated by 90° from one another with one lobe pointing towards the center of the benzene ring. These 4 positive lobes are separated by 4 negative lobes, which are offset from the positive lobes by 45°. The negative lobes point directly at the C—C bonds of the benzene ring. This pattern of accumulation and deficiency is readily explained by the same crystal field argument used earlier for the cobalt hexammine cation.

The most recent experimental deformation density maps [16], contrary to an earlier report [17], corroborates the positions of the maxima about the chromium atom. The experimental deformation density maps can be seen in Fig. 5. Accumulation peaks are found in both experimental and theoretical maps in the Cr—C and C—O bonds, although their relative peak heights are reversed. In addition, the oxygen lone pair density is much weaker in the experimental map. These differences can be attributed to a combination of effects. First, theoretical calculations tend to underestimate the electron density in the C—O bond due to truncation of the basis set. The oxygen lone pair density is weaker in the experimental maps due to the librational motion of the oxygen, sometimes called oxygen wag. The theoretical simulation of thermal motion does not totally account for this type of motion.

The main difference in the theoretical and experimental maps relates to this lack of C_{3v} symmetry in the experimental maps. The experimental deformation density map of a plane parallel to the plane of the benzene ring containing the chromium atom is shown in Fig. 5c. It can clearly be seen that the electron density in this plane does not exhibit the expected 3-fold symmetry. Even if ab initio calculations were performed with this slightly distorted geometry, this almost four-fold pattern of experimental electron density in this plane would not be reproduced. Structurally, the deviation of 3-fold symmetry is exhibited in the bond angles, most notably in the angles D—Cr—C(O) where D is the center of the benzene ring. The values for these angles of 123.7(1)°, 127.8(1)° and 127.8(1)° clearly show that the benzene ring is slightly bent towards one of the carbonyls. However, this small difference cannot be responsible for the observed deviation from C_{3v} symmetry seen in Fig. 5c.

In order to obtain a clearer picture of the nature of the interactions present in this molecule, two different fragment deformation density maps shown in Fig. 6, were calculated from the ab initio calculations. The first calculation was performed on a benzene ring and three carbon monoxide molecules in their equilibrium positions without the chromium atom. The electron of a free benzene and three carbon monoxide molecules was subtracted from the total electron density of the above arrangement in order to observe any interaction of the benzene ring with the

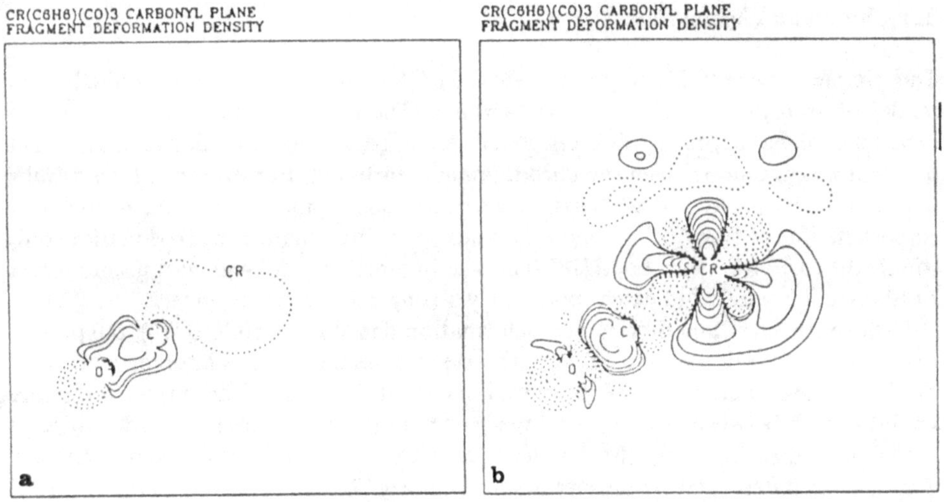

Fig. 6a, b. Fragment deformation density maps for $[(C_6H_6Cr(CO)_3]$: **(a)** Difference between the benzene ring and the carbonyls at their equilibrium positions and free benzene and carbon monoxide; **(b)** difference between the total molecular electron density and free benzene, three carbonyls and a lone chromium atom. The outermost contour has a value of 1/512 or .013 e $Å^{-3}$. Contours are geometric, differing by a factor of 2. (Reprinted with permission from [15] Copyright 1985, American Chemical Society)

carbonyls. However, in this difference map, the lack of any observable difference in the density of the benzene ring suggested that there is no interaction of the benzene ring with the carbonyl. There was a small difference in the density of the carbonyls, suggesting that the carbonyls interact with each other.

A second fragment deformation density was calculated by subtracting the density of free benzene and three carbon monoxide molecules and a lone, spherically symmetric, chromium atom from the total electron density of the molecule. In this map, the entire Cr—C bond region corresponds to a region of electron deficiency. At first, this seems in conflict with the large Cr—C accumulation observed in the standard deformation density but the buildup in the standard deformation density maps can be attributed to carbon lone-pair formation, not Cr—C bond formation.

The deficiency in the Cr—C bonding region in this fragment deformation density can be attributed to a combination of two effects: first, electron flow from the Cr *d* orbitals which point directly to the ligands to those which point between the ligands, the crystal field rearrangement, and second, donation of the carbon lone-pair density to the Cr atom. The donation of the carbon lone pair density to the Cr atom does not result in a zone of accumulation about the Cr since the density is donated to an already "depleted" orbital. But, if the electron deficient region between the Cr and C atoms is compared to the lobe of electron deficiency on the opposite side of the Cr atom, one can note fewer negative contours between the Cr and C atoms than on the opposite side. Other features of this fragment deformation include an elongation of charge accumulation perpendicular to the C—O bond axis indicating a Cr—CO interaction which leads to buildup in the carbonyls π region; a result of the π accepting nature of the carbonyl.

2.3 Chromium Hexacarbonyl

The simple octahedral compound, $Cr(CO)_6$, has been extensively studied as a model of transition metal-carbonyl bonding. These studies, both theoretical and experimental, have come to different conclusion as to, among other things, the amount of σ and π bonding between the chromium and carbonyl. For example, from results of scattered-wave Xα calculations, it was concluded that σ donation is the only important interaction for chromium-carbon bonding where π-back donation only affects the carbon-oxygen bond [18]. This was in direct conflict with various molecular orbital studies which indicated moderately strong π-back donation.

In the experimentally determined deformation density of $Cr(CO)_6$ [19], the pattern about the four independent Cr—C—O axes are similar to one another with most differences less than twice the estimated standard deviation. The largest difference ($\approx 3\sigma$) occurs between two carbon lone-pair regions, nevertheless, in order to get a better picture of bonding, the 4 independant Cr—C—O directions were averaged. This averaged deformation density is shown in Fig. 7.

The atomic charges of the Cr, C and O atoms were obtained by numerical integration of the averaged deformation density. They were found to be: $+0.15(\pm.12)$ for Cr, $+0.09(\pm.05)$ for C, and $-0.12(\pm.05)$ for O. When compared to previous quantum-chemically calculated values of the atomic charges [20], these values were found to

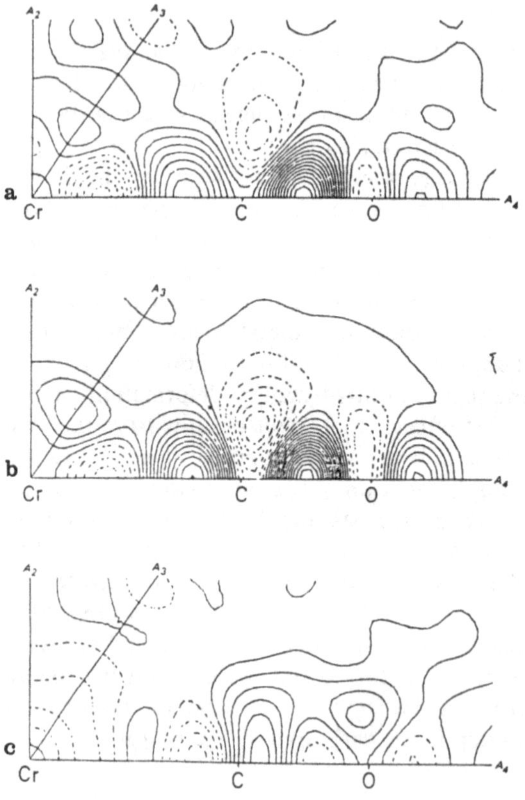

Fig. 7a–c. Average experimental deformation density for $Cr(CO)_6$: **(a)** average deformation density; **(b)** calculated deformation density, chromium atom in the configuration $3d(t_{2g})^{3.8}3d(e_g)^{1.2}4s$ plus carbon monoxide molecules; **(c)** average deformation density minus the molecular model density (**(a)**–**(b)**). A_2, A_3 and A_4 represent respectively two-, three-, and four-fold axes in the idealized octahedron. Countour interval: 0.05 e Å^{-3}. (Reprinted with permission from [19] Copyright 1976, American Chemical Society)

be relatively small, indicating that the net charge transfer between the chromium and carbonyl is relatively small.

The deformation density around the chromium atom showed maxima pointing towards the direction of the threefold axes of the idealized octahedron and minima pointing towards the bonds, as expected for octahedrally coordinated metals. The electronic configuration of the chromium atom was calculated from the best fit of the pattern of the deformation density about the chromium and this configuration was used in a molecular model of the charge density.

A molecular model of the charge density was computed by subtracting the aspherical chromium atom density and the density of a free carbon monoxide molecule from the averaged total density (see Fig. 7b). The resultant difference density showed deficiency in the σ region of the carbon monoxide compensated by accumulation in the π^* region as expected by σ-donation and π-back donation. Most of the region between Cr and C is negative except for a small accumulation zone at the midpoint of the bond. Values of σ and π electron charge transfers were obtained from the experimental map by computing the total occupancies for the 5σ and 2π orbitals of the carbon monoxide. They were calculated to be 1.6 and 0.4 electrons, respectively, which would indicate a charge transfer of 0.4 electrons for both σ and π bonding.

The theoretically determined deformation density for $Cr(CO)_6$ shows many of the same features (see Fig. 8) as the experimental map [21]. The comparison is quite good both qualitatively and quantitatively. In another theoretical study [22], the approach of a carbon monoxide ligand to a $Cr(CO)_5$ fragment was studied using deformation density maps. Two different difference densities were computed, the first was the atomic deformation density using the sum of the spherical atomic densities as

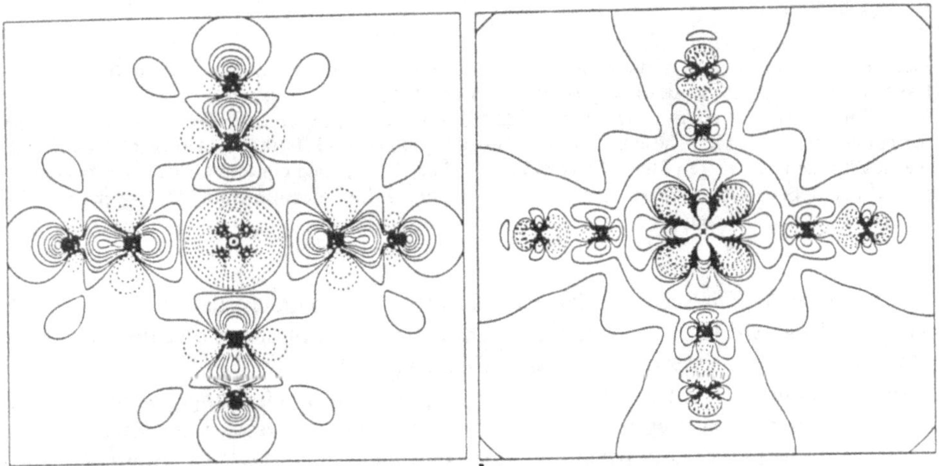

a b

Fig. 8a, b. Theoretical deformation density map from the LCAO-HFS wave function for $Cr(CO)_6$. Calculated from a double-zeta basis set plus *d* polarization: **(a)** Difference between $Cr(CO)_6$ and the spherically ayeraged atoms; **(b)** difference between $Cr(CO)_6$ and the $Cr(t_{2g})^6$ and the CO molecules. Contours: 0.00, $\pm$0.005, $\pm$0.01, $\pm$0.02, $\pm$0.03, $\pm$0.04, $\pm$0.05, $\pm$0.075, $\pm$0.10 e a.u.$^{-3}$; *solid lines*, positive and zero contours; *dashed lines*, negative contours. (Reprinted with permission from [21])

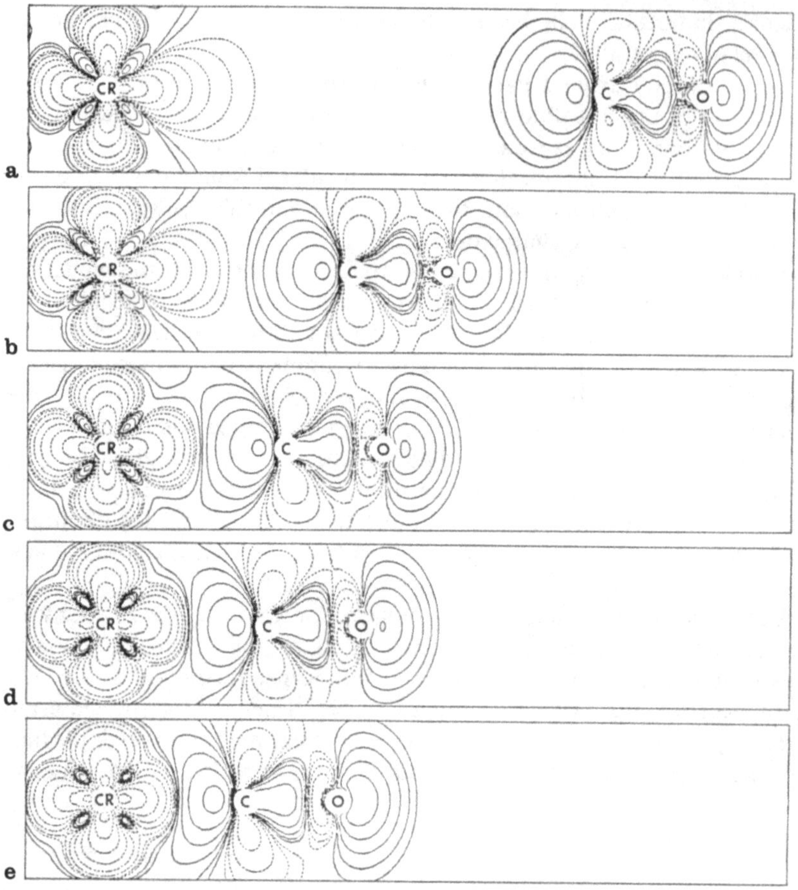

Fig. 9a–e. Atomic deformation density maps for the approach of a carbonyl to the $Cr(CO)_5$ fragment: **(a)** at infinite separation, plotted at 4 Å apart; **(b)** 1.0 Å longer than equilibrium distance; **(c)** 0.25 Å longer than equilibrium distance; **(d)** the equilibrium separation; **(e)** 0.25 Å shorter than the equilibrium distance. The contour intervals are geometric, differing by a factor of two, with the smallest contour representing ± 0.00390625 (2^{-8}) e a.u.$^{-3}$, *solid* contours denote electron gain, dashed contours denote electron density loss. Reprinted with permission from [22] Copyright 1983, American Chemical Society)

the promolecule. The second difference density was a fragment deformation density using the fragment $Cr(CO)_5$ and a free CO molecule as the promolecule. The series of deformation densities can be seen in Fig. 9 and 10.

In the atomic deformation density series (Fig. 9), only small changes were observed as the carbon monoxide approaches the $Cr(CO)_5$ fragment. Most of the changes were observed in the contours of the oxygen lone pair region and the C—O bond. Between the chromium and carbon, there is a deformation of C lone-pair density brought about by a shift of σ electron density from O towards C and the Cr—C bond region and the Cr d_σ loss. As the carbonyl approaches, the zero-density region moves from the carbon side of the Cr—C bond towards the midpoint of the bond, indicative of net movement of the "lone-pair" density into the interbond region. In this series of

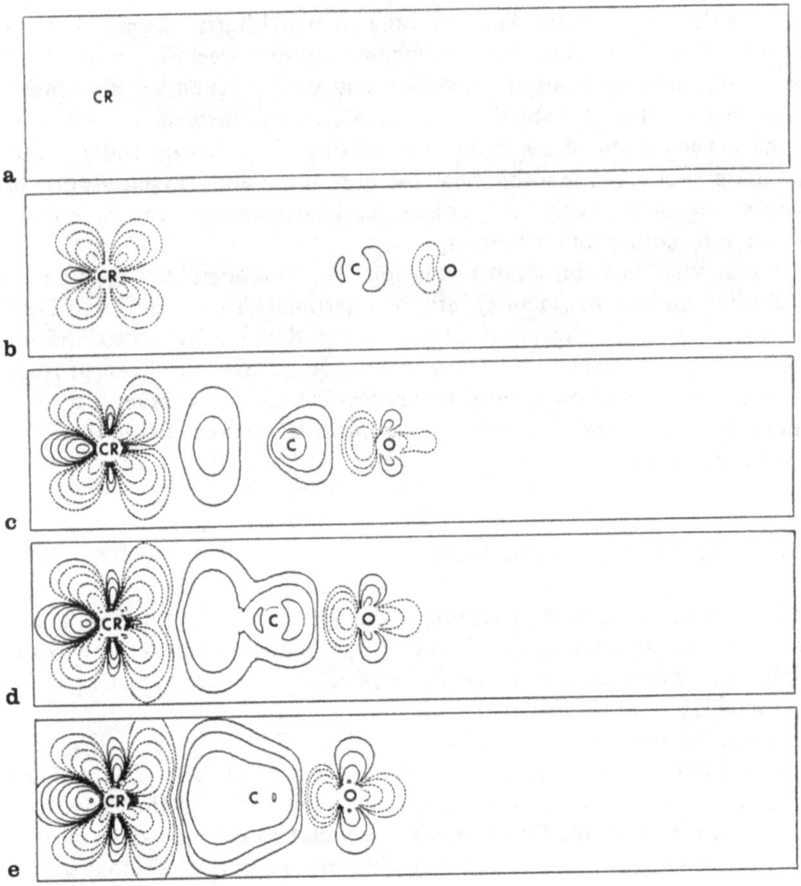

Fig. 10a–e. Fragment deformation density maps for the approach of a carbonyl to a $Cr(CO)_5$ fragment. The countering is the same as in Fig. 9. (Reprinted with permission from [22] Copyright 1983, American Chemical Society)

maps, the actual density change due to Cr—C bond formation is overwhelmed by "lone-pair" formation of the carbonyl and the atomic arrangement about the chromium atom.

The fragment deformation density maps, seen in Fig. 10, display the changes due to Cr—C bonding very clearly. At infinite separation, the fragment deformation density is zero everywhere. When the carbonyl comes to about 1 Å away from the equilibrium bond length (see Fig. 10b), changes in the carbonyl's σ framework occur in which the carbonyl ligand is beginning to act as a σ donor. Concurrently, the electron density around the chromium atom is responding to the incoming CO ligand. The Cr $3d_\pi$ orbitals interact strongly with the carbonyls of the $Cr(CO)_5$ fragment, donating more π-electron density to these carbonyls, as shown by the appearance of negative density contours in the Cr-d_π orbitals. As a result, these carbonyls become better σ donors in response to the extra π density donated to them, as evidenced by the accumulation in the d_σ orbitals of chromium. This accumulation is larger in the *trans* position than the *cis* positions. The incoming carbonyl ligand begins to act as a π acceptor at 0.25 Å away from the equilibrium position (see Fig. 10c) where the loss in the d_π orbital lobes pointing towards the incoming ligand becomes much larger than that in the other d_π orbital lobes. Concurrently, there is a buildup of π density on the oxygen atom of the incoming carbonyl. The density changes due to σ and π bonding is clearly separable near the chromium and oxygen atoms. In the Cr—C bonding region and near the carbon atom, the density difference is positive everywhere and due to both σ and π bonding.

Summing up, the changes upon bond formation are: 1) reorganization of CO σ density to put the carbon lone-pair density into the interbond Cr—C region and 2) an increase in π density on the CO ligand and loss of Cr d_π density. These descriptions indicate that π-bonding between Cr and CO actually incurs a transfer of charge from the chromium atom to the carbon monoxide whereas σ-bonding is more of a rearrangement of charge density and does not involve much transfer of charge from the carbonyl to the chromium atom.

2.4 Metall Carbene and Carbyne Complexes

Multiple bonds between a metal and carbon atoms has been the topic of much research. The experimental deformation density maps of two of the simpler metal carbene and carbyne complexes, $(OC)_5Cr{=}C(CH_3)OC_2H_5$ and $Cl(OC)_4Cr{\equiv}CCH_3$, have been reported [23].

The deformation density map of $(OC)_5Cr{=}C(CH_3)OC_2H_5$, shown in Fig. 11, has many features similar to the deformation density map of $Cr(CO)_6$. In the plane of the chromium and four carbonyl groups, there are accumulations in the Cr—C and C=O bonds. The maximum of the Cr—C bond accumulation is closer to the carbon atom, ranging from 1.20–1.35 Å away from the chromium atom. The arrow shown in Fig. 11a indicates a feature attributed to an intermolecular interaction between C3O3 and a carbonyl (C2O2) of a neighboring molecule.

Qualitatively, the deformation density associated with the Cr=C bond is very similar to that of the Cr—C(O) bonds. One difference is found in the position of the maximum of the Cr=C bond accumulation, which is 0.15 Å further away from the chromium atom (1.50 Å versus 1.35 Å for Cr—C(O)). However, the distance of this maximum to the carbon atom is the same as that in the Cr—C(O) bonds (0.55 Å). The smaller size of the Cr=C bond accumulation relative to the Cr—C(O) bond accumulation is attributed to a weaker Cr=C bond relative to the Cr—C(O) bond. This is also consistent with the longer Cr=C bond distance of 2.053 Å relative to the average Cr—C(O) bond distance of 1.905 Å. However, one must remember that most of this accumulation is due to carbon lone-pair formation and not Cr—C bonding.

The density in the carbene ligand shows large accumulations for the oxygen to carbene carbon bond and the carbene to methyl carbon bond. The oxygen-ethyl group bond peak is much weaker. In addition, a peak can be seen for a portion of the oxygen lone pair. About the chromium atom, there are accumulations in the threefold axes of the octahedron 0.6 Å from the chromium atom.

One of the surprising results of the low temperature x-ray structure determination of $Cl(CO)_4Cr{\equiv}CCH_3$ was that the $Cl(CO)_4Cr$ portion of the complex assumed

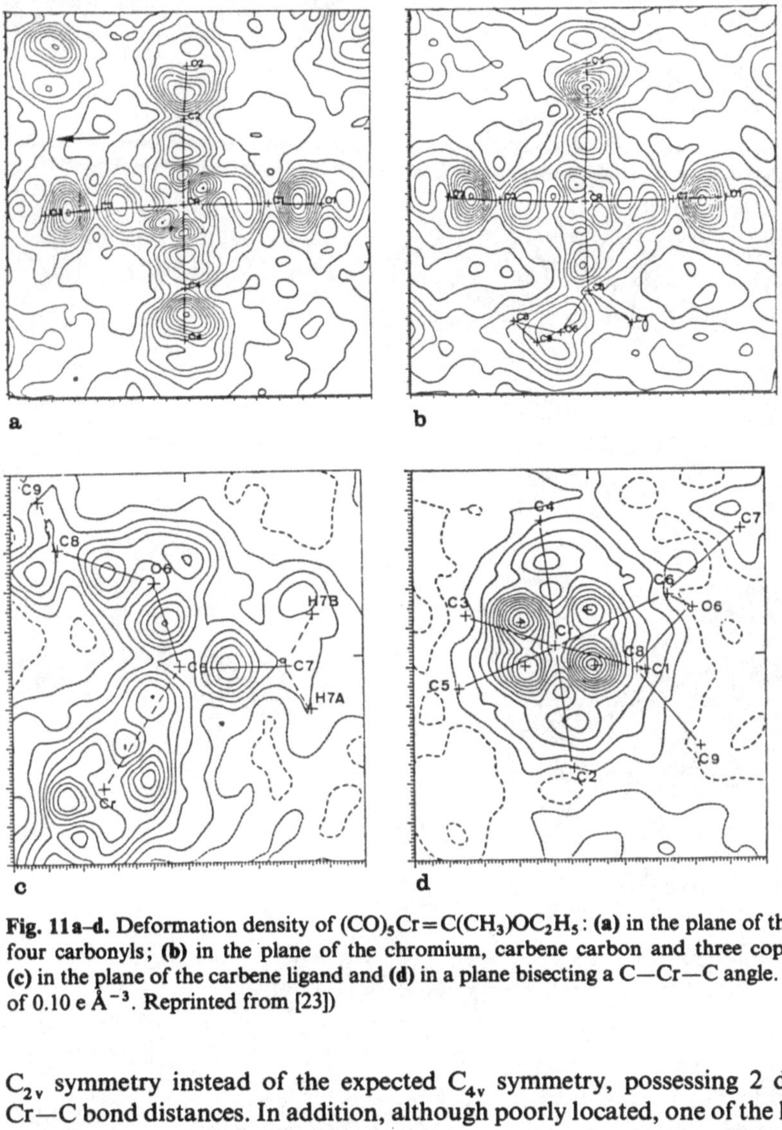

Fig. 11a–d. Deformation density of $(CO)_5Cr{=}C(CH_3)OC_2H_5$: **(a)** in the plane of the chromium and four carbonyls; **(b)** in the plane of the chromium, carbene carbon and three coplanar carbonyls; **(c)** in the plane of the carbene ligand and **(d)** in a plane bisecting a C—Cr—C angle. Contour interval of 0.10 e $Å^{-3}$. Reprinted from [23])

C_{2v} symmetry instead of the expected C_{4v} symmetry, possessing 2 distinct sets of Cr—C bond distances. In addition, although poorly located, one of the hydrogens was found to be clearly in the plane of the two carbonyls with the shorter Cr—C bond distances. These features are suggestive of a hyperconjugative effect involving one of the hydrogens of the methyl group and the chromium atom which would make the carbonyls inequivalent. This effect is shown below:

$$Cl(CO)_4Cr{\equiv}C{-}CH_3 \leftrightarrow$$

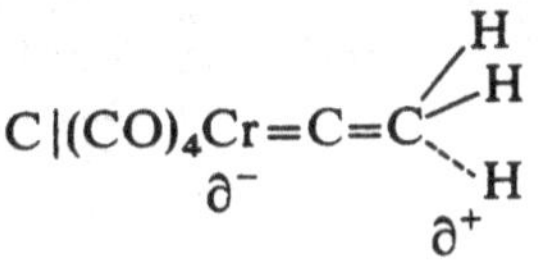

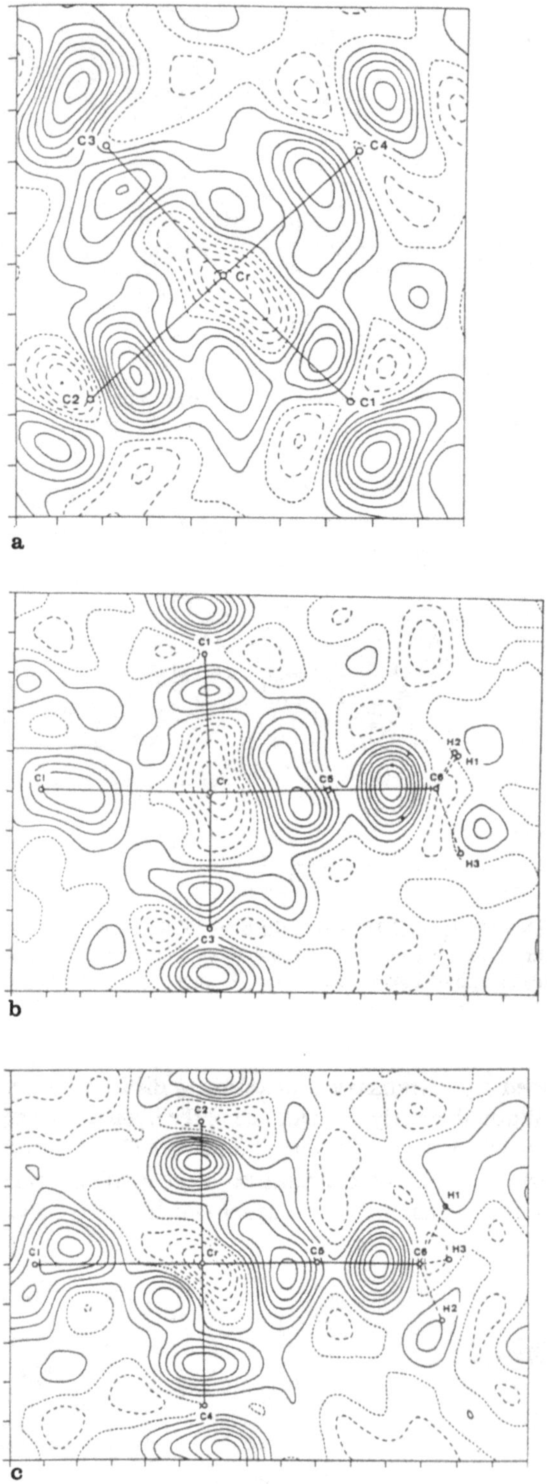

Fig. 12a–c. Deformation density of $Cl(CO)_4Cr{\equiv}CCH_3$: **(a)** in the plane of the chromium atom and the four carbonyl groups; **(b)** in the plane of the chromium, carbyne carbon and the two carbonyls with short Cr—C distances; **(c)** in the plane of the chromium, carbyne carbon and the carbonyls with long Cr—C distances. Contour interval of 0.1 e Å^{-3}. (Reprinted from [23])

The lowering of symmetry from 4-fold to 2-fold can clearly be seen in the deformation density maps shown in Fig. 12. There is a higher accumulation of density in the shorter Cr—C bonds. Also, the accumulation associated with the Cr≡C bond is quite extended towards the equatorial carbonyls as shown in Figures 11b and 11c. The accumulation between the carbyne carbon and the methyl carbon is also quite large. The last two observations seem to support the hyperconjugative effect causing the inequivalence of the carbonyls. Another possible cause of the lowering of symmetry could be crystal packing effects.

2.5 Metalloporphyrins

Another large class of organometallic compounds that have been studied extensively through the use of electron deformation densities is the metalloporphyrins. The electronic structure of metalloporphyrins is studied to serve as a model for biological processes involving heme proteins. Ideally, the deformation density maps should reveal the electronic ground state configuration of the central metal atom.

In almost all of the deformation density maps of the metalloporphyrins, the density about the porphyrin rings remains fairly consistent. In a typical deformation density ma of a tetraphenylporphyrin complex, shown in Fig. 13, density accumulations are found for all the C—C and C—N bonds. The accumulation peak for the C—C bond in the pyrrole ring opposite the nitrogen is larger than the other C—C ring bonds. Although π bonding is perpendicular to this plane, the shorter C—C bond length results in stronger accumulation in the σ bond, which is often taken as an indication of more double bond character in this bond. Additionally, the density peaks for the bonds of the pyrrole ring are located slightly off the C—C axis, outside of the ring, a result which can be attributed to ring strain.

Both experimental and theoretical deformation density maps have been studied for porphyrin complexes of a number of metals in differing oxidation and spin

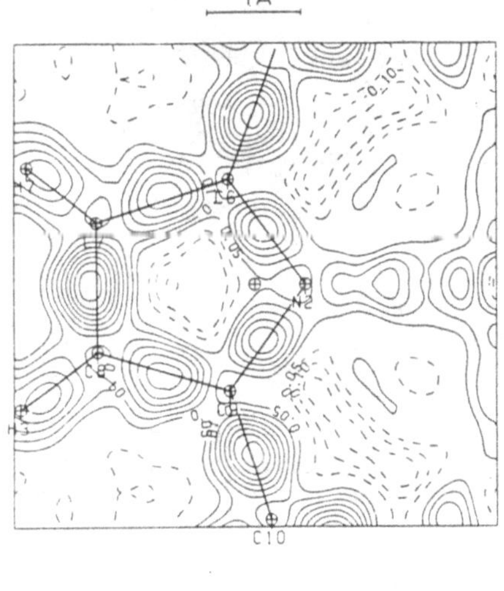

Fig. 13. Deformation density of a pyrrole ring of tetraphenylporphyrin after averaging over the four pyrrole rings of $[Fe(II)(THF)_2TPP]$. Contour interval of 0.05 e $Å^{-3}$. (Reprinted from [28])

states. Examples include: [Co(II)TPP] [24]; [Fe(III)OMeTPP] [25]; [Ni(II)(TMP)] [26]; [Fe(II)(Pc)] [27]; [Fe(II)(THF)$_2$TPP] [28]; [Fe(II)TPP] [29] and [Fe(III){meim}$_2$-OEP]ClO_4 [30]. The maps of [Fe(II)TPP] will be presented here as an example.

For [Fe(II)TPP], the electronic ground state configuration is a subject of some debate, with some experimental observations favoring the $^3A_{2g}$ from the $\{(xy)^2(xz, yz)^2(z^2)^2\}$ configuration [31] while other observations favor the ^3E_gA from the $\{(xy)^2(xz, yz)^3(z^2)^1\}$ configuration [32]. Theoretical ab initio SCF-CI calculations on iron (II) porphyrin has been reported on two levels [33]. The first CI level correlated all the Fe 3*d* electrons and found the found the $^3A_{2g}$ state lower in energy than the ^3E_g state by 600 cm^{-1}. The weight of the main configuration of the ^3E_g state $\{(xy)^2(xz, yz)^3(z^2)^1\}$ is 89%. However, the weight of another configuration of the same symmetry, designated as E_gB, $\{(xy)^1(xz, yz)^3(z^2)^2\}$, is found to be 10%. The weight of the main configuration of the $^3A_{2g}$ state is 99.5%. The second CI level correlated the Fe 3*d* electrons and the π electrons of the porphyrin ring. This calculation found the $^3A_{2g}$ state lower in energy than the ^3E_gA state by 2200 cm^{-1}. The weight of the main configuration for both state is found to be 79% with no other configurations contributing more than 1% for either state. The weight of the E_gB configuration in the ^3E_g state is now only 0.5%.

Deformation density maps of both electronic configurations (see Fig. 14) were plotted from this wave function. The main difference found in the deformation densities of the two states was found along the z axis with accumulation for the $^3A_{2g}$ state and deficiency for the ^3E_gA state. Due to its lower energy, the $^3A_{2g}$ state was predicted to be the electronic ground state configuration.

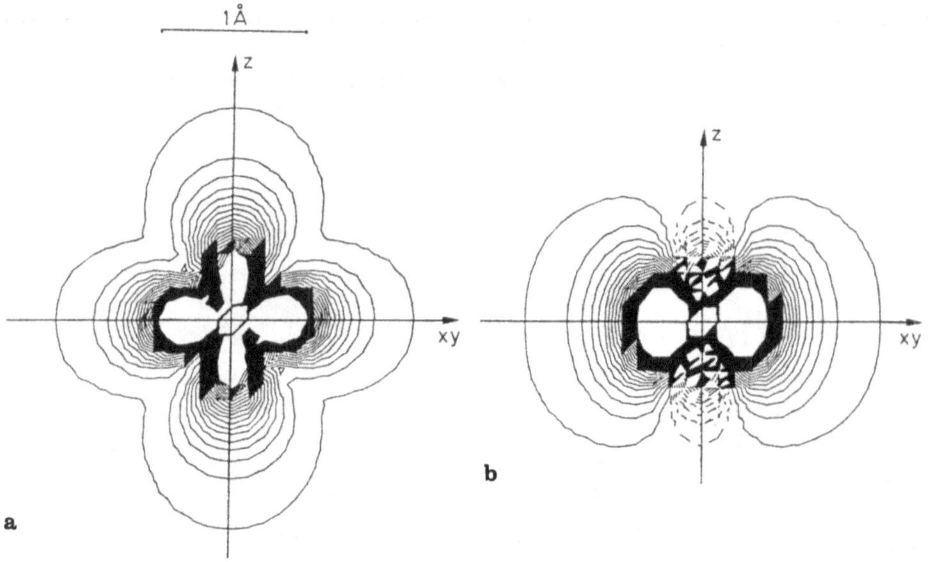

Fig. 14a, b. Computed deformation density maps from the ab initio SCF—CI wave function of iron(II)porphyrin: **(a)** the $^3A_{2g}$ state and **(b)** the ^3E_g state, in a plane perpendicular to the porphyrin ring and bisecting the N—Fe—N angle. Contour interval of 0.10 e Å^{-3}, negative contours are *dashed*. (Reprinted with permission from [33])

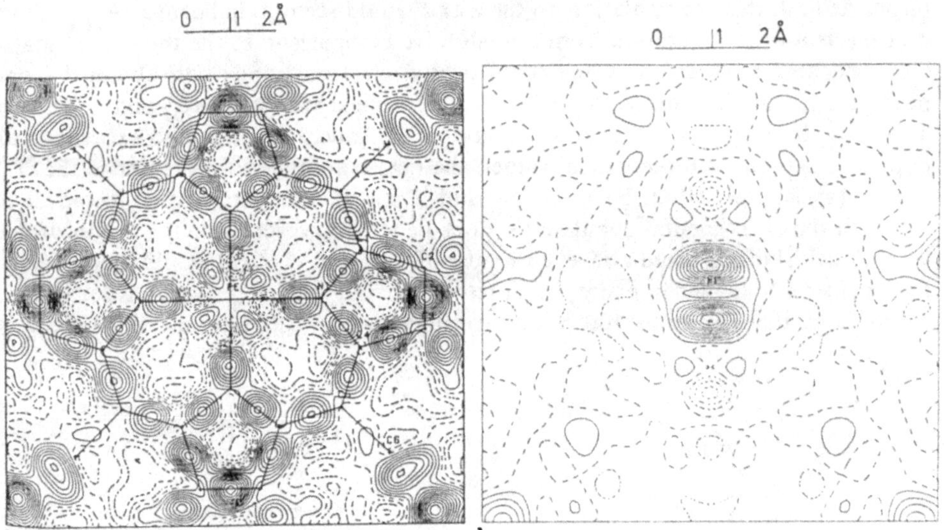

a b

Fig. 15a, b. Experimental deformation map of *meso*-[Fe(II)TPP]: **(a)** through the iron and the porphyrin plane. Contour interval of 0.05 e $Å^{-3}$; **(b)** through the plane bisecting N—Fe—N bonds, perpendicular to the average porphyrin plane. Contours at 0.10 e $Å^{-3}$. The Fe atom is indicated by a $\pm$. Zero and negative contours are *broken*. (Reprinted with permission from [29])

The experimental deformation density map from the crystal structure determination of meso-[Fe(II)TPP] [29] is shown in Fig. 15. In the map, the depopulation of the $d_{x^2-y^2}$ orbital is indicated by deficiency along the Fe—N bonds. Population of the d_{xy} orbital is indicated by accumulation in the bisector of the Fe—N bonds. There is also accumulation above and below the Fe atom perpendicular to the porphyrin

Table 2. Experimental and theoretical *d*-Orbital occupancies compared with those of a spherical Fe atom

Orbital	Experimental Fe(II)TPP	Theoretical		
		$^3A_{2g}$	^3E_g	Spherical
$d_{x^2-y^2}$	0.43 (6.0%)	0.18 (2.9%)	0.18 (2.9%)	1.2 (20%)
d_{z^2}	1.70 (23.7%)	1.94 (31.7%)	0.99 (16.1%)	1.2 (20%)
d_{xz}, d_{yz}	3.54 (49.3%)	1.98 (32.3%)	2.96 (48.1%)	2.4 (40%)
d_{xy}	1.50 (20.9%)	2.02 (33.0%)	2.02 (32.8%)	1.2 (20%)

d-Orbital Occupancies for Triplet State Configurations

Orbital	$^3A_{2g}$	^3E_gA	^3E_gB	$^3B_{2g}$
$d_{x^2-y^2}$	0	0	0	0
d_{z^2}	2 (33%)	1 (17%)	2 (33%)	1 (17%)
d_{xz}, d_{yz}	2 (33%)	3 (50%)	3 (50%)	4 (67%)
d_{xy}	2 (33%)	2 (33%)	1 (17%)	1 (17%)

plane. At first, this accumulation on the z axis would seem to indicate a $^3A_{2g}$ ground state electronic configuration from a qualitative comparison to the theoretical maps discussed above. However, when the *d* orbital electron population obtained from multipole parameters are compared to the theoretical values of $^3A_{2g}$ and ^3E_gA (see Table II), the (xy) populations are found to be much lower in the experimental populations than both theoretical values, the (xz, yz) population is closer to the ^3E_g and the (z^2) is intermediate between $^3A_{2g}$ and ^3E_g. A 40:60 mixture of ^3E_gA and ^3E_gB would fit the experimental populations much better. However, from the results of the SCF-CI calculations noted above, a mixture of 40:60 ^3E_gA:^3E_gB would theoretically be considerably higher in energy than either the $^3A_{2g}$ and ^3E_g states. According to the authors, these experimental results are suggestive of a ^3E_g ground state which is

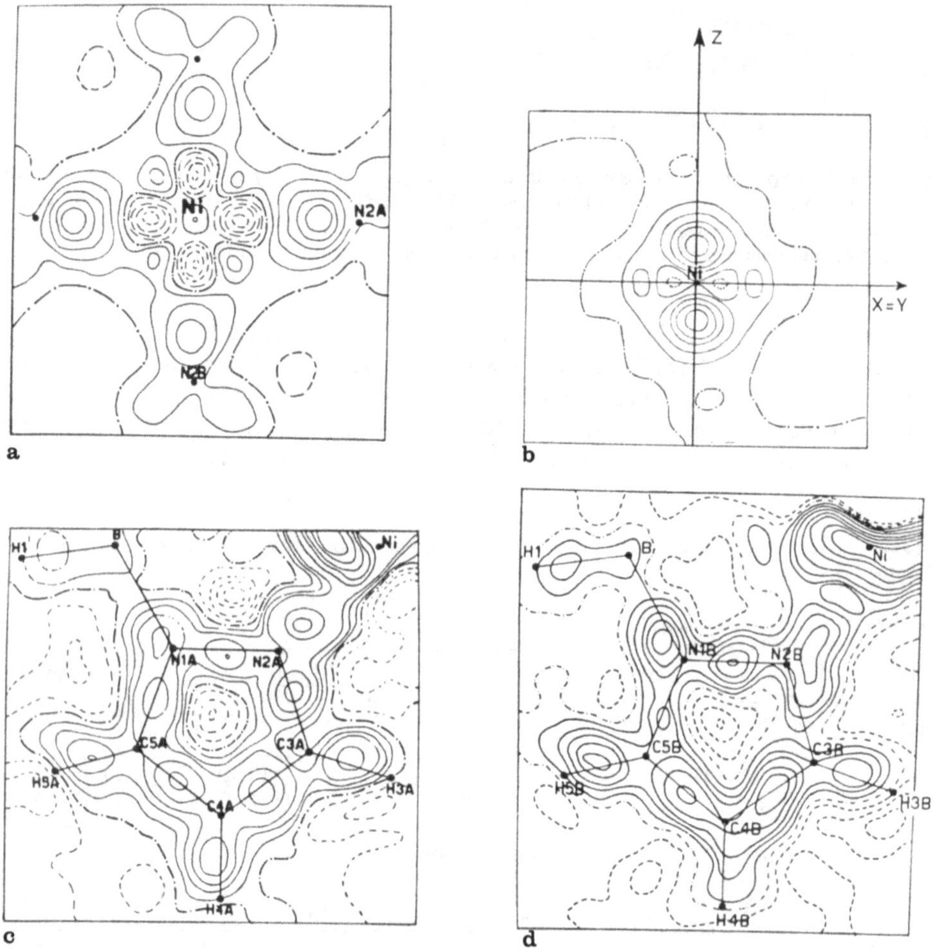

Fig. 16a–d. Experimental deformation density after multipole refinement of $[H_2B(C_3N_2H_3)_2]Ni(II)$: **(a)** in square NiN_4 plane; **(b)** in the plane perpendicular to the NiN_4 plane passing midway between the two pyrazolyl rings; **(c)** pyrazolyl ring A; **(d)** pyrazolyl ring B. Contours are at 0.1 e Å^{-3} intervals. (Reprinted with permission from Ref. [34]. Copyright 1987, American Chemical Society)

a mixture of several configurations with a smaller than 50% contribution of the ^3E_gA configuration.

The presence of ligands in the axial position will greatly influence the ground state electronic configuration at the iron. For example, for [Fe(II)Pc] the experimental deformation density [27] shows a deficiency in the z direction which rules out the $^3A_{2g}$ configuration. The pattern of accumulation around the iron atom points to the ^3E_gA ground state configuration. The reason for this seeming inconsistency can be found in the nonbonding interactions between neighboring molecules. The iron atom has a short nonbonding distance of 3.42 Å in the axial direction to a nitrogen atom of an adjacent molecule. This interaction is sufficient to cause a substantial change in the electronic ground state configuration.

More recently, the experimental deformation density of a related square planar nickel complex, bis(dihydrodi-1-pyrazolylborato)nickel(II) $\{[H_2B(C_3NH_3)_2]_2Ni\}$, has been reported [34]. This map, shown in Fig. 16, shows many of the same general features as the porphinato complexes. In the area around the nickel atom, there are accumulations in the xy direction bisecting the N—Ni—N angle and along the z axis perpendicular to the NiN_4 plane. In the Ni—N bonding region, there is an accumulation situated closer to the nitrogen which represents the nitrogen lone pair. The density of the ligands shows accumulations in all bonds. The accumulation of the B—N bond is shifted closer to the more electronegative nitrogen atom. The sizes of the accumulation peaks in the ligand are in the order: C—C > C—N ≃ N—N > C—H > N—B. There is no indication of alternating single- or double-bond character in the five-membered ring; this lack of alternation suggests electron delocalization about the ring.

3 Dinuclear Transition Metal Complexes

Deformation density maps of various dinuclear transition metal complexes have been studied to determine the nature of the metal-metal bond. However, for many covalently bonded dinuclear transition metal complexes, the buildup of charge between bonded metal atoms is not observed in either experimental nor theoretical maps. The reasons attributed to the lack of charge accumulation are similar to those discussed earlier in the introduction. Metal-metal bonds tend to involve the overlap of *d* orbitals over a relatively long bonding distance. The metal *d* orbitals tend to be quite compact in the region of the bond; thus, their weak overlap results in a diffuse accumulation.

Additionally, the choice of promolecule as the sum of sperhical ground state atoms, already a problem for molecules containing atoms with greater than halffilled 2*p* orbitals, is even less appropriate for the description of metal-metal bonds. For many of the dinuclear compounds, subtracting the density of properly chosen fragments produces a more chemically meaningful deformation density map.

3.1 Dinuclear Carbonyl Compounds

The dinuclear carbonyl compounds: [$Mn_2(CO)_{10}$], [$Fe_2(CO)_9$] and [$Co_2(CO)_8$] are a good series to study since they represent 3 different types of metal-metal bonds. According to electron counting rules, all three compounds should posess a single metal-metal bond. In [$Mn_2(CO)_{10}$], there is just an unsupported metal-metal bond between two square pyramidal $Mn(CO)_5$ fragments. The metal-metal bond in [$Fe_2(CO)_9$] is supported by three bridging carbonyls, whereas the metal-metal bond in [$Co_2(CO)_8$] is supported by two bridging carbonyls and has been described as "bent" in the direction of the "missing" bridging carbonyl (when compared to [$Fe_2(CO)_9$]).

In the crystal structure of [$Mn_2(CO)_{10}$], [35], it was observed that the equatorial carbonyls were slightly bent towards the other manganese atom. This was taken by some as evidence for semibridging character of these equatorial carbonyls [36]. However, this bend could also be caused by steric interaction between the carbonyls on the manganese, a proposal supported by the observed larger bend of the equatorial carbonyls towards the hydrogen atom in the structure of $HMn(CO)_5$ [37].

The experimentally determined deformation density of [$Mn_2(CO)_{10}$] [38], shown in Fig. 17, shows a slightly positive accumulation at the Mn—Mn midpoint which is

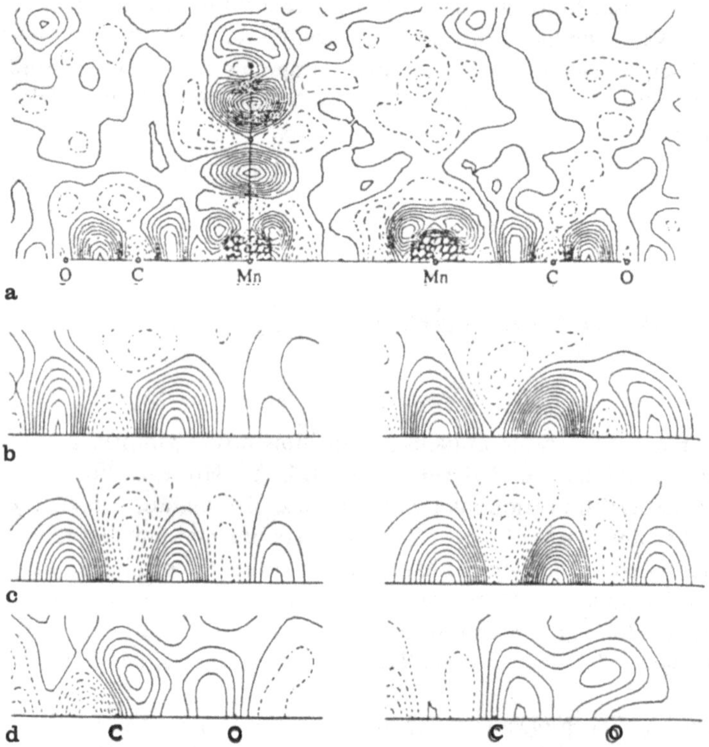

Fig. 17 a–d. Experimental deformation density of [$Mn_2(CO)_{10}$], from: **(a)** average deformation density along Mn—Mn bond; **(b)** Observed deformation density of the axial (*left*) and average equatorial carbonyl (*right*); **(c)** density calculated for carbon monoxide; **(d)** difference between **(b)** and **(c)**. Contour interval: 0.05 e Å^{-3}, negative contours are *dashed*. (Reprinted with permission from [38])

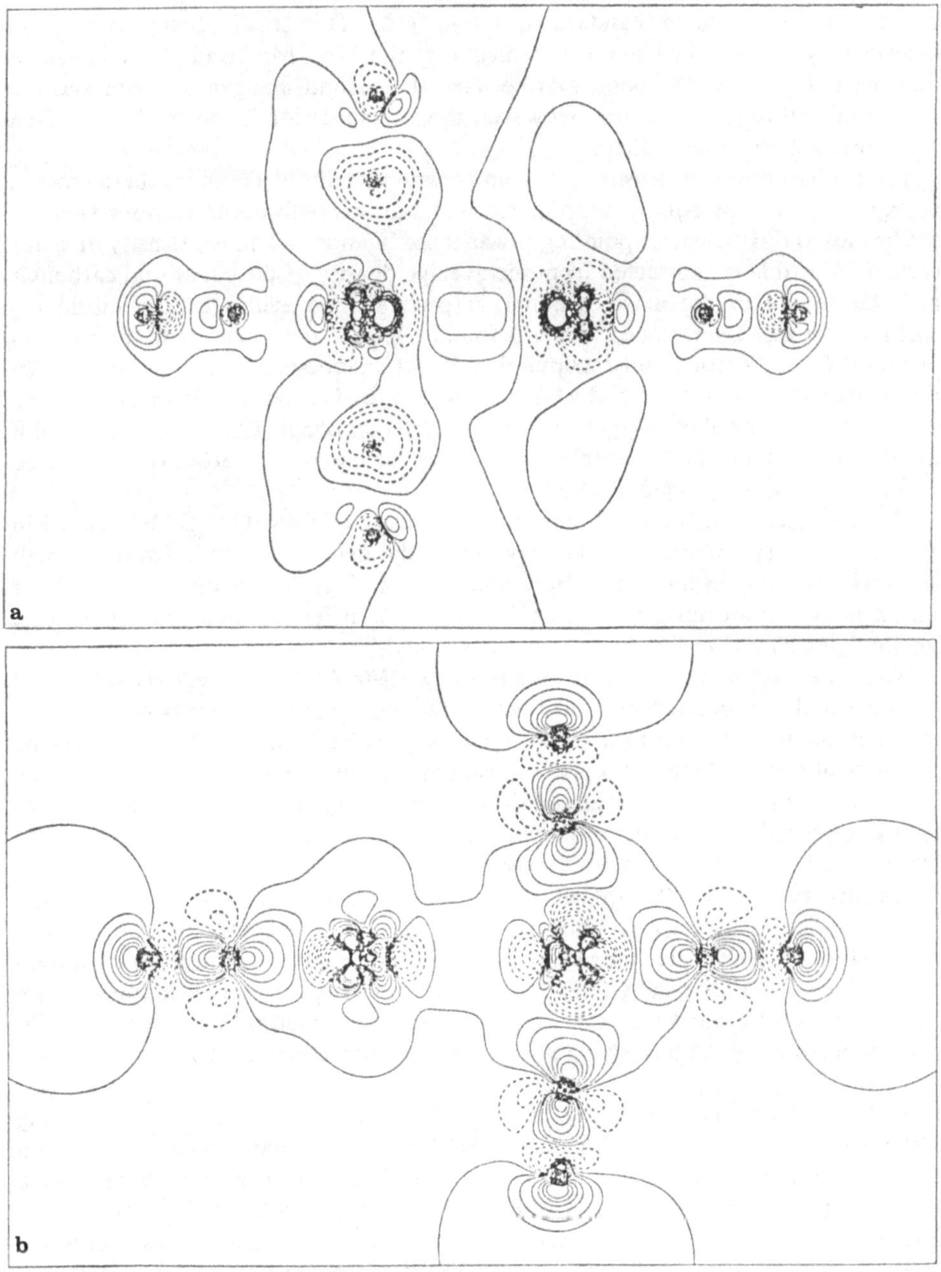

Fig. 18a, b. Theoretical deformation density maps from the LCAO-HFS wave function of $[Mn_2(CO)_{10}]$ from a double-zeta Slater-type basis set: **(a)** difference between $[Mn_2(CO)_{10}]$ and 2 $Mn(CO)_5$ fragments; **(b)** difference between $[Mn_2(CO)_{10}]$ and the Mn, C and O atoms, in the XZ plane. Contours (a.u.): 0.00, ±0.005, ±0.01, ±0.02, ±0.03, ±0.04, ±0.05, ±0.075, ±0.10 for (a); 0.00, ±0.01, ±0.02, ±0.04, ±0.06, ±0.08, ±0.10, ±0.15, ±0.20 for **(b)**, positive and zero contours are *solid*, negative contours are *dashed*. (Reprinted with permission from [39])

not above the estimated standard deviation (esd). This small accumulation zone extends very far into the plane perpendicular to the Mn—Mn bond. A second small maxima is found off the bond axis between the second manganese atom and the equatorial carbonyl on the first. However, this accumulation is too small to prove a direct interaction between them.

The configuration of density in the immediate vicinity of the manganese atom is as expected for octahedrally coordinated metal atoms with accumulations between the ligands and deficiencies pointing towards the ligands. When the density of a free carbon monoxide is subtracted from the average density of the equatorial carbonyls and the density of the axial carbonyl, respectively, no evidence of semibridging character for the equatorial carbonyls is found. There is accumulation in the π region about the C and O atoms and depopulation of carbon lone-pair region, features which are attributed to σ-donation and π^*-accepting nature of carbonyls. These effects are a bit larger for the axial carbonyl than the equatorial carbonyl. This is to be expected if the axial carbonyl is more strongly bonded than the equatorial carbonyl (as reflected in the shorter $M—C_{ax}$ bond lengths).

The theoretically determined deformation density of $[Mn_2(CO)_{10}]$ [39], shown in Fig. 18, is very similar to the experimentally determined map with a small accumulation zone in the Mn—Mn bonding region. The deformation density about the axial and equatorial carbonyls again show little difference giving no evidence of semibridging character for the equatorial carbonyls.

When the sum of the electron density of two $Mn(CO)_5$ fragments is subtracted from the toal molecular density, the accumulation in the Mn—Mn bonding region increases significantly, up to 0.10 $eÅ^{-3}$ (see Fig. 18b). Thus, the formation of the metal-metal bond in $[Mn_2(CO)_{10}]$ does result in accumulation in the bonding region. The lack of density in the standard deformation density is due to the attachment of the five carbonyls to the Mn atom, which causes a decrease of density, compared to the spherical atom, along the bond axis.

The structure of $[Fe_2(CO)_9]$ has D_{3h} symmetry with 3 bridging carbonyls [40]. Earlier studies have accepted a direct Fe—Fe bond for the compound. However, a recent Hartree-Fock-Slater molecular orbital calculation on the electronic configuration of $[Fe_2(CO)_9]$ has shown that the two $Fe(CO)_3$ fragments are held together not by a direct Fe—Fe bond but rather a $Fe—(CO)_{2\pi}—Fe$ bond [41]. More recent SCF calculations with a larger basis set corroborates the absence of a direct Fe—Fe bond [42].

In the descriptions of bonding in $[Fe_2(CO)_9]$, the 2π orbitals on the bridging carbonyls are split into two regions called $2\pi(\parallel)$ and $2\pi(\perp)$ relative to the metal-metal axis. The computed deformation density maps [39], shown in Fig. 19, display a very large $(CO)_{br}$-$2\pi(\parallel)$ population which is in agreement with the above theoretical formulation of the bonding. Along the Fe—Fe axis, there is a negative density difference which by itself, does not discount the possibility of a direct Fe—Fe bond. However, this observation and the extension of the CO-$2\pi(\parallel)$ population does differentiate the metal-metal bonding in $[Fe_2(CO)_9]$ from that in $[Mn_2(CO)_{10}]$. An experimental deformation density map of $[Fe_2(CO)_9]$ has been briefly mentioned [43] as having a positive accumulation in the Fe—Fe bonding region. However, this was a very preliminary report.

The structure of $[Co_2(CO)_8]$ is similar to $[Fe_2(CO)_9]$ with one of the bridging

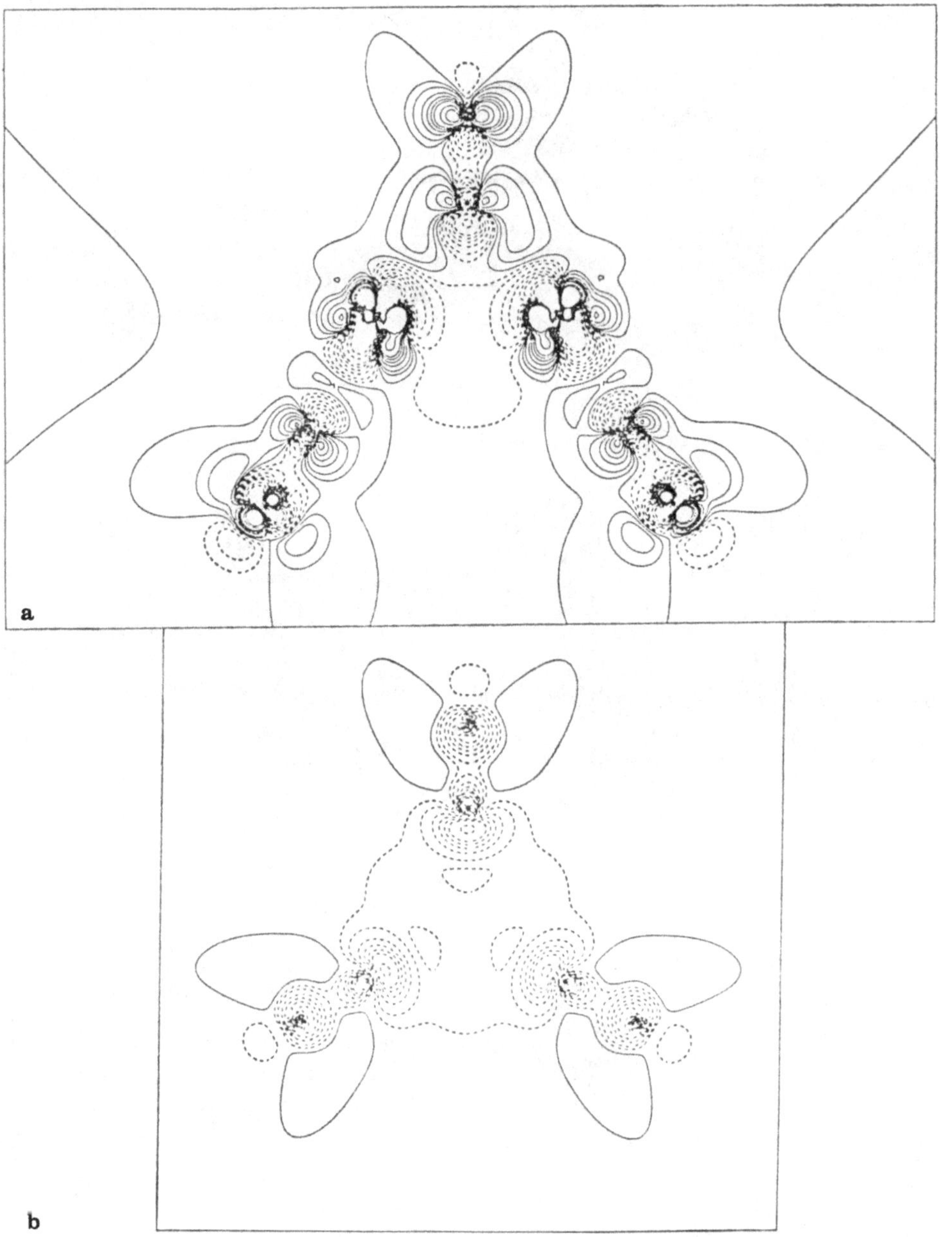

Fig. 19a, b. Theoretical deformation density map from the LCAO-HFS wave function of $[Fe_2(CO)_9]$ from a double-zeta Slater-type basis set. Difference between $[Fe_2(CO)_9]$ and the Fe atom and CO molecules: **(a)** in the **XZ** plane, **(b)** in the **XY** plane containing the three bridging CO's. Contours are the same as in Fig. 18a. (Reprinted with permission from [39])

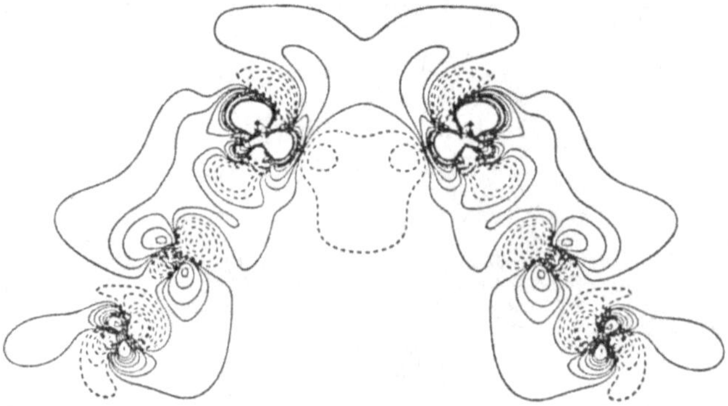

a

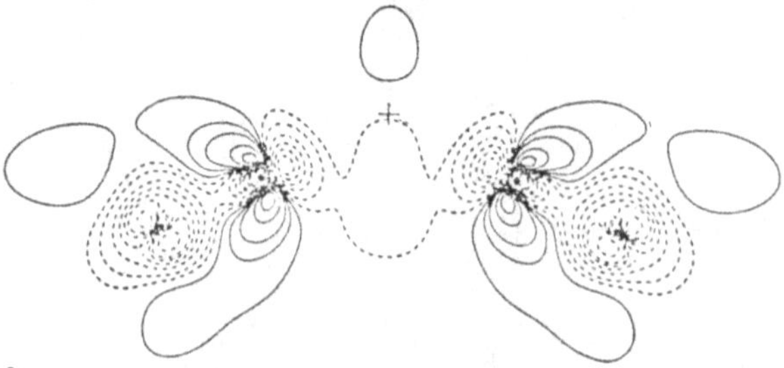

b

c

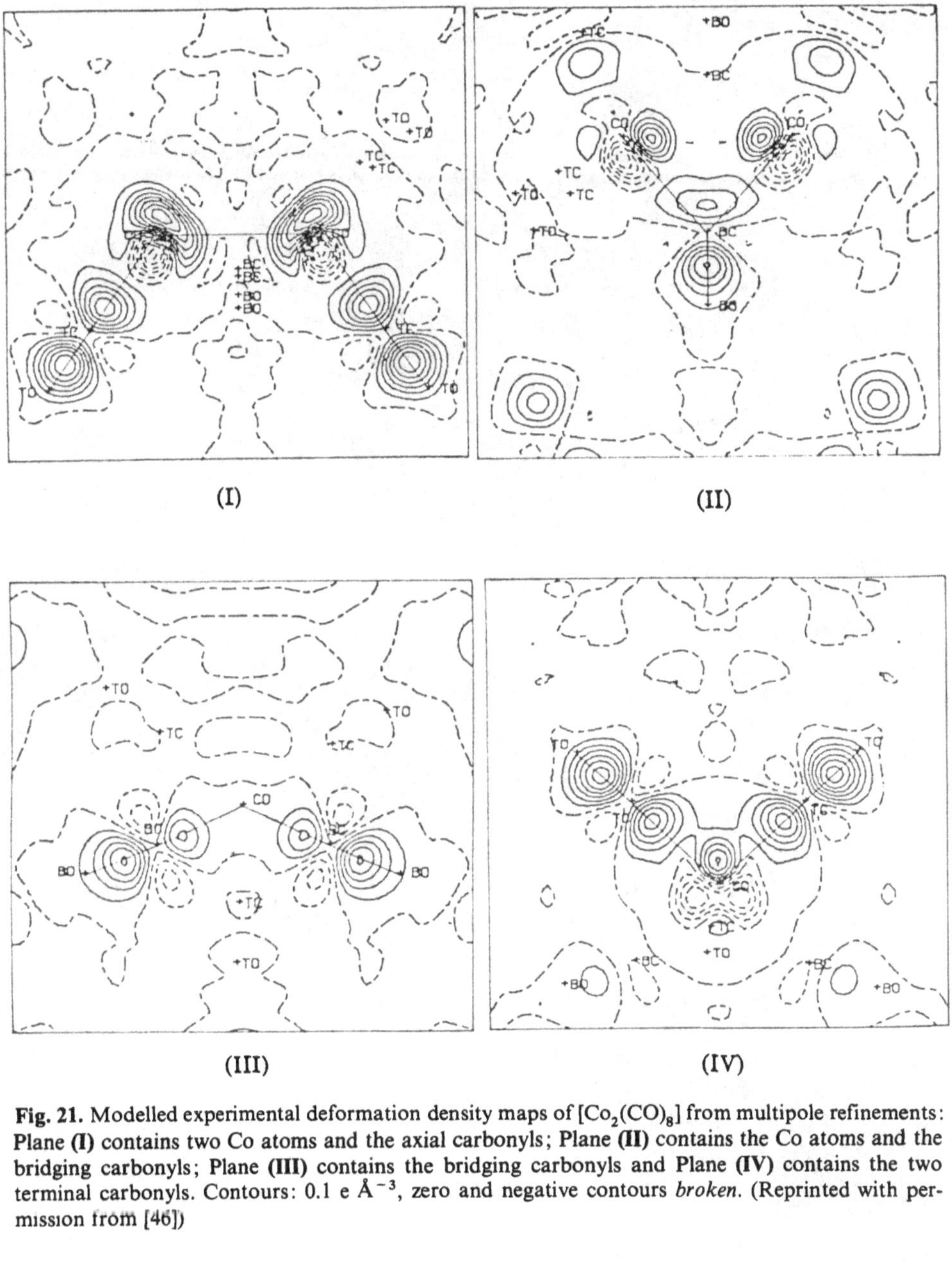

Fig. 21. Modelled experimental deformation density maps of $[Co_2(CO)_8]$ from multipole refinements: Plane **(I)** contains two Co atoms and the axial carbonyls; Plane **(II)** contains the Co atoms and the bridging carbonyls; Plane **(III)** contains the bridging carbonyls and Plane **(IV)** contains the two terminal carbonyls. Contours: 0.1 e $Å^{-3}$, zero and negative contours *broken*. (Reprinted with permission from [46])

◀

Fig. 20a–c. Theoretical deformation density maps of $[Co_2(CO)_8]$, density difference between $[Co_2(CO)_8]$ and the Co atom and the CO molecules: **(a)** in the XZ plane; **(b)** in the XY plane containing the bridging carbonyls (+ sign in the center indicates the Co—Co axis); **(c)** in the diagonal plane containing the two Co atoms and a bridging C atom. Contours are the same as in Fig. 18a. (Reprinted with permission from [39])

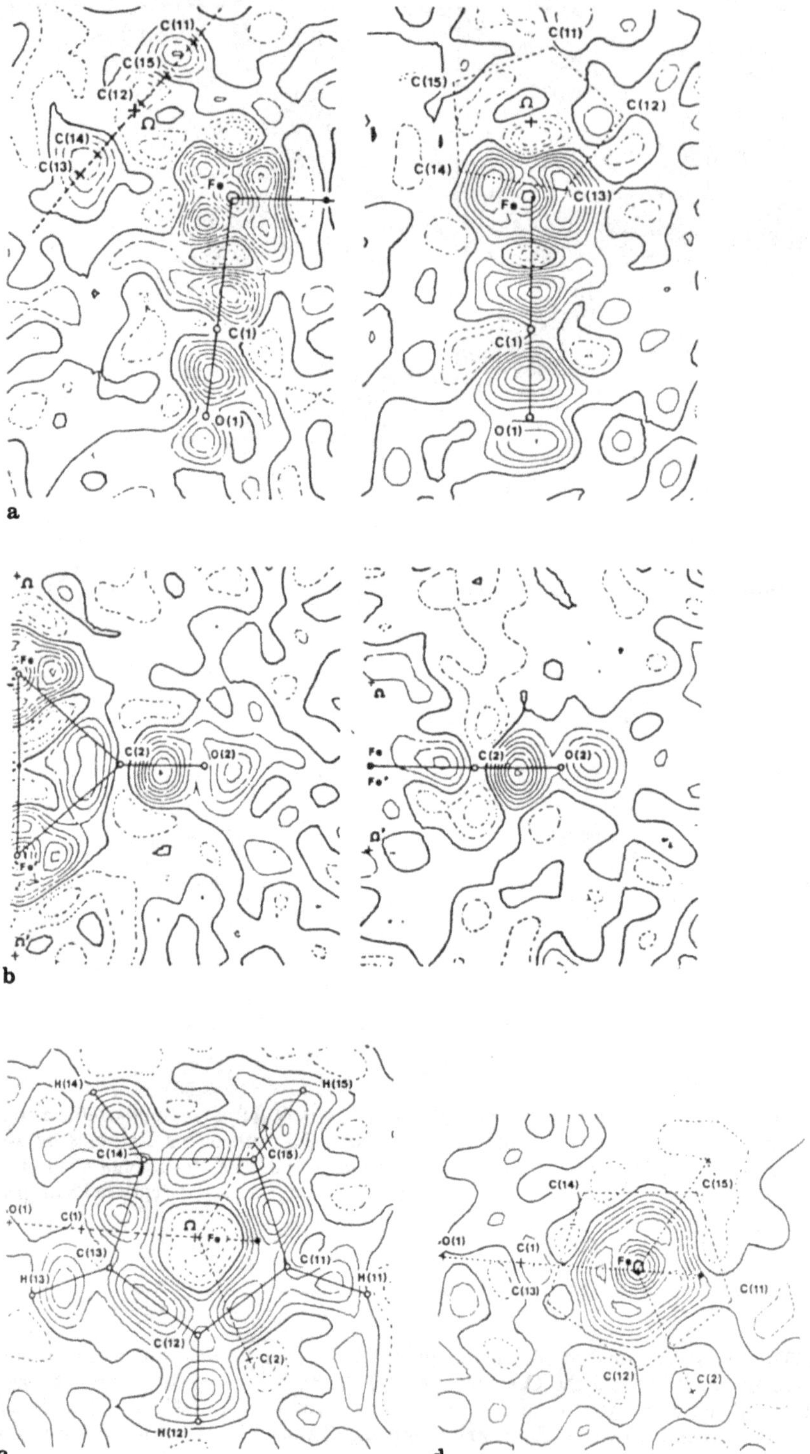

C(11)
C(15)
C(12)
C(14)
C(13)
Fe
C(1)
O(1)
a
Fe
C(2)
O(2)
Fe'
b
H(14)
H(15)
H(13)
H(11)
H(12)
C(2)
c
d

carbonyls missing [44]. Molecular orbital studies of the electron configuration of $[Co_2(CO)_8]$ seem to indicate the possibility of a bent Co—Co bond in the direction of the "missing" bridging carbonyl [45]. However, in the computed deformation density map, shown in Fig. 20, the very small accumulation in the bent bond region led to the conclusion that the bent bond could not be observed [39]. The density about the bridging carbonyls again shows a large population in the carbonyl $2\pi(\parallel)$ region. When compared to the deformation density map of $[Fe_2(CO)_9]$, a larger population in the carbonyl $2\pi(\perp)$ region is also found. There is no observable density accumulation on the Co—Co axis.

In the experimental deformation density map of $[Co_2(CO)_8]$ [46], (see Fig. 21) the electron density about the metal atoms is polarized towards the empty bridging site. The density at the back of the carbon atom of the bridging carbonyl is clearly elongated parallel to the Co—Co bond axis. One striking feature of the experimental deformation density is the presence of a large sharp peak at the center of symmetry of the complex. Since large sharp peaks are not what is expected for the overlap of metal d orbitals i.e., the absence of a peak in the theoretical map and the weak peak in the deformation density of $[Mn_2(CO)_{10}]$, this large peak was attributed to experimental error.

The compound $[CpFe(CO)_2]_2$ (Cp = C_5H_5) was one of the earliest dinuclear compounds studied with electron deformation density maps. As with the dinuclear carbonyls, electron counting rules require the formation of single Fe—Fe bond. Experimentally, the presence of a bond is also supported by the relatively short Fe—Fe distance and by the compound's diamagnetism.

However, in both the experimental [47] and theoretical [48] deformation density maps, shown in Fig. 22 and 23, respectively, little density accumulation is observed in the Fe—Fe bonding region. The lack of observable electron density in the metal-metal bonding region is consistent with the small negative Mulliken overlap population between the Fe atoms in SCF molecular orbital calculations on $[CpFe(CO)_2]_2$.

The experimental and theoretical deformation density maps qualitatively show similar features. The deformation density in part of the area about the iron atom shows features similar to other octahedrally coordinated metal atoms with accumulations in the three-fold axes of the octahedron (with one axis pointing towards the center of the cyclopentadiene ring, hereafter designated as Ω) and deficiencies pointing towards the carbonyl ligands. The pattern of density around the Fe-Ω axis, however, is different and tends toward cylindrical symmetry.

When the deformation density about the carbonyls is compared to that of free carbon monoxide, the bridging carbonyl is found to be altered from free carbon

---

◀

Fig. 22a–d. Experimental deformation density of $[(C_5H_5)Fe(CO)_2]_2$. The crystal structure determination of $[CpFe(CO)_2]$ has a center of symmetry at the center of the Fe—Fe' bond. **(a)** The two perpendicular planes which contain the terminal carbonyl C(1)O(1); **(b)** two perpendicular planes containing the bridged carbonyl C(2)O(2); **(c)** section containing the cyclopentadiene ligand and **(d)** a parallel plane through the iron nucleus. Contour intervals of 0.10 e $Å^{-3}$. *Dashed contours* for negative, *bold line* for zero. (Reprinted with permission from [47]. Copyright 1978, American Chemical Society)

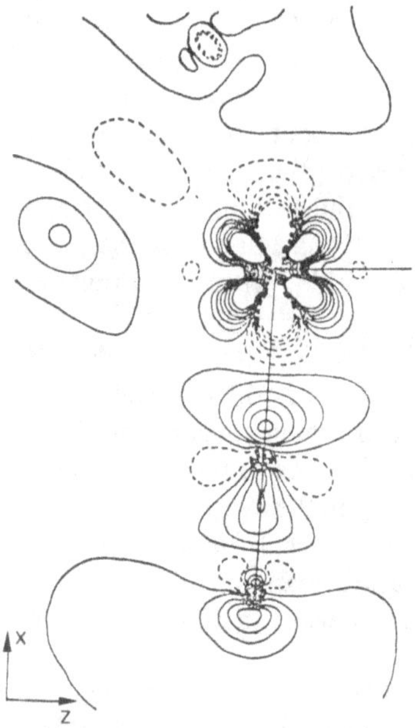

Fig. 23. Theoretical deformation density map from the ab initio SCF-HF wave function of $[(C_5H_5)Fe(CO)_2]_2$, in the plane containing the terminal carbonyl and the Fe—Fe line. Contour interval of 0.03 e $(au)^{-3}$. Deformation densities beyond ± 0.18 e $(au)^{-3}$ are not represented. *Dashed contours* are for negative densities, *solid lines* for zero and positive densities. (Reprinted with permission from [48]. Copyright 1978, American Chemical Society)

monoxide more than that of the terminal carbonyls. This is consistent with the observation that the bridging carbonyls are interacting with two Fe atoms at distances only slightly longer than the terminal carbonyl's Fe—C bond distance.

As in the dinuclear carbonyls with bridging carbonyls supporting the metal-metal bond, the actual existence of a direct Fe—Fe bond in $[CpFe(CO)_2]_2$ is still a matter of some debate. The lack of observable charge accumulation of the Fe—Fe axis and the elongation of density of the bridged carbonyl in both the theoretical and experimental deformation density maps seems to indicate a three-centered Fe-bridging carbonyl-Fe bond. However, once again, the lack of charge accumulation in the bonding region in the deformation densities is not sufficient proof to deny the existence of a direct Fe—Fe bond.

3.2 Dinuclear Acetylene Compounds

Deformation density maps of the dinuclear transition metal acetylene complexes $[(CpNi)_2(\mu\text{-}C_2H_2)]$ and $[Co_2(CO)_6(\mu\text{-}C_2R_2)]$ have been studied to give some insight in to the nature of the metal-metal bond in these complexes. As with $[Co_2(CO)_8]$, a "bent" metal-metal bond has been proposed for these complexes in early theoretical studies [49]. However, by analogy with the "straight" bond model for $[Co_2(CO)_8]$ [44], a straight bond has been proposed for these acetylene compounds. The study of the electron deformation density was undertaken in order to resolve this issue.

a

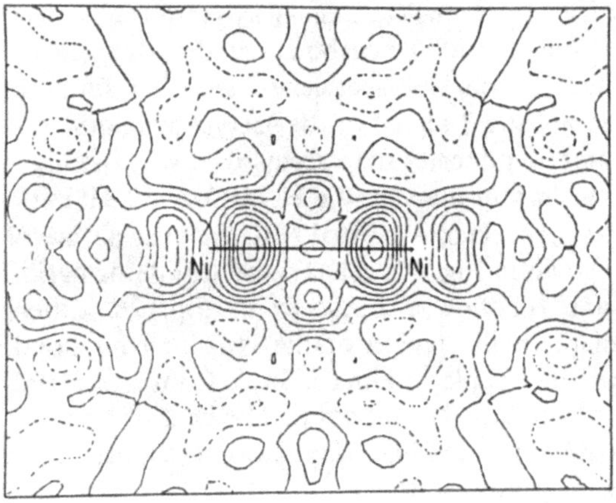

b

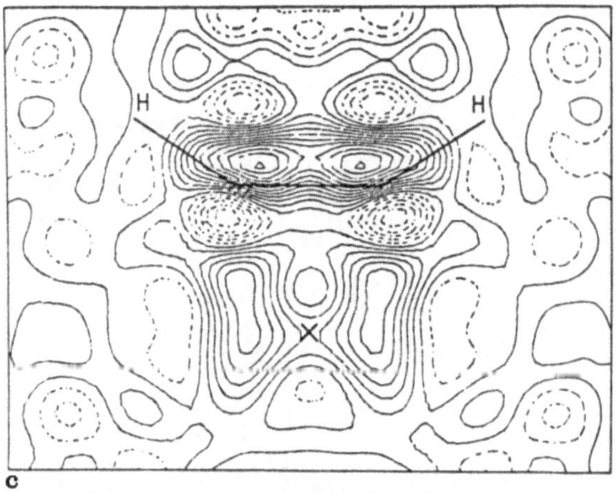

c

Fig. 24a–c. Experimental deformation density map of [(CpNi)$_2$(μ-C$_2$H$_2$)]: **(a)** In the plane containing the two Ni atoms and the midpoint of the acetylene C—C bond, designated by a *cross*. Crosses at the upper *left* and *right* indicate the centers of the cyclopentadienyl rings, **(b)** the plane containing the two Ni atoms parallel to the acetylene C—C bond; **(c)** the plane containing the acetylene carbons and the midpoint of the Ni—Ni line, designated by a *cross*. Contours at 0.1 e Å^{-3}, negative contours *dotted*. (Reprinted with permission from [50]. Copyright 1976, American Chemical Society)

The experimentally determined deformation density of [(CpNi)$_2$(μ-C_2H_2)] [50], shown in Fig. 24, shows substantial density between the two nickel atoms with two maxima about 0.5 Å from the each nickel atom. This accumulation of density along the Ni—Ni axis plus the lack of observable density in the region where the "bent" bond is expected, led the authors to the conclusion that the deformation density favored the straight bond model.

Other features of the experimental deformation density map included a pattern of three accumulation peaks near the nickel atom on the Ni-Ω (Ω = center of cyclopentadiene ring) axis, and towards the acetylenic carbons. This configuration of density about the nickel atoms is quite different than the commonly observed pattern for octahedrally coordinated metals. It may be best described as having trigonal symmetry with the Ni-Ω line as the symmetry axis.

Another feature of the experimental deformation density map is the shift of the density of the acetylenic C≡C bond off the internuclear axis away from the Ni—Ni bond. This shift can be attributed to a bending of the carbon bound hydrogens (to 148.4(5)°) to allow better Ni—C overlap. The main Ni-acetylene interaction found in the experimental map was concentrated in the π^* region of the acetylene ligand.

Even though the experimental deformation density map seems to favor a straight Ni—Ni bond, when the electron density contour map for the HOMO of the ab initio Hartree—Fock wavefunction of [(CpNi)$_2$(μ-C_2H_2)] is plotted [48], the bent nature of the Ni—Ni bond can clearly be seen (shown in Fig. 25). The theoretical deformation density map, obtained from this wave function and shown in Fig. 26, differs significantly from the experimental map. In the region of the Ni—Ni bent bond, there is a region of density deficit with a double minima 0.34 Å from the metal. This deficiency dereases and vanishes at the Ni—Ni midpoint. The lack of accumulation in the Ni—Ni region can be attributed to the high population (1.66 electrons) of each spherically averaged *d* orbital of ground state Ni.

Fig. 25. Electron density contour map for the HOMO of [(CpNi)$_2$(μ-C_2H_2)] in the plane containing two nickel atoms and the midpoint of the acetylene C—C bond. Contour interval of 0.01 e (au)$^{-3}$. (Reprinted with permission from [48]. Copyright 1978, American Chemical Society)

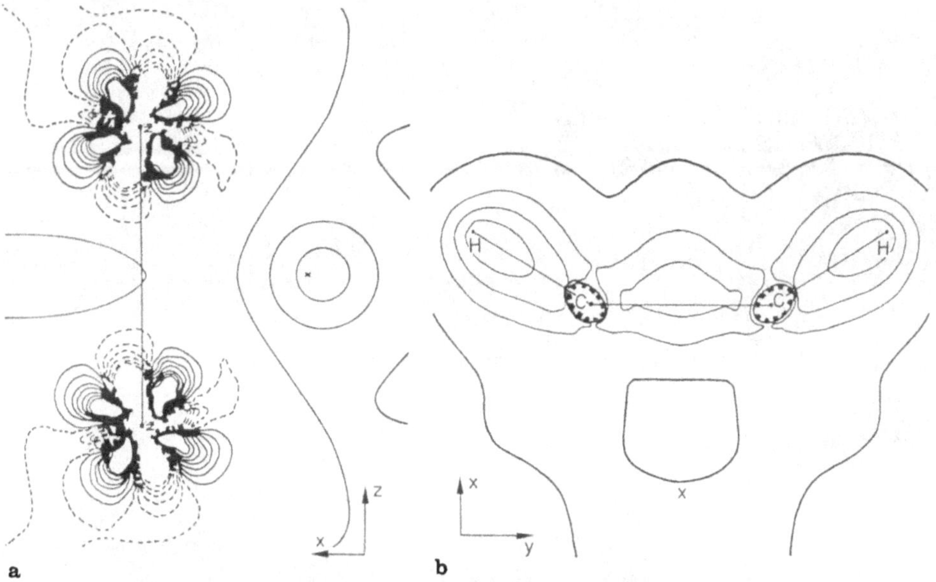

Fig. 26a, b. Theoretical deformation density map from the ab initio SCF-HF wavefunction of $[(CpNi)_2(\mu\text{-}C_2H_2)]$: **(a)** Plane containing two nickel atoms and the midpoint of the acetylene C—C bond; **(b)** Plane containing the acetylene ligand and the midpoint of the Ni—Ni line. Contour interval of 0.003 e $(au)^{-3}$ for both plots. Deformation densities beyond ± 0.18 e $(au)^{-3}$ not represented in **(a)**, negative densities beyond -0.09 e $(au)^{-3}$ not represented in **(b)**. Negative contours are *dashed*, positive and zero contours are *solid*. The zero contour in **(b)** is in *bold*. (Reprinted with permission from [48]. Copyright 1978, American Chemical Society)

The pattern of density about the Ni atom, contrary to the experimental map, is similar to other octahedral-like metal atoms with accumulations in the three-fold axes of the octahedron and deficiencies pointing towards the ligands. As in the experimental maps, the density about the acetylene C—C bond is again shifted off the internuclear axis away from the Ni—Ni bond. These discrepancies between theoretical and experimental deformation density maps were attributed to an artifact caused by the acentricity and partial disorder of the crystal structure of $[(CpNi)_2(\mu\text{-}C_2H_2)]$.

When the dicobalt complex, $[Co_2(CO)_6(\mu\text{-}C_2R_2)]$ was studied [51], the choice of R was chosen specifically to avoid the acentricity and disorder found for the dinickel complex. In addition, static deformation density maps were computed from multipole refinements in order to account for thermal smearing. Three models of multipole refinement were used to produce static maps, the first model included the 3*d* shell. The second model added the 4*s* orbital to the refinement. The third model added a 3*d* monopole to the refinement. The experimental studies were performed on $[Co_2(CO)_6(\mu\text{-}C_2\{t\text{-}Bu\}_2)]$. The theoretical maps were computed from the ab initio wave function of $[Co_2(CO)_6(\mu\text{-}C_2H_2)]$.

The experimental static deformation density map from the third multipole refinement is shown in Fig. 27. There is no observable accumulation in the Co—Co bonding region. Near the cobalt atom, there are two accumulation peaks. Between these peaks are large electron deficient regions. If a line is drawn on the density

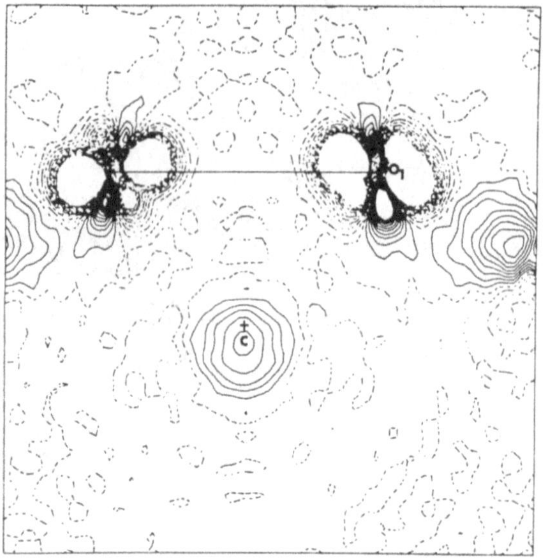

a

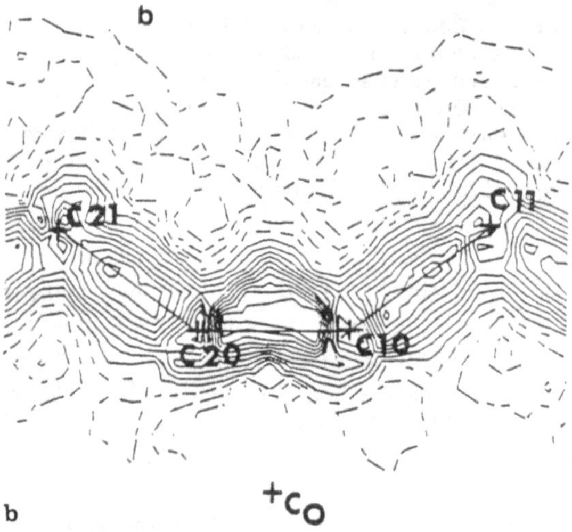

b

Fig. 27a, b. "Static" experimental deformation density map of $[Co_2(CO)_6(\mu\text{-}C_2\{t\text{-}Bu\}_2]$ from multipole refinement: **(a)** the plane containing the two metal atoms perpendicular to the acetylene C—C bond; **(b)** the plane containing the acetylene ligand perpendicular to the metal-metal line. Contours at 0.01 e $Å^{-3}$, *solid lines* for zero and positive contours, *dashed lines* for negative contours. (Reprinted with permission from [51]. Copyright 1986, American Chemical Society)

minima through the cobalt nuclei, an axis of minimum density which aproximates the "bent" bond region as found in the theoretical map can be seen. The static maps derived from the first two models reproduce the familiar octahedral pattern of density about the cobalt.

In the theoretical deformation density map of $[Co_2(CO)_6(\mu\text{-}C_2H_2)]$ (shown in Fig. 28), there is a small diffuse accumulation in the center of the Co—Co bent bond, but most of the metal-metal region is depopulated. The accumulation near the cobalt atom again resembles the familiar octahedral pattern. The lack of accumulation in the metal-metal bonding region is again attributed to the high (1.4 electrons) occupancy of each *d* orbital in a spherical Co atom and does not necessarily imply the lack of a Co—Co bond.

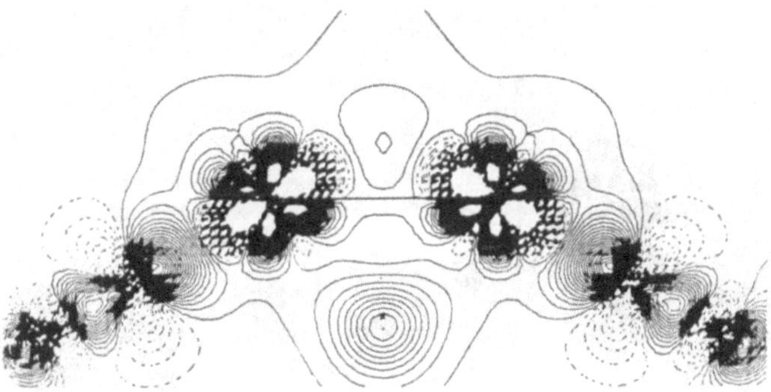

a

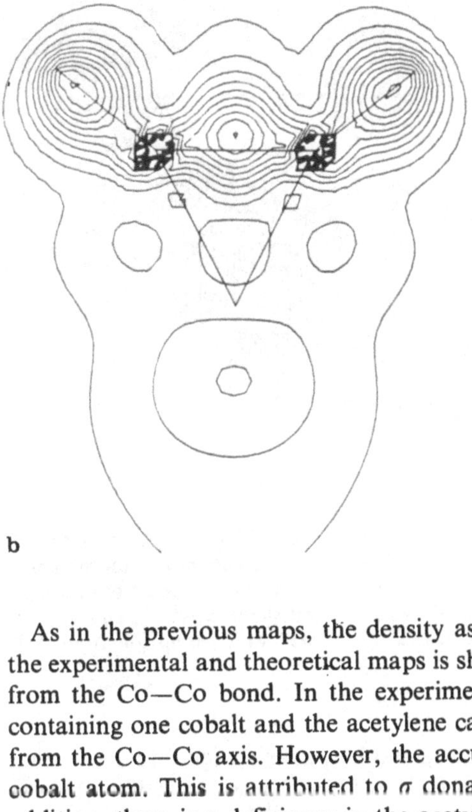

b

Fig. 28a, b. Theoretical deformation density map from the ab initio SCF-CI wave function of $[Co_2(CO)_6(\mu\text{-}C_2H_2)]$: (a) In the plane containing the two cobalts perpendicular to the acetylene C—C bond; (b) in the plane containing the acetylene ligand perpendicular to the metal-metal line. Countours at 0.05 e $Å^{-3}$, *solid lines* for zero and positive contours, negative contours are *dashed*. (Reprinted with permission from [51]. Copyright 1986, American Chemical Society)

As in the previous maps, the density associated with the acetylene bond in both the experimental and theoretical maps is shifted off the internuclear C—C axis, away from the Co—Co bond. In the experimental map, the accumulation in the plane containing one cobalt and the acetylene carbons is shifted from the C—C axis away from the Co—Co axis. However, the accumulation region is extended towards the cobalt atom. This is attributed to σ donation from the acetylene to the cobalt. In addition, there is a deficiency in the acetylenic carbon p_π region which is attributed to electron shift toward the center of the acetylene to form the acetylene π bond.

Overall, in all planes other than the plane containing the Co—Co line, there is general agreement between the static experimental and the theoretical maps. In the plane containing the Co—Co line, the static maps from the first two models are similar to the theoretical map while the static map from the third map is different near the cobalt atom. The striking difference of the experimental deformation density map of $[Co_2(CO)_6(\mu\text{-}C_2\{t\text{-Bu}\}_2]$ to that of $[(CpNi)_2(\mu\text{-}C_2H_2)]$, in addition to the similarity

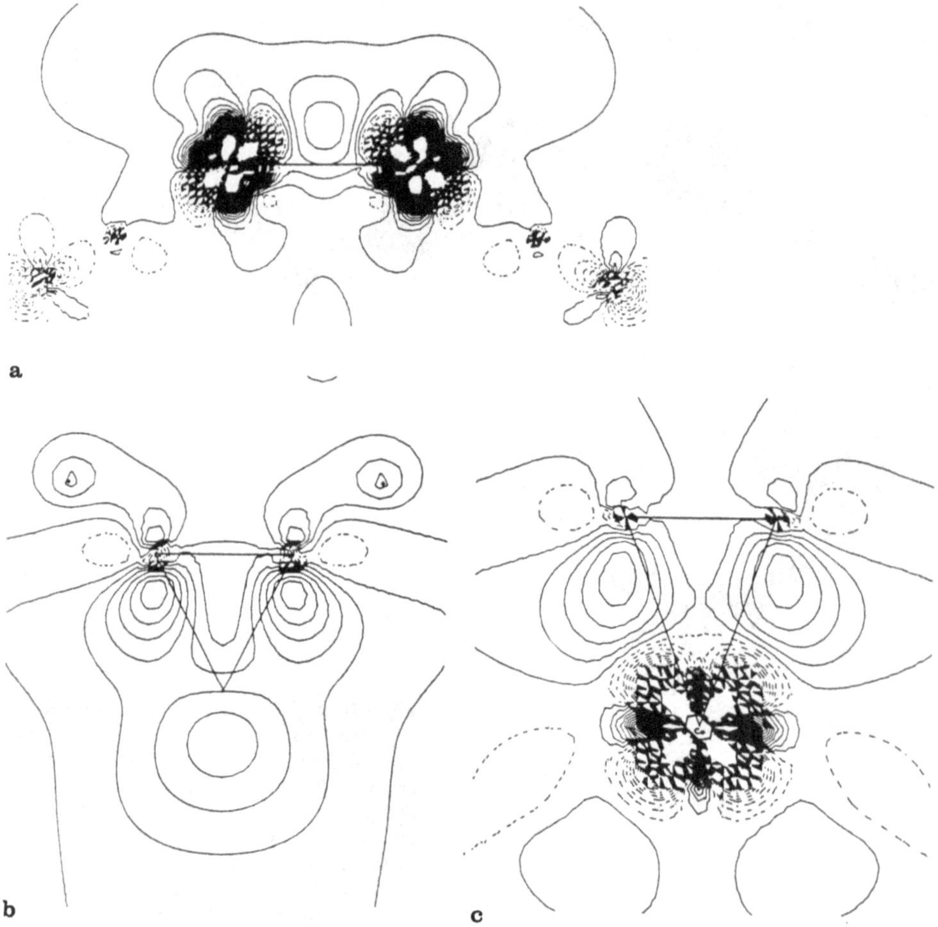

Fig. 29a–c. Theoretical fragment deformation density at CI level for $[Co_2(CO)_6(\mu\text{-}C_2H_2)]$ with regard to the superposition of 2 $Co(CO)_3$ + C_2H_2: **(a)** plane containing the two Co atoms and the midpoint of the acetylene C—C bond; **(b)** plane containing the two C atoms and the midpoint of Co—Co; **(c)** the plane containing the two C atoms and one cobalt atom. Contours at 0.05 e Å^{-3}, *solid lines* for zero and positive contours, negative contours are *dashed*. (Reprinted with permission from [51]. Copyright 1986, American Chemical Society)

of the theoretical maps of each compound to each other, adds credence to attributing the experimental density density accumulations in the Ni—Ni axis to artifacts due to crystal acentricity and disorder.

When the density from the ab initio SCF-CI wave function of two $Co(CO)_3$ fragments and a free acetylene molecule are subtracted from the total molecular electron density, a significant accumulation can be observed in the Co—Co bent bonding region (see Fig. 29). As noted earlier, the lack of electron density in the metal-metal bonding regions of the standard deformation densities of metal dimers has been attributed to rearrangement due to formation of metal carbonyl bonds and this fragment deformation density shows such an effect. In the region of the acetylene ligand, substantial accumulation peaks can be seen in the carbon π^* orbitals. The region of the acetylene π bond is electron deficient due to acetylene to metal donation. The important observation, however, is the accumulation in the Co—Co bent bond region, which indicates that the "bent" bond model for these complexes is essentially correct.

3.3 Dinuclear Compounds with Multiple Metal-Metal Bonds

A number of dinuclear metal complexes with multiple metal-metal bonds have been studied with the use of deformation density maps. Examples include: $[Mo_2\{N(CH_3)_2\}_4(CH_3)_2]$ [52], and Mo_2H_6 [53], both containing a metal-metal triple bond; $[Cr_2(tmp)_4]$ [52], $[Cr_2(O_2CCH_3)_4] \cdot 2\ H_2O$ [54], $[Cr_2(mhp)_4]$ [55], $[Mo_2(O_2CCH_3)_4]$ [56] and $[Mo_2(O_2CC_6H_5)_4]$ [52], all of which contain a metal-metal quadruple bond.

Of particular interest are the dichromium and dimolybdenum complexes with formal quadruple metal-metal bonds. This set of complexes generally consist of the two metals held together by four bidentate ligands, like the tetraacetates, $[M_2(O_2CCH_3)_4]$. Quadruply bonded dichromium and dimolybdenum compounds exhibit quite different chemical properties. For the quadruply bonded dichromium complexes, experimental chromium-chromium bond lengths vary over 0.7 Å. The dichromium tetraacetates always crystallize with some ligand, L, in the axial position. Experimentally, a correlation between the strength of the Cr-axial ligand interaction and the Cr—Cr bond length has been established [57]. However, theoretical ab initio calculations on anhydrous and hydrated chromium formate molecules concluded that axial water molecules had little effect on the strength of the Cr—Cr bond [58]. A number of quadruply bonded dichromium complexes with "supershort" Cr—Cr bond lengths (on the order of 1.85 Å) have been reported [59]. They posess a different bridging bidentate ligand which sterically blocks the axial site on chromium so that there are no axially coordinated ligands.

In contrast to the wide variance of chromium bond lengths in the dichromium compounds, quadruply bonded molybdenum complexes show comparatively little variance in Mo—Mo bond lengths with respect to axial coordination [60]. Furthermore, dimolybdenum tetraacetate will crystallize without axial ligands.

In the experimental deformation density study of $[Cr_2(O_2CCH_3)_4] \cdot 2\ H_2O$ [54], shown in Fig. 30, two density peaks located near the chromium atom in one of the planes of a pair of acetate ligands are not found in the chemically equivalent second plane. Differences can also be observed in the bisecting planes, 45° offset from the acetate ligands. Nevertheless, the deformation density map was averaged over the D_{4h} symmetry of the compound. The averaged experimental deformation density map, shown in Fig. 31, has density accumulations for all Cr—O, C—C and C—O bonds as well as for the oxygen lone pairs. However, in the Cr—Cr region, there is a lack of density accumulation in the σ bonding region. There is a region of accumulation off the bond axis which extends into both the π and σ bonding regions.

In contrast, the theoretically computed deformation density map from the limited CI wave function of $[Cr_2(O_2CH)_4]$ (shown in Fig. 32) has a diffuse accumulation peak about the center of the Cr—Cr bond. This accumulation extends into the π

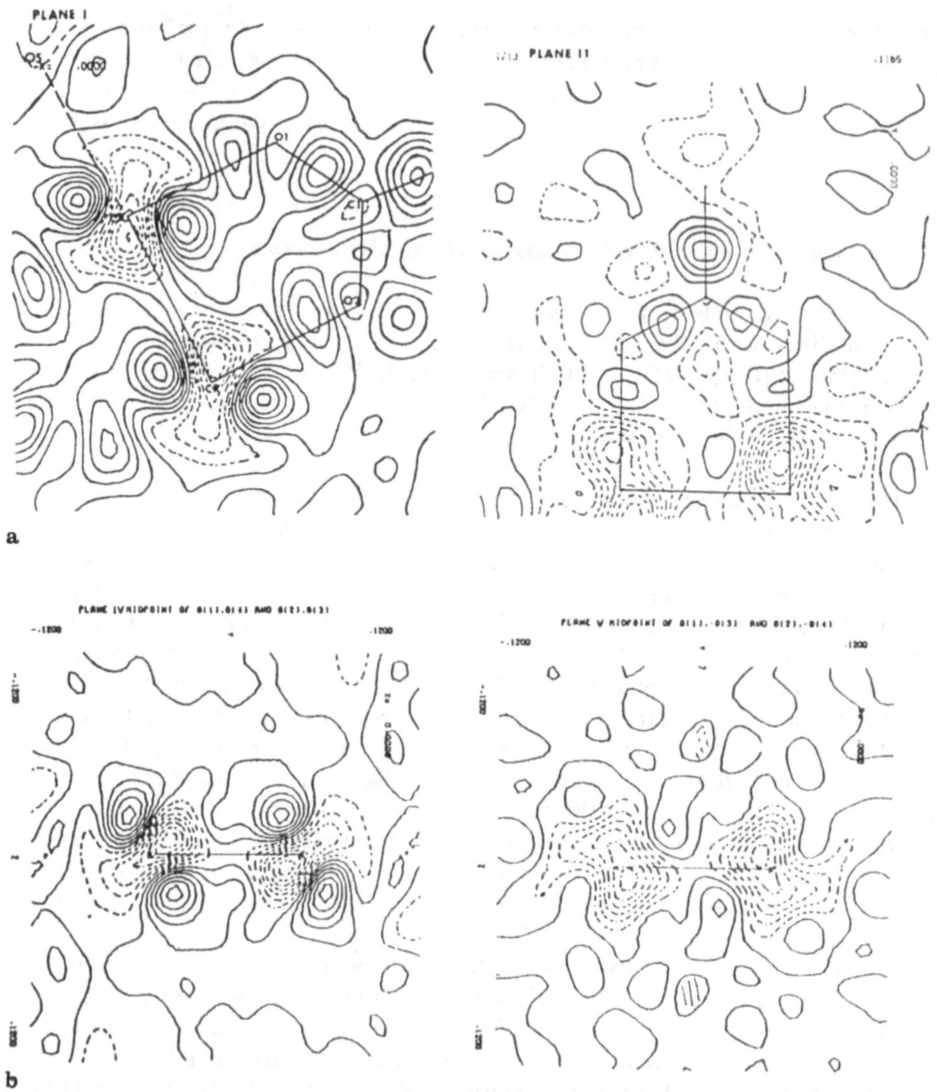

Fig. 30a, b. Experimental deformation density for $[Cr_2(O_2CCH_3)_4] \cdot 2\,H_2O$: **(a)** in the two planes containing the acetate ligands and the Cr atoms; **(b)** in the two planes bisecting the O—Cr—O angles. Contours are at 0.10 e $Å^{-3}$, and negative contours are *broken*. (Reprinted with permission from [54]. Copyright 1980, American Chemical Society)

and σ bonding regions. Initially, this difference between experimental and theoretical deformation densities in the σ-bonding region was attributed to the presence of axially situated water molecules in the experimental map. However, the theoretical and experimental deformation density maps of $[Mo_2(O_2CCH_3)_4]$, discussed next, show the same type of topological differences, without the presence of axial ligands in the experimental map.

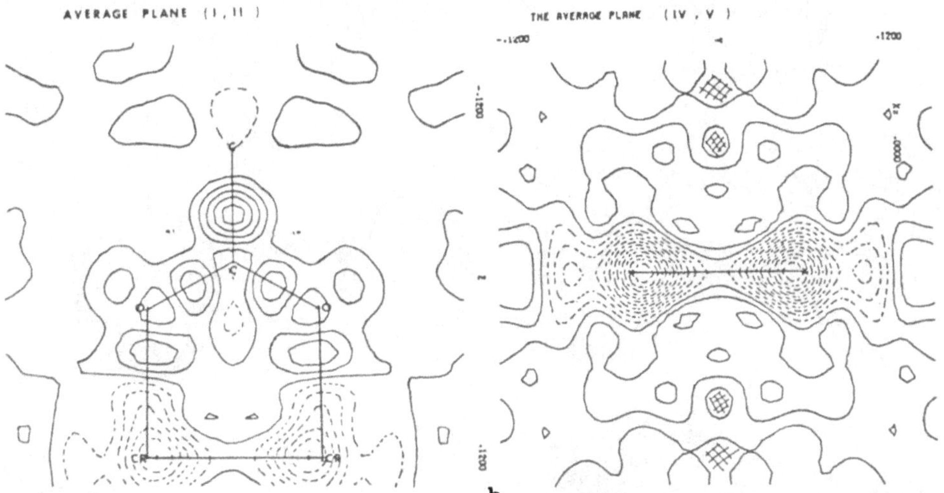

a b

Fig. 31 a, b. Experimental deformation density for $[Cr_2(O_2CCH_3)_4 \cdot 2\,H_2O]$: **(a)** Average of planes containing two Cr atoms and the acetate ligands; **(b)** Average of planes offset 45° from planes in **(a)** containing Cr atoms. Contour interval of 0.10 e Å^{-3} in **(a)**, 0.05 e Å^{-3} in **(b)**. Negative contours are *broken* in both maps. (Reprinted with permission from [54]. Copyright 1980, American Chemical Society)

The experimental deformation density map of $[Mo_2(O_2CCH_3)_4]$ [56], shown in Fig. 33, has a small accumulation at the center of the Mo—Mo bond. As in the experimental dichromium map above, the deformation density map was first averaged over the D_{4h} symmetry of the compound. This central point corresponds to a saddle

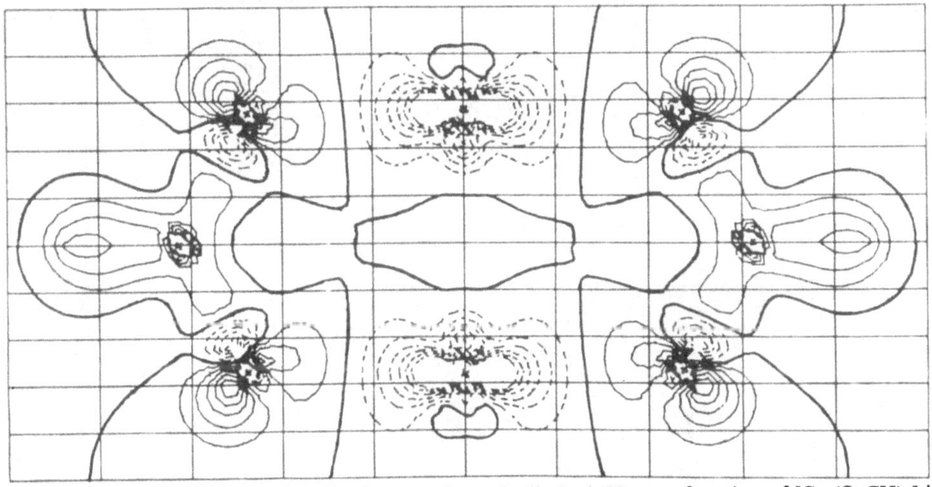

Fig. 32. Theoretical deformation density map from the limited CI wave function of $[Cr_2(O_2CH)_4]$ in a plane containing the Cr atoms and a formate ligand from a limited CI wave function. Contours are at 0.03 e (au)$^{-3}$. The *bold* line is the zero contour and negative contours are *broken*. (Reprinted with permission from [54]. Copyright 1980, American Chemical Society)

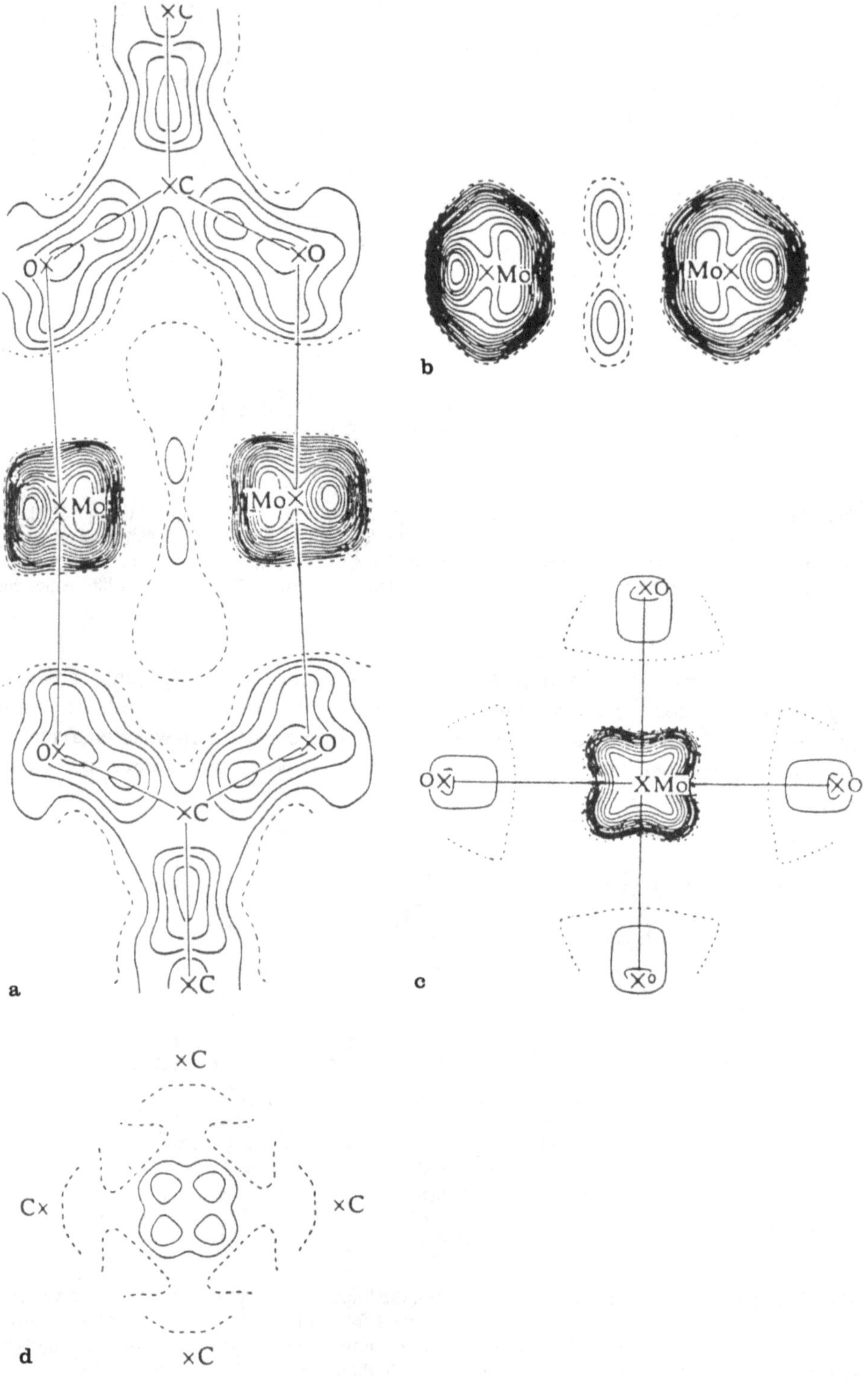
C
C
O
O
Mo
Mo
O
O
C
C
a
Mo
Mo
b
O
O
Mo
O
O
c
C
C
C
C
d

point between two accumulation zones symmetric to the Mo—Mo bond axis. In addition, abnormally large and sharp accumulation zones appear about the Mo atom. The appearance of these accumulation zones are in contrast to the large and diffuse nature of Mo 4*d* orbitals and are attributed to an artifact of scale factors and judged to be not chemically significant.

The theoretical map of $[Mo_2(O_2CH)_4]$, shown in Fig. 34, also has accumulation about the center of the Mo—Mo bond. The center of the bond corresponds to a peak of electron density in contrast to the experimental map. This difference in the pattern of the deformation density about the center of the Mo—Mo bond is similar to that found for the experimental and theoretical maps of the dichromium tetraacetates. The difference in the dichromium maps was attributed to the axially ligated water molecules in the experimental map. However, there are no axially ligated water molecules in the experimental map for the dimolybdenum complexes, therefore the difference must be due to some other effect.

Additionally, a feature of the theoretical map of the dimolybdenum complex that is not found in the theoretical map of the dichromium complexes is the presence of an accumulation zone on the external side of each Mo atom. This is masked in the experimental map by the extraneous sharp accumulation about each Mo atom. The presence of this axial accumulation could possibly be correlated to the relative insensitivity of the Mo—Mo bond to axial ligation when compared to the dichromium complexes.

In contrast to the exerpimental map of dichromium tetraacetate, the experimental deformation density map of $[Cr_2(mhp)_4] \cdot CH_2Cl_2$ [55] possessing a "supershort" Cr—Cr bond length of 1.879 Å (see Fig. 35), shows a large accumulation peak of 0.41 e $Å^{-3}$ at the bond midpoint. This accumulation extends into the π and σ bonding regions. This pattern about the center of the Cr—Cr bond contrasts to the two earlier experimental maps, where the center of the M—M bond corresponds to a saddle point between two larger accumulations off the bond axis. In contrast to the previous two complexes, this compound does not posess D_{4h} symmetry but D_{2d} symmetry. Therefore, the deformation density maps were averaged over the planes, even though there are still distortions in the two planes from D_{2d} symmetry.

As a comparison to the experimental map of $[Cr_2(mhp)_4]$, the computed deformation density map (shown in Fig. 36) from the ab initio SCF-CI wave function of $[Cr_2(H_2P(CH_2)_2)_4]$ was obtained. This complex is modelled from the compound $[Cr_2\{(CH_3)_2P(CH_2)_2\}_4]$, a compound with a similar "supershort" Cr—Cr bond distance (1.89 Å) [59b]. As in the previous experimental map, there is a large accumulation peak at the center of the Cr—Cr bond of 0.30 e $Å^{-3}$. Again, the accumulation zone extends into the π- and δ-bond regions.

◀

Fig. 33a–d. Experimental deformation density of $[Mo_2(O_2CCH_3)_4]$: **(a)** a section through the Mo—Mo bond and two acetate groups; **(b)** a section through the Mo—Mo bond bisecting the two planes of the adjacent acetate groups; **(c)** section through the midpoint of the Mo—Mo bond and perpendicular to it; **(d)** section through the plane formed by the Mo and the four ligating O atoms. Contour intervals of 0.1 e $Å^{-3}$, negative contours are *dotted*, zero contours are *chain dotted*. (Reprinted with permission from [56])

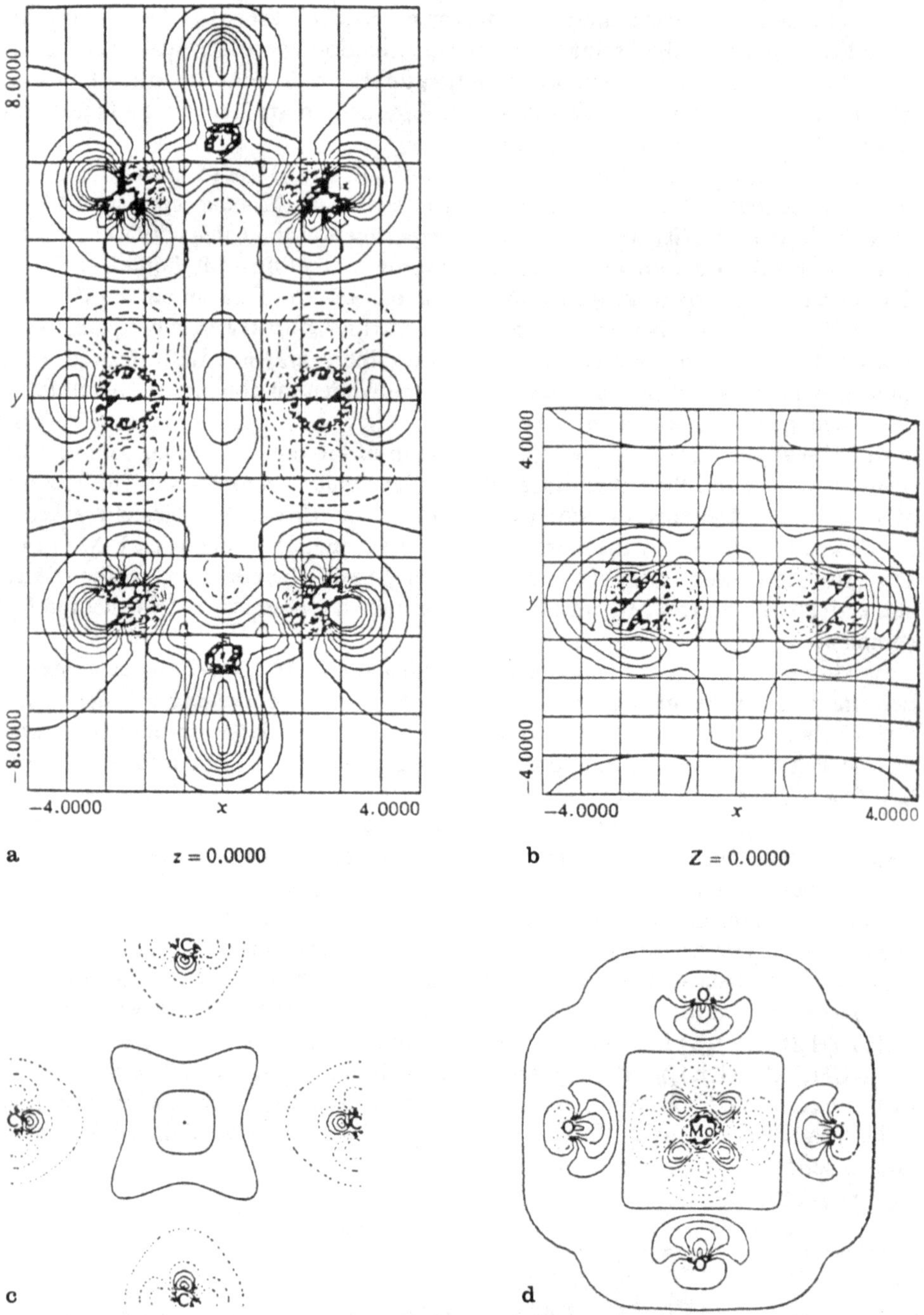

Fig. 34a–d. Theoretical deformation density map from the SCF-CI wave function of $[Mo_2(O_2CH)_4]$. The plots are in same order as in Fig. 33. Contour interval of 0.1 e Å^{-3}. (Reprinted with permission from [56])

a c

Fig. 35a–c. Average experimental deformation density of $[Cr_2(mhp)_4]$: **(a)** in the plane of a wing; **(b)** in the plane perpendicular to the Cr—Cr axis through the metal nucleus; **(c)** in the plane perpendicular to the Cr—Cr axis through the midpoint of the bond. Contours interval of 0.05 e $Å^{-3}$, negative contours are *dashed*. (Reprinted with permission from [55]. Copyright 1982, American Chemical Society)

The central accumulation zone in the theoretical map is shaped as a narrow disk which is squared in the xy directions. This pattern is seen in the two previous theoretical deformation density maps. The main difference between the maps is in the size of the central accumulation (0.1 e $Å^{-3}$ for $Cr_2(O_2CH)_4$; 0.15 e $Å^{-3}$ for $Mo_2(O_2CH)_4$ and 0.3 for $Cr_2[H_2P(CH_2)_2]_4$). In the experimental map of $[Cr_2(mhp)_4]$, the central accumulation zone in the experimental map is squared in the xy direction only at distances greater than 1 Å away from the bond axis. At closer distances, the accumulation is greater in the x and y directions than the xy directions. However, the difference is at most twice the value of the estimated standard deviation, therefore not significant.

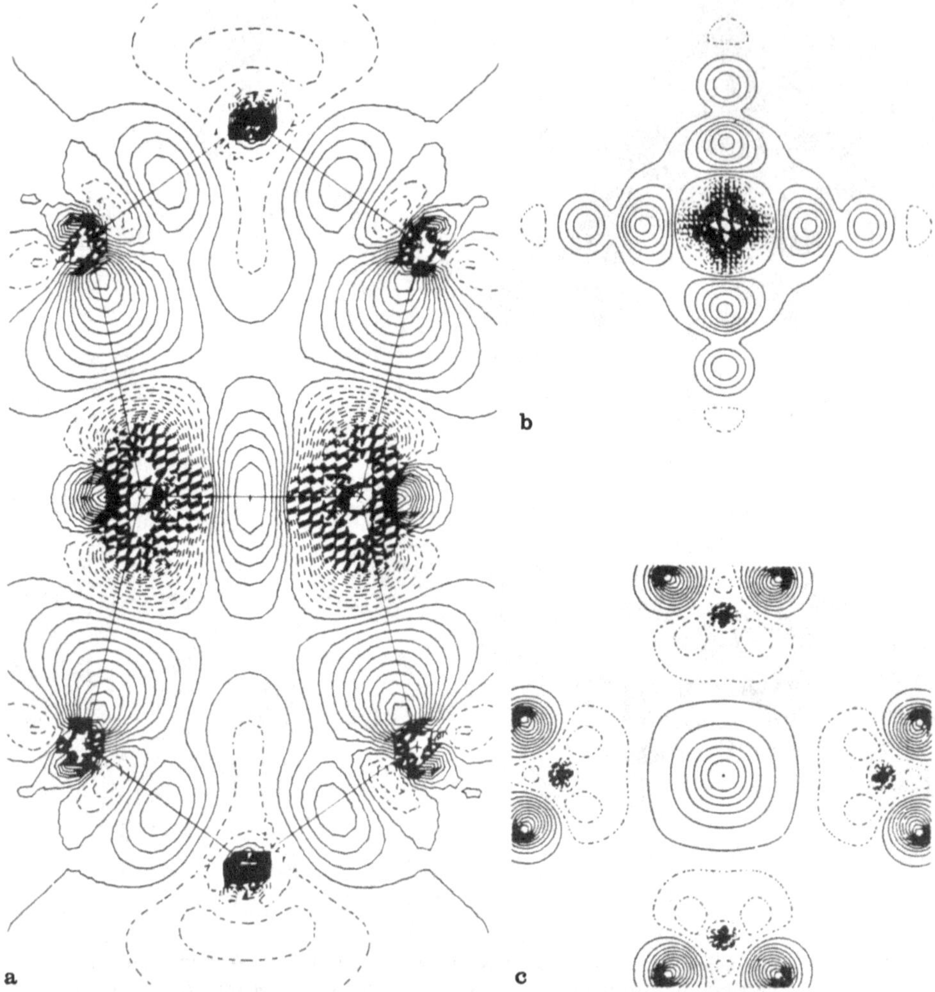

Fig. 36a–c. Theoretical deformation density from the SCF-CI wave function of $[Cr_2(H_2P(CH)_3)_4]$: **(a)** in the plane of a wing; **(b)** perpendicular to the Cr—Cr axis through the metal nucleus; **(c)** perpendicular to the Cr—Cr axis through the midpoint of the Cr—Cr bond. Contour interval of 0.05 e Å^{-3}. Negative contours are *dashed*. (Reprinted with permission from [55]. Copyright 1982, American Chemical Society)

4 Concluding Remarks

Through the use of deformation density maps, many of the features of metal-ligand and metal-metal bonding has been observed. However, in many of these studies rather puzzling features have been observed, such as the lack of accumulation in bonding regions. Many of these features may be attributed to three reasons listed in the introduction: 1) the rearrangement of the atom's valence shell in preparation for bonding; 2) the formation of bonds to other atoms resulting in loss of density along the bond direction and 3) the bond may be a relatively long bond between atoms with compact valence orbitals. Other features may be attributed to error in either the experimental or theoretical methods used to obtain the maps. In metal-ligand dative bonding, the problems of dividing the accumulation between that originally in the ligand lone pair and that due to metal-ligand bond formation still presents problems. Fragment analysis can assigst one in resolving both of these interpretative problems.

Although the accuracy of both measured and calculated maps has improved significantly, problems still remain in both methods. Experimental maps are sometimes plagued by "ghost" peaks near the heavy atoms. Additionally, lone pairs tend to disappear. Basis set quality is the chief obstacle to producing accurate calculated deformation density maps. Electron correlation can also be quite significant for long and weak bonds, such as some metal-metal bonds.

Comparison between theoretical and experimental maps is quite important as a tool to check on the accuracy of each method. Sometimes, conclusions based on just a theoretical or experimental map may by erroneous. A good case of this is in the experimental map of $[(CpNi)_2(\mu\text{-}C_2H_2)]$ noted earlier. However, comparison must be done with care. The effects of thermal smearing must be accounted for by either producing static experimental maps or dynamic theoretical maps. There is still room for improvement in both the experimental methods of accounting for thermal smearing.

Other systems which could be considered for future study could be trinuclear transition metal compounds. Trinuclear compounds present some different bonding arrangements which can not be found for dinuclear clusters such as triply-bridging ligands or different types of metal-metal bonds in the same cluster. Experimentally, the deformation density about the methylidyne group in $[Co_3(CO)_9(\mu\text{-}^3\text{-}CH)]$ has been studied [61]. Three different maps were produced with different promolecules for the CH ligand: 1) the sum of the spherical atoms; 2) the electron density of the $^2\Pi$ state of a CH fragment and 3) the density of the low-lying excited $^4\Sigma$ state of a CH fragment. The fragment deformation density from the $^2\Pi$ state showed excess density in the π region of the CH ligand and deficit in the σ region while that from the $^4\Sigma$ state showed the removal of the excess π density but excess in the σ region. These observations suggested that the electron density distribution of the CH ligand in the cluster was intermediate between these two states of the free CH fragment. A theoretical study of the cluster $[Co_3(CO)_9CCl]$ has been reported [62] in which the deformation density map in the plane of the three cobalts show features similar to $[Mn_2(CO)_{10}]$ with little accumulation in the Co—Co bond regions. However, as for $[Mn_2(CO)_{10}]$, when the appropriate fragments are subtracted from the total density (three $Co(CO)_3$ and CCl), accumulation in the Co—Co bond and the center of the tricobalt triangle can be seen.

However, the sheer size of these clusters will make theoretical calculations long and difficult to perform. Experimentally, obtaining crystals suitable for deformation density studies may prove to be quite difficult. Again, the choice of the promolecule may prove critical to produce chemically meaningful deformation density maps. Methods to intepret the electron density without the use of a reference through the Laplacian of the total density have been used to bypass the choice of a reference [63]. Perhaps this method, when applied to transition metal compounds will improve our understanding of the chemical bond in these systems.

5 References

1. For earlier reviews see: (a) Breitenstien M, Dannöhl H, Meyer H, Schweig A, Seeger R, Seeger U, Zittlau W (1983) International Reviews in Physical Chemistry 3: 355 (b) Coppens P, Hall MB (eds) (1982) Electron Distributions and the Chemical Bond, Plenum, New York
2. Cade PE (1972) Trans. Amer. Cryst. Assoc. 8: 1
3. Hall, MB (1982) In: Coppens P, Hall MB (eds) Electron distributions and the chemical bond, Plenum, New York, p 205
4. For examples see: (a) Hase H, Reitz H, Schweig A (1976) J. Chem. Phys. 39: 157 (b) Coppens P, Steverns ED (1977) Isr. J. Chem. 16: 175 (c) Stevens ED, Rys J, Coppens P (1977) Acta Crystallogr., Sect. A A33: 333 (d) see [15]
5. Kunze KL, Hall MB (1986) J. Am. Chem. Soc. 108: 5122
6. Newton MD, Switkes E, Lipscomb WN (1970) J. Chem. Phys. 53: 2645
7. Savarrault JM, Lehman MS (1980) J. Am. Chem. Soc. 102: 1298
8. Iwata M, Saito Y (1973) Acta Crystallogr. B29: 822
9. Kalman BL, Richardson JW (1971) J. Chem. Phys. 55: 4443
10. Maslen EN, Ridout SC (1987) Acta Crystallogr. B43: 352
11. Stevens ED, Coppens P (1979) Acta Crystallogr. A35: 536
12. Keulen E, Jellinek F (1966) J. Organomet. Chem. 4: 490
13. Bailey MF, Dahl LF (1965) Inorg. Chem. 4: 1314
14. Rees B, Coppens P (1972) J. Organomet. Chem. 42: C102
15. Kok RA, Hall MB (1985) J. Am. Chem. Soc. 107: 2599
16. Wang Y, Angermund K, Goddard R, Krüger C (1987) J. Am. Chem. Soc. 109: 587
17. Rees B, Coppens P (1973) Acta Crystalogr. B29: 2515
18. Johnson JB, Klemperer WG (1977) J. Am. Chem. Soc. 99: 7132
19. Rees B, Mitschler A (1976) J. Am. Chem. Soc. 98: 7918
20. (a) Cotton FA (1964) Inorg. Chem. 3: 702 (b) Dobson GR (1965) Inorg. Chem. 4: 1673
21. Heijser W, Baerends EJ, Ros P (1980) J. Molec. Struc. 63: 109
22. Sherwood DE, Hall MB (1983) Inorg. Chem. 22: 93
23. Krüger C, Goddard R, Claus KH (1983) Z. Naturforsch., B: Anorg. Chem. 38B: 1431
24. Stevens ED (1981) J. Am. Chem. Soc. 103: 5087
25. Lecomte C, Chadwick DL, Coppens P, Stevens ED (1983) Inorg. Chem. 22: 2982
26. Kutzler FW, Swepston PN, Berkovitch-Yellin Z, Ellis DE, Ibers JA (1983) J. Am. Chem. Soc. 105: 2996
27. Coppens P, Li L (1984) J. Chem. Phys. 81: 1983
28. Lecomte C, Blessing RH, Coppens P, Tabard A (1986) J. Am. Chem. Soc. 108: 6942
29. Tanaka K, Elkaim E, Li L, Jue ZN, Coppens P, Landrum J (1986) J. Chem. Phys. 84: 6969
30. Elkaim E, Tanaka K, Coppens P. Scheidt WR (1987) Acta Crystallogr. B43: 457
31. Goff H, La Mar GN, Reed CA (1977) J. Am. Chem. Soc. 99: 3641
32. Kitagawa T, Teraoka J (1979) Chem. Phys. Lett., 63: 443
33. Rohmer MM (1985) Chem. Phys. Lett., 116: 44
34. Clemente DA, Cingi-Biagini M (1987) Inorg. Chem. 26: 2350
35. Dahl LF, Rundle RE (1963) Acta Crystallogr. 16: 419
36. (a) Brown DA, Chambers WJ, Fitzpatrick NJ, Rawlinson RM (1971) J. Chem. Soc. A 720 (b) Bau R, Kirtley SW, Sorrell TN, Winarko S (1974) J. Am. Chem. Soc. 96: 988
37. La Placa SJ, Hamilton WC, Ibers JA, Davison A (1969) Inorg. Chem. 8: 1928
38. Martin M, Rees B, Mitschler A (1982) Acta Crystallogr. B38: 6
39. Heijser W, Baerends EJ, Ros P (1980) Farady Symposia of the Chemical Society 14: 211
40. Cotton FA, Troup JM (1974) J. Chem. Soc., Dalton Trans. 800
41. Lauher JW, Elias M, Summerville RH, Hoffmann R (1976) J. Am. Chem. Soc. 98: 3219
42. Bauschlicher CW (1986) J. Chem. Phys. 84: 872
43. Troup JM, Extine MW, Ziolo RF (1983) In: Coppens P, Hall MB (eds) Electron distributions and the chemical bond, Plenum, p 295
44. Summer CG, Klug HP, Alexander LE (1964) Acta Crystallogr. 17: 732
45. Thorn DL, Hoffmann R (1978) Inorg. Chem. 17: 126

46. Leung PC, Coppens P (1983) Acta Crystallogr. B39: 535
47. Mitschler A, Rees B, Lehmann MS (1978) J. Am. Chem. Soc. 100: 3390
48. Bénard M (1978) J. Am. Chem. Soc. 100: 7740
49. Teo BK, Hall MB, Fenske RF, Dahl LF (1975) Inorg. Chem. 14: 3103
50. Wang Y, Coppens P (1976) Inorg. Chem. 15: 1122
51. Baert F, Guelzim A, Poblet JM, Wiest R, Demuynck J, Bénard M (1986) Inorg. Chem. 25: 1830
52. Troup JM, Extine M, Ziolo RF (1983) In: Coppens P, Hall MB (eds) In: Electron distributions and the chemical bond. Plenum, p 288
53. Hall MB (1983) In: Coppens P, Hall MB (eds) Electron distributions and the chemical bond, Plenum, p 209
54. Bénard M, Coppens P, DeLucia ML, Stevens ED (1980) Inorg. Chem. 19: 1924
55. Mitschler A, Rees B, Wiest R, Bénard M (1982) J. Am. Chem. Soc. 104: 7501
56. Hino K, Saito Y, Bénard M (1981) Acta Crystallogr. B37: 2164
57. Cotton FA, Extine M, Rice GW (1978) J. Am. Chem. Soc. 17: 176
58. Bénard M (1978) J. Am. Chem. Soc. 100: 2354
59. (a) Cotton FA, Fanwick PE, Niswander RH, Sekutowski JC (1978) J. Am. Chem. Soc. 100: 4725
(b) Cotton FA, Hanson BE, Ilsley WH, Rice GW (1979) Inorg. Chem. 18: 2713
60. Cotton FA, Extine M, Gage LD (1978) Inorg. Chem. 17: 172
61. Coppens P, Coord Chem. Rev. (1985) 65: 285
62. Hall MB (1983) In: Coppens P, Hall MB (eds) Electron distributions and the chemical bond, Plenum, p 216

Fundamentals of Relativistic Effects in Chemistry

W. H. E. Schwarz*

Fachbereich Chemie, Universität Siegen, Postfach 10 12 40, D-5900 Siegen, FRG

Although at first sight the quantum theory of relativity looks quite academic to a chemist, it turned out during the 1970s that relativistic "corrections" are of significant importance to the chemistry of the majority of the heavy elements. The most important relativistic energy contributions are the mass-velocity effect, the Darwin effect, and the spin-orbit coupling. They change the energetic and spatial properties of the valence and outer core orbitals of many atoms by far more than a few percent. As a result, the structural and energetic properties of molecules may be quite different from what is expected in the conventional nonrelativistic framework. Both quantitative modifications and drastic qualitative changes may happen. Trends in the lower part of the periodic system are partly modified. The main emphasis is laid on the presentation of simple but nevertheless correct conceptual interpretation schemes. Depending on the theoretical framework, explanations of the relativistic effects can be given in rather different physical terms, which are complementary rather than contradictory. Several crude, but useful rules of thumb on relativistic effects are given.

* Visiting Professor at the Vrije Universiteit Amsterdam, and at the East China Normal University Shanghai.

1 Introduction

1.1 Some Experimental Evidence of Relativistic Effects in Chemistry

When analysing trends over the periodic system, the chemical relevance of some special mechanism, which must be operative in heavy atoms, becomes evident already at a qualitative level. Relativity stabilizes s^2 and s^2p^2 configurations and increases the bond strength of d and f shells.

While organic chemistry is a comparatively homogeneous field, the large number of different elements makes inorganic chemistry a science of rather diverse character. All the more, general and detailed classification schemes are needed to explain and predict trends at qualitative and semiquantitative levels. Usually this situation is mentioned in the preface of modern textbooks of general and inorganic chemistry, in order to justify the extensive amount of formalized theoretical concepts of "physical" nature, presented in many chapters. In this respect, these treatises differ significantly from their predecessors of the first half of this century, and from most organic chemistry textbooks, which often contain only qualitative "chemical" concepts.

One of the most important qualitative chemical concepts is the periodic system of elements invented 120 years ago. The atomic properties most basic for chemistry are the occupations, spatial extensions and energies of the outer valence shells. The first ionization potentials |E| of the atomic elements vary considerably between 3.9 eV for caesium and 24.6 eV for helium. We may represent |E| by:

$$|E| = 13.6\,\text{eV} \cdot Z_{eff}^2/n^2 \tag{1.1}$$

where Z_{eff} is the mean effective nuclear charge felt by the valence electron, and n is the principal quantum number. In general, |E| *decreases* monotonously when one descends in one group of the periodic system. Z_{eff} increases only slightly from one row to the next lower row, because the additional nuclear charge is efficiently screened by the additional core electrons. The increase of Z_{eff} is overcompensated by the decrease of $1/n^2$. Furthermore, there is an overall increase of |E|, when going from left to right within a row of the periodic table, i.e. from group 1 to group 18.

Pertinent curves are shown in Fig. 1. The shoulders in the curves reflect the structure of the atomic shells with 2, 8, 8, 18, 18, 32, 32 electrons in the shells. For the *lighter elements* these irregularities are due to the s–p sub-shell structure. |E| of group 13 (B, Al, Ga, ...) with configuration s^2p^1 is smaller than |E| of group 2 (Be, Mg, Ca, ...) with stable closed s^2-configuration. Similarly, |E| of group 16 (O, S, ...) with configuration s^2p^4 is smaller than |E| of group 15 (N, P, ...) with stable half-filled s^2p^3-configuration.

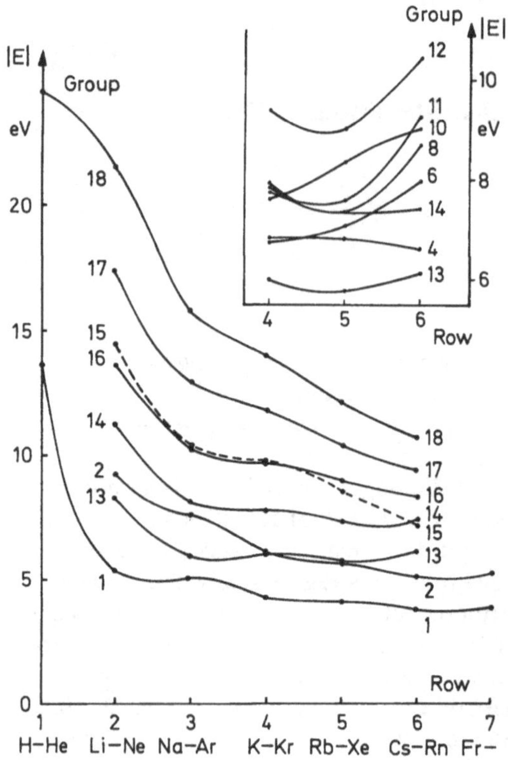

Fig. 1. First atomic ionization potentials |E| of the elements, in eV

However, the *heaviest elements* on the right side of Fig. 1 show different irregularities. The ionization potentials of the heaviest elements of groups 1 to 14 are nearly equal to, or are — in many cases — even larger, than those of their lighter predecessors. This indicates an additional stabilization of the s-vallence shells. Furthermore, |E| of the group 15 elements Sb and Bi is smaller than |E| of the corresponding group 16 elements Te and Po, and even smaller than the |E| of group 14 element Pb with configuration s^2p^2. It appears that s^2p^3 does no longer exhibit a special relative stability, but rather s^2p^2.

Therefore, we should not be surprised about further irregularities of chemical properties of heavy elements, which cannot be explained in the "nonrelativistic regime". In general, *higher valencies* are favored by the heavier elements. Examples are the pairs Ti/Zr, Fe/Ru, N/P or F/Cl. This trend is reversed, however, for the heaviest main-group elements, which are easily reduced to *lower valence* states (see Tab. 1). These data, too, indicate *special stability of the "inert s^2-pair", and of the s^2p^2-configuration in the heaviest elements.*

Table 1. Lower valencies of heavy main group elements

5th row 6th row	 Hg	 Tl	Sn Pb	Sb Bi	Te Po	 At
Low stable valencies	0, 1	1	2	3, 1	2	1, 3

In most cases the *bond strengths* of s/p-elements become weaker for the heavier homologues. This is understandable in the nonrelativistic regime. The valence orbitals become more diffuse and overlap less efficiently, and the bigger cores cause a larger closed-shell-repulsion.

Table 2. Melting points of transition metals (in K)

Group 3	Group 4	Group 6	Group 8	Group 10	Group 11	Group 12
Sc 1814	Ti 1933	Cr 2130	Fe 1809	Ni 1728	Cu 1337	Zn 693
Y 1795	Zr 2125	Mo 2883	Ru 2583	Pd 1827	Ag 1235	Cd 594
La 1194	Hf 2500	W 3683	Os 3318	Pt 2045	Au 1338	Hg 234

If we use the melting points of metallic elements as a rough measure of their bond strengths, then Tab. 2 indicates especially strong bonds of the heavy d/s-elements from Hf (group 4) to Au (group 11). Numerical calculations show [47] that this phenomenon can only partly be explained by the lanthanoid-contraction. Obviously, an additional mechanism *improves overlap and bonding properties of the transition metal d- and s-orbitals* with respect to what is expected in the nonrelativistic regime. Furthermore the "tungsten maximum" of a series of properties may be taken as evidence for the special stability of the s^2d^4-configuration.

Not only the heavy main-group and d-transition elements show anomalies, but *also the heavy f-elements*, the actinoids. While the lanthanoids are predominantly trivalent, the first six actinoids, Th to Am, predominantly exhibit variable, higher valencies indicating stronger bond-participation of 5f than of the 4f in the lanthanoids.

To summarize, the properties of the heaviest elements cannot completely be predicted by simply extrapolating the trends from the upper half of the periodic system. There is ample experimental evidence from all groups of the periodic system, even at the *qualitative level*, that in addition to the nonrelativistic shell effects, a mechanism is present in the heavy atoms, which stabilizes the s-AO and one single p-AO of the valence shell, and also modifies the valence-properties of the d- and f-levels.

In order to understand and to systematize the chemical properties of heavy elements, we therefore need additional theoretical tools and concepts, which both allow *quantitatively* accurate calculations and, what is even more important, a *qualitative* understanding and rationalization of the properties of the heavy elements. This tool is relativistic quantum mechanics. On the basis of this sophisticated physical theory and corresponding quantitative computational results, *new qualitative chemical concepts* can be derived.

1.2 Objects to be Covered

Despite the experimental findings, even until recently many chemists have believed that relativistic effects should not be important for the chemically relevant outer valence shells. It has originally been assumed that relativity must only be accounted for in the inner core shells of heavy atoms, where the electrons are moving "relativistically fast" in the near vicinity of high nuclear charge (distance from the nucleus $R \ll 1\,a_0$, $Z \gg 1$)*. In the *outer valence shell* ($R > 1\,a_0$) where the effective nuclear charge is $Z_{eff}(R) < 10$, the mean potential energy V, the mean kinetic energy T, and the mean velocity v, obtained from

$$V = -e^2 \cdot Z_{eff}(R)/(4\pi\varepsilon_0 R) \,; \qquad T = p^2/2m_e \,; \qquad v = p/m_e \quad (1.2)$$

are all of the order of one atomic unit*. $m_e = 1$ is the mass of the electron. The classical relativistic energy corrections are of order $(E^{nr})^2/(m_e c^2)$, where the non-relativistic energy E^{nr} stands for an expression containing V and T (see below) and c is the velocity of light. Because of the large value of c*, the naive expectation has been that relativistic energy corrections $\Delta^{rel}E$ in the valence shell are as small as of order $1/c^2 \sim 10^{-4}$ a.u. ~ 0.1 kJ/mol, which is evidently too small to explain the irregularities in the experimental facts mentioned above.

In the *second section* we will explain why relativistic effects in the *valence shells of heavy atoms* are in fact so large. To this end we will use semiclassical arguments, and in particular also approximate relativistic quantum mechanics. We will try to elucidate the *physical mechanism* in a transparent manner. So far, most *qualitative insight* into relativistic quantum chemistry has been gained from an analysis of the properties of heavy *atoms* [1, 20]. Therefore, it is most important to understand first *their* relativistic behavior. The largest part of this article is therefore devoted to atoms.

In the *third section* we will explain which consequences are to be expected for potential curves, geometries, and bond energies of *molecules*. We will focus on the *conceptual and interpretational aspects*. In order to avoid complications that are not necessary for a qualitative understanding of the fundamentals, we will confine ourselves to the Born-Oppenheimer and Hartree-Fock approximations. Nevertheless it must be kept in mind that avoided potential surface crossings are particularly frequent in heavy atom systems, and that spin-orbit coupling causes configuration mixing to become even more important than for light atoms, where only the interelectronic Coulomb interaction contributes.

Physical theories can be represented in *different mathematical forms* without changing the predictions of *physical quantities*. Furthermore the integrands of quantum mechanical expectation values can be recast into different forms without changing the integral value of the observable. Explanation schemes of physical phenomena are based on the interpretation of explicitly given mathematical formulae. *Different representations* of the mathematical formalism therefore yield different, *complementary explanation schemes*. Since this is often overlooked in interpretational work, for instance in the field of relativistic quantum chemistry, or in the fields of chemical and VB and MO concepts, it is still important to reiterate:

---

* Atomic units (a.u.) are defined by $e^2/4\pi\varepsilon_0 = m_e = h = 1$. The velocity of light is $c = 137$ a.u., the length unit is $1\,a_0 = 0.529$ Å, the energy unit is 1 ha = 27.21 eV, 1 eV = 96.5 kJ/mol. If not otherwise stated, atomic units are used throughout. Often factors $m_e = 1$ or $h = 1$ are omitted.

1) that one correct explanation does *not* rule out other (correct) ones,
2) that a *full* understanding of the physical principles requires the use of *several* complementary explanation schemes, and
3) that one must be cautious not to mix up incompatible conceptions of complementary schems.

The vast amount of numerical evidence of *relativistic effects in chemistry* has been reviewed recently by several authors [2–5]. We also mention Pyykkö's bibliography [6], which lists more than 3000 references up to 1985. We will not compete with these comprehensive works which put emphasis on chemically relevant numerical results. We will confine ourselves to conceptual aspects and summarize them in the final subsection. Each subsection will be preceeded by a short summary.

2 Relativistic Effects in Atomic Valence Shells

2.1 Semiclassical Estimates (Bohr's and Sommerfeld's Models)

In this section we will explain that, according to Bohr's and Sommerfeld's models for hydrogen-like ions [7], the relativistic increase of the electronic mass with velocity will cause the orbital energies to decrease and the orbital radii to contract by the relative amount of const. $(Z/c)^2 \cdot n^{-2}$ *for high angular momenta, and const.* $(Z/c)^2 \cdot n^{-1}$ *for low angular momenta. Z is the nuclear charge, c is the velocity of light* ($c = 137$ *a.u.), and n is the principal quantum number.*

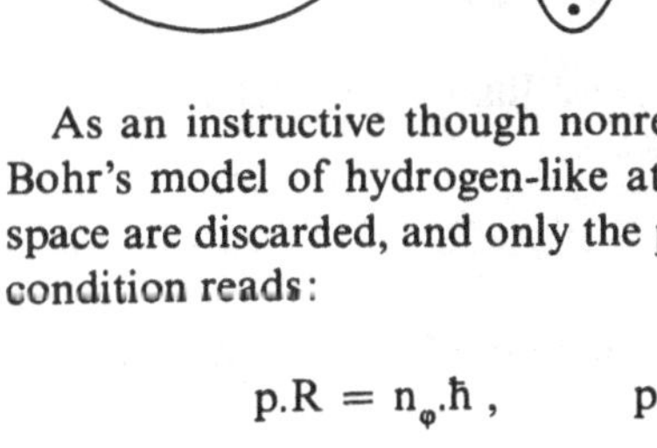

Fig. 2. Left side: Bohrs circular orbit for $n_\varphi = 3$ (3d); right side: Sommerfeld's penetrating orbit for $n = 3$, $n_r = 2$, $n_\varphi = 1$ (3s)

As an instructive though nonrealistic example, we will start the discussion with Bohr's model of hydrogen-like atoms. Here, two coordinates of three-dimensional space are discarded, and only the polar coordinate φ is considered. The quantization condition reads:

$$p.R = n_\varphi.\hbar\,, \qquad p = m_e.v\,, \tag{2.1}$$

where $n_\varphi = 1, 2, 3, \ldots$ is Bohr's (angular!) quantum number, v is the velocity, p the momentum, and p.R the angular momentum. A stationary state is characterized by equilibrium between Coulomb attraction and centrifugal "repulsion":

$$Z/R^2 = p^2/(m_e \cdot R)\,. \tag{2.2}$$

Solving for R and p, we obtain the nonrelativistic *values*:

$$R^{nr} = n_\varphi^2/(Z \cdot m_e)\,, \qquad p^{nr} = (Z \cdot m_e)/n_\varphi\,. \tag{2.3}$$

With the energy *relations*:

$$T = p^2/2m_e\,, \qquad V = -Z/r\,, \qquad E = T + V\,, \tag{2.4}$$

we obtain the nonrelativistic energy *values*:

$$\begin{aligned} T^{nr} &= (Z^2 \cdot m_e)/(2n_\varphi^2)\,, \qquad V^{nr} = -(Z^2 \cdot m_e)/n_\varphi^2\,, \\ E^{nr} &= -(Z^2 \cdot m_e)/(2n_\varphi^2)\,, \end{aligned} \tag{2.5}$$

which obey the (nonrelativistic) virial relation $V^{nr} = -2 \cdot T^{nr}$.

In the relativistic case, we have to replace the electronic rest-mass m_e in Eqs. (2.1) and (2.2) and in the following ones by the relativistic momentum mass or transversal inertial mass, resp. [48]:

$$m_{rel} = m_e/\sqrt{1 - v^2/c^2} = m_e \cdot (1 + v^2/2c^2 + 0(c^{-4})) > m_e\,. \tag{2.6}$$

The inrease of the relativistic mass with velocity is the so-called *mass-velocity effect* and, according to Eq. (2.3), causes a decrease of the radius of the relativistic orbit:

$$R^{rel} = R^{nr} \cdot \sqrt{1 - v^2/c^2} = R^{nr} \cdot (1 - v^2/2c^2 + 0(c^{-4})) < R^{nr}\,. \tag{2.7}$$

The *relative relativistic orbit-contraction* is given by

$$\delta^{rel}R := (R^{rel} - R^{nr})/R^{nr} \approx -0.5 \cdot (Z/n_\varphi c)^2\,. \tag{2.8}$$

Concerning energies, the nonrelativistic *kinetic energy expression* T has to be replaced by the corresponding relativistic one:

$$\tilde{T} = \sqrt{m_e^2 \cdot c^4 + p^2 \cdot c^2} - m_e c^2 \approx p^2/2m_e - p^4/(8m_e^3 c^2) < T\,, \tag{2.9}$$

which is *smaller than the nonrelativistic energy expression*. Inserting the relativistically reduced value R^{rel} and the increased value

$$p^{rel} = (Z \cdot m_{rel})/n_\varphi$$

into the expressions for $\tilde{T}$ and V, we obtain the relativistic kinetic, potential and total *energy values*

$$T^{rel} = (m_{rel} - m_e) \cdot c^2 \approx T^{nr} \cdot (1 + 3/4 \cdot (Z/cn_\varphi)^2 + \ldots) > T^{nr} \tag{2.10}$$

$$V^{rel} = -Z/R^{rel} \approx V^{nr} \cdot (1 + 1/2.(Z/cn_\varphi)^2 + \ldots) < V^{nr} \tag{2.11}$$

$$E^{rel} = E^{nr} \cdot (1 + 1/4 \cdot (Z/cn_\varphi)^2 + \ldots) < E^{nr}\,. \tag{2.12}$$

The increased relativistic mass results in an increased relativistic momentum, which overcompensates the decrease of the relativistic kinetic energy *expression*, so that the final relativistic kinetic energy *value* has increased. The contracted orbit results in a significant Coulomb energy lowering so that the *total energy is lowered.*

The relative relativistic energy-correction is of the same order $(Z/cn_\varphi)^2$ as the relativistic correction of the radii. That is, they increase with the squared nuclear charge felt by the electron, but they decrease quadratically with increasing quantum number n_φ. This corresponds to the naive expectation mentioned at the beginning of Sect. 1.2. These relations hold for Bohr's circular orbits, which correspond to quantum mechanical *orbitals of maximum angular momentum*, and Bohr's values approximate the corresponding exact quantum mechanical results.

However, penetrating orbitals of low angular momentum are the ones most important in chemistry. These can be approximated classically by Sommerfeld's elliptic orbits, where all three dimensions of space are accounted for in the quantization rules, giving rise to three quantum numbers n_r, n_φ, n_ϑ [7]. Instead of Eq. (2.12), Sommerfeld obtained:

$$\begin{aligned} E^{rel} &= m_e \cdot c^2 \cdot |1 + (Z/c)^2/(n_r + \sqrt{n_\varphi^2 - (Z/c)^2})^2|^{-1/2} - m_e c^2 \\ &\approx E^{nr} \cdot (1 + (Z/c)^2 \cdot (1/nn_\varphi - 3/4n^2)) \end{aligned} \tag{2.13}$$

where $n = n_r + n_\varphi$ is the principal quantum number. The lowest possible value $n_\varphi = 1$ yields the (exact) relativistic energy stabilization for s-orbitals:

$$\Delta^{rel}E = E^{rel} - E^{nr} \approx E^{nr} \cdot (1 - 3/4n) \cdot (Z^2/nc^2) \tag{2.14}$$

The corresponding relative orbit-contraction is approximately given by:

$$\delta^{rel}R = \Delta^{rel}R/R^{nr} \approx -(1 - 1/2n) \cdot (Z^2/nc^2) \tag{2.15}$$

This result agrees at the given order of n and c with the exact hydrogenic value, which has been discussed by Burke and Grant [8]. The leading term of the relativistic changes of energies and lengths is again proportional to $(Z/c)^2$. It will be called the first or lowest order term, while the nonrelativistic energy, wavefunction, etc. are called zeroth order quantities. Higher order terms become qualitatively important for the lowest row of the periodic system, but will not be discussed here.

While the relative relativistic stabilization and contraction of high angular momentum states decreases as $1/n^2$ (see Eqs. 2.8, 2.12), that of penetrating low angular momentum states decreases only as $1/n$ (see Eqs. 2.14, 2.15). This results because the mass-velocity correction $\sim(v/c)^2$ weighs the *inner parts* of excentric orbits more strongly.

In a many-electron atom the effective nuclear charge $Z_{eff}(R)$ increases up to the full value $Z_{nuc} \gg Z_{eff}^{Mean}$ for $R \to 0$ because of the decrease of shielding by the electrons. Therefore the relevant nuclear charge for the relativistic corrections, Z_{eff}^{rel}, will be larger than Z_{eff}^{Mean}, which describes the nonrelativistic expectation values. As will be discussed below, the relativistic effects in the valence shells of many-electron atoms are even further larger than is expected from this model.

The argument that R decreases due to the relativistic increase of m in Eq. (2.3) has not remained without opposition. It has been argued that according to Einstein's relation $m = E/c^2$, m decreases because of the relativistic energy decrease, Eqs. (2.12, 14). However, different masses are mixed up here. $m = E^{rel}/c^2$ is the inertial or gravitational mass of the stationary atomic system observed by a distant observer. A significant contribution to this m is the local potential energy of the electron. Note that by placing the atom in a large potential well, any change of m is possible without change of the atom itself. On the other hand, the orbital radius in Eq. (2.3) is determined by the curvature of the orbital. This depends on the momentum or transversal mass of the electron, Eq. (2.6), as measured in a reference frame at rest at the position, where the electron is instantaneously moving. In this reference frame the potential energy is always zero. Finally, we note that in the case of elliptical orbits also the longitudinal mass $m_e/\sqrt{1 - v^2/c^2}^3 > m_e$ plays a role.

2.2 Relativistic Quantum Mechanics

For those readers, who want to skip this more formal section, we summarize the main points. In order to derive the relativistic quantum mechanical corrections, and to obtain the basis for physical explanations, the Klein-Gordon equation for spin-zero particles and the Dirac equation for spin-1/2 particles are introduced. The equations are given in the standard representation and in the so-called free-particle representation. This will allow us to understand the physical origin of the Darwin effect.

The classical relativistic kinetic energy expression is a square root, and it is smaller than the nonrelativistic expression. A square root is two-valued (i.e. it has two signs). This leads to the concept of particle and antiparticle states. In the standard representation of the relativistic wave equations, the expression for the potential energy is the one known from nonrelativistic classical or quantum mechanics. For instance, the nuclear Coulomb potential is $-Z/R$*, where R is the distance of the electronic charge from the nucleus. The standard representation, however, is not diagonal in a basis of particle and antiparticle states, so that R cannot be interpreted as the position of a particle and its mass.* R_{charge} *and* $R_{particle}$ *are no longer identical.*

According to Heisenberg's uncertainty principle, position and momentum do not commute and cannot be defined or measured exactly at the same time. This causes the so-called kinetic zero-point energy. i.e. it keeps the kinetic energy of a localized system above a minimum value, and it keeps the average distance of the electron from the nucleus (and consequently the potential energy) above a minimum value. Measurement of mass and measurement of charge are different processes, which are no longer possible simultaneously in the relativistic regime. Consequently the nuclear Coulomb attraction energy is raised by an additional amount. This is the so-called Darwin effect [12].

There exists a unitary transformation from the standard relativistic (Dirac) representation, where $R = R_{charge}$*, to the relativistic Schrödinger-Pauli representation, where* $R = R_{particle}$*. It is important to note that all operators are represented by somewhat different expressions, which will give rise to different, complementary physical explanations [9, 53].*

Both nonrelativistic and relativistic quantum mechanics are compatible with the assumption of spin = 0 or spin = 1/2. However, a spin-orbit coupling energy emerges only in a relativistic theory.

The most important relativistic first-order effects are the mass-velocity energy (already discussed in Sect. 2.1), the Darwin energy, the spin-orbit energy, and the so-called indirect effects. (Higher order and two-particle effects will not be considered here. For further details see textbooks [11, 12].) It has often been stated that the Darwin effect is exceptional, as no classical explanation could be given for it. However, it can be explained in physical terms, if one has the appropriate theoretical formalism at hand.

We start with the classical relativistic energy expression:

$$(m_e \cdot c^2)^2 = (E - V)^2 - (\vec{p}c - \vec{A})^2 , \tag{2.16}$$

where V and $\vec{A}$ are the external electrostatic and magnetic potentials. In the following, we will assume $\vec{A} = 0$. Eq. (2.16) can be rewritten in three different, nearly equivalent forms:

$$\sqrt{m_e^2 \cdot c^4 + p^2c^2} = \tilde{E} - V \tag{2.17 RS}$$

$$m_e^2 \cdot c^4 + p^2c^2 = (\tilde{E} - V)^2 \tag{2.17 KG}$$

$$\beta m_e \cdot c^2 + \vec{\alpha}\vec{p}c = \tilde{E} - V . \tag{2.17 D}$$

Upon quantization, i.e. interpreting the variables as operators acting on wave-functions, for instance $\tilde{E} \rightarrow \hat{\varepsilon} = i \cdot \hbar \cdot \partial/\partial t$, the three Eqs. (2.17) are no longer equivalent. The *Relativistic Schrödinger* Eq. (2.17 RS) contains the classical relativistic kinetic energy (including the rest mass energy) explicitly on the left side (compare Eq. 2.9). The corresponding squared Eq. (2.17 KG) had already been analysed by Schrödinger before the invention of the nonrelativistic Schrödinger equation and had later been published by several authors, among others *Klein and Gordon* (see [9]). Interpreted as a scalar equation, it describes spin-zero particles. Introducing the so-called non-commuting four-by-four Dirac matrices $\vec{\alpha}$, β, the square-root can be solved explicitly in linear form, if $\vec{\alpha}.\vec{\alpha} = \underline{\underline{1}}$, $\beta.\beta = \underline{\underline{1}}$ and $\vec{\alpha}.\beta + \beta.\vec{\alpha} = \underline{\underline{0}}$ holds [10]. $\underline{\underline{0}}$ and $\underline{\underline{1}}$ are the zero and unit matrices. The corresponding *Dirac* Eq. (2.17D) describes spin-1/2 particles. In all three quantized equations, the operator $\hat{\varepsilon}$ has the eigenvalue ε for the stationary state $\Psi(r, t) = \psi(r) \cdot \exp(-i\varepsilon t)$. For particles the "frequency-factor" ε is positive (except in extremely attractive potentials where ε may even drop below zero) and is related to the energy E by $E + m_ec^2 = +\varepsilon$, while for antiparticles in general $\varepsilon < 0$ and $E + m_ec^2 = -\varepsilon$.

The differences between the quantized forms of Eqs. (2.17) become evident upon squaring the linear ones:

$$\left(m_e^2c^4 + p^2c^2 - (\varepsilon^{RS} - V)^2 + [\sqrt{m_e^2c^4 + p^2c^2},\ V(r)]_-\right) \cdot \psi^{RS} = 0 \tag{2.18RS}$$

$$\left(m_e^2c^4 + p^2c^2 - (\varepsilon^{KG} - V)^2\right) \cdot \psi^{KG} = 0 \tag{2.18KG}$$

$$\left(m_e^2c^4 + p^2c^2 - (\varepsilon^{D} - V)^2 + [\vec{\alpha}\vec{p}c,\ V(r)]_-\right) \cdot \psi^{D} = 0 \tag{2.18D}$$

Due to Heisenberg's uncertainty principle, position r and momentum p do not commute, $[r, p]_- = r \cdot p - p \cdot r = i \cdot \hbar$, which gives rise to nonvanishing commutators of the position dependent potential V(r) and terms containing p.

The four components of the *Dirac*-column vector ψ^D indicate that there are four "degress of freedom", spin-up and spin-down of electrons and positrons. The representation chosen in eqs. (2.17) and (2.18) for the *RS* and *KG* equations, however, have only one component, although these equations, too, describe (spin-zero) particles *and* antiparticles. Therefore, a two-component representation would be more satisfactory. The corresponding transformation has been given by Feshbach and Vilars for the Klein-Gordon equation [9]:

$$\begin{pmatrix} m_e \cdot c^2 + T + V - \varepsilon^{KG} & T \\ T & m_e \cdot c^2 + T - V + \varepsilon^{KG} \end{pmatrix} \cdot \begin{pmatrix} \psi_+^{KG} \\ \psi_-^{GK} \end{pmatrix} = 0 . \tag{2.19 KG}$$

For the RS-equation, see [13]. The standard Dirac equation can be written in a similar form:

$$\begin{pmatrix} m_e \cdot c^2 + V - \varepsilon^D & \pi \\ \pi & m_e \cdot c^2 - V + \varepsilon^D \end{pmatrix} \cdot \begin{pmatrix} \psi_+^D \\ i\psi_-^D \end{pmatrix} = 0 , \tag{2.19 D}$$

where $\pi = c \cdot \vec{\sigma} \cdot \vec{\nabla}$. The components of $\vec{\sigma}$ are Pauli's spin matrices, which are the off-diagonal two-by-two blocks of $\vec{\alpha}$. $\vec{s} = \vec{\sigma}/2$ represents the spin angular momentum operator of an electron. ψ_+^D and ψ_-^D are two-component spin-functions corresponding to the two upper and two lower components of ψ^D.

In the case of free particles, V = 0, Eqs. (2.19) can easily be diagonalized with the help of the transformations:

$$U^{KG} = ((m_e \cdot c^2 + T - \sigma_x \cdot T)/(m_e \cdot c^2 + T + \sigma_x \cdot T))^{1/4} , \tag{2.20}$$

$$U^D = ((m_e \cdot c^2 - \sigma_x \cdot \pi)/(m_e \cdot c^2 + \sigma_x \cdot \pi))^{1/4} ,$$

respectively. The resulting transformed Hamiltonian is formally similar for both cases:

$$\begin{pmatrix} +\sqrt{m_e^2c^4 + p^2c^2} - \varepsilon & 0 \\ 0 & +\sqrt{m_e^2c^4 + p^2c^2} + \varepsilon \end{pmatrix} \cdot \begin{pmatrix} \psi_+^{SP} \\ \psi_-^{SP} \end{pmatrix} = 0 . \tag{2.21}$$

The only difference is that in the Dirac case every term is to be multiplied by a 2×2-unit matrix. The "Schrödinger-Pauli" wavefunctions ψ_+^{SP} and ψ_-^{SP} have two components in the Dirac case, but only one in the KG case.

The advantage of the "Schrödinger-Pauli" representation (eq. 2.21) is that *free* particle and antiparticle states are now decoupled. *Particle states* with positive frequency factor ε and transformed eigenfunction $\begin{pmatrix} \psi_+^{SP} \\ 0 \end{pmatrix}$ are determined by the upper line of Eq. (2.21), and *antiparticle states* with negative frequency factor ε and transformed wavefunction $\begin{pmatrix} 0 \\ \psi_-^{SP} \end{pmatrix}$ are determined by the lower line.

If there is a potential $V \neq 0$ in the original Hamiltonian, U^+VU will appear in Eq. (2.21). This term contains "relativistic" diagonal corrections to V, and off-diagonal coupling terms between particle and antiparticle states. That is, the potential mixes free particle and free antiparticle states. The lower components of the wavefunction of a particle in a potential are no longer zero.

To summarize, different representations of operators and states may be used. In the "standard" representation (2.17) the potential operator is simply given by the classical expression V(r). Obviously, r is the *position of the electric charge* of the coupled particle-antiparticle system, and $|\psi^D|^2$ is the charge density. On the other hand, after linear transformation with U in particle-antiparticle space, $|\psi^{SP}|^2$ represents the particle density at *particle position* r. U is nonlocal in position space, that is U contains p, i.e. spatial derivatives (namely via T and π). Since momentum p and position r do not commute, the transformed operator for the charge position in this new representation is changed into:

$$\vec{r}^{tr} = U^+ \vec{r} U = \vec{r} + \beta[\vec{\alpha}p, r]_-/2mc + [\vec{\alpha}p,[\vec{r}, \vec{\alpha}\vec{p}]_-]_-/8m^2c^2 \ldots \qquad (2.22)$$

Particle and charge coordinates are different observables, which do not commute, because p and r do not commute. If one coordinate is defined exactly, the other one is smeared out by the order of magnitude of the Compton length $\lambda = \hbar/(m_e \cdot c)$. This holds for any relativistic system, be it described by the D-, KG-, or RS-equation.

2.3 Relativistic Energy Corrections of Lowest Order

If the relativistically corrected energy E^{rel} is expanded in powers of c^{-2}, $E^{rel} = E^{nr} + a_2/c^2 + \ldots$, the higher order terms are needed for light elements only if one wants to achieve very high spectroscopic accuracy. For the heaviest atoms, however, the higher order terms are even needed to achieve qualitatively correct results. Nevertheless, the main physical insight can be obtained by analysing just the relativistic correction of lowest order $\sim 1/c^2$. Corresponding expressions will be derived in this section for the kinetic velocity-mass, and potential Darwin and spin-orbit energies in both the standard Dirac and the Schrödinger-Pauli representations.

We will at first determine the total energy of an electron in both representations, using the relations:

$$H^{SP} = U^+H^DU\,; \qquad \Psi^{SP} = U^+\Psi^D\,; \qquad \Psi^D = U\Psi^{SP}\,. \qquad (2.23)$$

This gives rise to the following equivalent energy expressions:

$$\begin{aligned} E &= \langle \Psi^D |H^D| \Psi^D \rangle = \langle \Psi^{SP} |H^{SP}| \Psi^{SP} \rangle \qquad (2.24) \\ &= \langle U\Psi^{SP} |H^D| U\Psi^{SP} \rangle = \langle U^+ \Psi^D |H^{SP}| U^+ \Psi^D \rangle \end{aligned}$$

In the Dirac representation we have:

$$E = \langle \psi^D |(\beta - 1)\, m_e \cdot c^2 + c\vec{\alpha}p + V| \psi^D \rangle = T^D + V^D \qquad (2.25)$$

with the kinetic contribution:

$$T^D = -2m_ec^2 \cdot \langle \psi_-^D \mid \psi_-^D \rangle + c \cdot (\langle \psi_-^D \mid \vec{\sigma}\vec{p}\psi_+^D \rangle + \text{c.c.}) = T^\beta + T^\alpha \quad (2.26)$$

and the potential contribution:

$$V^D = \langle \psi_+^D \, |V| \, \psi_+^D \rangle + \langle \psi_-^D \, |V| \, \psi_-^D \rangle = V_+^D + V_-^D \quad (2.27)$$

c.c. is the complex conjugate matrix element.

2.3.1 Potential Energies

In order to clarify the meaning of these terms, we represent the upper and lower components of the Dirac function in terms of the upper components of the Schrödinger-Pauli function ψ_+^{SP}, where we will omit the index +. From Eq. (2.23) we obtain:

$$\begin{aligned} \psi_+^D &= \psi^{SP} - p^2\psi^{SP}/8m_e^2c^2 + O(c^{-4}) \quad (2.28) \\ \psi_-^D &= \vec{\sigma}\vec{p}\psi^{SP}/2m_ec + O(c^{-3}) \end{aligned}$$

which yields for the potential energies:

$$\begin{aligned} V_+^D &= \langle \psi^{SP} | \, V \, | \psi^{SP} \rangle - (\langle p^2\psi^{SP} | \, V \, | \psi^{SP} \rangle + \text{c.c.})/8m_e^2c^2 + O(c^{-4}) \\ &= V^{\text{“nr”}} + V_+^{Dar} \quad (2.29\,a) \\ V_-^D &= \langle \vec{\sigma}\vec{p}\psi^{SP} | \, V \, | \vec{\sigma}\vec{p}\psi^{SP} \rangle/4m_ec^2 + O(c^{-4}) = V_-^{Dar} + V^{SO} \quad (2.29\,b) \end{aligned}$$

Note that Eq. (2.29b) contains spin-independent (V_-^{Dar}) and spin-dependent terms (V^{SO}), because for the spin-matrices $\vec{\sigma}$ the following relation holds:

$$(\vec{\sigma} \cdot \vec{A})\,(\vec{\sigma} \cdot \vec{B}) = \vec{A} \cdot \vec{B} + i[\vec{A}, \vec{B}, \vec{\sigma}]$$

where $[\vec{A}, \vec{B}, \vec{\sigma}]$ is the cross-dot-product.

The first term $V^{\text{“nr”}}$ of Eq. (2.29a) would represent the nonrelativistic potential energy V^{nr} if the nonrelativistic function ψ^{nr} would occur instead of ψ^{SP}. The physical meaning of the terms of order c^{-2} in Eqs. (2.29a, b) becomes clear, if one applies the turn-over rule to the hermitean operators p^2 and $\vec{\sigma}\vec{p}$. (No problems are encountered in this step if the potential, and consequently the wavefunctions, are non-singular, as in the case for a real, extended nucleus.) One then obtains:

$$V^D = \langle \psi^{SP} \, | V - \vec{\nabla}^2 V/8m_e^2c^2 + (\vec{\nabla}V \times \vec{p}) \cdot \vec{s}/2m_e^2c^2 | \, \psi^{SP} \rangle + 0(c^{-4}) \quad (2.30)$$

$\vec{\nabla}^2 V$ may be expressed by the charge distribution ϱ, which corresponds to the potential V according to Poisson's equation:

$$\vec{\nabla}^2 V = -4\pi\varrho\,.$$

We note that V is the total potential operator in the Dirac representation [see Eq. (2.27)], whereas it is only the nonrelativistic contribution to the potential operator in the SP-representation [Eq. (2.30)]. In the latter representation the total potential operator contains also momentum and spin-dependent corrections of order $1/c^2$, which are already accounted for in $V^D = \langle\psi^D| V |\psi^D\rangle$.

The second term in Eq. (2.30) is the so-called Darwin correction, which is the sum of the second term of Eq. (2.29a), V_+^{Dar}, and of the spin-independent contribution of Eq. (2.29b), V_-^{Dar}. For an attractive nuclear Coulomb potential V, the Darwin term $+\pi \cdot \varrho/2m_e^2c^2$ is repulsive. It arises, because in the Schrödinger-Pauli representation the position of the charge of the electronic particle is "uncertain" by about λ, which reduces the value of the nonrelativistic nuclear Coulomb attraction in the vicinity of the nucleus. The Darwin-correction has no classical analogue in a similar sense as there is no classical analogue to the zero-point energy, which prevents the electron from collapsing into the nucleus. Both effects, which raise the expectation value of the potential energy, are consequences of Heisenberg's uncertainty relation.

The third contribution to Eq. (2.30) is the famous spin-orbit coupling. It can be interpreted as the interaction of the magnetic moment of the electron (associated with the electronic spin $\vec{s}$) and the magnetic field created by the "ring current" of an electron moving in an orbit with angular momentum $\vec{l}$ around the nucleus. It splits the $2l+1$-fold degenerate orbital level into two sublevels. Since opposite magnetic poles attract each other, "antiparallel" $\vec{l}$ and $\vec{s}$ (corresponding to toal angular momentum quantum number $j = l - s = l - 1/2$) lower l ones of the $2l + 1$ component states, while "parallel" $\vec{l}$ and $\vec{s}$ ($j = l + 1/2$) raises the remaining $l + 1$ components (see diagram for the case of p-AOs with $l = 1$).

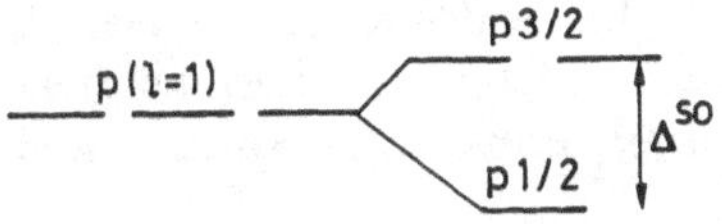

Fig. 3. Spin-orbit splitting of p-levels

A note is appropriate here concerning the meaning of the words "parallel" and "antiparallel". Two angular momentum vectors $\vec{l}$ and $\vec{l}'$ with angular momentum quantum numbers l,l' are called "parallel" (or "antiparallel"), if the sum of vectors $\vec{l} + \vec{l}' = \vec{j}$ corresponds to quantum numbers $j = l + l'$ (or $j = l - l'$, respectively). The angle γ between vectors $\vec{l}$ and $\vec{l}'$ has a sharply defined value:

$$\cos\gamma = \pm\sqrt{(l'/(l' + 1)) \cdot (l/(l + 1))^{\pm 1}}$$

We see that "parallel" angular momenta will always form an angle $\gamma_{par} = \text{acos}\,(l \cdot l'/(l + 1)/(l' + 1))^{+1/2} > 0$ with each other, while "antiparallel" angular momenta will only be antipyrallel with $\gamma_{antipar} = \text{acos}\,(-(l'/l \cdot (l + 1)/(l' + 1))^{1/2}) = 180°$, if $l = l'$. For example, for an electron with spin ($l' = 1/2$) in a p-orbital ($l = 1$), $\gamma_{par} = 66°$, and $\gamma_{antipar} = 145°$.

There are controversial statements in the literature as to whether the spin-orbit splitting is a relativistic effect. Levy-Leblond [14] has shown that both relativistic and nonrelativistic schemes offer similar possibilities to derive wave-equations without or with spin, although the need for deriving equations which transform according to multivalued spinor-representations seems more natural in the relativistic framework. Anyhow, the anomalous gyromagnetic ratio of spin-1/2 particles automatically energes also from nonrelativistic Galilei covariance. However, in a fully nonrelativistic theory the electron would only interact with an external magnetic field (Zeeman-effect). The lowest order contribution to the spin-orbit interaction in Eq. (2.30) has already been obtained by Thomas, treating the motion of a nonrelativistic electron in Maxwell's relativistically covariant electromagnetic field [15]. Higher-order corrections to the spin-orbit interaction require, however, a fully relativistic treatment.

2.3.2 Kinetic Energy

From Eqs. (2.26, 28) one obtains for the kinetic energy:

$$T^D = T^\beta + T^\alpha = -\langle \vec{\sigma}\vec{p}\psi^{SP} \mid \vec{\sigma}\vec{p}\psi^{SP}\rangle/2m_e + \langle \vec{\sigma}\vec{p}\psi^{SP} \mid \vec{\sigma}\vec{p}\psi^{SP}\rangle/m_e + 0(c^{-2})\,. \tag{2.31}$$

Applying again the turn-over rule, one recovers the nonrelativistic kinetic energy:

$$T^D = \langle \psi^{SP} | \, p^2/2m_e \, | \psi^{SP}\rangle + 0(c^{-2}) \approx T^{nr}\,. \tag{2.32}$$

Specifically, in zeroth order, the $\beta m_e c^2$-term of H^D contributes $T^\beta = E^{rel.} \approx E^{nr} = -T^{nr}$, while the $\vec{\alpha}\vec{p}c$-term contributes $T^\alpha \approx +2T^{nr}$. In order to obtain the lowest order relativistic corrections to the kinetic energy, higher order terms are needed in Eq. (2.28). Slightly different transformations U yield different relativistic corrections from $\beta m_e c^2$ and from $c\vec{\alpha}\vec{p}$, the sum of first order terms being constant and being equivalent to:

$$T^D = \langle \vec{p}\psi^{SP} \mid \vec{p}\psi^{SP}\rangle/2m_e - \langle p^2\psi^{SP} \mid p^2\psi^{SP}\rangle/(8m_e^3c^2) + 0(c^{-4})\,. \tag{2.33}$$

If we again apply the turn-over rule of hermitean operators, the kinetic energy in the Dirac picture expressed in terms of the SP-wavefunction, Eq. (2.33) is converted into:

$$T^D = \langle \psi^{SP} | \, p^2/2m_e - p^4/(8m_e^3c^2) \, | \psi^{SP}\rangle + 0(c^{-4})\,. \tag{2.34}$$

This is the total kinetic energy in the SP-representation. The second term of Eqs. (2.33, 34) represents the mass-velocity energy, already discussed in Sect. 2.1. As in the case of the potential expectation values (V^D from Eq. (2.27) = V^D from Eq. (2.30)), the *expectation values* for the kinetic energy are the same [T^D from Eq. (2.26) = T^D from Eq. (2.34)], although the *integrands* in the two schemes are different. For instance, the integrands in Eqs. (2.26, 31, 33) are positive definite since they are squared functions, while the integrands in Eqs. (2.32, 34) have nodes. For instance, $\psi^{SP}(p^2/2m_e)\,\psi^{SP}$ is positive in the classically allowed region, but is negative in the classically forbidden region.

2.4 The Paradoxical Role of the Mass-Velocity Correction

The paradoxical role of the mass-velocity correction had already been mentioned in Sect. 2.1. This paradox will be discussed further and brought in relation to well known paradoxa from other fields (covalent bond, Hund's rule, thermodynamics of stars).

We will discuss the relativistic energy of the 1s-state of a hydrogenlike ion with nuclear charge Z, using Eqs. (2.29/30) and (2.33/34). All terms of order c^{-4} and higher will be neglected. Then we may approximate ψ^{SP} by a single Slater-type orbital $\exp(-\zeta r)$. The following expressions for the matrix elements are obtained (in a.u.):

$$T = \langle p^2/2m_e \rangle = \zeta^2/2 \quad \text{nonrelativistic kinetic energy}$$
$$\Delta^{rel}T = \langle -p^4/8m_e^3c^2 \rangle = -5\zeta^4/8c^2 \quad \text{Mass-velocity correction} \tag{2.35}$$
$$V = \langle -Z/r \rangle = -Z\zeta \quad \text{nonrelativistic potential energy}$$
$$\Delta^{rel}V = \langle Z \cdot (p^2/r)/8m_e^3c^2 \rangle = Z\zeta^3/2c^2 \quad \text{Darwin correction.}$$

For s-states, the first order spin-orbit correction vanishes. Minimization of the non-relativistic energy expression T + V yields the optimal orbital exponent $\zeta_{opt}^{nr} = Z$, which in turn determines the nonrelativistic energy values:

$$T^{nr} = +Z^2/2\,, \quad V^{nr} = -Z^2\,, \quad E^{nr} = -Z^2/2\,. \tag{2.36}$$

If we insert ζ_{opt}^{nr} into the expression for the relativistic energy, we obtain:

$$T^{rel} = Z^2/2 - 5Z^4/8c^2\,, \quad V^{rel} = -Z^2 + Z^4/2c^2\,,$$
$$E^{rel} = -Z^2/2 - Z^4/8c\,, \tag{2.37a}$$

yielding the relativistic energy corrections:

$$\Delta^{rel}T = -5Z^4/8c^2\,, \quad \Delta^{rel}V = +Z^4/2c^2\,, \quad \Delta^{rel}E = -Z^4/8c^2\,. \tag{2.37b}$$

In this *first-order* treatment, the total energy has decreased, because the kinetic energy has relativistically decreased more than the Darwin term has increased the potential energy.

In the relativistic case we should use, however, the relativistically optimized orbital exponent $\zeta_{opt}^{rel} \approx Z(1 + (5 - 3)\, Z^2/2c^2)$. Due to the relativistically decreased "kinetic energy pressure", the orbital contracts, increasing ζ by $5Z^3/2c^2$, while the reduction of the nuclear attraction due to the Darwin effect causes a slight orbital expansion, corresponding to a decrease of ζ by $-3Z^3/2c^2$. The energy values now read, up to first order in c^{-2}:

$$\langle \hat{T}^{nr} \rangle = Z^2/2 + Z^4/c^2\,, \quad \langle \hat{V}^{nr} \rangle = -Z^2 - Z^4/c^2\,, \quad E^{nr} = -Z^2/2 \tag{2.38a}$$
$$\Delta_{dir}^{rel}T = -5Z^4/8c^2\,, \quad \Delta_{dir}^{rel}V = +Z^4/2c^2\,, \quad \Delta_{dir}^{rel}E = -Z^4/8c^2 \tag{2.38b}$$
$$T^{rel} = Z^2/2 + 3Z^4/8c^2\,, \quad V^{rel} = -Z^2 - Z^4/2c^2\,, \quad E^{rel} = -Z^2/2 - Z^4/8c^2\,. \tag{2.38c}$$

Here E^{nr} is the energy obtained with the nonrelativistic operator from the relativistically *relaxed* wavefunction. The basic reason for the relativistic energy decrease is the overall *decrease of the expression* for the relativistic energy due to the predominant mass-velocity correction. Thereby the balance of kinetic and potential energies is violated. $\langle \hat{T}^{nr} \rangle$ and $\hat{V}^{nr} \rangle$ change, until a new balance is reached. Since the nonrelativistic system was at a stationary point, $E^{nr} = \langle \hat{T}^{nr} \rangle + \langle \hat{V}^{nr} \rangle$ is not changed to first order, though. Therefore, the total energy can be predicted even in terms of the nonrelaxed wavefunction. However, nonstationary expectation values (e.g., $\langle \hat{T}^{nr} \rangle$ and $\langle \hat{V}^{nr} \rangle$, individually) are significantly modified by orbital relaxation. This so-called first-order "indirect" effect (Eq. 2.38a) may *partially compensate* the direct effect, as is predicted by LeChatelier-Braun's principle, or it may even *overcompensate* the direct changes [Eqs. (2.37b/38b)], as holds in the present case for T^{rel} and V^{rel}. This "non-LeChatelier like" behavior is quite common. Nevertheless, each example usually causes new astonishment. We will discuss below three typical cases.

We note that the indirect effects cancel to first order in the total energy. That is if we are interested in the total energy only, we need not discuss relativistic orbital relaxation, provided exact nonrelativistic functions are used for each state. Furthermore one should remember that the "virial relation" $T = -E = -V/2$ does not hold for the relativistic energies.

2.4.1 Similar Paradoxa

1) The covalent bond: If the (nonrelativistic) kinetic energy is written as:

$$T = \langle \psi | \, p^2/2m_e \, | \psi \rangle = \langle \psi | \, (\vec{\nabla}\psi/\psi)^2 \, | \psi \rangle / 2m_e \,, \tag{2.39}$$

$(\vec{\nabla}\psi/\psi)^2/2m_e$ may be interpreted as the positive definite kinetic energy density. In-phase overlap of atomic orbitals will reduce the slope of the molecular orbital and, consequently, of the kinetic energy density in the overlap region. At a saddlepoint of ψ, the kinetic energy density even becomes zero. Therefore, using an unrelaxed minimal AO-basis, the LCAO-MO-energy of a bonded system will be lower because of the decreased kinetic energy density functional. This causes a reorganization of the wavefunction, in general an overall orbital contraction, which drastically lowers the nuclear attraction energy. The accompanying indirect kinetic energy increase will significantly overcompensate the original direct kinetic energy decrease. Finally the kinetic energy has increased by the same amount as the total energy has decreased, thereby fulfilling the virial theorem. For details consult Ruedenberg [16] or the first chapter of the present volume by Kutzelnigg.

2) Hund's Rule: In general, the triplet state of a given electronic configuration a^1b^1 is lower in energy than the singlet state. The Hartree-Fock energy functional for the triplet contains a $-K$, instead of the $+K$ for the singlet, where K is the exchange potential operator. Therefore, using the unrelaxed singlet orbitals for the triplet, a direct decrease of the interelectronic repulsion energy by $-2K_{ab}$ results for the triplet. However, due to the reduced interelectronic repulsion, the orbitals contract in the triplet, thereby gaining a large amount of nuclear attraction energy.

Simultaneously the kinetic energy and the interelectronic repulsion energy increase. The indirect increase of the interelectronic repulsion overcompensates the direct decrease in most cases. So, the triplet state gets a *larger* interelectronic repulsion energy than the singlet, but its total energy is lower (because of the increased electron-nuclear attraction). For details see [17].

In all three cases (relativity, covalency, exchange stabilization), the driving force of the physical effect is the reduction of a specific positive term in the total energy functional. According to first order perturbation theory, the energy lowering can be estimated using the unperturbed orbitals. In this sense the energetic stabilization is due to a decrease of that specific energy term. Adjustment of the wavefunction to the new situation will then enlarge the negative electron-nuclear attraction and simultaneously compensate or overcompensate the original energy change. According to the virial theorem, which does not discriminate between different physical mechanisms, *any* stabilized system has a low energy "because" of low electron-nuclear attraction energy.

This "anti-LeChatelier" type behavior is not limited to quantum mechanical systems. Ruedenberg [16] has given an example from classical mechanics, and Kutzelnigg [19] one from economy. We add here one example from thermodynamics.

3) Evolution of stars: If a hot star emits radiation, it should cool down, and its constituent particles should loose kinetic energy. This causes a reduction of the internal temperature and pressure. The star then shrinks. Thereby it gains gravitational energy, which is converted into heat. Finally, according to the virial theorem, the internal kinetic energy or temperature has increased, while the total energy has decreased, due to radiation, by the same amount. For details see [18].

2.5 Magnitude of Relativistic Effects in Atomic Valence Shells

Relativistic calculations on many-electron atoms yield relativistic corrections which are quite different from those of hydrogenlike one-electron systems. The numerical results reveal remarkable trends. The magnitude of the relativistic corrections is so large as to cause drastic changes in the chemical properties of heavy elements. The data supply hints about the physical mechanism of relativistic effects in atoms, which will be elucidated in Sect. 2.6.

Desclaux [20] has carried out relativistic calculations on all neutral atoms at the Hartree-Fock level of approximation and has tabulated the relativistic orbital energies E^{rel} of all atomic orbitals. If the relative relativistic orbital energy correction is represented as:

$$\delta^{rel}E := (E^{rel} - E^{nr})/E^{nr} = a \cdot (Z_{nuc}/c)^2 , \qquad (2.40)$$

where Z_{nuc} is the charge of the bare nucleus, then for a given AO symmetry lj ($j = l \pm s$), the coefficient a varies comparatively little over the periodic table. In Tab. 3 the a-coefficients for the atomic valence orbitals are arranged in analogy to the elements in the periodic table. Most atoms have a doubly occupied s^2-valence shell. Singly occupied s^1-valence shells (as in groups 1 and 11) are indicated by a-values in italics. The a-parameters for the d- and f-valence shells are in italics, if the configuration differs from the standard $d^{n-2}s^2$ and $f^{n-2}s^2$ ones. The following observations can be made:

Table 3. Coefficient a for the relative relativistic orbital energy stabilization [see Eq. (2.40)], after Desclaux [20]

a) s-valence shell[1]

Row	1	2	3		4	5	6	7	8	9	10	11	12	13	14	15	16	17	18	Row
1	*.25*																		.2	1
2	*.17*	.20	.32											.32	.40	.44	.48	.51	.53	2
3	*.20*	.20	.29											.29	.34	.37	.39	.40	.42	3
4	*.20*	.19	.27		.30	.33	*.53*	.37	.38	.40	.41	*.68*	.44	.45	.44	.44	.44	.44	.44	4
5	*.22*	.20	.33		.38	*.60*	*.62*	.45	*.64*	*.65*		*.66*	.51	.50	.48	.47	.46	.45	.45	5
6	*.23*	.21	.31	(4f)	.50	.55	.59	.62	.65	.68	*.94*	*.97*	.75	.70	.66	.63	.60	.59	.57	6
7	*.31*	.28	.44	(5f)																7

b) d-valence shell average[2] c) p-valence shell average

Row	3	4	5	6	7	8	9	10	11	13	14	15	16	17	18	Row
2										.18	.08	.00	—.07	—.14	—.20	2
3										.10	.01	—.05	—.10	—.15	—.18	3
4	—.99	—.82	—.73	*—.57*	—.62	—.58	—.55	—.53	*—.40*	.01	—.07	—.13	—.16	—.20	—.23	4
5	—.89 (4f)	—.73	*—.51*	*—.47*	—.53	*—.40*	*—.39*	*—.10*	*—.34*	—.03	—.11	—.16	—.20	—.23	—.25	5
6	—.70/—.83[3] (5f)	—.69	—.60	—.55	—.51	—.48	—.45	*—.32*	*—.31*	—.11	—.20	—.25	—.30	—.33	—.36	6

d) p-valence shell spin-orbit splitting

Group						
13	14	15	16	17	18	Row
.22	.37	.52	.69	.85	1.01	2
.29	.40	.49	.58	.67	.74	3
.34	.45	.55	.62	.69	.75	4
.39	.49	.57	.64	.70	.75	5
.52	.64	.74	.82	.87	.93	6

e) f-valence shell average[4]

	Group													
	3/1	3/2	3/3	3/4	3/5	3/6	3/7	3/8	3/9	3/10	3/11	3/12	3/13	3/14
Lanth.	*−1.29*	−1.66	−1.53	−1.44	−1.37	−1.31	*−.97*	−1.24	−1.21	−1.19	−1.17	−1.16	−1.14	*−.86*
Act.		*−1.15*	*−1.05*	*− .97*	−1.09	−1.04	*−.83*	− .95	− .92	− .89	− .86			

1 s^2 shells; s^1 shells are given in italics
2 $d^{n-2}s^2$ shells; $d^{n-1}s^1$ (d^n) are given in italics
3 La/Lu
4 $f^{n-2}s^2$ shells; $f^{n-3}d^1s^2$ shells are given in italics

1. *Dependence on principal quantum number n.* While the relativistic stabilization of hydrogenic AOs decreases with increasing n [compare Eqs. (2.12/14)] this does not hold generally for the orbitals of neutral many-electron atoms. For s-AOs, for instance, there is, on the whole, a significant increase of a when going down in the groups. d- and f-AOs are destabilized, with a slight decrease of the absolute value of the (negative) a-parameter, that is also an increase of a with increasing n. Only for p-AOs, a decreases.

2. *Dependence on nuclear charge.* Z_{eff} for valence electrons increases somewhat when going down in a group, or from left to right in a row. In Sect. 2.1 we expected $\delta^{rel}E$ to increase more strongly than $\sim Z_{eff}^2/n$. The empirically found increase is proportional to the squared, totally unscreened nuclear charge, Z_{nuc}^2, and is unexpectedly strong. The numerically determined energy changes (e.g., +30% for Au6s, −50% for actinoide 5f) are large enough to trace the large qualitative changes mentioned in Sect. 1.1 back to relativistic effects.

3. *Shell effects.* The variation of a over the periodic table is not monotonous. a_s (the index s refers to s-AOs) increases in a row when the d- or f-population increases. It exhibits a maximum in group 11. a_s is significantly larger for $d^{n-1}s^1$ configurations than for $d^{n-2}s^2$ configurations. Similarly the a_f-parameter for f-AOs increases if an f-electron is transferred to the d-shell: $a_f(s^2d^1f^{n-3}) > a_f(s^2d^0f^{n-2})$.

4. *Relativistic stabilization and destabilization.* Only the s-level is consistently stabilized by relativity. The slight relativistic stabilization of the p-level average for the lightest elements changes into destabilization, as the number of electrons increases from the left top to the right bottom of the periodic system. The dominant effect for p-AOs is spin-orbit coupling, which splits the three degenerate p-levels up into a stabilized p1/2 level and two destabilized p3/2 levels (see Fig. 3). The splitting increases up to several eV in the right bottom of the periodic table. The d- and f-shells are split due to spin-orbit interaction by a smaller fraction, and their averages are considerably *destabilized* throughout.

2.6 Spatial Origin of the Relativistic Effects

In order to understand the physical mechanism of relativistic effects in atoms, it is important to know in which domains of space the relativistic hamiltonian contributes most strongly to the energy. The spatial valence shell determines the nonrelativistic energy; the nearest neighborhood of the nucleus contributes to the direct relativistic energy; the outer atomic core region is important for the "indirect" relativistic energy due to relativistic orbital changes. In addition to the well-known "indirect destabilization" of d- and f-AOs, s- and p-AOs are also subject to indirect stabilization. The interplay of stabilization and destabilization effects explains the "gold maximum".

As usual, the term "shell" denotes one-electron orbital states of specified energy. In the following the term "spatial shell" denotes the spatial region of the outer main maximum of the corresponding orbital. So, the atomic volume is divided up into the innermost spatial K-shell, surrounded in an onion-skin like manner by the spatial L-, M-, N-, ... shells. The largest part of the density of a 1s-AO is found in the spatial K-shell. The 2s-AO has its outer main maximum in the spatial L-shell and its inner core wiggle in the spatial K-shell. The 2p-AO density is mainly located in the spatial L-shell, the inner tail of the p-AO extending into the spatial K-shell. The innermost and outermost spatial shells of a nl-AO are the spatial $l+1$st and nth shells, respectively. About 90% of the total density of a nl-AO is contained in the nth spatial shell. Each innermore spatial shell contains about one order of magnitude less density, so that the innermost core-tails appear to be rather unimportant, at least concerning the density.

For an orbital ψ with orbital energy eigenvalue E of the Fock Hamiltonian $\hat{F}$, we obtain the expectation value of E by integrating over the whole domain:

$$E := \lim_{R\to\infty} E(R) = \lim_{R\to\infty} \int_0^R \psi^* \cdot \hat{F}\psi \cdot d\tau = \lim_{R\to\infty} (E \cdot P(R)) . \tag{2.41}$$

R is the distance from the nucleus, and:

$$P(R) = \int_0^R \psi^* \cdot \psi \cdot d\tau$$

is the radially integrated orbital density with the appropriate normalization condition $\lim_{R\to\infty} P(R) = 1$.

In Fig. 4a the functions E(R) and P(R), and in addition the kinetic and potential energy contributions T(R) and V(R) to E(R), are shown for the uranium 7s-AO [21]. The dominant contributions to P (largest increase of P(R)), and accordingly to E, result from the Q-valence shell, and to a minor extent from the outermost P-core shell. However, the innermost shells contribute large energy amounts to T and V individually. For valence orbitals we have $|E| \lesssim 1$ a.u., and inside the atomic cores, where $R < 1$ a.u., we have $|V| > Z$ a.u. It then follows from:

$$\hat{F}\psi = (\hat{T} + \hat{V})\,\psi = E\psi \tag{2.42}$$

that, because $|V| \gg |E|$ in the core, T and V must cancel approximately:

$$\hat{T}\psi \approx -\hat{V}\psi . \tag{2.43}$$

Consequently, the virial ratio V/T is -2 for orbitals in the pure Coulomb field of a bare nucleus, but is approximately -1 for orbitals in the field of an atomic many-electron core, which effectively shields the nuclear attraction in the spatial valence shell.

2.6.1 Direct Relativistic Contributions

The innermost spatial shells are even more important for the relativistic energy contributions than for the *individual* nonrelativistic energy contributions. In the Schrödinger-Pauli representation, the kinetic mass-velocity operator is $\sim\hat{p}^4 \sim \hat{T}^2$, which is very large near the nucleus, namely $\sim\hat{V}^2$ according to Eq. (2.43). The Darwin potential operator is $\vec{\nabla}^2 V$ [see Eq. (2.30)]. The electronic contribution to

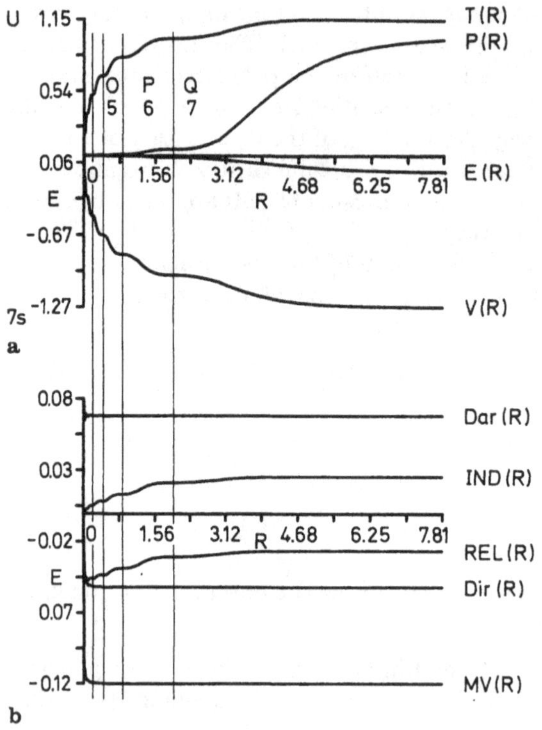

Fig. 4a, b. Radially integrated contributions to orbital energy and density of the uranium 7s valence orbital (in a.u.), after [21].
a nonrelativistic values, **b** relativistic corrections in 1st order perturbation theory.
This legend refers also to Figs. 5 and 6.
Dar = Darwin energy, Dir = direct relativistic energy contribution (Darwin + mass-velocity energy, without spin-orbit contribution which vanishes for s-AOs), E = nonrelativistic energy, IND = indirect relativistic energy correction, IND(—5d) = indirect effect without contribution from the 5d-shell, MV = mass-velocity energy, P = orbital density, REL = total relativistic spin-averaged energy (DIR + IND), SO = total spin-orbit splitting, SOel = electronic contribution to the spin-orbit splitting, T = nonrelativistic kinetic energy, V = potential energy (nuclear attraction and interelectronic Coulomb and exchange interaction)

$\vec{\nabla}^2 V$ is negligible in comparison to the nuclear contribution. The latter is a pointlike δ-potential at the position of a pointlike nucleus, and it is $\pi \cdot \varrho_{nuc}/2c^2$ for a spatially extended nucleus, where ϱ_{nuc} is the charge density *inside* the nucleus, $\varrho_{nuc} > 10^{12}$ a.u. The resulting Darwin potential inside the nuclear radius R_{nuc} of about $\sqrt[3]{M}/2c^2 \sim 10^{-4}$ a.u. is $V^{Dar} \sim (3Z/M) \cdot c^4 \sim 1/3 \times 10^9$ a.u. (M = nuclear mass number).

In the Dirac representation, which was used by the authors of Figs. 4–6 [21], the mass-velocity and Darwin operators are less localized than in the Schrödinger-Pauli picture. Still, they are significant only in the innermost spatial shells, namely in the spatial K-shell for valence s-AOs, in the spatial K- and L-shells for valence p-AOs, in the spatial L- and M-shells for valence d-AOs, etc. (Note, that the

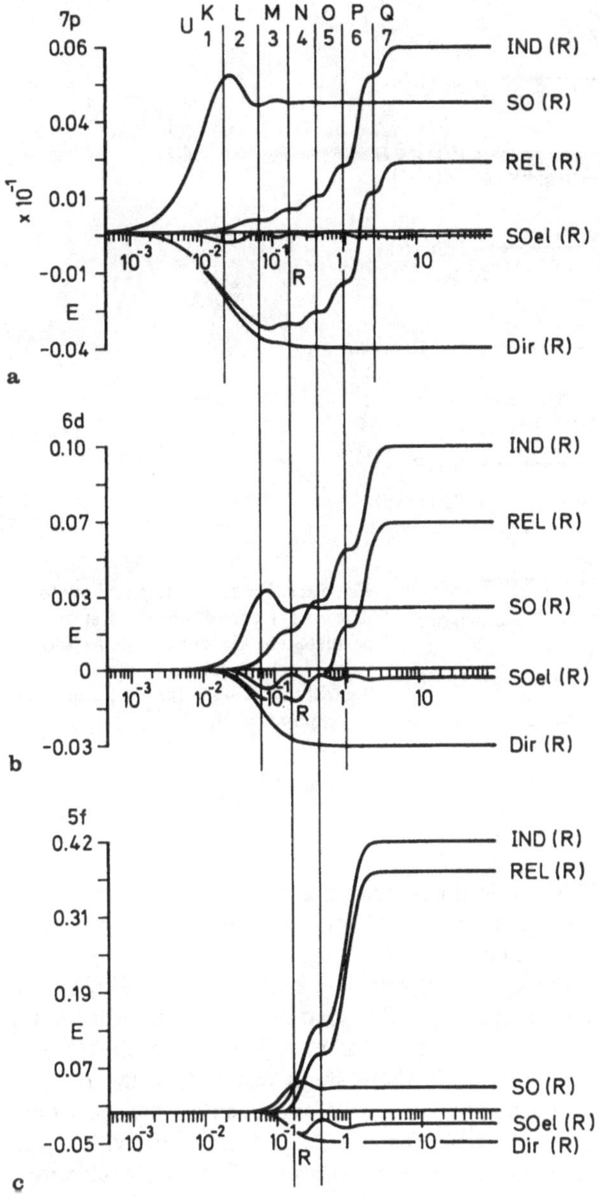

Fig. 5a–c. Radially integrated relativistic energy corrections (in first order perturbation theory) for the 7p, 6d, 5f valence orbitals of uranium over a longarithmic R-scale (in a.u.) after [21]. See caption of Fig. 4

U6d-AO extends only with the inner tail of the innermost wiggle into the L-shell). The described behavior is shown in Figs. 5 and 6, where the relativistic energy contributions to the U7p, U6d, U5f and Au6s, Au6p valence orbitals are plotted over a logarithmic R-scale.

The relevant relativistic correction to the potential is the Darwin term in the case of s-AOs, and the spin-orbit term in the case of p-, d-, and f-AOs. The

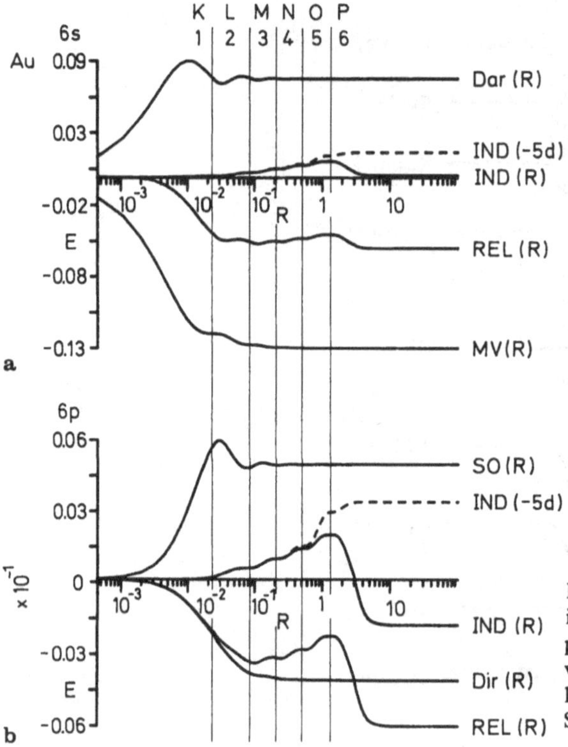

Fig. 6 a, b. Radially integrated relativistic energy corrections (in first order perturbation theory) for the 6s and 6p valence orbitals of gold over a logarithmic R-scale (in a.u.) after [21]. See caption of Fig. 4

spin-orbit term vanishes for s-AOs, while the nuclear Darwin term vanishes for p-, d-, and f-AOs. The electronic contribution to the Darwin term is negligibly small and is not shown separately in the figures.

The electronic contribution to the spin-orbit splitting, that is the shielding of the nuclear field by the other electrons (the so-called two-electron spin-other orbit terms), is small. Since the spin-orbit coupling of a nl-AO arises in the lth and $l+1^{st}$ spatial shell, only core electrons inside these shells can reduce the nuclear spin-orbit coupling by shielding the nuclear charge. Consequently, the electronic reduction of spin-orbit coupling is negligible for p-AOs and increases with l (see Fig. 5). It decreases *relatively*, if the nuclear charge increases. While the inner electrons may reduce the nuclear spin-orbit coupling by one third in the first transition series, the electronic effect is already comparatively small for the d shell of the 7th row (see Fig. 5b).

The Dar(R) and SO(R) curves behave very similarly (compare Figs. 4–6). They increase at small R-values and then exhibit a slight decrease. This decrease is *not* due to core-electronic shielding (which is much weaker), but is due to the oscillatory behavior of the integrands in the Dirac representation. The SO(R)-curves are monotonic in the Schrödinger representation, the physical matrix element values SO(∞) of course being identical in both representations.

2.6.2 Indirect Relativistic Contributions

If we treat a system nonrelativistically and relativistically in the same representation, so that the terms in the Hamiltonians and the wavefunctions have the same physical meaning, we may write:

$$(H^{nr} - E^{nr})\,\psi^{nr} = 0\,, \qquad (H^{rel} - E^{rel})\,\psi^{rel} = 0 \tag{2.44}$$

$$\Delta^{rel}H = H^{rel} - H^{nr}\,, \qquad \Delta^{rel}\psi = \psi^{rel} - \psi^{nr}\,, \qquad \Delta^{rel}E = E^{rel} - E^{nr} \tag{2.45}$$

$$\begin{aligned}\Delta^{rel}E &= \langle\psi^{nr}|\,\Delta^{rel}H\,|\psi^{nr}\rangle + (\langle\psi^{nr}|\,H^{nr}\,|\Delta^{rel}\psi\rangle + \text{c. c.}) + 0(\Delta^{rel})^2 \\ &= \Delta^{rel}_{dir}E + \Delta^{rel}_{ind}E\,.\end{aligned} \tag{2.46}$$

If for a given ψ^{nr}, Eq. (2.44) is sufficiently accurately fulfilled (for details see [22], the second term in Eq. (2.46) vanishes. That is, performing *fully optimized all-electron calculations*, $\Delta^{rel}H$ and $\Delta^{rel}E$ consist only of the first "direct" relativistic correction term (for core *and* valence electrons) discussed in Sect. 2.3.

This does not hold, however, if the nonrelativistic wavefunction is not appropriately optimized, for instance, if the same set of *frozen core* orbitals is used for several different valence states. The frozen-core approach seems especially appropriate in heavy atom quantum chemistry, where we have a large number of atomic core orbitals, which vary only very slightly from molecule to molecule. Then the second "indirect" energy contribution does *not* vanish exactly.

Instead of frozen core-all electron calculations one may prefer a *valence only approach*. In the latter case, H^{nr} is an effective valence Hamiltonian containing core electron potential terms. Consequently, $\Delta^{rel}H$ will consist of the direct terms for the valence electrons, *and* of additional indirect terms, which take into account that the valence electrons move in the field of relativistically modified core orbitals ($\Delta^{rel}\psi_{core}$).

Let us discuss a simplified example explicitly. We apply the Hartree approximation, i.e. we neglect both exchange and correlation effects. The Hamiltonians for the core system be:

$$h_c^{nr}\,, \qquad h_c^{rel} = h_c^{nr} + \Delta^{rel}h_c$$

and for the core + valence system be:

$$H^{nr} = h_c^{nr} + h_v^{nr} + g_{cv}\,, \qquad H^{rel} = H^{nr} + \Delta^{rel}h_c + \Delta^{rel}h_v\,.$$

g_{cv} describes the core-valence interelectronic interaction, and relativistic corrections to it are neglected for the moment. The wavefunctions for the core system be:

$$\varphi_c^{nr}\,, \qquad \varphi_c^{rel} = \varphi_c^{nr} + \Delta^{rel}\varphi_c\,,$$

and those for the total system be:

$$\varphi_{c'}^{nr}\cdot\varphi_v^{nr}\,, \qquad \varphi_{c'}^{rel}\cdot\varphi_v^{rel}\,,$$

where the difference of the core functions φ_c and $\varphi_{c'}$ for the core and core + valence systems, respectively, has been accounted for. The relativistic correction to the valence energy is then:

$$\begin{aligned}\Delta^{rel}E_v := &(E^{rel}_{c'+v} - E^{nr}_{c'+v}) - (E^{rel}_c - E^{nr}_c) = (E^{rel}_{c'+v} - E^{rel}_c) - (E^{nr}_{c'+v} - E^{nr}_c)\\ = &\langle\varphi^{rn}_v|\,\Delta^{rel}h_v\,|\varphi^{nr}_v\rangle + (\langle\varphi^{nr}_{c'}|\,h^{nr}_c + J_v\,|\Delta^{rel}\varphi_{c'}\rangle + \text{c. c.})\\ &+ (\langle\varphi^{nr}_v|\,h^{nr}_v + J_{c'}\,|\Delta^{rel}\varphi_v\rangle + \text{c. c.})\\ &+ (\langle\varphi^{nr}_c|\,\Delta^{rel}h_c\,|\varphi^{nr}_{c'} - \varphi^{nr}_{c'}\rangle + \text{c. c.})\\ &+ 0(\Delta^{rel})^2\end{aligned} \tag{2.47}$$

J_v and $J_{c'}$ are the potentials created by the valence and core electrons, respectively.

In the *fully optimized all-electron* case, where both $\varphi_{c'}$ and φ_v are optimized, the second and third terms vanish, but $\varphi^{nr}_{c'} \neq \varphi^{nr}_c$. Then

$$\Delta^{rel}E_v = \langle\varphi^{rn}_v|\,\Delta^{rel}h_v\,|\varphi^{nr}_v\rangle + (\langle\varphi^{nr}_{c'}|\,\Delta^{rel}h_c\,|\varphi^{nr}_{c'} - \varphi^{nr}_c\rangle + \text{c. c.}) + 0(\Delta^{rel})^2 \tag{2.48}$$

In the *all-electron frozen-core* approach, $\varphi^{nr}_{c'} = \varphi^{nr}_c$, and only φ^{nr}_v is optimized, i.e. φ^{nr}_v is an eigenfunction of $h^{nr}_v + J_c$, and φ^{nr}_c is an eigenfunction of h^{nr}_c. Then

$$\Delta^{rel}E_v = \langle\varphi^{nr}_v|\,\Delta^{rel}h_v\,|\varphi^{nr}_v\rangle + (\langle\varphi^{nr}_c|\,J_v\,|\Delta^{rel}\varphi_c\rangle + \text{c. c.}) + 0(\Delta^{rel})^2 \tag{2.49}$$

Finally, the *valence-only* approach, where only φ_v appears, is obtained, if the (first order) identity:

$$\langle\varphi^{nr}_c|\,J_v\,|\Delta^{rel}\varphi_c\rangle + \text{c. c.} = \langle\varphi^{nr}_v|\,\Delta^{rel}J_c\,|\varphi^{nr}_v\rangle \tag{2.50}$$

is inserted in Eq. (2.47). One then obtains:

$$\Delta^{rel}E_v = \langle\varphi^{nr}_v|\,\Delta^{rel}h_v + \Delta^{rel}J_c\,|\varphi^{nr}_v\rangle + 0(\Delta^{rel})^2\,. \tag{2.51}$$

$\Delta^{rel}J_c$ is the change of the core-potential due to the relativistic changes of the core electrons and causes the indirect relativistic effect.

According to Eqs. (2.48), (2.49), and (2.51), three physically different interpretations can be given to $\Delta^{rel}E_v$. $\Delta^{rel}E_v$ is defined in Eq. (2.47) as a "double difference". The first difference is that between the relativistic and nonrelativistic behavior, denoted by Δ^{rel} ($\Delta^{rel}\varphi$ and $\Delta^{rel}h$). The second difference is that between the core and the core + valence systems, the corresponding differences of the wavefunction and the Hamiltonian being ($\varphi_{c'} - \varphi_c$), and (h_v, J_v), respectively.

The first term in all three expressions of Eqs. (2.48, 49, 51) is the same. It contains the "double-difference hamiltonian" $\Delta^{rel}h_v$, the direct relativistic correction of the Hamiltonian with respect to the additional valence electrons. The second terms differ. Equation (2.48) contains the relativistic correction to the Hamiltonian, and the valence shell correction to the core wavefunction. Equation (2.49) contains the relativistic correction to the core wave function, and the valence shell contribution to the electronic interaction Hamiltonian. Finally, Eq. (2.51) contains the relativistic correction to the core contribution of the electronic interaction Hamiltonian.

So, the three different expressions for the same quantity give rise to three different physical interpretations of the origin of relativistic effects. In addition to the direct relativistic energy of the valence electrons, the total relativistic correction to the valence energy can be attributed

a) either to the change of the direct relativistic energy of the core, which is changed by the valence electrons [Eq. (2.48)], or

b) to the change of the nonrelativistic valence-electronic Coulomb repulsion of the core, changed by relativity [Eq. (2.49)], or

c) to the relativistic change of the core-electronic Coulomb potential [Eq. (2.51)].

It does not make a difference, whether the atom is divided into core and valence shells, or into one particle and all the remaining ones. In the latter case $\Delta^{rel}E_v$ is the relativistic correction to the *ionization potential.* The first scheme based on Eq. (2.48) corresponds to the relativistic SCF approach, while the last scheme based on Eq. (2.51) corresponds to the relativistic Coupled Hartree-Fock approach and Koopmans' theorem. A detailed analysis of relativistic corrections to ionization potentials has been presented in section 2 of Ref. [24]. (See also Sect. 2.7.2 below).

Using Eq. (2.51), the authors of Ref. [21] have calculated the radial behavior of the indirect relativistic energy correction:

$$\mathrm{IND}(R) = \int_0^R \varphi_v^* \cdot \Delta^{rel}J_c \cdot \varphi_v \cdot d\tau \qquad (2.52)$$

While the three direct effects (mass-velocity, Darwin, spin-orbit) originate in the immediate vicinity of the nucleus, also the outer spatial shells contribute to the indirect effect (see Figs. 4–6). Core orbitals, which *overlap* with valence orbitals (at least with their inner tails), will alter the effective nuclear charge when they become relativistically modified. Core shells *inside* the innermost core-wiggle of the valence orbital (i.e. the K shell for p-AOs, the K and L shells for d-AOs, etc.) will not contribute to the indirect effect.

In the literature (e.g. [1, 23]) the indirect effect is in general identified with energetic *destabilization* of "nonpenetrating" valence orbitals (d- and f-AOs) due to the increased nuclear shielding by relativistically contracted core orbitals. Since, however, only s- and p1/2-core orbitals contract relativistically, while d- and f-core orbitals expand [20], it is evident that the s- and p1/2-orbitals contribute to indirect destabilization, while the d- and f-orbitals contribute to *indirect stabilization.*

d- and f-valence orbitals, which overlap strongly with the relativistically contracted outer s-p-core shells, undergo significant indirect destabilization (i.e. negative contribution to the *a*-parameter). This effect by far exceeds the direct stabilization, which is comparatively small, because d- and f-orbitals have very low density in the vicinity of the nucleus (see Fig. 5 and Tab. 3). An interesting observation is that the indirect destabilization is partially reduced, if the number of d-electrons is increased from $d^{n-2}s^2$ to $d^{n-1}s^1$ (see the italic values in Tab. 3a/b).

If s- and p-valence orbitals overlap mainly with other s- and p-orbitals, as occurs at the beginning and at the end of a row, the destabilization is again the main indirect effect (see Fig. 4b, 5a and Tab. 3c). However, if a d- or f-shell becomes highly populated in the vicinity of the s- or p-valence shells, the concomitant *indirect stabilization* compensates or even overcompensates the indirect destabilization. The strong stabilizing effect of the 5d shell on the Au6s and especially on the diffuse, sensitive Au6p orbital is demonstrated in Fig. 6.

The cancellation of indirect stabilization and destabilization for Au6s has special chemical relevance. It explains the so-called "gold maximum", namely the extraordinary large relativistic stabilization of the s-valence AO in group 11 and in Pt, resulting in large chemical effects [2]. The "gold maximum" is due to the indirect stabilization by the relativistically expanded d-shell. A nearly vanishing total indirect effect for Au6s, and only a small indirect destabilization for Tl6p had already been found by Grant et al. [23]. However, as shown in Ref. [21], it was premature to conclude that the indirect effect is in general negligible for s- and small for p-valence orbitals. The latter assertion is already at variance with the early extensive data set of Desclaux [20]. Although the direct stabilization increases with increase of nuclear charge, the a-parameter decreases from Au6s to Rn6s in the 6th row of Tab. 3a. or from Tl6p to Rn6p in the 6th row of Tab. 3c. It indicates significant indirect destabilization for s- and p-valence AOs of elements, which are not "near the d-block".

2.7 Further Paradoxa

As already mentioned in Sect. 2.4, the values of the individual energy contributions are very different, depending on whether they are calculated within perturbation theory from the zeroth order state-function; or using the fully optimized statefunction. However, the value of the total energy is approximately the same in both approaches. The significance of using approximate models for the development of physical understanding is discussed. Inferences on the mechanism of the direct relativistic effects, and on the calculation of relativistic corrections to ionization energies are drawn.

2.7.1 Reexamination of the Spatial Origin of Relativistic Effects

On the basis of Eq. (2.41) we concluded that the total nonrelativistic energy of a valence orbital originates in the spatial valence shell, while according to Eq. (2.43) kinetic and potential energies, though individually large, cancel each other effectively in the core. This should not only hold for the nonrelativistic energy contributions, but likewise for the relativistically corrected ones. On the other hand, we have found in Sect. 2.6 that the direct relativistic energy-correction is created in the neighborhood of the nucleus. This apparent antinomy has the same logical structure as the paradoxa mentioned in Sect. 2.4. Before we examine it, some general philosophical comments seem appropriate.

When we want to analyse the exact theoretical solution for a given physical system, we may use general theorems like the virial theorem, the variational principle, Eqs. (2.42, 43), etc. Our conclusions are general ones, which do *not* distinguish between qualitatively different physical situations. For instance, the exact wavefunction of any bound molecule will tell as that the bond energy is due to potential energy lowering, which is partially compensated by kinetic energy increase. This holds irrespective of covalent or ionic character of the bond. We do not obtain an answer, *why* a physical phenomenon occurs. We should know, for instance, why overlap, exchange, or relativity causes the exact wavefunction to develop in such a manner that the final physical effect, which is a small difference of large numbers (see, e.g. Fig. 4a), displays a specific qualitative character.

In order to approach an understanding of the physical mechanism, we must split up the physical phenomenon artificially and nonuniquely into two "nonphysical" parts. The first part should be describable by a simple model, which is transparent enough to exhibit the physical change (energy decrease, orbital contraction, etc.) in a directly evident manner. The second part of the physical phenomenon consists of the transition from the model to the exact, real system, and is as complex as reality itself. Appropriate splittings of reality and appropriate models are those ones, where one can prove or at least expect with justification that the contribution of the second part to the total effect under consideration is small. In such cases the effect predicted by the model (e.g., the total energy change) approximates the physically measurable quantity, while individual contributions like potential and kinetic energies may be changed drastically in the second part. Furthermore, an appropriate model should also allow to predict these latter changes qualitatively on the basis of general theorems (variational principle, virial theorem).

It is not uncommon in the field of chemical bonding, both in textbooks and in original papers, to reject explanations which are based on models that violate a special theorem (e.g., the virial theorem), and to accept *only* those schemes which fulfil special theorems and which do not artificially split up the physical phenomenon into parts. However, this latter standpoint will not accelerate the progress of physical understanding.

Accordingly, we have presented and discussed Figs. 4–6, which were obtained within the framework of first order perturbation theory. This violates the virial theorem. We have learned that the physical mechanism of the direct relativistic effects acts in the vicinity of the nucleus, and we have achieved an intuitive

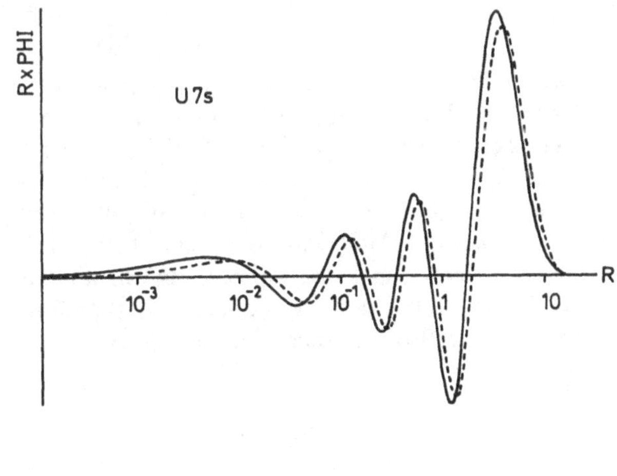

Fig. 7. Nonrelativistic Schrödinger and relativistic Schrödinger-Pauli radial wavefunction $R \cdot \varphi(R)$ for the U7s orbital over logarithmic R-scale

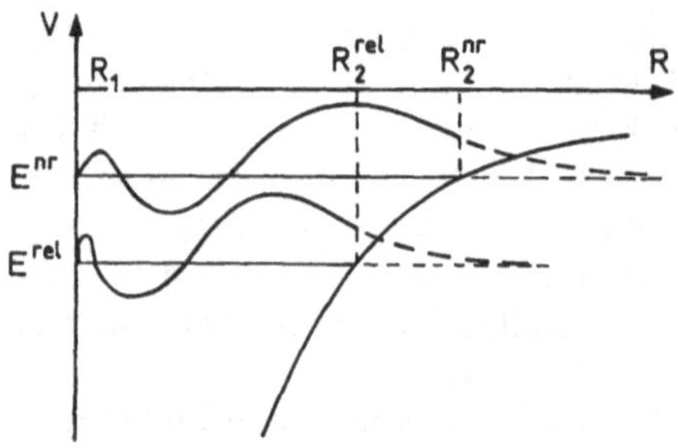

Fig. 8. Change of wavefunction and energy due to relativity, schematically

understanding of the signs and magnitudes of contributions from different shells of different atoms.

In a fully optimized relativistic calculation, the wavefunction will change (see Fig. 7), so that finally the (relativistic) virial theorem is fulfilled. The kinetic energy ($T^{nr} + \Delta^{rel}T$) will have increased for relativistically stabilized states, though originally $\Delta^{rel}T$ was negative (as already discussed with the help of a simple variational model in Sect. 2.4).

Using the semiclassical model, the relativistic mechanism can be described as follows. The nonrelativistic wavelength of a matter-wave with energy E in a potential V is:

$$\lambda^{nr} = 2/(2m_e(E^{nr} - V))^{1/2} . \tag{2.53a}$$

From Eq. (2.17) one obtains:

$$\lambda^{rel} \approx \lambda^{nr}/(1 + (E^{nr} - V)/(2m_e c^2) + (\Delta^{rel}E - \Delta^{rel}V)/(E^{nr} - V))^{1/2} . \tag{2.53b}$$

The relativistic change of λ is appreciable in the direct vicinity of the nucleus, where no longer $|V| \ll 2m_e c^2$, and from where the direct relativistic energy correction stems in first order perturbation theory.

The boundary conditions for the wavefunction lead to the quantum condition (see Fig. 8):

$$2\int_{R_1}^{R_2} dR/\lambda = n - \gamma_1 - \gamma_2 , \tag{2.54}$$

where R_1, R_2 are the classical boundary points ("turning points"), and γ_1, γ_2 account for quantum mechanical phase shifts at the classical end points (for instance penetration of the wavefunction into the classically forbidden region $R > R_2$). The main effect of relativity, the change of λ at $R \approx 0$ on the left side of Eq. (2.54), corresponds to an increased γ_1-value on the right side of Eq. (2.54). We may say that "relativity sucks the tail of the valence orbital into the nuclear region". The corresponding phase-shift at R_1 propagates into the valence region, all the nodes being shifted inwards (see Fig. 7). The outer bondary condition is therefore met at a smaller value $R_2^{rel} < R_2^{nr}$, causing a shift of the energy according to:

$$E^{rel} = V(R_2^{rel}) < V(R_2^{nr}) = E^{nr} \tag{2.55}$$

(see Fig. 8). The relaxed relativistic wavefunction will guarantee effective cancellation of V and T in the core region, the total energy (including the relativistic correction) now originating mainly in the valence region.

Since the total nonrelativistic energy originates in the valence region, and since in the latter approach also the relativistic energy change originates in the valence region, we have a sound basis for valence-only calculations (both of the semiempirical or the pseudopotential type), which do not take any details of the inner core into account. The requirement on the valence-only Hamiltonian is, that the relativistic phase-shift of the wavefunction at the outer core boundary should be recovered. This can be achieved by shifting the repulsive branch of the pseudopotential appropriately. Then the inner core region is no longer of significance.

2.7.2 Spin-orbit Splitting of Ionization Potentials

The Hartree-Fock approximation to valence ionization potentials $|E|$ suffers from significant correlation errors, so that the comparison of relativistic Hartree-Fock and accurate $|E|$'s is problematic. Correlation corrections cancel, however, to a large extent for the spin-orbit splitting of nonrelativistically degenerate orbital levels, provided the spin-orbit coupling is not as large as to give rise to severe configuration mixing by itself (compare, e.g. Ref. [25]).

Table 4. Spin-orbit splitting of ionization potentials (in eV)

		Schrödinger-Pauli approach				Dirac approach		
Atom	ionized level	$\langle\Delta^{so}h\rangle$	ionic reorg.	ΔE^{so}	exp. ΔE^{so}	ΔE^{so}	ionic reorg.	$\Delta^{so}E^{D}$
Ne	2p	0.085	+0.009	0.094	0.097	0.094	−0.030	0.124
Pd	4d	0.398	+0.040	0.438	0.439	0.424	−0.141	0.565
Xe	5p	0.989	+0.121	1.11	1.306	1.326	−0.130	1.436

If we divide the electronic system of an atom into one particle, to be ionized, and the remainder, then $\Delta^{rel}E_v$ from the first order theory of Sect. 2.6.2 can be identified with the relativistic correction of the ionization potential. Taking into account that to *first order*, the spin-orbit energy of closed shells vanishes, and that the radial matrix elements of the ionized orbital n and its non-ionized spin-orbit mate n′ are identical, the first order Eq. (2.48) can be rewritten as

$$\Delta^{so}E_n = \langle\varphi_n^{nr}|\,\Delta^{so}h\,|\varphi_n^{nr}\rangle + (\langle\varphi_n^{nr}|\,\Delta^{so}h\,|\Delta^{ion}\varphi_n^{nr}\rangle + \text{c. c.} \tag{2.56}$$

φ_n^{nr} is the degenerate orbital in the neutral system before ionization, and $\Delta^{ion}\varphi_n^{nr}$ is the change of that orbital due to ionization. The first term of Eq. (2.56) is the spin-orbit splitting according to Koopmans' theorem, where the ionization energy is described with the help of canonical orbitals of the nonionized system. The second term describes the energy change due to reorganization of the remaining orbitals in the ion. The two terms are given in Tab. 4 under heading "Schrödinger-Pauli approach" ($\langle \Delta^{so}h \rangle$, and ionic reorg., respectively). Since in general orbitals contract after ionization (see Tab. 5), the ionic reorganization increases the spin-orbit splitting. The total effect (ΔE^{so}) in the third column of Tab. 4 agrees reasonably well with the experimental values for the lighter elements, where higher order effects and spin-orbit configuration-mixing are not yet important.

Table 5. Orbital radii (in 10^{-3} a_0)

		$\langle r \rangle$ Schrödinger			$(\langle r \rangle_{j=l+1/2} - \langle r \rangle_{j=l-1/2})$ Dirac		
Atom	level	neutral atom	positive ion	diff. %	neutral atom	positive ion	diff. %
Ne	2p	965	875	−9.3	2.69	1.90	−30
Pd	4d	1550	1440	−7.1	36.3	22.7	−38
Xe	5p	2315	2197	−5.1	110.	88.5	−20

If on the other hand, fully relativistic orbitals from Dirac-Fock calculations are used instead of nonrelativistic orbitals, ΔE^{so} is, according to Koopmans' theorem, approximately given by the splitting of the corresponding Dirac orbital energies, $\Delta^{so}E^{D}$ in the last column of Tab. 4. While Koopmans' spin-orbit splittings are too low at the Schrödinger-Pauli level, they are too large at the Dirac level. The ionic reorganization effect (next to the last column of Tab. 4) is now negative, yielding good agreement with experiment for most elements.

Of course, the orbitals contract after ionization also in the Dirac case, thereby increasing the relativistic energy corrections. However, as shown in Tab. 5, the more extended upper $j = l + 1/2$-orbital contracts more than the already smaller $j = l - 1/2$-orbital, so that the spin-orbit difference becomes indeed reduced by ionic orbital relaxation. Further details are given in Ref. [25]. See also Ref. [26].

3 Relativistic Effects in Molecules

3.1 Relativistic Change of Bond-Lengths

s- and p1/2-AOs contract relativistically; most bonds are relativistically contracted, too. Before 1980 the naive argument was popular that, since covalent bonds are "due" to AO-overlap, AO-contraction will cause bond-length contraction. However, this argument is too simple. After the "Dutch revolution" in 1980 [2, 30, 31], the argument was popular that, since bond-contraction — as most other relativistic effects — can be estimated by first order perturbation theory, using zeroth-order nonrelativistic uncontracted orbitals, bond contraction is not due to orbital contraction. This argument is too simple, as well. An analysis of the phenomenon of relativistic bond-length changes will be presented on the basis of double perturbation theory.

The most important basic properties of molecules are their energies, their geometrical structures and their force constants. Although there exists a large body of numerical results from more or less sophisticated computational work in the fields of molecular and solid state chemistry, the understanding of relativistic effects in chemistry is yet less developed than in atomic physics.

Some effort has already been spent on the determination and analysis of relativistic changes of structural parameters. A table in Pyykkö's recent review article [2] on explicit studies of relativistic bond-length changes contains approximately 400 entries. All but a few exceptions show relativistic *bond length concentration* of the order of

$$\delta^{rel}R_e := (R_e^{rel} - R_e^{nr})/R_e^{nr} \approx -a_R \cdot (Z/c)^2 . \tag{3.1}$$

The coefficient a_R depends on the electronic valence configuration. a_R has a maximum of ~ 0.25 in group 11 (gold maximum) and is very small for the first and last groups of the periodic table [2]. A few cases of *bond length expansion* are also known. The largest ones occur in TlH^+ and Tl_2 [2].

In order to elucidate the mechanism of relativistic bond length changes, we expand the molecular energy E in a power series of R and r. R is the internuclear distance. It is appropriate to shift the origin of the length scale to the nonrelativistic internuclear equilibrium distance, so that R is the *deviation* of the internuclear separation from the nonrelativistic bond length, $R_e^{nr} := 0$. $r \sim c^{-2}$ is the relativity parameter. The nonrelativistic case corresponds to $c \to \infty$ and $r = 0$, the relativistic case with $c = 137$ a.u. is identified with $r = 1$. The bond length R_e is determined by:

$$(\partial_E/\partial R)_{R=R_e} = 0 .$$

From the Taylor series:

$$E = E(R=0) + (\partial E/\partial R) \cdot R + (\partial^2 E/\partial R^2) \cdot R^2/2 + (\partial^3 E/\partial R^3) \cdot R^3/6 + \dots \tag{3.2}$$

we obtain for the relativistic bond-length change $\Delta^{rel}R_e = R_e^{rel} - R_e^{nr} = R^{rel}$ (note that $R_e^{nr} := 0$):

$$\Delta^{rel}R_e = -E'/E'' * (1 + (E''' \cdot E')/(2E''E'') + \dots) \tag{3.3}$$

where $E'' = \partial^2 E^{rel}/\partial R^2 = k > 0$ is the (relativistic) force constant. The relevant term, which determines whether relativistic bond length contraction ($\Delta^{rel}R_e < 0$) or expansion ($\Delta^{rel}R_e > 0$) occurs, is $E' = \partial E^{rel}/\partial R$. Taking into account that $(\partial E^{nr}/\partial R) = 0$ at $R = R_e^{nr} := 0$, and expanding E also in powers of r yields in first order:

$$\Delta^{rel}R_e \approx -k^{-1} \cdot (\partial^2 E^{rel}/\partial R\, \partial r)_{R=0,r=1} , \tag{3.4}$$

where we have used:

$$\Delta^{rel}E/\partial R \approx \partial^2 E^{rel}/\partial R\, \partial r\ . \tag{3.5}$$

We will understand the behavior of $\Delta^{rel}R_e$ if we understand the behavior of the second derivative in Eq. (3.5).

Because of the normalization condition $\langle\psi \mid \psi\rangle \equiv 1$, the following relations hold for any variable x:

$$\partial\langle\psi \mid \psi\rangle/\partial x = \langle\partial\psi/\partial x \mid \psi\rangle + \text{c.c.} = 0 \tag{3.6}$$

$$\begin{aligned}\partial E/\partial x = \partial\langle\psi|\, H\, |\psi\rangle/\partial x &= \langle\partial\psi/\partial x \mid H\, |\psi\rangle + \text{c.c.} + \langle\psi|\, \partial H/\partial x\, |\psi\rangle \\ &= \langle\psi|\, \partial H/\partial x\, |\psi\rangle\ .\end{aligned} \tag{3.7}$$

Equations (3.7) is the well-known Hellmann-Feynman theorem. By identifying either R or r with x, the mixed second derivative can be written in two equivalent forms:

$$\partial^2 E^{rel}/\partial r\, \partial R = \partial\langle\psi|\partial H^{rel}/\partial r\, |\psi\rangle/\partial R \tag{3.8a}$$

or:

$$\partial^2 E^{rel}/\partial R\, \partial r = \partial\rangle\psi|\, \partial H^{rel}/\partial R\, |\psi\rangle/\partial r\ . \tag{3.8b}$$

Thus it is clear from the beginning that $\Delta^{rel}R_e$ *can be explained along two different lines*, either with the help of the relativistic correction to the *Hamiltonian*, $\partial H^{rel}/\partial r = \Delta^{rel}H$, or with the help of the (relativistic) Hellmann-Feynman *force*, $f^{rel}_{HF} = \partial H^{rel}/\partial R$:

$$\Delta^{rel}R_e = -k^{-1} \cdot \partial(\Delta^{rel}E)/\partial R \tag{3.9a}$$

$$\Delta^{rel}R_e = -k^{-1} \cdot \partial\langle f^{rel}_{HF}\rangle/\partial r\ . \tag{3.9b}$$

During the last decade, several groups have analysed relativistic bond length changes according to Eqs. (3.8a, 3.9a) [3, 27–31]. Since consistent results have been obtained, it has been reiterated that consequently other interpretation schemes do not apply. Schwarz [24, 33], and recently Pyykkö [2], too, have stressed, however, the possibility of different complementary interpretation schemes.

3.1.1 Explanation Scheme *a*

Using scheme (a) in actual calculations, one has to determine the relativistic energy correction $\Delta^{rel}E$ at different internuclear distances R. $\Delta^{rel}E$ can be calculated approximately — using first order perturbation theory, where only the nonrelativistic wavefunctions are needed at several different R-values; or $\Delta^{rel}E$ can be calculated accurately — with the help of relativistically corrected wavefunctions. Consequently, it is possible to explain relativistic bond length contractions with or without considering explicitly the relativistic orbital contractions. In this scheme, changes of bond lengths and changes of orbital radii are both a "direct" consequence of the relativistic Hamiltonian. In this scheme, bond length contraction is not a direct result of relativistic AO-contraction. Relativistically corrected orbitals are needed in this scheme only for higher order contributions, which may become important for the heaviest elements.

Writing Eq. (3.9a) down explicitly:

$$\Delta^{rel}R_e = -k^{-1} \cdot (\langle\psi|\, \partial^2 H/\partial r\, \partial R\, |\psi\rangle + (\langle\psi|\, \partial H/\partial r\, |\partial\psi/\partial R\rangle + \text{c.c.})) \tag{3.10a}$$

we have split the relevant matrix element up into two terms, so that we get *two contributions to the relativistic bond-length change*:

1) the expectation value of the distance-change of the relativistic correction-operator, $\partial^2 H/\partial r\, \partial R = \partial\Delta^{rel}H/\partial R$, determined with the nonrelativistic density $|\psi\rangle\langle\psi|$, and

2) the expectation value of the relativistic correction-operator $\partial H/\partial r = \Delta^{rel}H$, determined with the distance-change of the density $\partial(|\psi\rangle\langle\psi|)/\partial R$.

The sum of the two contributions determines the "observable" $\Delta^{rel}R_e$.

The two terms may be evaluated individually in different ways: a) using either nonrelativistic or relativistic wavefunctions, b) using the Dirac or the Schrödinger-Pauli representation for the operator and the wavefunction, and c) expressing ψ in terms of the Dirac or Schrödinger-Pauli functions according to Eqs. (2.23, 24). So we can write down many different expressions for the two terms, which may give rise to different numerical values for them, and which may lead to different explanations in physical terms. The sum of two corresponding terms, of course, will always lead to the same result for the physical quantity $\Delta^{rel}R_e$ (provided no further approximations are made).

A) In the all-electron Dirac representation, V depends only on R, and $\vec{\alpha}\vec{p}c$ and βmc^2 depend only on r. Therefore:

$$\partial^2 H^D/\partial R\, \partial r \equiv 0 \tag{3.11}$$

and *only the second term* in Eq. (3.10a) contributions. The relativistic correction operator is:

$$\partial H^D/\partial r = -\beta' \cdot m_e c^2 - \vec{\alpha}\vec{p}c/2 \tag{3.12}$$

where $\beta' = \beta - 1$. If we insert the *fully relativistic* function ψ^D, so that the relativistic energy correction under the integral stems from the valence region, one may speculate that

$$\Delta^{rel}R_e = -k^{-1} \cdot \text{tr}(\partial H^D/\partial r) \cdot (\partial\, |\psi\rangle\langle\psi|/\partial R) \tag{3.13}$$

will gain its *dominant contributions in the valence region*, too. The phase shift, which originates at the nucleus due to the direct relativistic effect, and also in the whole core region due to the "indirect" relativistic change of the Hartree-Fock potential, will modify the atomic *and* the molecular orbitals in a *similar* manner. However, a detailed investigation of the relation between AO contraction and bond length contraction has notyet been carried out.

If we insert the *zeroth-order* nonrelativistic function $U \cdot \psi^{Schr}$ according to Eq. (2.23) into Eq. (3.13), where ψ^{Schr} is the *non*relativistic function in the Schrödinger representation, one may speculate that in this first-order perturbation theoretical approximation of Eq. (3.13) the *dominant contributions* to the integral come from the *region near the nucleus.* Orthogonalization of a valence AO to the occupied orbitals of an adjacent atom* leads to additional "core tails" on that atom, which usually increase the density at the nucleus, and thereby increase the direct *relativistic stabilization at shorter bond lengths.*

B) In the Pauli representation, we may write:

$$\partial H^{SP}/\partial r = h^{vm} + h^{Dar} + h^{SO} + h^{IND} \tag{3.14}$$

where h^{vm} is the mass-velocity operator from Eq. (2.34), and h^{Dar} and h^{SO} are the Darwin and spin-orbit operators from Eq. (2.30). In frozen-core calculations, the valence hamiltonian contains, in addition, h^{IND}, the indirect term from Eq. (2.51). Since h^{Dar}, h^{SO} and h^{IND} depend on the position of the nucleus, we have now:

$$\partial^2 H^{SP}/\partial r\, \partial R = \partial(h^{Dar} + h^{SO} + h^{IND})/\partial R \neq 0\,, \tag{3.15}$$

contrary to Eq. (3.11) of scheme (A). The matrix-element of the nuclear Darwin contribution is:

$$\langle \psi^{SP} |\, \partial h^{Dar}/\partial R\, | \psi^{SP} \rangle = (\pi Z/4m_e^2c^2)\,(\partial \varrho/\partial R_+)\,(\partial \varrho/\partial R_-) \tag{3.16}$$

where $\partial \varrho/\partial R_\pm$ means the slopes of the density at the nucleus on the same and opposite sides with respect to the other, bonded atom. The last factor in Eq. (3.16) represents the dipolar asymmetry of the charge distribution around the nucleus. If the density decreases faster on the back side than on the bond side, as happens frequently, the Darwin term will contribute to bond expansion. This is consistent with the destabilizing influence of the Darwin term in atoms. (The previous statement of the present author in Ref. [33] that the matrix element in Eq. (3.16) vanishes, was not correct).

Except [54], no investigations on the numerical significance of the Darwin term in Eq. (3.16) and of the corresponding spin-orbit and indirect terms seem to be available. One may speculate that h^{IND}, which is destabilizing and AO-expanding in most cases in the atoms, will also contribute to bond-expansion in most cases in molecules.

The *dominant term* in Eq. (3.10a) is most probably still *the second one.* In first-order perturbation theory, the mass-velocity effect $\sim -p^4 \sim -T^2$ will gain its

---

* This Pauli exclusion effect results in an *overall decrease* of valence electron density of atom A in the core region of atom B, but yet in local density *increases in special areas,* for instance around the nucleus.

bond-contracting influence in the *nuclear region* so that $\Delta^{rel}R_e$ appears to be determined by the innermost tails of the valence orbitals. If ψ increases in the nuclear region with decreasing R (e.g. because of the orthogonality condition, as mentioned above), the dominance of the stabilizing mass-velocity term in Eq. (3.14) will result in bond-length contraction [30, 31]. However, no detailed numerical investigations are available at present.

C) The pseudopotential approach. In general, the orthogonality condition increases the curvature of the valence orbital, when the bond-length decreases. Thereby the kinetic energy T^{nr} is increased. This effect is known as the "Pauli", "exchange" or "Besetzungsverbot" repulsion between the valence orbital and the occupied core orbitals, and it contributes significantly to the repulsive branch of the molecular potential curve. Those contributions, which determine R_e^{nr}, originate from the *outer core* shell. The effect is simulated by the repulsive pseudopotential in the effective core potential approach [34]. In this approach, the relativistically corrected Hamiltonian is written as [35, 36]:

$$H^{rel} = H^{nr} + \Delta^{rel}V_c \tag{3.17}$$

where $\Delta^{rel}V_c$ modifies the repulsive branch of the atomic pseudopotential in the outer core region. We rewrite Eq. (3.10a):

$$H^{rel} = H^{nr} + \Delta^{rel}V_c \tag{3.17}$$

where $\Delta^{rel}V_c$ modifies the repulsive branch of the atomic pseudopotential in the outer core region. We rewrite Eq. (3.10a):

$$\Delta^{rel}R_e = -k^{-1} \cdot \int dr^3 \cdot (\partial\Delta^{rel}V_c/\partial R \cdot \varrho + \Delta^{rel}V_c \cdot \partial\varrho/\partial R)\,. \tag{3.18}$$

In this approach, the first term is no longer small. According to results by Collignon et al. [37], partly cited in [33], and according to the thorough theoretical analysis by Schwarz [33] of numerical data by Christiansen and Ermler [5, 27], *the first term is now the dominant one*, while the second term is small and sometimes even of the opposite sign. The largest contributions to the first term originate in the *outer-core/inner-valence region.*

We note that pseudopotentials are not uniquely defined. Different pseudopotential approaches, which yield very similar *total* molecular energies and bond lengths, will not necessarily yield similar results for the individual contributions, into which the total effect is being broken up. This is indeed the case for the two pseudopotential approaches used by Collignon, and Christiansen, resp. [33].

To summarize, the relativistic bond-length change *may be* explained as due to the change of the relativistic energy correction with change of internuclear distance. This "effect" may be split up into two parts. The first one is due to the change of $\Delta^{rel}H$ with bond length. The second one is due to the change of ψ or ϱ with bond length. Which one of the two terms is dominant, and which regions of space dominantly contribute, depends on the type of the wave-equation (Dirac, Schrödinger-Pauli, frozen-core, pseudopotential), and whether relativistic, or zeroth-order nonrelativistic wavefunctions are used to determine $\Delta^{rel}R_e$ accurately, or approximately to first order, respectively.

The explanation of physical effects in physical terms depends partly on the physical system itself, and partly on the theoretical scheme chosen. Therefore, the general form of explanations should always (and not only in the field of relativistic quantum mechanics) read: "Within the framework of the theoretical scheme chosen the phenomenon under discussion is due to . . .". Different correct explanation schemes are possible, they are complementary. Of course not every explanation is correct; one can formulate many seemingly sensible explanations which are incorrect.

3.1.2 Explanation Scheme *b*

The explicit form of Eq. (3.9b) is:

$$\Delta^{rel}R_e = -k^{-1}(\langle\psi|\,\partial^2H/\partial R\,\partial r\,|\psi\rangle + (\langle\psi|\,\partial H^{nr}/\partial R\,|\,\partial\psi/\partial r\rangle + \text{c.c.}))\,. \tag{3.10b}$$

The *first terms* of Eqs. (3.10a and b) are identical. The first term of Eq. (3.10b) is interpreted, however, differently from that of Eq. (3.10a) as being due to the relativistic change ($\partial/\partial r$) of the Hellmann-Feynman force operator $\partial H/\partial R$. As discussed in Sect. 3.1.1, this term vanishes in the Dirac approach. It is bond expanding in most cases in the Schrödinger-Pauli approach, and it is the dominant bond contracting one in the pseudopotential approach. In frozen-core calculations the first term contains a contribution due to the indirect effect.

The *second term* of Eq. (3.10b) is the Hellmann-Feynman force exerted by the relativistic density change on the nuclei:

$$\int dr^3 \cdot f^{nr}_{HF} \cdot \Delta^{rel}\varrho\,. \tag{3.18}$$

A) In all-electron calculations, we have:

$$f^{rel}_{HF} = \partial H^{nr}_{el}/\partial R = -Z\cdot(z_i/r_i^3) \tag{3.19}$$

where z_i, r_i are the electronic coordinates with respect to the nucleus (charge Z), the bond direction being the z-axis. The nuclear contribution to the Hellmann-Feynman force has been omitted, since it is not changed relativistically. The difference between the Dirac and Schrödinger-Pauli representations is here (to first order) due to the difference of ψ^D and ψ^{SP}. The Hellmann-Feynman force of Eq. (3.19) heavily weighs the *region around the nucleus*. Small changes of the dipolar contribution of the density around the nucleus will have a noticable effect on the force. One can show with the help of Eq. (2.23) that:

$$\psi^{SP}_+ \approx \psi^D_+ + (4m_ec - 2V/c)^{-1}\,\vec{\sigma}\vec{p}\psi^D_-\,. \tag{3.20}$$

If ψ^D_- (the lower two components of the Dirac wavefunction) has components of even angular momentum, then ψ^{SP}_+ will obtain some odd contribution, giving rise to an additional Hellmann-Feynman force contribution in the SP representation.

B) In frozen-core and pseudopotential calculations, H^{nr} contains the (effective) core potential, which yields an additional contribution to the effective Hellmann-Feynman force. Since the pseudopotential is rather steep at the *outer core boundary*, large contributions from the "pseudo-force" will arise.

The second factor under the integral in Eq. (3.18) is the relativistic change of the *molecular* electron density. It is common practice to split up the total molecular density into contributions from the spherically averaged unperturbed *atomic* ground-states, and an additional so-called *difference* density:

$$\varrho_{mol} = \sum_{at} \varrho_{at} + \varrho_{dif} \tag{3.21}$$

Similarly we split $\Delta^{rel}\varrho_{mol}$ up:

$$\Delta^{rel}\varrho_{mol} = \sum_{at} \Delta^{rel}\varrho_{at} + \Delta^{rel}\varrho_{dif} \tag{3.22}$$

so that the second term of Eq. (3.10b), i.e. Eq. (3.18), now consists of two contributions, an atomic one (at) and a specific molecular one (dif). Inserting Eq. (3.19) and the atomic contribution of Eq. (3.22) into Eq. (3.18), one obtains [33] the atomic contribution to the second matrix element of Eq. (3.10b):

$$\Delta^{rel}_{2nd,\,at}\, R_e = +Z/(k \cdot \bar{R}_e^2) \cdot \int_{\bar{R}_e}^{\infty} \Delta^{rel}\varrho_{at}(r) \cdot dr^3 \tag{3.23}$$

where $\bar{R}_e$ is the total bond length. If the atomic density contracts in the outer valence region, then $\Delta^{rel}\varrho_{at} < 0$ in the outer region. Then the integral is negative, and the contribution of Eq. (3.23) to the total $\Delta^{rel}R_e$ is negative. *Atomic orbital contraction contributes to bond length contraction*, as anticipated intuitively before 1980 [39]. However, it should be noted, that 1) Eq. (3.23) is only *one* contribution, and that 2) there exists *no* simple linear relation between $\Delta^{rel}R_e$ and the relativistic AO contraction $\Delta^{rel}R_i$. The assumption that $\Delta^{rel}R_e$ is given numerically by the values $\Delta^{rel}R_i$ of the different AOs i, weighed by the AO-populations p_i [40], is unfounded.

So far, no investigation of the genuine molecular contribution to the second matrix element:

$$\Delta^{rel}_{2nd,\,dif} R_e = -k^{-1} \cdot \int f_{HE} \cdot \Delta^{rel}\varrho_{diff} \cdot dr^3 \tag{3.24}$$

has been published. Two effects may be especially important. First, the orthogonality requirement of the valence AO on the relativistically modified core orbitals of the adjacent atom will contribute to $\Delta^{rel}\varrho_{diff}$. Second, the appropriate state of an atom with open p- or d-valence shell to undergo bond formation is not the spherically averaged mixed state, but a state with oriented, differently populated AOs. Further speculations may be found in Refs. [24, 33].

To summarize, the relativistic bond-length change may, alternatively to scheme (a) (Sect. 3.1.1), be explained as due to the change of the Hellmann-Feynman force because of relativity. This "effect" may also be split up into two parts. The *first* one is due to the relativistic change of the Hellmann-Feynman force operator $\Delta^{rel}f_{HF}$. This term is numerically identical with the first term of approach (a), provided one works in the same operator representation with the same wavefunction. The *second* term is due to the change of the Hellmann-Feynman force caused by the relativistic modification of the molecular orbitals. The total molecular orbital change "consists" of the relativistic atomic orbital change, which causes a corresponding bond length change, and of a genuine molecular orbital change, which has so far not been investigated in detail.

3.2 Spin-orbit Coupling in Molecules

p-bonds are destabilised by spin-orbit coupling. Spin-orbit coupling is very large in atoms from the right bottom of the periodic table. Bond angles are only weakly influenced by spin-orbit coupling.

The discussion of spin-orbit effects in molecules is more complicated than that of the mass-velocity, Darwin and indirect effects, because the spin-orbit coupling operator is of more complicated nature. Namely, it is a nonlocal differential operator, which destroys nonrelativistic symmetry. However, a few general rules can be obtained rather easily.

3.2.1 Spin-orbit Coupling and Bond Energies

The spin-orbit splittings of the atomic p-valence shells are presented in Tab. 6. They increase strongly from left to right in each row. Within one column, the splitting increases approximately proportional to Z^2_{nuc}.

Table 6. Valence-p-orbital spin-orbit splitting in kJ/mol [20]

	Group												
Row	1	2	3	...	11	12	13	14	15	16	17	18	Row
2	0	0	0				0	1	2	3	5	7	2
3	0	1	1				1	3	5	7	11	17	3
4	1	2	3		3	7	10	17	27	33	46	64	4
5	3	7	10		11	21	26	41	63	82	96	125	5
6	7	15	20	...	46	77	92	133	200	245	305	370	6
7	20	43	50	...									

In light atoms, the spin-orbit splitting is small in comparison to both the electron-interaction between the valence electrons and the chemical interaction between adjacent atoms (i.e. $\sigma_g - \sigma_u$ — MO splitting). Therefore the nonrelativistic MO-scheme, using conventional so-called LS/ΛS coupling, is a reasonable starting point. Then one may take spin-orbit coupling into account by first-order perturbation theory.

In the "strongly relativistic limit", one assumes the spin-orbit splitting to be large in comparison to both electron-interaction and chemical interaction. Here, one starts with the relativistic MS-scheme (MS = molecular spinor), using jj/$\omega\omega$ coupling of atomic spinors.

However since all three interactions are in fact of similar magnitude in the valence shell of heavy molecules, only a complicated relativistic configuration mixing scheme will yield satisfactory numerical results [3, 4, 46]. Nevertheless, the limiting cases are helpful for a qualitative understanding.

A potential energy curve has two ends. At the separated atoms end the spin-orbit coupling splits the nonrelativistically degenerate states and will always stabilize the atomic ground states. In the molecule, spin-orbit coupling is partially quenched by the non-spherically symmetric molecular field. Therefore, one expects bond destabilization and bond length expansion by spin-orbit coupling [25, 38]. A typical example is Tl_2 in Fig. 9 ([32], see also [38]). Further ones are reviewed in Refs. [3, 4].

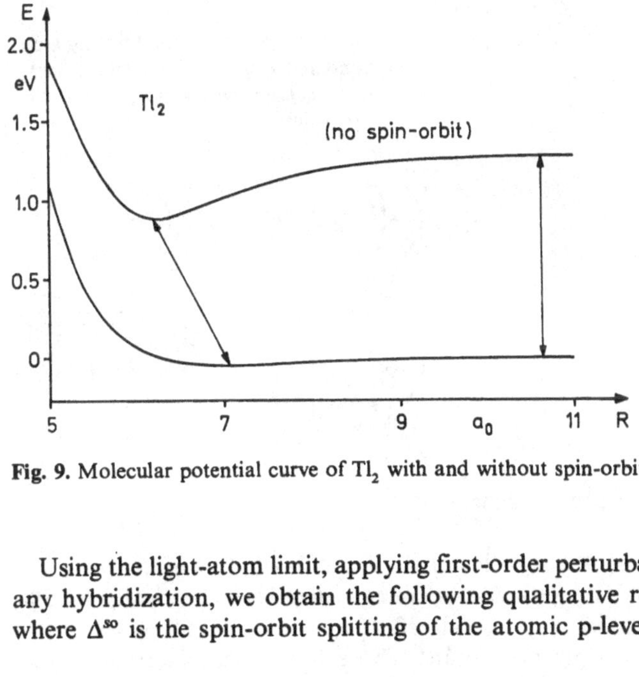

Fig. 9. Molecular potential curve of Tl_2 with and without spin-orbit interaction, after [32]

Using the light-atom limit, applying first-order perturbation theory, and neglecting any hybridization, we obtain the following qualitative result for $p\sigma$ and $p\pi$ bonds, where Δ^{so} is the spin-orbit splitting of the atomic p-level (see Fig. 3):

Table 7. Spin-orbit contribution of different $p\sigma$- and $p\pi$-MOs to the bond energy (lowest order approximation)

Type of MO	First order SO energy of MO	Correlates with AO	First order SO energy of AO	SO contribution to bond energy
$p\sigma 1/2$	0	p1/2	$-2\Delta^{so}/3$	$-2\Delta^{so}/3$
$p\pi 1/2$	$-\Delta^{so}/3$	p1/2	$-2\Delta^{so}/3$	$-\Delta^{so}/3$
$p\pi 3/2$	$+\Delta^{so}/3$	p3/2	$+\Delta^{so}/3$	0
$p\pi^* 1/2$	$-\Delta^{so}/3$	p3/2	$+\Delta^{so}/3$	$+2\Delta^{so}/3$
$p\pi^* 3/2$	$+\Delta^{so}/3$	p3/2	$+\Delta^{so}/3$	0
$p\sigma^* 1/2$	0	p3/2	$+\Delta^{so}/3$	$+\Delta^{so}/3$

Bonding orbitals are in general destabilized, and antibonding orbitals are stabilized by spin-orbit coupling. A similar result is obtained in the "strongly relativistic limit" [2, 41, 42]. Spin-orbit interaction couples the p_x-, p_y, p_z-AOs with specified spin and phase factors. For example, a p1/2-atomic spinor reads $p_x \cdot \beta + i \cdot p_y \beta + p_z \cdot \alpha$. As shown in Fig. 10, two p1/2-spinors can only form a pσ-bond together with a pπ^*-antibond, or a pπ-bond together with a pσ^*-antibond. It should be noted, however, that hybridization and second-order configuration mixing may modify these simple rules (see e.g. [38]).

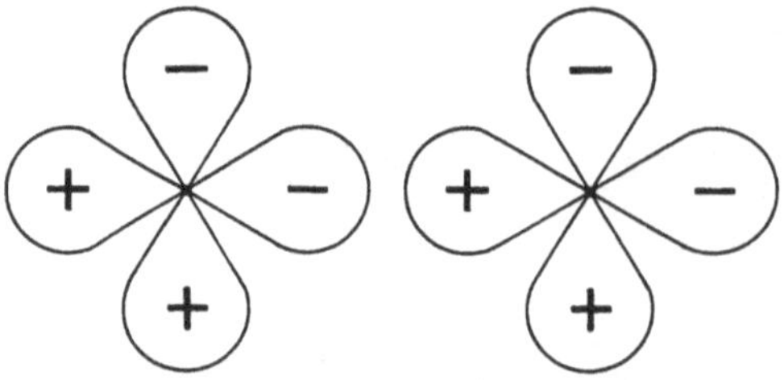

Fig. 10. Two p1/2-AOs, forming a pσ-bond and a pπ^*-antibond (right side), forming a pπ-bond and a pσ^*-antibond (left side)

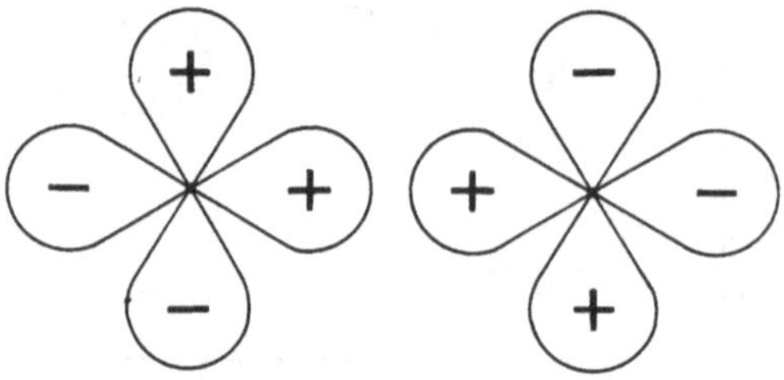

3.2.2 Spin-orbit Splitting and Bond Angles

Bond stretching force constants are significantly larger than bending force constants. Therefore, one may expect that spin-orbit coupling, which can drastically modify bond energies and bond lengths, may also significantly change bond angles.

Nonrelativistic p-AOs are triply degenerate. Those linear combinations, which are orthogonal in function space and allow for maximum overlap with neighbouring atoms, are the p_x, p_y, p_z ones. They are orthogonal in real space. Consequently the atoms of the right side of the periodic system, which form pσ-bonds, tend to form bond angles of 90°. Spin-orbit coupling removes the orbital degeneracy. The p3/2 spinors have less pronounced directional properties, their quadrupole moments being only half as large as those of the p_x-, p_y-, p_z-AOs. The p1/2 spinor has even a spherically symmetric density.

A series of pseudopotential-SCF and Hückel-MO results by Collignon [37] are presented in Tab. 8. Further results from the literature are reviewed by Pyykkö [2]. Although the spin-orbit splitting is of the order of 10^2 kJ/Mol in the heavy main-group elements, its influence on the bond angles α is still small, and only of the order of the numerical accuracy of the methods.

Table 8. Change of bond angles by spin-orbit coupling (SO)

Molecule	Method[a]	Angle	Calcul. angle	Angle change due to SO
$PbCl_2$	RPP	Cl-Pb-Cl	98.2°	−0.4°
	REX		93.9°	−0.0°
PbH_2	RPP	H-Pb-H	92.2°	+0.6°
	REX		91.7°	+1.0°
Pb_2H_4	RPP	H-Pb-H	97.5°[b]	+0.4°
		Pb-Pb-H_2	123.5°[b]	−0.4°
		Pb-Pb-H	111.3°[b]	−0.3°
$^{114}EH_2$	REX	H-E-H	90.3°	+2.3°
TeH_2	REX	H-Te-H	99.5°[b]	−0.0°
PoH_2	REX	H-Po-H	95.0°[b]	−0.5°

[a] RPP: relativistic SCF pseudopotential method including spin-dependent terms [38]
REX: relativistic extended Hückel method [43]
[b] similar values have been obtained in Ref. [44]

These somewhat unexpected results can be explained as follows. Real basis functions such as p_x, p_y, p_z have exactly zero angular momentum with respect to their symmetry axis. Consequently, the spin-orbit operator will not change the diagonal energy expectation values, but will only contribute off-diagonal elements in this basis. A real basis is adapted to the molecular symmetry, which is lower than in the atom due to the electrostatic fields of the ligands. As long as the spin-orbit coupling is not the largest interaction in the valence shell, it will only cause second order corrections in the molecules. That is, $\Delta^{rel}\alpha \sim (\Delta^{so})^2 \sim O(c^{-4})$, which is very small.

Very recently [52] the first drastic change of a bond angle by spin-orbit interaction has been found for an excited state. The A_1 component of the $^3B_1$ state of PbH_2 ($\alpha = 120°$) is energetically destabilized by spin-orbit induced configuration mixing with the $^1A_1$ ground state ($\alpha = 91°$). Thereby the bond angle of $^3B_1(A_1)$ is reduced by 10° to $\alpha = 110°$. However, this big change has not been reconfimed by [37].

First-order changes of bond angles are to be expected, if Δ^{so} is larger than both the valence shell exchange and resonance parameters and if, in addition, the molecular electron configuration is sensitive to spin-orbit coupling. For instance, consider a triatomic system like CH_2 or NO_2, which is bent "because" of the Teller-Renner effect, i.e. in linear geometry the degenerate π level is partially filled. In the nonrelativistic case, the degeneracy can only be broken by bending, while in the relativistic case, spin-orbit coupling will do the job. In such cases, bond angle increases may occur. So far, no clear example has been found.

3.3 Relativistic Changes of Bond Energies

Both relativistic stabilizations and destabilizations of bonds are frequent. A systematic scheme of bond energy increases and decreases is presented in Tab. 9.

In Sect. 3.2.1 we have discussed the first order influence of spin-orbit coupling on covalent bonds. In addition, the bond energies may also be modified by the mass-velocity, Darwin, indirect and higher order effects. Furthermore, we have to distinguish between covalent and ionic binding. In the ionic case it also makes a difference whether the anion or the cation is the heavy one, i.e. significantly influenced by relativity. While in most cases bond lengths are relativistically shortened and bond angles are only weakly influenced, relativistic changes of bond energies exhibit a more complex behavior. Extending ideas of Pyykkö [2] we have tried to systematize the bond energy increases and decreases in Tab. 9. The original references to the examples given in Tab. 9 can be found in Refs. [2–6] (see also Refs. [51, 52]).

Table 9. Relativistic bond energy (BE) changes

Dominant relativistic effect	Decisive AO	Covalent bonding	Ionic bonding	
			heavy cation	heavy anion
Mass-velocity (+ Darwin) stabilization	s	BE increase (Hg_2^{+2})	BE decrease (Au^+Cl^-)	BE increase (Cs^+Au^-)
Spin-orbit stabilization	p1/2, d3/2	BE decrease (Tl_2)	BE decrease (Tl^+Cl^-)	(BE increase) (Tl^-M^+, ?)
Spin-orbit destabilization	p3/2, d5/2	BE decrease (Te_2, Pt_2)	BE increase ($Pt^{+2}(Cl^-)_2$)	BE decrease (Na^+I^-)
Indirect destabilization	d, f	BE increase (W_2)	BE increase ($Au^{+3}(Cl^-)_3$	(BE decrease)

3.3.1 Mass-velocity and Darwin effect

The most prominent relativistic changes of s-AOs are contraction and stabilization, caused by the mass-velocity effect (and partially compensated by the Darwin effect). *Covalent bonds*, if mainly due to s-AO interaction, are significantly stabilized (and contracted, as usual). Typical examples are Au_2 and isoelectronic Hg_2^{+2}. Relativity is responsible for the extraordinary stability of the latter species and for a large part of the chemistry of mercury.

The bond energy of *ionic compounds* is the larger, the smaller the energy of charge transfer:

decrease of ionization potential
increase of electron affinity } increase of ionic bond energy

increase of ionization potential
decrease of electron affinity } decrease of ionic bond energy

Relativity stabilizes s-AOs and therefore increases the corresponding s-IP and s-EA. Consequently, the bond energy of ionic Au^+ compounds, e.g. AuCl, is decreased because of the relativistic increase of the Au6s IP, while the bond energy of Au^- compounds, e.g. CsAu, is increased because of the relativistic increase of the Au6s EA. It is important to note that despite the different bond energy changes, the bond lengths are relativistically contracted in all these examples.

3.3.2 Spin-orbit Coupling

The most prominent relativistic effect of p-AOs is the spin-orbit splitting. Atoms in groups 13 and 14 have one or two electrons in the stabilized p1/2 AO while atoms from the following groups have also electrons in the destabilized p3/2 AOs.

Concerning *covalent bonds*, relativity decreases the bond energy. The effect is rather large in molecules from groups 13 and 14 (Tl_2, Pb_2), compare Fig. 9 in Sect. 3.2.1 and Refs. [4, 38]. For molecules from the next groups (Bi_2, Te_2, I_2), higher MOs become occupied which are more or less stabilized by relativity. However, the total effect of all occupied MOs is still destabilizing (compare Tab. 7).

The spin-orbit coupling in the d-shell has similar effects. It weakens the bond energies of transition metal dimers such as Pt_2.

Spin-orbit splitting increases IP and EA of p1/2 and d3/2 levels and decreases IP and EA of p3/2 and d5/2 levels. Consequently, *ionic compounds* of group 13 and 14 cations (e.g. TlCl) or compounds of anions from the higher groups (e.g. NaI) are relativistically destabilized.

On the other hand, ionic compounds from group 13 anions and from higher group cations, for instance Cs^+Tl^- or Bi^+F^-, are predicted to be relativistically stabilized. The same holds for the corresponding d-compounds, e.g. $Pt^{+2}Cl_2^-$.

3.3.3 Indirect AO-destabilization

The indirect destabilization and expansion is the dominant relativistic effect for d and f AOs. This effect improves the overlap properties of these "inner" valence AOs with the surrounding ligand orbitals, thereby increasing the covalent bond strength of different transition metal compounds, for instance W_2 or PtH_2. In the case of platinum, the bond strengthening is a combined effect of indirect d-AO destabilization and s-AO stabilization. While Pd has a $d^{10}s^0$ closed shell ground state, the relativistically reduced d-s gap results in a d^9s^1 open shell ground state for Pt. Both singly occupied AOs can form stable covalent bonds, especially since relativistic expansion of the compact d and contraction of the diffuse s improves the overlap properties of both AOs [52].

The indirect d-AO destabilization should also contribute to the bond strength of transition metal cations. Relativistic stabilizations have been reported, for instance, for Th(IV) and U(VI) compounds (e.g. by about 200 kJ/mol for $ThHCl_3$) [50]. However, some counter-examples are also known. The actinocenes $M^{+4}(C_8H_3^{-2})_2$, which are relativistically destabilized [49], show that these oversimplified rules do not always work. Though, they give at least some vage impression why relativistic bond energy changes exhibit such a diversity.

3.4 Summary and Conclusions

A "nonrelativistic valence electron" in an atom feels only the effective nuclear charge Z_{eff} because of the effective screening of the nuclear charge Z_{nuc} by the core electrons. The physical mechanism of relativity works in the vicinity of the nucleus. Therefore, the relativistic corrections for outer valence electrons are still determined by the unscreened nuclear charge Z_{nuc}. Relativistic effects in the valence shells scale approximately as $(Z_{nuc}/c)^2$.

In first-order perturbation theory, the relativistic *energy* correction emerges in the vicinity of the nucleus. The total energy decreases, because the relativistic kinetic energy expression is smaller. Relativistically optimized wavefunctions undergo a phase shift near the nucleus, which propagates out into the valence shell. This results in a spatial shift of the orbital density maxima. Although there is no theoretical basis, it has been reiterated in the literature that the relativistic contraction of outer orbitals is "only" due to the orthogonality constraint on the inner core orbitals, which are "directly" modified by the relativistic Hamiltonian. Writing the relativistic change of orbital φ_i as [21]:

$$\Delta^{rel}\varphi_i = \varphi_i^{rel} - \varphi_i^{nr} = \sum_j |\varphi_j^{rel}\rangle \cdot (\delta_{ji} - \langle\varphi_j^{rel} | \varphi_i^{nr}\rangle) \tag{3.25}$$

one can examine the importance of the orthogonality constraint on different relativistic orbitals φ_j^{rel}. Numerical results [21, 23] show, that the Schmidt-orthogonalization overlap-integrals on the inner core-AOs $j < i$ [Eq. (3.25)] are negligibly small. Valence orbitals are *directly* influenced by the action of the relativistic Hamiltonian on their inner tails. Thereby the infinite number of *higher* orbitals φ_j^{rel}, $j > i$ are mixed in.

The relativistic change of the energy expectation value of a *relativistically optimized* valence orbital originates in the outer valence shell. The decrease of the kinetic energy *expression* results via orbital relaxation in an increase of the kinetic energy expectation *value*, which is overcompensated by the potential energy decrease in the valence shell. *Valence*-density-faithful valence-only approaches are therefore capable of simulating relativistic effects, although the *inner* core part of those pseudo-valence orbitals does not play any role.

There are three basic relativistic effects. Relativistic kinematics increases the effective mass and decreases the kinetic energy expression. This mass-velocity effect stabilizes and contracts atomic orbitals and molecular bonds. It is modified by momentum-dependent potential energy corrections. These are the Darwin and spin-orbit terms.

If only a subsystem is investigated (for instance the valence shell, or a single electron) the so-called indirect effect emerges. To lowest order, this is the change of the classical (nonrelativistic) electron-repulsion energy caused by the relativistically modified charge distribution of the core. Inner s- and p-shells are relativistically contracted and shield the nucleus more effectively. Therefore all valence orbitals are *indirectly destabilized*, especially the d- and f-AOs, but *also the s- and p-AOs*. Inner d- and f-shells are indirectly expanded. Therefore they cause indirect valence orbital *stabilization*, especially for s- and p-AOs of the central groups of the periodic table. In addition to the *nonrelativistic* d-block- and lanthanoide-contraction, there is a similar *relativistic* d-block- and lanthanoide-contraction.

Neglecting for the moment the variation of the relativistic AO-effects along a row of the periodic system (which is due to the variation of the population of the other shells), we can give very crude estimates of the magnitude (up to ± 50–$70\,\%$) of the relativistic AO-energy and AO-radius changes (see Tab. 10).

These AO-changes are related to changes of bond-energies and bond lengths. The connection is complicated, because it also depends on differences of electronegativities and radii of the bonding atoms. Details have been discussed in Refs. [2, 24]. In

Table 10. Crude estimates for relativistic AO-changes

$\Delta^{rel}E/E^{rel} \approx 0.6 \cdot (Z/c)^2$	$\Delta^{rel}R/R^{rel} \approx -1/3 \cdot (Z/c)^2$	for s-AOs
$0.3 \cdot (Z/c)^2$	$-1/3 \cdot (Z/c)^2$	p1/2-AOs
$-0.3 \cdot (Z/c)^2$	0	p3/2-AOs
$-0.6 \cdot (Z/c)^2$	$+1/4 \cdot (Z/c)^2$	d-AOs
$-1.2 \cdot (Z/c)^2$	$+1/3 \cdot (Z/c)^2$	f-AOs

many cases, s- and p1/2-AO contraction and stabilization "leads" to bond contraction and stabilization, while p-AO spin-orbit splitting tends to expand and destabilize covalent bonds.

The most transparent relativistic effects are the changes of molecular structure. Relativistic bond angle changes, especially those due to spin-orbit coupling, seem to be small in general, while large bond length changes have been "observed".

The relativistic bond length changes can be explained either by discussing energies or forces. In any case there are two contributions, the first being an expectation value of the zeroth-order wavefunction (nonrelativistic wavefunction at the nonrelativistic bond length). The second contribution depends on a first-order wavefunction. The magnitude of the two terms individually depends on the theoretical framework chosen, only the sum of them being uniquely defined. This is similar as in the case of magnetic properties, where only the sum of dia- and para-magnetic contributions is uniquely defined. Different theoretical approaches attribute the relativistic effects to different spatial regions. The situation is similar as with optical transition moments, where the dipole-length formula weights the outer spatial regions, and the velocity and especially the acceleration formula weights the inner spatial regions.

Concerning the energy and force concepts of binding, the *first* contribution to $\Delta^{rel}R_e$ contains either the change of $\Delta^{rel}H$ with bond length, $d(\Delta^{rel}H)/dR$, or the change of the Hellmann-Feynman force $f = dH/dR$ due to relativity, $\Delta^{rel}(dH/dR)$. The *second* contribution contains either the change of the *molecular* wavefunction with bond length, $d\psi/dR$, or the relativistic *molecular* orbital change, $\Delta^{rel}\psi$, in combination with $\Delta^{rel}H$, and $f = dH/dR$, respectively. In both cases there exists no *direct* relation between relativistic AO changes and bond length changes, as first pointed out in [30], but there exist complicated relations between $\Delta^{rel}R$, $\Delta^{rel}\psi$ and $\Delta^{rel}H$ [24].

Concerning the relativistic changes of molecular energies, or even of reactivities, the theoretical understanding is less advanced. A large number of case studies on individual molecules have been reviewed recently (see especially [2–6]). So far only the first steps in developing general interpretation schemes as that of Tab. 9 have been made.

Acknowledgement The author is grateful to G. Collignon, R. Jaquet, W. Kutzelnigg and P. Pyykkö for critically reading the manuscript and for many helpful comments. He thanks K. Balasubramanian, P. A. Christiansen, W. Ermler and P. Pyykkö for preprints, and J. E. Baerends, G. Collignon and E. van Wezenbeek for results prior to publication. This work has benefitted very much from the author's stays at the Free University Amsterdam and at the East China Normal University Shanghai. He thanks Prof. Baerends and Prof. Pan, respectively, for their warm hospitality; furthermore, he thanks for financial support by these two Universities, by DAAD and by Fonds der Chemischen Industrie.

4 References

1. Pykkö P (1978) Adv Quantum Chem 11: 353; see also Fricke B (1975) Struct Bonding 21: 89
2. Pykkö P (1988) Chem Rev 88: 563
3. Ermler WC, Ross RB, Christiansen PA (1988) Adv Quantum Chem 19: 139
4. Balasubramanian K, Pitzer KS (1987) Adv Chem Phys 57: 287
5. Christiansen PA, Ermler WC, Pitzer KS (1985) Ann Rev Phys Chem 36: 407
6. Pyykkö P (1986) Relativistic theory of atoms and molecules, Lecture notes in chemistry 41. Springer-Verlag, Berlin
7. Sommerfeld A (1919) Atombau und Spektrallinien, Vieweg, Braunschweig
8. Burke V, Grant IP (1967) Proc Phys Soc 90: 297
9. Feshbach H, Villars F (1958) Rev Mod Phys 30: 24
10. Dirac PAM (1928) Proc Roy Soc A 117: 610
11. Sakurai JJ (1967, 1984) Advanced quantum mechanics. Benjamin, Menlo Park
12. Moss RE (1973) Advanced molecular quantum mechanics. Chapman and Hall, London
13. Kissel-Phillip M, Schwarz WHE (1988) Phys Rev A 38: 6027
14. Levy-Leblond JM (1967) Commun Math Phys 6: 286
15. Thomas LH (1927) Phil Mag 3: 1
16. Ruedenberg K (1962) Rev Mod Phys 34: 326
17. Katriel J, Pauncz R (1977) Adv Quant Chem 10: 143
18. Tayler RJ (1970) The stars: Their structure and evolution. Wykeham, London
19. Kutzelnigg W (1978) Einführung in die Theoretische Chemie. Verlag Chemie, Weinheim
20. Deslaux JP (1973) At Data Nucl Data Tabs 12: 311
21. Schwarz WHE, van Wezenbeek E, Baerends EJ, Snijders J, Schwerdtfeger P (1988) J Phys B 22: 1515
22. Epstein ST (1974) The variation method in quantum theory. Academic Press, New York
23. Rose SJ, Grant IP, Pyper NC (1978) J Phys B 11: 1171
24. Schwarz WHE, Chu SY, Mark F (1983) Mol Phys 50: 603
25. Mark F, Marian C, Schwarz WHE (1984) Mol Phys 53: 535
26. Pyper NC, Rose SJ, Grant IP (1982) J Phys B 15: 1319
27. Christiansen PA, Ermler WC (1985) Mol Phys 55: 1109
28. Katriel J, Feller D, Davidson ER (1984) Int J Quantum Chem 26: 489
29. Martin RL (1983) J Phys Chem 87: 750
30. Snijders JG, Pyykkö P (1980) Chem Phys Let 75: 5
    Ziegler T, Snijders JG, Baerends EJ (1980) Chem Phys Let 75: 1
31. Pyykkö P, Snijders JG, Baerends EJ (1981) Chem Phys Let 83: 432
32. Christiansen PA, Pitzer KS (1981) J Chem Phys 74: 1162
33. Schwarz WHE (1987) Phys Scripta 36: 403
34. Kahn LR, Baybutt P, Truhlar DG (1976) J Chem Phys 65: 3826
35. Hafner P (1976) PhD Dissertation, University of Bonn
    Hafner P, Schwarz WHE (1978) J Phys B 11: 217
36. Lee YS (1978) Diss Abstr Int B 39: 772
    Christiansen PA, Lee YS, Pitzer KS (1979) J Chem Phys 71: 4445
37. Collignon G (1983) unpublished results; (1989) Ph. D. thesis, University, Siegen
    Schwarz WHE, Collignon G, Chu SY, Mark F (1985) 5th Intern Congr Quant Chem Montreal, Abstract 29-A
38. Hafner P, Habitz P, Ishikawa Y, Wechsel-Trakowski E, Schwarz WHE (1981) Chem Phys Let 80: 311
39. Pyykkö P, Desclaux JP (1979) Acc Chem Res 12: 276
40. Walch SP, Bauschlicher CW, Siegbahn PEM, Portridge J (1982) Chem Phys Let 92: 54
41. Pitzer KS (1975) J Chem Phys 63: 1032, Chem Comm 760
42. Pyper NC (1980) Chem Phys Let 73: 385, 74: 554
43. Pyykkö P (1989) Semi-empirical relativistic molecular structure calcularions, in: Methods in computational chemistry, Vol 2, Wilson S (ed) Plenum, London
44. Pyykkö P, Lohr LL (1981) Inorg Chem 20: 1950
45. Gleghorn JT, Hammond NDA (1984) Chem Phys Let 105: 621

46. Esser M, Butscher W, Schwarz WHE (1981) Chem Phys Let 77: 359
Esser M (1984) Int J Quantum Chem 26: 313
47. Bagus PS, Lee YS, Pitzer KS (1975) Chem Phys Let 33: 408
48. Joos G (1959) Lehrbuch der Theoretischen Physik, § 2.6.9, Akademische Verlagsgesellschaft, Frankfurt am Main
Landau LD, Lifschitz EM (1977) Klassische Feldtheorie, § 9, Akademie-Verlag, Berlin
49. Boerrigter PM (1987) Thesis, Vrije Universiteit, Amsterdam
50. Ziegler T (1988) ACS symposium series, Toronto
51. Schwerdtfeger P, Dolg M, Schwarz WHE, Bowmaker GA, Boyd PSW (1989) J Chem Phys 91: 1762
52. Balasubramanian K (1988) J Chem Phys 88: 6955, 89: 5731
53. Foldy LL, Wouthuysen SA (1950) Phys Rev 78: 29
54. Rutkowski A, Schwarz WHE (1989) Theoret. Chim. Acta in press

FSC
www.fsc.org
MIX
Papier | Fördert
gute Waldnutzung
FSC® C083411

Kolibri 360 GmbH
Ferdinand-Jühlke-Straße 7
99095 Erfurt, Deutschland
produktsicherheit@kolibri360.de

www.ingramcontent.com/pod-product-compliance
Ingram Content Group UK Ltd.
Pitfield, Milton Keynes, MK11 3LW, UK
UKHW021942290926
8980IPUK00016B/1763

9 783039 113095